TABLE OF ATOMIC WEIGHTS 1971

Scaled to the relative atomic mass, $A_r(^{12}C) = 12$

The values of $A_r(E)$ given here apply to elements as they exist in materials of terrestrial o
artificial elements. When used with due regard to the footnotes they are considered relia
digit, or ±3 if that digit is subscript.

Alphabetical Order in English

Name	Symbol	Atomic number	Atomic weight
Mercury	Hg	80	200.5_9
Molybdenum	Mo	42	95.9_4
Neodymium	Nd	60	144.2_4
Neon	Ne	10	20.17_9[c]
Neptunium	Np	93	237.0482[f]
Nickel	Ni	28	58.7_1
Niobium	Nb	41	92.9064[a]
Nitrogen	N	7	14.0067[b,c]
Nobelium	No	102	—
Osmium	Os	76	190.2
Oxygen	O	8	15.999_4[b,c,d]
Palladium	Pd	46	106.4
Phosphorus	P	15	30.97376[a]
Platinum	Pt	78	195.0_9
Plutonium	Pu	94	—
Polonium	Po	84	—
Potassium	K	19	39.09_8
Praseodymium	Pr	59	140.9077[a]
Promethium	Pm	61	—
Protactinium	Pa	91	231.0359[f]
Radium	Ra	88	226.0254[f,g]
Radon	Rn	86	—
Rhenium	Re	75	186.2
Rhodium	Rh	45	102.9055[a]
Rubidium	Rb	37	85.467_8[c]
Ruthenium	Ru	44	101.0_7
Samarium	Sm	62	150.4
Scandium	Sc	21	44.9559[a]
Selenium	Se	34	78.9_6
Silicon	Si	14	28.08_6[d]
Silver	Ag	47	107.868[c]
Sodium	Na	11	22.98977[a]
Strontium	Sr	38	87.62[g]
Sulfur	S	16	32.06[d]
Tantalum	Ta	73	180.947_9[b]
Technetium	Tc	43	—
Tellurium	Te	52	127.6_0
Terbium	Tb	65	158.9254[a]
Thallium	Tl	81	204.3_7
Thorium	Th	90	232.0381[f]
Thulium	Tm	69	168.9342[a]
Tin	Sn	50	118.6_9
Titanium	Ti	22	47.9_0
Tungsten	W	74	183.8_5
Uranium	U	92	238.029[b,c,e]
Vanadium	V	23	50.941_4[b,c]
Wolfram	W	74	183.8_5
Xenon	Xe	54	131.30
Ytterbium	Yb	70	173.0_4
Yttrium	Y	39	88.9059[a]
Zinc	Zn	30	65.38
Zirconium	Zr	40	91.22

[a] Element with only one stable nuclide.

[b] Element with one predominant isotope (about 99 to 100 per cent abundance); errors in abundance determinations have a correspondingly small effect on the confidence in the value of $A_r(E)$.

[c] Element for which the value of $A_r(E)$ derives its reliability from calibrated measurements (i.e. from comparisons with synthetic mixtures of known isotopic composition).

[d] Element for which known variations in isotopic abundance in terrestrial material prevent a more precise atomic weight being given, $A_r(E)$ values should be applicable to any 'normal' material.

[e] Element for which values of A_r may be found in commercially available products that differ from the tabulated value of $A_r(E)$ because of inadvertent or undisclosed changes of isotopic composition.

[f] Element for which the value of A_r is that of the most commonly available long-lived nuclide.

[g] Element for which geological specimens are known in which the element has an anomalous isotopic composition.

PRINCIPLES OF CHEMISTRY

A SERIES OF BOOKS IN CHEMISTRY

James W. Cobble, *Editor*

PRINCIPLES OF CHEMISTRY

Loren G. Hepler
University of Lethbridge
Lethbridge, Alberta, Canada

Wayne L. Smith
Colby College
Waterville, Maine

Macmillan Publishing Co., Inc.
New York
Collier Macmillan Publishers
London

Printed in the United States of America

Portions of the material in this book have been adapted from *Chemical Principles,* by Loren G. Hepler, © 1964 by Blaisdell Publishing Company

MACMILLAN PUBLISHING CO., INC.
866 THIRD AVENUE, NEW YORK, NEW YORK 10022

COLLIER-MACMILLAN CANADA, LTD.

Library of Congress Cataloging in Publication Data

Hepler, Loren G.
Principles of chemistry.

(A Series of books in chemistry)
Includes bibliographies.
1. Chemistry. I. Smith, Wayne L., joint author.
II. Title. [DNLM: 1. Chemistry. QD31.2 H529p 1974]
QD31.2.H46 540 73-5282
ISBN 0-02-353680-2

Printing: 1 2 3 4 5 6 7 8 Year: 5 6 7 8 9 0

PREFACE

This textbook has been written primarily for students who are taking a first course in university or college chemistry and who have taken a high school course in chemistry. We assume that students who use this book will have some acquaintance with the vocabulary and some experience with various aspects of chemistry. But because we have learned by experience that students come to our classes with widely different backgrounds, we have not assumed that the students have any particular knowledge or skills. We have, therefore, included discussions of such "elementary" topics as use of exponentials and logarithms, balancing equations, and the meaning of Avogadro's number along with more "advanced" topics such as the kinetic theory of gases, bonding theory, thermodynamics, electrochemistry, and kinetics.

Most teachers believe that students learn chemistry best by solving problems, and most students feel that problem solving is the most difficult part of the study of chemistry. There is no easy road to success as a problem solver, but we have tried to smooth the road as much as possible in several different ways. First, this book contains a large number of problems with a substantial number of fully worked examples. Further, we have included both easy and difficult problems so that students are led along reasonably gently. And finally, we have provided numerous specific applications of general principles to particular problems. Answers to many problems are given at the back of the book.

Several comments about our use of mathematics are in order. Exponential notation for large and small numbers, logarithms, and simple algebra are used in many parts of this book. The principles on which these uses are based are reviewed in Chapter 1. Although algebraic manipulations frequently appear obvious to experienced scientists, they are not usually obvious to beginners. We have, therefore, tried to include the "missing steps" in derivations and solutions to example problems, which accounts for the substantial number of mathematical expressions that appear in many parts of this book.

In Chapters 11, 12, 13, 16, 21, and 22 we make use of simple calculus, mostly in connection with derivation of thermodynamic and rate equations. Many of these derivations are preceded by or accompanied by less rigorous algebraic or purely verbal discussions so that the meaning of the derived equations can be understood without calculus. Our experience has been that most students can quickly learn enough calculus to follow derivations, but that more skill and experience are required to use the same operations in solving problems. Thus we feel that formal training in calculus is not necessary for use and understanding of this book. For those students who want to learn quickly some highly selected calculus or who want to review pertinent material from a formal course, we have included such material in Appendix I.

We have included discussion of such topics as fundamental constants, units (including SI units), and dimensions in Chapter 1, rather than in appendices. Our hope is that students will see these useful tables and make use of them throughout their study of chemistry.

We have settled on the order of presentation following Chapter 1 for several good reasons, but not with the expectation that our order is "best." First, we have tried to progress (with lots of ups and downs) from material that is partly familiar to material that is largely unfamiliar to most students. Second, we have also tried to progress (again with lots of ups and downs) from material that students will find reasonably easy to material that most students will find difficult. Finally, we have tried to organize and present our material so that teachers and students can choose their own order within wide limits.

We have resisted the temptation to write a chapter on "chemistry and society" or "chemistry and the environment." Instead, we have attempted to integrate discussions of such topics as corrosion, water and air pollution, heavy metal poisoning, and energy sources with our discussions of chemical principles.

Many people have helped us in many ways and it is a pleasure to thank them all. First, we express our gratitude to our colleagues in the Departments of Chemistry of the University of Lethbridge and Colby College who have patiently answered our questions and who have provided the favorable atmosphere in which we wrote this book. We also thank all of the following for the suggestions and helpful comments that they have provided at various times: Stan Angrist, Gordon Atkinson, John Bauman, Jerry Bell, Gary Bertrand, Al Caretto, Al Colter, Frank Ellison, Paul Fugassi, Bob Graham, Truman Kohman, Bob Kurland, Doug Maier, Gil Mains, R. A. Matheson, Bill O'Hara, Evans Reid, Paul Wenaas, Earl Woolley, and Claus Wulff. L. G. H. also thanks the faculty of the Department of Chemistry, University of Otago, Dunedin, New Zealand, for their hospitality while part of this book was written.

L. G. H.
W. L. S.

CONTENTS

1

INTRODUCTION

Language of Chemistry

A common definition of chemistry a few years ago was "the science that treats the composition of substances and the transformations they undergo." Various chemists and students have semifacetiously defined chemistry as "what chemists do." But these statements and all other concise definitions are too restrictive. Many nonchemists do some chemistry in the course of their work and also make use of chemistry done by chemists. Further, many chemists do work that fits customary definitions of other disciplines—and all the while happily think of themselves as chemists doing chemistry.

In addition to the traditional subdivisions within chemistry, we now have biochemistry, geochemistry, nuclear chemistry, electrochemistry, oceanographical chemistry, and many other adjectives and adjectival prefixes that modify chemistry. There are also such thriving disciplines as chemical physics and chemical metallurgy. We must conclude that modern chemistry is inextricably mixed with other sciences and that no particular discipline can be studied or practiced without some regard for other fields of science. This mixing of traditional disciplines may be illustrated by considering briefly some areas in which chemists now work.

Many chemists are currently concerned with fuel cells that permit the direct conversion of chemical energy to electrical energy, but mechanical engineers do similar work in this field. Other chemists are much concerned with the thermal, magnetic, and electrical properties of solids—a field that also attracts the interests of physicists, metallurgists, geologists, and electrical engineers. Still other chemists and biologists share an interest in proteins and other molecules that make up living matter.

We must conclude that it is impossible to define chemistry so that the definition would include all of modern chemistry and exclude all other fields. Nevertheless, we do have a good idea as to what constitutes the central area of modern chemistry and are now able to see many of the relationships between chemistry and other fields of science and engineering. Therefore we can proceed with an introduction to the scientific principles of modern chemistry and with selected applications of these principles to other disciplines.

The language of chemistry (and other sciences) is partly the same as the language of ordinary conversation. But the language of science is different in that we make considerable use of symbols and abbreviations with specific meanings. We also use numbers, and units of some sort are associated with nearly all of these numbers. An important part of the language of science is mathematics.

The language of chemistry contains such terms as Ca, Cl_2, and H_2O. Sometimes these letters are used as abbreviations for the substances calcium, chlorine, and water. More commonly, however, these letters are used as symbols to represent particular quantities of these substances. For instance, these letters might represent an atom of calcium, a molecule of chlorine consisting of two atoms bound together, or a molecule of water consisting of two atoms of hydrogen and one of oxygen.

Another important use of these letters is as symbols to represent a particular large number of atoms of calcium or molecules of chlorine and water. The counting unit commonly used in chemistry is the **mole,** which consists of 6×10^{23} atoms or molecules and is of sufficient importance to warrant considerable attention later in this book.

Water and many other substances exist in different physical states, depending on the temperature and pressure. For example, we may have solid, liquid, or gaseous water. Chemists commonly write $H_2O(c)$, $H_2O(liq)$, and $H_2O(g)$ to represent crystalline water (ice), liquid water, and gaseous water (steam or water vapor). Similarly, Ca(c), $CH_3OH(liq)$, and $Cl_2(g)$ represent solid (crystalline) calcium, liquid methyl alcohol (wood alcohol), and gaseous chlorine. Some substances exist in more than one solid form. For instance, carbon can be graphite, diamond, or in an amorphous noncrystalline form. We write C(gr), C(diamond), and C(amorph) to represent carbon in these forms. Similarly, we write S(rh) and S(mono) to represent solid sulfur in the rhombic and monoclinic forms.

Many substances dissolve in water to form solutions that are discussed in detail later in this book. To indicate that a substance is dissolved in water (in aqueous solution), we write (aq) after the symbol for the substance. Thus $CH_3OH(aq)$ represents methyl alcohol dissolved in water. We might also write NaCl(aq) to represent ordinary table salt dissolved in water, but usually we write $Na^+(aq)$ + $Cl^-(aq)$ to indicate that aqueous sodium chloride is dissociated into charged particles called ions. In general, we write chemical symbols, formulas, and reaction equations as realistically as possible and in a fashion that conveys as much information as possible.

In writing chemical equations to represent chemical reactions, we use the symbols $\rightarrow$ and $\rightleftharpoons$. For example, we might write

$$NH_3(g) + HCl(g) \rightarrow NH_4Cl(c)$$

for the reaction of gaseous ammonia with gaseous hydrogen chloride to form crystalline ammonium chloride. Or we might write

$$NH_4Cl(c) \rightarrow NH_3(g) + HCl(g)$$

when we are concerned with the reaction that occurs on heating solid ammonium chloride. Finally, we might also write

$$NH_4Cl(c) \rightleftharpoons NH_3(g) + HCl(g)$$

when we are concerned with both the forward and reverse reactions, as in the study of chemical equilibria.

Physical processes are also usefully represented by equations. We write

$$H_2O(c) \rightarrow H_2O(liq)$$

for the melting of ice, but we write

$$H_2O(c) \rightleftharpoons H_2O(liq)$$

when we are concerned with both the melting of ice and the freezing of liquid water or when we are concerned with ice and water in equilibrium as at 0°C and one atmosphere (atm) pressure.

Mathematics of Chemistry

Numerical calculations are an important part of chemistry. These calculations involve the use of logarithms, determination of square roots and higher roots, arithmetical operations with both very large and very small numbers, and the solution of algebraic equations. All of these subjects are discussed later in this section and illustrated with worked examples.

Algebraic rules for handling exponentials are the following:

1. To multiply exponentials having the same base, add exponents.
2. To divide exponentials having the same base, subtract the exponent of the divisor from the exponent of the dividend.
3. To raise an exponential to a power, multiply the exponent by the power.
4. To obtain the nth root of an exponential, divide the exponent by n.
5. An exponential x^n equals $1/x^{-n}$; conversely, $x^{-n} = 1/x^n$.

The uses of these rules are illustrated in the following examples.

Example Problem 1.1. Evaluate the following expression:

$$\frac{(3^{1.5})^2(3^2)(x^{14})(x^{-2})}{(3x^5)^2(x^2)}$$

Application of rule 1 gives $(x^{14})(x^{-2}) = (x^{12})$. Rule 3 gives us $(3^{1.5})^2 = (3^3)$ and $(3x^5)^2 = (9x^{10})$. Substitution of these results in the expression above gives

$$\frac{(3^3)(3^2)(x^{12})}{(9x^{10})(x^2)}$$

Now application of rule 1 gives (3^5x^{12}) for the numerator and $9x^{12}$ or 3^2x^{12} for the denominator of this expression. Rule 3 then leads to $(3^3) = 27$ for the desired answer. ■†

Example Problem 1.2. Evaluate the following expression:

$$\frac{10^{3.167} \times (10^{2.664})^{0.5}}{(10^{1.080})^{1.5}}$$

We reduce the above expression to

$$\frac{10^{3.167} \times 10^{1.332}}{10^{1.620}}$$

and then with rules 1 and 2 obtain

$$10^{2.879}$$

Handy tables (called logarithm tables) permit us to convert the exponential to the more convenient number 757 as illustrated later. ■

Before proceeding with detailed discussion of the use of logarithms, we consider the use of exponentials as a convenient means of expressing both very large and very small numbers. The number 5918 may be expressed as 5.918×1000 or as 5.918×10^3. Similarly, 602,000,000,000,000,000,000,000 is conveniently expressed as 6.02×10^{23}. The small number 0.000453 may be expressed as 4.53/10,000, as $4.53/10^4$, or as 4.53×10^{-4}. Arithmetical operations with large and small numbers expressed as exponentials are illustrated in Example Problems 1.3 and 1.4.

Example Problem 1.3. Evaluate the expression

$$\frac{(3.8 \times 10^{21})(2.6 \times 10^{-18})}{(4.9 \times 10^5)^{0.5}}$$

The numerator of this expression is found to be 9.88×10^3. The denominator could be given as $\sqrt{4.9} \times 10^{2.5}$, but in general it is more convenient to work with integral exponents. We therefore express 4.9×10^5 as 49×10^4 (the even exponent yields an integer when multiplied by 0.5 or divided by 2) and then obtain 7.0×10^2 for the value of the denominator of the above expression. Combination of our results so far yields

$$\frac{9.88 \times 10^3}{7.0 \times 10^2}$$

and thence $1.4 \times 10^1 = 14$ for the desired answer. ■

Example Problem 1.4. Evaluate the following expression:

$$\frac{[(3.12 \times 10^{-4}) - (5.6 \times 10^{-5})](1.27 \times 10^3)}{(4.22 \times 10^{-2})^2(2.50 \times 10^{-3})^{0.5}}$$

We first express 5.6×10^{-5} as 0.56×10^{-4}, which is then subtracted from 3.12×10^{-4} to yield 2.56×10^{-4}. Multiplying 2.56×10^{-4} by 1.27×10^3 gives 3.25×10^{-1} for the numerator of the expression above.

†The symbol ■ signals the end of example problem.

Squaring 4.22×10^{-2} gives 17.8×10^{-4} or 1.78×10^{-3}. We express 2.50×10^{-3} as 25.0×10^{-4}, and then take the square root to obtain 5.0×10^{-2}, which is multiplied by 1.78×10^{-3} to give 8.90×10^{-5} for the denominator.

Dividing the numerator by the denominator gives $0.365 \times 10^4 = 3650$ for the desired answer. ■

Multiplication, division, determination of roots, and raising numbers to specified powers are arithmetical operations that are conveniently carried out with the aid of logarithms. The common logarithm of a number is the power to which 10 must be raised to equal the number under consideration. Thus the log of 1 is 0, the log of 10 is 1, the log of 100 is 2 and (by rule 5) the log of 0.01 is −2. In general, the log of 10^n is n.

Logarithms of numbers that are not integral powers of 10 are ordinarily obtained from log tables or a slide rule. Since tables of logarithms list the exponents to which 10 must be raised to yield numbers between 1 and 10, we find it convenient to express all numbers as numbers between 1 and 10 times the appropriate 10^n. Thus we write 462 as 4.62×10^2 and write 0.0000387 as 3.87×10^{-5}. The log of 4.62 is listed in log tables as 0.6646 and the log of 10^2 is 2. According to rule 1 for handling exponentials, we add 0.6646 to 2 to obtain 2.6646 as the desired log of 462. Similarly, the log of 3.87 is 0.5877 and the log of 10^{-5} is −5. Adding 0.5877 to −5 gives −4.4123 as the desired log of 3.87×10^{-5}.

An antilog is the number to which a logarithm corresponds. For example, the antilog of 2 is 100. To find the antilog of 2.6646, we see in a log table that 4.62 is the antilog of 0.6646 and already know that 100 is the antilog of 2. Since $2.6646 = 2 + 0.6646$, we multiply 4.62×100 to obtain 462 as the desired antilog of 2.6646.

To find the antilog of a negative log, we rewrite the negative log in the form of a positive number between 0 and 1 minus the appropriate integer. Thus we write −4.4123 as $0.5877 - 5$. The antilog of 0.5877 is 3.87 and of −5 is 10^{-5}, so the antilog of −4.4123 is 3.87×10^{-5}.

Example Problem 1.5. Use logarithms to evaluate the following expression:

$$\left[\frac{(5.12 \times 10^{-2}) \times (6.42 \times 10^{4})}{3.62 \times 10^{-3}}\right]^{\frac{3}{5}}$$

We proceed as follows, adding and subtracting logarithms (exponents of 10) to multiply and divide, and multiplying a log by an exponent to determine a root.

$$\begin{aligned}
\log(5.12 \times 10^{-2}) &= 0.7093 - 2 = & -1.2907 \\
\log(6.42 \times 10^{4}) &= & \underline{+4.8075} \\
&\text{Sum} & +3.5168 \\
\log(3.62 \times 10^{-3}) &= 0.5587 - 3 = & \underline{-2.4413} \\
&\text{Difference} & +5.9581 \\
\tfrac{3}{5} \times 5.9581 &= 0.6 \times 5.9581 = 3.57486 &
\end{aligned}$$

The antilog of 3.5749 is 3.76×10^3, and we have the desired answer. ■

Most calculations with logarithms are those involving ordinary or base 10 logarithms in which the log of a number is the power to which 10 must be raised to give that number. It is possible, however, to set up a system of logarithms based on any number. For a variety of purposes, so-called natural logarithms to the base $e = 2.71828\ldots$ are of fundamental importance. These natural logarithms are usually represented by **ln** rather than the **log** commonly used for ordinary logarithms based on 10. The relationship between natural and common logarithms is

$$\ln x = 2.303 \log x$$

Example Problem 1.6. Find the natural logarithm of 3.68×10^{-6}.

An ordinary log table shows that $\log 3.68 = 0.5658$. Thus we find that $\log (3.68 \times 10^{-6}) = 0.5658 - 6 = -5.4342$. Multiplying this last value by 2.303 gives us $\ln (3.68 \times 10^{-6}) = -12.5150$ as the desired natural logarithm. ■

Example Problem 1.7. We have $\ln x = -5.4302$. What is the value of x?

First, we divide -5.4302 by 2.303 to obtain $\log x = -2.3579$, which can also be written as $\log x = 0.6421 - 3$. The antilog of 0.6421 is 4.386 and the desired value of x is 4.386×10^{-3}. ■

We commonly apply mathematics to deducing a general equation from appropriate specific information. For example, given the information that the temperature at which ice melts or water freezes (ice and water in equilibrium) is 0°C or 32°F and that the normal boiling point (temperature at which the vapor pressure equals one standard atmosphere) is 100°C or 212°F, we can deduce a general relationship between temperatures on the Fahrenheit and Celsius (commonly called centigrade) scales.

Finding a desired general relationship from specific data is often facilitated by making an appropriate drawing, as in Figure 1.1. We see from the figure that

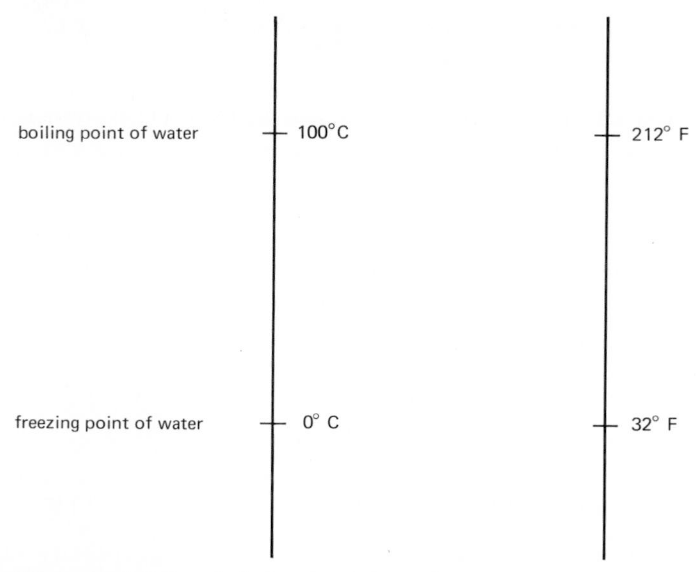

Figure 1.1. Celsius and Fahrenheit temperature scales.

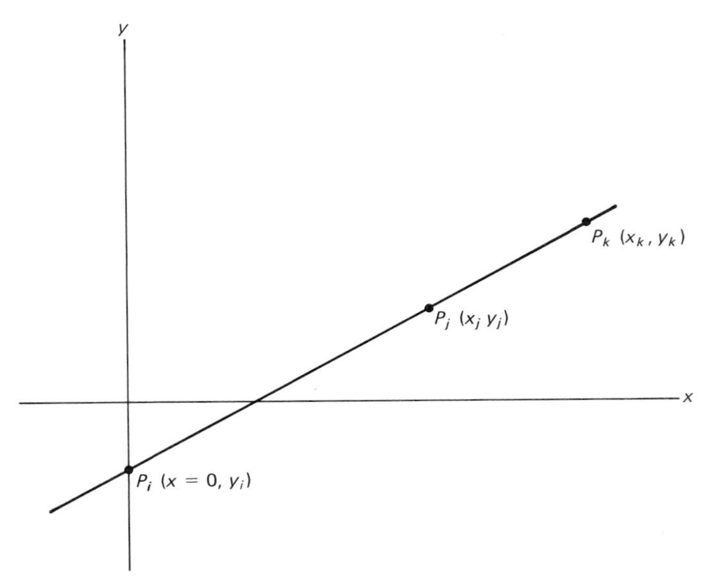

Figure 1.2. Graph of a straight line of slope m and intercept b, as given by the equation $y = mx + b$. The slope m of the line is given by $(y_k - y_j)/(x_k - x_j)$ and the intercept on the y axis has value $b = y_i$.

the temperature interval between the freezing and boiling points of water is 100°C or 212° − 32° = 180°F and conclude that the Fahrenheit degree is $\frac{100}{180} = \frac{5}{9}$ as large as the Celsius degree.

Taking account of the different zero points on the two scales, we write

$$\mathrm{F} = \tfrac{9}{5}\mathrm{C} + 32°$$

or

$$\mathrm{C} = \tfrac{5}{9}(\mathrm{F} - 32°).$$

The validity of either of these equations may be confirmed by inserting a selected value of the temperature on one scale and determining whether the correct temperature on the other scale is calculated. For example, setting the Celsius temperature equal to 100°, we calculate

$$\mathrm{F} = \tfrac{9}{5}100° + 32° = 212°$$

to verify the first equation above.

Equations of the form

$$y = mx + b \tag{1.1}$$

represent a straight line of slope m and intercept b, as illustrated in Figure 1.2. In Figure 1.3 we have a graph showing the relationship between the Celsius and Fahrenheit temperature scales.

Equations of the form

$$ax^2 + bx + c = 0 \tag{1.2}$$

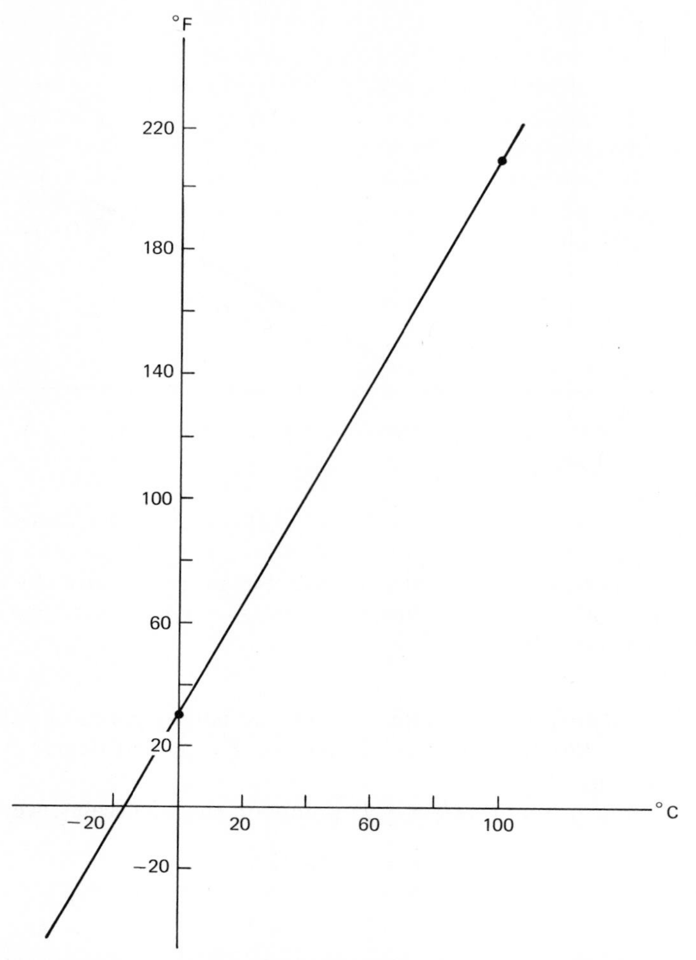

Figure 1.3. A graph showing the linear relationship between temperatures on the Celsius and Fahrenheit scales.

The slope is $(212° - 32°)/(100° - 0°) = \frac{9}{5}$ and the intercept on the vertical axis is 32°. Thus we see from $y = mx + b$ that an equation for the straight line is $F = \frac{9}{5}C + 32°$, as previously given.

are called quadratic equations and can be solved by means of

$$x = \frac{-b \pm (b^2 - 4ac)^{\frac{1}{2}}}{2a} \tag{1.3}$$

When quadratic equations occur in some form different from the standard form given above, algebraic rearrangement is required before the standard solution equation can be applied. Generally only one of the two possible values of x is physically reasonable for a given problem. In chemistry the nature of the problem or the system under consideration often suggests some approximation that converts a quadratic or even higher order equation to one that is easily solved for the unknown by algebraic rearrangement.

Calculus is first used in this book in Chapter 11. The mathematical operations involved are simple and easily learned from the brief discussion that appears in Appendix I and examples later in the book. With most of our calculus derivations there are parallel verbal descriptions so that the meaning of the derived equations should be readily grasped, even by those without formal training in calculus.

It is frequently desired to express a set of experimental data by means of an equation. For example, thermodynamic considerations described in Chapter 13 lead us to an equation of the form

$$\log P = \frac{A}{T} + B \tag{1.4}$$

for expressing vapor pressure P as a function of temperature T. Even without the theoretical principles leading to the convenient equation (1.4), we could still express P as a function of T by using series expressions such as

$$\log P = a + bT + cT^2 + dT^3 + \cdots$$

In addition to their usefulness as a means of summarizing information by means of an equation, series are also useful for various approximate or limiting calculations. Some important series expressions are listed in Table 1.1 and discussed below.

We shall find particular use for the series in Table 1.1 in obtaining approximations to e^x and $\ln(1 + x)$ when x is very small. In such instances we need consider only the leading terms in the series (because terms involving x^2, x^3, and higher powers of x are negligibly small) and can write

$$e^x \cong 1 + x$$

and

$$\ln(1 + x) \cong x$$

as excellent approximations. Similarly, for x near 1, we can write

$$\ln x \cong x - 1$$

as an excellent approximation.

Table 1.1. Infinite Series

$e^x = 1 + x + \frac{x^2}{2!} + \frac{x^3}{3!} + \cdots$	
$\ln(1 + x) = x - \frac{x^2}{2} + \frac{x^3}{3} - \frac{x^4}{4} + \cdots$	$-1 < x \leqq 1$
$\ln x = (x - 1) - \frac{(x-1)^2}{2} + \frac{(x-1)^3}{3} - \cdots$	$0 < x \leqq 2$

Dimensions, Units, and Fundamental Constants

In most of the calculations and equations of chemistry, the numbers we use have units associated with them and represent definite chemical or physical quantities. Thus when we mean to express a length or a mass, we write 10 centimeters (cm) or 5 grams (g) rather than merely 10 or 5. Units (such as g, cm, cal, etc.) have been included throughout most of the calculations illustrated in this book. In other calculations, insertion of units has been left as an exercise for the reader.

It is both possible and useful to express most dimensions as appropriate combinations of a few selected dimensions, which might be called fundamental dimensions. The choice of particular dimensions as "fundamental" is largely arbitrary and may be made on the basis of convenience, simplicity, or what seems most "reasonable." Choices of fundamental dimensions that are convenient and commonly used in much scientific work are given in following tables.

Dimensions of mechanical quantities are usefully expressed in terms of mass, length, and time, which are represented by M, L, and T in Table 1.2.

Table 1.2. Dimensions of Mechanical Quantities

Acceleration	LT^{-2}
Density	ML^{-3}
Energy and Work	ML^2T^{-2}
Force	MLT^{-2}
Frequency	T^{-1}
Momentum (angular)	ML^2T^{-1}
Momentum (linear)	MLT^{-1}
Power	ML^2T^{-3}
Pressure	$ML^{-1}T^{-2}$
Velocity (angular)	T^{-1}
Velocity (linear)	LT^{-1}

Dimensions of electrical quantities (in the esu system) can also be expressed in terms of mass, length, and time as shown in Table 1.3.

Along with dimensions, we must have units. For example, the quantity of length (a fundamental dimension) may be expressed in terms of units such as inches, feet, miles, light years, millimeters, centimeters, meters, kilometers, and so on. We begin by presenting mostly "traditional" units and then consider SI units at the end of this chapter.

The **dyne** and the **erg** are the fundamental units of force and energy in the **cgs** (centimeter, gram, second) system. Since the dimensions of force are MLT^{-2}, the units describing the dyne are g cm sec^{-2} in the **cgs** system. Similarly, the dimensions of energy are ML^2T^{-2} and the erg (one dyne cm) is expressed in terms of g cm^2 sec^{-2}. The **joule** which is equal to 10^7 ergs, is another commonly used unit of energy. Energy in terms of joules is often determined electrically as (volts)^2 (ohms)$^{-1}$ (seconds) or (amperes)^2 (ohms) (seconds). A rate of energy transfer of

Table 1.3. Dimensions of Electrical Quantities (esu)

Capacity	L
Charge (quantity of electricity)	$M^{\frac{1}{2}}L^{\frac{3}{2}}T^{-1}$
Current	$M^{\frac{1}{2}}L^{\frac{3}{2}}T^{-2}$
Electrical potential	$M^{\frac{1}{2}}L^{\frac{1}{2}}T^{-1}$
Resistance	$L^{-1}T$

Table 1.4. Relationships Between Energy Units

1 joule (J) = 1 watt-second = 10^7 ergs
1 calorie (cal) = 4.1840 J = 3.966×10^{-3} British thermal units (Btu)
1 kilocalorie (kcal) = 1000 cal = 3.966 Btu
1 liter atmosphere = 101.33 J = 24.218 cal
1 kilowatt-hour = 3412 Btu = 860.4 kcal
1 electron-volt = 1.602×10^{-12} erg
1 electron-volt molecule^{-1} = 23.062 kcal mole^{-1}

one joule per second corresponds to the important unit of power called the **watt,** which is also equal to one volt-ampere. Therefore one watt-second (W-sec) equals one J of energy. Other units of energy that are commonly used are the kilowatt hour and the British thermal unit (Btu).

It may be verified by combining the dimensions in Table 1.2 that any product of pressure times volume is an energy. Energies associated with the work of expansion or compression are commonly expressed in terms of liter atmospheres or other PV products.

Energies involved in experiments with small numbers of atoms or molecules are often expressed in terms of electron-volts; one electron-volt is the energy an electron gains when it is accelerated by a potential difference of one volt.

Relationships between the various energy units are summarized in Table 1.4.

Relationships between common units of length, mass, volume, and pressure are summarized in Table 1.5.

Table 1.5. Conversion Factors for Length, Mass, Volume, and Pressure

Length:	1 m	= 100 cm = 1000 mm
	1 cm	= 10^8 Angstroms (Å)
	1 inch	= 2.540 cm
Mass:	1 kg	= 1000 g = 2.205 lb
Volume:	1 liter	= 1000 ml = 1000.028 cm^3 = 0.2642 gal
	1 gal	= 3.785 liters
Pressure:	1 atm	= 760 mm Hg = 1.013 bars
	1 bar	= 10^6 dyne cm^{-2}
	1 mm Hg	= 1 torr

Values of the fundamental constants that are important in chemistry are listed in Table 1.6. Some of these constants (especially R, the gas constant) are given in several systems of units and are useful as conversion factors as illustrated in Example Problem 1.8.

Example Problem 1.8. The energy change associated with a certain change of volume at constant pressure is calculated from $w = P\Delta V$ as 9.00 liter atm. Use values of R from Table 1.6 to convert this energy to calories.

We write

$$9.00 \text{ liter atm} \times \frac{1.987 \text{ cal deg}^{-1} \text{ mole}^{-1}}{0.082 \text{ liter atm deg}^{-1} \text{ mole}^{-1}} = 218 \text{ cal} \quad \blacksquare$$

Table 1.6. Fundamental Constants

c	Speed of light	2.997925×10^{10} cm sec^{-1}
h	Planck's constant	6.6256×10^{-27} erg sec
e	Electronic charge	4.80298×10^{-10} esu
		1.60210×10^{-19} coulomb
m_e	Rest mass of electron	9.1091×10^{-28} g
m_p	Rest mass of proton	1.6725×10^{-24} g
N	Avogadro's number	6.02252×10^{23} molecules mole^{-1}
k	Boltzmann's constant	1.38054×10^{-16} erg deg^{-1} molecule^{-1}
R	Gas constant	1.9872 cal deg^{-1} mole^{-1}
		8.3143 joule deg^{-1} mole^{-1}
		82.055 cm^3 atm deg^{-1} mole^{-1}
		0.082055 liter atm deg^{-1} mole^{-1}
$\mathfrak{F}$	Faraday constant	96,487 coulombs (mole of electrons)$^{-1}$
		23,060 cal volt^{-1} (mole of electrons)$^{-1}$
g	Standard gravitational constant	980.655 cm sec^{-2}
	Ice point (0°C)	273.15°K
	Triple point of H_2O	273.16°K

Physics

As indicated in the first section of this chapter, the borders between chemistry and other sciences are vague. They depend on the point of view and also shift with time as scientific interests and technological applications change. It is therefore impossible to develop an understanding of chemistry without making use of physics. In this section we summarize by means of equations some of the results of physics that are of particular use in the development of the principles of chemistry.

Speed is the distance traveled per unit time without specifying direction, and velocity is the distance traveled per unit time in a given direction. In the specific case of constant speed or velocity, either may be calculated as a ratio of distance to time as in

$$v = \frac{d}{t} \tag{1.5}$$

Both speed and velocity have dimensions of length divided by time and are commonly expressed in terms of centimeters per second (cm sec^{-1}) in scientific work.

The rate of change of velocity is the acceleration, which has dimensions of length divided by time squared, and in the case of constant acceleration can be calculated as

$$a = \frac{v_2 - v_1}{t_2 - t_1} \tag{1.6}$$

Mechanical force may be calculated as the product of mass times acceleration as in

$$f = ma \tag{1.7}$$

Force therefore has dimensions of mass times length divided by time squared. With mass in grams, length in centimeters, and time in seconds (cgs system), the unit of force is the dyne. The force due to gravity is given by the mass of the object times the gravitational constant g:

$$f = mg \tag{1.8}$$

The gravitational constant therefore has dimensions of length divided by time squared.

When action of a constant force on an object results in a linear displacement of the object over a certain distance, the work done is given by

$$w = fd \tag{1.9}$$

Work therefore has dimensions of mass times length squared divided by time squared. The dimensions and units of work are the same as those of energy.

Pressure P is defined as force per unit area A:

$$P = \frac{f}{A} \tag{1.10}$$

Multiplying both sides of $f = PA$ by distance d and recognizing that $fd = w$ and that area times distance represents a change in volume gives us

$$w = P(V_2 - V_1) \tag{1.11}$$

where V_2 and V_1 represent final and initial volumes. This equation is also concisely written as

$$w = P\Delta V \tag{1.12}$$

in which we use ΔV to represent the change in volume given by $V_2 - V_1$. This same result might also have been obtained by recognizing that the dimensions of pressure are mass divided by length and time squared ($ML^{-1}T^{-2}$) and that multiplying pressure times volume (length cubed, L^3) yields a quantity with dimensions of work (ML^2T^{-2}).

Since energy is a measure of the ability to do work, the dimensions of energy and work are the same. The energy an object has because of motion of its mass is called kinetic energy (KE) and is calculated as

$$\text{KE} = \tfrac{1}{2}mv^2 \tag{1.13}$$

The reader should verify from this equation that the dimensions of work and energy are indeed the same.

Potential energy is a consequence of the position of an object and depends on the location of the object with respect to other relevant objects. For example, the potential energy of an airplane or of a satellite depends on its distance from the earth. Potential energy of importance in chemistry arises from the attraction and repulsion of electrical charges.

The force exerted on charged bodies separated by a distance d and having charges q_1 and q_2 is given by Coulomb's law as

$$f = \frac{q_1 q_2}{\varepsilon d^2} \tag{1.14}$$

in which ε represents the dielectric constant of the medium between the charged bodies. If the charges on the bodies are of opposite sign, the force is attractive and f is negative. Common units of charge are the electrostatic unit (esu), the charge of one electron, and the coulomb.

A group of moving electric charges is called an electric current. Flow of one coulomb of charge per second is one ampere, the most common unit of current. A constant current i may be defined by the equation

$$i = \frac{q}{t} \tag{1.15}$$

and the total charge transferred by a constant current in time t is given by

$$q = it \tag{1.16}$$

The electric current carried by a conductor is given by Ohm's law:

$$i = \frac{V}{R} \tag{1.17}$$

in which V and R represent electrical potential and resistance (usually expressed in volts and ohms).

Electrical power may be calculated as Vi, i^2R, or V^2/R. Energy is power times time, so electrical energy is calculated as Vit, i^2Rt, or V^2t/R. Since $it = q$, electrical energy is also given by Vq.

A simple capacitor (condenser) consists of two parallel metallic plates separated a certain distance and is a device for storing electrical energy. The charge q that can be stored on a capacitor is proportional to the potential difference V between the plates as indicated by

$$q = CV \tag{1.18}$$

where the proportionality constant C is called the capacitance. The ratio of capacitance C of a given capacitor with some material of interest filling the space between the plates to the capacitance C_0 of that same capacitor with the space between the plates evacuated is the dielectric constant ε of the material of interest, as shown by

$$\varepsilon = \frac{C}{C_0} \tag{1.19}$$

SI Units and Symbols

There have been many attempts to introduce or impose some sort of order and uniformity into scientific units and symbols. In the past, all such attempts have foundered on the diversities of science and scientists. But as a result of extensive organizational work and international cooperation, we are slowly approaching a consistent set of units and symbols to be used (with some variations) by all scientists in all countries. This International System of Units (SI) was first defined and given some official status by the General Conference on Weights and Measures in 1960.

The SI units and symbols represent an excellent compromise between the needs and uses of many different kinds of scientists, engineers, and others who use the language of science. Unfortunately, some of the SI units and symbols are less convenient for use in connection with chemistry than older "traditional" units and symbols. Further, most readers of this book are likely to be more familiar with certain traditional units and symbols (for example, calories and °K) than with corresponding SI units and symbols (joules and K). Because our principal concern has been to explain principles and applications as clearly as possible, we have chosen to write much of this book with traditional units and symbols so that the readers can concentrate on new ideas rather than on unfamiliar units and symbols. It must, however, be recognized that the SI units and symbols are being more and more widely used and that it is impractical to continue for long without becoming familiar with them. Therefore this section of the book contains a summary of SI units and symbols that readers may use for future reference. Also, at the ends of several chapters we have included problems that involve conversions between traditional and SI units.

The basic SI units and their symbols are summarized in Table 1.7. Certain units that are not part of the SI but are accepted for use with SI units are listed in Table 1.8. The basic SI units listed in Table 1.7 lead to a number of commonly used derived SI units; some of these are listed in Table 1.9.

Table 1.7. Basic SI Units and Symbols

Physical quantity	*Name of unit*	*Symbol*
length	meter	m
mass	kilogram	kg
time	second	s
thermodynamic temperature	kelvin	K
electric current	ampere	A
amount of substance	mole	mol

Because it is impossible to devise any set of units that will be of convenient size for all uses and users, it is common to make use of various prefixes to indicate multiples and fractions. Familiar examples are the kilogram and the milligram. A summary of prefixes used with SI (and other) units is given in Table 1.10.

Students, beginning scientists, and experienced scientists are mostly alike in being unable or unwilling to remember the names, symbols, and definitions of all units and the relationships between these units. Nor is there any need to remember all of this information. It is, however, necessary to know where to find such information and to know how to make use of it. It is likely, therefore, that

Table 1.8. Units Accepted for Use with SI

Unit	*Symbol*
minute (of time)	min
hour	h
liter (defined as $10^{-3}m^3$)	l

Table 1.9. Names and Symbols for Certain Derived SI Units

Physical quantity	*Name of SI unit*	*Symbol*	*Definition of SI unit*
force	newton	N	$kg \cdot m \cdot s^{-2}$
pressure*	pascal	Pa	$kg \cdot m^{-1} \cdot s^{-2} = N \cdot m^{-2}$
energy	joule	J	$kg \cdot m^2 \cdot s^{-2}$
power	watt	W	$kg \cdot m^2 \cdot s^{-3} = J \cdot s^{-1}$
electric charge	coulomb	C	$A \cdot s$
electric potential	volt	V	$kg \cdot m^2\ s^{-3} \cdot A^{-1} = J \cdot A^{-1} \cdot s^{-1}$
electric capacitance	farad	F	$A^2 \cdot s^4 \cdot kg^{-1} \cdot m^{-2} = A \cdot s \cdot V^{-1}$
frequency	hertz	Hz	s^{-1} (cycle per second)

*Pressure is often expressed in SI units in terms of kilonewtons per square meter. The relationship between these SI units and the familiar atmosphere is 1 atm = 101.325 $kn \cdot m^{-2}$

Table 1.10. Prefixes for Fractions and Multiples of SI Units

	Prefix	*Symbol*
Fraction		
10^{-1}	deci	d
10^{-2}	centi	c
10^{-3}	milli	m
10^{-6}	micro	μ
10^{-9}	nano	n
10^{-12}	pico	p
10^{-15}	femto	f
10^{-18}	atto	a
Multiple		
10	deka	da
10^2	hecto	h
10^3	kilo	k
10^6	mega	M
10^9	giga	G
10^{12}	tera	T

readers of this book will often refer to this chapter while reading or using later chapters. In particular, information about SI units is summarized in Tables 1.7–1.9, with information about "traditional" units summarized in Tables 1.4–1.6. Related useful information about dimensions is summarized in Tables 1.2 and 1.3, while a summary of common prefixes to indicate multiples or fractions is given in Table 1.10.

References

Beiser, A.: *Essential Math for the Sciences: Analytic Geometry and Calculus.* McGraw-Hill Book Co. Inc., New York, 1969 (paperback).

Butler, J. A. and D. G. Bobrow: *The Calculus of Chemistry.* W. A. Benjamin, Inc., New York, 1966 (paperback).

Daniels, F.: *Mathematical Preparation for Elementary Physical Chemistry.* McGraw-Hill Book Co. Inc., New York, 1928 (paperback).

Kleppner, D. and N. Ramsay: *Quick Calculus.* John Wiley and Sons, Inc., New York, 1965 (paperback).
Masterton, W. L., and E. J. Slowinski: *Mathematical Preparation for General Chemistry.* W. B. Saunders Co., Philadelphia, 1970 (paperback).
Mellor, J. W.: *Higher Mathematics for Students of Chemistry and Physics,* 4th ed. Dover Publications, Inc., New York, 1955 (paperback).
Paul, M. A.: International system of units (SI). *Chemistry,* **45**:14 (Oct. 1972).

Problems

1. Find the common and natural logarithms of the following:
(a) 3740 (b) 0.000489 (c) 832.1×10^{-6} (d) 6.49×10^{7}

2. Find the antilogs of the following common logarithms:
(a) 2.6037 (b) −8.4195

3. Find the antilogs of the following natural logarithms:
(a) −2.0910 (b) 23.456

4. Use logarithms to find the cube root of 94.25.

5. Evaluate each of the following:
(a) $e^{4.56}$ (b) $e^{-11.3}$ (c) $10^{-6.66}$ (d) $10^{9.004}$

6. Evaluate each of the following expressions:
(a) $4.81 \times 10^{-8} \times (5.11 \times 10^{6})^{4.5}$
(b) $(7.84 \times 10^{-6})^{-0.667}$
(c) $(10^{4.334})/(10^{-0.478})$

7. Use the quadratic formula to evaluate the unknown x in the equation

$$1.8 \times 10^{-5} = \frac{(x)^2}{1.5 - x}$$

8. Suppose that general chemical knowledge tells us that the unknown x in the equation in problem 7 has a very small positive value. Use this qualitative information to simplify the equation so that it may be solved directly for x. Insert the value of x so obtained in the original equation to verify that it is indeed a satisfactory solution.

9. Use the quadratic formula to evaluate the unknown x in the equation

$$2 \times 10^{-9} = (x)(0.99 + x)$$

10. Suppose that general chemical knowledge tells us that the unknown x in the equation in problem 9 has a very small positive value. Use this qualitative information to simplify the equation so that it may be solved directly for x. Insert the value of x so obtained in the original equation to verify that it is indeed a satisfactory solution.

11. Consideration of chemical equilibrium in a certain system leads to the equation

$$2 \times 10^{-14} = (2x)^2(0.20 + x)$$

Using the information that the chemically significant value of x is small and positive, evaluate x and show that the value so obtained satisfies the equation above.

12. Chemical considerations tell us that the unknown x in the following equation is a small positive number:

$$2 \times 10^{-5} = \frac{x^2}{(0.00010 - x)}$$

Because we cannot be sure that x is small enough to make $0.00010 - x \cong 0.00010$ an adequate approximation, we must either use the quadratic formula or some method of approximation followed by a check of the adequacy of the approximation, and then further approximation, until a satisfactory answer is obtained. Use the latter method to determine the small positive value of x that is desired.

13. Consideration of chemical equilibrium data for a system in which several reactions occur leads to the following equations:

$$A = \frac{K_1K_2}{B} + K_1$$

$$A = \frac{K_1K_2}{B} + K_1 + K_1K_3B$$

In these equations K_1, K_2, and K_3 are unknowns that we wish to evaluate and A and B represent experimental data. A graph of A against $1/B$ should yield a straight line whose slope and intercept will permit evaluation of K_1 and K_2. Explain. Show how the second equation can be rearranged so that a complicated function of A, B, K_1, and K_2 can be plotted against B to yield a straight line through the origin with slope equal to K_3.

14. Theoretical work indicates that vapor pressures should vary with temperature in accord with an equation of the form

$$P = Ae^{-B/T}$$

Show that this equation is consistent with a straight line graph when $\log P$ is plotted against $1/T$. Show how to evaluate the important constant B from the graph.

15. Linear momentum is defined as mass times velocity, as in

$$p = mv$$

Derive a relationship between kinetic energy, momentum, and mass.

16. Show by consideration of dimensions that angular momentum can be expressed in terms of multiples of Planck's constant.

17. It was stated in the text that electrical energy can be calculated in terms of Vit, i^2Rt, or V^2t/R. Use Ohm's law to convert from any one of these expressions to the other two. Then show that any one of these expressions has the dimensions of energy.

18. A generating station converts thermal energy to electrical energy with an efficiency of 27%. How much thermal energy, expressed in Btu and kilocalories, is required to produce 1000 kilowatt-hours of electrical energy?

19. Express the following physical constants in SI units: (a) speed of light in vacuum, (b) gas constant, (c) Planck's constant, (d) Faraday constant, (e) gravitational constant.

20. Express:

(a) $J \times Hz$ in terms of A and V

(b) $F \times V$ in terms of C

(c) $V \div N$ in terms of m and C

(d) $J \div (kg \cdot m^2 \cdot s^{-2} \cdot A^{-2})$ in terms of A

21. Calculate the work involved in compressing an ideal gas from a volume of 10.0 to 2.0 liters at a constant external pressure of 4.0 Pa.

22. In SI units calculate the kinetic energy of an electron moving at a velocity of $2.0 \times 10^7\ m \cdot s^{-1}$. Compare this value to that of a 140 g baseball thrown at a velocity of $30\ m \cdot s^{-1}$. The rest mass of an electron is 9.1×10^{-31} kg.

2

ATOMS

Introduction

It is common knowledge today that scientists believe ordinary matter is made up of atoms. Much of modern science consists of the search for new knowledge about atoms or the application of existing knowledge of atoms. Although most chemical interest is centered on the ways that atoms combine with other atoms and on the properties of various combinations of atoms, we shall begin by devoting some attention to individual atoms in order to build a foundation for later study of groups of atoms.

From the time of the ancient Greeks until the seventeenth century, the concept of atoms as the fundamental building blocks of matter was entirely philosophical. This situation began to change when natural philosophers related their thoughts and speculations about the nature of matter to the observable properties of matter and began to do experiments. Results of a great number of experiments carried out after about 1700 provide much indirect evidence in support of the atomic theory. This indirect evidence of the reality of atoms comes from various empirical (means derived from experiment) laws concerning the composition of chemical compounds, from the consistent atomistic interpretation of chemical reactions, from Faraday's laws of electrolysis, and from the kinetic theory of gases.

Direct evidence of the existence of atoms was first obtained shortly before 1900 and soon much new understanding followed. For instance, it gradually became known that atoms are composed of smaller particles. For students of chemistry the most important of these subatomic particles are the electron, the proton, and the neutron.

The Electron

In England about 1830 Michael Faraday carried out many quantitative investigations to determine the behavior of matter when an electric current passes through it. Many years later G. J. Stoney concluded from the results of Faraday's experiments that electricity probably exists in discrete units, and in 1891 he suggested the name **electron** for the postulated fundamental unit of electricity. About this same time J. J. Thomson at Cambridge University carried out a series of experiments that provided direct evidence of electrons and yielded information about some of their properties.

In the latter half of the nineteenth century many physicists carried out experiments on the conduction of electricity through gases. A schematic diagram of a typical discharge tube used for such experiments is shown in Figure 2.1. The apparatus consists of a glass tube fitted with electrodes and openings through which gases can be admitted or removed. Gases at ordinary pressure are poor conductors of electricity, and a very high voltage is required to initiate current flow. As some of the gas is pumped out, thereby reducing the pressure in the tube, the conductivity is increased and light is emitted. If neon is the gas in the tube, one has a neon light.

At still lower pressures the tiny amount of gas remaining in the tube ceases to emit light, but the glass of the tube glows or fluoresces. Both William Crookes and Jean Perrin showed that the glow was caused by bombardment of the glass by high speed "rays" emitted by the cathode and called cathode rays.

Thomson then carried out experiments with the apparatus illustrated in Figure 2.2, which permitted investigation of the deflection of cathode rays by electric and magnetic fields. He concluded from the results of these experiments that the cathode rays consisted of negatively charged particles with mass much less than that of any atom. Although earlier work had been done with cathode rays and the measurements of Faraday had led Stoney to postulate the existence of electrons, the detailed investigations by Thomson provided the first convincing experimental evidence for negative particles that are much lighter than atoms. Thus Thomson is generally given credit for discovery of the electron.

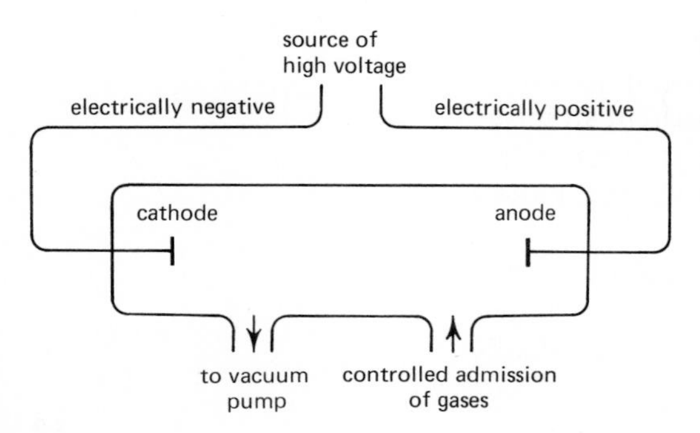

Figure 2.1. Schematic illustration of a simple discharge tube that was used for investigation of conduction of electricity through gases.

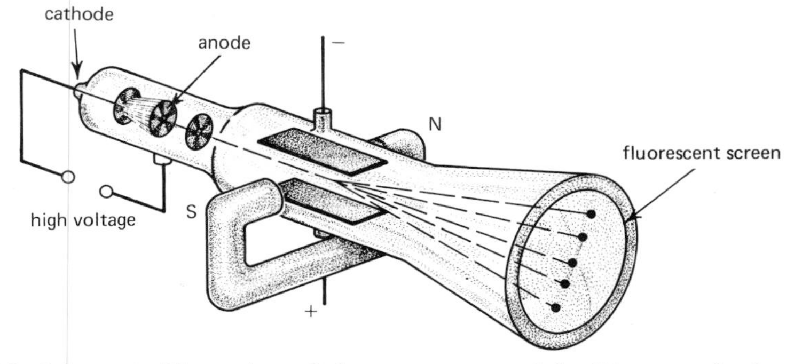

Figure 2.2. Schematic illustration of the apparatus used by Thomson for investigation of the ratio of electric charge to mass of the electron by means of deflection of cathode rays in electric and magnetic fields.

Some of the electrons emitted by the cathode pass through the hole in the anode and then through the hole in the plate to the right of the anode. In the absence of external fields, these electrons continue in straight lines until they hit the fluorescent screen. The direction of displacement of the beam of electrons when subjected to external fields shows that the electrons are negatively charged. Quantitative study of these deflections of the beam permits calculation of the ratio of electric charge to mass of the electron.

A few years after discovery of the electron by Thomson, R. A. Millikan, an American physicist, began experiments that led to the first direct and accurate evaluation of the charge of the electron. A simplified version of the apparatus that he used is illustrated in Figure 2.3. Millikan obtained small charged drops of oil by spraying oil into the air space above two parallel condenser plates. Some

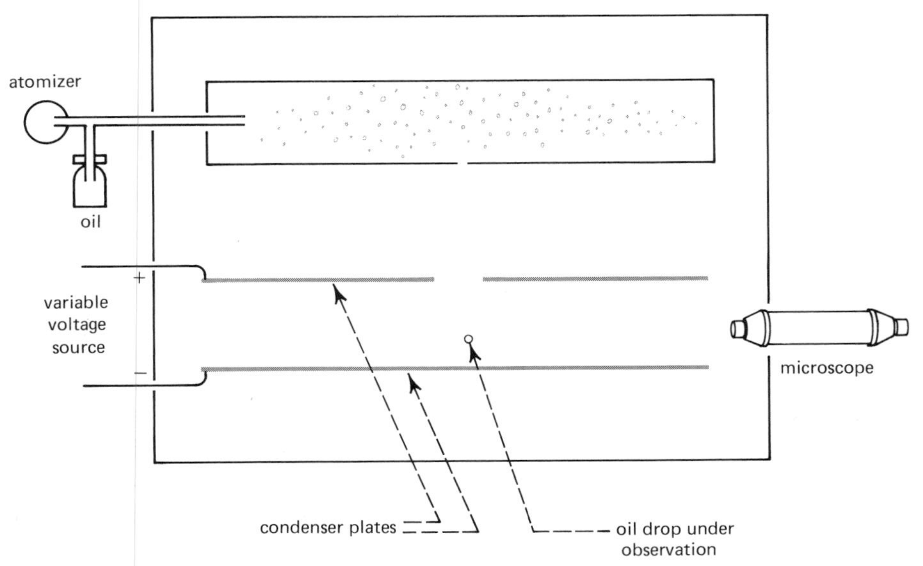

Figure 2.3. Simplified schematic diagram of the oil-drop apparatus used by Millikan in determining the charge of the electron. The oil drops become charged either as a result of friction or by contact with air that has been ionized by x rays.

of the droplets fell through the hole in the upper plate and were then observed individually through the microscope. The rate at which the oil drop falls in the earth's gravitational field permits accurate calculation of the mass of the drop. Then the condenser plates are connected to the voltage source, and the rate of fall or rise of the same drop in the resulting electric field is observed. This rate and the previously calculated mass of the drop permit calculation of the electric charge on the drop.

Millikan reasoned that the charge on each oil drop must result from attachment of an integral number of electrons to it. The smallest charge that he ever found on a drop was 1.6×10^{-19} coulomb, and he always found the charge to be some integral multiple of 1.6×10^{-19} coulomb. Millikan therefore identified this amount of charge with the charge of a single electron.

Thomson and others were investigating the charge/mass ratio of the electron even before Millikan's determination of the charge of the electron. Combination of these results permitted calculation of the mass of the electron. A value for the charge/mass ratio of the electron as determined by the Thomson method is 1.76×10^8 coulombs per gram. From this value and 1.6×10^{-19} coulombs for the charge of an electron, the rest mass of the electron is calculated as

$$m_e = \frac{1.6 \times 10^{-19} \text{ coulomb}}{1.76 \times 10^8 \text{ coulomb g}^{-1}} = 9.1 \times 10^{-28} \text{ g}$$

Since it had been known for some time from chemical evidence that the mass of a hydrogen atom is about 1.67×10^{-24} g, it was clear that the lightest atom was about 1840 times as heavy as the electron.

The Proton

Discovery of the electron naturally led to attempts to identify the fundamental positively charged particle corresponding, at least in magnitude of charge, to the negative electron. Careful study of the charge/mass ratio of *positive* rays in discharge tubes showed that this ratio depended on what gas was in the tube. Since the largest charge/mass (smallest mass/charge) ratio was obtained when hydrogen was in the tube, the fundamental particle of positive charge was deduced to be a hydrogen atom from which an electron has been removed. This deduction was suggested partly by the results of chemical experiments carried out before 1900. The name assigned to this fundamental positively charged particle was **proton.**

The mass of the hydrogen atom can be calculated from results of chemical experiments or from the charge/mass ratio determined from deflection of positive rays. For this latter calculation it is necessary to realize that the charge of the proton is exactly equal in magnitude but opposite in sign to the charge of the electron. Further, when it is realized that a hydrogen atom is composed of one proton and one much lighter electron, it becomes apparent that the mass of a proton is very nearly equal to the mass of a hydrogen atom.

The mass of the proton is 1.67×10^{-24} g, and its charge is $+1.6 \times 10^{-19}$ coulomb.

The Neutron

Several scientists suggested about 1920 that, to account for isotopes, there must be another subatomic particle with nearly the same mass as the proton but with no net electric charge. In 1932 James Chadwick proved the existence of these particles called *neutrons*. Neutrons are important in many nuclear reactions and are components of all atoms heavier than hydrogen.

The Nuclear Atom

The experiments of Thomson, Millikan, Ernest Rutherford, and others made it clear in the early years of the twentieth century that atoms were made up of protons and electrons. Other work on scattering of light and beams of electrons by atoms provided estimates of the total number of electrons in various atoms. It was also clear that neutral atoms must contain an equal number of protons and electrons. Several suggestions were advanced for the arrangement of these particles in atoms, but no relevant experiments were performed until 1910.

Hans Geiger had come from Germany to the University of Manchester where he worked with Rutherford and developed the instrument later called the **Geiger counter.** Rutherford suggested that Geiger and Ernest Marsden, a student, should investigate the scattering of alpha (α) particles by thin foils of metals. (It is sufficient for the present to know that an alpha particle is an atom of helium that is missing two electrons and is therefore positively charged and about four times as heavy as an atom of hydrogen.)

The alpha particle scattering experiments of Rutherford, Geiger, and Marsden are illustrated in Figure 2.4. Most of the alpha particles passed straight through the foil, but a small number were deflected from their straight path through angles ranging from a few degrees to nearly 180°. Several years later, Rutherford wrote:

> . . . It was quite the most incredible event that has ever happened to me in my life. It was almost as incredible as if you had fired a 15-inch shell at a piece of tissue paper and it came back and hit you.
>
> On consideration I realized that this scattering backwards must be the result of a single collision and when I made calculations it was impossible to get anything of that order of magnitude unless you took a system in which the greater part of the mass of the atom was concentrated in a minute nucleus.

In 1911 Rutherford presented his model of the nuclear atom. His scattering experiments could be explained if most of the mass of an atom is concentrated in what he called the nucleus and if this nucleus is positively charged. An appropriate number of electrons in the space surrounding the nucleus maintains the net electrical neutrality of the atom. Calculations based on his measurements showed that the nuclei of several elements are about 10^{-13} cm in diameter. Since it was already known that atoms as a whole are about 10^{-8} cm in diameter, it was recognized that an atom should be pictured as an extremely small and dense nucleus with much lighter electrons scattered around the nucleus and with a relatively large amount of empty space between the electrons.

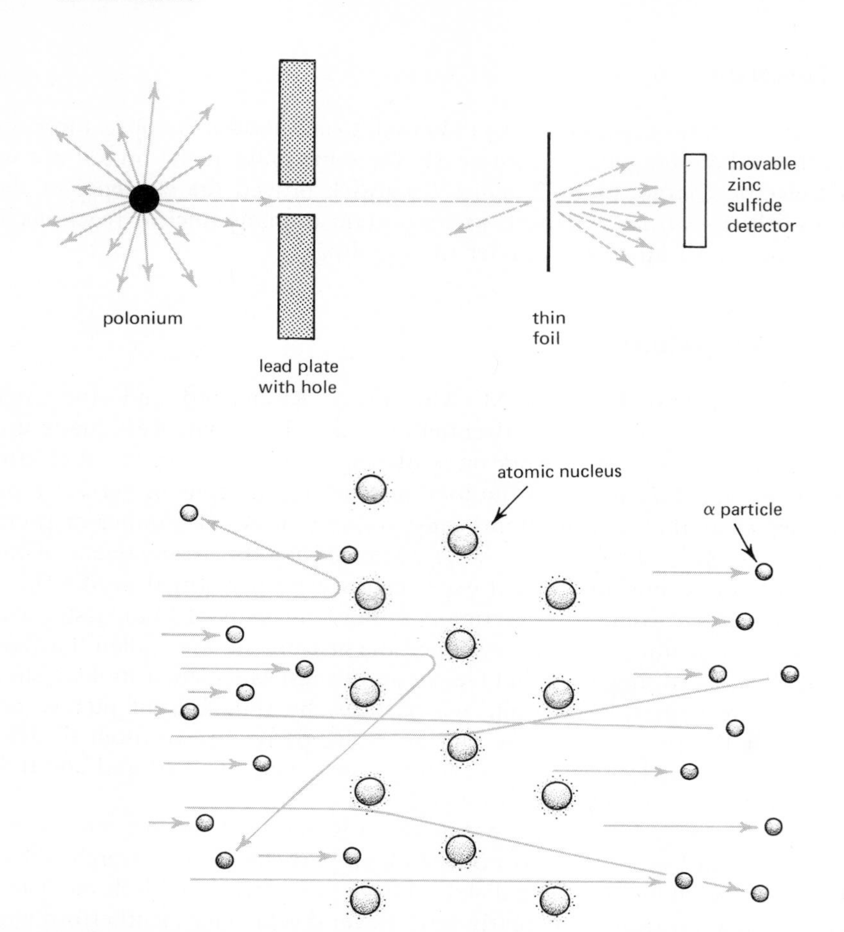

Figure 2.4. The upper diagram is a schematic illustration of the alpha particle scattering experiments carried out by Rutherford, Geiger, and Marsden. The lower diagram illustrates how the Rutherford model of the atom accounts for the scattering of some of the alpha particles through large angles.

Isotopes

It has already been mentioned that analysis of the charge/mass ratio of positive rays led to the realization that the proton is a fundamental unit of positive electricity. J. J. Thomson, A. J. Dempster, F. W. Aston, and others subsequently refined the technique of studying these rays with an instrument called a mass spectrograph, which operates on the same principles as the instrument pictured in Figure 2.2. Masses of many atoms have been determined by mass spectrograph and are used for calculation of what we call atomic weights. But before we consider atomic weights, we must take up the question of whether all atoms of the same element have the same mass.

About 1803 John Dalton concluded from the available chemical evidence that matter consisted of minute particles that could not be divided or changed into other particles. He also concluded that all atoms of any element are identical to each other in mass and other properties. No direct and conclusive evidence contrary to Dalton's proposal of equal masses for all atoms of the same element was found until 1913 when J. J. Thomson investigated neon in his mass spectrograph and found neon atoms of two distinctly different masses.

Atoms of the same element but of different mass are called **isotopes.** Isotopes of hydrogen, oxygen, uranium, and several other elements are of particular importance in chemistry.

The word "element" has just been used in a special way. It is necessary to consider carefully what is meant by "atoms of the same element" and by "chemical element." Further, it is necessary to understand what is meant when we say that a given atom is an atom of some particular element or that some substance, such as iron or carbon, is an elementary substance. The search for satisfactory understanding of statements about elements and kinds of atoms played an important role in the development of chemistry and atomic physics.

Scientists in the nineteenth century thought that all atoms of the same element had the same chemical and physical properties, including mass. The discovery of isotopes clearly showed that this belief was, at least partly, mistaken. A somewhat more satisfactory statement would be that atoms of the same element are almost identical with respect to chemical and most physical properties but might have different masses. In some ways this statement is a satisfactory way of defining what we mean by element; in other ways it is unsatisfactory.

For example, three isotopes of hydrogen are now known. These three isotopes have *relative* masses of 1, 2, and 3 and are sufficiently different from one another in some ways that they commonly are given different names. The lightest isotope, which is the most abundant, is called hydrogen, but the heavier isotopes are often called **deuterium** and **tritium** rather than hydrogen. Deuterium is also called **heavy hydrogen.** Pure hydrogen and pure deuterium are significantly unlike each other in some ways, as are compounds of hydrogen and deuterium. For example, ordinary water containing hydrogen and **heavy water** containing deuterium affect living organisms quite differently. Their boiling and freezing points are also different. This practice of using different names for different isotopes of the same element is applied only to hydrogen.

The statement "same with respect to chemical properties" could possibly result in classifying hydrogen and deuterium as different elements. Because minor differences in the chemical behavior of isotopes of other elements have also been observed, we are forced to refrain from defining an element solely in terms of chemical behavior. It is necessary to consider further what is known about atomic structure and the relationship of atomic structure to chemical and physical properties.

Atomic Number and Elements

Rutherford's experiments showed that most of the mass of atoms is concentrated in the positively charged nucleus. Further information about the nucleus was obtained in 1913 and 1914 by H. G. J. Moseley who studied the x rays emitted by a large number of metals as they were bombarded by high speed electrons. He discovered that the frequency, or wavelength, of these x rays could be correlated with what he called the **atomic number.** By atomic number Moseley meant the net positive charge of the nucleus in terms of the charge of one proton. Moseley's work made it possible to assign atomic numbers to many elements, and chemical evidence made possible the assignment of atomic numbers to the rest of the elements known at that time.

We now define an element by stating that all atoms of the same element have the same atomic number. Atoms of a given element may or may not all have the same mass. The significance of the definition of an element in terms of atomic number becomes clear as some of the details of nuclear structure are considered and as the connection between atomic structure and chemical behavior is studied. We shall see that the atomic number of an element is equal to the number of protons in the nucleus of each atom of that element.

Atomic Weights

Chemists first began to be interested in the masses of single atoms about the time of Dalton. At that time and for many years after, it was not possible to observe atoms one at a time; therefore chemists found ways of deducing masses of atoms from the results of measurements on large numbers of atoms. In the next chapter we will return to a discussion of these investigations.

The mass spectrographs of Thomson and other early investigators had sufficient resolving power to permit isotopes of several light elements to be distinguished from each other, but did not permit precise determination of atomic masses. Subsequent work by A. J. Dempster and F. W. Aston led to such improvements that by 1926 Aston's instrument was used for the determination of masses to 1 part in 10,000. At that time atomic masses were determined mass spectrographically relative to the mass of the most abundant isotope of oxygen, which was adopted as a reference. Oxygen was first chosen as a mass reference primarily because of the chemical investigations of atomic weights in the preceding hundred years.

The mass of a macroscopic object is commonly determined by "weighing" it on a balance. If the weights on one pan of the balance are gram weights, the mass of the object is determined in grams. Similarly, mass spectrographic studies led to determination of atomic masses in terms of the oxygen atom. Results of mass spectrographic investigations might have been reported in grams, especially if the mass of the oxygen atom had been sufficiently well known. Instead, however, a new mass unit was defined. This defined unit of mass, the **atomic mass unit,** was set exactly equal to one sixteenth of the mass of the oxygen atom.

When the atomic mass unit was first defined, it was thought that there was only one isotope of oxygen and that atomic masses determined mass spectrographically and those determined chemically would be identical. The discovery of other isotopes of oxygen by W. F. Giauque and H. L. Johnston in 1929 complicated this situation. Mass spectroscopists used an atomic mass unit defined as one sixteenth of the mass of the most abundant isotope of oxygen. The mass of an atom of the most abundant isotope of oxygen was therefore exactly 16 of these defined atomic mass units.

Since chemists determined atomic masses in terms of the mass of a large number of atoms, definition of the atomic mass unit in terms of a single isotope was slightly different from the chemical definition. Chemists defined the atomic mass unit as exactly one sixteenth of the average mass of naturally occurring oxygen. This difference in definitions led to a small difference between the atomic masses determined chemically and those determined mass spectrographically. Although small, this difference in atomic weight scales caused increasing difficulties, especially in the borderland between chemistry and physics, until 1961 when a new atomic weight scale based on carbon was adopted for both chemistry and physics. One atomic mass unit is now defined as exactly one twelfth of the mass of an atom of the most abundant isotope of carbon. Numerical consequences of this revision are small—negligible for many purposes but of sufficient importance for some purposes to justify the change.

Masses and natural abundances of atoms of many elements have been determined mass spectrographically with high accuracy, and many of the atomic weights now used for chemical calculations are based on these atomic masses.

Atomic masses, whether expressed in grams or the more usual atomic mass units, are useful for consideration of quantities of matter containing huge numbers of atoms. For example, atoms of the only naturally occurring isotope of manganese have been found to have mass of 54.9380 atomic mass units. We therefore conclude that an atom of this isotope of manganese has a mass that is 54.9380/12.0000 times the mass of an atom of the most abundant isotope of carbon. Further, we know that if we have 54.9380 g of manganese, we have the same number of manganese atoms as there are carbon atoms in 12.0000 g of carbon (containing only the most abundant isotope).

Example Problem 2.1. It is known that there are 6.02×10^{23} carbon atoms in 12.0 g of carbon. Use this information with that in the paragraph above to calculate the number of manganese atoms in 35.0 g of manganese.

We solve this problem by first calculating the mass of a single atom of carbon to be

$$\frac{12.0 \text{ g}}{6.02 \times 10^{23} \text{ atoms}} = 1.99 \times 10^{-23} \text{ g atom}^{-1}$$

The mass of a single atom of manganese is therefore

$$\frac{54.9}{12.0} \times 1.99 \times 10^{-23} = 9.12 \times 10^{-23} \text{ g atom}^{-1}$$

We now divide 35.0 g of manganese by this mass to obtain

$$\frac{35.0 \text{ g}}{9.12 \times 10^{-23} \text{ g atom}^{-1}} = 3.83 \times 10^{23} \text{ atoms}$$

in 35.0 g of manganese. ■

When we are dealing with elements having two or more naturally occurring isotopes, it is frequently desirable to calculate and use an average atomic mass that is commonly called an **atomic weight.** The meaning of average atomic mass or of atomic weight is illustrated by calculation of this quantity for chlorine.

There are two naturally occurring isotopes of chlorine, which have atomic masses of 34.962 and 36.966 atomic mass units. The natural abundances of these isotopes are 75.4 and 24.6%, respectively. As the first step in calculating the average atomic mass or atomic weight of chlorine we calculate the mass of 1000 (or any other convenient number) chlorine atoms to be

$$(754 \times 34.962) + (246 \times 36.966) = 35453 \text{ atomic mass units}$$

The desired average atomic mass is found to be

$$35453/1000 = 35.453 \text{ atomic mass units}$$

We shall later identify 35.453 g of chlorine as being 6.02×10^{23} atoms.

Nuclear Composition

We now put together our information about subatomic particles, atomic numbers, and atomic masses in order to understand some of what is known about the composition of atomic nuclei.

The most abundant isotope of carbon has a defined mass of 12.0000 atomic mass units and has atomic number 6. This isotope is represented by $^{12}_{6}C$, where the subscript 6 indicates the atomic number and the superscript 12 represents the mass number of the isotope. This mass number is the integer closest to the atomic mass expressed in atomic mass units. Although this symbol is redundant in that the 6 and the C both indicate carbon, it is commonly used because it is a convenient way of summarizing important information.

Carbon is assigned the atomic number six because each carbon nucleus has a net positive charge equal to that of six protons. An ordinary neutral atom of carbon therefore has six electrons surrounding the nucleus to account for electroneutrality. For present purposes the mass of the proton is taken to be 1.0 atomic mass unit (actually, it is very slightly more) and the mass of one or even six electrons is ignored because of being much smaller than one. So we see that the total mass of six protons and six electrons is very nearly six atomic mass units. Since the mass of the carbon atom under consideration is twelve atomic mass units, there must be more particles in the nucleus to account for the remaining mass. This extra mass is provided by six neutrons, each with atomic mass of about 1.0 atomic mass unit and no net charge.

The atomic number of chlorine is 17. We know, therefore, that atoms of $^{35}_{17}Cl$ and of $^{37}_{17}Cl$ have 17 protons in their nuclei and 17 electrons outside of their nuclei. Atoms of each isotope must have enough neutrons in their nuclei to make the total mass equal to the proper value. Hence there are 18 neutrons in $^{35}_{17}Cl$ and 20 neutrons in $^{37}_{17}Cl$ nuclei.

Note that the mass number is always equal to the total number of protons and neutrons in the nucleus.

Since the masses of the proton (1.007825 atomic mass units) and the neutron (1.008665 atomic mass units) are known, the total mass of the component protons and neutrons of any nucleus may be calculated and compared to the actual mass. It is found that the nuclear masses are lighter than the calculated masses. It is this mass difference that accounts for the stability of nuclei and also for the tremendous energies that can be obtained in nuclear fission and fusion processes. As nearly everyone now knows, these energies and masses may be related by means of Einstein's famous $E = mc^2$.

Chemical reaction behavior is almost entirely determined by the number and arrangement of electrons; therefore, we find that isotopes of the same element have nearly identical chemical properties. Isotopes of uranium, principally $^{235}_{92}U$ and $^{238}_{92}U$, are so chemically similar that no practical method of separation based on chemical reactions has been developed. On the other hand, chemical differences between $^{1}_{1}H$ (ordinary hydrogen) and $^{2}_{1}H$ (heavy hydrogen or deuterium, often given the symbol $^{2}_{1}D$) are sufficient to permit chemical separation of these isotopes from each other. The chemical differences between hydrogen and deuterium are significant in many reactions.

In general there are larger chemical differences between isotopes of a light element than there are between isotopes of a heavier element because the per cent mass difference must be large, as it is for isotopes of light elements, for mass to affect chemical properties significantly.

Radioactivity and Atomic Structure

After the discovery of x rays by Wilhelm Roentgen in 1895, a search began for other sources and kinds of radiation. Henri Becquerel was the first to find and recognize what we call radioactivity. He discovered that uranium salts emit a radiation sufficiently penetrating to affect a photographic plate after passing through paper and thin sheets of metal. Many other investigations of the radiation given off by uranium and other heavy elements were carried out by Rutherford, Frederick Soddy, the Curies, and others in the years following Becquerel's discovery in 1896.

Within a few years of the initial discovery, three kinds of radiation were recognized. These are called alpha rays (or alpha particles), beta rays (or beta particles), and gamma rays. Alpha particles are helium nuclei moving at high speeds. Beta particles are electrons moving at high speeds. Gamma radiation is light of short wavelength and high energy.

Alpha particles are helium nuclei with mass of four atomic mass units. Each helium nucleus has a net positive charge of two proton units, has atomic number two, and can become an atom of helium by capturing two electrons. These statements about alpha particles are all based on the results of carefully conducted experiments. For example, it has been possible to measure directly the total charge of a known large number of alpha particles and thence to calculate the charge of one alpha particle. Measurements of the deflections of beams of alpha particles in electric and magnetic fields confirm that their charge is positive and that their mass is 4 atomic mass units. The helium that is formed when alpha particles capture electrons has been collected, identified, and measured. There is no reasonable doubt that alpha particles are helium nuclei.

Now consider the consequences of emission of an alpha particle from the nucleus of a radioactive atom. First, it is known that the positively charged alpha particle must come from the nucleus of the emitting atom because all of the positive charge and most of the mass of any atom is in the nucleus. We know that an atom of $^{238}_{92}U$ contains 92 protons and $238 - 92 = 146$ neutrons in its nucleus, and we also know that a helium nucleus is made up of two protons and two neutrons. Thus we deduce that, when an atom of $^{238}_{92}U$ emits an alpha particle, the atom that remains will contain 92 − 2 = 90 protons and 146 − 2 = 144 neutrons. This atom must be an atom of the element with atomic number 90, which is thorium, and must have mass number 90 + 144 = 234. All of this information is summarized by

$$^{238}_{92}U \rightarrow \, ^{4}_{2}He + \, ^{234}_{90}Th$$

or by the equivalent

$$^{238}_{92}U \rightarrow \alpha + \, ^{234}_{90}Th$$

The rate of decomposition of a group of $^{238}_{92}U$ atoms remains to be considered. Rates of radioactive decay processes are generally characterized by what is called a half-life. This **half-life** is the time that it takes for half of a sample of the radioactive material to undergo the process that makes the material radioactive. The half-life for $^{238}_{92}U$ decay by alpha emission is 4.5×10^9 years.

Half-lives of various radioactive substances range from only a fraction of a second to more than 10^9 years. Half-lives and rates of radioactive decay processes are statistical and are significant only when large numbers of atoms are involved. Detailed numerical calculations are considered in Chapter 22 with the rates of chemical reactions.

Interpretation of the emission of beta particles (symbolized by β^-) is less direct than interpretation of alpha emission. It might be expected that the electrons constituting beta radiation come from the cloud of electrons outside the nucleus, but this supposition is false as illustrated by the beta decay of $^{14}_{6}C$ to $^{14}_{7}N$. The $^{14}_{6}C$ nucleus contains six protons and eight neutrons, while the $^{14}_{7}N$ nucleus that is formed contains seven protons and seven neutrons. Hence we deduce that one neutron in the $^{14}_{6}C$ nucleus has been transformed into a proton, which remains in the nucleus, and an electron, which is emitted and called a beta particle. We summarize this process by

$$^{14}_{6}C \rightarrow \, ^{14}_{7}N + \beta^-$$

In connection with the conversion of a neutron in a nucleus to a proton, which remains in the nucleus, and an electron, which is ejected, it is interesting to note that *free* neutrons decay or disintegrate to yield protons and electrons. The half-life for free neutron disintegration is 13 minutes.

Gamma rays usually accompany alpha and beta radiation. These gamma rays are like x rays in that they are light waves of extremely short wavelength and high energy. They originate in energy changes accompanying rearrangement of the particles in the nucleus after emission of an alpha or beta particle.

A process that is in some respects the reverse of β^- decay is also known. Certain nuclei are able to capture an electron, thereby converting one proton in the nucleus to a neutron and forming an atom with atomic number one less than that of the parent. In 1938 Louis Alvarez first found evidence for this mode of nuclear transformation, which is called **electron capture** or *K* capture. This latter name is intended to show that the electron is captured from the "*K* shell" of electrons surrounding the nucleus. The rearrangements of electrons about the nucleus after this electron capture ordinarily result in emission of x rays.

Nuclear Reactions

In 1917 Rutherford began the experiments that led in 1919 to the discovery of the first man-made nuclear reaction. Although positively charged alpha particles are repelled by the positively charged nucleus, Rutherford reasoned that it might be possible for alpha particles to penetrate the nuclei of atoms with lower atomic numbers. By bombarding nitrogen with alpha particles from a sample of radium, he produced oxygen by means of the nuclear reaction

$$^{14}_{7}\text{N} + ^{4}_{2}\text{He} \rightarrow ^{17}_{8}\text{O} + ^{1}_{1}\text{H}$$

and detected the emitted protons.

Because the alpha particles available from naturally occurring radioactive materials have a limited range of energies, relatively few nuclear reactions could be carried out in this way. Subsequent development of high energy particle accelerators such as the cyclotron have made it possible to produce a large number of nuclear reactions. These accelerators make available high energy beams of protons, deuterons, alpha particles, and even heavier nuclei so that many nuclear reactions are now known. The first such reaction was observed by John Cockcroft and Ernest Walton in 1932 in England when they bombarded lithium with protons accelerated in a machine they had designed and built for this purpose. The reaction was

$$^{7}_{3}\text{Li} + ^{1}_{1}\text{H} \rightarrow 2\,^{4}_{2}\text{He}$$

One of the difficulties in carrying out nuclear reactions by bombardment of a "reactant" sample with protons, alpha particles, or other positively charged particles is in securing high enough energies to enable the bombarding particle to overcome the coulombic repulsion of the positively charged nucleus. Shortly after Chadwick's discovery of the neutron in 1932, Enrico Fermi in Rome realized that these particles might easily penetrate atomic nuclei and thereby cause nuclear reactions. Within a short time, Fermi had produced a wealth of new radioactive substances by means of reactions utilizing neutron bombardment.

Neutrons used for these nuclear reactions were produced by bombardment of light elements with alpha particles. Typical reactions yielding neutrons (represented by $^{1}_{0}n$—zero atomic number and mass of one atomic mass unit) are the following:

$$^{11}_{5}\text{B} + ^{4}_{2}\text{He} \rightarrow ^{14}_{7}\text{N} + ^{1}_{0}n$$

$$^{9}_{4}\text{Be} + ^{4}_{2}\text{He} \rightarrow ^{12}_{6}\text{C} + ^{1}_{0}n$$

Three important examples of nuclear reactions involving neutrons are shown in equations (2.1) through (2.3).

$$^{14}_{7}\text{N} + ^{1}_{0}n \rightarrow ^{14}_{6}\text{C} + ^{1}_{1}\text{H} \qquad (2.1)$$

$$^{59}_{27}\text{Co} + ^{1}_{0}n \rightarrow ^{60}_{27}\text{Co} \qquad (2.2)$$

$$^{6}_{3}\text{Li} + ^{1}_{0}n \rightarrow ^{3}_{1}\text{H} + ^{4}_{2}\text{He} \qquad (2.3)$$

Reaction (2.1) accounts for the cosmic ray induced production of ^{14}C in the atmosphere and makes possible the method of ^{14}C dating proposed and developed by W. F. Libby. Neutron bombardment in nuclear reactors is now used to produce large quantities of ^{14}C, which decays by β^- emission to ^{14}N with a half-life of 5720 years, for use as a radioactive tracer.

Reaction (2.2) yields the radioactive ^{60}Co (half-life for β^- decay is 5.26 years, accompanied by emission of gamma rays) that now has important uses in medicine. Production of ^{60}Co in the "cobalt bomb" is also possible, but fortunately has not yet been done.

Reaction (2.3) is useful for production of tritium (^3_1H), which is used as a radioactive tracer and in thermonuclear fusion.

Nuclear Fission

Bombardment of many nuclei with neutrons causes the formation of an unstable species containing the captured neutron. This unstable species then emits a β^- particle and becomes an atom of an element with atomic number one greater than the original element. Thus when Fermi found that the products formed by bombarding uranium with neutrons emitted β^- particles, he quite reasonably concluded that he had produced elements with atomic numbers greater than 92. These experiments and Fermi's conclusions attracted much attention and were the source of several controversies that were difficult to resolve because of the very small amounts of radioactive substances involved. Within a few years, however, experiments by Fermi and especially by Otto Hahn and Lise Meitner showed that the radioactivity was certainly not derived from elements with atomic numbers between those of lead and uranium, thus providing indirect support for Fermi's original interpretation.

Ida Noddack in Germany pointed out that insufficient attention had been paid to the possibility that much lighter atoms might have been formed as a result of splitting the uranium nucleus by the bombarding neutrons. These suggestions were not well received because of the difficulties associated with the chemical analysis of such small samples and the belief that such atom splitting was contrary to the laws of physics.

After Irene Joliot-Curie and Paul Savitch in France had come close to identifying radioactive lanthanum in the products arising from the bombardment of uranium by neutrons, Hahn and Fritz Strassmann succeeded in identifying barium in these products and reported their startling results in January, 1939. In the months that followed, Hahn and Strassmann succeeded in identifying several other products of splitting uranium nuclei.

Lise Meitner had fled from Germany to Sweden in 1938 when Hitler occupied her native Austria. She learned of the latest results of Hahn and Strassmann during the Christmas vacation of 1938, which she was spending with her nephew, Otto Frisch. Together they introduced the term "nuclear fission" for this process and showed that it was, in fact, consistent with Bohr's theory of the nucleus. Within a month Frisch had demonstrated that tremendous energies were liberated during fission, a result that was soon substantiated in several laboratories in the United States and also in France by Joliot. Then the liberation of neutrons, which had been suggested by both Hahn and Fermi, was confirmed in the laboratories of two groups in the United States and one in France.

Once the facts of fission and neutron liberation were established, it became clear that a chain reaction yielding huge quantities of energy might be possible. Much work was devoted toward that end and by 1945 the first nuclear fission bombs had been made, tested, and used in World War II. With the end of the war came increased interest in constructive uses of the energy from nuclear fission, and efforts to use nuclear reactors for generation of electrical energy were undertaken. Increasing numbers of people in several countries now receive useful electrical energy from this source, which will become more and more important as our reserves of fossil fuels diminish.

As a result of fission bombs and the fusion bombs that they trigger, it is now clear that we humans have the power to destroy ourselves. But with the aid of nuclear power and other technological advantages made possible by modern science, we also have the power to make a better life than was even dreamed of in the past. If we choose the better life, actual attainment of this desirable end will involve many contributions from chemistry as well as other sciences and such fields as economics and politics. Increased production of food through better fertilizers and insecticides and decreased production of people through birth control pills are examples of the obvious contributions of chemistry and chemical technology. We must also recognize the dangers associated with nuclear reactors, persistent insecticides, fertilizer run-off, etc. Technological blessings are rarely unmixed.

References

Choppin, G. R.: *Nuclei and Radioactivity*. W. A. Benjamin, Inc., New York, 1964 (paperback).

Hahn, O.: The discovery of fission. *Sci. Amer., 198,* 76 (Feb. 1958).

Harvey, B. G.: *Nuclear Chemistry*. Prentice-Hall, Englewood Cliffs, N.J., 1965 (paperback).

Katz, J. J.: The biology of heavy water. *Sci. Amer., 203,* 106 (July, 1960).

Libby, W. F.: *Radiocarbon Dating*. University of Chicago Press, 1965 (paperback).

Perlman, I., and G. T. Seaborg: The synthetic elements. *Sci. Amer. 182,* 38 (April, 1950).

Romer, A. (ed. and commentary): *The Discovery of Radioactivity and Transmutation.* Dover Publications, Inc., New York, 1964 (paperback).

Seaborg, G. T.: *Man-made Transuranium Elements.* Prentice-Hall, Englewood Cliffs, N.J., 1963 (paperback).

Seaborg, G. T., and A. R. Frisch: The synthetic elements III. *Sci. Amer. 208,* 68 (April, 1963).

Seaborg, G. T. and A. Ghiorso: The synthetic elements II. *Sci. Amer. 195,* 66 (Dec. 1956).

Problems

1. Calculate the ratio of nuclear volume to atomic volume for a nucleus of diameter 1.5×10^{-13} cm and for an atom of diameter 1.2×10^{-8} cm.
2. How many protons and neutrons are in nuclei of the following:
   (a) $^{39}_{19}K$ (b) $^{197}_{79}Au$ (c) $^{24}_{11}Na$ (d) $^{141}_{56}Ba$
3. Two naturally occurring isotopes of lithium have masses of 6.0151 and 7.0160 atomic mass units and natural abundances of 7.98 and 92.02% respectively. Calculate the average atomic mass of lithium.
4. Three naturally occurring isotopes of neon have masses of 19.9924, 20.9938, and

21.9914 atomic mass units and natural abundances of 90.92, 0.26, and 8.82%, respectively. Calculate the average atomic mass of neon.

5. The product of radioactive decay of $^{60}_{27}Co$ has been found to be $^{60}_{28}Ni$. What particles are emitted by $^{60}_{27}Co$?

6. Atoms of $^{24}_{11}Na$ decay by β^- emission. Atoms of what element and mass number are formed?

7. What atom is formed as a product of the radioactive decay of an atom of $^{226}_{88}Ra$ by alpha particle emission?

8. If 1.0 g of $^{226}_{88}Ra$ were to disintegrate entirely by alpha decay, how much helium (expressed in grams) would be formed?

9. Fill in the blank in the following equation that represents a typical fission of $^{235}_{92}U$.

$$^{235}_{92}U + ^{1}_{0}n \rightarrow ^{141}_{56}Ba + 3^{1}_{0}n + ______ + \text{energy}$$

10. Fill in the blanks in the following nuclear reaction equations:

$$^{14}_{7}N + ^{1}_{0}n \rightarrow ______ + ^{1}_{1}H$$

$$^{241}_{95}Am + ^{4}_{2}He \rightarrow ______ + 2^{1}_{0}n$$

$$^{214}_{82}Pb \rightarrow ______ + \beta^-$$

$$^{14}_{7}N + ______ \rightarrow ^{17}_{8}O + ^{1}_{1}H$$

11. An atom of $^{238}_{92}U$ captures a neutron (possibly in a nuclear reactor) and becomes ______. This atom then decays by β^- emission to become ______, which also decays by β^- emission to become ______.

12. Nuclei that contain relatively too many protons and too few neutrons approach stability by capturing an orbital electron as in *K* capture. For example, $^{48}_{23}V$ may capture an electron to become $^{48}_{22}Ti$. It is also possible for such nuclei to approach stability by emission of positrons (positively charged electrons) represented by β^+. Write an equation to represent the decay of $^{22}_{11}Na$ by positron (β^+) emission.

13. The atomic mass unit is just as "good" as the gram or pound, but because we have no laboratory balances for convenient comparison of unknown masses with the mass of a single atom of $^{12}_{6}C$, it is often necessary to know and use a relationship between atomic mass units and grams. The knowledge that there are 6.02×10^{23} carbon atoms in 12.0 g of carbon is the usual basis for such calculations. What mass in grams corresponds to one atomic mass unit? What is the mass in grams of one atom of $^{19}_{9}F$? How many atoms of phosphorus are there in 18.5 g of $^{31}_{15}P$? Note that these calculations can be done conveniently in terms of a counting unit of 6.02×10^{23} items, which is called a mole and is discussed extensively in later chapters.

14. Calculate the charge to mass ratio of a proton and compare it to that of an $^{16}_{8}O$ atom with two outer electrons removed.

15. What is the atomic weight of an electron on the ^{12}C scale?

16. (a) What is the average mass of a mercury atom in grams? (b) The density of liquid mercury is 13.5 g cm^{-3}. Calculate the atomic volume and maximum radius of a mercury atom, assuming that atoms are spheres.

17. An average penny weighs about 3.1 g. Calculate the energy in ergs available if the mass of a penny could be converted to energy. Compare this value with 40 kcal, which is a rough average of the energy that can be obtained by combustion of 3 g of a "typical" fuel.

3

GASES

Introduction

Gases differ markedly from liquids and solids in many ways. For example, the volume of a sample of gas can be changed considerably by changing either the pressure or the temperature. In contrast, volumes of solids and liquids are little changed by changes in pressure or temperature. Gases tend to fill completely any space that is available to them and also adapt easily to changes in the shape of any container. Liquids and solids do not readily expand to fill a large container or contract to fit into a small one, and solids do not easily adjust to changes in shape of a container. Diffusion of one gas into another is noticeably faster than diffusion of one liquid into another or of one solid into another.

Another important contrast between gases and condensed phases is that all gases show great similarities in many properties, whereas liquids and solids exhibit considerable individuality with regard to these same properties. For example, all gases at the same temperature and pressure have about the same coefficients of compressibility and thermal expansion, whereas solids and liquids vary considerably.

Ideal Gas Laws

All gases consist of real molecules and atoms that are frequently confined in real containers and measured by real instruments. It is the behavior of real gases that is important to chemistry and other sciences. Nevertheless, it is convenient and useful to define what we call an **ideal gas,** with properties that are described by simple equations. These ideal gas equations are very useful for describing the behavior of real gases under many conditions where the real gas behavior is in close agreement with that given by the simple ideal gas equations. Even for conditions where the ideal gas equations are clearly inadequate, they still offer a useful start toward an accurate description.

In all of the following discussion of gases, it is important to distinguish between real gas behavior, ideal gas behavior as described by the ideal gas equations, and human explanations of observed properties of gases.

The first experiments on gases of interest to us were described by Robert Boyle in 1662. In these experiments he determined how the volume of a given sample of air varied with pressure. The results of Boyle's experiments can be summarized by stating that the volume of a sample of gas is inversely proportional to the pressure, as indicated by

$$V = \frac{\text{constant}}{P} \tag{3.1}$$

or by

$$PV = \text{constant} \tag{3.2}$$

A few years after Boyle's experiments, Mariotte in France discovered that these equations are valid only at constant temperature. Still later these equations were found to be valid only if the pressure of the gas is low. At high pressures discrepancies between actual behavior and the behavior predicted by equation (3.1) or (3.2) become large, and no simple equation adequately relates pressure to volume. The graph in Figure 3.1 shows that Boyle's law describes the behavior of real gases at low pressures, although deviations are often large at high pressures.

We now say that Boyle's law as expressed in words or by equations (3.1) or (3.2) is an ideal gas law that is exactly valid only in the limit of low pressure.

In 1699 Guillaume Amontons, who had done important work in developing

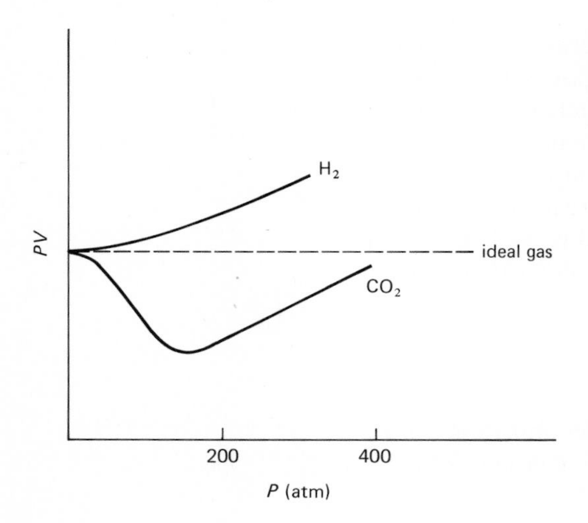

Figure 3.1. Boyle's law is represented by the dashed line in the graph. Real gas behavior is in close agreement with Boyle's law at low pressures.

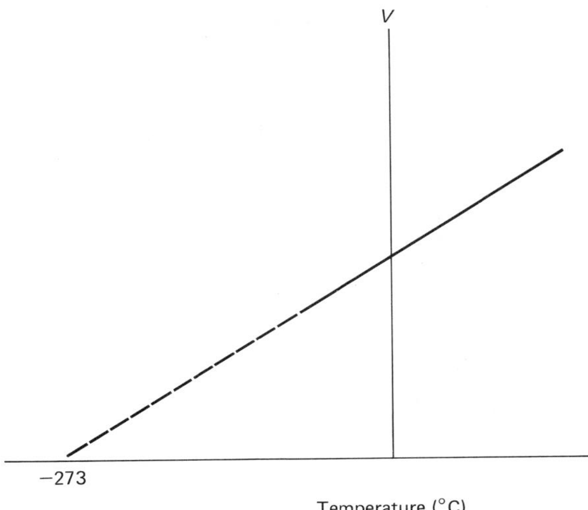

Figure 3.2. Graph of the volume of a given mass of gas at constant pressure as a function of temperature.

The dashed line is an extrapolation to zero volume at −273°C. Measurements on gases cannot be carried to such low temperatures because even helium condenses to a liquid at a temperature higher than −273°C.

thermometers, studied the effect of temperature on the volumes of several gases and showed that the volume of each gas changed by the same fraction of the original volume for a given change in temperature. As a result of this work, Amontons postulated a kind of absolute zero of temperature—an idea that became firmly established a century and a half later.

Amontons' results were verified and made more accurate by Jacques Charles in 1787. Then in 1802 Joseph Gay-Lussac extended these experiments and accurately determined the amount of expansion per degree for several gases. He found that the increase in volume of a given quantity of gas (at constant pressure and initially at 0°C) was about 1/270 of the initial volume for every degree rise in the temperature. For example, 100.0 cm^3 of gas at 0°C would occupy a volume of $100.0 + (100.0/270) = 100.4$ cm^3 at 1°C and the original pressure.

Later investigations have shown that an equation based on Gay-Lussac's observations is valid only for gases at constant low pressure and that the factor 1/270 should be 1/273.15. This information is summarized by

$$V = V_0 + \frac{tV_0}{273.15} = V_0\left(1 + \frac{t}{273.15}\right)$$

or

$$V = V_0\left(\frac{273.15 + t}{273.15}\right) \tag{3.3}$$

where V_0 represents the volume of a given mass of gas at 0°C and V represents the volume of the same mass of gas at the same pressure but at the temperature t°C.

Equation (3.3) suggests that if a gas were to be cooled to −273.15°C, its volume V would become zero. This is illustrated in Figure 3.2. No such phenomenon can actually be observed because all gases condense to form liquids or solids at temperatures above −273.15°C.

By defining a new temperature T in terms of the Celsius temperature t as

$$T = t + 273.15 \tag{3.4}$$

we can simplify equation (3.3) to equation (3.5).

$$V = \left(\frac{V_0}{273.15}\right)T. \tag{3.5}$$

Because $V_0/273.15$ is a constant for any given mass of gas at any constant pressure, equation (3.5) may also be written as

$$V = \text{constant} \times T \tag{3.6}$$

Equation (3.6) is now the most common way of expressing Gay-Lussac's law.

The temperature T defined by equation (3.4) and used in equations (3.5) and (3.6) is today called the **absolute** or **Kelvin** temperature and is widely used in scientific work.

The pressure-volume-temperature (PVT) behavior of real gases is accurately described by equations (3.2) and (3.6) only when the gas pressure is low. At high pressures real gas behavior deviates from the behavior predicted by these equations. We say that equations (3.2) and (3.6) are ideal gas equations and are valid for real gases only in the limit of low pressure where real gas and ideal gas behavior coincide. Ideal gas equations are useful for calculations concerning real gases at low and moderate pressures because differences between real and ideal gas behavior are generally small except at high pressures.

Combined Ideal Gas Equations

Boyle's law as expressed by equation (3.2) can be used for calculating changes in volume caused by changes in pressure at constant temperature. Gay-Lussac's law as expressed by equation (3.6) can be used for calculating changes in volume caused by changes in temperature at constant pressure. These equations can be combined for calculating the changes in volume caused by changes in both pressure and temperature or can be inverted for calculating the pressure or temperature changes required to cause certain changes in volume. Several calculations with these equations are illustrated in the following examples.

Example Problem 3.1. We have a given mass of gas that occupies 200 cm^3 when the pressure of the gas is 1.0 atm and we want to know what the pressure must be to make the volume 160 cm^3 with the gas still at the original temperature of 273°K.

On the basis of Boyle's law as expressed by equation (3.2) we write

$$1.0 \text{ atm} \times 200 \text{ cm}^3 = \text{constant} = 200 \text{ cm}^3 \text{ atm}$$

Because Boyle's law is obeyed at other low pressures, we also write

$$P \times 160 \text{ cm}^3 = 200 \text{ cm}^3 \text{ atm}$$

and solve for P to find

$$P = 200 \text{ cm}^3 \text{ atm}/160 \text{ cm}^3 = 1.25 \text{ atm}$$

as the desired pressure. ■

Boyle's law can be put into another form that is convenient for use in solving problems like Example Problem 1. Let P_1 and V_1 represent the initial pressure and volume of a sample of gas at some temperature. Then from equation (3.2) we have

$$P_1V_1 = \text{constant}$$

Similarly for P_2 and V_2, which represent the final pressure and volume, we have

$$P_2V_2 = \text{constant}$$

Combination of these two equations gives

$$P_1V_1 = P_2V_2 \tag{3.7}$$

Equation (3.7) leads to an easy solution for Example Problem 3.1.

The reasoning that led from equation (3.2) to equation (3.7) can be applied to equation (3.6) after it has been rearranged to

$$\frac{V}{T} = \text{constant}$$

to give

$$\frac{V_1}{T_1} = \frac{V_2}{T_2} \tag{3.8}$$

Example Problem 3.2. The gas considered in Example Problem 3.1 is heated at constant pressure of 1.25 atm from its initial temperature of 273°K to a final temperature of 323°K. What is the final volume?

We see from Example Problem 3.1 that the volume occupied by the gas at the initial temperature 273°K is 160 cm^3. Insertion of these data in equation (3.8) for V_1 and T_1 with $T_2 = 323$°K gives

$$\frac{160 \text{ cm}^3}{273\,^\circ\text{K}} = \frac{V_2}{323\,^\circ\text{K}}$$

This equation is solved to give a final volume $V_2 = 189$ cm^3. (Note that absolute temperatures are required in this calculation.) ■

Boyle's law says that V is proportional to $1/P$ at constant temperature, and Gay-Lussac's law says that V is proportional to T at constant pressure. It is useful to combine these statements for calculations concerning systems that are not restricted to constant temperature or pressure. We may state the desired combination of Boyle's law with Gay-Lussac's law by saying that V is proportional to T/P. Thus we write

$$V = \text{constant} \times (T/P)$$

and rearrange to obtain

$$PV/T = \text{constant} \tag{3.9}$$

Because the constant depends only on the mass of gas, by the same reasoning that led from (3.2) to (3.7) we also obtain

$$\frac{P_1V_1}{T_1} = \frac{P_2V_2}{T_2} \tag{3.10}$$

Example Problem 3.3. A particular mass of gas occupies 200 cm³ when the pressure of the gas is 1.0 atm and the temperature is 273°K. What volume will this gas occupy when the pressure is 1.25 atm and the temperature is 323°K?

We let P_1, V_1, and T_1 in equation (3.10) represent the initial conditions while P_2, V_2, and T_2 represent the final conditions. Thus we have

$$\frac{1.0 \text{ atm} \times 200 \text{ cm}^3}{273\,^\circ\text{K}} = \frac{1.25 \text{ atm} \times V_2}{323\,^\circ\text{K}}$$

This equation is solved to give a final volume $V_2 = 189 \text{ cm}^3$. Note that this calculation is for the same total change as in Example Problems 3.1 and 3.2. ■

Kinetic Molecular Theory of Gases

It has often proven useful in science to devise mental pictures, sometimes highly simplified, of real systems. The kinetic molecular theory of gases is a mathematical representation of one such mental picture.

The eminent mathematician Daniel Bernoulli took the first step toward the kinetic theory in 1738. Unfortunately, at that time the scientific world was unready for either an atomic theory or for statistical calculations. Over a century later in 1856 August Kronig, a German schoolmaster, improved Bernoulli's work and presented it in a receptive scientific climate. Within a few years several of the greatest men in theoretical physics (Rudolf Clausius, James Clerk Maxwell, and Ludwig Boltzmann) had made important contributions to the subject and in the process did pioneering work in the related fields of thermodynamics and statistical mechanics.

We shall now describe a mental picture of an ideal gas and then develop a simple mathematical representation of that picture.

1. An ideal gas consists of tiny particles called molecules, which are so small and far apart (on the average) that the total volume of all the molecules is negligible compared to the empty space in the system.
2. The molecules move in straight lines except when they collide with each other or with the walls of the container. All collisions are perfectly elastic, which means that no kinetic energy is converted into internal excitation of the molecules.
3. Forces between molecules are negligible except during collisions.
4. The average kinetic energy of all the molecules is proportional to the absolute temperature.

Before we make use of this model, it is well to consider how the ideal gas compares with a real gas. The large compressibilities of real gases show that they are mostly empty space, in agreement with the first statement above. As the total pressure is lowered, the total volume of all the molecules in a given mass of gas is an increasingly smaller fraction of the total volume; therefore, we expect that ideal gas equations derived from this model will apply most accurately to real gases when the pressure is low.

The rapid diffusion of real gases into one another provides evidence that molecules of real gases move rapidly. Several experimental investigations have provided data on the velocities of gas molecules and have confirmed that their motion is linear.

The third statement about our ideal gas model is applicable to real gases if the pressure is low, so that the molecules will, on the average, be far apart. It is also applicable if the temperature is high, so that the molecules are moving sufficiently rapidly (large kinetic energy) that the effect of intermolecular forces is negligible. Since the measured PVT relations of real gases at low pressures do not depend on the nature of the gas being investigated, we may infer that interactions between the molecules are unimportant under these conditions.

All of the preceding discussion is summarized by stating that our imaginary ideal gas is very much like a real gas at low pressure and in some ways is like a real gas at high temperature even if the pressure is not low.

It has been observed many times that equations (3.2) and (3.6) based on Boyle's law and on Gay-Lussac's law satisfactorily represent the behavior of real gases at low pressures. Therefore proper mathematical treatment of the ideal gas we have described should lead to these equations. We might also hope that further mathematical treatment would lead to other equations that will be in agreement with other experimental results for real gases. Some of the most impressive triumphs of the kinetic theory in the last third of the nineteenth century resulted from theoretical predictions that were *later* verified quantitatively in the laboratory.

Consider a container in which there are $\mathcal{N}$ molecules, each having mass m. In a real gas these molecules are moving in all directions with a great variety of speeds. A rigorous treatment of this system of chaotically moving molecules confined in a container of any shape is too complicated for this book, but we may arrive at the same results by way of the simpler treatment given on the following pages.

Our simplified treatment begins with consideration of a cubical box of side length L containing $\mathcal{N}$ molecules, each of mass m. We imagine that one third of the molecules are moving in the x direction, one third are moving in the y direction, and one third in the z direction. Thus each molecule is moving parallel with four walls and is colliding with only two walls. We also imagine that all of the molecules are moving with the same speed, represented by s. If we remember that pressure is defined as force per unit area and that force is defined as the change in momentum per unit time, we have all of the information that we need to derive a mathematical relationship between P, V, T, $\mathcal{N}$, m, and s.

We begin by considering the change in momentum associated with collision of one molecule with one wall. The momentum of the molecule is ms in one direction (considered +) before the collision and ms in the other direction (considered −) after the elastic collision with the wall. The change in momentum is therefore $2ms$.

Since we express speed in terms of centimeters per second (cm sec^{-1}) we know that a molecule with speed s travels s cm in one sec. The molecule under consideration therefore makes s/L collisions in one sec with the two walls confining it and collides with one of these walls half as many times—that is, $s/2L$ times per second.

The force exerted by this one molecule on one wall is the change of momentum per collision multiplied by the number of collisions per second, which gives us

$$f \text{ (one molecule, one wall)} = 2ms\left(\frac{s}{2L}\right) = ms^2/L.$$

The pressure exerted by this one molecule is the force divided by the area of the wall, which gives

$$P \text{ (one molecule)} = (ms^2/L)/L^2 = ms^2/L^3 = ms^2/V$$

where $V = L^3$ is the volume of the box. The total pressure on a wall is the pressure exerted by one molecule times the number of molecules that collide with that wall. We therefore multiply the single molecule pressure above by $\mathcal{N}/3$ to obtain

$$P = \frac{\mathcal{N}}{3}\left(\frac{ms^2}{V}\right)$$

We both multiply and divide the right-hand side of this equation by 2 and then multiply both sides of the equation by V to obtain

$$PV = \frac{2\,\mathcal{N}}{3}\left(\frac{ms^2}{2}\right) \tag{3.11}$$

The fourth descriptive statement of our ideal gas was that the average kinetic energy ($ms^2/2$) is proportional to the absolute temperature. We may therefore write equation (3.11) as

$$PV = \frac{2\,\mathcal{N} \times \text{constant} \times T}{3} \tag{3.12}$$

Remembering that $\mathcal{N}$ (the number of molecules) is a constant for a particular experiment, we see that equation (3.12) is equivalent to equation (3.9) that resulted from combination of Boyle's law with Gay-Lussac's law. We therefore conclude that our kinetic theory of the ideal gas leads to a PVT equation that is in accord with measurements on real gases at low pressures.

It should be noted that more rigorous treatment (no assumptions about cubical boxes, direction of motion of molecules, or uniformity of speed) is more complicated, but leads to an equation of the same form as (3.12).

Avogadro's Law and Atomic Weights

The kinetic molecular theory equations of the preceding section involve $\mathcal{N}$, the number of molecules in the sample of gas under consideration. By rearranging equation (3.12) we obtain

$$\mathcal{N} = \frac{3\,PV}{2 \times \text{constant} \times T} \tag{3.13}$$

This equation shows that all samples of any gas at the same temperature and (low) pressure and occupying the same volume must contain the same number of molecules. This statement, without the kinetic molecular theory justification for it, is Avogadro's law substantially as first proposed by Amedeo Avogadro in 1811.

Before we proceed with discussion of the use and significance of Avogadro's law, let us try to imagine ourselves in the early days of the nineteenth century so that we might follow part of the development of atomic theory up to the time of Rutherford, Thomson, and other atomic scientists of the first quarter of the twentieth century.

In 1799 the **law of constant proportions** was first stated by Joseph Proust, a French chemist working in Madrid. His analytical data showed that all samples of the same pure compound contain the constituent elements in the same proportions.

The **law of simple multiple proportions** was first stated in 1803 by John Dalton, who at the same time was formulating his atomic theory. Dalton's law of multiple proportions states that if two elements combine to form more than one compound, the weights of one element that combine with a fixed weight of the other are in ratios of small whole numbers. The validity of this law was substantiated by chemical analysis of a number of compounds. It is likely that Dalton was led to his atomic theory at least partly by the experimentally verified law of simple multiple proportions—certainly this law is a logical consequence of the atomic theory.

The efforts of Dalton and his contemporaries to assign atomic weights to all the elements known at that time were frustrated by their inability to assign correct formulas to compounds. For instance, they knew that water is 88.8% oxygen and 11.2% hydrogen, but they had no way of knowing that there are twice as many atoms of hydrogen as oxygen in a molecule of water or that both oxygen and hydrogen molecules contain two atoms.

In connection with his work on the analysis of air, Gay-Lussac undertook investigations of volume relations in reactions between various gases and in 1808 published his **law of combining volumes.** According to this law, the volumes (measured at the same pressure and temperature) of the gases consumed or produced in chemical reactions are in ratios of small integers. Although Gay-Lussac stated that his work provided support for Dalton's atomic theory, Dalton and many other scientists of the time were slow to see the connection.

In 1811 Avogadro advanced the hypothesis that equal volumes of all gases at the same pressure and temperature contain the same number of molecules. On the basis of this hypothesis, Avogadro was able to account for all of the experimental facts concerning combining volumes of gases and relative densities of gases. Further, he was able to show that hydrogen and oxygen must be diatomic H_2 and O_2 rather than H and O. Avogadro also offered proof that formulas for water and nitric oxide are H_2O and NO rather than HO and N_2O_2 as commonly thought at that time. Finally, Avogadro's hypothesis was consistent with Dalton's atomic theory and with Gay-Lussac's experimentally verified law of combining volumes.

Avogadro's hypothesis was almost entirely ignored or rejected by his contemporaries, although his reasoning appears clear today and his experimental support was convincing then. For nearly 50 years, Ampere was apparently the only scientist of note to accept Avogadro's ideas.

In 1858 Stanislao Cannizzaro revived Avogadro's hypothesis that equal volumes of gases at the same pressure and temperature contain equal numbers of mole-

Table 3.1.

Substance	*Mass of gas in 22.4 liters at 0°C and 1.0 atm, g*	*Composition of gas, g*
Hydrogen	2.0	2.0 H
Oxygen	32.0	32.0 O
Nitrogen	28.0	28.0 N
Chlorine	71.0	71.0 Cl
Water	18.0*	2.0 H and 16.0 O
Hydrogen chloride	36.5	1.0 H and 35.5 Cl
Ammonia	17.0	3.0 H and 14.0 N
Nitrous oxide	44.0	28 N and 16 O
Nitric oxide	30.0	14 N and 16 O
Carbon monoxide	28.0	12 C and 16 O
Carbon dioxide	44.0	12 C and 32 O

*Actually determined at higher temperature and then calculated for these conditions, using equation (3.10).

cules. He both provided better justification for the hypothesis than had Avogadro and applied it usefully to the vexing problems of atomic weights and molecular formulas.

Rather than become involved with the reasoning followed by Avogadro and Cannizzaro, we shall accept the validity of Avogadro's law on the basis of equation (3.13) and proceed with applications. Let us imagine that we have a 22.4 liter container that can be filled with various gases at 0°C and 1.0 atm. (Note that we can choose any size for the container, but this choice proves convenient.) The difference in weight of the empty container and the container with gas gives the mass of gas. Further, let us assume that we know the composition of each gas in terms of the percentage of each constituent element and therefore can calculate the mass of each element in the gas weighed in the 22.4 liter container. These data, which were also available to Cannizzaro, are given in Table 3.1.

In a sufficient number of compounds of any element, probably at least one compound will have only one atom of that element per molecule. Other compounds might have two, three, or other integral numbers of atoms of that element per molecule. This statement is almost equivalent to Dalton's law of simple multiple proportions. The existence of Table 3.1 is itself a confirmation of the law of constant composition, because if this law were not valid it would be impossible to give a single list of compositions for a group of compounds.

We now set about deducing some relative atomic weights from the data in Table 3.1. Because we are interested in relative weights (at least for now), we may choose any reference atomic weight that we want. In order to be consistent with chemical practice prior to 1961, let us choose exactly 16 atomic mass units as the atomic weight of oxygen. All of the substances containing oxygen that are listed in Table 3.1 contain 16.0 g or some integral multiple of 16.0 g of oxygen in 22.4 liters at 0°C and 1.0 atm. Not one compound contains less than 16.0 g of oxygen under these conditions. We therefore conclude that molecules of water, nitrous oxide, and nitric oxide each contain one atom of oxygen, while molecules of elementary oxygen and carbon dioxide each contain two atoms of oxygen. *These deductions are all based on the idea that there are equal numbers of molecules in equal volumes of all gases at the same pressure and temperature.*

In exactly the same way we deduce that each molecule of hydrogen chloride contains one atom of hydrogen, while each molecule of water and ammonia contains two and three atoms of hydrogen. The data in Table 3.1 also show that each molecule of ammonia and nitric oxide contains one atom of nitrogen, while each molecule of elementary nitrogen and nitrous oxide contains two nitrogen atoms. We also can deduce that each molecule of hydrogen chloride contains one atom of chlorine and that each molecule of elementary chlorine contains two atoms of chlorine.

We put all of this information together to write the following formulas for substances listed in Table 3.1: H_2, O_2, N_2, Cl_2, H_2O, HCl, NH_3, N_2O, and NO. Similar data for other compounds of carbon also permit us to establish that molecules of carbon monoxide and carbon dioxide contain one atom each of carbon so that their formulas are CO and CO_2. Since we have chosen a value for the atomic weight of oxygen, the masses listed in Table 3.1 can be used with these formulas to establish the familiar atomic weights of the other elements.

All of the gas samples considered in Table 3.1 contain the same number of molecules. This important number is called **Avogadro's number** and is assigned the symbol N. There are N molecules of O_2 in 22.4 liters at 0°C and 1.0 atm, and the mass of these N molecules of O_2 is 32.0 g. Because each molecule of oxygen contains two atoms of oxygen, there are $2N$ atoms in 32.0 g of O_2. We can calculate a value of N if we know the mass of a single atom of oxygen, or we can determine a value for N in some other way and then calculate the mass of a single atom of oxygen.

Since atoms and molecules are extremely small and so many of them are involved in ordinary chemical processes, the large number represented by N is a convenient counting unit in chemistry just as a dozen is a convenient counting unit in a bakery. We now define a collection of N objects (usually atoms, molecules, or ions) as being one **mole.** Thus N molecules of oxygen are one mole of O_2 molecules and contain two moles of O atoms. We also define the mass (expressed in grams) of N objects as the atomic weight or molecular weight. Thus the atomic weight of oxygen (atoms) is 16.0 g $mole^{-1}$ and the molecular weight of oxygen (molecules) is 32.0 g $mole^{-1}$.

With the discovery of the less abundant isotopes of oxygen, $^{17}_{8}O$ and $^{18}_{8}O$, it was recognized that so-called chemical and physical atomic weight scales should be slightly different. Physical atomic weights were determined by comparing the mass of oxygen ions taken one at a time with ions of other elements also taken one at a time. Thus the mass spectroscopist compared masses of various isotopes to the mass of $^{16}_{8}O$. But the chemist who made use of mass and volume data for gases (or weight relations in certain chemical reactions) ultimately compared his mass of interest to the mass of the same number of oxygen atoms or molecules, all having the distributions of isotopes that occur in nature. Because naturally occurring oxygen is mostly $^{16}_{8}O$, the discrepancy was small—amounting to about 2 parts in 10,000 or 0.02%.

In 1961 a new atomic weight scale based on $^{12}_{6}C$ was adopted for both chemistry and physics. One atomic mass unit is now defined as exactly one twelfth of the mass of one atom of $^{12}_{6}C$, and the atomic weight of $^{12}_{6}C$ is therefore exactly 12 atomic mass units. Twelve grams of $^{12}_{6}C$ is called one mole of this isotope of carbon and consists of Avogadro's number of atoms. On this scale the atomic weight of the naturally occurring mixture of isotopes of carbon is 12.011 atomic mass units per average atom of carbon or 12.011 g mole^{-1} of carbon consisting of the naturally occurring mixture of isotopes. The atomic weight of the naturally occurring mixture of oxygen isotopes is 15.9994 atomic mass units per atom or 15.9994 g mole^{-1} of O atoms. We shall find that atomic and molecular weights expressed in terms of grams per mole are especially useful in chemistry.

After consideration of the magnitude of Avogadro's number, we shall turn to some chemical calculations involving atomic and molecular weights in connection with properties of gases.

Avogadro's Number

Avogadro's number, N, is the number of molecules of any gas in 22.4 liters of that gas at 0°C and 1.0 atm. Its principal importance in chemistry is as a counting unit that is used to define the mole, which has mass expressed in grams equal to the appropriate sum of atomic weights. There are many ways to determine this number, some of which are discussed here.

One method for calculation of Avogadro's number makes use of data for the electron and proton. Combination of Millikan's value for the charge of the electron with Thomson's value for the charge/mass ratio gives a value for the mass of the electron. Recognizing that the charge of the proton is of equal magnitude to that of the electron permits us to calculate the mass of the proton from Thomson's value of its charge/mass ratio. Then adding the mass of the proton to the mass of the electron gives us the mass of a single atom of hydrogen. Consideration of data for gaseous hydrogen and oxygen shows that the atomic weight of hydrogen is 1.0 g mole^{-1} (compared to 16.0 g mole^{-1} for oxygen). Thus we can calculate Avogadro's number from

$$\frac{\text{g mole}^{-1}}{N \text{ atoms mole}^{-1}} = \text{g atom}^{-1} \quad \text{or} \quad N \text{ atoms mole}^{-1} = \frac{\text{g mole}^{-1}}{\text{g atom}^{-1}}$$

It is also possible to calculate values of Avogadro's number by application of Einstein's theory of Brownian motion to data for small particles suspended in a liquid. Another approach involves relation of the viscosity of a gas to the kinetic theory of gases. Still another method involves actual counting of single molecules as alpha particles and measurement of the resulting helium as in problem 15 at the end of this chapter. These methods do not lead to a highly accurate value for N, but do provide strong support for the meaning that we attach to Avogadro's number.

Another method for calculation of Avogadro's number makes use of results of x-ray investigations of crystalline substances. We illustrate this method with a simple calculation applied to sodium chloride.

Study of the scattering of x rays by crystalline sodium chloride has led to determination of the distance between adjacent ions and the arrangement of the

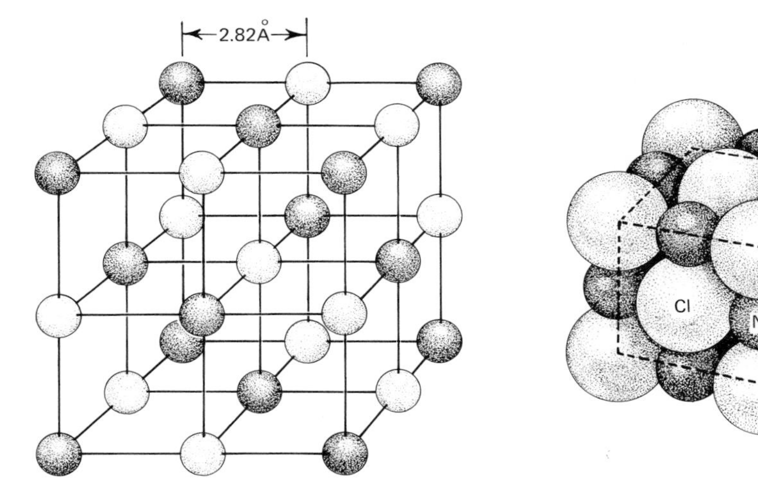

Figure 3.3. Illustration of the sodium chloride crystal lattice, in which dark and light circles represent sodium and chloride ions, respectively. The distance between centers of ions is 2.82 Å $= 2.82 \times 10^{-8}$ cm. The illustration on the left most clearly shows the arrangement of ions, but the illustration on the right is more realistic.

ions in the crystal as shown in Figure 3.3. The density of crystalline sodium chloride has also been determined to be 2.164 g cm^{-3}.

We now use atomic weights of sodium and chlorine, which have been determined in a variety of ways without making use of Avogadro's number, to calculate that the mass of one mole of NaCl is 58.443 g. Although it is common to say that the molecular weight of NaCl is 58.443 g mole^{-1}, it should be recognized that there are no NaCl molecules in a crystal of NaCl and that what is meant is that 58.443 g is the mass of one mole (Avogadro's number) of sodium ions plus the mass of one mole (Avogadro's number) of chloride ions.

We now put all of this information together for the desired calculation of Avogadro's number. First, we calculate from the information given and density = mass/volume that the volume of one mole of NaCl is

$$V = \frac{58.443 \text{ g mole}^{-1}}{2.164 \text{ g cm}^{-3}} = 27.00 \text{ cm}^3 \text{ mole}^{-1}$$

The length of a side of a cube with $V = 27.00 \text{ cm}^3$ is 3.00 cm. Along each 3.00-cm edge there are $3.00/(2.82 \times 10^{-8}) = 1.064 \times 10^8$ ions and the whole cube contains $(1.064 \times 10^8)^3 = 12.04 \times 10^{23}$ ions. Because half of these are sodium ions and half are chloride ions, we have 6.02×10^{23} ions of each kind in 58.443 g of sodium chloride. Recalling that 58.443 g of NaCl is one mole of sodium chloride, we see that the value of Avogadro's number is 6.02×10^{23}.

The best value for Avogadro's number from all sources is 6.02252×10^{23}, which is usually shortened to 6.02×10^{23} for calculations in this book.

We now summarize and illustrate once again what we mean by **mole.** When we say that we have one mole of oxygen, we mean that we have $N = 6.02 \times 10^{23}$ molecules of oxygen, each consisting of two atoms of oxygen and having total mass of $2 \times 16 = 32$ g mole^{-1}. One mole of gaseous methane (CH_4) consists of Avogadro's number of carbon atoms and four times Avogadro's number of hydrogen atoms, having a total mass of $12.0 + 4(1.0) = 16$ g mole^{-1}.

One mole of a solid or liquid substance also contains Avogadro's number of molecules. In some substances the atomic and molecular units are identical, as in sodium and other metals. Naphthalene, on the other hand, consists of molecules for which we write the formula $C_{10}H_8$ to indicate that each molecule contains ten carbon atoms and eight hydrogen atoms. The molecular weight of naphthalene is $(10 \times 12.0) + (8 \times 1.0) = 128$ g mole^{-1}. This mass of naphthalene consists of Avogadro's number of naphthalene molecules.

Since a crystal of sodium chloride or other ionic substance consists of discrete ions, there is no molecular unit analogous to that found in a crystal of naphthalene. Usually we arbitrarily write the simplest formula for such ionic substances and consider the appropriate sum of atomic weights to be the molecular weight. Thus the molecular weight of NaCl is $23.0 + 35.5 = 58.5$ g mole^{-1}. This mass of sodium chloride contains Avogadro's number (one mole) of sodium ions and the same number (one mole) of chloride ions. Similarly, the molecular weight of sodium sulfate, Na_2SO_4, is $(2 \times 23.0) + 32.0 + (4 \times 16.0) = 142.0$ g mole^{-1}. Each mole of sodium sulfate contains $2N$ sodium ions and N atoms of sulfur combined with $4N$ atoms of oxygen to form N sulfate ions.

A General Ideal Gas Equation

From the number of molecules $\mathcal{N}$ in a particular sample and Avogadro's number N we can calculate the number of moles in that sample as

$$\frac{\mathcal{N} \text{ molecules}}{N \text{ molecules mole}^{-1}} = n \text{ moles}$$

We now multiply and divide the right side of equation (3.12) from the kinetic theory of gases by N to obtain

$$PV = \frac{2\,\mathcal{N} \times \text{constant} \times TN}{3N}$$

Recognizing that $n = \mathcal{N}/N$ gives us

$$PV = n\left(\frac{2 \times \text{constant} \times N}{3}\right)T$$

Finally, we represent $(2 \times \text{constant} \times N)/3$ by R to obtain the very useful general ideal gas equation

$$PV = nRT \tag{3.14}$$

in which the constant R is called the gas constant and n represents the number of moles of gas.

One way to obtain a numerical value for R to use for subsequent calculations with this equation is to solve for R as

$$R = PV/nT$$

From previous discussion, we know that 1.0 mole of gas occupies 22.4 liters at 1.0 atm and 0°C = 273°K. Substitution into the equation for R gives

$$R = \frac{1.0 \text{ atm} \times 22.4 \text{ liters}}{1.0 \text{ mole} \times 273 \text{ deg}} = 0.082 \text{ liter atm deg}^{-1} \text{ mole}^{-1}$$

Example Problem 3.4. Consider the determination of the molecular formula and molecular weight of grain alcohol. Because grain alcohol is a liquid with low vapor pressure at 0°C, it is impossible to do experiments on gaseous alcohol at one atm at that temperature. We must therefore make use of PVT and mass data at some other temperature and pressure. We must also give attention to transforming per cent composition data derived from chemical analyses into the desired chemical formula.

Some experimental data for grain alcohol follow:

1. Chemical analyses indicate that grain alcohol is 52.2% carbon (by weight), 13.0% hydrogen, and 34.8% oxygen.
2. 3.00 liters of alcohol vapor at 100°C and 700 mm Hg pressure weigh 4.15 g.

We begin by solving the general ideal gas equation (3.14) for n to obtain

$$n = PV/RT$$

Use of this equation for numerical calculation requires consideration of the units in which the experimental data are expressed.

We first express the given pressure in terms of atm as

$$P = \frac{700 \text{ mm Hg}}{760 \text{ mm Hg atm}^{-1}} = 0.922 \text{ atm}$$

The absolute temperature on the Kelvin scale is given by

$$T = 273 + 100 = 373\text{°K}$$

The volume is already expressed in liters and we have $R = 0.082$ liter atm deg^{-1} mole^{-1}. Substituting in the equation for n gives

$$n = \frac{0.922 \text{ atm} \times 3.00 \text{ liters}}{0.0820 \text{ liter atm deg}^{-1} \text{ mole}^{-1} \times 373 \text{ deg}} = 0.0904 \text{ mole}$$

This result tells us that 4.15 g of alcohol are 0.0904 mole, so the molecular weight (mass of one mole) is

$$M = 4.15 \text{ g}/0.0904 \text{ mole} = 46.0 \text{ g mole}^{-1}$$

Now we use the per cent composition data with this molecular weight to calculate the amounts of carbon, hydrogen, and oxygen in one mole of alcohol as follows:

$$\begin{aligned} 0.522 \times 46.0 &= 24.0 \text{ g carbon} \\ 0.130 \times 46.0 &= 6.0 \text{ g hydrogen} \\ 0.348 \times 46.0 &= 16.0 \text{ g oxygen} \end{aligned}$$

We divide each of these weights by the corresponding atomic weight to obtain the number of moles of carbon, hydrogen, and oxygen in one mole of alcohol as follows:

$$\begin{aligned} 24.0 \text{ g}/12.0 \text{ g mole}^{-1} &= 2.0 \text{ moles carbon} \\ 6.0 \text{ g}/1.0 \text{ g mole}^{-1} &= 6.0 \text{ moles hydrogen} \\ 16.0 \text{ g}/16.0 \text{ g mole}^{-1} &= 1.0 \text{ mole oxygen} \end{aligned}$$

The formula for grain alcohol is therefore written as C_2H_6O.

We do not have sufficient information in this problem to permit us to go from the molecular formula C_2H_6O to a structural formula that will show how the atoms are arranged in molecules of grain alcohol. Other kinds of evidence discussed later in this book show that the structure of grain alcohol may be represented by

```
    H  H
    |  |
 H—C—C—O—H  ■
    |  |
    H  H
```

Sometimes it is desirable to calculate the density of a gas under some specified conditions, or we might want to calculate the molecular weight of a gas from an experimentally determined density. Although both of these calculations can be made with equations already given, they can be done more conveniently after writing the general ideal gas equation $PV = nRT$ in a different form.

The number of moles of a substance is the mass of that substance divided by its molecular weight. Representing mass by m and molecular weight by M, we have

$$n = m \text{ g}/M \text{ g mole}^{-1} = \text{moles}$$

Substitution of m/M for n in $PV = nRT$ gives

$$PV = (m/M)RT$$

Next we multiply both sides of this equation by M and divide both sides by V to obtain

$$PM = (m/V)RT$$

Density is defined as mass per unit volume, as shown by the equation $d = m/V$; therefore we substitute d for m/V in the equation above to obtain

$$PM = dRT$$

Simple rearrangement now gives two directly useful equations:

$$M = dRT/P \qquad (3.15)$$

$$d = PM/RT \qquad (3.16)$$

Here it is well to recognize that it is better to be able to derive these equations from $PV = nRT$ than to memorize them.

Equation (3.15) can be applied to the solution of Example Problem 3.4 if we recognize that the given mass and volume permit calculation of the density of grain alcohol vapor under the stated conditions.

Example Problem 3.5. Calculate the density of gaseous methane at 45 °C and 620 mm Hg pressure.

Equation (3.16) is used for this calculation. We have $T = 273 + 45 = 318\,°\text{K}$ and $P = 620 \text{ mm Hg}/760 \text{ mm Hg atm}^{-1} = 0.817 \text{ atm}$. The molecular weight of methane is calculated from its formula CH_4 to be $12.0 + 4(1.0) = 16.0 \text{ g mole}^{-1}$. Putting these data into equation (3.16) gives

$$d = \frac{0.817 \text{ atm} \times 16.0 \text{ g mole}^{-1}}{0.082 \text{ liter atm deg}^{-1} \text{ mole}^{-1} \times 318 \text{ deg}} = 0.50 \text{ g liter}^{-1} \quad ■$$

Mixtures of Gases

In 1801 John Dalton discovered by experiment that the total pressure exerted by a mixture of gases is equal to the sum of the partial pressures of the various gases in the mixture, where the **partial pressure** is defined as the pressure a gas would exert if it were alone in the container. We may summarize Dalton's law of partial pressures by the equation

$$P_t = P_1 + P_2 + P_3 + \cdots \tag{3.17}$$

where P_t represents the total pressure and the numbered P's represent the partial pressures of the various gases in the mixture. Another way to write this equation is

$$P_t = \Sigma P_i \tag{3.18}$$

where the ΣP_i means to add or sum all of the individual partial pressures. Dalton's law of partial pressures is another ideal gas law that is exactly valid only in the limit of low pressure but is closely obeyed by most gases at moderate pressures.

The first and third statements (page 40) that describe the ideal gas account for the low pressure validity of Dalton's law of partial pressures. Because the molecules occupy a negligible fraction of the total volume of the container, and because they do not interact appreciably with one another, each molecule strikes the walls with the same frequency and the same force it would if no other molecules were present. Each gas therefore exerts a pressure that is independent of other gases that may be present. Hence Dalton's law is in accord with and should be predicted by the kinetic theory.

We find it useful to have an equation that relates the partial pressure of one gas in a mixture to the total pressure and the composition of the mixture. Taking each component of the mixture to be ideal, we can write

$$P_1 = n_1RT/V, \quad P_2 = n_2RT/V, \quad \cdots$$

and substitute these equations into (3.17) to obtain

$$P_t = (n_1 + n_2 + \cdots)(RT/V) = (\Sigma n_i)(RT/V), \tag{3.19}$$

where the Σ indicates that the various n_i are to be summed.

Now suppose that we are especially interested in the gas represented by subscript 1. We rearrange $P_1V = n_1RT$ to $RT/V = P_1/n_1$ and substitute in (3.19) to obtain

$$P_t = (\Sigma n_i)(P_1/n_1)$$

Solving for the desired P_1 gives

$$P_1 = \frac{n_1}{\Sigma n_i} P_t \tag{3.20}$$

This equation permits us to calculate the partial pressure of gas 1 from knowledge of the total pressure, the number of moles of gas 1 in the system, and the total number of moles of gas in the system.

It is useful to write equation (3.20) in general terms as

$$P_i = \frac{n_i}{\Sigma n_i} P_t \tag{3.21}$$

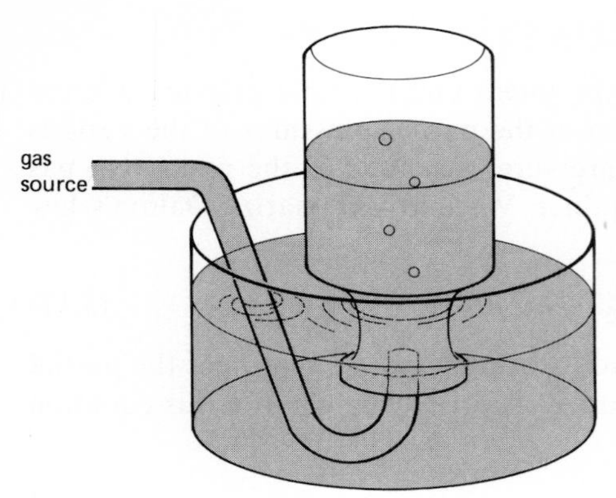

Figure 3.4. Schematic illustration of a simple arrangement for collecting and measuring the amount of gas. It is necessary to correct for the vapor pressure of the liquid and to assume (or prove by experiment) that the gas does not react with or dissolve in the liquid enough to cause a significant error.

The expression $n_1/\Sigma n_i$ is commonly called the mole fraction of component 1 and the general expression $n_i/\Sigma n_i$ is commonly called the mole fraction of component i. It is common to use X to represent the mole fraction and a subscript to indicate the component to which this X refers. Thus we can write equation (3.21) as

$$P_i = X_i P_t \quad \text{or} \quad X_i = P_i/P_T \tag{3.22}$$

Example Problem 3.6. Suppose that hydrogen is collected over water at 25 °C in an apparatus like that pictured in Figure 3.4. How much H_2 (expressed in moles) has been collected when the volume of gas is 223 ml under atmospheric pressure of 74.2 cm Hg?

Dalton's law of partial pressures tells us that the total pressure of the gas, which must be equal to the atmospheric pressure, is given by the sum of the partial pressures. Therefore we have

$$P_{\text{atm}} = P_{\text{H}_2} + P_{\text{H}_2\text{O}}$$

We know that $P_{\text{atm}} = 74.2$ cm Hg and find in Appendix VI that the vapor pressure of water at 25 °C is 23.8 mm Hg (= 2.38 cm Hg). We can calculate

$$P_{\text{H}_2} = P_{\text{atm}} - P_{\text{H}_2\text{O}} = 74.2 - 2.4 = 71.8 \text{ cm Hg}$$

This pressure is conveniently used in subsequent calculations when expressed in atmospheres as 71.8 cm Hg/76.0 cm Hg atm^{-1} = 0.945 atm.

Solving the ideal gas equation for n, we find

$$n = \frac{PV}{RT} = \frac{0.945 \text{ atm} \times 0.223 \text{ liter}}{0.0820 \text{ liter atm deg}^{-1} \text{ mole}^{-1} \times 298 \text{ deg}} = 0.00863 \text{ mole H}_2 \quad \blacksquare$$

Real Gases

Much effort has been directed at finding PVT equations (called equations of state) that agree better with PVT data at high pressures than does the ideal gas equation. To illustrate the nature of real gas behavior, we introduce the compressibility factor represented by Z and defined as

$$Z = PV/nRT$$

This equation can be used more conveniently if we recognize that V/n is the volume per mole or the volume of one mole or the **molar volume,** which we can represent by $\bar{V}$ and thus obtain

$$Z = P\bar{V}/RT \tag{3.23}$$

For the ideal gas $Z = 1.0$ and is independent of both pressure and temperature. But for real gases Z depends on both pressure and temperature, as illustrated by the graphs in Figure 3.5.

How can we modify the ideal gas equation $PV = nRT$ so that the result will provide a better description of real gas behavior? First we note that for an ideal gas

$$\frac{V}{n} = \bar{V} = \frac{RT}{P}$$

This equation predicts that $\bar{V}$ approaches zero as the temperature gets very small or the pressure gets very large. We know that $\bar{V}$ for a real gas cannot go to zero, but must approach a nearly constant value that is related to the actual volume of the molecules. Thus we write

$$\bar{V} = b + \frac{RT}{P}$$

as a first improvement on the ideal gas equation. Multiplication of this equation by P/RT gives

$$\frac{P\bar{V}}{RT} = Z = 1 + \frac{b}{RT}P \tag{3.24}$$

Equation (3.24) represents a straight line of slope b when Z is plotted against P and therefore accounts fairly well for the behavior of H_2 at 273°K and N_2 at 670°K illustrated in Figure 3.5.

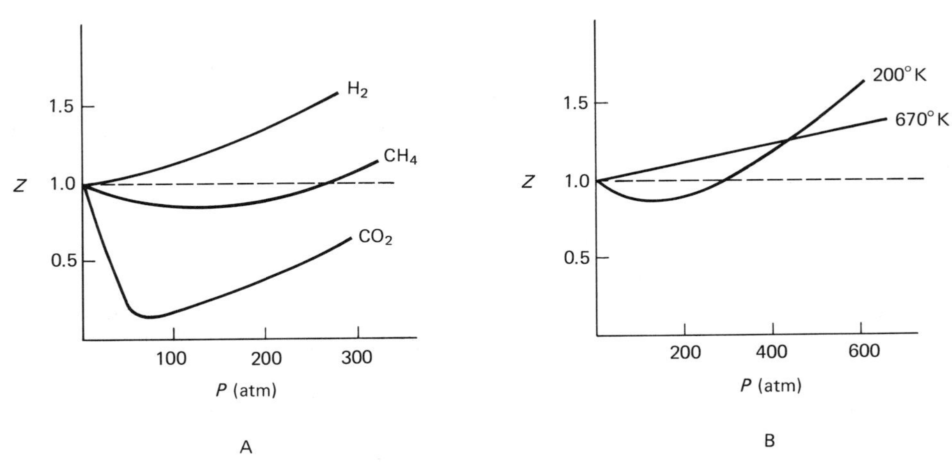

Figure 3.5. Graph A shows deviations from ideal gas behavior (dashed line) for gases at 273°K, while graph B shows the effect of temperature on these deviations for nitrogen.

A further improvement can be made by allowing for the attractions between molecules of a real gas. These attractive forces account for the liquefaction of gases as the pressure is increased or the temperature is lowered. The effect of these attractive forces is to make the actual pressure of a real gas lower than that of an ideal gas in the same volume at the same temperature. The magnitude of this effect clearly increases as $\bar{V}$ decreases and the gas molecules are crowded closer together, which can be expressed by setting the effect of intermolecular attractions proportional to $1/\bar{V}^2$, as first deduced by van der Waals.

Now we rearrange equation (3.24) to

$$P = \frac{RT}{\bar{V} - b}$$

and remember that the effect of intermolecular forces is to decrease the pressure. We therefore subtract $a/\bar{V}^2$ from the right side to obtain

$$P = \frac{RT}{\bar{V} - b} - \frac{a}{\bar{V}^2} \tag{3.25}$$

This equation is called the van der Waals equation in honor of the man who originated it in 1873.

To show how the van der Waals equation accounts for deviations from ideal gas behavior we multiply equation (3.25) through by $\bar{V}/RT$ to obtain

$$\frac{P\bar{V}}{RT} = Z = \frac{\bar{V}}{\bar{V} - b} - \frac{a}{RT\bar{V}}$$

At high temperatures the term on the far right becomes relatively small and we obtain

$$Z \cong \frac{\bar{V}}{\bar{V} - b} > 1 \qquad \text{(high temp. approx.)}$$

At lower temperatures the term on the far right becomes larger and cannot be neglected, but $\bar{V}/(\bar{V} - b) \cong 1$ at moderate pressures, so we obtain

$$Z \cong 1 - \frac{a}{RT\bar{V}} < 1 \qquad \text{(low temp. approx.)}$$

We see that the van der Waals equation accounts for $Z > 1$ at high temperatures and $Z < 1$ at low temperatures. High and low in this sense are taken relative to the temperature of condensation of the gas to a liquid or solid. Thus 273°K is a high temperature for hydrogen but a low temperature for carbon dioxide (see Figure 3.5A).

Although the van der Waals equation is only fairly good for quantitative representation of real gas behavior at high pressures or low temperatures, it works well at moderate pressures and temperatures and is especially interesting because it can also account for some of the properties of liquids and liquid-gas equilibria.

Molecular Speeds, Diffusion, and Heat Capacity

It has already been shown that the kinetic theory of gases accounts for Avogadro's law and leads to the useful equation $PV = nRT$. In this section we illustrate two more applications of the kinetic theory.

We begin by writing equation (3.11) specifically for Avogadro's number of molecules so that $\mathcal{N}$ becomes N, giving us

$$PV = \frac{2N}{3}\left[\frac{ms^2}{2}\right] \tag{3.11a}$$

For one mole of gas we know that $PV = RT$, and we can represent the average kinetic energy $ms^2/2$ by $\mathscr{E}$ to obtain

$$RT = \frac{2N}{3}\left[\frac{ms^2}{2}\right] = \frac{2N\mathscr{E}}{3} \tag{3.26}$$

We solve equation (3.26) for s^2 to obtain

$$s^2 = \frac{3RT}{Nm} \tag{3.27}$$

Taking square roots of both sides of this equation gives us an equation that permits calculation of the average speed of the molecules in a gas:

$$s = \left(\frac{3RT}{Nm}\right)^{\frac{1}{2}} \tag{3.28}$$

Equation (3.28) shows that the average velocity of molecules in a gas is proportional to the square root of the temperature and inversely proportional to the square root of the mass. Light molecules collide more frequently with the walls of the container than do heavier molecules, but the heavier molecules undergo a greater change in momentum per collision. These two factors cancel each other so that gas pressure is independent of the mass of the molecules.

Numerical calculations with equations (3.27) and (3.28) are most conveniently done after making either of two substitutions. Avogadro's number times the mass of a single molecule (Nm) is the mass of one mole of molecules, which is the molecular weight represented by M. Thus $3RT/Nm = 3RT/M$. Or we may replace R/N by another constant k, called the Boltzmann constant. Thus $3RT/Nm = 3kT/m$. Since R is the gas constant per mole of gas, $R/N = k$ is the gas constant for one molecule of gas. The molar gas constant R is expressed in terms of energy per degree per mole, while the molecular gas constant k is expressed in terms of energy per degree per molecule. We now write equation (3.28) as

$$s = \left(\frac{3RT}{M}\right)^{\frac{1}{2}} = \left(\frac{3kT}{m}\right)^{\frac{1}{2}} \tag{3.29}$$

Before proceeding with numerical calculations and applications of equations (3.27) through (3.29), it is necessary to consider what we mean by average speed. More detailed treatment than is presently appropriate shows that the s^2 that appears in the various equations we have used is actually the average of squared speeds of the molecules. We illustrate this kind of average with calculations for an imaginary system of three molecules having speeds of 4, 6, and 8. The average value of s^2 for this system is computed by summing the squares of the speeds and dividing by the number of molecules. Thus the average $s^2 = (4^2 + 6^2 + 8^2)/3 = 38.7$, and we take the square root to obtain the average $s = 6.2$. This average speed that we have denoted by s is commonly called the root mean square speed to distinguish it from the ordinary average speed. For our particular simple example, the root mean square speed is 6.2, while the ordinary average speed is seen to be $(4 + 6 + 8)/3 = 6.0$. Although important for some purposes, the difference between the various ways of averaging is not large enough to require further attention here.

It has long been known that gases will leak (diffuse) through a small hole in a container. Many years ago Thomas Graham carried out quantitative studies of the diffusion of gases through porous plugs, and by about 1830 he had established that the rate of diffusion is inversely proportional to the square root of the molecular weight of the gas. This statement, now called Graham's law, is commonly summarized by the equation

$$\frac{\text{rate of diffusion of gas 1}}{\text{rate of diffusion of gas 2}} = \left(\frac{M_2}{M_1}\right)^{\frac{1}{2}} \tag{3.30}$$

Since the rate of diffusion is proportional to the average speed of the molecules, we see that equation (3.29) from the kinetic theory leads to Graham's law as expressed by equation (3.30).

Several important separation processes are based on the mass dependence of rate of diffusion. For example, ^{235}U is separated from ^{238}U by means of a process based on the difference in rates of diffusion of gaseous $^{235}UF_6$ and $^{238}UF_6$. Because the difference in diffusion rates for these closely similar substances is small, the actual separation process requires a complicated series of porous barriers and arrangements for recycling.

An equation that permits calculation of the average kinetic energy of a molecule of gas is obtained by solving (3.26) for $\mathscr{E}$:

$$\mathscr{E} = \frac{3}{2}\frac{RT}{N} = \frac{3}{2}kT \tag{3.31}$$

The total kinetic energy of all of the molecules in a mole of gas is given by the average energy times the number of molecules; therefore we can obtain this energy, represented by $E = N\mathscr{E}$, as

$$E = \tfrac{3}{2}RT \tag{3.32}$$

Heating a gas results in increasing the energy of the gas and can cause the gas to do work on its surroundings. In the special case where the gas is confined in a constant volume container, no work can be done on the surroundings and all of the added heat results in an increase in energy of the gas. Thus for constant volume processes we may equate energy changes with amount of heat transferred. Equation (3.32) permits calculation of the energy change (and thus also the amount of heat transferred in a constant volume process) when the temperature increases from T_1 to T_2 as

$$E_2 - E_1 = \tfrac{3}{2}R(T_2 - T_1)$$

Remembering that the heat capacity may be defined as the heat required to increase the temperature of a mole of substance by 1°, we have

$$C_v = \tfrac{3}{2}R \tag{3.33}$$

In this equation we have written C_v to emphasize that we are referring to a heat capacity that applies to constant volume conditions. A heat capacity determined at constant pressure rather than at constant volume would have a larger value because part of the heat transferred would cause work to be done in the expansion of the gas against the atmosphere.

Before we can compare equation (3.33) with the results of measurements, we must consider the limitations of the equation. First, it is restricted to gases under those PVT conditions for which the ideal gas model and equations are satisfactory. Second, we must recognize that our derivation was concerned only with kinetic energy. Thus we cannot consider gases for which energies of rotation and vibration of the molecules are important. These considerations limit the validity of equation (3.33) to monatomic gases (no rotational or vibrational energies possible for these) at low densities.

Equation (3.33) leads to

$$C_v = 1.5 \times 1.987 \text{ cal deg}^{-1} \text{ mole}^{-1} = 2.98 \text{ cal deg}^{-1} \text{ mole}^{-1}$$

for the constant volume heat capacity due to translational motion and the resulting kinetic energy of the molecules. Heat capacities have been measured at constant volume for the monatomic gases He, Ne, Ar, Kr, Xe, and Hg and found to be in excellent agreement with this value.

Example Problem 3.7. Calculate the average speed of oxygen molecules at normal room temperature of ~25°C.

From equation (3.29) we have

$$s = \left(\frac{3RT}{M}\right)^{\frac{1}{2}} = \left(\frac{3 \times 8.3 \times 10^7 \text{ ergs deg}^{-1} \text{ mole}^{-1} \times 298 \text{ deg}}{32 \text{ g mole}^{-1}}\right)^{\frac{1}{2}}$$

In this equation we have expressed R in the cgs system (see Chapter 1) and the temperature on the Kelvin scale. After doing the indicated arithmetic and cancelling some of the units, we have

$$s = 4.8 \times 10^4 \text{ (g cm}^2 \text{ sec}^{-2})^{\frac{1}{2}} \text{ g}^{-\frac{1}{2}}$$

in which we have made use of the knowledge that the dimensions of the erg (dyne cm) are g cm^2 sec^{-2}. Again cancelling units, we obtain the desired average speed as

$$s = 4.8 \times 10^4 \text{ cm sec}^{-1}$$

This same result could have been obtained less conveniently from

$$s = \left(\frac{3kT}{m}\right)^{\frac{1}{2}}$$

For this calculation we use $k = 1.38 \times 10^{-16}$ erg deg^{-1} molecule^{-1} and calculate the mass of a single molecule of O_2 as 32.0 g mole^{-1}/6.02 × 10^{23} molecules mole^{-1}. ■

Example Problem 3.8. It requires 7.50 cal to increase the temperature of 10.0 g of monatomic gas by 10°C at constant volume. Calculate the molecular weight of the gas.

Recognizing that it would require 7.50/10 = 0.750 cal to cause the same temperature change for 1.0 g of gas permits us to calculate the heat capacity for 1.0 g (sometimes called the specific heat) as

$$0.750/10.0 = 0.0750 \text{ cal deg}^{-1} \text{ g}^{-1}$$

We equate this specific heat times molecular weight with $C_v = 2.98$ cal deg^{-1} mole^{-1} from the kinetic theory applied to a monatomic gas and obtain

$$M \text{ g mole}^{-1} \times 0.075 \text{ cal deg}^{-1} \text{ g}^{-1} = 2.98 \text{ cal deg}^{-1} \text{ mole}^{-1}$$

and then

$$M = \frac{2.98}{0.075} = 39.8 \text{ g mole}^{-1}$$

for the desired molecular weight, which permits us to identify the gas as argon. ■

References

Feifer, N.: The relationship between Avogadro's principle and the law of Gay-Lussac. *J. Chem. Educ.*, **43**:411 (1966).

Hawthorne, R. M., Jr.: Avogadro's number: early values by Loschmidt and others. *J. Chem. Educ.*, **47**:751 (1970).

Hildebrand, J. H.: *An Introduction to Molecular Kinetic Theory.* Reinhold Publishing Corp., New York, 1963 (paperback).

Nash, L. K.: *Stoichiometry.* Addison-Wesley Publishing Co., Inc., Reading, Mass., 1966 (paperback).

Ruckstahl, A.: Thomas Graham's study of the diffusion of gases. *J. Chem. Educ.*, **28**:594 (1951).

Swinbourne, E. S.: The van der Waals gas equation. *J. Chem. Educ.*, **32**:366 (1955).

Problems

1. How many atoms are there in 4.8 g of carbon?
2. What is the mass (in grams) of one atom of carbon?
3. How many grams of water would you weigh out to obtain 2.5 moles of water?
4. What volume does 1.0 mole of any gas occupy at 0.50 atm and 335°K?
5. A sample of gas originally at 560 mm Hg pressure and 20°C occupied a volume of 16.1 liters. What volume does this gas occupy after it is heated to 345°K and the pressure is increased to 1.5 atm?
6. At 0°C and 1.0 atm there are $N = 6.02 \times 10^{23}$ atoms of argon in 22.4 liters. The radius of an atom of argon is approximately 1.8×10^{-8} cm. Calculate the volume of N spherical argon atoms. What percentage of the 22.4 liters is actually occupied by N argon atoms?
7. The density of a gas at 0°C and 1.0 atm is 0.002856 g cm^{-3}. What is the molecular weight of this gas?
8. What is the density of propane (C_3H_8) gas at 35°C and 0.50 atm?
9. A compound of carbon and hydrogen contains 82.8% carbon and 17.2% hydrogen.

The density of the gaseous compound is 1.66 g $liter^{-1}$ at 665 mm Hg and 100°C. What is the molecular formula of this compound?

10. A chemist uses a pump to produce a vacuum in which the pressure is 10^{-6} mm Hg at 25°C. How many molecules remain in such a vacuum with total volume of 1.0 liter?

11. Assuming that 20% of the molecules in air are O_2 and that 80% are N_2, calculate the density of air at 25°C and 740 mm Hg.

12. Why do different gases have different efficiencies in conducting heat away from hot bodies?

13. Both He and H_2 have been used to fill balloons designed to carry people or instruments aloft. What are the relative efficiencies of these two gases for this purpose?

14. What is the total pressure exerted by a mixture of 4.00 g of H_2 and 5.00 g of He in a 30-liter container at 20°C?

15. Radium emits alpha particles at the rate of 1.16×10^{18} particles per year per gram of radium. The helium that results from the alpha particles emitted in six months by 2.5 g of radium occupies a volume of 0.054 cm^3 at 1.0 atm and 0°C. It is also known that 4.0 g of He occupy 22.4 liters under these same conditions. Use these data to calculate a value for Avogadro's number. Because alpha particles can be counted individually, the agreement of this value with values obtained in other ways provides strong support for Avogadro's law and the meaning that we attach to Avogadro's number.

16. Since $\bar{V} = V/n$, we may write equation (3.24) in the form

$$PV = nRT + nbP$$

Making use of $n = m/M$ and $d = m/V$, derive an equation that relates the density to the molecular weight. Suppose that densities of a gas are determined at several moderate pressures, all at the same known temperature. Derive the appropriate equation and illustrate its use in graphically determining M and b from the slope and intercept of a straight line.

17. On the assumption that the kinetic theory of gases is applicable to tiny bodies of any kind under conditions comparable to a gas at low pressure, calculate the average speed of smoke particles with mass of 10^{-13} g at 25°C.

18. How many calories are required to heat 1.0 mole of neon from 300°K to 400°K at constant volume?

19. Calculate the average velocity of a molecule of ammonia (NH_3) at 400°K.

20. The total pressure in a mixture containing twice as many molecules of H_2 as He is 800 mm Hg. What is the partial pressure of H_2?

21. It has been reported that the density of gaseous sulfur is 3.29 g $liter^{-1}$ at 600°C and 700 mm Hg. Calculate the molecular weight of sulfur vapor under these conditions and give the formula of the principal species in the sulfur vapor.

22. How hot must a sample of O_2 be to make the average speed of the O_2 molecules equal to the average speed of neon atoms at 25°C?

23. Two moles of an ideal gas initially at 1.0 atm and 100°C are heated to 200°C with the volume held constant. What is the final pressure?

24. Calculate the ratio of rates of diffusion of $^{235}UF_6$ and $^{238}UF_6$ and compare this ratio with a similar ratio for H_2 and D_2.

25. A tube that is 100 cm long has small holes in each end so that HCl can diffuse in from one end while NH_3 can diffuse in from the other end. How far from the HCl inlet will solid NH_4Cl first appear as a result of reaction of the diffusing gases when diffusion begins from each end at the same time?

26. A sample of H_2 is collected over water in an apparatus like that pictured in Figure 3.4. This sample of gas occupies 124 cm^3 at 30°C when the pressure is 750 mm Hg. What volume would this same sample of H_2 have occupied had it been collected under the same conditions over a liquid with completely negligible vapor pressure?

27. At what temperature is the average speed (root mean square speed) of H_2 molecules equal to 9.5×10^4 cm sec^{-1}?

28. A 1.0-liter flask was filled with gas at 20°C and 1.0 atm. When gas was allowed to escape at constant temperature until the pressure was reduced to 0.813 atm, the weight of gas in the flask was found to decrease by 0.050 g. Calculate the molecular weight of the gas.

4

SOLUTIONS

Introduction

We considered a special kind of solution without calling it a solution when we stated and used Dalton's law of partial pressures for a mixture (solution) of ideal gases. Although it is more customary to consider sweetening syrup to be a solution of sugar in water than air to be a solution of oxygen in nitrogen or vice versa, both are true solutions. Before we proceed with detailed discussion of liquid solutions, it is desirable to define and illustrate some general terms.

A homogeneous system is one whose properties are the same in all parts, while a heterogeneous system is one that consists of two or more distinct homogeneous regions. These homogeneous regions are called phases. We say that a sample of liquid water is homogeneous and is therefore a single phase. Liquid water at 0°C in contact with ice at 0°C is a heterogeneous system of two phases, regardless of whether the ice is in one large lump or several small lumps. If we drop one gold nugget or many gold nuggets into the ice-water system, we then have a three phase system.

A solution is a single homogeneous region called a phase, consisting of two or more species not readily converted into one another. Thus a system consisting of sugar dissolved in water is a single phase system of two components and is called an aqueous solution. On the other hand, a piece of cast iron is not a single phase solution but is a two phase system consisting of small homogeneous regions of iron and carbon.

It is common to designate one of the components of a solution as the solvent and other components as solutes. The solvent is usually the component that is present in greatest amount, but it must be remembered that solvent and solute are merely convenient labels that are sometimes attached to components of some solutions. We ordinarily say that water is the solvent and sugar is the solute in a solution made up of these two components. Similarly, since we are accustomed to thinking of water as solvent rather than as solute, we say that water is the solvent and alcohol the solute in a solution made of these two components. A solution that is 95% water and 5% alcohol would be described as a dilute solution of alcohol in water, and a solution that is 5% water and 95% alcohol might be described as a concentrated solution of alcohol in water. But we could equally well label alcohol the solvent and water the solute, which would lead us to describe the first solution as a concentrated solution of water in alcohol and the second as a dilute solution of water in alcohol.

Concentration

When we deal with solutions, we are often concerned with the amounts of these solutions. But since the properties of the solutions depend on the relative amounts of the components, we are frequently more concerned with the composition of a solution than with the amount of it. Even in those cases in which the total amount of solution is important, it is still usually necessary to consider the proportions of the various components.

The relative amounts of the components in solutions may be expressed in terms of mass fraction or mass per cent (commonly called weight per cent) or volume fraction or volume per cent. More commonly, however, scientists use the terms mole fraction, molality, or molarity as a means of expressing relative amounts of components of solutions.

The **mole fraction** is the number of moles of the substance of interest divided by the total number of moles in the solution. We let X_1 represent the mole fraction of component 1 and $n_1, n_2, \cdots$ represent the numbers of moles of components designated 1, 2, $\cdots$ and write

$$X_1 = \frac{n_1}{n_1 + n_2 + \cdots} = \frac{n_1}{\Sigma n_i}$$

Similar expressions may also be written for the mole fractions of the other components as summarized by the general equation

$$X_i = \frac{n_i}{\Sigma n_i} \tag{4.1}$$

The sum of the mole fractions of all the components adds up to unity, as illustrated in equation (4.2) for a two-component solution:

$$X_1 + X_2 = \frac{n_1}{n_1 + n_2} + \frac{n_2}{n_1 + n_2} = \frac{n_1 + n_2}{n_1 + n_2} = 1 \tag{4.2}$$

It is customary to designate the solvent as component 1 and the solute as component 2 in two component solutions, but it should be remembered that this assignment and choice of names is entirely a matter of convention.

The **molality** of a solution is the number of moles of solute per kilogram (1000 g) of solvent and is represented by m. A defining equation for the molality is

$$m = \frac{\text{moles of solute}}{\text{kg of solvent}} \tag{4.3}$$

Except when chemical equilibria (such as dissociation) are involved, the molality of a solution does not depend on either the temperature or the pressure. Since molality is defined in terms of masses of components, which can be measured very accurately, this method of expressing relative proportions of components is commonly used for very accurate work.

It is often more convenient to measure volumes of solutions, thus making it desirable to have a similar expression in terms of volume of solution rather than mass of solvent. The "volume based" **molarity** of a solution (sometimes specifically called the *concentration*) is the number of moles of solute per liter of solution and is represented by M. A defining equation for the molarity is

$$M = \frac{\text{moles of solute}}{\text{liters of solution}} \tag{4.4}$$

The molarity of a solution depends on both the temperature and pressure.

Because the density of water is very nearly 1.000 g ml^{-1}, the molalities and molarities of dilute aqueous solutions are not very different.

Example Problem 4.1. Enough water is added to 8.00 g of NaCl to make 100 g of solution. The density of this solution is 1.054 g ml^{-1} at 25°C. Calculate the molality and molarity of NaCl in this solution. What will be the effects of increasing temperature or increasing pressure on these quantities?

We first divide the mass of NaCl by the molecular weight to find that we have

$$\frac{8.00 \text{ g}}{58.4 \text{ g mole}^{-1}} = 0.137 \text{ mole}$$

of NaCl in the solution. There are $100 - 8 = 92$ g of water in the solution. This mass of water is also expressed as 0.092 kg.

The molality is calculated from equation (4.3) as follows:

$$m = \frac{0.137 \text{ mole}}{0.092 \text{ kg}} = 1.49 \text{ mole NaCl/kg of } H_2O$$

This solution is described as being 1.49 molal (also abbreviated m) in NaCl.

The volume of 100 g of solution is calculated from the given density as

$$V = \frac{100 \text{ g}}{1.054 \text{ g ml}^{-1}} = 94.8 \text{ ml} = 0.0948 \text{ liter}$$

Then the molarity is calculated from equation (4.4) as

$$M = \frac{0.137 \text{ mole}}{0.0948 \text{ liter}} = 1.45 \text{ moles NaCl/liter of solution}$$

Thus we also describe this solution as being 1.45 molar (M) in NaCl (at 25°C and 1.0 atm).

Note that the molarity and molality differ by less than 3% and that the difference will be even smaller for more dilute solutions for which the density is closer to 1.000 g ml^{-1}.

The molality of this solution is not changed by increasing temperature or pressure, since masses of both components remain unchanged.

Increasing temperature causes the solution to expand, thereby making the volume of solution greater than 94.8 ml and the molarity less than 1.45 moles/liter of solution. Increasing pressure compresses the solution into a volume smaller than 94.8 ml and makes the molarity greater than 1.45 moles/liter of solution. ■

Example Problem 4.2. Water is added to 8.00 g of NaCl to make a solution with volume of 600 ml. What is the molarity of the solution?

It was calculated in the preceding problem that 8.00 g of NaCl is 0.137 mole of NaCl. By equation (4.4) we have

$$M = \frac{0.137 \text{ mole}}{0.600 \text{ liter}} = 0.228 \text{ mole NaCl/liter of solution.}$$

Thus this solution is said to be 0.228 M or 0.228 molar.

To calculate the molality of this solution, we must know either the mass of water added or the density of the solution. By comparison with the preceding problem we predict that the molality differs from 0.228 by less than 3%. ■

Example Problem 4.3. What is the molarity of a solution that is made by diluting 100 ml of a 1.20 molar solution to 350 ml?

We begin by rearranging and using equation (4.4) as follows:

$$\text{moles of solute} = M \times \text{liters of solution}$$

Now, for the original solution we have

$$\text{moles of solute} = (1.20 \text{ moles liter}^{-1}) \times (0.100 \text{ liter})$$
$$= 0.120 \text{ mole of solute}$$

Since we have the same *amount* of solute in the final solution as we had in the initial solution, we find from equation (4.4) that

$$M = \frac{0.120 \text{ mole}}{0.350 \text{ liter}} = 0.343 \text{ mole liter}^{-1}$$

for the final solution. ■

Example Problem 4.4. A sulfuric acid solution is 88.0% (by mass) H_2SO_4 and has density 1.802 g ml^{-1} at room temperature. Calculate the molarity and the molality of this solution.

We know from the composition that 100 g of solution contains 88.0 g of H_2SO_4 and 12.0 g of water. From the atomic weights we calculate that the molecular weight of H_2SO_4 is 98.1 g $mole^{-1}$ and that we have 88.0 g/98.1 g $mole^{-1}$ = 0.897 mole of H_2SO_4.

The molality of the solution is calculated from equation (4.3) as

$$m = \frac{0.897 \text{ mole}}{0.012 \text{ kg } H_2O} = 74.8 \text{ moles per kg } H_2O$$

As the first step in calculating the molarity we obtain the volume of 100 g of solution as

$$V = \frac{\text{mass}}{\text{density}} = \frac{100 \text{ g}}{1.802 \text{ g ml}^{-1}} = 55.5 \text{ ml} = 0.0555 \text{ liter}$$

Then the molarity is calculated from equation (4.4) as

$$M = \frac{0.897 \text{ mole}}{0.0555 \text{ liter}} = 16.2 \text{ mole liter}^{-1} \quad \blacksquare$$

Digression on Liquids

Before proceeding with our treatment of solutions, it is useful to consider some of the properties of liquids that are relevant to their behavior as solvents in solutions.

The small compressibilities of liquids as compared to gases is consistent with a molecular picture in which the molecules in liquids are close together, leaving little free space between them. Because liquids readily adapt to the shape of any container, we conclude that the molecules of a liquid are not held in definite positions with respect to one another, as they are in solids, but are able to slide past one another relatively easily. Since such properties as coefficients of compressibility and thermal expansion are independent of the particular gas under consideration at low pressures, we conclude that intermolecular forces do not contribute importantly to the properties of gases at low pressures. On the other hand, these same properties for liquids and solids depend markedly on the particular solid or liquid under consideration. We must conclude that intermolecular forces contribute importantly to the properties of both solids and liquids.

In many ways it is appropriate to describe a liquid as intermediate between a solid and a gas. In terms of most physical properties, such as compressibilities and densities, liquids are more similar to solids than to gases under ordinary conditions. In other respects, however, liquids are properly regarded as more gas-like than solid-like. For instance, solids are characterized by long-range order in the arrangement of their molecules, while such long-range order is absent from both liquids and gases.

We have already seen how the kinetic molecular theory leads to a simple equation of state ($PV = nRT$) that describes the behavior of gases at low and moderate pressures. The kinetic theory has also been applied successfully to calculation of such properties as molecular speeds, heat capacities, and viscosities of gases. Analogous calculations have been carried out for solids, in which the atoms are pictured as confined to certain points (called lattice points) about which they vibrate in the crystal.

The equation of state developed by van der Waals has been applied by van der Waals and others to liquids and is reasonably successful in several respects. The van der Waals approach might be described as one in which a liquid is regarded as dense or compressed gas. Others have approached the problem from the opposite point of view, in which the liquid is regarded as a sort of broken down solid. Each approach has its own advantages and disadvantages. In recent years several combinations of these two approaches have also contributed to understanding liquids. A partial alphabetical list of those who have contributed to the molecular theory of liquids (Alder, Barker, Bernal, Dahler, Eyring, Frenkel, Hamann, Hirschfelder, Kirkwood, Lennard-Jones, Pople, Rice, . . .) looks like an extract from an international *Who's Who* in physical science.

Although considerable progress in the theory of liquids has been made and more can be expected, we still must rely mainly on experiment for information

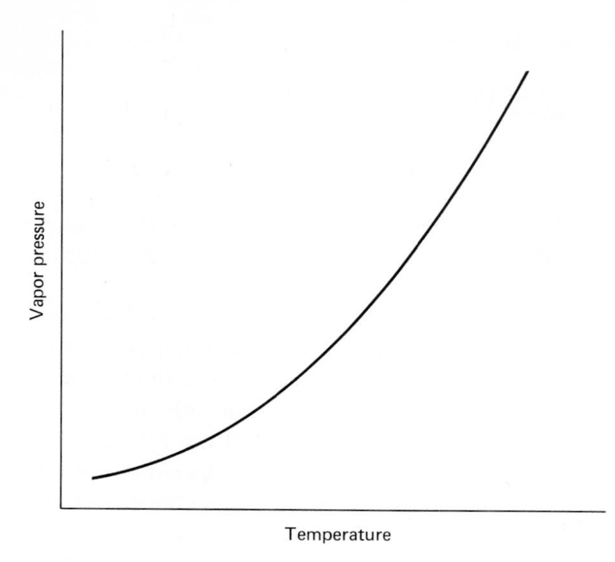

Figure 4.1. Graph of the vapor pressure of a liquid against temperature.

It has been found that vapor pressure data are summarized by equations of the form $\log P = A + (B/T)$ so that graphs of $\log P$ against $1/T$ (T is the absolute temperature, commonly on the Kelvin scale) give straight lines.

about most of the liquids that are of chemical, biological, or industrial importance. In the following paragraphs we shall define some of the properties of liquids that are of immediate interest and later consider still other properties.

The normal freezing point of a liquid is the temperature at which the liquid and solid phases coexist at equilibrium under a pressure of one atm. Freezing points at other pressures are generally slightly different from the normal freezing point.

The vapor pressures of liquids increase with increasing temperatures, as shown in Figure 4.1. Liquids boil at the temperature at which the vapor pressure is equal to the prevailing atmospheric pressure. Boiling points therefore depend on the atmospheric pressure. The boiling point is raised as the atmospheric pressure is raised and lowered as the atmospheric pressure is lowered. The standard or normal boiling point is the temperature at which the vapor pressure of the liquid is one standard atmosphere, 760 mm Hg.

The normal boiling point of water is 100°C, which means that the vapor pressure of water is 760 mm Hg at 100°C. At a pressure of 526 mm Hg, the observed boiling point of water is 90°C, and at a pressure of 10 mm Hg the observed boiling point is only 22°C. The lowering of boiling points as pressure is lowered makes the use of pressure cookers necessary at high altitudes where the boiling point of water is too low for satisfactory cooking. Pressure cookers or their industrial equivalents are also used at low altitudes when water is to be heated to a temperature substantially higher than the normal boiling point.

Properties of Solutions

As previously indicated, we have no theory of the liquid state that is generally successful in relating molecular properties to ordinary macroscopic properties of liquids. The situation is much better with respect to solutions, however, in the sense that we now understand many of the properties of solutions in relation to the properties of the components of the solution. For instance, if we know the vapor pressures of two liquids at a given temperature, we are often able to calculate accurately the partial pressures of each component over solutions made by mixing these two liquids.

As a result of both experimental and theoretical investigations, we are able to make several generalizations about solutions in terms of the properties of the pure components. For example, the partial pressure of any component of a solution is always less than the vapor pressure of the pure material at the same temperature. Figure 4.2 illustrates one experimental verification of this generalization.

The boiling point of a solution is the temperature at which the total vapor pressure of the solution equals the atmospheric pressure. The total vapor pressure of a solution is given by the sum of the partial pressures of all of the components of the solution. In the special case of a two component solution in which the solute has negligible vapor pressure, the vapor pressure of the solution is just equal to the vapor pressure of the solvent. Because the vapor pressure of any component of any solution is always less than the vapor pressure of that same pure substance at the same temperature, we conclude that the boiling point of a solution of a *nonvolatile* solute is always higher than the boiling point of the pure solvent. This elevation of the boiling point is a consequence of the lowering of the solvent vapor pressure: the solution must be heated to a higher temperature than the pure solvent to raise the vapor pressure to atmospheric pressure, as illustrated in Figure 4.3.

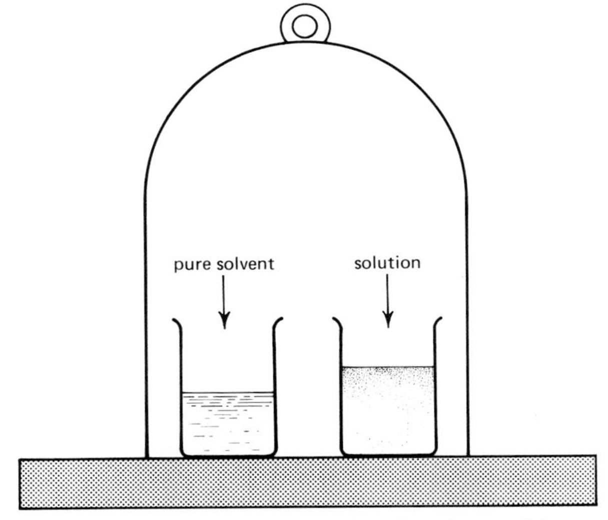

Figure 4.2. Illustration of an apparatus for showing that the partial pressure of a solvent over a solution is lower than the vapor pressure of the pure solvent at the same temperature.

As time passes, the solvent evaporates from the pure solvent and condenses in the solution. The transfer of solvent is easily observed because of resulting changes of liquid level in the beakers and occurs because the vapor pressure of the pure solvent is greater than the vapor pressure of the solvent in the solution. All of the solvent can eventually evaporate from the beaker on the left and condense in the beaker on the right.

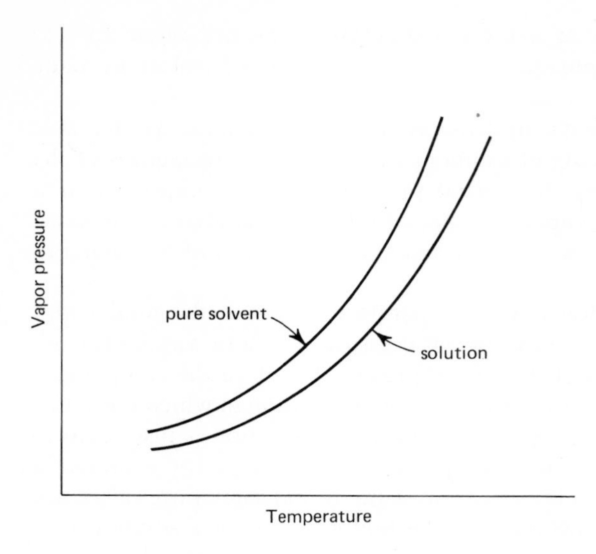

Figure 4.3. Graph of the vapor pressure of some pure substance against temperature with a similar graph of the vapor pressure (or partial pressure) of that same substance over a solution in which that substance is the solvent.

The partial pressure of solvent is always less than the vapor pressure of the pure solvent at the same temperature. It is therefore necessary to heat a solution to a higher temperature than the pure solvent to reach a solvent partial pressure of one atm. In the case of nonvolatile solutes, the boiling point of a solution is therefore higher than the boiling point of the pure solvent.

The spontaneous formation and existence of stable solutions implies that the solvent in such solutions would rather exist in the solution state than in the pure state. In other words, the solvent molecules have a greater tendency to stay in solution than they have to stay in the pure state at the same temperature. Hence solvent molecules have less tendency to escape from solution than they have to escape from the pure solvent at the same temperature. The vapor pressure of a liquid is a direct measure of the escaping tendency of the molecules in a pure solvent, just as the partial pressure of the solvent is a direct measure of the escaping tendency of the solvent molecules from the solution. Application of the idea of escaping tendencies of components of solutions was pioneered by G. N. Lewis, who made many other important contributions to chemistry.

Because both the vapor pressure of a pure solvent and the partial pressure of solvent over a solution decrease as the temperature is lowered, we know that escaping tendency of a solvent (pure or in solution) is lowered by decreasing temperature. At the freezing point the escaping tendencies of the substance common to both phases, solid and liquid, must be exactly equal in order for equilibrium to exist between the two phases. It has been verified for many substances that the vapor pressures of solid and liquid do become equal at their melting or freezing points.

The effect of a solute on the freezing point of a solution from which it is possible to freeze the pure solvent is illustrated in Figure 4.4. This illustration shows that, in order for the escaping tendency of the solvent in a solution to be equal to the escaping tendency of the pure solid solvent, the equilibrium freezing temperature of the solution must be lower than the normal freezing temperature of the pure solvent.

At a pressure of one atmosphere, ice and water are in equilibrium at 0°C: attempts to raise or lower the temperature by heating or cooling result only in melting some of the ice or freezing some of the water as long as both ice and water are present. But the equilibrium temperature is easily lowered by adding some substance that will dissolve in the water to form a solution. This melting point lowering explains the use of salt-water-ice mixtures for making ice cream and explains how salt can cause ice on the roads to melt when the temperature is lower than 0 °C.

It is observed experimentally that the magnitudes of freezing point lowerings and boiling point elevations are roughly proportional to the concentration of solute. It therefore appears that an important effect of the solute is to decrease the probability of escape of a solvent molecule from the solution as compared to the escape from pure solvent at the same temperature. This probability or statistical effect is the only important one for some solutions and is easily calculated for some of these solutions. Such solutions are called **ideal solutions.** Because the proper theoretical background for making and using these statistical calculations has not been developed in this book, we shall proceed with a discussion of ideal solutions in terms of equations that can be deduced directly from the results of experiments.

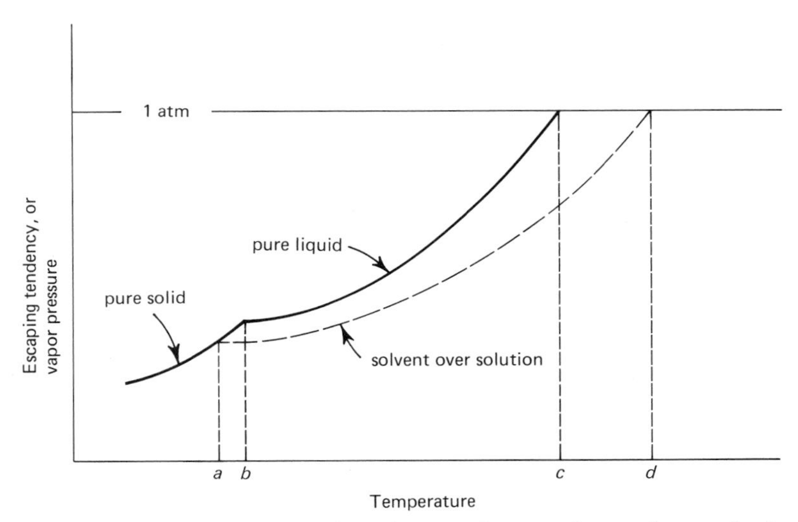

Figure 4.4. Illustration of the effect of a solute on the escaping tendency of solvent and consequent lowering of freezing point and elevation of boiling point.

Temperature *b*, where escaping tendencies of pure solid and pure liquid are equal, is the normal freezing point. The escaping tendency of the liquid solvent is lowered by solute so that the freezing point of the solution is lowered to temperature *a*. Temperature *c*, where the vapor pressure of the pure liquid equals 1 atm, is the normal boiling point of the pure solvent. Because a solute lowers the escaping tendency of the solvent, the temperature must be increased to temperature *d* to make the vapor pressure of solvent over the solution equal 1 atm. If the solute is nonvolatile, so that only the solvent contributes to the total vapor pressure, temperature *d* will be the boiling point of the solution.

Ideal Solutions and Raoult's Law

Although it has been known for hundreds of years that solutes lower the freezing temperature of water and in 1822 Faraday observed the elevation of boiling points by nonvolatile solutes, there were few quantitative measurements and these few referred almost entirely to complicated systems of salts dissolved in water. Several people recognized that elevation of the boiling point by nonvolatile solutes was due to lowering of the vapor pressure of the solvent, and a few measurements made about 1850 had indicated that the amount of lowering of vapor pressure was proportional to the amount of added solute, but neither data nor theory justified any sort of generalization.

The investigations that led to the useful relations that presently concern us were carried out by Francois Marie Raoult in France. He made careful measurements of the vapor pressure of solvent over a variety of solutions and was able to recognize a pattern that led to what we now call Raoult's law. This law says that the partial pressure of a component of a solution is equal to the mole fraction of that component in the solution times the vapor pressure of that same pure substance at the same temperature. All this is summarized conveniently by

$$P_i = X_i P_i^0, \tag{4.5}$$

in which P_i and X_i represent the partial pressure and the mole fraction of the i component of the solution and P_i^0 represents the vapor pressure of the same *pure* component at the same temperature as the solution.

We define an **ideal solution** as one for which Raoult's law, as in equation (4.5), describes the vapor pressures of all components over the entire range of concentration at all pressures and temperatures. In general we expect and find that solutions in which molecules of all components are similar are best described by Raoult's law.

The properties of many solutions are not accurately described by Raoult's law. These deviations from ideal behavior are important and are discussed later. But it is now well established on both experimental and theoretical grounds that Raoult's law is always accurate for the principal component of any solution that is sufficiently dilute. Since we usually call the principal component the solvent, we say that the solvent in a dilute solution always follows Raoult's law. Thus even for nonideal solutions Raoult's law is valid for any component in the limit of small concentration of other components. This conclusion is important because it permits the application of Raoult's law to the determination of molecular weights without regard to whether solutions are ideal, provided that they are dilute.

Example Problem 4.5. The vapor pressure of pure benzene, C_6H_6, is 74.66 mm Hg at 20°C. The partial pressure of benzene over a solution made up of 100 g of benzene and 2.11 g of a compound of unknown molecular weight is 73.94 mm Hg at the same temperature. Calculate the molecular weight of the solute.

Since the solution is dilute with respect to the unknown solute, Raoult's law applies to the solvent benzene. Thus we write equation (4.5) as

$$P_1 = X_1 P_1^0$$

where we have used the subscript 1 for the solvent. This equation is solved for X_1 as follows:

$$X_1 = P_1/P_1^0 = 73.94 \text{ mm Hg}/77.66 \text{ mm Hg} = 0.9903$$

The mole fraction of benzene that is represented by X_1 is the number of moles of benzene in the solution divided by the total number of moles of benzene plus solute in the solution as indicated by

$$X_1 = \frac{n_1}{n_1 + n_2}$$

There are 100 g/78.1 g mole^{-1} = 1.280 moles of benzene in the solution and 2.11/MW$_2$ moles of solute, where MW$_2$ represents the unknown molecular weight of the solute. We now write

$$X_1 = 0.9903 = \frac{1.280}{1.280 + (2.11/\text{MW}_2)}$$

and solve to find MW$_2$ = 168 g mole^{-1}. ■

As illustrated in Example Problem 4.5, it is possible to calculate the molecular weight of solute from the vapor pressures of pure solvent and of solvent over a dilute solution containing known masses of solute and solvent. A little arithmetic will verify that extremely accurate data are needed for the calculation of molecular weights by this method. As a result of this requirement, methods have been developed for measuring the difference in vapor pressures of the pure solvent and a solution at the same temperature. Compared to the direct measurement of vapor pressures with the required accuracy, these measurements of the vapor pressure difference are easily carried out.

To make use of differential vapor pressure data, we derive an appropriate equation based on Raoult's law. We begin by writing Raoult's law for the solvent and substituting $1 - X_2$ for X_1 from equation (4.2) to obtain

$$P_1 = X_1 P_1^0 = (1 - X_2) P_1^0$$

Carrying out the multiplication gives

$$P_1 = P_1^0 - X_2 P_1^0$$

which is solved for X_2 to obtain

$$X_2 = \frac{P_1^0 - P_1}{P_1^0} = \frac{n_2}{n_1 + n_2} \tag{4.6}$$

Now we can proceed to make use of differential vapor pressure data $(P_1^0 - P_1)$ by way of equation (4.6) to obtain the molecular weight of solute as illustrated later in Example Problem 4.6. But, as with many other problems, we can make our arithmetic easier by doing some algebra first.

We begin by solving equation (4.6) for n_2 as follows:

$$n_2 = X_2(n_1 + n_2) = X_2 n_1 + X_2 n_2$$

$$n_2 - n_2 X_2 = n_2(1 - X_2) = X_2 n_1$$

$$n_2 = \frac{X_2 n_1}{1 - X_2}$$

Equation (4.2) shows that $1 - X_2 = X_1$ so that we also have

$$n_2 = \frac{X_2 n_1}{X_1}$$

Now we put $X_2 = (P_1^0 - P_1)/P_1^0$ from (4.6) and $X_1 = P_1/P_1^0$ from Raoult's law into this equation to obtain

$$n_2 = \left(\frac{P_1^0 - P_1}{P_1^0}\right)\left(\frac{P_1^0}{P_1}\right)n_1$$

and thence

$$n_2 = \left(\frac{P_1^0 - P_1}{P_1}\right)n_1$$

Next, we express n_1 and n_2 in terms of masses (G) and molecular weights (MW) to obtain

$$\frac{G_2}{\mathrm{MW}_2} = \left(\frac{P_1^0 - P_1}{P_1}\right)\left(\frac{G_1}{\mathrm{MW}_1}\right)$$

and finally

$$\mathrm{MW}_2 = \frac{G_2 P_1(\mathrm{MW}_1)}{G_1(P_1^0 - P_1)} \tag{4.6a}$$

In the usual case it is necessary to calculate P_1 from the already known P_1^0 and the measured $(P_1^0 - P_1)$ as $P_1 = P_1^0 - (P_1^0 - P_1)$.

Measurements of the sort under discussion are almost always made on dilute solutions because it is only for these solutions that we can be sure that Raoult's law accurately describes the vapor pressure of solvent. It is therefore of interest to consider approximations in equations (4.6) and (4.6a) for this special case of dilute solutions. Since a dilute solution is one for which n_2 is very much smaller than n_1, we may approximate equation (4.6) by

$$\frac{P_1^0 - P_1}{P_1^0} \cong \frac{n_2}{n_1}$$

Again expressing n_1 and n_2 in terms of G and MW, we obtain

$$\frac{P_1^0 - P_1}{P_1^0} \cong \frac{(G_2/\mathrm{MW}_2)}{(G_1/\mathrm{MW}_1)}$$

and solve for the desired molecular weight of solute as

$$\mathrm{MW}_2 \cong \frac{G_2 P_1^0(\mathrm{MW}_1)}{G_1(P_1^0 - P_1)} \tag{4.6b}$$

Another approach to this convenient approximation is by way of equation (4.6a). In the special case of a dilute solution, the vapor pressure lowering is very small. That is, $(P_1^0 - P_1)$ is small compared to P_1^0 and we therefore have $P_1^0 \cong P_1$. So we substitute P_1^0 for P_1 in the numerator of (4.6a) and obtain the approximation given above as (4.6b).

Use of equation (4.6) and the following equations is illustrated in Example Problem 4.6.

Example Problem 4.6. A differential vapor pressure measurement on the solution described in Example Problem 4.5 led to $P_1^0 - P_1 = 0.715$ mm Hg. Calculate the molecular weight of the solute.

PROCEDURE 1. We use equation (4.6) with the given information to obtain

$$\frac{0.715 \text{ mm Hg}}{74.66 \text{ mm Hg}} = X_2 = 0.00958$$

Further use of this result with the right side of (4.6) gives

$$0.00958 = \frac{n_2}{1.280 + n_2}$$

which is solved for $n_2 = 0.0124$ mole of solute. Since

$$0.0124 \text{ mole} = \frac{2.11 \text{ g}}{\text{MW}_2 \text{ g mole}^{-1}}$$

we have $\text{MW}_2 = 170 \text{ g mole}^{-1}$ for the desired molecular weight.

PROCEDURE 2. We can obtain the same result with less effort by making use of equation (4.6a). We calculate $P_1 = P_1^0 - (P_1^0 - P_1) = 74.66 - 0.72 = 73.94$ mm Hg and insert this value with given information in equation (4.6a) to obtain

$$\begin{aligned}\text{MW}_2 &= \frac{(2.11 \text{ g})(73.94 \text{ mm Hg})(78.1 \text{ g mole}^{-1})}{(100 \text{ g})(0.715 \text{ mm Hg})}\\ &= 170 \text{ g mole}^{-1}\end{aligned}$$

PROCEDURE 3. With still less arithmetic we can make use of our given information in the approximate equation (4.6) as follows:

$$\begin{aligned}\text{MW}_2 &\cong \frac{(2.11 \text{ g})(74.66 \text{ mm Hg})(78.1 \text{ g mole}^{-1})}{(100 \text{ g})(0.715 \text{ mm Hg})}\\ &\cong 172 \text{ g mole}^{-1}\end{aligned}$$

It should be recognized that the error (2 g mole^{-1}) associated with the approximations leading to (4.6b) is entirely acceptable in many cases. For example, if chemical analysis has established that the empirical or simplest formula of the solute is C_6H_{13}, it is only necessary to obtain the molecular weight accurately enough to distinguish between (C_6H_{13}), $(C_6H_{13})_2$, $(C_6H_{13})_3$, etc. ■

Freezing and Boiling Points of Solutions

Figure 4.4 illustrates that the freezing point of a solution is lower than that of the pure solvent. As a solution is cooled below the normal freezing point of the solvent, solid solvent will freeze out at some temperature. The temperature at which the first solid solvent appears is called the freezing point of the solution. As cooling continues, more solvent freezes out and the remaining solution becomes more concentrated. Since the freezing point lowering becomes greater as the concentration increases, the temperature continues to drop. This behavior of a solution is different from that of a pure substance because a pure substance remains at its normal freezing point until all of the liquid is frozen. This difference

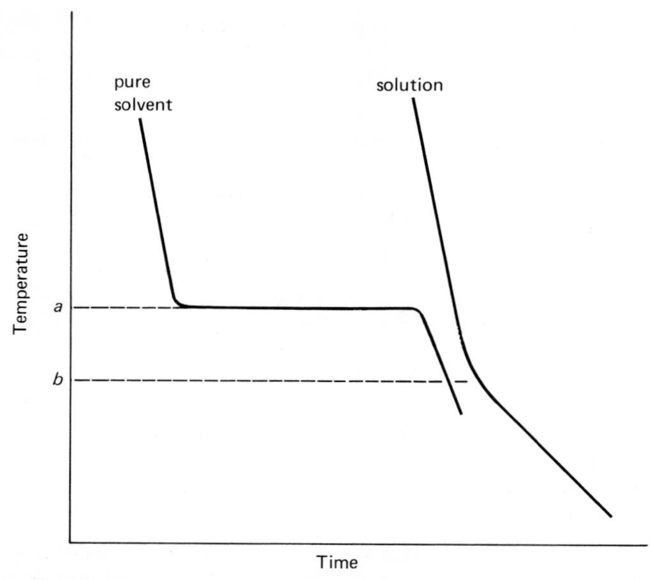

Figure 4.5. These "cooling curves" are obtained when a liquid sample is allowed to cool slowly with an approximately constant rate of heat loss.

The freezing point of the pure solvent is indicated by *a*, and the lower freezing point of the solution is indicated by *b*, which is the temperature at which the first solid solvent appears. For the pure solvent, the temperature remains constant at freezing point *a* as long as both liquid and solid are present. The temperature can fall below *a* only when all of the liquid has frozen to become solid. For the solution there is no constant freezing point because the concentration of solute in solution increases as solid solvent freezes out—thus causing the freezing point to continue to fall.

in behavior between a pure substance and a solution, which is illustrated in Figure 4.5, is the basis for a widely used test for purity.

Many experiments have shown that the difference between the freezing point of a dilute solution and the freezing point of the pure solvent is proportional to the concentration of the solute. We represent the temperature difference by ΔT and express the proportionality to concentration with the equation

$$\Delta T = k_f \times (\text{concentration}) \tag{4.7}$$

Figure 4.4 shows that the freezing point lowering is proportional to the lowering of the escaping tendency of the pure solute. Since Raoult's law, which is applicable to the solvent in all dilute solutions, tells us that the vapor pressure lowering of the solvent is proportional to the concentration of the solute (equation 4.6), we see that the freezing point lowering for dilute solutions should be proportional to the solute concentration as already expressed in equation (4.7).

Thermodynamic treatment that is beyond the scope of this book permits evaluation of the constant k_f for a particular solvent from knowledge of the heat required to melt a known quantity of the pure solvent. It is also possible to evaluate k_f from measured values of ΔT obtained for solutions of known concentrations. The k_f values obtained in these independent ways have been found to be in excellent agreement.

The numerical value of k_f for a particular solvent depends on the units in which concentration is expressed. Since we are going to be concerned mostly with aqueous solutions and often will express the concentration in terms of molality, we write the general equation (4.7) as

$$\Delta T = \left(\frac{1.86\,°\mathrm{C} \times \mathrm{kg\ of\ H_2O}}{\mathrm{moles\ of\ solute}}\right) \times m \tag{4.8}$$

The units for k_f must be degrees times the reciprocal of the units of concentration. Since this equation is specific for *dilute aqueous solutions* and we have already seen that $m \cong M$ for such solutions, we can also write

$$\Delta T \cong 1.86\ M \tag{4.9}$$

which is necessarily limited to dilute aqueous solutions.

Example Problem 4.7. Calculate the freezing point of a solution made by adding 3.0 moles of ethylene glycol ($HOCH_2CH_2OH$, commonly used as antifreeze) to 1.2 kg of water.

From equation (4.3) we have

$$m = \frac{3.0\ \mathrm{moles}}{1.2\ \mathrm{kg}} = 2.5\ \mathrm{molal}$$

Substitution of this molality in equation (4.8) gives

$$\Delta T = 1.86 \times 2.5 = 4.66°$$

for the difference between the freezing points of pure water and the solution. The freezing point of the solution is therefore calculated to be $-4.66\,°\mathrm{C}$.

As mentioned previously, equations (4.7) and (4.8) are limiting equations in the same sense that ideal gas equations are limiting equations. The ideal gas equations are exactly valid in the limit of low pressure, while these freezing point lowering equations are exactly valid in the limit of low concentration of solute. We should therefore expect that the calculation above for a 2.5 molal solution is only approximate and report the calculated freezing point as about $-4.7\,°\mathrm{C}$. ■

Example Problem 4.8. The solution formed by dissolving 10.0 g of glucose in 100 g of water freezes at $-1.03\,°\mathrm{C}$. Chemical analysis has established that glucose consists of 40.0% (by mass) carbon, 6.7% hydrogen, and 53.3% oxygen. Use these data to determine the molecular weight and formula of glucose.

We rearrange equation (4.8) and calculate the molality of the solution to be

$$m = \frac{\Delta T}{1.86} = \frac{1.03}{1.86} = 0.554\ \mathrm{molal}$$

Using this result in equation (4.3) we calculate the number of moles of glucose in this solution as

$$\begin{aligned}\mathrm{moles\ of\ solute} &= m \times \mathrm{kg\ of\ solvent}\\ &= 0.554\ \mathrm{mole\ kg^{-1}} \times 0.100\ \mathrm{kg}\\ &= 0.0554\ \mathrm{mole}\end{aligned}$$

The molecular weight of glucose is therefore

$$\frac{10 \text{ g}}{0.0554 \text{ mole}} = 180 \text{ g mole}^{-1}$$

Now we use the per cent composition data with appropriate atomic weights and this molecular weight to calculate the amounts of carbon, hydrogen, and oxygen in one mole of glucose as follows:

$$0.400 \times 180 = 72 \text{ g carbon}$$
$$0.067 \times 180 = 12 \text{ g hydrogen}$$
$$0.533 \times 180 = 96 \text{ g oxygen}$$

We divide each of these weights by the corresponding atomic weight to obtain the number of moles of carbon, hydrogen, and oxygen in one mole of glucose as:

$$72 \text{ g}/12 \text{ g mole}^{-1} = 6 \text{ moles C}$$
$$12 \text{ g}/1.0 \text{ g mole}^{-1} = 12 \text{ moles H}$$
$$96 \text{ g}/16 \text{ g mole}^{-1} = 6 \text{ moles O}$$

The formula for glucose is therefore $C_6H_{12}O_6$. ■

As already stated in connection with Figure 4.4 and the subsequent discussion of Raoult's law, the vapor pressure of solvent over a solution is always lower than the vapor pressure of the pure solvent at the same temperature. In the case of nonvolatile solutes that contribute negligibly to the total vapor pressure, the boiling point of a solution is therefore *higher* than that of the pure solvent.

Many measured boiling points for solutions have confirmed the thermodynamic prediction that the difference between the boiling points of solutions and pure solvent is proportional to the concentration for all dilute solutions of nonvolatile solutes. The value of the proportionality constant, which depends on the solvent and the units used to express concentration, can be evaluated either from measured boiling point elevations or from knowledge of the heat of vaporization of the pure solvent.

A specific equation for the difference between the boiling points of water and a dilute aqueous solution of nonvolatile solute is

$$\Delta T = k_b m \qquad (4.10)$$

where k_b has the value 0.51 °C × kg H_2O/moles solute.

Osmotic Pressure

Some properties of dilute solutions, called **colligative** properties, depend on the relative numbers of solute and solvent particles but not upon what the solute particles are. Three of these colligative properties, freezing point lowering, vapor pressure lowering, and boiling point elevation, have already been discussed. The fourth colligative property, the **osmotic pressure,** is directly related to the lower escaping tendency of the solvent in solution as compared to the pure solvent at the same temperature.

We may illustrate the phenomenon of osmosis and the resulting osmotic pressure by means of the experiment pictured in Figure 4.6. The end of a glass tube

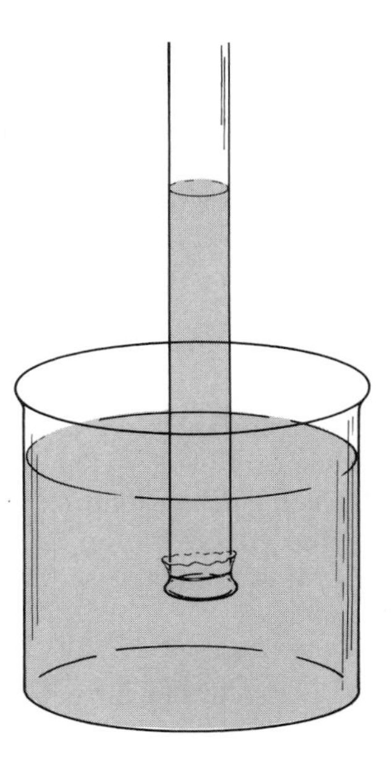

Figure 4.6. Illustration of osmosis, in which the osmotic pressure is indicated by the height to which the solution in the tube rises above the level of water in the beaker.

is covered with a piece of animal membrane or parchment and placed in a beaker of water. Then a sugar solution is poured into the tube until the levels of sugar solution and water are equal. It is observed that the level of liquid in the tube slowly climbs as water diffuses through the membrane into the solution. After the level has reached a certain height that depends on the temperature and concentration of the solution, the system is at equilibrium and no further change is observed.

The semipermeable membranes used in osmotic pressure studies permit the passage of molecules of only one component of a solution, generally the solvent. Now suppose that pure water and some aqueous solution are separated by a membrane that is permeable only to water. Since the escaping tendency of water in the solution is less than the escaping tendency of pure water, water passes through the membrane from the pure water side to the solution side. The osmotic pressure is equal to the hydrostatic pressure that is sufficient to stop this flow of water.

It has been observed experimentally and may be deduced from Raoult's law that the osmotic pressure for a dilute solution is proportional to the concentration of solute. Thus it is possible to use osmotic pressure data for calculation of molecular weights. This method, except in the special case of solute molecules having extremely large molecular weights, ordinarily offers no advantages (and often offers disadvantages) over the freezing point lowering method already discussed.

The molecular weight of hemoglobin was first deduced from osmotic pressure data to be about 68,000 g mole^{-1}. Much of our knowledge of the molecular weights of large molecules of biological and industrial importance comes from the results of osmotic pressure studies, to which we return in Chapter 21.

Henry's Law

Some solutions behave according to Raoult's law over the whole range of composition and are called ideal solutions. Other solutions do not behave according to Raoult's law over the whole range of composition and are called real or nonideal solutions. Even for these nonideal solutions, we have seen that the vapor pressure of solvent follows Raoult's law when the solution is dilute. Henry's law, which is applicable to the solute in dilute solution, is complementary to Raoult's law in that either can be derived from the other by means of the laws of thermodynamics.

Consider a solution so dilute that each solute molecule is surrounded only by solvent molecules. The solute molecules are therefore in a uniform environment. In such a dilute solution the escaping tendency of solute molecules is proportional to the concentration of the solute. Since the partial pressure is a measure of escaping tendency, we write

$$P_2 = kX_2 \tag{4.11}$$

for the solute in dilute solution. This equation is one way of expressing Henry's law. Except in special cases, we have no way of predicting the numerical value of the constant k in equation (4.11). One special case for which we can predict the value of k is for an ideal solution that (by definition) follows Raoult's law. Combination of equation (4.11) with the Raoult's law expression for the solute ($P_2 = X_2P_2^0$) gives $k = P_2^0$ for this special case.

For dilute solutions the mole fraction is proportional to the molality. We can therefore write equation (4.11) as

$$P_2 = km \tag{4.12}$$

or as

$$P_2 = kM \tag{4.13}$$

The value of k appropriate to these last two equations is not the same as that for equation (4.11) above. In general, it is necessary to use experimental data to evaluate the Henry's law constant to be used in any of these equations.

Henry's law was discovered in 1801 by William Henry, who was investigating the solubilities of gases in water. As a result of many later investigations, it is now well established that Henry's law is applicable to the solute in all dilute solutions while Raoult's law applies to the solvent. These laws are valid for some solutions over wide ranges of concentrations, but it is important to recognize that they are applicable to all solutions in the limit of small concentrations of one component.

Nonideal Solutions

Now we turn to consideration of nonideal solutions with properties described by Raoult's law and Henry's law only in the limit of small concentrations. As a guide to what can be expected from these nonideal solutions, we first consider briefly ideal solutions of liquids we label A and B. From Raoult's law we write

$$P_A = X_AP_A^0 \tag{4.14}$$

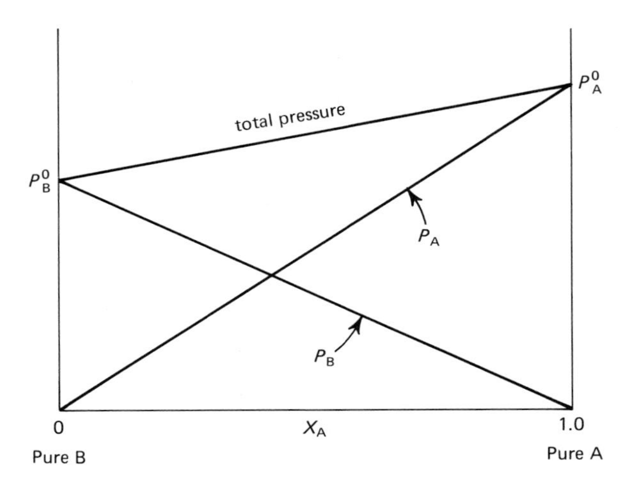

Figure 4.7. Graphs of total pressure and partial pressures over ideal solutions. The slopes and intercepts of the lines are in accord with equations (4.14), (4.15), and (4.16).

and

$$P_B = X_B P^0_B \tag{4.14a}$$

for the partial pressures of the two components. Because it is useful to display both P_A and P_B on the same graph, we substitute $1 - X_A$ for X_B (see equation 4.2) in equation (4.14a) and obtain

$$P_B = (1 - X_A)P^0_B$$

and then

$$P_B = P^0_B - P^0_B X_A \tag{4.15}$$

A graph in which both P_A and P_B are plotted abainst X_A is shown in Figure 4.7. The slopes and intercepts of these lines are in accord with equations (4.14) and (4.15).

The total vapor pressure over the solution is also shown in Figure 4.7. This total pressure P_t is the sum of the partial pressures of A and B over the solution and is given by

$$P_t = P_A + P_B = X_A P^0_A + (P^0_B - X_A P^0_B).$$

After rearranging, we have

$$P_t = (P^0_A - P^0_B)X_A + P^0_B \tag{4.16}$$

This equation represents a straight line of slope $(P^0_A - P^0_B)$ and intercepts P^0_B at $X_A = 0$ and P^0_A at $X_A = 1.0$ when P_t is plotted against X_A as in Figure 4.7.

Now we indicate in another graph the properties of solutions of A and B that are not ideal. For pure A (at $X_A = 1.0$) we know that $P_A = P^0_A$ and that $P_B = 0$. Similarly, for pure B (at $X_A = 0$) we know that $P_B = P^0_B$ and that $P_A = 0$. This information, which fixes the end points of the lines to be drawn, is supplemented by the knowledge that the solvent obeys Raoult's law and the solute obeys Henry's law in dilute solution.

Thus Raoult's law is applicable to P_A in the region near $X_A = 1.0$ and to P_B in the region near $X_A = 0$. Since the proportionality constants in Raoult's law equations are known to be the vapor pressures of the pure components, we know

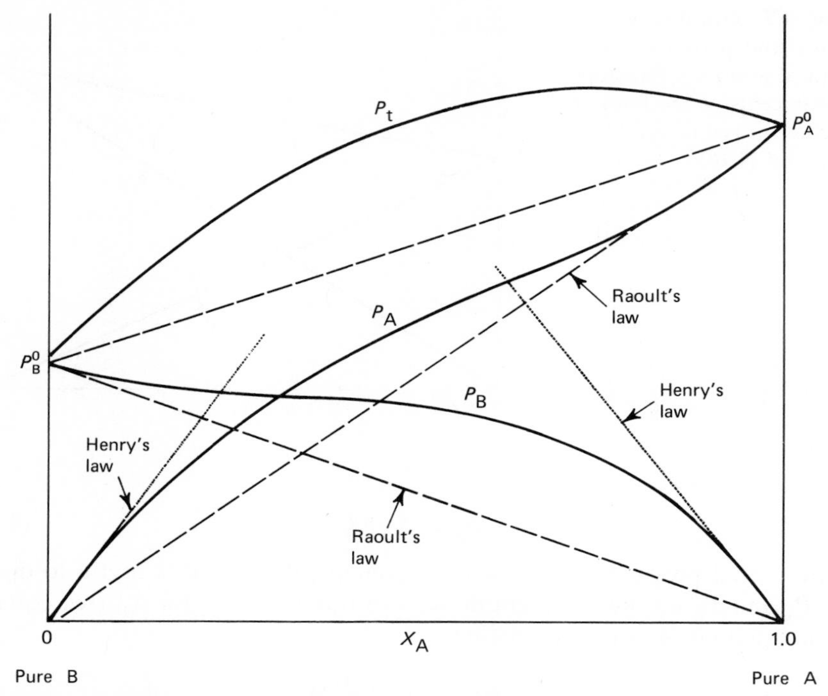

Figure 4.8. Graphs of partial pressures of A and B and total pressure over nonideal solutions. Henry's law (dotted lines) is followed by the components present in low concentrations, and Raoult's law (dashed lines) is followed by components with mole fraction near unity. Solutions (as in this illustration) that have pressures greater than the ideal solution values are said to exhibit positive deviations while those with pressures lower than the ideal values are said to exhibit negative deviations.

the slope of each partial pressure line in the limit at one end of the concentration range. Further, Henry's law is applicable to P_A in the region near $X_A = 0$ and also is applicable to P_B in the region near $X_A = 1.0$. Although we know that the partial pressures of solutes must vary linearly with composition in these ranges, we do not necessarily know the slopes. All of this information is summarized in Figure 4.8, where data for a typical nonideal system are displayed.

Electrolyte Solutions

Comparisons of solutions of sugar in water with solutions of sodium chloride in water disclose several differences. Possibly the property that is most strikingly different is the electrical conductivity. Pure water and solutions of sugar in water are poor conductors of electricity. On the other hand, aqueous solutions of sodium chloride are good conductors of electricity. Substances (like sodium chloride) that form highly conducting solutions in water are called **electrolytes,** while substances (like sugar) that form practically nonconducting solutions are called **nonelectrolytes.**

Although such eminent scientists as Michael Faraday, Alexander Williamson, Wilhelm Hittorf, Rudolph Clausius, and Friedrich Kohlrausch had investigated the electrical conductivities of electrolyte solutions, it remained for graduate student Svante Arrhenius to carry out some of the most relevant experiments and especially to recognize the meaning of the results of those experiments.

Arrhenius pointed out (1883–1887) that the results of a great variety of experiments could be understood if one assumed that electrolytes are dissociated into charged particles called ions. He assumed that electrolytes are completely dissociated into ions when the solution is infinitely dilute and that the extent or per cent of dissociation gradually decreases as the solutions become more concentrated. Arrhenius, van't Hoff, and Ostwald showed that these ideas could account for many of the available data on conductivities, vapor pressures, boiling points, and freezing points of solutions. In this chapter we are especially concerned with freezing points of electrolyte solutions; in Chapter 16 we consider conductivities.

For convenient comparison of freezing points of aqueous solutions we rearrange equation (4.8) to

$$\frac{\Delta T}{1.86\ m} = 1.0 \tag{4.17}$$

According to this equation, a graph of $\Delta T/1.86\ m$ versus m should have intercept 1.0 at $m = 0$ and should be horizontal (zero slope) in the limit of small m. The behavior of nonelectrolytes is in agreement with equation (4.17), as can be seen by plotting the typical data given in Table 4.1.

The situation is markedly different when we consider freezing point lowering data for dilute solutions of electrolytes as in Table 4.2. The values of $\Delta T/1.86\ m$ are not 1.0 for dilute solutions and are, in fact, approaching various integers larger than one. Part of this difficulty is easily explained if we consider the meaning of *molality* in equations (4.8) and (4.17). A sugar solution that is 1.0 m is one that contains 1.0 mole (Avogadro's number) of sugar molecules (solute particles) per 1000 g of water. But the total molality of solute particles in a solution containing 1.0 mole of sodium chloride is 2.0 if the sodium chloride is completely dissociated into sodium ions and chloride ions.

On the basis of this reasoning van't Hoff suggested rewriting equations (4.8) and (4.17) to take dissociation into account by introducing a numerical factor i, which represents the number of particles into which each solute molecule

Table 4.1. Freezing Point Lowering Data for Aqueous Nonelectrolytes

n-Propanol ($CH_3CH_2CH_2OH$)		*Sucrose* ($C_{12}H_{22}O_{11}$)	
m	$\Delta T/1.86\ m$	m	$\Delta T/1.86\ m$
0.00	1.00*	0.00	1.00*
0.01	1.000	0.01	1.001
0.05	0.995	0.05	1.006
0.10	0.990	0.10	1.011
0.20	0.986	0.20	1.023
0.50	0.978	0.50	1.052
1.00	0.961	1.00	1.108

* Extrapolated

Table 4.2. Freezing Point Lowering Data for Aqueous Electrolytes

	KCl	*HCl*	*KOH*	K_2SO_4	$CoCl_2$	$K_3Fe(CN)_6$
m	*i*	*i*	*i*	*i*	*i*	*i*
0.0*	2.00	2.00	2.00	3.00	3.00	4.00
0.001	1.98	1.98	—	2.84	2.91	3.82
0.005	1.96	1.95	1.99	2.77	2.80	3.51
0.010	1.94	1.94	1.98	2.70	2.75	3.31
0.050	1.88	1.90	1.93	2.45	2.64	3.01
0.100	1.85	1.89	1.87	2.32	2.62	2.85
0.500	1.78	—	—	2.17	—	—

*All values of i at $m = 0$ are extrapolated from the experimental data given here. Note that these extrapolated values for $\Delta T/1.86\ m = i$ are in accord with the idea that these electrolytes are completely dissociated in the limit of small concentration.

dissociates. Thus we have

$$\Delta T = 1.86\ im \tag{4.18}$$

and

$$\frac{\Delta T}{1.86\ m} = i \tag{4.19}$$

where i is the numerical factor mentioned above that is sometimes called the van't Hoff factor. This factor is unity for nonelectrolytes and is larger than unity for electrolytes. The data in Table 4.2 show that i approaches 2.0 for electrolytes such as KCl, HCl, and KOH. The limiting value of i is seen to be 3.0 for salts such as K_2SO_4 and $CoCl_2$ and to be 4.0 for $K_3Fe(CN)_6$, which dissociates into K^+ and $Fe(CN)_6^{3-}$ ions. As expected for nonelectrolytes, the data in Table 4.1 are consistent with $i = 1.0$ for these solutes, which do not dissociate.

Comparison of the data in Tables 4.1 and 4.2 shows that deviations from the simple freezing point lowering equation are much greater for electrolytes than for nonelectrolytes at corresponding concentrations. Arrhenius tried to explain these deviations by saying that salts are completely dissociated in the limit of small concentration while in more concentrated solutions only some of the solute molecules are dissociated, which can account for values of i decreasing with increasing concentration. This same idea also accounts qualitatively for changes in electrical conductivity with changes in concentration.

It was soon discovered, however, that this explanation in terms of dissociation of only some of the solute molecules could not account quantitatively for the freezing point lowerings (also vapor pressure lowerings, boiling point elevations, and conductivities) of electrolyte solutions. First, the i values obtained from freezing point data were found to be in conflict with calculations involving equilibrium constants, which we will consider in Chapter 14. Second, i values for many electrolytes were found to become quite large for concentrated solutions. For example, the i value for 2.0 m $CaCl_2$ is 4.2. One might expect from the formula that each mole of $CaCl_2$ dissociates into one mole of calcium ions and two moles of chloride ions. Freezing point data support the predicted value of $i = 3.0$ for the limit of zero concentration where we expect complete dissociation. Incomplete dissociation might account for i values less than 3.0 for solutions of finite concentration, but cannot possibly account for i values larger than 3.0.

It now seems certain that the Arrhenius interpretation of complete dissociation in the limit of small concentration is correct, but the electrostatic interactions of the charged ions have a marked effect and are the principal cause of deviations from the behavior predicted by equation (4.19). Because of the electrical attraction of oppositely charged ions and the mutual repulsion of like charged ions, the ions are not as independent of each other as the neutral solute molecules in a nonelectrolyte solution. Hence we find that deviations from ideal solution behavior are greater for electrolytes than for nonelectrolytes.

In 1923 Peter Debye and Erich Hückel were able to account quantitatively for the observed freezing points (and some other properties) of dilute solutions of electrolytes by means of their theory of electrostatic interactions of ions. Inherent in the Debye-Hückel theory is the assumption that many electrolytes are completely dissociated in all reasonably dilute solutions. There is now ample experimental evidence to support this assumption for many dilute solutions. It is also certain that even concentrated solutions of many electrolytes contain, at most, only a few per cent of undissociated molecules or associated ions.

There is at present no satisfactory theory to account quantitatively for the freezing points, conductivities, and other properties of concentrated solutions of electrolytes. It is certain that electrostatic interactions between ions are important and so complicated for concentrated solutions that the Debye-Hückel theory is unable to account for them. Electrostatic attractions between oppositely charged ions can cause these ions to associate to form what are called ion pairs or sometimes complex ions. For example, magnesium ions (Mg^{2+}) and sulfate ions (SO_4^{2-}) can associate to form magnesium sulfate ion pairs, or ferric ions (Fe^{3+}) can associate with chloride ions (Cl^-) to form such species as $FeCl^{2+}$, $FeCl_2^+$, etc. Ample experimental data are available to verify that this sort of association of ions is generally most common for highly charged ions and increases with increasing concentration.

It is largely arbitrary whether we consider an electrolyte in concentrated solution to be incompletely dissociated or to be completely dissociated but with some of the ions paired. If we choose to look at concentrated solutions from the first of these viewpoints, we see that there was considerable truth in Arrhenius' original hypothesis of complete dissociation in the limit of small concentration but incomplete dissociation in more concentrated solutions: Arrhenius merely overestimated the importance of incomplete dissociation and underestimated the importance of nonideal behavior due to electrostatic interactions between the charged ions.

We may summarize by saying that we regard *many* electrolytes as being completely dissociated into ions in all dilute aqueous solutions. Deviations from ideal behavior in dilute solutions are adequately accounted for by the Debye-Hückel theory. Detailed information about concentrated solutions of these electrolytes can, at present, be found only by experiment.

Weak Electrolytes

Our concern so far has been with solutes that are not dissociated at all into ions and also with those solutes that are completely dissociated in dilute aqueous solutions. Such substances have been called nonelectrolytes and electrolytes, but we shall henceforth speak of the latter as strong electrolytes. However, a great many important solutes fall into neither class. These "in between" substances, such as acetic acid, are called *weak* electrolytes. Weak electrolytes are said to be slightly dissociated, which means that only a small fraction of the solute molecules are dissociated into ions. One of the first and most impressive triumphs of the original Arrhenius theory was that it permitted Arrhenius, Ostwald, and van't Hoff to account very accurately for the properties of dilute solutions of a variety of weak electrolytes.

As might be expected, the electrical conductivity of a weak electrolyte solution is much less than the conductivity of a strong electrolyte solution of the same concentration because there are fewer ions in the weak electrolyte solution to carry the electric current. Results of conductivity measurements have been used in calculating the fraction of molecules in weak electrolyte solutions that are dissociated into ions. There are also other methods, some of which are discussed in later chapters, for investigating the dissociation of weak electrolytes.

Solubility

Suppose that we slowly add potassium bromide (KBr) to 1000 g of water that is maintained at 25 °C. At first the added salt dissolves readily, but after 677 g have been added, the solution is saturated and no more KBr will dissolve. The solution is said to be **saturated** when it is in contact with excess undissolved solute and has reached a state of equilibrium. In this state of equilibrium, the escaping tendency of the dissolved potassium bromide is equal to the escaping tendency of solid potassium bromide.

The equilibrium state is not a state in which there is no change. Rather, it is a state in which the concentration of potassium bromide in solution does not change with time, although there is continual exchange of potassium bromide between the solution and the solid. This state of dynamic equilibrium (demonstrated by experiments with radioactive tracers) is established when the rate of solution of KBr is equal to the rate of precipitation so that there is no *net* change in the system.

As stated above, a saturated solution is one in which the dissolved solute is in equilibrium with excess undissolved solute. After a solution is saturated, addition of more solute causes no change in the solution because it already contains all the solute that it can dissolve at the specified temperature.

The solubility of most substances in water increases as the temperature increases. A solution that is saturated at a low temperature may therefore dissolve all the excess solid solute and become unsaturated as the temperature is increased. On the other hand, a solution that is unsaturated at a high temperature may become saturated and yield a solid precipitate as the temperature is lowered.

Now let us return to consideration of a saturated solution prepared by adding a little more than 677 g of KBr to 1000 g of water at 25 °C. Because the solubility of KBr increases as the temperature increases, this solution will dissolve more KBr as the temperature is raised. After all of the solid KBr is dissolved, the solution is no longer saturated. As the hot solution containing more than 677 g of dissolved KBr per 1000 g of water is cooled to 25 °C, KBr precipitates out of solution until the solution again contains 677 g of dissolved KBr.

If a solution that is nearly saturated at a high temperature is cooled very carefully without shaking or stirring, it may happen that no KBr will precipitate from solution. Thus it may be possible to obtain a solution that contains more than 677 g of dissolved KBr per 1000 g of water at 25 °C. The solution is said to be **supersaturated.** A supersaturated solution is neither stable nor in equilibrium. A stray particle of dust, a mechanical shock, stirring, or adding a crystal of solute can cause a supersaturated solution to become saturated by precipitating all of the excess solute that it contains.

Solubility data at several temperatures for some inorganic salts are shown in Figure 4.9 in terms of grams of solute per 1000 g of water. Division of any of these solubilities by the molecular weight of the solute gives the number of moles dissolved per 1000 g of water, which is the molality of the saturated solution. It is necessary to know the density to calculate the molarity of the saturated solution.

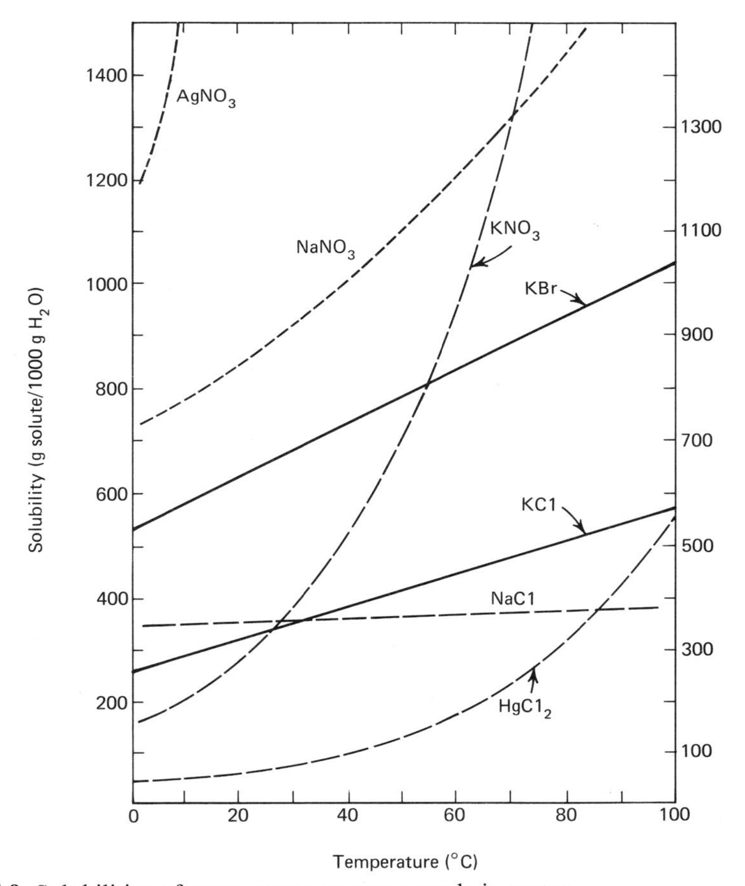

Figure 4.9. Solubilities of some common compounds in water.

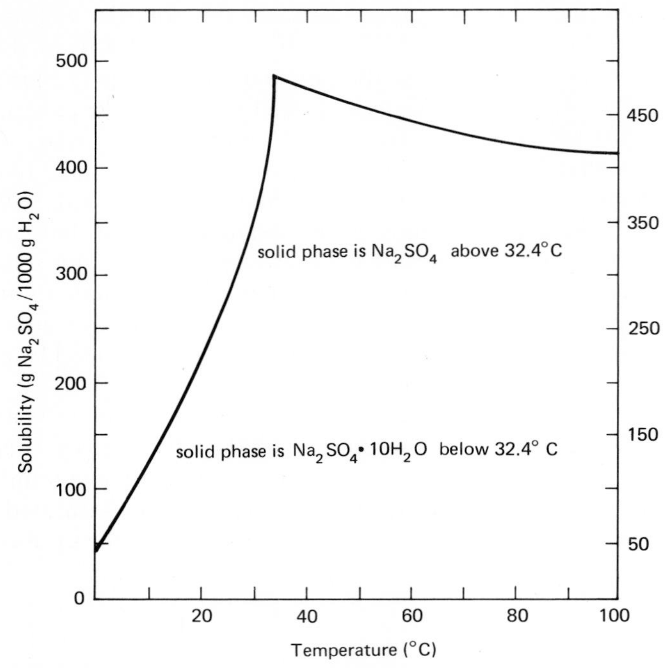

Figure 4.10. Solubility of sodium sulfate in water at several temperatures.

Solubility data for sodium sulfate (Na_2SO_4) are shown in Figure 4.10. To understand the difference between the solubility behavior of Na_2SO_4 and the salts cited in Figure 4.9 it is necessary to consider the solid phase in equilibrium with the saturated solution. At temperatures up to 32.4°C, the solid phase is sodium

Table 4.3. Summary of Solubilities in Water

Generalizations	*Comments*
Most nitrates and acetates are quite soluble.	There are no common exceptions. Silver acetate is the least soluble of the common nitrates and acetates.
Most chlorides, bromides, and iodides are quite soluble.	The following chlorides, bromides, and iodides are only slightly soluble: AgCl, AgBr, AgI, $PbCl_2$, $PbBr_2$, PbI_2, Hg_2Cl_2, Hg_2Br_2, and Hg_2I_2.
Most sulfates are quite soluble.	$PbSO_4$, $SrSO_4$, and $BaSO_4$ are only slightly soluble.
Most carbonates and sulfites are slightly soluble.	Carbonates and sulfites of the alkali metals and ammonium ion are moderately soluble.
Most sulfides are slightly soluble.	Sulfides of the alkali metals and ammonium ion are quite soluble. Cr_2S_3, Al_2S_3, and sulfides of the alkaline earth metals are decomposed by water.
Most hydroxides are slightly soluble.	Hydroxides of the alkali metals are quite soluble. Hydroxides of Ca^{2+}, Sr^{2+}, and Ba^{2+} are moderately soluble.

sulfate decahydrate ($Na_2SO_4 \cdot 10H_2O$) and the appropriate line in the figure indicates the solubility of this substance. Above the transition temperature of 32.4°C, the solid phase in equilibrium with saturated solution is anhydrous Na_2SO_4, and the appropriate line in the figure refers to the solubility of this compound. Saturated solution, solid Na_2SO_4, and solid $Na_2SO_4 \cdot 10H_2O$ exist in equilibrium with each other at the transition temperature of 32.4°C.

A useful qualitative summary of solubilities of inorganic compounds in water is given in Table 4.3.

References

Bernal, J. D.: The structure of liquids. *Sci. Amer.*, **203:**124 (Aug. 1960).

Dreisbach, D.: *Liquids and Solutions.* Houghton Mifflin Company, Boston, 1966 (paperback, a discussion of theories of liquids and solutions with numerous excerpts from classic research papers.)

Harris, F. E., and L. K. Nash: A Raoult's law experiment for the general chemistry course. *J. Chem. Educ.*, **32:**575 (1955).

Hildebrand, J. H.: *An Introduction to Molecular Kinetic Theory.* Reinhold Publishing Corp., New York, 1963 (paperback).

Jones, H. C.: *The Modern Theory of Solutions.* Harper, New York, 1899 (English translations of papers by van't Hoff, Raoult, and Arrhenius).

King, E. J.: *Qualitative Analysis and Electrolytic Solutions.* Harcourt, Brace and Co., New York, 1959.

McGlashan, M. L.: Deviations from Raoult's law. *J. Chem. Educ.*, **40:**516 (1963).

Nash, L. K.: *Stoichiometry.* Addison-Wesley Publishing Co., Inc., Reading, Mass., 1966 (paperback).

Snyder, A. E.: Desalting water by freezing. *Sci. Amer.*, **207:**41 (Dec. 1962).

Problems

1. How many grams of calcium chloride are required for making 500 ml of 1.4 molar solution?
2. What is the molality of a solution made by adding 6.0 g of ethyl alcohol to 150 g of water?
3. A solution is made by dissolving 1.0 g of NaCl in 50.0 g of water. The density of this solution is 1.0125 g ml^{-1} at 25°C. Calculate both the molality and the molarity of the solution.
4. Raoult's law is applicable to all solutions of benzene in toluene (or toluene in benzene). The vapor pressures of pure benzene and pure toluene are 119 mm Hg and 37 mm Hg at 30°C, respectively. Calculate the total pressure and the partial pressure of each component at 30°C above a solution in which the mole fraction of benzene is 0.35 and that of toluene is 0.65.
5. What is the vapor pressure of water over a solution at 30°C that contains 2.0 g of urea (NH_2CONH_2) and 60 g of water?
6. How many grams of methyl alcohol, CH_3OH, are required to lower the freezing point of 100 g of water by 0.30°?
7. A sample of impure water is found to freeze at a temperature 0.031° lower than the freezing point of pure water. What is the molality of the impurity?
8. A solution prepared by adding 2.95 g of nonelectrolyte to 75 g of water is found to freeze at −1.56°C. What is the molecular weight of the solute?

9. It is observed that red corpuscles that are placed in pure water swell and finally burst. Explain in terms of osmosis and escaping tendency.

10. Nonvolatile solutes lower the freezing point and raise the boiling point of water. Ethyl alcohol (normal boiling point = 78 °C) lowers the freezing point of water and also *lowers* the boiling point since the solution formed by mixing water and alcohol boils at a lower temperature than does pure water. Explain.

11. The vapor pressure of a dilute aqueous solution (containing a nonvolatile solute) is 0.211 mm Hg lower than that of pure water at the same temperature. Calculate the molality of the solution and the boiling point of this solution. We have $k_b = 0.51° \times$ kg of H_2O/moles of solute.

12. Use the vapor pressure data in Appendix VI to estimate the vapor pressure of water at 160 °C. See the discussion accompanying Figure 4.1 for a hint.

13. Concentrations of commercially available solutions are often expressed in terms of weight or mass per cent. Such data, with densities at room temperature (~25 °C) are given below for some solutions. Calculate both molality and molarity for each of these solutions.

(a) Ammonia: 58.6% NH_3; $d = 0.90$ g ml^{-1}.
(b) Hydrochloric acid: 36.0% HCl; $d = 1.19$ g ml^{-1}.
(c) Acetic acid: 99.5% CH_3COOH; $d = 1.06$ g ml^{-1}.

14. One hundred proof ethyl alcohol has density 0.9344 g ml^{-1} at room temperature and consists of 42.5 mass per cent CH_3CH_2OH. Calculate the molality, molarity, and mole fraction of alcohol in a 100-proof solution.

15. A stock solution of hydrochloric acid is 1.00 *M*. How would you use some of this solution to make 100 ml of solution that is 0.15 *M*?

16. A dilute aqueous solution of nonelectrolyte freezes at −1.21 °C. If the solute is nonvolatile, the boiling point of the solution is how much greater than that of pure water? (See problem 11.)

17. The freezing point of a solution of 2.11 g of naphthalene ($C_{10}H_8$) in 100 g of benzene is 0.845° lower than the freezing point of pure benzene. Calculate the molal freezing point lowering constant for benzene (with units).

18. The normal melting point of sodium nitrate is 307 °C. A variety of evidence suggests that liquid sodium nitrate is best pictured as being completely dissociated into Na^+ and NO_3^- ions. It has been found that a 0.10 molal solution of NaCl in $NaNO_3$ has a freezing point that is 1.58° lower than that of pure $NaNO_3$, while a 0.10 molal solution of KCl freezes at a temperature that is $2 \times 1.58° = 3.16°$ lower than the freezing point of pure solvent. These data are consistent with the idea that NaCl dissolved in $NaNO_3$ is completely dissociated into Na^+ ions (not foreign particles from the point of view of the solvent) and Cl^- ions (foreign or solute particles from the point of view of the solvent) while KCl is completely dissociated into K^+ and Cl^- ions that are both to be regarded as solute particles. Predict the freezing point lowering of a 0.10 molal solution of $BaCl_2$ in $NaNO_3$. How can you explain the observation that the freezing point lowering of a 0.10 molal solution of $CdCl_2$ in $NaNO_3$ is significantly smaller than that of a 0.10 molal solution of $BaCl_2$?

19. The boiling point elevation constant for carbon disulfide (CS_2) is 2.34 °C × kg CS_2/moles solute. A solution that consists of 8.53 g of sulfur dissolved in 150 ml of CS_2 has a boiling point that is 0.411° higher than that of pure CS_2. The density of CS_2 is 1.263 g ml^{-1}. What is the molecular weight and formula of sulfur dissolved in carbon disulfide?

20. Two beakers containing aqueous solutions of sucrose are placed under a bell jar and allowed to stand until equilibrium is attained. One beaker initially contains 5.0 g of sucrose ($C_{12}H_{22}O_{11}$) dissolved in 100 g of water and the other initially contains 7.5 g of sucrose in 125 g of water. How much water will be in each beaker at equilibrium?

21. Some mothballs are analyzed and found to contain 49.0 mass % carbon, 2.7 mass % hydrogen, and 48.3 mass % chlorine. Measurements of colligative properties show that the molecular weight of the principal constituent is 147 g $mole^{-1}$. What is the molecular formula for this substance?

22. A 2.00 g sample of naphthalene ($C_{10}H_8$) in 100 g of a volatile solvent has a boiling point 0.40° higher than the boiling point of the pure solvent. In a similar experiment, 2.00 g of compound X of unknown molecular weight in 100 g of the same solvent raises the boiling point by 0.55°. Calculate the molecular weight of compound X.

23. Solutions of methyl alcohol (CH_3OH) and ethyl alcohol (CH_3CH_2OH) are nearly ideal. The vapor pressures of methyl alcohol and ethyl alcohol are 625 mm Hg and 353 mm Hg at 60°C. Calculate the total vapor pressure and the composition of the vapor at 60°C over a solution that is made by mixing 50 g of one alcohol with 50 g of the other.

24. The vapor pressure of diethyl ether ($CH_3CH_2OCH_2CH_3$) is lowered by 16.2 mm Hg at 20°C when 10.0 g of nonvolatile solute is dissolved in 100 g of ether. The vapor pressure of pure ether is 442.2 mm Hg at this temperature. Calculate the molecular weight of the solute.

25. An aqueous solution of nonvolatile solute freezes at −1.52°C. Calculate the normal boiling point of this solution and the vapor pressure at 30°C. (See problem 11.)

5

WATER

Introduction

A great many chemical substances are essential to life on earth and there are many more substances that are nearly essential to our civilization. Of all these substances, water is justifiably considered to be the most important in the study of chemistry and other sciences. This is due to its abundance on earth, its unusual physical properties, its characteristics as solvent for many other compounds, its biological properties, and its unusual structures in the liquid and solid states.

In these days of spectacular scientific and technological developments that have made it possible for man to destroy his civilization and possibly himself, it is important and interesting to consider the consequences of other scientific and technical advances that are quietly altering the course of history. Many of these advances that have been made or that are being developed are concerned with the chemistry of water. Before proceeding with detailed scientific discussion of some of the chemical and physical properties of water, we briefly consider a few of the economic, social, and political problems associated with water in North America and some other parts of the world.

Thousands of years ago men discovered that salt could be obtained by evaporating away ocean water. Even today this primitive process is important to some people. But for most of the world it is the water, rather than the salt, that is of greatest importance. One of the principal methods used to desalinate ocean water is boiling, followed by condensation of the distillate, all carried out in ways designed to minimize energy consumption and thus cost.

In North America *as a whole* there is no water shortage now, but there certainly are many important problems associated with water, First, there are serious local shortages of water. Second, and possibly more important, is the problem of the quality of the water near almost every urban or industrial center. In connection with this second problem, it is appropriate to mention in a textbook on chemistry that the chemical industry is one of the worst offenders in polluting our water supplies and also one of the leading contributors to methods of alleviating the problem.

In a general way, problems concerning water in the United States are typical of the rest of the world, although there are important differences as well as similarities. There is certainly no shortage of fresh water in the world as a whole, as the following figures show. Annual rainfall on planet earth amounts to about 1.3×10^{17} gallons of water. Of this huge amount, about 3.5×10^{16} gallons fall on the land area during one year, or about 1×10^{14} gallons per day. Since the total population on earth is about 3 billion people, the daily rainfall on land averages almost 30,000 gallons of fresh water per person. The average per capita use of water in the United States and Canada amounts to about 1650 gallons per day (about 750 gallons for agriculture, 750 gallons for industrial use, and 150 gallons for household use), and average use in most of the rest of the world is significantly less. Thus the average per capita rainfall is well over 20 times the average per capita use of fresh water.

More water is used in agriculture than in all other human activities put together, and irrigation is by far the largest controlled use of water. Most of the water for irrigation comes from rivers, but only about 3% of the total river flow is used for this purpose. Thus it might be said that there is great waste of much of our water resources.

Since we have plenty of water in the world as a whole, why can't we solve all of our problems merely by using the available water efficiently and transporting water to those areas that are afflicted with an inadequate supply? First, it is certainly true that some of our problems can be alleviated by using the available water more efficiently, but this involves many complicated technical, economic, and political problems that are associated with the water treatment that is a necessary part of water reuse. A few of the problems of local shortages have been solved by transporting water. Some two thousand years ago magnificent aqueducts were built to carry water to Rome, and huge quantities of water are now piped from northern to southern California. But transporting water may be so expensive (often 5 to 15 cents per 1000 gallons per 100 miles) that it is often an impractical solution.

Further, considerable waste of water is inevitable. Many of our rivers naturally flow through areas having plentiful rainfall where irrigation is not needed. In many areas rainfall is seasonal or so erratic that drought or flood is the usual condition.

Various methods for desalination of ocean water are now known and hold considerable promise for largely solving most water shortage problems for several coastal areas. But it should be noted that there is presently no indication that desalination of ocean water can be made cheap enough to provide a practical source of water for large scale irrigation. Because the costs of some methods of desalination are almost directly proportional to salt concentration, these methods offer attractive possibilities for the reclamation of brackish water (not nearly as salty as ocean water) in many areas.

Closely related to the problems of desalination of ocean water and brackish water are the problems of handling water pollution. To the extent that industrial waste consists of inorganic electrolytes, the problems are very similar. But concentrations of inorganic electrolytes in "ordinary" polluted water are considerably less than in ocean water and are often only a minor factor in water treatment. Polluted water often contains undesirably large concentrations of dissolved and suspended organic materials, ranging from detergents and oils to sewage. Considerable progress has been made in developing chemical, electrical, and biological treatments of such polluted waters, but many of the processes are regarded as uneconomical by those who may have to pay for them. Little progress has been made in the United States or Canada toward political solution of these economic problems that have their origin in technology, although Germany and a few other countries appear to have achieved reasonably fair and workable solutions to problems of river water pollution.

Other approaches to more efficient utilization of available water have already yielded important benefits and promise more for the future. About half of the water presently provided for irrigation is lost in transport and only about half of the water that reaches the fields is actually used by the growing plants. New mulching methods are being applied to reduce evaporation from soils. Loss of water by seepage from irrigation canals can be lowered by development of improved inexpensive lining (plastics?) for the canals. Considerable progress has already been made, particularly in Australia, in retarding evaporation of standing water by means of a thin film of an organic chemical on the water surface.

One last aspect of water purification deserves mention. Salt and water are by no means the only useful substances obtained from ocean water. Both magnesium and bromine are now extracted in large quantities from ocean water, and it is possible that other elements will be obtained similarly. The possibility of combining processes that yield useful chemicals with those that yield fresh water is appealing, as is the possibility of combining a fresh water production facility with a nuclear reactor for generation of electric power.

An important problem involving quite a lot of chemistry that is related to problems already discussed is the use of salty water without removing the salts. For instance, coastal areas could conceivably use ocean water for many sanitation and industrial purposes. One present barrier to such use is the extreme corrosiveness of ocean water, which might be circumvented by means of chemical retardation of corrosion or by use of plastic or glass in contact with such waters.

We now proceed with a detailed consideration of some of the properties of water and its solutions. The scientific principles that are the foundation for several of the approaches to water purification and related problems will be discussed later.

The H_2O Molecule

Chemical analysis of water establishes that the empirical formula is $(H_2O)_n$, and determinations of the molecular weight by way of density measurements on gaseous water establish that the molecular weight is 18.0 g mole^{-1}. Thus the molecular formula for water is known to be the familiar H_2O.

One of the earliest methods of learning about the shapes of molecules and the distribution of electric charges within molecules involved determination of the **dipole moment.** As our first illustration of the meaning of dipole moment, consider two charges, $+\delta$ and $-\delta$, attached to the ends of a stick of length l. This "molecule" is electrically neutral, but because of its separated charges it interacts electrically with external electric fields. These fields might originate in the plates of a capacitor to which a potential has been applied or might be due to separated charges in nearby molecules. The dipole moment μ defined by $\mu = \delta \times l$ is the molecular quantity that is of particular interest and is seen to have dimensions of charge times distance. Some molecules have $\mu = 0$ and are said to be nonpolar. Other molecules for which $\mu \neq 0$ are said to be polar and commonly have dipole moments of the order of 10^{-18} esu-cm. It is now customary to define 10^{-18} esu-cm as one debye unit, represented by D, in honor of Peter Debye.

We now turn to consideration of the dipole moment of the H_2O molecule. As described in Chapter 1, an electric condenser or capacitor commonly consists of two parallel metallic plates and is a device for storing electrical energy. The dielectric constant of any material is defined as

$$\varepsilon = \frac{C}{C_0}$$

in which C represents the capacitance when the material of interest is between the plates and C_0 represents the capacitance of the same capacitor when the space between the plates is evacuated. The relationship between the dielectric constant of a gas and the dipole moment of the gas molecules that was derived by Debye permits evaluation of the dipole moment of the water molecule from dielectric constant data for gaseous water (steam). In this fashion it has been found that $\mu = 1.85$ D for the H_2O molecule.

Several hypothetical arrangements of the atoms in a molecule of H_2O are pictured in Figure 5.1. The structure might be H—H—O linear or nonlinear or H—O—H linear or nonlinear. A molecule with the symmetrical H—O—H linear structure would have zero dipole moment and can be ruled out as a possibility. On the basis of dipole moment alone it is impossible to decide definitely between the other three possible structures, but results of a variety of other experiments

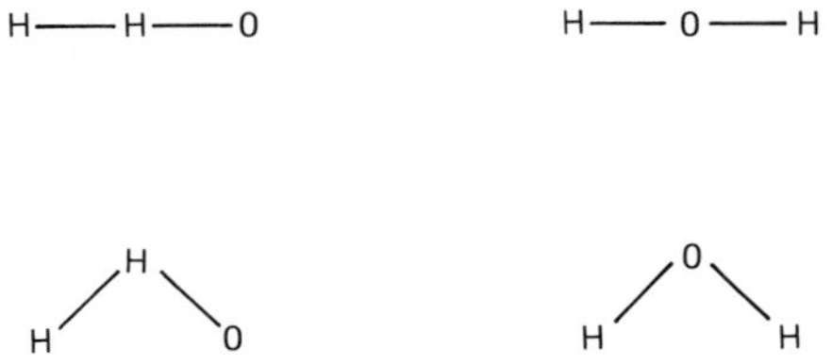

Figure 5.1. Four hypothetical structures of the H_2O molecule.

H—H—O H—O—H

H H O H O H

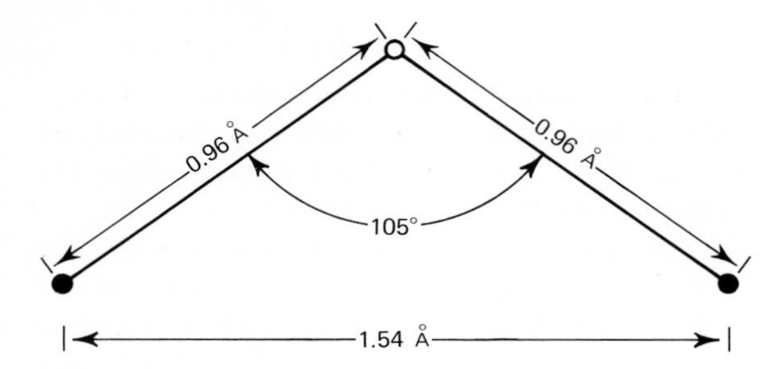

Figure 5.2. Illustration of the geometry of the H_2O molecule. The hollow circle represents the center (nucleus) of the oxygen atom and the solid circles represent the centers (nuclei) of the hydrogen atoms.

(mostly spectroscopic) show that the water molecule has the nonlinear H—O—H structure pictured in more detail in Figure 5.2.

The distribution of electrons and the nature of the bonding in the H_2O molecule are important problems to which we will return in Chapter 8. Now we can only say that the oxygen end of the water molecule bears an excess of negative charge while the hydrogen ends bear slight excesses of positive charge corresponding to a deficiency of negative charge. Since the dipole moment is a *product* of charge times distance, we are unable to specify definitely the magnitudes of the excess charges (say $-\delta$ near the oxygen and $+\delta/2$ near each hydrogen) or their distances of separation.

Although the structure and properties of gaseous water are interesting and important for many purposes, we shall devote most of our attention to liquid and solid water. It is when many water molecules come together that the unusual properties of this familiar substance become most striking.

Some Properties of Liquid Water

The behavior of polar molecules in the electric field between the parallel plates of a charged condenser is illustrated in Figure 5.3, which shows that the dipoles tend to line up so that negative ends are nearest the positive plate and positive ends nearest the negative plate. This alignment partially neutralizes the electric field due to the charges on the condenser plates and permits more charges to

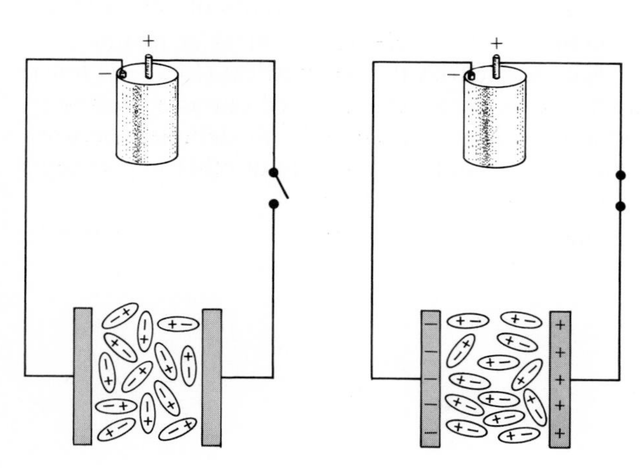

Figure 5.3. Schematic diagrams of polar molecules randomly oriented in absence of external field (on the left) and partially oriented by the electric field due to charges on condenser plates (on the right).

be stored on the plates. Therefore the capacitance of the condenser is greater than it would be with the space between the plates evacuated. The substance between the plates has dielectric constant greater than unity.

We might expect that the ability of molecules in a liquid to neutralize charges on the condenser plates would be roughly proportional to the dipole moments of the molecules. This proportionality cannot be entirely simple because alignment of the dipoles is determined not only by the field arising from the charges on the condenser plates but also by the interactions of the dipoles with one another, which may differ considerably from one liquid to another. Dielectric constants for a number of liquids and corresponding dipole moments of the molecules are listed in Table 5.1. The data for the compounds listed above the line in Table 5.1 exhibit the expected correlation between dipole moment of the single molecules and dielectric constants of the liquids, as illustrated by the crosses in Figure 5.4.

The dielectric constants of all of the liquids listed below the line in Table 5.1 and represented by circles in Figure 5.4 are larger than might be expected on the basis of dipole moments of single molecules. These unexpectedly high dielectric constants can be explained in terms of association of molecules with one another in the liquid state to form constantly changing and shifting aggregates of molecules that have been called "flickering clusters." The effective dipole moments of these aggregates may be considerably greater than the dipole moments of the individual molecules. Thus, if we could plot the unknown dipole moments of average aggregates on our graph in Figure 5.4, we might expect to preserve the general rough proportionality between dipole moment and dielectric constant.

Further evidence in support of the idea of association of molecules in water can be obtained from consideration of the boiling point of water in relation to boiling points of some other liquids. As a prelude to discussing the significance

Table 5.1. Dielectric Constants of Liquids at 25°C and Dipole Moments of Single Molecules

Substance	*Dielectric constant*	*Dipole moment, D*
CCl_4 (carbon tetrachloride)	2.2	0.00
C_6H_6 (benzene)	2.3	0.00
PCl_3	3.4	0.78
$CHCl_3$ (chloroform)	4.8	1.01
CH_3OCH_3 (dimethyl ether)	5.0	1.30
CH_3I (iodomethane)	7.0	1.6
SO_2	14.0	1.6
CH_3Cl (chloromethane)	12.8	1.87
- - - - - - - - - -	- - - -	- - - -
NH_3 (ammonia)	16.9	1.46
$CH_3CH_2CH_2CH_2OH$ (butyl alcohol)	17.1	1.67
$CH_3CH_2CH_2OH$ (propyl alcohol)	20.1	1.68
CH_3CH_2OH (ethyl alcohol)	24.3	1.69
CH_3OH (methyl alcohol)	32.6	1.70
H_2O	78.5	1.85

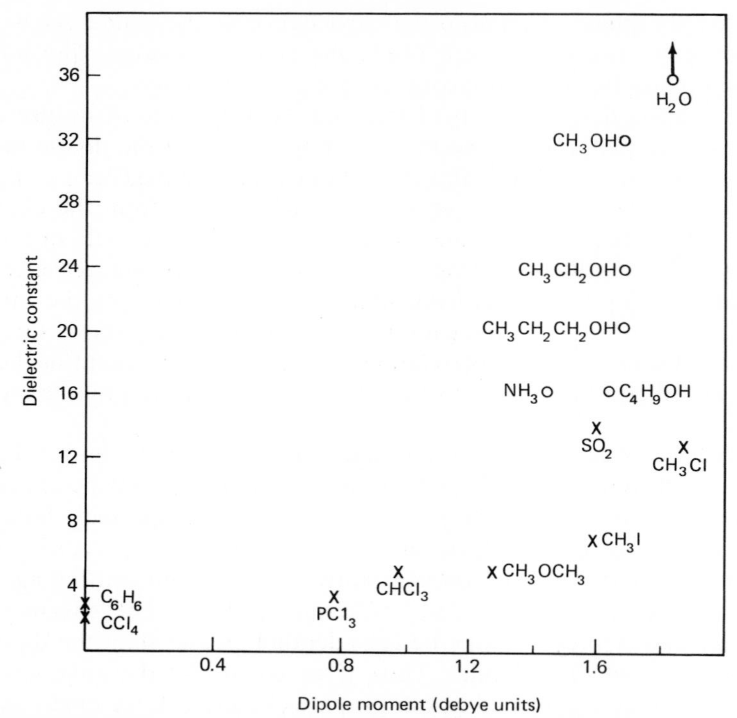

Figure 5.4. Graph of dielectric constants of liquids against dipole moments of molecules. Substances in which hydrogen bonding is important are represented by circles while other substances are represented by crosses.

of boiling points it is necessary to consider the forces that are responsible for the existence of nonionic liquids. Suppose that we carry out an imaginary experiment in which the temperature of a gas is gradually lowered while the pressure is held constant. Experience tells us that the gas will condense to form a liquid at a temperature that is characteristic of the substance and the particular constant pressure chosen for the experiment. This condensation to a liquid occurs for substances such as HCl that are made up of polar molecules and also occurs for nonpolar substances such as methane, carbon tetrachloride, and even for the so-called noble gases. Existence of these latter substances in the liquid state is convincing evidence that nonpolar molecules do attract one another.

Attractive forces between nonpolar molecules were postulated many years ago by van der Waals in Holland. Then some thirty years passed before Fritz London applied quantum theory to obtain an explanation for these weak but important forces between molecules. A simple picture of the origin of these forces can be obtained now from consideration of two atoms close together.

According to quantum theory, the electrons in atoms are constantly moving about. Thus each atom is an oscillating system of electric charges and has at any instant a small electric moment that induces an opposite moment in nearby atoms and results in a small attractive force. These attractive forces are often called van der Waals forces and are roughly proportional in strength to the number of electrons per atom or molecule.

We can now make a qualitative prediction about the relationship between boiling points of nonpolar substances and the total number of electrons per molecule. Because molecules that contain many electrons are attracted more strongly to one another than are molecules containing fewer electrons, more energy is required to pull the "many electron" molecules away from one another than is required to pull the "few electron" molecules away from one another. Hence we must heat a liquid made up of "many electron" molecules to a higher temperature (put in more thermal energy) to make it boil than we would to boil a liquid made up of "few electron" molecules. The data plotted in Figure 5.5 show the predicted relationship between number of electrons per molecule and boiling point for a variety of nonpolar substances. It is interesting to note that the constant a of the van der Waals gas equation (see page 54) is also roughly proportional to the number of electrons per molecule for nonpolar substances.

Total attractive forces between polar molecules are stronger than the weak van der Waals forces between nonpolar molecules. In addition to the van der Waals forces, polar molecules also experience another attractive force that is due to interactions between the permanent electric dipoles. We therefore predict that boiling points of polar substances will be higher than those of nonpolar substances containing the same numbers of electrons per molecule, and we find that this prediction is supported by many known boiling points.

Boiling points of a number of substances that are made up of polar molecules are plotted against numbers of electrons per molecule in Figure 5.6. All of these boiling points are higher than the boiling points shown in Figure 5.5 for nonpolar

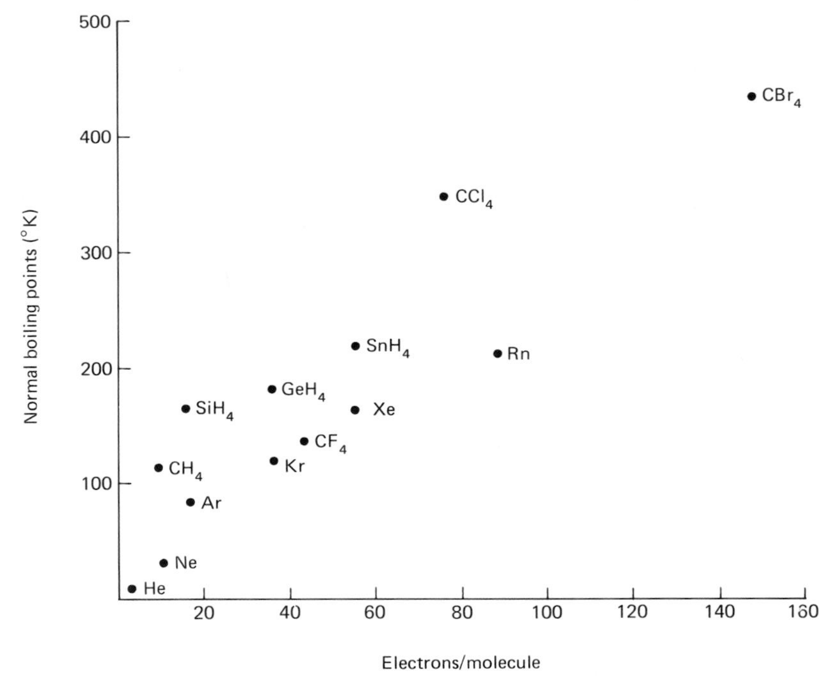

Figure 5.5. A graph of the normal boiling points of some liquids (nonpolar molecules) against the numbers of electrons per molecule.

molecules with the same numbers of electrons per molecule. As already mentioned, these higher boiling points for polar substances are expected because polar molecules are attracted to each other by dipolar forces in addition to the van der Waals force that is the only attractive force between nonpolar molecules.

We might estimate from Figure 5.6 that an imaginary "normal" water would have a boiling point of about 210°K (−63°C) as compared to the observed boiling point of real water that is 163° higher. The explanation for this unexpectedly high boiling point of water involves an attractive force that is unimportant for H_2S, H_2Se, and H_2Te. In 1920 Wendell M. Latimer and Worth H. Rodebush recognized the general importance of this extra attractive force and were the first to provide an explanation for it. They deduced that what we now call **hydrogen bonds** are formed between adjacent molecules of water. In order to be boiled, any liquid must be heated to some temperature high enough that the molecules have sufficient energy to escape from the attractive forces holding them together in the liquid state. This boiling temperature is higher than "normal" for water because it takes extra energy to break the hydrogen bonds between adjacent molecules.

It has also been shown that the existence of hydrogen bonds is consistent with many other properties of water. For example, hydrogen bonding between adjacent molecules accounts for the formation of molecular aggregates that were suggested by the very high dielectric constant of water.

Comparisons of boiling points and other properties of series of related compounds show that hydrogen bonds of sufficient strength to be of general importance occur only in compounds containing fluorine, oxygen, nitrogen, and chlorine. In general, fluorine forms the strongest hydrogen bonds, oxygen the next strongest, nitrogen the next strongest, and chlorine the weakest.

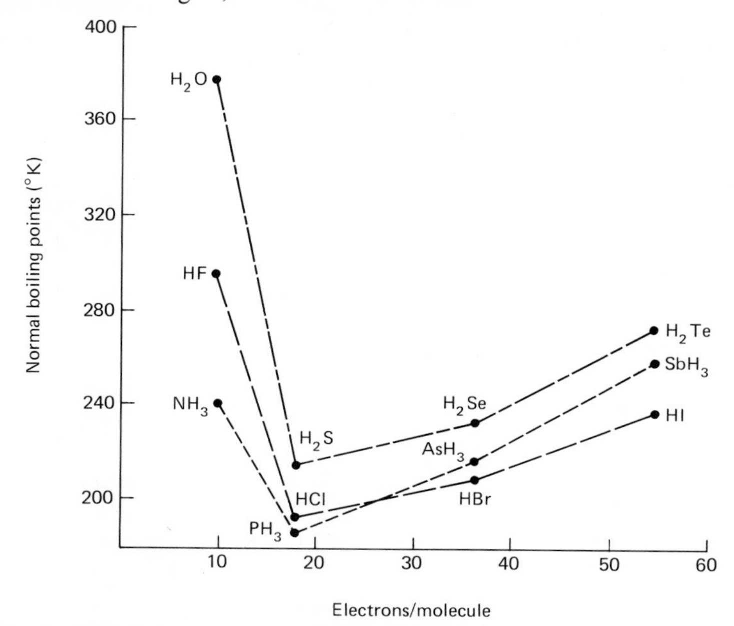

Figure 5.6. A graph of the normal boiling points of some liquids that consist of polar molecules against the numbers of electrons per molecule.

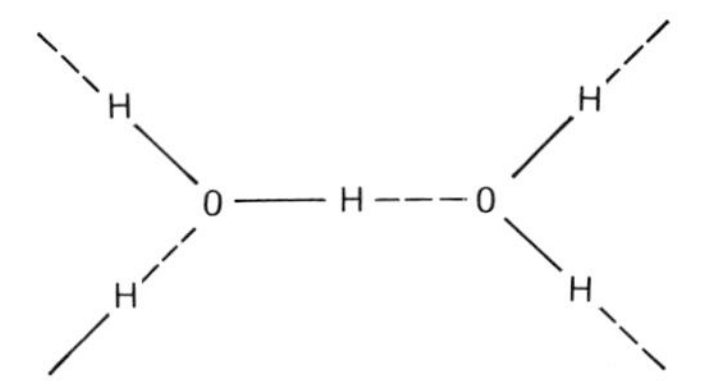

Figure 5.7. Illustration of hydrogen bonding in water and ice.
The solid lines represent "ordinary" chemical bonds that bind atoms of hydrogen and oxygen together in H_2O, while the dashed lines represent the hydrogen bonds that connect adjacent molecules. There are also attractive forces due to electric dipoles of the molecules and to van der Waals forces, but it is primarily the hydrogen bonding that accounts for the unusual and important properties of water. It should be noted that the distinction between "ordinary" bonds connecting H and O and hydrogen bonds connecting H and O is often much less definite than indicated by this illustration, as will be discussed later in some detail.

Explanation of hydrogen bonds (and other chemical bonds) is based on the results of quantum theory. Later chapters are devoted to the foundations of quantum theory and to application of the theory to chemical bonding of several sorts, including hydrogen bonding. For the present, however, we can describe hydrogen bonding in terms of Figure 5.7, which illustrates such bonding in water and ice. In this picture, the solid lines represent the "ordinary" chemical bonds that bind atoms of hydrogen and oxygen together in H_2O and the broken lines represent the hydrogen bonds that connect adjacent molecules to each other. In some substances the hydrogen atom is definitely connected to one oxygen atom by an "ordinary" chemical bond and to another by a hydrogen bond. In other substances, such as water, the distinction between "ordinary" O—H bonding and hydrogen bonding represented by O---H is indistinct.

Structures of Ice and Water

Investigation of the diffraction of x rays has led to determination of the arrangement of the oxygen atoms in ice. Each oxygen atom is surrounded tetrahedrally by four other oxygen atoms, as indicated in Figure 5.8 for a group of five atoms and in Figure 5.9 for a larger number of atoms. This is a very open structure and accounts for the low density of ice (remember that water expands on freezing, corresponding to a decrease in density).

The distance between an oxygen atom and any one of its four nearest neighbors in ice is 2.76 Å. Hydrogen atoms *might* be midway (1.38 Å) between neighboring oxygen atoms, but a variety of evidence shows that each hydrogen atom is closer to the oxygen atom on one side than to the oxygen atom on the other. The arrangement is random as far as permitted by the requirement that each oxygen atom must have two "close" and two "distant" hydrogen atoms on the lines drawn to the four surrounding oxygen atoms.

The arrangements of the molecules in water (and other liquids) are matters of great current scientific interest. Considerable progress has been made in gathering structural data and relating information about structures to observable macroscopic properties such as densities, dielectric constants, heat capacities, and

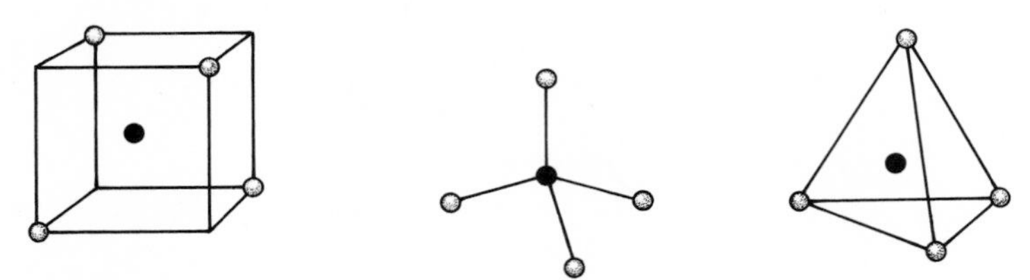

Figure 5.8. Three illustrations of the tetrahedral arrangement of oxygen atoms in ice. In each of these illustrations four oxygen atoms represented by open circles are arranged tetrahedrally around a central oxygen atom represented by a filled circle. The illustration on the left clearly shows the openness of the structure of ice, since only four of the eight corners of the cube are occupied.

The closest distance between oxygen atoms is 2.76 Å and the angles formed by straight lines from ○ to ● to ○ are 109°.

Although it is not shown in these illustrations, each oxygen atom that is represented by an open circle is also in the center of a tetrahedral array of four other oxygen atoms. The hydrogen atoms in ice are located on straight lines drawn between neighboring oxygen atoms, as shown in Figure 5.9.

absorption of sound. On the other hand, some of the available information that is relevant to structures of liquids is of uncertain reliability or subject to conflicting but more or less equally reasonable interpretations. Further, we still have no adequate experimental methods for obtaining some of the data that would probably prove most useful in developing the desired molecular understanding.

As might be expected from the above generalities, we have a considerable body of experimental data for liquids and a lesser body of derived structural information. Because water and aqueous solutions are very important and have been investigated in many ways, we have more information about them than we have about other systems. Although a complete review of either the data or the derived conclusions would require more pages than this entire book, we can briefly summarize some of the most important experimental results and some structural features of water that can account for these observations.

As already mentioned, its high dielectric constant indicates that water is what we call an associated liquid in which the molecules are connected to each other in such fashion that there are aggregates or clusters that consist of fairly large numbers of H_2O units. In Chapter 12 we will also see that the entropy of vaporization provides strong support for this picture of water as an associated liquid.

Since there seems to be overwhelming evidence that there are clusters or aggregates of H_2O units in liquid water, we must consider questions about the structures and average lifetimes of these aggregates and investigate the possibility that some of the molecules in liquid water are not tied up in these clusters. Again we find that dielectric constant data are helpful. In general, dielectric constants can be measured with a capacitor in a direct current circuit or in an alternating current (of any frequency) circuit. It is the static or *d.c.* dielectric constant to which we have referred earlier in this chapter that is useful for later calculations of coulombic forces. Measurements with high frequencies have provided evidence that the average lifetimes of the aggregates in water are from 10^{-10} to 10^{-11} seconds and partly accounts for Henry Frank's suggestion that these aggregates be described as "flickering clusters." Although this average lifetime of a flickering cluster is very short in terms of our everyday experience, it is 100 to 1000 times longer than the period of a molecular vibration and seems long enough to make these clusters meaningful physical entities.

Consideration of the densities of ice and liquid water provides evidence for the structures of these flickering clusters and also compels us to recognize that there is almost certainly another important component in the structure of liquid water. The density of liquid water is 0.9998 g cm^{-3} at 0°C, while that of ice is 0.917 g cm^{-3} at the same temperature. Thus the volume of a given mass of

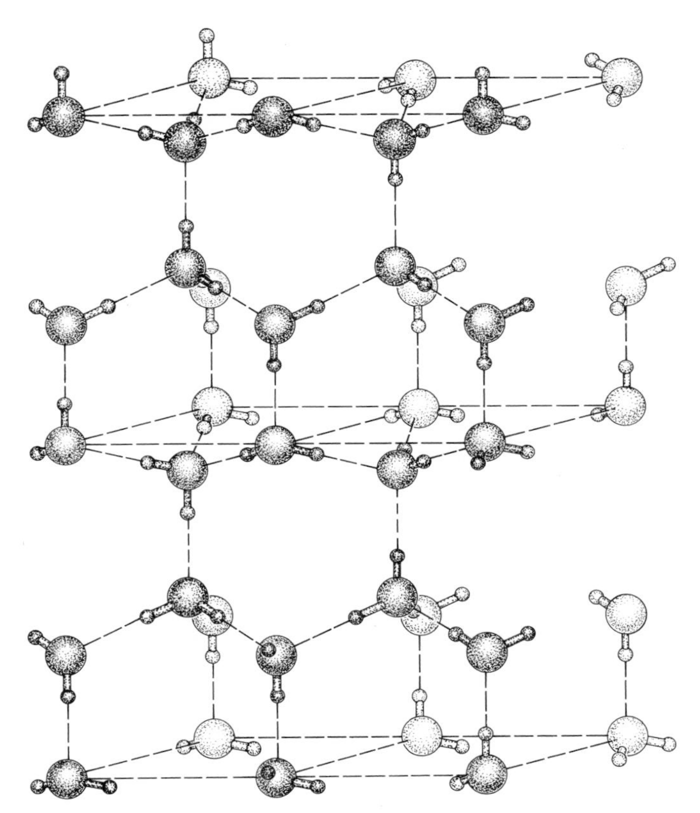

Figure 5.9. Illustration of the structure of ice.

Each hydrogen atom is located on a line between two nearest neighbor oxygen atoms. The hydrogen atoms are not located midway between oxygen atoms but are 1.01 Å from the closest oxygen atom and 1.75 Å from the more distant oxygen atom. Each oxygen may be regarded as bonded to two hydrogen atoms by "ordinary" chemical bonds and to two hydrogen atoms by weaker forces that we call hydrogen bonds. Linus Pauling has carried out a statistical analysis of the number of arrangements of the hydrogen atoms that are consistent with this picture and has shown that the results of his calculation are in agreement with the thermodynamic properties of ice and with the neutron diffraction experiments carried out by S. W. Peterson and H. A. Levy.

ice is 9% greater than that of the same mass of liquid water. The densities and volumes are illustrated in Figure 5.10. In contrast to the H_2O system, most other substances have greater density in the solid state than in the liquid state and contract, rather than expand, on freezing.

The consequences of the relationship between densities of ice and water are of considerable importance to life on earth. One illustration of this importance is provided by an experiment carried out over a century and a half ago by Count Rumford, who fastened a lump of ice in the bottom of a vessel, which he then filled with water. He found that the water in the top of the vessel could be heated and actually boiled while the ice melted only very slowly. Were it not for the anomalous densities of ice and water, the coldest water in our streams, lakes, and oceans would sink to the bottom and there freeze to form ice that would remain on the bottom. Once formed, this ice on the bottom would not be easily melted because warm water would remain at the surface. Year after year, the ice would increase in winter and largely persist through the summer until eventually all or most water would be frozen to ice. But because ice actually has density less than that of water, ice forms only on the surface of a body of water and is readily melted in warm weather.

Although the expansion of water on freezing acts for the benefit of most life on earth, it can be detrimental because freezing the water in living organisms results in expansion that bursts or seriously damages the cells.

Before turning to a molecular interpretation of the relative densities of ice and water, we also consider the unusual variation of the density of liquid water with temperature. The density of water *increases* from 0°C to 4°C, passes through a maximum, and then decreases as the temperature is raised. These densities and corresponding volumes are illustrated in Figure 5.11.

We now turn to a description of the structure of liquid water in terms that are consistent with the picture of an associated liquid already gained from other considerations. One description, which accounts for the expansion on freezing and the density maximum at 4°C, centers on the idea that liquid water can be regarded as a mixture of species. Possibly the simplest generally satisfactory picture is one similar to that first set forth in 1892 by Wilhelm Roentgen. He

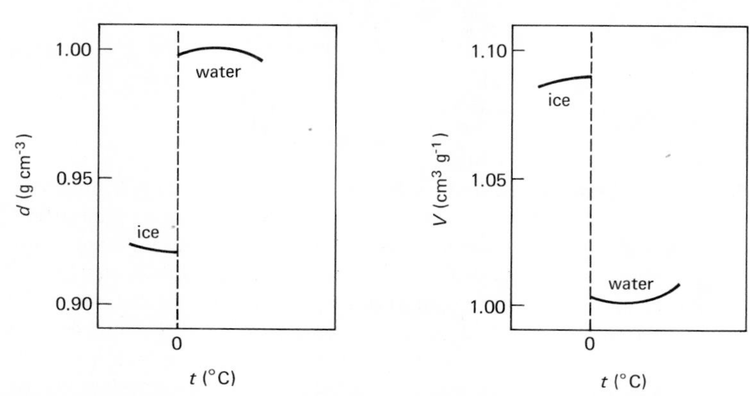

Figure 5.10. Illustrations of the unusual relationships between the densities and volumes of water and ice.

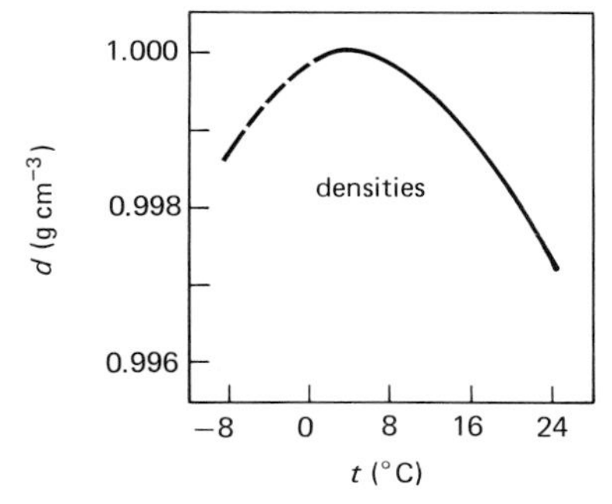

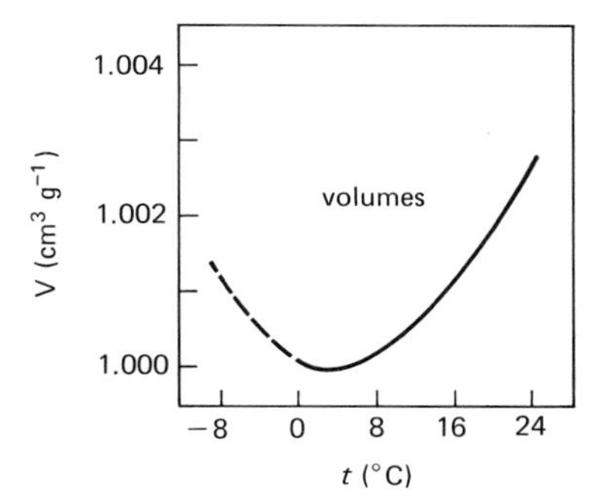

Figure 5.11. Illustrations of the unusual variation of density and volume of liquid water with temperature. The dashed lines refer to supercooled water.

considered water to be an equilibrium mixture of two kinds of arrangements of H_2O molecules. One species was supposed to be a bulky arrangement with low density like ice, while the other was supposed to be a more compact arrangement with greater density. Freezing corresponds to conversion of a liquid mixture of bulky and compact species to a solid that consists entirely of the bulky species—thereby explaining the observed expansion on freezing. By assuming that both species expand on heating and that the proportion of compact species increases with increasing temperature, it is possible to account for the maximum density of water and the unusually large heat capacity of water.

Since the time of Roentgen, many attempts by such outstanding scientists as J. D. Bernal, R. H. Fowler, A. Eucken, John A. Pople, John Lennard-Jones, Henry Eyring, Henry Frank, Harold A. Scheraga, and Linus Pauling have met with varied success. Several of these efforts have involved a picture of water as a mixture of more or less distinguishable species of varying compactness and other properties. Another approach is based on a picture of liquid water as a sort of fully bonded giant molecule in which the hydrogen bonds have considerable freedom to bend. This bending bond model can account for many of the properties of water as well as the mixture of species models. At present, we can conclude that we have lots of information about the structure of liquid water but still no clear picture of details of the structure. Later we will return to consideration of the structure of water in connection with thermal properties of water and also in connection with certain chemical reactions.

Ions in Aqueous Solution

The significance of dielectric constant to solutions containing ions is most easily approached in terms of Coulomb's law, which could have been used originally to define the dielectric constant. According to Coulomb's law, the force f between two particles separated by distance d and bearing charges q_1 and q_2 is given by the equation

$$f = \frac{q_1 q_2}{\varepsilon d} \tag{5.1}$$

where ε represents the dielectric constant of the medium surrounding the particles. If the charges are of opposite sign, the force is attractive and f is negative. Equation (5.1) shows that the force between two charges a given distance apart decreases as the dielectric constant of the medium increases. If we start with two oppositely charged particles close together, the work required to pull them apart decreases as the dielectric constant of the medium increases.

We can now begin to see why liquids that have high dielectric constants are generally the best solvents for electrolytes. Suppose that we could start with two separated ions, one Na^+ and one Cl^-, in a variety of solvents. Equation (5.1) tells us that the attractive force tending to pull these ions together is larger when the ions are in a solvent having low dielectric constant than when they are in a solvent having high dielectric constant. Thus ions are best able to stay apart in solvents having high dielectric constants, which partly accounts for the great solvent power of water for ionic substances.

A related effect that helps to account for the stability of aqueous solutions of electrolytes is the formation of *hydrates* of ions. A *hydrated* ion is an aggregate of ion and one or more neighboring water molecules. The electric field around every ion in solution tends to orient the surrounding water molecules and to attract one end of each of these molecules. These attractions may be so strong that they result in what we call hydrated ions. Because the electric field around an ion is proportional to the charge on the ion and inversely proportional to the square of the radius of the ion, we expect small and highly charged ions to be most strongly hydrated and also expect larger ions with a charge of only one electron unit to be least strongly hydrated. Since most monatomic positive ions are smaller than most monatomic negative ions, hydration is often more important for positively charged ions than for negatively charged ions. Figure 5.12 shows schematic illustrations of hydration of positive and negative ions.

The number of water molecules that can be accommodated in the first layer around an ion is often called the **coordination number** or **hydration number** of that ion. For some ions, such as Cr^{3+}, the hydration number is well established from appropriate measurements on solutions. For some other ions, hydration numbers are inferred from information about solids. The forces between ions and water molecules that cause hydration are sufficiently strong that many ions retain water molecules around them when crystals are precipitated from aqueous solu-

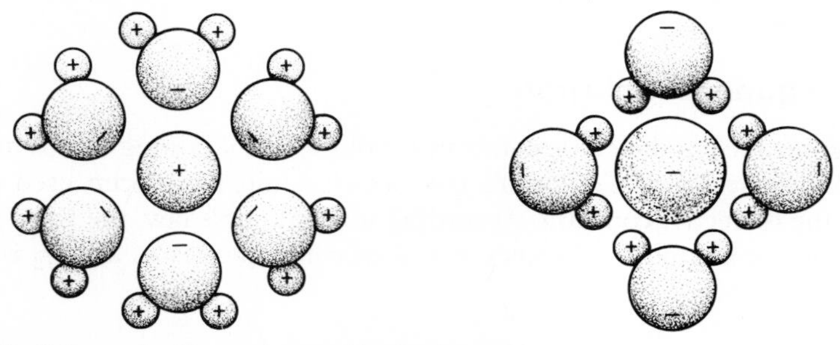

Figure 5.12. Schematic illustrations of hydration of ions.

The oxygen (negative) end of the water dipole is attracted to positive ions while the hydrogen (positive) end of the water dipole is attracted to negative ions.

tion. For example, $MgCl_2 \cdot 6H_2O(c)$ may be obtained by evaporation of a solution of magnesium chloride. Since each magnesium ion in the crystal is surrounded by six water molecules, it is reasonable to *assume* that the hydration number of magnesium ion in solution is also six.

The forces that attract the six water molecules in the hydration sphere to a Cr^{3+} ion do not entirely disappear outside that first layer of water. Thus the $Cr(H_2O)_6^{3+}$ entity formed by the ion and six water molecules interacts with neighboring water molecules. Similarly, the $Mg(H_2O)_6^{2+}$ that we have assumed to exist in solution also interacts with its neighboring water molecules. For $Cr(H_2O)_6^{3+}$, we are able to distinguish between the primary hydration layer of six water molecules and the unknown number of water molecules outside the first layer that are close enough to be appreciably influenced by the charge of the ion. For $Mg(H_2O)_6^{2+}$, however, we are currently unable to distinguish clearly between water molecules in the first layer and those in more distant layers that are also influenced by the charge of the ion. Thus we often write $Mg^{2+}(aq)$ for the hydrated ion of magnesium and make no attempt to describe the details of hydration.

One or more of the water molecules surrounding an ion can often be replaced by other chemical species. For example, one of the water molecules in violet $Cr(H_2O)_6^{3+}$ can be replaced by Cl^- to give green $[Cr(H_2O)_5Cl]^{2+}$. It is also possible for certain species to replace two water molecules each. For example, ethylenediamine of formula $H_2NCH_2CH_2NH_2$ can react with $Cr(H_2O)_6^{3+}$ to form a species represented by $[Cr(en)_2(H_2O)_2]^{3+}$, in which en is used as an abbreviation for ethylenediamine. Structures of these and related compounds are considered in more detail in Chapter 20.

Acids and Bases: Ionization of Water

The small electrical conductivity of very pure water indicates that there are a few ions present. Many measurements on solutions confirm the presence of ions and give information about the numbers and charges of these ions. The presence of these ions and their charges are accounted for by the dissociation or ionization of water, which is most simply represented by

$$H_2O \rightleftharpoons H^+ + OH^- \quad (5.2)$$

Only a very small fraction of the water molecules undergo this ionization reaction to form hydrogen ions and hydroxide ions, represented by H^+ and OH^-. At 25°C the concentrations of hydrogen ions and hydroxide ions are each 1×10^{-7} *m*. The double arrow symbol ($\rightleftharpoons$) is used to indicate that this ionization is an equilibrium process, as discussed and thoroughly illustrated in Chapters 13 and 14.

Both hydrogen ions and hydroxide ions are hydrated in aqueous solution and might be represented by $H(H_2O)_x^+$ and $OH(H_2O)_y^-$ in which x and y indicate appropriate numbers of water molecules. Some evidence indicates that the principal hydrogen ion species in aqueous solution is a monohydrate that is represented by $H(H_2O)^+$ or more often by H_3O^+ and called a **hydronium ion.** Thus it is more realistic, although more complicated, to rewrite the simple equation (5.2) as

$$(1 + y)H_2O + H_2O \rightleftharpoons H_3O^+ + OH(H_2O)_y^- \quad (5.3)$$

We are still writing $OH(H_2O)_y^-$ because we do not know the hydration number of the hydroxide ion.

Since water is an associated liquid in which most molecules are attached to other molecules by means of hydrogen bonds, it is appropriate to say that each H_2O in liquid water is hydrated. Further, each H_3O^+ in solution is certainly interacting strongly with some or all of the water molecules adjacent to it and might also be regarded as hydrated. In fact, there is evidence that each H_3O^+ entity in solution is hydrated to about the same extent as each H_2O in pure water. Thus equations such as (5.3) are at best only a partial representation of the total ionization reaction.

Equations such as (5.2), in which hydration is ignored, are adequate for some purposes. More often, however, we want to take general notice of hydration without specifying details and do so by writing (aq) after each ion or other dissolved species to indicate that it is in aqueous solution and is hydrated in some fashion. Thus we commonly write

$$H_2O(\text{liq}) \rightleftharpoons H^+(\text{aq}) + OH^-(\text{aq}) \tag{5.4}$$

Because many of the things that we want to know and that can be determined from results of measurements are related to the numbers or concentrations of ions rather than details of hydration, equations such as (5.4) are generally all that we need.

In a few cases, especially when we are interested in certain properties of acids, it is customary to indicate the primary hydration of hydrogen ions and write

$$H_2O + H_2O \rightleftharpoons H_3O^+(\text{aq}) + OH^-(\text{aq}) \tag{5.5}$$

There are many substances that dissolve in water and thereby change the concentrations of hydrogen ions and hydroxide ions in the water. Substances that increase the concentration of hydrogen (or hydronium) ions are called acids; substances that increase the concentration of hydroxide ions are called bases or alkaline substances.

The concentrations of hydrogen ions and hydroxide ions in aqueous solution are related in a way that can be described by a simple equation. At 25°C the product of the concentrations of hydrogen ions and hydroxide ions is 1×10^{-14}, as indicated by

$$[H^+][OH^-] = 1 \times 10^{-14} \tag{5.6}$$

In this equation we have used [ ] to indicate concentrations in units of molarity or molality. It is appropriate to use either molality or molarity because this equation is only exactly valid for such dilute solutions that molality and molarity are nearly identical.

It should be realized that the $[H^+]$ and $[OH^-]$ in equation (5.6) above represent the concentrations of hydrated hydrogen ions and hydrated hydroxide ions as they really exist in solution, whether or not we know the details of their hydration.

Equation (5.6) is satisfied by the concentrations of hydrogen ions and hydroxide ions in pure water where both have been determined to be $1 \times 10^{-7}\,M$. The hydrogen ion concentration in acidic solution is greater than $1 \times 10^{-7}\,M$, and equation (5.6) tells us that the concentration of hydroxide ions must therefore be less than 1×10^{-7} *M* in acidic solution. Similarly, the concentration of hydroxide ions is greater than 1×10^{-7} *M* in alkaline or basic solutions, and equation (5.6) tells us that the concentration of hydrogen ions is less than 1×10^{-7} *M* in these solutions.

Although acidic solutions contain both hydrogen ions and hydroxide ions, the concentrations of hydrogen ions are generally very much larger than concentrations of hydroxide ions. For example, consider a solution in which the concentration of hydrogen ions is 0.1 *M*. We calculate from equation (5.6) that the concentration of hydroxide ions is only 1×10^{-13} *M* in this solution. It is therefore ordinarily proper to ignore the hydroxide ions and consider only the much more numerous hydrogen ions in acidic solutions. Similarly, in basic or alkaline solution we have both hydrogen ions and hydroxide ions, but only the more numerous hydroxide ions need ordinarily be considered.

Hydrogen chloride is well known to be an acidic substance, and solutions of HCl in water are commonly called hydrochloric acid solutions. Gaseous HCl molecules split to form hydrogen ions and chloride ions in aqueous solution as shown by

$$HCl(g) \rightarrow H^+(aq) + Cl^-(aq) \tag{5.7}$$

which might also be written as

$$HCl(g) + H_2O(liq) \rightarrow H_3O^+(aq) + Cl^-(aq) \tag{5.8}$$

In equation (5.7) we have only generally indicated the hydration of the ions, while in equation (5.8) we have specifically indicated the hydration of hydrogen ions (protons) to hydronium ions. In both equations the hydration of the chloride ions is indicated only generally.

Sodium hydroxide is a well known base. It readily dissolves in water as indicated by

$$NaOH(c) \rightarrow Na^+(aq) + OH^-(aq) \tag{5.9}$$

in which we have only generally indicated the hydration of both sodium ions and hydroxide ions. Sodium and some other metals are called **alkali metals** because these metals, as well as hydroxides and oxides of these metals, react with or dissolve in water to form alkaline solutions. Balanced equations for the reactions of sodium metal and sodium oxide with water are given to illustrate these reactions:

$$Na(c) + H_2O(liq) \rightarrow Na^+(aq) + OH^-(aq) + \tfrac{1}{2}\,H_2(g) \tag{5.10}$$

$$Na_2O(c) + H_2O(liq) \rightarrow 2\,Na^+(aq) + 2\,OH^-(aq) \tag{5.11}$$

Because reaction (5.10) is one in which considerable heat is evolved, the hydrogen that is formed is often hot enough to react explosively with oxygen from the air.

The data in Table 4.2 show that aqueous hydrogen chloride is completely ionized in dilute solution and is therefore called a strong electrolyte or, more specifically, a strong acid. Similar data show that aqueous sodium hydroxide is also a strong electrolyte, and we therefore call sodium hydroxide a strong base. Several other strong acids and strong bases are known and commonly used in the laboratory and in industry.

A few salts are incompletely dissociated even in dilute solution and are therefore called weak electrolytes. Similarly, there are a considerable number of weak electrolytes among acids and bases. We call these substances weak acids and weak bases.

Acetic acid is a common weak acid. Even in dilute solution not all of the dissolved acetic acid molecules are split into hydrogen ions and acetate ions. In fact, most of the acetic acid molecules dissolved in water remain as acetic acid molecules while only a small fraction split to form ions. This partial or fractional ionization is illustrated by stating that the process or reaction represented by the equation

$$CH_3CO_2H(aq) \rightleftharpoons H^+(aq) + CH_3CO_2^-(aq) \qquad (5.12)$$

or by

$$CH_3CO_2H(aq) + H_2O(liq) \rightleftharpoons H_3O^+(aq) + CH_3CO_2^-(aq) \qquad (5.13)$$

is incomplete or proceeds only a small fraction to the right. In Chapter 14 we will find it useful to apply the concept of chemical equilibrium and the equilibrium constant to calculate some of the properties of acetic acid solutions. Here we point out only that equation (5.6) is one example of an equilibrium constant expression.

One of the characteristics of acids and bases, whether strong or weak, is that they react with one another to form salts and water. For example, the reaction of hydrochloric acid (strong electrolyte) with sodium hydroxide (strong electrolyte) to form sodium chloride (strong electrolyte) and water is represented by

$$H^+(aq) + Cl^-(aq) + Na^+(aq) + OH^-(aq) \rightleftharpoons H_2O(liq) + Na^+(aq) + Cl^-(aq) \qquad (5.14)$$

Although sodium chloride can be obtained from the solution by evaporating the water, the salt exists as separated aqueous ions in solution and should be so represented in the reaction equation.

When the specific salt that is formed in a neutralization reaction is irrelevant, equations such as (5.14) can be simplified by striking out the ions that appear on both sides, which leaves

$$H^+(aq) + OH^-(aq) \rightleftharpoons H_2O(liq) \qquad (5.15)$$

Equation (5.15) is the reverse of equation (5.4) that we wrote earlier to represent the ionization of water. Because only a small fraction of water molecules are split into hydrogen ions and hydroxide ions, we expect that these ions have sufficient affinity for each other that they will react with each other almost completely to form water. Thus we say that reaction (5.15) proceeds almost completely to the right and that reaction (5.4) proceeds only very slightly to the right as it is written.

Weak acids and bases can also be involved in neutralization reactions as illustrated by the following equation for acetic acid and sodium hydroxide:

$$CH_3CO_2H(aq) + Na^+(aq) + OH^-(aq) \rightleftharpoons H_2O(liq) + CH_3CO_2^-(aq) + Na^+(aq) \qquad (5.16)$$

We have written $CH_3CO_2H(aq)$ because acetic acid is a weak electrolyte that exists largely in the molecular rather than ionic form. On the other hand, we have written $Na^+(aq) + OH^-(aq)$ and $Na^+(aq) + CH_3CO_2^-(aq)$ because both sodium hydroxide and sodium acetate are strong electrolytes that exist as largely independent ions in solution.

For many purposes it is immaterial whether the reaction above is carried out with NaOH or some other strong base (such as KOH), and it is appropriate to cancel the $Na^+(aq)$ from each side of the reaction equation.

The preceding discussion of acids and bases in terms of hydrated hydrogen ions and hydroxide ions is limited to aqueous systems and thus is not applicable to the chemistry of liquid ammonia or other nonaqueous solvent systems. Further, the discussion we have so far given is not entirely satisfactory for understanding the basicity of solutions of sodium sulfide, sodium acetate, or ammonia in water. We must also consider the acidity of aqueous solutions of ferric chloride ($FeCl_3$) and other substances that do not even contain hydrogen and thus cannot directly add hydrogen ions to the solution.

In 1923 J. N. Bronsted in Denmark and T. M. Lowry in England proposed that an acid be defined as a substance that is capable of giving a proton to another substance and that a base be defined as a substance that accepts a proton. Thus an acid is a proton donor and a base is a proton acceptor in the Bronsted-Lowry system.

Hydrogen chloride is an acid because it donates a proton to a water molecule to form a (hydrated) hydronium ion. Water, the proton acceptor, is a base in this reaction. Similarly, acetic acid is an acid because it also can donate a proton to a solvent water molecule to form a hydronium ion. Again, water is the base.

Acetate ions and chloride ions are bases because they can accept protons to form hydrogen chloride and acetic acid molecules. Water is a stronger base than chloride ion because water wins the competition with chloride ions for the donated proton, as shown by the absence of hydrogen chloride molecules in aqueous solutions. Acetate ion, however, is a stronger base than water because it largely wins in the competition for protons, as indicated by the small extent of reaction (5.13).

When sodium acetate is added to water, the resulting aqueous acetate ions, being bases, "want" protons. Water molecules have protons to lose. We have shown above that acetate ions are more basic than water molecules, so we can predict that some of the acetate ions will be able to take protons away from some of the water molecules. The water molecules that have lost protons become hydroxide ions, and we say that the solution is basic or alkaline because the concentration of hydrated hydroxide ions is greater than $1 \times 10^{-7}\,M$ and the concentration of hydrated hydrogen ions is therefore less than $1 \times 10^{-7}\,M$. In this reaction the proton-donating water acts as an acid.

In later chapters we will return to quantitative consideration of acid-base equilibria in water and also qualitative discussion of acid-base behavior in nonaqueous solvents.

References

Bernal, J. D.: The structure of liquids. *Sci. Amer.*, **203:**124 (Aug. 1960).

Buswell, A. M., and W. H. Rodebush: Water. *Sci. Amer.*, **194:**76 (April 1956).

Chalmers, B.: How water freezes. *Sci. Amer.*, **200:**114 (Feb. 1959).

Clever, H. L.: The hydrated hydronium ion. *J. Chem. Educ.*, **40**:637 (1963).
Davis, K. S., and J. A. Day: *Water: The Mirror of Science*. Doubleday and Co., Inc., Garden City, N.Y., 1961 (paperback).
Dorsey, N. E.: *Properties of Ordinary Water-Substance*. Reinhold Publishing Corp., New York, 1940.
Eisenberg, D., and W. Kauzmann: *The Structure and Properties of Water*. Oxford University Press, Oxford, 1969.
Forbes, G. S.: Water: some interpretations more or less recent. *J. Chem. Educ.*, **18**:18 (1941).
Henderson, L. J.: *The Fitness of the Environment*. Beacon Press, Boston, 1958 (paperback). This classic "inquiry into the biological significance of the properties of matter," which was originally published in 1913, contains an interesting account of the properties of water.
Katz, J. J.: The biology of heavy water. *Sci. Amer.*, **203**:106 (July 1960).
Kavanau, J. L.: *Water and Solute-Water Interactions*. Holden-Day, Inc., San Francisco, 1964.
Knight, C., and N. Knight: Snow crystals. *Sci. Amer.*, **228**:100 (Jan. 1973).
Pauling, L.: *The Nature of the Chemical Bond*, 3rd ed. Cornell University Press, Ithaca, N.Y., 1960. Chapter 12 in this pioneering book by a great modern scientist is especially relevant to topics discussed in the preceding pages.
Penman, H. L.: The water cycle. *Sci. Amer.*, **223**:98 (Sept. 1970).
Pimentel, G. C., and A. L. McClellan: *The Hydrogen Bond*. W. H. Freeman and Co., San Francisco, 1960.
Revelle, R.: Water. *Sci. Amer.*, **209**:54 (Sept. 1963).
Snyder, A. E.: Desalting water by freezing. *Sci. Amer.*, **207**:41 (Dec. 1962).

Problems

1. Suggest an explanation for the observed decrease in dielectric constant of liquid water as the temperature is increased.
2. Compare the attractive force between a Na^+ ion and a Cl^- ion that are 20 Å apart in water with the attractive force between the same ions separated by the same distance in ethyl alcohol. Make the same comparison for a Mg^{2+} ion and a SO_4^{2-} ion. What do these comparisons *suggest* about the relative solubilities of NaCl and $MgSO_4$ in water and in alcohol?
3. Many salts may be precipitated from aqueous solution by adding alcohol. Suggest an explanation.
4. Heat is absorbed when $Na_2SO_4 \cdot 10H_2O(c)$ dissolves in water, but heat is evolved when anhydrous $Na_2SO_4(c)$ is dissolved in water. Suggest an explanation. (Heat effects associated with chemical processes are discussed in detail in several later chapters starting with Chapter 11.)
5. Why is it reasonable to say that HCl ionizes when it dissolves in water but incorrect to say that NaCl ionizes when it dissolves?
6. Hydrogen peroxide has dielectric constant comparable to that of liquid water. What can you conclude about the structure of a single molecule of H_2O_2 and about the nature of molecular interactions in liquid hydrogen peroxide?
7. How many H_2O molecules are in 1.0 kg of water? Given that the concentrations of both hydrogen ions and hydroxide ions are 1.0×10^{-7} molal in pure water, calculate the fraction of H_2O molecules that are ionized.
8. Calculate the change in volume associated with freezing one mole of water at 0°C

to ice at the same temperature. Densities of water and ice at 0°C are 0.99987 and 0.915 g cm^{-3}, respectively.

9. Although Kr and HBr molecules have 36 electrons each, the normal boiling point of Kr is 121°K whereas that of HBr is 206°K. Suggest an explanation for the large difference in normal boiling points.

10. Estimate the normal boiling point of the imaginary "ideal" liquid ammonia that would have no hydrogen bonds.

11. One popular water conservation measure is to place a brick in the flush tank of a toilet. A normal brick will displace about 0.25 gallons of water. Assuming that the average person flushes a toilet six times a day, how much water would be saved in a year in a city of 100,000 if all flush tanks had bricks in them?

12. Dimethyl ether, $(CH_3)_2O$, is completely miscible with water, whereas dimethyl sulfide, $(CH_3)_2S$, is only slightly soluble in water. Suggest an explanation for the difference in solubilities.

13. Densities of water and ice at 0°C are 0.9999 and 0.915 g cm^{-3}, respectively. Calculate the fraction of an iceberg that will show above the water line (in a fresh water lake) at 0°C. What fraction of an iceberg would show above the water line in sea water of density 1.025 g cm^{-3}? (*Hint:* It is necessary to consider Archimedes' principle.)

6

ELECTRONS IN ATOMS

Introduction

A large part of chemistry is concerned with reactions between atoms or groups of atoms. Atomic structure is therefore of great importance in understanding chemical reactions. Starting with the Rutherford picture of the atom as a dense positively charged nucleus surrounded by electrons, many scientists concluded that the number, arrangement, and energies of electrons are the most important factors influencing the interactions between atoms. Through the efforts of Bohr, de Broglie, Schrodinger, Heisenberg, Dirac, Pauli, Lewis, Pauling, and many others, we now feel that we have a reasonable understanding of the electronic structure of atoms and of the relationship of this structure to chemistry. Before we undertake a detailed discussion of atomic structure, we should first consider the discontinuous nature of matter and energy in the subatomic world in terms of conflicts with classical mechanics and the newer concepts that have led to our present picture of the atom.

Origins of Quantum Theory

All heated objects emit electromagnetic radiation. When an iron rod is heated, the color gradually changes from gray to red to dull orange to white, indicating that higher frequencies (shorter wavelengths) become relatively more important in

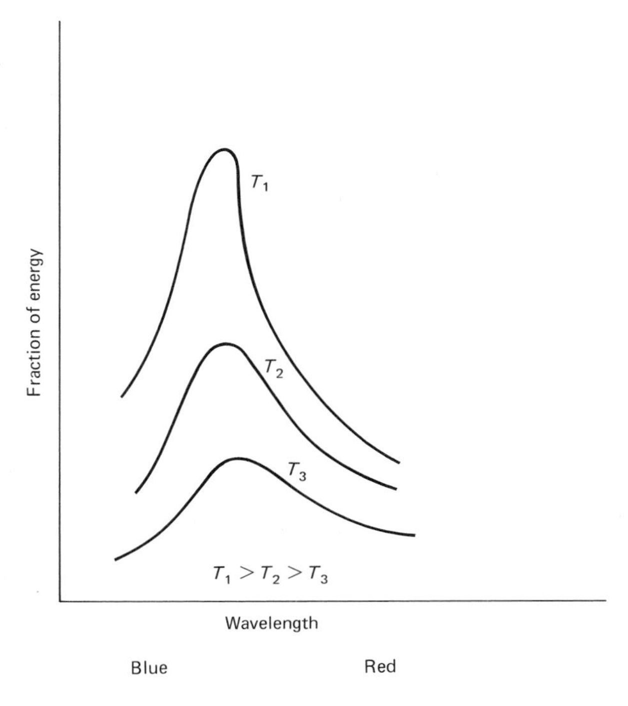

Figure 6.1. Distribution of radiation from a black body at three temperatures. Planck developed his quantum theory to explain the shapes of these curves.

the radiation as the temperature increases. Figure 6.1 shows some curves representing the observed distribution of energy as a function of wavelength in such radiation.

The electromagnetic spectrum, illustrated in Figure 6.2, can be conveniently expressed in terms of wavelength λ (Greek letter lambda) or frequency ν (Greek letter nu). These quantities are inversely proportional to each other. Wavelengths of light are expressed in any of several units of length; for the present we use

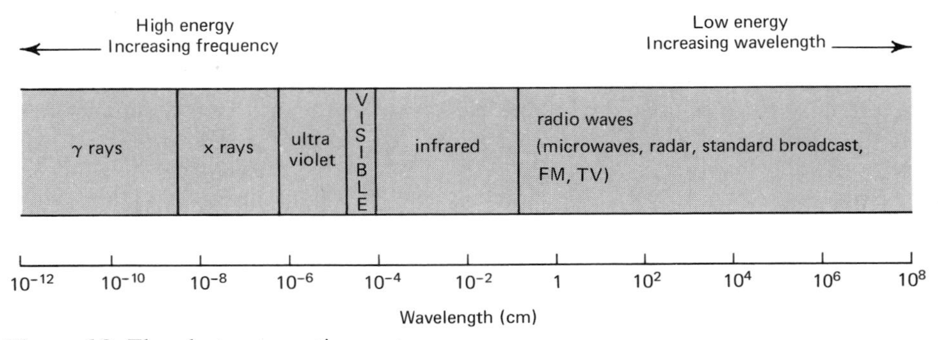

Figure 6.2. The electromagnetic spectrum.

The boundaries between any two regions of the electromagnetic spectrum are no more definite than between any two colors of the visible spectrum. Various regions of the spectrum actually overlap, and naming depends both on wavelength and the origin of the waves. In all cases the velocity of the radiation (in vacuum) will be 3.0×10^{10} cm/sec.

centimeters. Frequency is the number of waves or oscillations per unit time; we might use waves per second, but more commonly write sec^{-1}, although Hz (hertz) is gaining increasing acceptance for this unit. Velocity is length per unit time; we use $cm\ sec^{-1}$. Remembering that c represents the velocity of light, we have

$$c = \lambda\nu \qquad (6.1)$$

Example Problem 6.1. Calculate the frequency of light that has a wavelength of 5.0×10^{-5} cm.

We use equation (6.1) and the knowledge that the velocity of light in vacuum is $3.0 \times 10^{10}\ cm\ sec^{-1}$. We first rearrange (6.1) and then put in the proper numbers (*with units*) for c and λ to obtain the desired frequency as

$$\nu = \frac{c}{\lambda} = \frac{3.0 \times 10^{10}\ \text{cm sec}^{-1}}{5.0 \times 10^{-5}\ \text{cm}} = 6.0 \times 10^{14}\ \text{sec}^{-1}$$

From Figure 6.2 we note that this light is in the ultraviolet region of the spectrum. ■

Prior to 1900 several physicists had made unsuccessful attempts to explain the curves shown in Figure 6.1. Implicit in the classical physics used by these investigators is the assumption that energy can be absorbed or emitted *continuously* in arbitrarily small amounts. This breakdown of classical mechanical principles when applied to radiation was viewed with dismay by the physicists of the time; they termed it the "ultraviolet catastrophe."

The way out of this predicament was suggested in 1900 by Max Planck, a professor of physics at the University of Berlin. Planck discarded some of the precepts of classical mechanics and showed that experiment and theory could be brought into agreement by assuming that light energy could be absorbed or emitted only in very small discrete units called **quanta:** hence the name **quantum theory.**

The success of Planck's quantum theory of radiation was partly responsible for Einstein's proposal in 1905 that light is also propagated through space in definite quanta called **photons.** The energy, E, of these photons is given by

$$E = h\nu \qquad (6.2)$$

where h is the Planck constant. The value of this constant is 6.62×10^{-27} erg sec. The dimensions of energy times time constitute a quantity called action. Note that the energy of a photon is not fixed, but depends on the frequency of the light.

Another problem for classical physicists was the **photoelectric effect.** A beam of light falling on a metal surface can cause electrons to be emitted from the metal, as represented in Figure 6.3. This phenomenon is the basis for present-day photoelectric cells (electric eyes). The velocities of the emitted electrons have been determined experimentally. For any given metal the *maximum* velocity of these electrons depends only on the *frequency* of incident light. On the other hand, classical electromagnetic theory leads to the prediction that the energy of photoelectrons should be independent of frequency but should vary with intensity, that is, on the amount of light. Again, classical physics fails to account for experimental results.

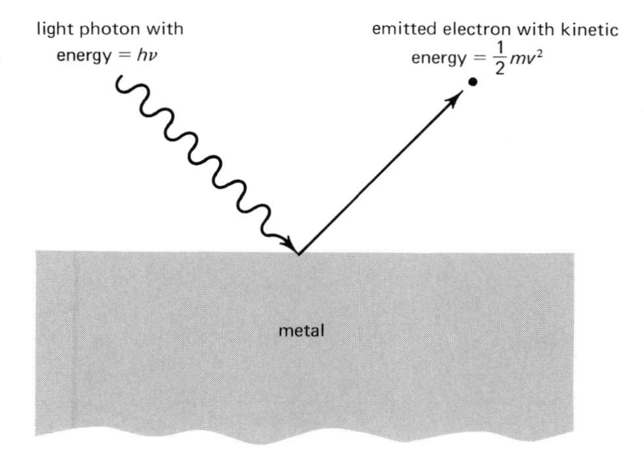

Figure 6.3. Schematic representation of the photoelectric effect.

Now let us consider Einstein's explanation of the photoelectric effect, for which he originally postulated that light is propagated in quanta. A photon of frequency ν and energy $h\nu$ is absorbed by a metal. Some or all of this energy is used in ejecting an electron from the metal. The emitted electron will have maximum energy and velocity when all of the absorbed light energy is used in ejecting this photoelectron. It will have some lesser energy and velocity when part of the absorbed light energy goes into the vibrations of the metal atoms and, therefore, is not used to eject the photoelectron.

Before proceeding with interpretation of the photoelectric effect, it is necessary to review the relationship between the velocity of a moving body and the kinetic energy of the moving body. As stated in Chapter 1, one of the fundamental equations of elementary physics is

$$\mathrm{KE} = \tfrac{1}{2}mv^2 \tag{6.3}$$

where KE, m, and v represent kinetic energy, mass, and velocity, respectively.

Einstein put all this knowledge together and derived a relationship between the maximum velocity of emitted electrons and the frequency of the incident light. The total energy absorbed by the metal is $h\nu$; the energy "used up" is the kinetic energy of the electron plus the energy necessary to get the electron out of the metal, called the **work function** of the metal and given the symbol WF. These statements are summarized by the equation

$$h\nu = \mathrm{KE}_{\max} + \mathrm{WF} \tag{6.4}$$

where $\mathrm{KE}_{\max}$ represents the kinetic energy of the emitted electrons with the maximum velocity.

If Einstein's ideas about the photoelectric effect are correct, *no* electrons should be emitted when the frequency of the light is so low that $h\nu$ is less than WF. It is observed experimentally that light of frequency lower than what is called the threshold frequency, although it may be very intense, does not cause electrons to be emitted.

At frequencies greater than this threshold frequency, the maximum velocity or kinetic energy of the emitted electrons is related to the frequency as predicted

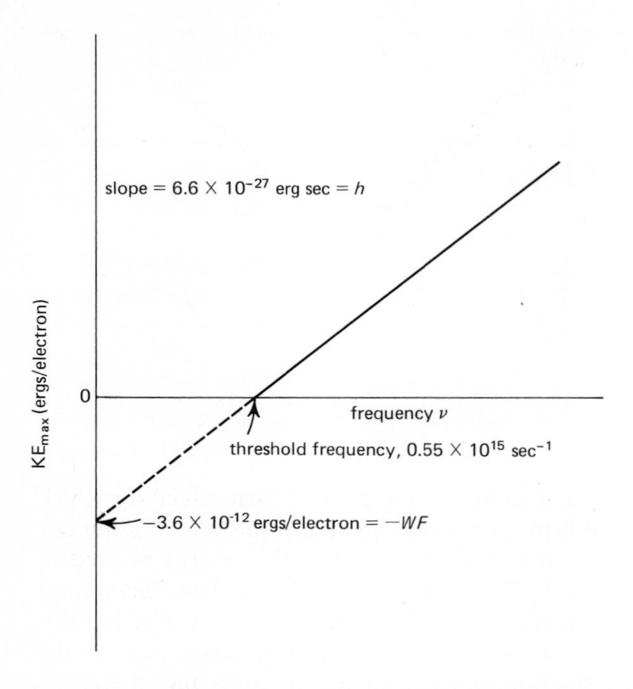

Figure 6.4. Graph of maximum kinetic energy of photoelectrons from potassium against frequency of absorbed light. The dashed line indicates extrapolation to obtain the intercept on the KE_{max} axis.

by Einstein. This relationship is best seen by rearranging equation (6.4) to the form

$$KE_{max} = h\nu - WF \tag{6.5}$$

We know that (see Chapter 1) $y = mx + b$ is the equation of a straight line of slope m and intercept b on the y axis when y is plotted against x. A graph of KE_{max} against ν should therefore give a straight line of slope h and intercept $-$WF. A graph of this sort for potassium is shown in Figure 6.4, and work functions for several metals are listed in Table 6.1. All of these work functions came from graphs like that shown for potassium in Figure 6.4. Each of these graphs also gives an independent value for h, all of which are the same within experimental uncertainty.

Now consider the units used for kinetic energy, Planck's constant, and the work function (see Chapter 1). Because mass is expressed in grams and velocity in centimeters per second, we calculate kinetic energy in terms of (g)(cm sec^{-1})^2 and

Table 6.1. Work Functions of Metals

Metal	*WF (ergs/electron)*
Li	3.6×10^{-12}
Na	3.7×10^{-12}
K	3.6×10^{-12}
Ag	7.5×10^{-12}
Zn	5.5×10^{-12}
Hg	7.2×10^{-12}
Pt	10.0×10^{-12}

use the term erg for energy expressed in these units. The units for $h\nu$ and WF must also be ergs if these energies are to be additive and, because ν is expressed in sec^{-1}, h must be in units of erg sec.

The value of h determined from photoelectric data is in agreement with the value deduced earlier by Planck. Between 1900 and 1912 the theories of Planck and Einstein and the quantity of experimental data gathered to test the validity of these theories clearly showed that energy, at least under some conditions, is not continuous but comes in discrete bunches or packets called quanta.

Another important advance in quantum theory came in 1913 when Niels Bohr, a Danish physicist, first applied quantum theory to energies of electrons in atoms.

Bohr's Theory of the Hydrogen Atom

The alpha particle scattering experiments of Rutherford, discussed in Chapter 2, led to the miniature solar system model of the atom. In 1912 it was generally believed that an atom was composed largely of empty space with negative electrons revolving at relatively great distances about a small, very dense, positive nucleus. However, this model presents another problem for classical electromagnetic theory. To illustrate the difficulties let us consider the hydrogen atom in some detail.

We will start with the generally accepted picture of the hydrogen atom in 1912 as consisting of a proton nucleus with one electron moving around or located near the proton. The positive proton and negative electron should exert an attractive force for each other. If the electron were stationary, it would be pulled into the nucleus. Therefore we must assume that the electron is in motion, which counteracts the pull of the proton.

If classical laws of motion were applicable on an atomic scale, we would expect the electron at first to revolve about the proton in the same way that planets move about the sun. The electron's orbit could be of any size, determined by the energy of the system. But according to classical theories of electricity and magnetism, a charged particle moving in a circle should emit light; and the frequency of this light should equal the frequency of revolution of the electron around the proton. Since energy is lost as light is emitted, the electron would move closer to the nucleus. Its frequency of revolution would increase, thus increasing the frequency of the emitted light. The orbital radius would gradually decrease and the electron would spiral into the nucleus. Thus classical mechanics predicts the complete collapse of atoms. Even in 1912 it was clear that something must be wrong with this analysis. Atoms do not collapse nor do atomic spectra consist of light of all wavelengths. Rather, excited atoms emit radiation of only certain well defined wavelengths, a line spectrum.

Before continuing with Bohr's proposals to solve these difficulties, we must briefly consider atomic spectra. All substances absorb or emit light under proper conditions. It is well known that salts of certain metals, when heated to high temperatures in a flame, impart a characteristic color to the flame. Thus sodium salts impart a yellow color to flame, lead salts impart a light blue color, and calcium salts give a pale red color. In addition, when an electrical discharge is passed through a gas at low pressures the gas emits a characteristically colored glow discharge. The orange-red discharge of neon is all too familiar in commercial neon signs. Mercury vapor lamps show a greenish-blue glow discharge. These flame colors or glow discharges may be characterized more completely by dispersing the emitted light with a prism so that various wavelengths are spread out into a spectral pattern. An appropriate experimental arrangement is shown schematically in Figure 6.5. With visible white light as the source, a continuous spectrum is obtained. With the heated salt or glow discharge of a gas as the source, a line spectrum results.

In 1913, Bohr proposed a theory that accounted for the existence of line spectra and also explained why atoms do not collapse. Applying the quantum ideas of Planck and Einstein, Bohr suggested that the total energy (kinetic plus potential) of an electron in an atom is **quantized,** that is, has only certain restricted values. Expressed another way, this means that the electron in a hydrogen atom can move around the atom only in certain definite orbits, frequently called stationary states or energy levels, and while doing so undergoes no energy change. The only way the electron can change its energy is to shift from one discrete energy level to another. The emission spectra we have been discussing thus involve transition of an electron from a higher to a lower energy state with emission of a quantum of energy $h\nu$. If no lower energy levels are available, there can be no transition. For this reason, atoms do not collapse.

In order to test these assumptions, Bohr applied them to the problem of calculating some of the properties of a hydrogen atom. He started with a mental picture of a hydrogen atom based on the Rutherford nuclear atom as illustrated in Figure 6.6.

Bohr assumed that the electron moves with velocity v in a circular path or orbit of radius r around the proton. Newton's equation, $f = ma$, where f, m, and a represent force, mass, and acceleration, furnished the starting point for Bohr's derivation. The electrostatic attractive force between the oppositely charged proton and electron is given by Ze^2/r^2 from Coulomb's law, where e represents the magnitude of the charge of the electron and Z is the atomic number. The centripetal acceleration is v^2/r. Substitution of Ze^2/r^2 for f and v^2/r for a in $f = ma$ gives

$$\frac{Ze^2}{r^2} = \frac{mv^2}{r} \tag{6.6}$$

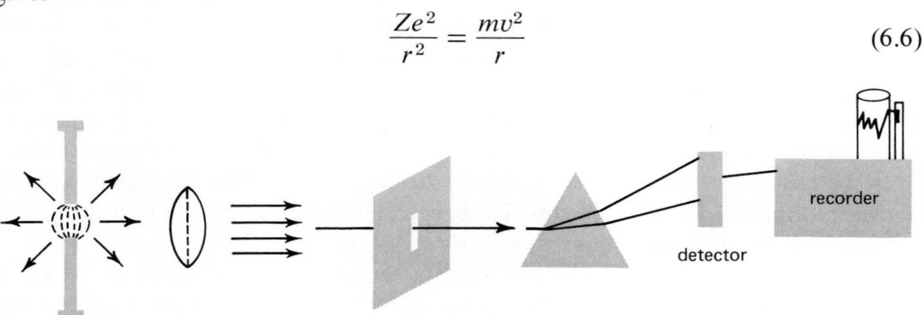

Figure 6.5. Schematic diagram of spectrograph for emission spectroscopy.

The light source can be a flame, an electric arc, or merely the substance under investigation heated to a high temperature.

For absorption spectroscopy a heated filament may be used as a light source. The substance to be investigated is placed between the defining slit and the prism (or a ruled grating).

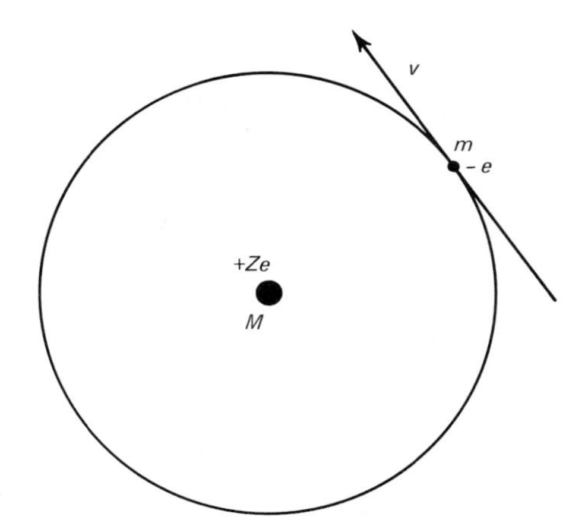

Figure 6.6. Schematic illustration of hydrogen atom according to Bohr.

The electron of charge $-e$ and mass m is moving with velocity v in a circular path or orbit of radius r around the proton, which has mass M and charge $+Ze$. In this case Z, the atomic number, is unity.

Bohr then introduced his assumption that the electron could not move around in just any orbit by stating that the angular momentum, mvr, of the electron must be some integral multiple of $h/2\pi$ as expressed by the equation

$$\text{angular momentum} = mvr = nh/2\pi \tag{6.7}$$

In this equation and the other Bohr equations to follow, n represents any positive integer and is called a **quantum number.**

Before we combine equations (6.6) and (6.7), justification of Bohr's assumption should be considered. Although in 1913 it was impossible for Bohr to justify his assumption rigorously, he could rationalize it, partly on the grounds that quantization of energies of some atomic processes had been shown by Planck and Einstein to be necessary. Once the idea was accepted that only certain energy states are allowed in atoms, it became clear that Planck's constant should be included somewhere in the equations describing the electrons in atoms. In 1913 the best evidence for the validity of Bohr's assumption was the success of the derived equations.

It is interesting to speculate on the thoughts which led Bohr to propose the quantum condition. Perhaps he used reasoning like the following. Planck's constant, h, has dimensions of energy times time. Since energy has dimensions of mass times velocity squared, and velocity is distance per unit time, we may write

$$h = (\text{energy})(\text{time}) = (\text{mass})(\text{distance})^2/\text{time}.$$

Recalling that momentum is mass times velocity or mass times distance per unit time we see that Planck's constant has dimensions of momentum times distance:

$$h = (\text{momentum})(\text{distance})$$

This analysis could have led Bohr to reason that the quantum condition he desired should be introduced by requiring that the electron's momentum times the circumference of its circular orbit should be some integral multiple of h. Knowing that the circumference of a circle of radius r is $2\pi r$, he could write

$$(\text{momentum})(\text{circumference}) = mv2\pi r = nh$$

where n is an integer. Division of this equation by 2π gives

$$\text{angular momentum} = mvr = nh/2\pi$$

which is seen to be identical with equation (6.7) used to introduce the quantum condition.

Equation (6.7) may be solved for v and then v^2 as shown by

$$v = \frac{nh}{2\pi mr} \tag{6.8}$$

and

$$v^2 = \frac{n^2h^2}{4\pi^2m^2r^2} \tag{6.9}$$

The right hand side of (6.9) is then substituted for v^2 in (6.6), which is solved for r to give

$$r = \frac{n^2h^2}{4\pi^2mZe^2} \tag{6.10}$$

The radius of the smallest and hence the lowest energy orbit in a hydrogen atom is calculated from (6.10) by setting the quantum number n equal to unity and putting in the numerical values for h, π, m, and e with proper attention to units. These are: $h = 6.62 \times 10^{-27}$ erg sec, $\pi = 3.14$, $m = 9.1 \times 10^{-28}$ g, and $e = 4.8 \times 10^{-10}$ esu. The abbreviation **esu** stands for electrostatic unit of charge.

From (6.10) we have

$$r = \frac{(1)^2(6.62 \times 10^{-27}\text{ erg sec})^2}{(4)(3.14)^2(9.1 \times 10^{-28}\text{ g})(1)(4.8 \times 10^{-10}\text{ esu})^2}.$$

This equation is simplified by doing all of the indicated arithmetic and by expressing the erg and esu in terms of the fundamental units of mass, length, and time. Because dimensions of the erg are g cm^2 sec^{-2} and those of the esu are g$^{\frac{1}{2}}$ cm$^{\frac{3}{2}}$ sec^{-1},

$$r = \frac{0.53 \times 10^{-8}(\text{g cm}^2\text{ sec}^{-2})^2(\text{sec}^2)}{(\text{g})(\text{g}^{\frac{1}{2}}\text{ cm}^{\frac{3}{2}}\text{ sec}^{-1})^2} = 0.53 \times 10^{-8}\text{ cm}$$

for the radius of the electron's orbit in the most stable state of the hydrogen atom.

Because centimeters are an inconvenient unit of length for discussions of atoms, it has become customary to use another unit, the **angstrom,** defined as 10^{-8} cm and given the symbol Å. The radius of the smallest electron orbit is therefore 0.53 Å. It is known from a variety of experiments that atoms have diameters of the order of 1 or 2 Å, so it is clear that the Bohr theory is at least close to the truth in this respect.

An important feature of the Bohr theory is that it can be used to calculate the energies of the various allowed electron orbits or, as they are now called, energy levels. When an electron is excited from a given energy level to some higher energy level, light of frequency ν is absorbed such that $h\nu$ is equal to the energy difference between the two levels. Similarly, when an electron moves from a given energy level to a lower energy level, light of frequency ν is emitted such that

$h\nu$ equals the energy difference of the two levels. The general equation is

$$\Delta E = |E_f - E_i| = h\nu \tag{6.11}$$

where E_f and E_i represent the energies of the final and initial states, respectively. We take the absolute value (indicated by vertical bars) of the energy difference to ensure that the energy of the photon is positive.

We may also calculate the total energy E of the electron as the sum of the kinetic energy $mv^2/2$, resulting from motion, and the potential energy $-Ze^2/r$, attributable to position in the nuclear force field:

$$E = \frac{mv^2}{2} - \frac{Ze^2}{r} \tag{6.12}$$

The potential energy is a relative value that depends on the state we define as having zero potential energy. It is convenient to define the system to have a potential energy of zero when the electron is infinitely far from the nucleus. Rearranging equation (6.6), we obtain

$$mv^2 = \frac{Ze^2}{r} \tag{6.13}$$

and substitute into (6.12) to find that

$$E = \frac{Ze^2}{2r} - \frac{Ze^2}{r} = -\frac{Ze^2}{2r} \tag{6.14}$$

Substituting the value of r from equation (6.10) into equation (6.14) yields

$$E = -\frac{2\pi^2 m Z^2 e^4}{n^2 h^2} \tag{6.15}$$

When the electron is in the lowest energy level ($n = 1$), the atom is said to be in its **ground state,** the state of maximum stability. With the electron in any higher energy level, the atom is in an **excited state.**

Example Problem 6.2. Calculate the energy of the first excited state of the unipositive helium ion, He^+, which is formed by removing an electron from a neutral helium atom.

We use equation (6.15), noting that $Z = 2$ and $n = 2$.

$$E = \frac{2(3.14)^2(9.11 \times 10^{-28}\ \text{g})(2)^2(4.80 \times 10^{-10}\ \text{esu})^4}{(2)^2(6.62 \times 10^{-27}\ \text{erg sec})^2}$$
$$= -2.17 \times 10^{-11}\ \text{erg}$$

Energies of electronic states are commonly expressed in terms of electron volts per atom (eV atom^{-1}) or kilocalories per mole (kcal mole^{-1}). Using the conversion factors from Chapter 1 (1 eV atom^{-1} = 1.60×10^{-12} erg atom^{-1} = 23.1 kcal mole^{-1}), we find that

$$E = -13.6\ \text{eV atom}^{-1} = -314\ \text{kcal mole}^{-1}$$

Note that this energy is based on $E = 0$ for the helium nucleus and the electron infinitely far apart. ■

The most impressive triumph of the Bohr model was the precision with which it predicted the spectral lines for the hydrogen atom. Rearranging equation (6.11) and combining it with (6.15), the energy of the photon emitted on transition from energy level n_i to energy level n_f is

$$h\nu = E_i - E_f = \frac{2\pi^2 mZ^2e^4}{h^2}\left(\frac{1}{n_f^2} - \frac{1}{n_i^2}\right) \qquad n_i > n_f \tag{6.16}$$

where n_i and n_f represent the initial and final quantum numbers, respectively. Spectroscopists generally express energies in terms of wave numbers, $\bar{\nu}$, defined as $1/\lambda$. Since $\nu = c/\lambda = c\bar{\nu}$,

$$\bar{\nu} = \frac{2\pi^2 mZ^2e^4}{ch^3}\left(\frac{1}{n_f^2} - \frac{1}{n_i^2}\right) \tag{6.17}$$

Comparison of equation (6.17) with

$$\bar{\nu} = R_{\mathrm{H}}\left(\frac{1}{n_f^2} - \frac{1}{n_i^2}\right) \tag{6.18}$$

obtained earlier from experimental investigations of the spectrum of hydrogen yields a valuable result. Prior to 1913 there was no theoretical explanation for the empirically obtained constant R_{H}, called the Rydberg constant. The experimentally determined value for R_{H} is 109,677.60 cm^{-1}. The calculated value for R_{H} obtained from

$$R_{\mathrm{H}} = \frac{2\pi^2 mZ^2e^4}{ch^3} \tag{6.19}$$

is 109,737 cm^{-1}. In this calculation the assumption is made that the mass of the nucleus is stationary, which is valid only if the nucleus has infinite mass. For a nucleus of finite mass, both the electron and nucleus must revolve about a common center of mass. Replacing m, the mass of the electron, with the reduced mass μ defined by

$$\mu = \frac{m_e m_p}{m_e + m_p} \tag{6.20}$$

yields a calculated value for R_{H} of 109,677 cm^{-1}. This remarkable agreement between theoretical and experimental values constituted a major victory for the Bohr model of atomic structure.

A schematic representation of the line spectrum of the hydrogen atom is shown in Figure 6.7. The different series are named after the men who first observed them.

A more detailed investigation of the fine structure in the atomic spectrum of hydrogen, with subsequent attempts to explain it, led to modifications of the Bohr theory. Perhaps the most significant was the introduction by Arnold Sommerfeld of elliptical orbits and the accompanying necessity for a second quantum number. In spite of several important modifications between 1913 and 1925, even the revised Bohr model could not account for the spectra of systems more complex than the hydrogen atom. Its failure to account for the relative intensities of spectral lines, the *ad hoc* introduction of various quantum numbers, and its inability to describe polyelectronic systems indicated some fundamental flaws in the Bohr theory.

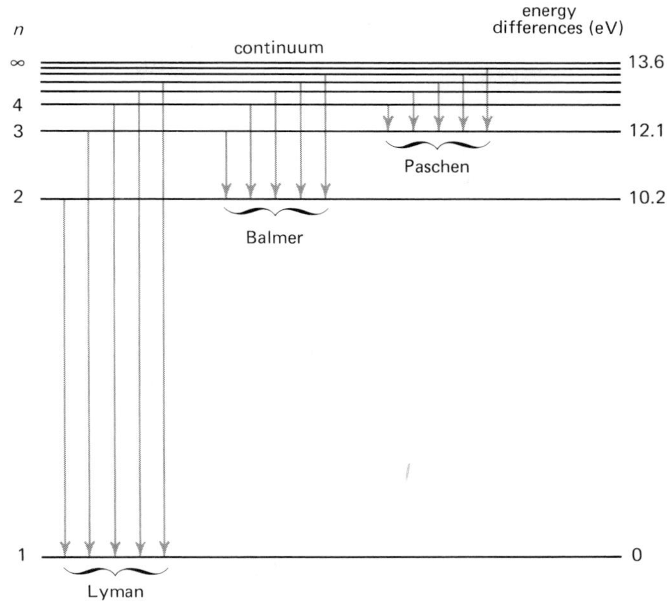

Figure 6.7. Energy levels of the hydrogen atom.

The energy changes for electrons dropping from an excited state to a lower energy state are determined by the energy levels of the initial and final states, which are in turn dependent on the principle quantum number. The difference in energy between the lowest (most stable) level ($n = 1$) and the series limit ($n = \infty$, $r = \infty$) is called the ionization energy.

We accept theories as long as they are consistent with experimental observations; when additional facts that contradict a theory are discovered, the theory must be revised or discarded. Although Bohr's theory of atomic structure must be rejected as inadequate in detail, we should not belittle its contribution. His postulates provided the necessary break with classical mechanics and set the stage for the more satisfying explanations we now consider. It is altogether fitting that Bohr received the Nobel Prize in Physics in 1922.

Through the Bohr treatment of the atom we have noted that integers appear as necessary parts of the equations. There is another well-established branch of physics in which integers appear naturally and only certain solutions of equations are allowed, namely, classical wave theory. For example, the audio frequencies of a banjo string are not continuous but are limited to certain frequencies called fundamentals and overtones. Perhaps the necessity of quantum conditions is related to a fundamental wave nature associated with all particles.

Matter Waves and the Uncertainty Principle

To describe adequately the behavior of light, it was necessary to discard certain clear-cut classical ideas and admit the difficult concept of wave-particle duality. This necessity created concern, but not nearly as much as the suggestion in 1924 by a young French physicist, Louis de Broglie, that this wave-particle duality applied to *all* matter. Our personal perceptions of radiant energy are only indirect, but we pick up sticks and stones and the assignment of wave nature to them inevitably seems strange.

The quantum theory of light described the energy of photons (light "particles") by the expression $E = h\nu$. But de Broglie realized that a particle theory contained nothing that even permitted definition of a frequency ν. However, Einstein's now famous equation, $E = mc^2$, relating energy, mass, and the velocity of light had been proposed in 1905 in his special theory of relativity. By equating these two expressions for energy we obtain

$$mc^2 = h\nu \tag{6.21}$$

where m may be considered as representing the "equivalent mass" of a photon of frequency ν or the actual mass of an "equivalent particle." Dividing each side of equation (6.21) by c gives

$$mc = \frac{h\nu}{c} = \frac{h}{\lambda} \tag{6.22}$$

where mc is the momentum of a photon or an equivalent particle (note from equation 6.1 that $\nu/c = 1/\lambda$). We also represent the momentum mc by p and write equation (6.22) as

$$p = \frac{h}{\lambda} \tag{6.23}$$

Equation (6.23) relates the momentum of a photon considered as a *particle* with the wavelength of the photon considered as a *wave*.

De Broglie suggested that this particle-wave relationship applied not only to radiant energy, but to *all* particles. Substituting the particle mass times particle velocity v for p in equation (6.23) yields the expression

$$\lambda = \frac{h}{p} = \frac{h}{mv} \tag{6.24}$$

where λ, the particle wavelength, is often referred to as the **de Broglie wavelength.**

De Broglie's theory and the resulting equations were regarded as wild speculation by the skeptical. There was at first no remarkable agreement between theoretical equations and experimental results as there had been in 1913 to support Bohr's theory. But in 1927 Davisson and Germer found that electrons are scattered by crystals and produce diffraction patterns similar to x-ray diffraction patterns, thus verifying that electrons do indeed have wave properties. In addition, the experimental wavelengths of moving electrons were in complete agreement with the values calculated from the de Broglie equation (6.24). More recently, wave properties have also been observed for such particles as neutrons and helium atoms.

Example Problem 6.3. Calculate the de Broglie wavelengths associated with an electron that has been accelerated through a 100 volt potential difference and a 56 g ball traveling at 100 miles per hour (a tennis ball accelerated by Arthur Ashe's serve).

In both cases we use equation (6.24) with $h = 6.62 \times 10^{-27}$ erg sec. For the electron in question we can first calculate the velocity by rearranging the kinetic energy expression (6.3) to yield

$$\begin{aligned} v &= \left(\frac{2\text{KE}}{m}\right)^{\frac{1}{2}} \\ &= \left(\frac{2(100\text{eV})(1.60 \times 10^{-12}\text{ erg eV}^{-1})}{(9.1 \times 10^{-28}\text{ g})}\right)^{\frac{1}{2}} \\ &= 5.9 \times 10^{8}\text{ cm sec}^{-1} \end{aligned}$$

Remember that an erg has dimensions of g cm^2 sec^{-2}. Thus the wavelength of the electron is

$$\lambda = \frac{(6.62 \times 10^{-27}\text{ erg sec})}{(9.1 \times 10^{-28}\text{ g})(5.9 \times 10^{8}\text{ cm sec}^{-1})} = 1.2 \times 10^{-8}\text{ cm} = 1.2\text{ Å}$$

This wavelength of an electron is about the same as most atomic radii and also about the same as the wavelength of some x rays. Most significantly, electron wavelengths are about the same as the internuclear distances in the crystal lattices that provided the basis for the diffraction experiments of Davisson and Germer.

The velocity of the tennis ball is

$$\begin{aligned} v &= \frac{(100\text{ mi hr}^{-1})(5280\text{ ft mi}^{-1})(12\text{ in. ft}^{-1})(2.54\text{ cm in.}^{-1})}{(60\text{ min hr}^{-1})(60\text{ sec min}^{-1})} \\ &= 4.5 \times 10^{3}\text{ cm sec}^{-1} \end{aligned}$$

The de Broglie wavelength of this tennis ball would be

$$\lambda = \frac{(6.62 \times 10^{-27}\text{ erg sec})}{(56\text{ g})(4.5 \times 10^{3}\text{ cm sec}^{-1})} = 2.6 \times 10^{-32}\text{ cm} = 2.6 \times 10^{-24}\text{ Å}$$

The wavelength of the ball (and all other macroscopic particles) is so short that it has no observable significance. ■

The de Broglie relationship (6.24) is simply related to the Bohr quantum condition for the electron in a hydrogen atom. Consider an electron moving in a circular orbit around a proton, just as Bohr pictured a hydrogen atom. According to equation (6.24) this electron has a wavelength that is determined by its velocity. As shown in Figure 6.8, the circumference of the orbit must be an integral multiple

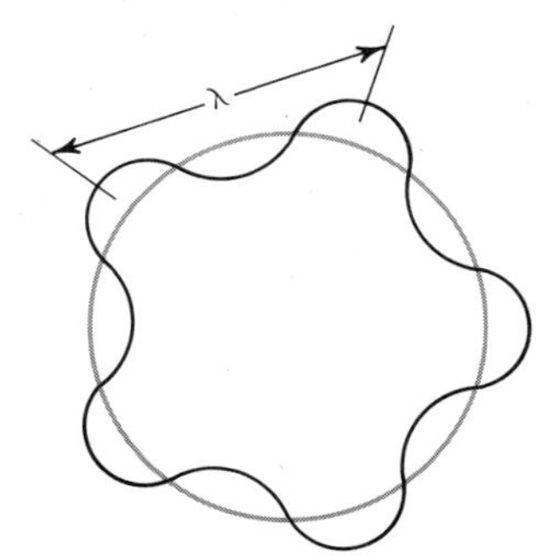

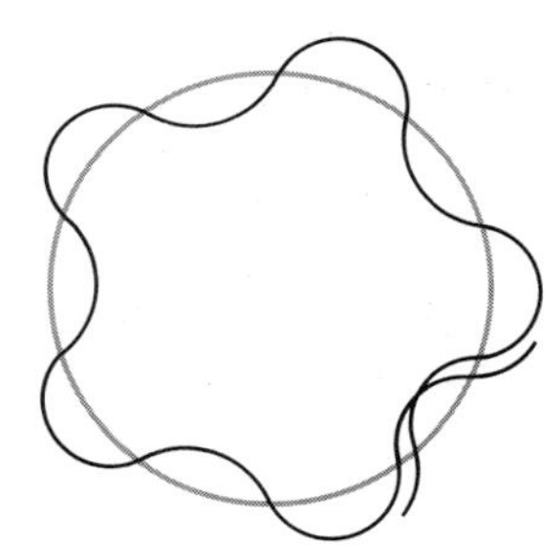

Figure 6.8. Drawing representing electron waves in a circular orbit. The drawing on the left shows a wave for which $2\pi r = 5\lambda$; the drawing on the right shows a wave of slightly different wave length that is not permitted because it would destroy itself by interference. Hence the requirement arises that $2\pi r = n\lambda$ as in equation (6.25).

of the wavelength of the electron in order to avoid destructive interference (analogous to destructive interference in optics); hence we write

$$2\pi r = n\lambda \tag{6.25}$$

Substitution of h/mv for λ in (6.25) gives

$$2\pi r = nh/mv$$

from which we obtain

$$mvr = nh/2\pi$$

The last equation is the Bohr quantum condition as introduced earlier in this chapter by equation (6.7), which reinforces our belief in a connection between the necessity for quantum conditions and the wave nature of matter.

One of the most important consequences of the dual nature of matter is the uncertainty principle postulated by Werner Heisenberg in 1927. This principle represents a complete break with classical physical thought and drastically affects how we must think about electrons in atoms and molecules. The Heisenberg uncertainty principle states that it is impossible for us to know *precisely and simultaneously* the position and momentum of a particle. Heisenberg showed that the lower limit of this uncertainty is Planck's constant divided by 4π as indicated in the expression

$$\Delta x \, \Delta p_x \geq h/4\pi \tag{6.26}$$

In this expression Δx is the uncertainty in position and Δp_x is the uncertainty in momentum in the x direction. There are other equivalent ways of stating this principle. For example, we could have $\Delta y \, \Delta p_y \geq h/4\pi$ or $\Delta E \, \Delta t \geq h/4\pi$. The latter expression, relating uncertainties in energy and time, is useful in estimating the sharpness of spectral lines.

Heisenberg's mathematical research showed that the uncertainty principle is a logical and necessary consequence of attributing wave properties to matter that we ordinarily think of as consisting of particles. It is also possible to consider this problem qualitatively by conducting a "thought experiment" in which we try to see and thereby locate an electron. From optics we know that it is necessary to use light with a wavelength shorter than the size of a particle to observe it accurately. Therefore we must try to see and locate accurately a very small particle (such as an electron) with very short wavelength light (such as γ rays or x rays).

Even neglecting experimental problems associated with designing and operating a γ-ray or x-ray microscope, we have now a fundamental difficulty. Because γ rays and x rays have very short wavelengths, they also have very high frequencies, which means that their photon energies ($E = h\nu$) are also very high. Collision of a high energy photon with a small particle considerably alters the velocity of the small particle, thereby causing a considerable uncertainty in the momentum of the particle. We might decrease this latter uncertainty by using light of lower frequency, but the longer wavelength light would lead to reduced resolution in locating the particle. Thus we see that we cannot simultaneously determine position and momentum with unlimited accuracy, as already stated more precisely in equation (6.26). Heisenberg's uncertainty principle is not something that can be overcome or avoided by way of technical improvements in our instruments; it is inherent in the nature of our measuring processes.

The uncertainty principle is important only in the subatomic world of very light particles. In our ordinary macroscopic world, in which we are concerned with relatively heavy particles, measurements of position and momentum are not sufficiently precise to be limited by the uncertainty principle.

The difficulties inherent in the Rutherford-Bohr model of the atom as a miniature solar system may now be partly obvious. In this model it was implicitly assumed that both the position and momentum of each electron could be defined at any given moment. Such a precise designation is contrary to experimental reality, as expressed by the uncertainty principle. From this principle we conclude that it is only possible to locate electrons in atoms in a statistical way. We should therefore speak in terms of the probability of finding an electron at any given position.

From the contributions of de Broglie and Heisenberg it was apparent that a new system of mechanics was necessary to describe events in the subatomic world. Because an electron has wave properties, its behavior might be described by a wave equation. We thus come to the contributions of the Austrian physicist Erwin Schrodinger and the realm of wave mechanics.

The Schrodinger Wave Equation and Quantum Numbers

In 1926, Schrodinger suggested that the behavior of electrons in atoms could be described by a second order differential equation similar to that used in classical wave mechanics to describe the waves of a vibrating violin string. He was thus able to relate the energies of atomic systems to wave properties. Schrodinger's system of wave mechanics has been justified in the best possible way—by many experimental verifications. The time-independent Schrodinger equation, which is dependent only on the three dimensions of space, may be written in operator form as

$$H\Psi = E\Psi \tag{6.27}$$

where H, the Hamiltonian operator, is

$$H = -\frac{h^2}{8\pi^2 m}\left(\frac{\partial^2\Psi}{\partial x^2} + \frac{\partial^2\Psi}{\partial y^2} + \frac{\partial^2\Psi}{\partial z^2}\right) + V \tag{6.28}$$

An operator is a symbol for a mathematical procedure that changes one function into another. In these equations m is the mass of the electron, h is Planck's constant, V the potential energy (a function of the position of the particle), and E the total energy (also a function of position).

To obtain solutions for the wave function Ψ involves complicated mathematical procedures; therefore, we shall limit ourselves to largely qualitative discussion of the results. Exact solutions for the wave equation have been obtained only for very simple systems, such as the hydrogen atom. For systems of more than one electron each variable in the Schrodinger equation depends upon the position of each of the electrons and only approximate solutions have been obtained, although some approximations have been carried to high accuracy. Many approximate solutions are sufficiently close to reality that they account for observed phenomena, and we shall be content to use the results of these approximate solutions in this book.

When the Schrodinger equation is applied to real systems, physically reasonable results can be obtained only if the energy is restricted to certain discrete values related to one another by integers. Thus quantization and quantum numbers are a necessary consequence of the wave nature of electrons and do not have to be introduced arbitrarily, as was done in the Bohr theory.

A question that frequently troubles students relates to a physical interpretation for Ψ. Unfortunately, while the wave mechanics of Schrodinger are satisfying mathematically and yield "answers" in agreement with experiment, they do not contain the pleasing pictorial ideas and explicit detail of classical mechanics or the Bohr atom. The wave function, Ψ, a particular solution of the Schrodinger wave equation, mathematically describes an electron in a given energy state. By itself Ψ has no physical meaning, but by analogy to classical wave mechanics it is often called a probability amplitude. The square of Ψ, however, does have an important physical interpretation. Ψ^2 may be considered a probability distribution and is proportional to the probability of finding the electron in any given volume element; however, Ψ^2 cannot tell us where the electron is actually located at any given instant. This indefinite picture is, of course, consistent with and required by Heisenberg's uncertainty principle.

The concept of electron probability can be related to a useful picture of atomic orbitals by considering a hypothetical experiment. Suppose we take instantaneous pictures of an electron in the lowest energy state of a hydrogen atom. Three-dimensional photography would enable us to assign coordinates x, y, z to the electron and plot it as a point in a three-dimensional diagram. Because the electron is in rapid motion, a photograph taken a fraction of a second later would show the electron in a new position, and provide another point on our diagram. After several million photographs we would have an array of dots that would resemble a cloud, dense in regions where there are many points crowded together, diffuse in regions where there are few points. In dense regions of the cloud there would be a high probability of finding the electron.

The cloud picture may then be considered as a map of probability density. A cross section of such a cloud for a spherical orbital is shown in Figure 6.9. Since there is a finite probability of finding the electron anywhere in space, as illustrated in Figure 6.10, the atom cannot be given a definite radius. Instead

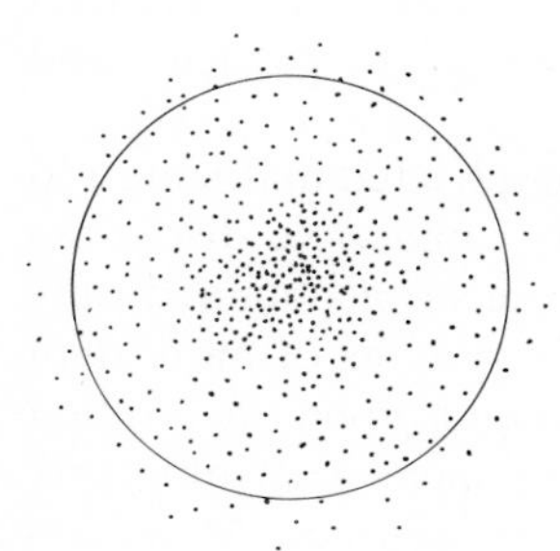

Figure 6.9. Cross section of a spherical charge cloud representation or probability density pattern for the lowest energy state of a hydrogen atom. The boundary contour encloses an arbitrary fraction of the total electron density, often 95%.

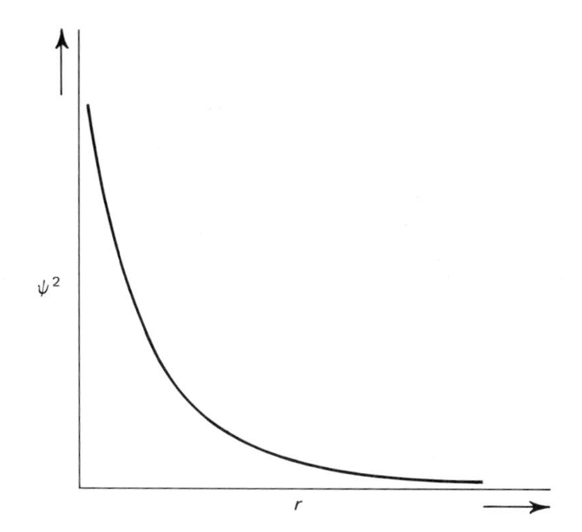

Figure 6.10. Probability distribution for an electron in the 1s orbital of hydrogen. Here r represents radial distance from the nucleus.

In principle, values of ψ^2 never reach zero, and there is a finite probability of finding the electron anywhere in the universe. But 99% of the probability is within a sphere of radius 2.2 Å from the nucleus.

we have a fuzzy electron cloud with no definite boundaries. We therefore draw an arbitrary boundary contour to enclose most of the electron density, usually 95%.

Our electron density picture thus has a shape represented by the circular cross section of the boundary surface in Figure 6.9. This boundary surface, which encloses most of the electron density for a given energy state, is termed an **orbital.** An orbital may be represented by a mathematical function, Ψ, or by a probability density representation, Ψ^2. Ψ may have positive or negative values, but Ψ^2 must always be positive. As we shall see after discussing quantum numbers, different shapes for charge cloud distributions can be drawn for other energy states of the hydrogen atom electron.

Integers appear naturally when the Schrodinger equation is applied to the hydrogen atom. Four quantum numbers are sufficient to characterize the properties of the single electron in the hydrogen atom and permit an explanation of the atom's energy level scheme obtained from its atomic spectrum. Although repulsive interactions between electrons affect the energy level spacings in many-electron atoms, there are clear relationships to the hydrogen atom case. If allowances are made for electron-electron repulsions, an energy level scheme consistent with observed atomic spectra of many-electron atoms is obtained by using quantum numbers and orbitals similar to those of the hydrogen atom.

The **principal quantum number, *n*,** analogous to the previously discussed Bohr quantum number, gives a measure of the energy of an electron in an orbital. Because of the uncertainty principle and the statistical interpretation of wave functions, we cannot consider electrons in atoms to be moving in definite orbits. Nevertheless, the principal quantum number is a measure of the mean radial distance of the electron density from the nucleus. On the average, electrons with a given principal quantum number will be closer to the nucleus and have lower energies (that is, more negative energies) than will electrons having a larger value of n. The principal quantum number can take only the integral values 1, 2, 3, . . . , ∞.

Table 6.2 Allowed Quantum Numbers for Electrons in Atoms

n:	*positive integers; 1, 2, 3, . . . , ∞ (not zero)*
l:	*positive integers less than* n *(including zero)*
m_l:	*positive and negative integers from* $+l$ *to* $-l$ *(including zero)*
m_s:	$+\frac{1}{2}$ *and* $-\frac{1}{2}$

The quantum number l is called the **azimuthal** or **angular momentum quantum number.** Although the classical concept of angular momentum is not defined in wave mechanics because we no longer visualize the electron as a discrete body, l may be considered as determining the shape of the electron cloud. Energies of electrons are also dependent on the value of l, since angular motion implies an angular kinetic energy, which is limited by the total energy of the electron. Values for the angular momentum quantum number are restricted by mathematical solutions of the Schrodinger equation and also by experimental results to integers from zero to $n - 1$, inclusive. For example, if $n = 3$, l might be 0, 1, or 2.

The **magnetic quantum number,** m_l, indicates how the orbital angular momentum is oriented relative to some fixed direction. The introduction of an external magnetic or electric field most conveniently provides the arbitrary reference axis. Electron energies are less dependent on m_l than on n and l. Theory and experiment restrict the m_l quantum number values to integers between $+l$ and $-l$, including zero. Thus, when $l = 2$, m_l may be $+2$, $+1$, 0, -1, or -2.

These three quantum numbers, n, l, and m_l, explicitly determine the energy of an electron orbital. To indicate completely the state of an electron in an orbital, it is necessary to define a fourth quantum number, m_s, the **spin quantum number.** The necessity for this quantum number was first deduced from very closely spaced lines in the spectra of certain atoms. In this connection it is convenient to picture the electron as spinning on its own axis so that it has an inherent magnetic moment. Since the spin can be in either of two possible directions (such as clockwise or counterclockwise), there are two possible values of the spin quantum number. Thus we say that m_s can have values of $+\frac{1}{2}$ or $-\frac{1}{2}$. Theoretical justification for this quantum number was obtained in 1928 when Paul Dirac developed a wave equation that allowed for relativity effects.

The allowed values for the four quantum numbers are summarized in Table 6.2. In Chapter 7 we will consider the relationships of these quantum numbers to the arrangements and energies of electrons in atoms of some of the lighter elements.

So far we have said little about the shapes of the electron clouds or the most probable region for locating an electron in space, meaning the region of maximum probability density. We will now consider these facets of our knowledge.

Representations of Atomic Orbitals

As a preliminary to verbal and pictorial description of atomic orbitals, it is useful to summarize a convenient and traditional means of identifying atomic orbitals in terms of their n and l quantum numbers. The system we use had its origin in the early study of spectral lines. Many of the observed lines were classified as *sharp, principal, diffuse,* or *fundamental,* and the initial letters s, p, d, and f were used to label them. Quantum theory has led to identification of the energy levels associated with these lines and the same letters s, p, d, and f have come into common use to indicate the numerical values of the l quantum numbers associated with the various levels as follows:

Value of azimuthal quantum number l	*Identifying letter*
0	s
1	p
2	d
3	f

Thus we represent the atomic orbital with quantum numbers $n = 1$ and $l = 0$ by the number-letter combination 1s. We have no other orbitals with $n = 1$ because in this case l is restricted to zero. The orbital with $n = 2$ and $l = 0$ is identified by the combination 2s, whereas the other orbital with $n = 2$, which has $l = 1$, is identified by the combination 2p. In the case of $n = 3$, we may have orbitals with $l = 0$, $l = 1$, and $l = 2$. We identify these orbitals with combinations 3s, 3p, and 3d. When $n = 4$, we have atomic orbitals with $l = 0$, $l = 1$, $l = 2$, and $l = 3$, and identify these orbitals with 4s, 4p, 4d, and 4f.

Earlier in this chapter we presented the concept of electron probability density with boundary contours to represent an orbital. In Figures 6.9 and 6.10 we were referring to the spherical charge cloud of a 1s orbital ($n = 1$, $l = 0$) of hydrogen. The 1s orbital corresponds to the lowest energy or ground state of this atom. Figure 6.10, in which the probability of finding the electron in a given volume in space is plotted against the distance from the nucleus, is also included in Figure 6.11. A more informative picture can be obtained if we divide the volume surrounding the nucleus into spherical shells, much like the layers of an onion.

The volume of each of these concentric shells of thickness dr is equal to $4\pi r^2 dr$ and will increase as the distance from the nucleus increases, as shown in Figure

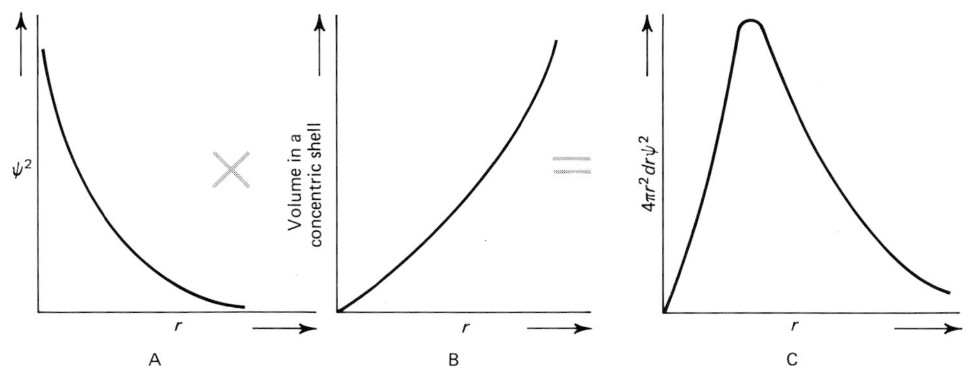

Figure 6.11. A. Probability density as a function of distance from nucleus. **B.** Total volume of concentric shells as a function of distance from nucleus. **C.** Radial probability distribution pattern for 1s orbital of hydrogen, giving probability of finding electron at given distance from nucleus.

6.11B. To obtain the probability of locating an electron in such a shell, we must multiply the probability that the electron is in a volume element times the volume of the shell. The resulting function, $\Psi^2 4\pi r^2 dr$, called the **radial probability distribution,** is shown for the 1*s* orbital of hydrogen in Figure 6.11C. Figure 6.11C clearly illustrates why the maximum probability is not at the origin (where $r = 0$), even though ψ^2 is a maximum there, because the center is a mathematical point of zero volume. The maximum probability is at $r = 0.529$ Å, which is also the value obtained for the radius of the first Bohr orbit.

There is the important distinction, however, that the Bohr theory confined the electron to a fixed orbit. The wave mechanical picture gives the electron a maximum probability of being at a distance from the nucleus equal to the Bohr radius, but also gives a finite probability of the electron being closer to the nucleus or farther from the nucleus. More significant is the conclusion that most of the charge (~90%) is contained in a sphere with radius about three times that of the maximum probability shell. Although the orbital does not have definite boundaries, the probability of finding the electron rapidly decreases towards zero as the distance from the nucleus is greatly increased.

The radial distribution function plotted against distance from the nucleus for a 2*s* orbital ($n = 2$, $l = 0$) is shown in Figure 6.12A and a cross section of the probability distribution in Figure 6.12B. In both diagrams we notice a nodal region. As in classical mechanics, a node is a point where the wave function is equal to zero, corresponding to a minimum amplitude. Although the 2*s* orbital also has a spherical contour surface containing 95% of the total probability density of the electron, there is a nodal spherical shell inside this sphere where there is zero probability of finding the electron. The question: "How does the electron get across this nodal space?" is meaningless. We cannot properly talk about an electron moving in some definite path; we are discussing wave behavior. It is equally meaningless to ask how the standing wave of a vibrating violin string gets across the node in its first overtone. The two concentric spheres of the 2*s* orbital do not constitute two independent waves but are two parts of the same wave.

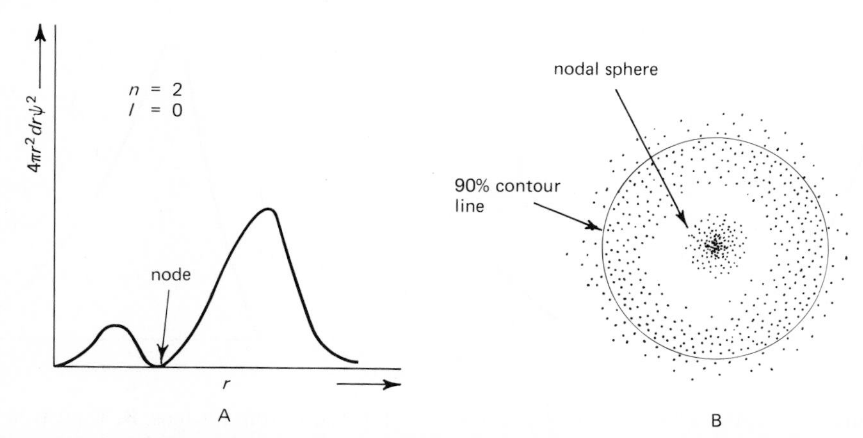

Figure 6.12. A. Radial distribution function for a 2*s* orbital of hydrogen. **B.** Cross section of probability density pattern for spherical 2*s* orbital. In the nodal sphere, there is a zero probability density.

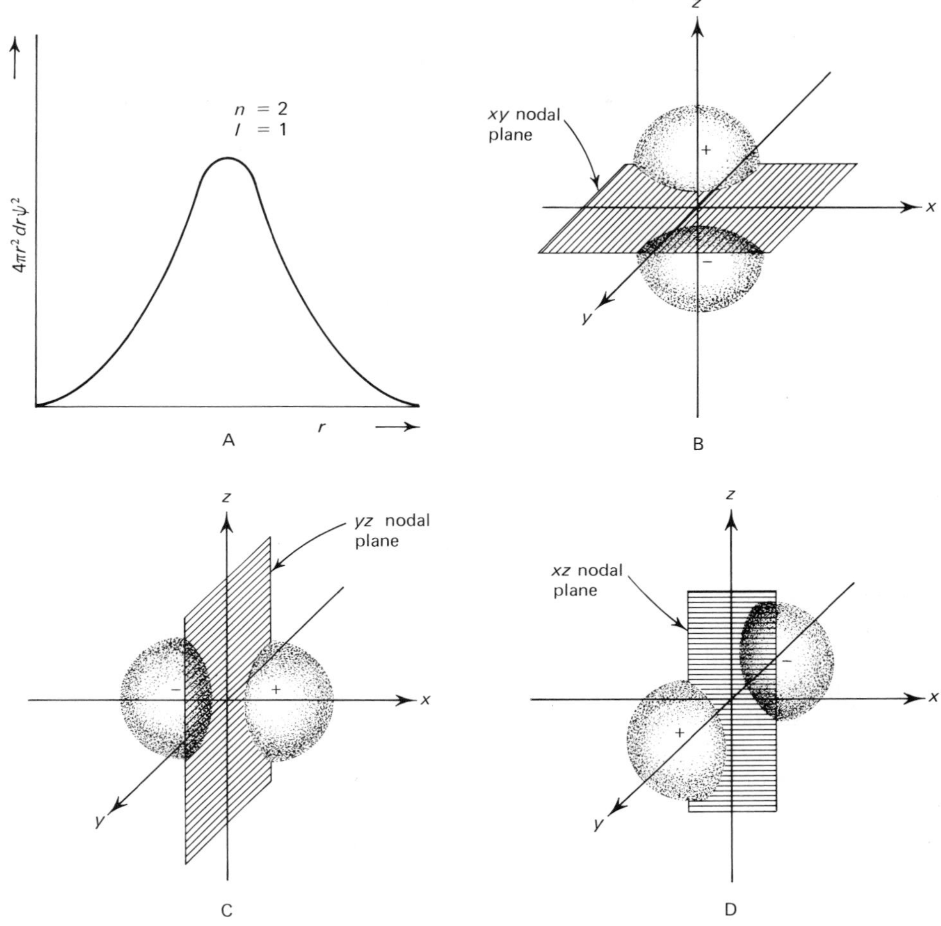

Figure 6.13. A. Radial distribution function for a $2p$ electron of hydrogen. **B.** Boundary contour for $2p_z$ orbital. **C.** Boundary contour for $2p_x$ orbital. **D.** Boundary contour for $2p_y$ orbital.

A more satisfying answer to this problem of the nodes is obtained from the Dirac relativistic treatment of quantum mechanics. An important consequence of this more correct, but also more mathematically complicated, relativistic approach is the disappearance of the regions of zero probability. In brief, the nodes of the nonrelativistic Schrodinger approach are replaced by regions where the probability density, although very small, is nevertheless finite.

For the $n = 2$ quantum level, there is another type of orbital, which is not spherical but rather has directional properties. There are three of these p orbitals because, when $l = 1$, m_l may have the values -1, 0, or $+1$. A characteristic of p orbitals is that they contain a nodal plane. For convenience, we orient these p orbitals along the principal coordinate axes and designate them as $2p_x$, $2p_y$, and $2p_z$. These three equivalent p orbitals differ only in their orientation in space. The radial probability distribution function is plotted against r in Figure 6.13A. The spatial contours for the $2p$ orbitals are shown in Figure 6.13B, C, D. It will

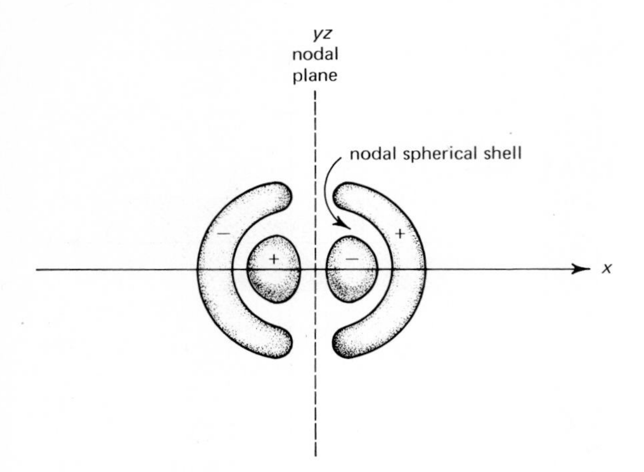

Figure 6.14. Cross section through $3p_x$ orbital showing both the nodal *yz* plane and a nodal sphere. The solid lines represent boundary contours enclosing about 90% of the total electron density.

later be important to remember that the signs on the lobes of the *p* orbitals refer to the mathematical sign of the wave function, Ψ, and are not related to electrical charge. It is also important to realize that a combination of the three $2p$ orbitals will occupy a spherical domain in space.

In the next quantum level ($n = 3$), there are three possible values for l and three types of orbitals. The $3s$ orbital ($l = 0$) is similar to the $2s$ orbital, but there are two nodal shells in the spherical boundary contour. The three $3p$ orbitals ($l = 1$) are similar to the $2p$ orbitals but have additional nodal surfaces, as represented for the $3p_x$ orbital in Figure 6.14. However, the outer part of the $3p$ orbital looks like a $2p$ orbital, and it is customary to represent all *p* orbitals by the three spatial diagrams shown in Figure 6.13.

The spatial contours for the five $3d$ orbitals ($l = 2$) are more complicated and will be presented in Chapter 20. Orbitals with l values greater than 2 are much more complicated and will not be discussed in this book.

Before closing our discussion, we should clarify our means of presentation of orbitals. Most of our pictorial representations are boundary surfaces of constant probability density. They are therefore plots of constant Ψ^2, and the mathematical sign must always be positive. The signs that we write on the lobes of these pictorial representations correspond to those of Ψ, which may be either positive or negative. Our patterns are thus dual-purpose drawings. The probability patterns, related to Ψ^2, are most useful pictorially, but it is convenient to include the mathematical signs from the orbital pattern, related to Ψ, for later discussions of chemical bonding.

References

Adamson, A. W.: Domain representations of orbitals. *J. Chem. Educ.*, **42:**140 (1965).

Berry, R. S.: Atomic orbitals. *J. Chem. Educ.*, **43:**283 (1966) (an Advisory Council on College Chemistry resource paper).

Cohen, I., and T. Bustard: Atomic orbitals: limitations and variations. *J. Chem. Educ.*, **43:**187 (1966).

Darrow, K. K.: The quantum theory. *Sci. Amer.* **186:**47 (March 1952).

Gamow, G.: *Mr. Tompkins in Wonderland.* Cambridge University Press, Cambridge, 1965 (paperback).
Gamov, G.: *Thirty Years That Shook Physics.* Doubleday and Co., Inc., Garden City, N.Y., 1966 (paperback).
Hoffmann, B.: *The Strange Story of the Quantum.* Dover Publications, New York, 1959 (paperback).
Johnson, R. C., and R. R. Retlew: Shapes of atoms. *J. Chem. Educ.*, **42:**145 (1965).
Lambert, F. L.: Atomic orbitals from wave patterns. *Chemistry,* **41:**10 (Feb. 1968); **41:**8 (Mar. 1968).
Ogryzlo, E. A.: Atomic orbitals and classical motion. *J. Chem. Educ.*, **45:**80 (1968).
Pimentel, G. C., and R. D. Spratley: *Chemical Bonding Clarified Through Quantum Mechanics.* Holden-Day, Inc., San Francisco, 1969 (paperback).
Powell, R. E.: Relativistic quantum chemistry: the electrons and the nodes. *J. Chem. Educ.*, **45:**558 (1968).
Sebera, D. K.: *Electronic Structure and Chemical Bonding.* Blaisdell Publishing Co., New York, 1964 (paperback).

Problems

1. What is the frequency of light with wavelength 8.0×10^{-5} cm? In what region of the electromagnetic spectrum would this light fall?
2. Light in the red region of the visible spectrum has a frequency of 4.6×10^{14} sec^{-1}. Calculate the wavelength in angstroms.
3. Calculate the frequency and the energy per photon for electromagnetic radiation with wavelength (a) 3.0×10^{5} cm (long radio wave), (b) 3.0×10^{-3} cm (far infrared), (c) 1500 Å (ultraviolet), and (d) 0.0010 Å (γ ray).
4. Calculate the energy in ergs per photon and in kilocalories per mole of photons for light with a wavelength of 2500 Å. In what region of the electromagnetic spectrum does this light belong?
5. Using the work function value given in Table 6.1, calculate the threshold frequency and wavelength for emission of electrons from silver.
6. What is the maximum velocity of electrons ejected from potassium by light of frequency 1.2×10^{15} sec^{-1}?
7. To initiate many chemical reactions it is necessary to supply energy in the range of 10 to 100 kcal mole^{-1}. Give the limits for this range in angstroms, wave numbers, and frequency. To what region of the electromagnetic spectrum do these limits correspond?
8. Calculate the velocity and kinetic energy of the electron (in atomic hydrogen) in the Bohr orbit with $n = 1$, and in the Bohr orbit with $n = 2$.
9. Calculate the wavelengths (in angstroms) of the first two lines in the Balmer series (see Figure 6.7) for atomic hydrogen.
10. Calculate the ionization energy of hydrogen. The ionization energy is the amount of energy necessary to remove an electron from the influence of the nucleus (a transition from $n = 1$ to $n = \infty$).
11. Suppose that a hydrogen atom has been bombarded with radiation with the result that the electron has been excited to the $n = 3$ quantum level. Calculate the ionization energy for this excited hydrogen atom.
12. Calculate the radius and the velocity for an electron in the Bohr orbit with $n = 3$ for the hydrogen-like ion, Be^{3+}.
13. (a) Calculate the "simple" Rydberg constant for the deuterium atom.
    (b) Calculate the reduced mass of the deuterium atom.
    (c) Using the reduced mass correction, calculate the Rydberg constant for deuterium.

14. Calculate the frequency (in wave numbers) for the lowest energy transition in the Lyman series ($n = 2$ to $n = 1$) of the unipositive helium ion, He^+.

15. What must be the velocity of an electron if it is to have a wavelength of 4.0×10^{-8} cm?

16. What must be the velocities of an electron and of a proton to have a wavelength equivalent to that of an x ray of frequency 3.0×10^{18} sec^{-1}?

17. For the Li^{2+} ion, calculate the wavelength of the lowest energy transition ($n = 4$ to $n = 3$) in the Paschen series.

18. An average hydrogen molecule at 200°C has a velocity of 2.4×10^5 cm sec^{-1}. Calculate the de Broglie wavelength of this particle.

19. Electrons in a particular experiment are accelerated by a potential difference of 10,000 volts. Calculate the de Broglie wavelength of these electrons.

20. (a) What is the minimum uncertainty in the instantaneous position of a baseball (according to the uncertainty principle of Heisenberg rather than to a poor batter) of mass 140 g and velocity 3000 cm sec^{-1}? Assume that it is possible to measure both the mass and the velocity of the ball with an accuracy of 1 part in a million.
(b) Determine the minimum uncertainty in the instantaneous position of an electron moving at the same velocity as the baseball in part (a). Assume that the momentum of the electron can be measured with the same degree of accuracy.

21. Calculate the volume of spherical shells of thickness 0.020 Å at distances of 0.25, 0.50, 0.75, and 1.00 Å from the origin and make a graph of your results.

22. The maximum in a graph of *fraction of solar energy* against *wavelength* as in Figure 6.1 occurs at about 5000 Å. According to Planck's theory of radiation from a "black body," the wavelength at which the emission of energy is a maximum is given by $\lambda_{max} = ch/4.97kT$. Calculate the approximate surface temperature of the sun, assuming that it radiates as a "black body."

23. In the Lyman emission series for atomic hydrogen (see Figure 6.7) a line is observed that has $\bar{\nu} = 97{,}492.2$ cm^{-1}. Calculate the quantum number of the initial state.

24. The threshold wavelength for photoemission of electrons from lithium is about 5200Å. What is the velocity of an electron emitted as a result of absorption of a photon with a wavelength of 3000Å?

25. A bullet with a mass of 0.067 oz was fired with a velocity of 720 mi hr^{-1}. Calculate the wavelength of this bullet in angstroms.

26. In order to evaluate Planck's constant, Millikan [*Phys. Rev.*, **7**:355 (1916)] performed photoelectric effect experiments on several different metals; in the process he reaffirmed Einstein's interpretation of this phenomenon. He made the assumption that the kinetic energy of the photoelectron must be equal to the charge on the electron times the potential difference just necessary to prevent its ejection. From the results of his experiments, Millikan plotted volts versus frequency of the bombarding light. The slope of the volt versus frequency line for sodium was 4.13×10^{-15}. From this information calculate Planck's constant. (*Hint:* From elementary calculus we know that, if $V \times e = h\nu - WF$, then the slope of a volts versus frequency line, $d(\text{volts})/d\nu$, must be equal to h/e.)

7

ELECTRONIC CONFIGURATIONS AND PERIODIC PROPERTIES

Introduction

The energies and arrangements of electrons in atoms are important to an understanding of the chemical interactions of atoms with one another. In the previous chapter we discussed the theories that led to our present day picture of electrons in atoms. This chapter is concerned with the electronic configurations of representative elements that have only *ns* and *np* electrons in their valence shells. The transition metals, in which the $(n - 1)d$ electrons must also be considered, will be discussed in Chapter 20.

The periodic table is a convenient and useful means of classifying much information about elements. In it the elements are arranged in order of increasing atomic numbers and, hence, in order of increasing numbers of electrons surrounding the nucleus. More important to chemistry, however, is the arrangement of elements in vertical groups. These groups help make possible the systematization of the chemistry of families of elements and eliminate the necessity of remembering many details of the chemistry of all the elements.

Various fundamental chemical and physical properties of atoms show regular periodic trends among the elements. Of particular importance in chemistry are those properties affected by atomic size and the energies involved on addition or removal of electrons. A knowledge of atomic radii, ionization energies, and electron affinities can be very useful when attempting to predict the formation or reactions of chemical compounds. Another property with many applications in chemistry is electronegativity.

Electronic Configuration of Atoms

By the term "configuration of an atom" we mean the distribution of electrons among the various orbitals. Before proceeding to use the concept of quantum numbers obtained from the Schrodinger equation to describe the electronic structure of atoms, we must briefly discuss three additional principles. Although our methods and conclusions are only approximate for atoms more complicated than hydrogen, we use these principles with confidence because the results are consistent with the data obtained from a huge number of experiments.

The **Pauli exclusion principle,** suggested by German physicist Wolfgang Pauli in 1925, states that no two electrons in the same atom can have the same set of four quantum numbers. The exclusion principle is based both on theoretical considerations and experimental evidence and is roughly analogous to the classical principle that two bodies cannot be in the same place at the same time.

The principle quantum number, n, is the most important quantum number in determining the energy of an electron in an atom. It is therefore convenient to consider each integral value of n as representing a "shell" or level of allowed energies analogous to the orbits associated with the Bohr theory. Thus $n = 1$ is associated with the level of lowest energy, $n = 2$ with the level of next lowest energy, and so on.

Each of the energy levels denoted by a particular value of n may be further divided into sublevels on the basis of the l quantum number. For historical reasons the various levels are sometimes denoted by letters *K*, *L*, *M*, *N*. . . . Thus we have the following levels and sublevels:

K level: 1*s* only
L level: 2*s* and 2*p*
M level: 3*s*, 3*p*, and 3*d*
N level: 4*s*, 4*p*, 4*d*, and 4*f*

Other sublevels such as 2*d* or 3*f* cannot exist because these combinations of quantum numbers are not permitted by the rules already given in Chapter 6. The maximum number of electrons that can be in any level or sublevel is determined by the allowed quantum numbers and the Pauli exclusion principle.

Electrons with given values of n, l, and m_l, are said to be in a particular **orbital.** Since there are only two possible values for m_s, each orbital can hold just two electrons. For example, the *N* level ($n = 4$) consists of 4*s*, 4*p*, 4*d*, and 4*f* sublevels. In this level we have one 4*s* orbital, three 4*p* orbitals, five 4*d* orbitals, and seven 4*f* orbitals. The maximum number of electrons in each orbital is two, and the maximum number of electrons in the $n = 4$ level is 32, which can also be represented as $2n^2$.

The second principle we should consider in describing the electronic arrangement of atoms is the **aufbau** (building up) **principle.** The basic assumption is that in the ground state the electrons will be in their lowest possible energy levels. The approximate order of sublevels, in terms of energy, for elements through krypton ($Z = 36$) is shown in Figure 7.1.

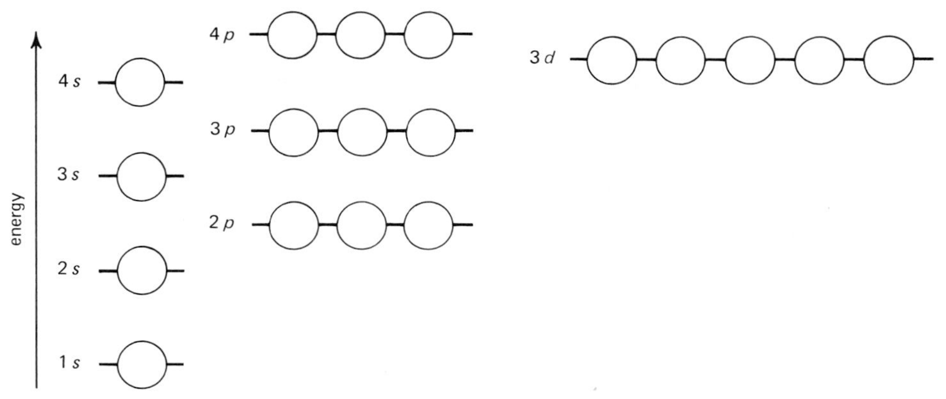

Figure 7.1. Energy level diagram for atomic orbitals. For many-electron atoms the relative order of energy levels is shifted somewhat due to electronic interactions.

Each orbital represented by a circle can hold two electrons, which will have identical quantum numbers except for m_s. There are three $2p$ orbitals corresponding to the three values of m_l that are allowed when $n = 2$ and $l = 1$. Similarly, there are three $3p$ orbitals corresponding to three values of m_l that are allowed when $n = 3$ and $l = 1$. There are five $3d$ orbitals corresponding to the five values of m_l that are allowed when $l = 2$.

The third principle necessary to avoid ambiguities in our aufbau process for many-electron atoms involves the filling of orbitals of equivalent energy. A simplified statement of **Hund's rule** is that electron spins remain "unpaired" if possible. Two electrons that have different m_s values are said to be "spin-paired." If both spins are "parallel," meaning the same m_s values, the electrons are "unpaired." The effect of Hund's rule will be illustrated when we consider specific elements. It should be understood that this "pairing" terminology applies only to electrons with the same n and l quantum numbers.

Although it is useful and convenient to classify electrons according to their energies into levels and sublevels, we should add a note of caution. The use of quantum numbers and the principles for filling oribitals provide a theoretical framework for a system of bookkeeping; the main justification for this framework is that the predicted results are consistent with experimental facts. The everyday meanings of the words "level" and "sublevel" (or "shell" and "subshell") cannot be applied to electrons in atoms. When we say that two electrons are in the same level we do not mean (and cannot because of the uncertainty principle) that these two electrons are circling the nucleus in identical orbits. Rather, we mean that the two electrons have the same principal quantum number n. If they have the same quantum number l they also have the same energy. If they have different values of l, they have slightly different energies, although similar enough that it is useful for some purposes to group them together.

A hydrogen atom has one electron, and we want to characterize that electron by means of quantum numbers. We are presently interested in the lowest energy state, for which $n = 1$. When $n = 1$, l can only be 0 and, therefore, $m_l = 0$. The spin quantum number, m_s, can be either $+\frac{1}{2}$ or $-\frac{1}{2}$. This information can be summarized by writing $1s^1$ where 1 indicates that we are concerned with the energy level with $n = 1$. The s indicates that $l = 0$, and the superscript 1 indicates that there is one electron in this level. Superscript numbers used in this manner to indicate numbers of electrons should not be confused with exponents.

Now let us suppose that we can build an atom of helium by adding the proper nuclear particles (in this case one proton and two neutrons) and one electron to an atom of hydrogen. Since we are concerned with the lowest electronic energy state, we want n for this second electron to be as small as possible. It has already been shown for $n = 1$ that $l = 0$, $m_l = 0$, and m_s can be either $+\frac{1}{2}$ or $-\frac{1}{2}$. Therefore one electron in an atom of helium can have $n = 1$, $l = 0$, $m_l = 0$, and $m_s = +\frac{1}{2}$, while the other electron has $n = 1$, $l = 0$, $m_l = 0$, and $m_s = -\frac{1}{2}$. These electrons differ only in the m_s quantum number and have the same energy. Note that the Pauli exclusion principle prohibits both electrons from having all four identical quantum numbers. The electronic configuration of helium is represented concisely by $1s^2$ where 1 again indicates that $n = 1$, s indicates that $l = 0$, and superscript 2 indicates that there are two electrons.

The next element that we come to in our imaginary building process is lithium with three electrons. The third electron cannot have $n = 1$ because all the values of l, m_l, and m_s that are allowed when $n = 1$ have been used by the first two electrons. The third electron must have $n = 2$ and can have $l = 0$, $m_l = 0$ and $m_s = +\frac{1}{2}$ or $-\frac{1}{2}$, as is summarized by writing $1s^22s^1$ for the electronic configuration of lithium. The $1s^2$ has the meaning previously described for helium, and the $2s^1$ means that there is one electron with $n = 2$ and $l = 0$. Similarly, the electronic configuration of beryllium, which has four electrons, is represented by $1s^22s^2$. Note that we use the $2s$ orbital ($l = 0$) rather than a $2p$ orbital ($l = 1$) for lithium and beryllium, because the $2s$ orbital is lower in energy than any of the $2p$ orbitals (see Figure 7.1).

Atoms of boron contain five electrons. Four of these electrons are characterized by $1s^22s^2$, and we can deduce the proper characterization of the fifth by consideration of the allowed quantum numbers. Because the allowed values of m_l and m_s for $n = 2$ and $l = 0$ have been used, it is necessary to consider the next highest energy level, which is characterized by $n = 2$ and $l = 1$. When $l = 1$, the allowed values for m_l are $+1$, 0, and -1; and for each of these values of m_l there are two allowed spin quantum numbers. For the fifth electron in boron, we choose $n = 2$, $l = 1$, $m_l = -1$, and $m_s = -\frac{1}{2}$ and write $1s^22s^22p^1$ as a concise description of the electronic configuration. The $2p^1$ indicates that there is one electron with $n = 2$ and $l = 1$.

It should be noted that $m_l = -1$ and $m_s = -\frac{1}{2}$ have been chosen arbitrarily. We could as well have chosen $m_l = 0$ or $+1$ and $m_s = +\frac{1}{2}$, but it is customary to choose the smallest (-1 rather than 0 or $+1$, and $-\frac{1}{2}$ rather than $+\frac{1}{2}$) quantum numbers first.

Atoms of carbon contain six electrons. It is possible for the sixth electron in carbon to have $m_l = -1$ (the same as the fifth electron) and $m_s = +\frac{1}{2}$ (not the same as the fifth electron), or it is possible to have m_l be one of the other allowed values. To handle this situation we resort to Hund's rule. Two electrons in the same orbital will necessarily spend more time in the same region in space (in essence, closer together) than two electrons in different orbitals. Thus the electrostatic repulsion energy is higher for the former case (spins paired) than for the latter case (parallel or unpaired spins). We therefore describe the sixth electron

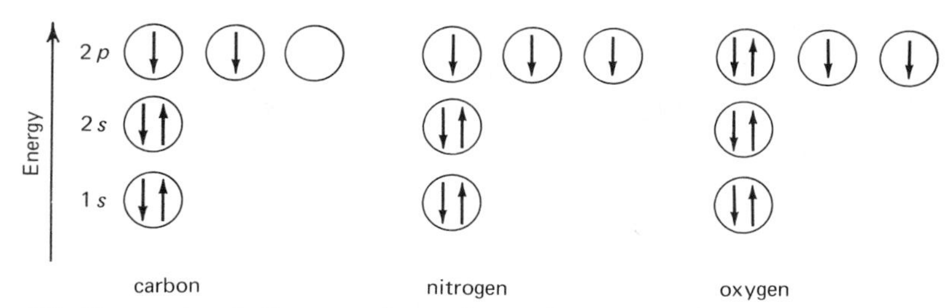

Figure 7.2. Energy level diagrams for carbon, nitrogen, and oxygen. Arrows pointing downward indicate $m_s = -\frac{1}{2}$, and arrows pointing upward indicate $m_s = +\frac{1}{2}$. Note the pairing of electrons in orbitals, and compare with Tables 7.1 and 7.2 and the preceding discussion.

in carbon with the quantum numbers $n = 2$, $l = 1$, $m_l = 0$ (rather than -1 as for the fifth electron), and $m_s = -\frac{1}{2}$. The electronic configuration of carbon is summarized by $1s^2 2s^2 2p^2$.

By analogy to carbon we can now correctly predict that the highest energy electron in nitrogen has quantum numbers $m_l = +1$ and $m_s = -\frac{1}{2}$ and write the electronic configuration as $1s^2 2s^2 2p^3$. Similarly, the eighth electron in oxygen has $m_l = -1$ and $m_s = +\frac{1}{2}$, so that two of the $2p$ electrons have their spins antiparallel or paired. The electronic configuration of oxygen is $1s^2 2s^2 2p^4$. The ninth electron in fluorine has $m_l = 0$ and $m_s = +\frac{1}{2}$. The electronic configuration of fluorine is written $1s^2 2s^2 2p^5$. The tenth electron in neon has $m_l = +1$ and $m_s = +\frac{1}{2}$, and we write $1s^2 2s^2 2p^6$ for the electronic configuration of neon. All six of the $2p$ electrons are spin paired in neon, as are the two $1s$ and the two $2s$ electrons.

Figure 7.2 shows energy level diagrams for carbon, nitrogen, and oxygen. Using Figure 7.1 as a guide, similar representations can be made for all elements through krypton ($Z = 36$). There are some irregularities in the filling of d orbitals in the transition metals, which will be discussed in Chapter 20. The present chapter and immediately succeeding chapters are concerned mostly with electronic configurations involving s and p orbitals.

Following the aufbau rules we have just discussed, it is possible to assign quantum numbers for the highest energy electron in any atom. Table 7.1 lists the quantum number for this "last" electron in all atoms through argon ($Z = 18$), and the electronic configurations of these same atoms are listed in Table 7.2.

Table 7.1. Quantum Numbers of Highest Energy Electrons in Atoms

Atom	n	l	m_l	m_s	*Atom*	n	l	m_l	m_s
H	1	0	0	$-\frac{1}{2}$	Ne	2	1	$+1$	$+\frac{1}{2}$
He	1	0	0	$+\frac{1}{2}$	Na	3	0	0	$-\frac{1}{2}$
Li	2	0	0	$-\frac{1}{2}$	Mg	3	0	0	$+\frac{1}{2}$
Be	2	0	0	$+\frac{1}{2}$	Al	3	1	-1	$-\frac{1}{2}$
B	2	1	-1	$-\frac{1}{2}$	Si	3	1	0	$-\frac{1}{2}$
C	2	1	0	$-\frac{1}{2}$	P	3	1	$+1$	$-\frac{1}{2}$
N	2	1	$+1$	$-\frac{1}{2}$	S	3	1	-1	$+\frac{1}{2}$
O	2	1	-1	$+\frac{1}{2}$	Cl	3	1	0	$+\frac{1}{2}$
F	2	1	0	$+\frac{1}{2}$	Ar	3	1	$+1$	$+\frac{1}{2}$

Table 7.2. Electronic Configurations of Atoms

Atom	*Atomic number*	*Configuration*	*Atom*	*Atomic number*	*Configuration*
H	1	$1s^1$	Ne	10	$1s^22s^2p^6$
He	2	$1s^2$	Na	11	[Ne] $3s^1$
Li	3	$1s^22s^1$	Mg	12	[Ne] $3s^2$
Be	4	$1s^22s^2$	Al	13	[Ne] $3s^23p^1$
B	5	$1s^22s^22p^1$	Si	14	[Ne] $3s^23p^2$
C	6	$1s^22s^22p^2$	P	15	[Ne] $3s^23p^3$
N	7	$1s^22s^22p^3$	S	16	[Ne] $3s^23p^4$
O	8	$1s^22s^22p^4$	Cl	17	[Ne] $3s^23p^5$
F	9	$1s^22s^22p^5$	Ar	18	[Ne] $3s^23p^6$

Figure 7.1 gives an indication of problems that we will meet when we consider the electronic configuration of atoms with more than 18 electrons. In accord with our earlier statement that the principal quantum number, n, is most important in determining the energy of an electron, we might expect that the lowest energy level available to the nineteenth electron would be a $3d$ orbital with $n = 3$, $l = 2$, $m_l = -2$, and $m_s = -\frac{1}{2}$. But the situation in an atom containing more than one electron is complicated by electronic interactions. Thus we cannot be entirely confident of any such specific predictions until there is some experimental verification. In fact, the energies of the $n = 3$ and $n = 4$ levels overlap. The $3s$ and $3p$ sublevels are at lower energies than any of the sublevels of the $n = 4$ level, but the lowest of the $n = 4$ sublevels (the $4s$ sublevel) is at a lower energy than the highest of the $n = 3$ sublevels. The energy dependence of these atomic orbitals as a function of atomic number will be considered in Chapter 20.

Atomic Structure and the Periodic Table

Using the phrase "the periodic table" is somewhat misleading. Literally hundreds of versions of the periodic table of elements have been devised, many of which predated modern theories of atomic structure. The familiar "long form" periodic table is a modification of that of Russian chemist Dmitri Mendeleef, although a German chemist, Lothar Meyer, working independently published a similar table a few months later. Both tables, presented in 1869, were based on increasing atomic weights, but with allowances for grouping elements according to chemical behavior. This last reservation was significant because there are three inversions of atomic weight values in the modern periodic table (argon-potassium, cobalt-nickel, and tellurium-iodine). For example, despite the fact that argon has a slightly greater atomic weight than potassium, the chemistry of argon clearly places it in the noble gas family, whereas potassium is similar to the other alkali metals.

Of considerable interest and importance is the fact that Mendeleef left vacant positions in his proposed table to allow for undiscovered elements. Later, in 1871, Mendeleef published a more comprehensive treatment which he called the **periodic law:** the properties of elements depend on the structure of the atom and vary in a systematic way. Utilizing periodic character, Mendeleef predicted the properties of three unknown elements that he named *eka-boron, eka-aluminum,* and *eka-silicon.* When later discovered, the properties of scandium, gallium, and germanium were remarkably similar to the predictions of Mendeleef.

The atomic weight inversions indicated that an order based on increasing atomic weight was not the best way to arrange elements in the periodic table; therefore, elements were assigned arbitrary atomic numbers. Elements in the periodic table were then arranged in order of increasing atomic number from 1 to 92 (now 105). Justification for this arrangement came in 1913 from the work of Moseley, who showed that the frequencies of certain x rays followed atomic number and not atomic weight.

Horizontal rows in the table are called **periods** and vertical sequences are called **groups,** although common names for some of these groups are often used. A better understanding of the part that electronic structure plays in the properties of elements produces a corresponding understanding of periodic behavior. In terms of our present model, we can see that the group IA elements, the alkali metals, are characterized by an outer electronic configuration of ns^1. The chemical properties of the group IIA elements, the alkaline earth metals, are largely determined by their outer electronic configuration of ns^2. Similarly, the outer configuration of ns^2np^5 largely determines the chemistry of the halogens (group VIIA), whereas all the noble gases have the ns^2np^6 outer configuration.

It might be useful to close this section by showing how wave mechanical predictions substantiate the view that it is the outer electrons, the "valence shell," of an element that primarily determine its chemical behavior. In Figure 7.3 are shown the electronic radial distribution curves for argon and the unipositive potassium ion. Note that the *K*, *L*, and *M* shells are quite distinct and that the

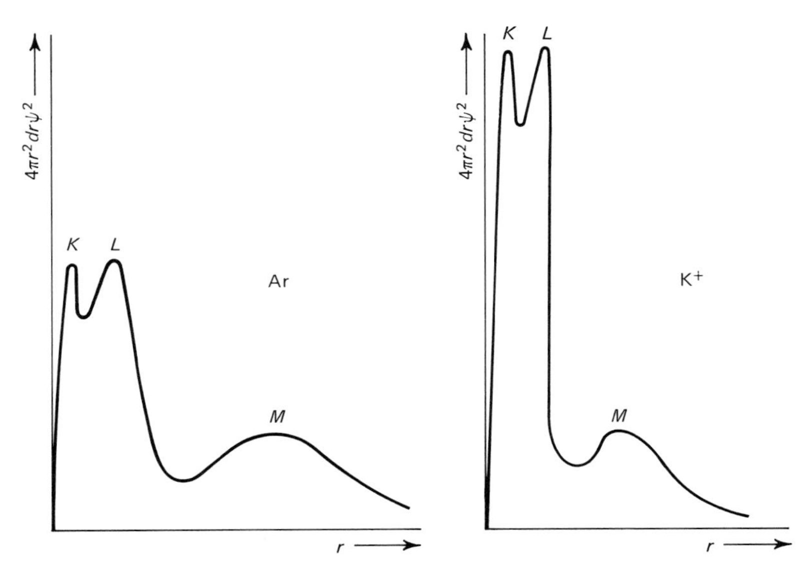

Figure 7.3. Radial distribution curves showing electron density as a function of distance from the nucleus for argon and the unipositive potassium ion. The total area under the curves is proportional to the total number of electrons, 18 in both cases.

outer part of the atom is dominated by the outer shell, here the *M* shell. Clearly, the size and chemistry of an atom is determined mainly by the relatively diffuse outer shell. The electronic shells are closer to the nucleus for potassium than for argon because the former has a larger nuclear charge.

Atomic Radii

The diffuse nature of the charge cloud makes it difficult to determine precisely the effective radius of an atom. As a result different kinds of radii are defined. For the purposes of this book we will only be concerned with covalent radii and ionic radii. We are principally interested in trends in the periodic table and all types of radii exhibit similar relationships. For the present we will consider covalent radii; ionic radii are treated in Chapter 21.

Despite the uncertain size of atoms, distances between nuclei can be determined precisely. The covalent radius of an atom is defined as one half the distance between two nuclei held together by a single covalent bond. Most of the lighter atoms form simple diatomic molecules, such as H_2, Li_2, and Cl_2, and the calculation is relatively simple. For instance, the internuclear distance in the H_2 molecule is 0.74 Å; we therefore assign a value of 0.37 Å as the covalent radius of a hydrogen atom. In some cases, such as N_2 and O_2, the bonding is more complicated and other reference molecules must be chosen. The values selected are averages from a large series of molecules chosen to give a consistent and additive set. The results for some of the light elements are shown in Figure 7.4.

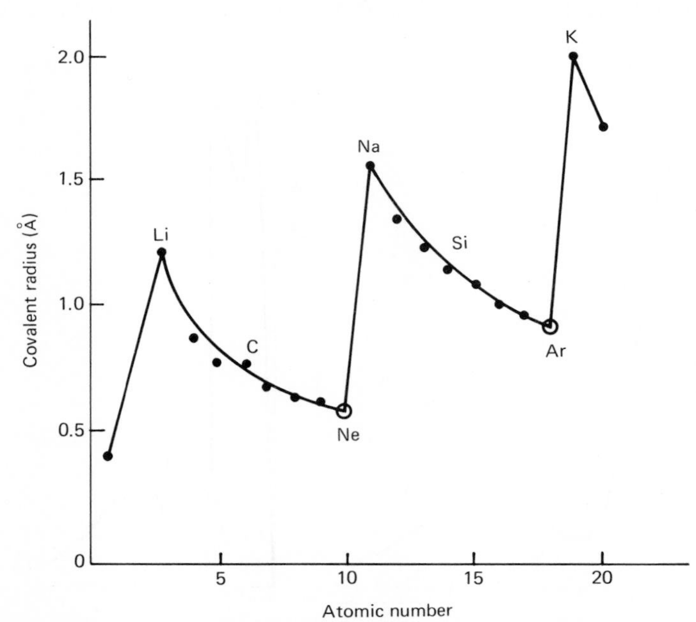

Figure 7.4. Single-bond covalent radii for the first 20 elements of the periodic table. The numbers are averages of values for several molecules containing the particular atom. The values for the noble gases (Z = 10 and 18) are extrapolated values.

In general, covalent radii increase with increasing atomic number in a given group in the periodic table. This may be explained using the alkali metals as an example. The valence electron in lithium is a $2s$ electron; in sodium, a $3s$ electron; in potassium, a $4s$ electron; and so on. Because the effective orbital radius increases as the principal quantum number increases, the effective size of the atom also increases as we go down a group in the periodic table.

A more striking trend in covalent radii may be noticed in horizontal rows (see Figure 7.4). Starting with an alkali metal there is a gradual decrease in radius across a period, reaching a minimum with the halogen, which is in turn much smaller than the following alkali metal. This trend may also be explained by considering the outer orbital arrangement of the atoms. Proceeding from lithium to fluorine in the aufbau process, we are adding electrons to the $n = 2$ quantum level, the $2s$ and $2p$ orbitals. At the same time the nuclear charge is increasing. Since electrons in the same quantum level are less effective than inner electrons in shielding each other from the nucleus, there is an increasingly greater attraction between the nucleus and the electrons. The net result is a shrinkage in effective size with increasing atomic number in going across a period.

From the above discussion we predict that the covalent radius of neon is less than that of fluorine. Since no chemical compounds containing the lighter noble gases have yet been obtained, there is no experimental verification of this prediction. Compounds of xenon, however, are known (Chapter 17). Consistent with the above arguments, the covalent radius of xenon (1.30 Å) is less than that of iodine (1.33 Å).

Ionization Energy

When one or more electrons are detached from or added to an atom, a charged species called an **ion** is formed. Because many chemical reactions involve electron transfers, energy changes involved in adding or removing electrons from atoms or ions are important in understanding chemical behavior.

The ions of lithium and sodium formed by removing one electron are represented by Li^+ and Na^+; the superscript + indicates that each ion has a net positive charge of one electron unit compared to a neutral atom. Similarly, we write Be^{2+} and Al^{3+} for the ions formed by removing two and three electrons, respectively, from each neutral atom of beryllium and aluminum. Negative ions are formed by addition of one or more electrons to a neutral atom. Thus we write Cl^- and S^{2-} to represent the ions formed by adding one electron to a neutral chlorine atom and two electrons to a neutral sulfur atom.

The minimum energy required to remove an electron from a gaseous atom is called the ionization energy (sometimes ionization potential), and may be determined from atomic spectra or by mass spectroscopy. The **first ionization energy, IP_1,** is the energy necessary to remove one electron from a neutral gaseous atom to form a unipositive ion. The **second ionization energy, IP_2,** is the energy required to remove one more electron from the unipositive ion. Third and higher ionization energies are similarly defined. Ionization energies are generally expressed in electron volts (eV) units (1 eV atom^{-1} = 23.1 kcal mole^{-1}). Some ionization energies for the first 20 elements are given in Table 7.3. Note that ionization

Table 7.3. Some Ionization Energies (eV units)

Atomic number	*Element*	IP_1	IP_2	IP_3
1	H	13.6		
2	He	24.6	54.4	
3	Li	5.4	75.6	122.4
4	Be	9.3	18.2	153.9
5	B	8.3	25.1	37.9
6	C	11.3	24.4	47.9
7	N	14.5	29.6	47.6
8	O	13.6	35.1	55.1
9	F	17.4	35.0	62.6
10	Ne	21.6	41.0	63.4
11	Na	5.1	47.3	71.7
12	Mg	7.6	15.0	80.1
13	Al	6.0	18.8	28.4
14	Si	8.1	16.3	33.5
15	P	11.0	19.7	30.2
16	S	10.4	23.4	35.0
17	Cl	13.0	23.8	39.9
18	Ar	15.8	27.6	40.7
19	K	4.3	31.8	46.5
20	Ca	6.1	11.9	51.0

requires an input of energy and has a positive sign; this is consistent with usual thermodynamic notation.

For a given atom the ionization energies always increase in order $IP_1 < IP_2 < IP_3 < \cdots < IP_n$. This is understandable on the basis of increasing coulombic attraction and decreasing size as electrons are removed. As electrons are removed from an atom, the effective nuclear charge experienced by the outer electrons increases due to decreased shielding by the remaining electrons. Therefore, the effective size of the atom decreases, and the energy necessary to remove an outer electron increases. For example, the ionization energies for N, N^+, and N^{3+} increase in the order $IP_1 = 14.5$, $IP_2 = 29.6$, and $IP_4 = 77.5$ eV, while the effective radius decreases in the order 0.92, 0.25, and 0.16 Å.

Ionization energies are one of the few fundamental properties of an isolated atom that can be measured directly. Therefore periodic trends and apparent anomalies are of great interest in chemistry. The variation of first ionization energies with atomic number is shown in Figure 7.5.

Inspection of either Table 7.3 or Figure 7.5 shows that less energy is required to remove one electron from an alkali metal, with an outer electronic configuration of ns^1, than from any other element in a given period. Although there are irregularities, the first ionization energy generally increases across a period, reaching a maximum with the noble gases, which are characterized by an outer electronic configuration of ns^2np^6.

The outer electronic configuration of an alkali metal ion (Li^+, Na^+, and so on) is the same as that of the preceding noble gas, and we might therefore suspect that the electronic configurations of the noble gases are particularly stable. Consistent with this suspicion, we note that the sum of IP_1 and IP_2 for beryllium (giving a helium configuration) is less than the corresponding sum for either lithium or boron. Similarly, we expect the removal of three electrons from boron or aluminum will require less energy than for any neighboring elements in the periodic table. Since these predictions are in accord with the experimentally observed ionization energies and with quantum mechanical calculations, our suspicion concerning the special stability of noble gas electronic configurations seems justified.

The steady (slightly irregular) increase in ionization energies from alkali metal to noble gas in a given period is due to the steady increase in the effective nuclear charge with increasing atomic number. In "building" from sodium to argon the electrons are all added to 3*s* and 3*p* orbitals, that is, to the valence shell, and do not shield each other effectively from the increasing nuclear charge.

By reasoning already described we deduce that C^{4+} and Si^{4+} ought to be more stable than other ions of the same charge that do not have noble gas electronic configurations. These predictions, and others for such ions as N^{5+}, are supported by ionization energies but are of little chemical importance because atoms of carbon, silicon, and nitrogen generally do not react with other atoms by losing four or five electrons each. Except for the transition elements, compounds con-

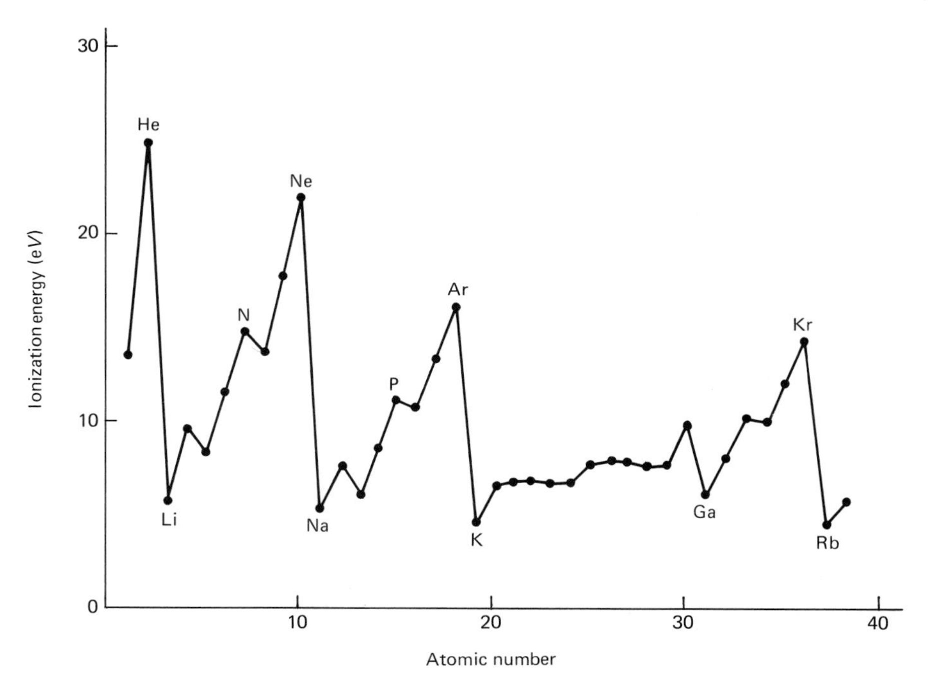

Figure 7.5. First ionization energies of the elements as a function of atomic number. Note that the alkali metal has the minimum and the noble gas the maximum ionization energy in a given row. Also note the extra stability toward ionization of atoms with filled or half-filled subshells compared to the immediately preceding or succeeding elements. The first ionization energies of the transition metals are all very close together.

taining ions with a charge greater than +3 electron units are uncommon, and monatomic ions with a charge greater than +4 electron units are unknown in stable chemical compounds.

On the basis of the conclusions that we have reached about the stabilities of various ions, we predict correctly that lithium and sodium (and the other alkali metals) form compounds in which each atom of the alkali metal has lost one electron and exists as a unipositive ion. We also predict correctly that the alkaline earth metals readily lose two electrons and exist as bipositive ions. It is reasoning of this sort that enables us to understand and remember the formulas of many chemical compounds.

We have already noted that a filled energy level (noble gas) configuration is a particularly stable one. Further inspection of Figure 7.5 shows that atoms with filled or half-filled sublevels have higher first ionization energies than might otherwise be expected. For example, beryllium ($2s^2$) and nitrogen ($2s^22p^3$) have slightly larger ionization energies than boron and oxygen, respectively. Similarly, zinc ($3d^{10}4s^2$) has a larger ionization energy than gallium ($3d^{10}4s^24p^1$). This special stability is generally considered to be due to the spherically symmetrical charge cloud of a half-filled or completely filled sublevel.

Another general trend may be noticed by inspecting the first ionization energies for the alkali metals (see Figure 7.5). These ionization energies become progressively smaller in going from lithium to rubidium. Again this trend may be explained by considering atomic radii. The $2s$ electron of lithium (atomic radius = 1.23 Å) experiences a much stronger attraction to the lithium nucleus than does the $5s$ electron of rubidium to the rubidium nucleus (atomic radius = 2.11 Å). This trend is continued through to cesium, which has the smallest first ionization energy of any common element. Ionization energies of elements in other groups also decrease in going from smaller to larger atoms.

The existence of definite ionization energies for the various elements is support for the fundamental postulate that electrons in atoms are in definite, well-defined energy states or levels.

Electron Affinity

Neutral atoms may also accept or capture additional electrons to form negative ions. The **electron affinity, EA,** is defined as the energy released when a neutral gaseous atom accepts an electron to form a negative ion. When energy is released, the sign is positive; when energy is required, the sign is negative. This sign convention is opposite that for ionization energies (and usual thermodynamic practice), but it is well established and we shall not change it here.

Unfortunately, electron affinities are difficult to measure experimentally. Values must usually be obtained by indirect methods, and the accuracy is low. Some of the better known electron affinities are given in Table 7.4.

Halogen atoms, which can gain the rare gas configuration by accepting one electron, have the highest electron affinities. There is currently no simple explanation for the fact that the EA of chlorine is greater than that of fluorine. The values for the other halogens are in the order expected on the basis of atomic

Table 7.4. Electron Affinities for Some Common Elements (eV units)

H	0.75	O	1.47
Li	0.58	F	3.52
Be	−0.19	Na	0.78
B	0.33	Cl	3.69
C	1.12	Br	3.44
N	−0.27	I	3.14

size. Metals, in which the outer electrons are rather loosely bound, have very small or even negative electron affinities.

Electronegativity

A less fundamental but more chemically useful concept is that of electronegativity, first introduced by Pauling in 1932. Electronegativity is a term used to describe the tendency of an atom *in a molecule* to attract electrons in the molecule to itself. It is not the same as electron affinity, which refers to free atoms and free electrons.

In molecules like F_2 or H_2 there is no tendency for the bonding electrons to drift in one direction or the other. There is "equal sharing" of the bonding electron pair. However, in a molecule containing different atoms, such as HF, there is an "unequal sharing" with the bonding electrons being pulled toward the atom of greatest electronegativity.

It is reasonable to assume that the energy necessary to break the bond in HF might be approximately equal to the mean value of the bond energies for the hydrogen and fluorine molecules. However, the bond energy of an AB molecule is generally found to be greater than the geometric mean of the bond energies of the A_2 and B_2 molecules, that is

$$E_{\mathrm{A-B}} = (E_{\mathrm{A-A}} \times E_{\mathrm{B-B}})^{\frac{1}{2}} + \Delta_{\mathrm{A-B}}$$

Pauling found that this "extra" bond energy, $\Delta_{\mathrm{A-B}}$, could be related to the difference in electronegativity of the two atoms by the expression

$$\mathrm{EN_A} - \mathrm{EN_B} = 0.208\ (\Delta_{\mathrm{A-B}})^{\frac{1}{2}}$$

The 0.208 factor converts the kilocalorie per mole units of bond energies to electron volts in which we express electronegativities. The square root of $\Delta_{\mathrm{A-B}}$ is used simply because it leads to the most consistent and additive set of electronegativity values.

The electronegativity difference between hydrogen and fluorine can be evaluated from bond energy data (see Chapter 11). The bond energy values for H_2, F_2, and HF are 104, 38, and 136 kcal mole^{-1}, respectively. The geometric mean of the H_2 and F_2 values, $(104 \times 38)^{\frac{1}{2}} = 63$, is significantly less than $E_{\mathrm{H-F}}$. The "extra" bond energy, $\Delta_{\mathrm{A-B}}$, is 73 kcal mole^{-1}. Taking the square root and multiplying by the conversion factor yields $(0.208)(8.55) = 1.78$.

Table 7.5. Electronegativity Values for Some Common Elements

Alkali metals		*Alkaline earth metals*		*Halogens*		*Other elements*	
Li	0.98	Be	1.57	F	3.98	H	2.20
Na	0.93	Mg	1.31	Cl	3.16	C	2.55
K	0.82	Ca	1.00	Br	2.96	N	3.04
Rb	0.82	Sr	0.95	I	2.66	O	3.44
Cs	0.79	Ba	0.89			S	2.58
						P	2.19
						B	2.04

Only differences are obtained by this method, and therefore some reference point must be chosen. The most electronegative element, fluorine, was originally assigned the arbitrary value of 4.0. Table 7.5 gives improved electronegativity values for some common elements, calculated by the Pauling method.

Consistent with ionization energies and electron affinities, the halogens are the most electronegative elements; the alkali metals are the least electronegative. In a general way, electronegativities are related to size. In any family, the smallest atom generally has the highest electronegativity. This correlation between size and electronegativity may also be noted in any horizontal row, for example, chlorine atoms are both smaller and more electronegative than sodium atoms.

Among several other electronegativity scales which have been suggested since 1932 we might mention that of R. S. Mulliken. Mulliken showed by theoretical arguments that the electronegativity of an atom should be equal to the average of its first ionization energy and its electron affinity. Physically this seems quite reasonable, but a severe limitation is the lack of reliable electron affinity data. After some adjustments this method gives values which are generally consistent with Pauling values.

Although much debate centers around electronegativities (for example, their theoretical significance, the best way to obtain them, values for groups of atoms) they remain useful in chemistry. They are particularly useful when considering the polarity of bonds (Chapter 8), in estimating approximate bond energies, and predicting thermal stabilities.

References

Little, E. J., Jr., and M. M. Jones: A complete table of electronegativities. *J. Chem. Educ.*, **37**:231 (1960).

Pritchard, H. O.: The determination of electron affinities. *Chem. Rev.*, **52**:529 (1953).

Pritchard, H. O., and H. A. Skinner: The concept of electronegativity. *Chem. Rev.*, **55**:745 (1955).

Redfern, J. P., and J. E. Salmon: Periodic classification of the elements. *J. Chem. Educ.*, **39**:41 (1962).

Rich, R.: *Periodic Correlations.* W. A. Benjamin, Inc., New York, 1965.

Sanderson, R. T.: A rational periodic table. *J. Chem. Educ.*, **41**:187 (1964).

Seaborg, G. T.: Prospects for further considerable extension of the periodic table. *J. Chem. Educ.*, **46**:626 (1969).

Sebera, D. K.: *Electronic Structure and Chemical Bonding.* Blaisdell Publishing Co., New York, 1964 (paperback).
Sisler, H. H.: *Electronic Structure, Properties, and the Periodic Law.* Reinhold Publishing Corp., New York, 1963 (paperback).
Spice, J. E.: *Chemical Binding and Structure.* Macmillan Co., New York, 1964 (paperback).
van Spronsen, J. W.: The prehistory of the periodic system of the elements. *J. Chem. Educ.*, **36**:565 (1959).

Problems

1. Following the aufbau principle, Pauli exclusion principle, and Hund's rule, list all four quantum numbers for the "last" electron in the following atoms and ions: Li^+, P, Ca^{2+}, Al, Cl^-, and Ar.
2. What is the maximum number of electrons that may have a principle quantum number of 2, 3, and 4 in a given atom? Devise a mathematical formula that gives the maximum number of electrons for any value of n.
3. Write the electronic configurations for the following ions: O^{2-}, Si^{4+}, K^+.
4. Use energy level diagrams similar to those in Figure 7.2 to show the electronic distribution in the valence shell for the following atoms and ions: N^{3-}, Al, Cl^-, S, and Cl.
5. In 1829, the German chemist Dobereiner observed several triads of elements with similar chemical properties. In each case the atomic weight of one element in the triad was close to the average of the other two. Compare the atomic weight values, most common ions, and general chemical properties for the following triads: Cl, Br, I; Li, Na, K. These triad relationships proved not to be significant but did encourage other classification attempts.
6. Plot the melting points (°C) versus the periodic group (or family) for the following elements. Is the plot perfectly periodic?

Element	*Melting point* (°C)	*Element*	*Melting point* (°C)
B	2030	Al	660
C	3730	Si	1410
N	−210	P	44
O	−219	S	119

Now add the values for the next row of the period table and predict the value for Ge (germanium).

Element	*Melting point* (°C)
Ga	30
As	817
Se	217

It was by way of plots like these that Mendeleef was able to predict with amazing accuracy the chemical and physical properties of germanium, which he called eka-silicon.

7. Given the following information about chlorides of silicon and tin, predict the corresponding values for the tetrachloride of germanium and compare with experimental values in a *Handbook of Chemistry and Physics* or other reference work.

Formula	$SiCl_4$	$SnCl_4$
Boiling point	57.6°C	114°C
Density of liquid	1.50 g cm^{-3}	2.23 g cm^{-3}

8. Heats of solution of the various noble gases in water are as follows: He, −0.4; Ne, −1.1; Ar, −2.9; Kr, −3.7 kcal $mole^{-1}$. The negative sign indicates that the solution reactions are exothermic (evolve heat). Plot these heats of solution versus the period number for these four elements and then predict the heat of solution of xenon. (The actual value is −4.2 kcal $mole^{-1}$.) Suggest an explanation for the fact that the energy liberated when a noble gas is dissolved in water increases with increasing atomic size of the noble gas. In several later chapters we return to further interpretation of energies of various physical processes and chemical reactions.

9. Francium ($Z = 87$) is not found in nature because all of its isotopes are radioactive and have short half-lives. By noting its position in the periodic table, predict a value for its melting point, formulas of its compounds with chlorine and oxygen, its relative ability to conduct electricity, and its valence shell electronic configuration.

10. Seaborg [*J. Chem. Educ.*, **46**:626 (1969)] has suggested that it should be possible to extend the periodic table to a considersble extent. Consider the possible element with atomic number 118. For this element predict reasonable values for atomic weight, covalent radius, first ionization energy, and suggest its chemical behavior. Give the basis for your reasoning in each case.

11. Plot the melting points and boiling points versus the period number for the group V hydrides.

	Melting point (°C)	*Boiling point* (°C)
PH_3	−132	−87
AsH_3	−116	−62
SbH_3	−88	−17

From your plot predict the melting and boiling points for ammonia and compare them with the experimental values. What does this suggest about the concept of periodicity in predicting physical properties?

12. In each of the following sets pick the atom with the largest radius and the one with the smallest radius. (a) Cl, Br, I; (b) Se, Kr, Sr; (c) K, Ga, Br; (d) B, P, Br; (e) Al, Ge, Sb.

13. Suggest an explanation for the following experimental fact: the decrease in radius for the seven elements from lithium to fluorine is 0.62 Å, whereas the decrease for the ten elements from scandium to zinc is only 0.13 Å.

14. Given the bond lengths for the following homonuclear diatomic molecules:

H_2, 0.74 Å Cl_2, 1.99 Å
F_2, 1.42 Å Br_2, 2.28 Å

Calculate the bond lengths for the following: HF, HCl, HBr, ClF, BrF, and BrCl.

15. Predict the relative radii for the following isoelectronic species: H^-, He, Li^+, Be^{2+}. Briefly explain your reasoning.

16. In each of the following sets pick the atom with the highest first ionization energy. (a) Li, Be, B; (b) P, S, Cl; (c) Cl, Br, I; (d) Sr, Ba, Ra; (e) I, Xe, Cs; (f) N, As, Bi.

17. In each of the following sets pick the atom with the highest second ionization energy. (a) Na, Mg, Al; (b) Li, Na, K; (c) Cl, Br, I; (d) Br, Kr, Rb.

18. Which factor is more important in determining the ionization energy of an element in a family, the radius or the nuclear charge? Why?

19. (a) Molecules as well as atoms exhibit distinct ionization energies. The first ionization energy of molecular oxygen is 282 kcal $mole^{-1}$, whereas IP_1 for the noble gas xenon is 280 kcal $mole^{-1}$. Write an equation for each process.

(b) Oxygen reacts with platinum hexafluoride (PtF_6) vapor to form the compound dioxygenyl hexafluoroplatinate (V), $[O_2^+(PtF_6)^-]$. Do you think that it would be feasible to attempt the preparation of the corresponding compound, $[Xe^+(PtF_6)^-]$? Explain your reasoning.

20. Would it require more energy to remove the valence electron from a neutral gaseous potassium atom or from a gaseous lithium atom in which the valance electron had previously been promoted to the 4*s* orbital? Explain your reasoning.

21. Calculate the ionization energy for the helium ion, He^+.
Hint: Ionization involves an electronic transition from $n = 1$ to $n = \infty$.

22. When gaseous diboron tetrafluoride, B_2F_4, is put into a mass spectrometer a peak corresponding to the $B_2F_4^+$ ion is first observed at 12.07 ± 0.01 eV. Write an equation for this process and convert the energy to kilocalories per mole and kilojoules per mole.

23. Ionization energies are rather easy to obtain experimentally, whereas electron affinities are very difficult to obtain directly. Suggest an explanation.

24. The first electron affinity for oxygen is 1.47 eV and the second electron affinity is -7.3 eV. Write equations for both processes; calculate the energy involved in the overall process of $O + 2e^- \rightarrow O^{2-}$; and explain the significance of the signs for each step and for the overall process.

25. Show that the arithmetic mean of two bond energies E_{A-A} and E_{A-B} must be greater than or equal to the geometric mean of these energies.

26. The bond energies for Cl_2, I_2, and ICl are 58, 36, and 51 kcal $mole^{-1}$, respectively. Calculate the electronegativity difference between iodine and chlorine. What else is needed to establish electronegativity values for these elements?

27. From the information in Tables 7.3 and 7.4, calculate the electronegativity on the Mulliken scale for fluorine, oxygen, and nitrogen. How do these values compare with the Pauling values?

28. Because all its isotopes have relatively short half-lives, the chemistry of astatine (At) has not been extensively studied. Predict the formula of its compounds with hydrogen and cesium. Would the values for the melting point, first ionization energy, electron affinity, and electronegativity of elemental astatine be greater than or less than those for iodine?

8

BONDING AND STRUCTURE OF CHEMICAL COMPOUNDS

Introduction

In this chapter we will consider some of the current theories that are used by chemists to explain bonding between atoms in discrete molecules. In doing so we make the tacit assumption that molecules can be described in terms of individual atoms. Among the questions that any bonding theory should answer are the following: Why do atoms come together to form molecules? What factors are important in formation of ionic compounds or covalent compounds? Why do only certain combinations of atoms form stable molecules, such as F_2 and H_2O, and not other combinations? Why do molecules with the same number of atoms have different shapes, for example, linear CO_2 and bent H_2O?

When atoms come together and react to form a stable aggregate, the electronic configurations of all the atoms involved in the reaction are affected. We shall begin our discussion with a brief consideration of ionic bonding, the forces that hold together atoms with valence electrons in dissimilar energy states. The rest of the chapter will be devoted to bonding in covalent molecules where the energy states of the valence electrons in the bonded atoms are similar.

It is convenient to view ionic bonds as being formed by a complete transfer of electrons from one atom to another and covalent bonds as a sharing of electrons between atoms. We should realize that these divisions are artificial and that many borderline examples can be cited. In all cases the word **bond** refers to the forces that hold two atoms together.

Ionic Crystals

When atoms with valence electrons in very dissimilar energy states come together, electrons may be transferred from the atoms of one element to the atoms of the other. Such a transfer of electrons results in formation of positive and negative ions. Large numbers of these oppositely charged ions can be held together by nondirectional electrostatic forces, and they constitute an ionic crystal. The total energy of the ions in the ionic crystal will be lower than that of the individual atoms. Low energy corresponds to stability.

A consideration of the energies involved tells us which elements are expected to react with other elements to form ionic substances. The formation of crystalline sodium chloride from gaseous atoms of sodium and chlorine may be taken as an example. It is convenient for energy considerations to picture crystal formation as occurring in a series of simple steps. The first step in this reaction sequence can be taken to be the ionization of sodium atoms. This step requires energy equal to the ionization energy of each sodium atom times the number of atoms involved. The second step is the addition of the electrons taken from the sodium atoms to an equal number of chlorine atoms. The energy released in the second step is determined by the electron affinity of the chlorine atoms. The third step is the combination of equal numbers of sodium ions and chloride ions to form a crystal of sodium chloride. The energy released in this third step is dependent on both the size and charge of the ions and on the arrangement of these ions in the crystal. It is called the **lattice energy,** U, and will be discussed more fully in Chapter 21. The overall process may be summarized by the following equations:

$$\begin{array}{llr}
(1)\ \mathrm{Na(g)} \rightarrow \mathrm{Na^+(g)} + e^- & \mathrm{Energy} = \mathrm{IP_1} & (8.1)\\
(2)\ \mathrm{Cl(g)} + e^- \rightarrow \mathrm{Cl^-(g)} & \mathrm{Energy} = \mathrm{EA} & (8.2)\\
(3)\ \mathrm{Na^+(g)} + \mathrm{Cl^-(g)} \rightarrow \mathrm{NaCl(c)} & \mathrm{Energy} = \mathrm{U} & (8.3)\\
\hline
\quad\ \mathrm{Na(g)} + \mathrm{Cl(g)} \rightarrow \mathrm{NaCl(c)} & \mathrm{Energy} = \mathrm{IP_1} + \mathrm{EA} + \mathrm{U} & (8.4)
\end{array}$$

As shown in our later discussions of thermodynamics, spontaneous reactions of the type represented by equation (8.4) are generally **exothermic,** meaning that energy is liberated by the reaction. Therefore the sum of the energies released in the second and third steps must be greater than the energy required in the first step.

The formation of an ionic bond is most likely when one element has a low ionization energy and the other element has a high electron affinity. Ionization energies are smallest for atoms of those elements that can attain a noble gas electronic configuration by losing one electron per atom. Electron affinities are largest for those elements that can attain a noble gas electronic configuration by gaining one electron per atom. Hence we expect the alkali metals to react with halogens to form stable ionic crystals, a prediction that is verified experimentally.

Since an alkali metal ion has a positive charge of one electron unit, it is customary to say that it is in an **oxidation state** of +1 in its ionic compounds.

Similarly, the halide ions have a negative charge of one electron unit and are said to be in a -1 oxidation state. Sodium chloride, therefore, contains equal numbers of Na^+ ions and Cl^- ions and is represented by the formula NaCl.

Atoms of the alkaline earth metals have the lowest ionization energies for loss of two electrons per atom. The energy required for double ionization of these atoms is larger than that required for single ionization of either the alkali metals or the alkaline earth metals. However, the energy released in the third step, in which positive and negative ions come together to form a crystal, is much greater for doubly charged ions than for singly charged ions. The net result is that formation of ionic compounds with the alkaline earth metals in $+2$ oxidation states is more energetically favorable (lower energy state) than formation of compounds with the alkaline earth metals in a $+1$ oxidation state. Calcium is said to be in a $+2$ oxidation state because it has a positive charge of two electron units. When a calcium atom loses two electrons, which can then be captured by two chlorine atoms, all three atoms attain the noble gas configuration of argon. Calcium chloride is electrically neutral, so its formula must be $CaCl_2$.

Most solid metal oxides are ionic crystals. This may seem surprising when one considers that the electron affinity of oxygen to form the doubly negative oxide ion (acquisition of two electrons) is negative, namely -7.3 eV. For example, magnesium oxide (MgO) is a stable ionic compound despite the fact that energy must be supplied to form both the Mg^{2+} and O^{2-} ions. The explanation is simply that the energy liberated in the third step, when these doubly charged ions come together, more than compensates for the energy required in the first two steps.

As a general rule we expect ionic compounds to be formed from atoms of elements with widely different electronegativities. This is consistent with the earlier statement that ionic compounds are usually formed by atoms of elements with low ionization energies reacting with atoms of elements with high electron affinities. The electronegativity rule is easier to apply, however, since electronegativity values are more readily available than electron affinities. Unfortunately, various attempts to relate electronegativity differences with quantitative measures of ionic bonding have not really been successful.

Like sodium chloride, most ionic compounds at room temperature are brittle white solids that dissolve in water to form electrically conducting solutions. Another characteristic property of ionic compounds is that they melt at relatively high temperatures; this property results from the strength of ionic bonds. The resultant fused salts are also electrically conducting, indicating the presence of mobile ions. The properties of the ionic substances formed from various elements are discussed more fully in several later chapters.

Lewis Theory

The properties of sodium chloride and many other substances are adequately explained on the basis of aggregation of ions formed by transfer of electrons from atoms of one element to atoms of another element. There are many elements, however, for which electron transfer seems unlikely because of either large ionization energies or small electron affinities. Further, there are many substances that do not have the characteristic physical and chemical properties associated with ionic materials. Substances such as ammonia, water, and methane melt at relatively low temperatures and are poor conductors of electricity in the liquid state. Instead of a transfer of charge from one atom to another, we believe that the bonding in such substances, called **covalent bonding,** involves a sharing of electrons between the two atoms.

In 1916, G. N. Lewis developed a covalent bonding theory based on valence shell octets. This Lewis approach was the first reasonably successful attempt to explain bonding and stoichiometry in terms of outer electronic configurations. Lewis began by noting the special stability of noble gas electronic configurations and suggested that atoms form bonds by losing, gaining, or sharing enough electrons to achieve an octet of electrons in their valence shells. Bonds would be ionic if electrons are transferred, covalent if they are shared.

When writing Lewis (electron dot) structures, each valence electron is represented by a dot. Only the electrons with the largest value of n, the principal quantum number, need be considered for most representative elements. If we utilize our present knowledge of quantum theory, the reason for the importance of the octet becomes apparent. For an electronic configuration of ns^2np^6, each of the ns,np_x, np_y, and np_z orbitals can hold two electrons, making a total of eight.

The transfer of electrons in the formation of an ionic compound like sodium chloride can be represented as

$$\mathrm{Na}\cdot + \cdot\ddot{\underset{\cdot\cdot}{\mathrm{Cl}}}: \rightarrow \mathrm{Na}^+ \ :\ddot{\underset{\cdot\cdot}{\mathrm{Cl}}}:^-$$

By losing one electron, sodium atoms achieve the electronic configuration of neon; by gaining one electron, chlorine atoms take on the argon configuration.

With ionic compounds we do not ordinarily speak of molecules, because in the solid state each positive ion is attracted to several negative ions and each negative ion is attracted to several positive ions. The oppositely charged ions attract each other by electrostatic, or coulombic, forces. However, discrete molecules of sodium chloride and some other substances that we ordinarily think of as ionic do exist as gases at high temperatures. Thus it is proper to speak of molecules of sodium chloride in the gas phase even though crystalline sodium chloride does not contain discrete molecules.

Now consider the example of molecular fluorine (F_2). Experimental evidence shows that fluorine ordinarily exists in the form of diatomic molecules and does not exhibit the properties characteristic of ionic substances. Since the electronegativities of the two fluorine atoms are identical, sharing electrons, as suggested by Lewis, seems plausible. Diatomic fluorine may be represented by

$$:\ddot{\underset{\cdot\cdot}{\mathrm{F}}}:\ddot{\underset{\cdot\cdot}{\mathrm{F}}}:$$

which indicates that each fluorine atom has achieved the neon configuration as a result of sharing a pair of electrons. We can similarly represent the diatomic molecules of the other halogens, such as Cl_2 or Br_2. In diatomic hydrogen, represented by H:H, each atom achieves a helium configuration by sharing the electron pair. Diatomic molecules in which both atoms are the same are called **homonuclear** molecules.

Heteronuclear molecules, in which the atoms are not identical, can also be represented by similar Lewis "electron dot" structures. If the electronegativities of the elements are not too different, sharing electron pairs to form bonds seems reasonable. For example, one atom of hydrogen can unite with one atom of chlorine to form HCl; two atoms of hydrogen can unite with one atom of oxygen to form H_2O; or three atoms of hydrogen can combine with one atom of nitrogen to form NH_3. In each case the atoms achieve a noble gas configuration by sharing electron pairs. The formulas for hydrogen chloride, water, and ammonia are represented in the Lewis notation as

$$\mathrm{H:\underset{\cdot\cdot}{\overset{\cdot\cdot}{Cl}}:} \qquad \mathrm{H:\underset{\cdot\cdot}{\overset{\cdot\cdot}{O}}:H} \qquad \mathrm{H:\underset{\cdot\cdot}{\overset{\overset{\displaystyle H}{\cdot\cdot}}{N}}:H}$$

Note that Lewis structures give no indication of the true geometry of the molecule. They do, however, indicate the connections between atoms within a molecule correctly and the existence of unshared pairs of electrons, called **lone pairs.** Knowledge of the presence of lone pairs often enables chemists to predict correctly the geometry and chemical behavior of simple molecules.

For many molecules we picture atoms as sharing more than one electron pair. For example, each atom in carbon dioxide shares two pairs of electrons and thereby attains the neon configuration, as represented by

$$\mathrm{:\overset{\cdot\cdot}{O}::C::\overset{\cdot\cdot}{O}:}$$

In carbon monoxide and diatomic nitrogen there is a sharing of three electron pairs:

$$\mathrm{:C:::O:} \qquad \mathrm{:N:::N:}$$

A dash is frequently used to represent a shared electron pair. In this notation the symbols for carbon dioxide and carbon monoxide become

$$\mathrm{:\overset{\cdot\cdot}{O}{=}C{=}\overset{\cdot\cdot}{O}:} \quad \text{and} \quad \mathrm{:C{\equiv}O:}$$

We refer to a carbon-oxygen double bond in carbon dioxide and to a carbon-oxygen triple bond in carbon monoxide. Often, particularly in organic chemistry the nonbonding electrons are omitted entirely, and only the dashes that represent bonding pairs are shown. There are as many "dash" as "dot" chemists, and students should be familiar with both notations.

As a general rule, bond lengths decrease and bond energies increase from single to double to triple bonds. As an example, consider the following data for carbon-oxygen compounds:

Molecule	$H_3C{-}O{-}H$	$O{=}C{=}O$	$C{\equiv}O$
Carbon-oxygen bond length (Å)	1.43	1.16	1.13
Carbon-oxygen bond energy (kcal $mole^{-1}$)	85.5	192	256

The bond energy given here for carbon dioxide is an average value for the two carbon-oxygen bonds in this molecule.

Polarity of Covalent Bonds

In homonuclear diatomic molecules such as H_2 or Cl_2 the bonding electrons are shared equally between the bonded atoms. These molecules, electrically symmetrical as well as neutral, are said to be **nonpolar.** In the case of HCl, the hydrogen and chlorine atoms do not share equally the electron pair that bonds them. A molecule of HCl is electrically neutral, but it is not electrically symmetrical. The more electronegative chlorine atom exerts a greater pull on the bonding electron pair than does the hydrogen atom. Thus, the hydrogen end of the molecule is positive and the chlorine end negative, as represented by $\overset{+\!\!\longrightarrow}{\mathrm{HCl}}$. Because the centers of positive and negative electricity are not the same, the bond between hydrogen and chlorine is called **polar** and hydrogen chloride is called a polar molecule.

As another example of a polar covalent bond consider the bond between chlorine and fluorine in the molecule ClF. Both chlorine and fluorine atoms have *p* orbitals that are singly occupied. Overlap of these *p* orbitals of the two atoms can occur to produce a single covalent bond. Because chlorine and fluorine have different electronegativities, the bond is polar. In this case, fluorine is the negative end of the molecule and chlorine the positive end, as represented by $\overset{+\!\!\longrightarrow}{\mathrm{ClF}}$.

Diatomic molecules that contain polar bonds are said to have a dipole moment, which is a measure of the electrical asymmetry of a molecule. (See Chapter 5 for another discussion of dipole moment.) The presence of polar bonds in a polyatomic molecule does not necessarily mean that the molecule as a whole has a nonzero dipole moment. Because dipole moments are vector quantities, the magnitude of the dipole moment depends on the geometry of the molecule.

As examples we consider carbon dioxide and water. Oxygen is more electronegative than carbon so that carbon dioxide has two polar bonds. But because carbon dioxide is a linear molecule, represented by $\overset{\longleftarrow\!\!+}{\mathrm{O{=}C}}\overset{+\!\!\longrightarrow}{\mathrm{{=}O}}$, there is no *net* dipole. Water also has two polar bonds, as pictured in Chapter 5. The fact that water has a net dipole moment shows that the molecule must be nonlinear. The electrical asymmetries of water and of other polar substances have considerable effect on their chemical and physical properties.

Another question arises: At what point does a covalent bond become so polar that we consider it to be ionic? The answer must be that there is no sharp distinction between ionic forces, or bonds, and polar covalent bonds. Bonds between atoms with different electronegativities are polar. If the electronegativity difference, and hence the polarity, is sufficiently large, the bond may be classed as ionic. In many cases, however, we can only state that the bond is either highly polar *or* ionic.

Resonance

Lewis electron dot formulas are simple and useful for depicting the bonding in most ionic and covalent compounds. There are difficulties, however. For some molecules, such as PCl_5 and SF_6, the central atom must be assigned more than an octet of electrons. For other species, no single electron dot formula can be drawn that accounts satisfactorily for the observed properties.

This latter problem is encountered with sulfur dioxide, SO_2. From a variety of experiments we know that the sulfur atom is located between the two oxygen atoms. Because SO_2 has a high dipole moment, we conclude that the molecule must be nonlinear. Both sulfur and oxygen have six outer-shell electrons; there are thus a total of 18 valence electrons. Possible electron dot formulas that show eight electrons around each atom are

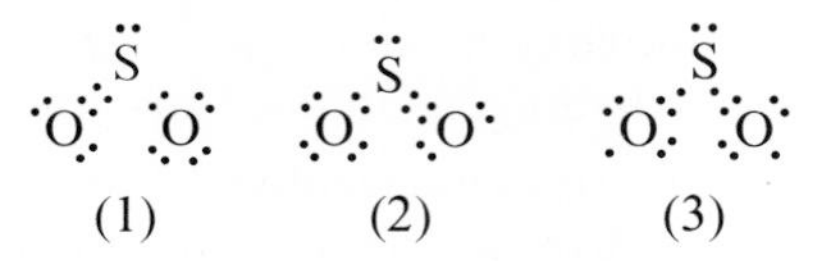

Formula 3 is rejected because it contains unpaired electrons. Substances that contain unpaired electrons are attracted to magnets and are called **paramagnetic.** If all electrons are paired so that magnetic properties of the spinning electrons cancel, the substance is **diamagnetic.** Diamagnetic substances are weakly repelled by magnets. Diamagnetism occurs in all matter, even though it may be obscured in paramagnetic substances. Because SO_2 is diamagnetic, we know that it has no unpaired electrons.

Formulas 1 and 2 indicate that the SO_2 molecule should have one double (short) sulfur-oxygen bond and one single (long) sulfur-oxygen bond. Experiments show that both bonds in SO_2 are exactly the same length. We are thus unable to write a single electron dot formula for SO_2.

A way out of this difficulty was suggested by Pauling and is termed **resonance theory.** The actual electronic distribution in the SO_2 molecule, described by a combination of formulas 1 and 2, is said to be a **resonance hybrid.** Because it may lead to the erroneous impression that the molecule flips back and forth between the two structures, the choice of the word "resonance" is unfortunate. The molecule has a single structure with a fixed set of properties. The difficulties arise because of our limitations in describing structures with pencil and paper.

A story of uncertain origin that has been devised to clarify what is meant by resonance is the following: A knight of the round table saw a rhinoceros on one of his journeys. He described it to a fellow knight as a cross, or hybrid, between a unicorn and a dragon. The unicorn and dragon are imaginary creatures just as our electron dot formulas are imaginary pictures of the SO_2 molecule. But the rhinoceros and the electronic structure of SO_2 are real. The knight described the real animal in terms that were easy for him to use. We describe the bonding in SO_2 in terms that are easy for us to use.

Resonance represents an attempt to patch up valence bond theory to account for certain molecules such as SO_2. In molecular orbital theory, in which the electrons are pictured as belonging to the molecule as a whole, this problem does not arise.

Another alternative to resonance structures, the **double quartet** approach, has been suggested recently by J. W. Linnett. Instead of an octet of valence electrons, one pictures two interlocking quartets differing in electron spin. The four electrons in each quartet are arranged tetrahedrally, which minimizes electron-electron repulsions by keeping electrons as far apart as possible. In the illustrative drawing of this representation of the neon atom, we use hollow circles to represent electrons

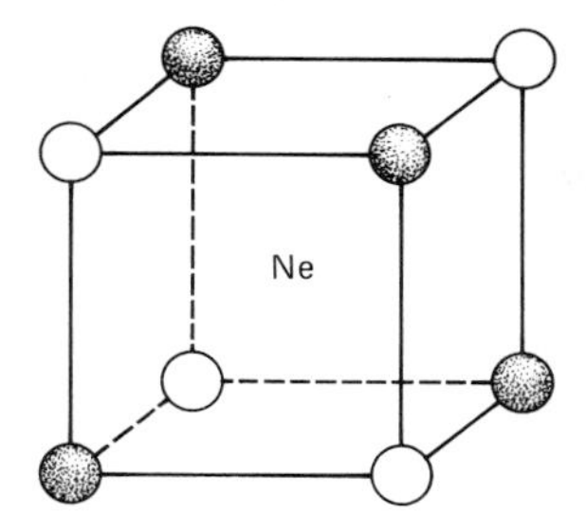

of one spin and solid circles to represent electrons of the other spin. These two quartets of electrons with opposite spins may be considered independently of each other. In forming bonds, electron sharing occurs to complete each quartet.

Single bonds are formed by having two tetrahedra share a common corner, as illustrated for diatomic fluorine. Each atom of fluorine has seven electrons and the molecule has seven electrons of each spin type, as represented by

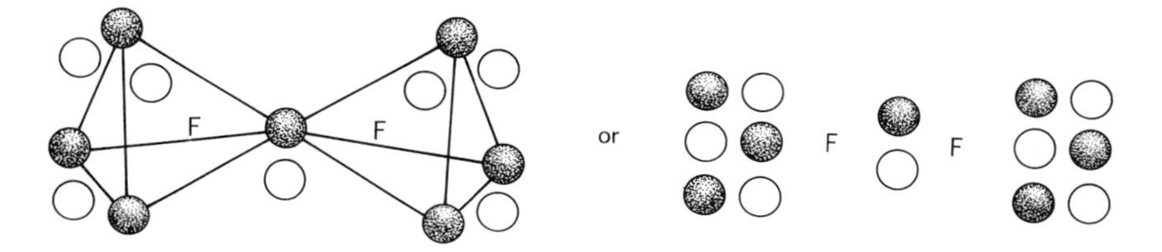

There is no net electron spin because the two sets just cancel.

To form double bonds, the two tetrahedra of the same spin on different atoms share a side. The net result is to place four electrons in the internuclear region. Triple bonds are formed by tetrahedra sharing a face, thus placing six electrons between the nuclei.

Further illustrations of Linnett's double quartet approach are given later in connection with discussion of bonding in certain molecules and also in references cited at the end of this chapter.

Electrons in Molecules

In an isolated atom an electron is attracted to the nucleus and repelled by all the other electrons. When two atoms come together there are additional attractions and repulsions resulting from the interaction of the nucleus and electrons of one atom with those of the other. If the net result is a larger attraction, an electronic rearrangement must have occurred to give a more stable state. In other words, the molecular state must be one of lower energy than that of the isolated atoms.

Figure 8.1 illustrates the change in potential energy that accompanies bond formation in a covalent molecule. When the atoms are far apart there is no interaction between them, and the potential energy is set at zero. As the two atoms approach each other, attractive terms become predominant and the potential energy goes through a minimum. The position of this minimum corresponds to the internuclear distance for the most stable state of the molecule, that is, the

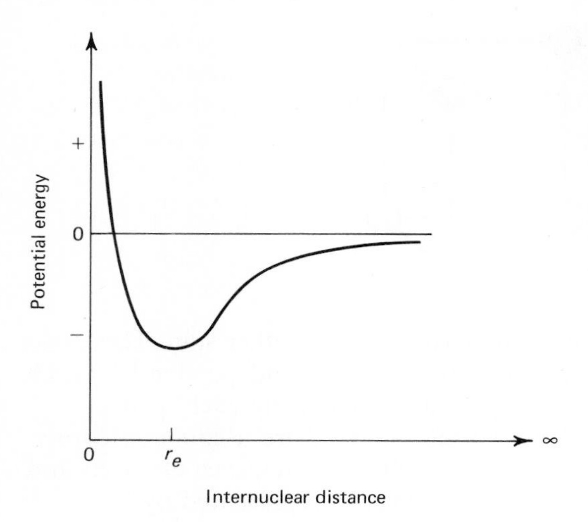

Figure 8.1. Change in potential energy as two atoms approach each other. By convention, the potential energy is zero at infinite separation. The minimum potential energy corresponds to r_e, the equilibrium bond distance for the stable molecule.

equilibrium bond length. As the atoms get closer together, electron-electron and nucleus-nucleus repulsion cause the potential energy to increase rapidly.

There are two general theories that attempt to explain this phenomenon of bond formation. In the **molecular orbital** approach that we will be considering soon, one assumes that a molecule is much like a polynuclear atom. The emphasis is placed on the valence electrons. The nuclei (or nuclei plus inner electrons) are brought into position. The valence electrons of all the atoms are then allotted to molecular orbitals, which spatially cover the entire molecule. The concept of chemical bonds between individual atoms is abandoned. In the **valence bond** method the molecule is treated as a collection of atoms held together by definite bonds. The emphasis is placed on the individual atoms, including nuclei plus inner and valence electrons. These complete atoms are brought together and allowed to interact using the orbitals of the isolated atoms. Valence bond theory may be considered a more rigorous version of the simple "electron dot" approach.

It is important to realize that both methods are approximations and that both describe limiting conditions. As corrections are applied, the two methods tend to converge. The valence bond approach is visually simpler and more consistent with traditional ideas about atoms and molecules. The molecular orbital approach often gives more satisfying answers but is sometimes difficult both visually and mathematically. A chemist generally chooses the theory that seems more appropriate to his particular problem.

Hybrid Orbitals and Shapes of Molecules

The arrangement of ions with respect to one another in most ionic crystals is determined by the charges and sizes of both the positive and negative ions. The arrangement of atoms in covalent molecules is not as simply explained. Molecules containing two atoms are necessarily linear, but those with three or more atoms present complications. One approach to this problem is to use valence bond theory and one additional principle, the overlap criterion of bond strength.

The most reliable criterion for the strength of a given bond is the energy released when the bond is formed. Unfortunately, these energies are often difficult to predict or obtain experimentally. Sometimes a more easily applied criterion is the **principle of maximum overlap:** As the overlap of two atomic orbitals increases, the bond strength between the atoms increases. In a qualitative way this makes sense. The more two orbitals overlap, the more the bonding electrons are concentrated between the two nuclei. In this position the bonding electrons should minimize repulsions between nuclei and maximize the attractions between themselves and both nuclei. If the spatial orientation of the bonding orbitals is known, the geometry of the molecule is readily determined.

To illustrate, we consider the hydrides of the first row elements: carbon, nitrogen, and oxygen. Methane, CH_4, as shown in Figure 8.2, has a tetrahedral shape with the carbon at the center of the tetrahedron and the four hydrogens at the corners. The experimentally observed angles for the H—C—H bonds are 109.5°. The electronic configuration of an isolated carbon atom is $1s^22s^22p^2$, whereas hydrogen is $1s^1$. To form four equivalent C—H bonds, we evidently need to use all the carbon $2s$ and $2p$ orbitals. We might "promote" one of the $2s$ electrons to give a configuration of $1s^22s^12p_x^12p_y^12p_z^1$. The energy released during bond formation more than compensates for this "promotional energy." Since s orbitals are spherical and p orbitals are directional, the resultant C—H bonds would not all be equivalent. We therefore mathematically mix the $2s$ and three $2p$ orbitals of carbon to obtain a set of four equivalent hybrid orbitals. These orbitals, designated sp^3, consist of equal parts of each of the four atomic orbitals used to form them. The superscript indicates the number of atomic orbitals used in forming the hybrid orbitals, not the electron population. The sp^3 hybrid orbitals are directed toward the corners of a regular tetrahedron. Overlap of the carbon sp^3 orbitals and the hydrogen $1s$ orbitals leads to the observed tetrahedral shape of the CH_4 molecule.

In the ammonia molecule, NH_3, the observed H—N—H bond angles are 106.6°. The electronic configuration of an isolated nitrogen atom is $1s^22s^22p^3$. Overlap of the hydrogen $1s$ orbital with the $2p_x$, $2p_y$, and $2p_z$ orbitals of nitrogen would produce three equivalent bonds but give bond angles of 90°. Proton-proton repulsions would logically lead to some widening of the bond angles but probably not by as much as 16.6°. It seems more reasonable to assume a basic tetrahedral shape for the NH_3 molecule also. The five valence electrons of nitrogen can be distributed so that there are two in one of the sp^3 orbitals, and one in each of

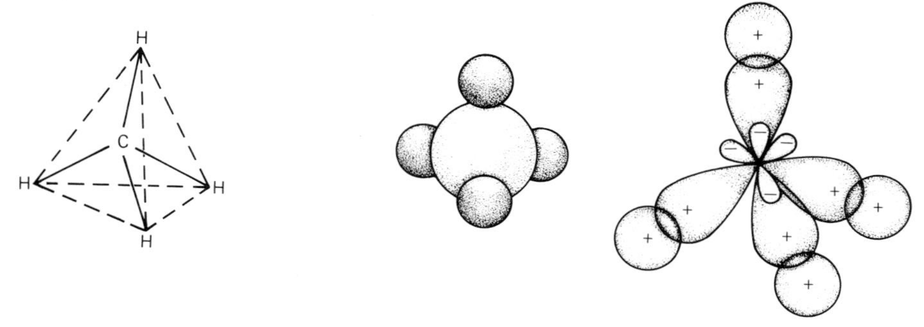

Figure 8.2. Tetrahedral CH_4 molecule.

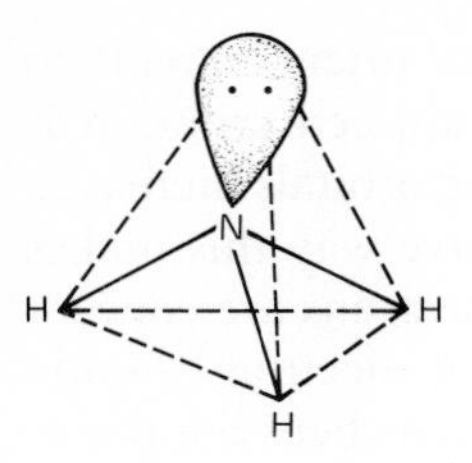

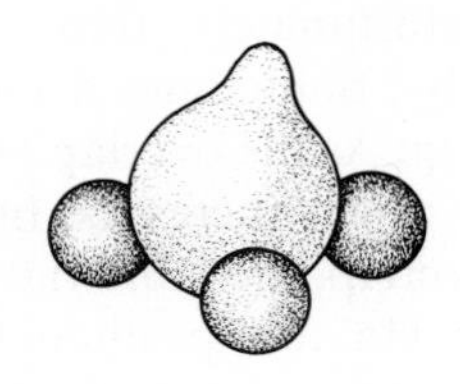

Figure 8.3. Trigonal-pyramidal NH_3 molecule.

the other three sp^3 orbitals. Overlap of these three singly occupied sp^3 orbitals of nitrogen with each of the hydrogen $1s$ orbitals would yield the observed shape of the molecule, as shown in Figure 8.3. The small deviation of the H—N—H angle from the tetrahedral angle can be explained through lone pair–bonding pair repulsions. The existence of a lone pair of electrons in a directional orbital is consistent with much of the chemistry of NH_3 and other related nitrogen compounds.

In the water molecule the bond angle is 104.5°. An isolated oxygen atom has an electronic configuration of $1s^22s^22p^4$. As in the NH_3 case we might assume that the bonding in H_2O involves use of pure p orbitals with proton-proton repulsions causing the deviation from 90°. It seems more plausible, however, to assume sp^3 hybrid orbitals leading to two bonding pairs and two lone pairs, as shown in Figure 8.4. The deviation from the tetrahedral angle of 109.5° would result from bonding pair–lone pair repulsions. The chemistry of water and its ability to form two hydrogen bonds per molecule is consistent with the presence of two lone pairs in directional orbitals. Lone pair electrons in directed orbitals are believed to make sizable contributions to the dipole moments of molecules such as NH_3 and H_2O.

The concept of hybrid orbitals is also useful in understanding the shapes of more complicated molecules. The designations and geometric arrangements of some other hybrid orbitals are summarized in Table 8.1. The use of these hybrid orbitals will be considered in connection with organic molecules (Chapter 19) and transition metal complexes (Chapter 20).

A qualitative, but very useful, theory for predicting molecular shapes has recently been proposed by R. J. Gillespie. The basic idea is that the shape of a molecule is determined primarily by repulsive interactions between electron pairs in the valence shell of the central atom. It is called the **valence-shell electron-pair repulsion** theory, abbreviated **VSEPR.** Molecular formulas are written in the form AX_mE_n where A is the central atom, X is any atom or group (called a ligand) attached to the central atom, and E is a lone pair. The $m + n$ electron pairs are located as far apart as possible in order to minimize repulsions. Figure 8.5 shows the preferred structures for several common types of molecules.

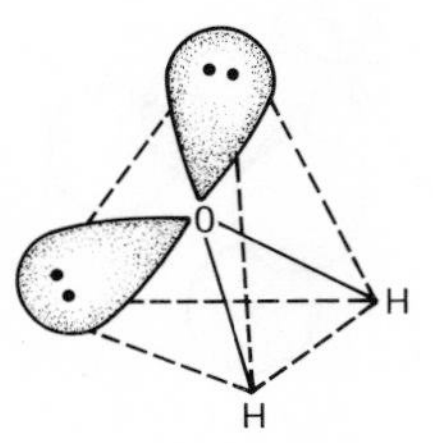

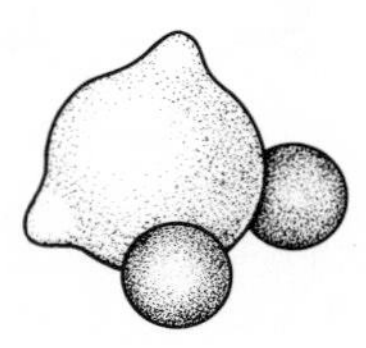

Figure 8.4. Bent H_2O molecule.

Table 8.1. Hybrid Orbitals

Designation	*Geometry*	*Bond angle*	*Example*
sp	linear	180°	$BeCl_2$
sp^2	trigonal	120°	BCl_3
sp^3	tetrahedral	109.5°	CH_4
dsp^2	square planar	90°	$PtCl_4^{2-}$
d^2sp^3	octahedral	90°	SF_6
dsp^3	trigonal bipyramidal	90° and 120°	PF_5

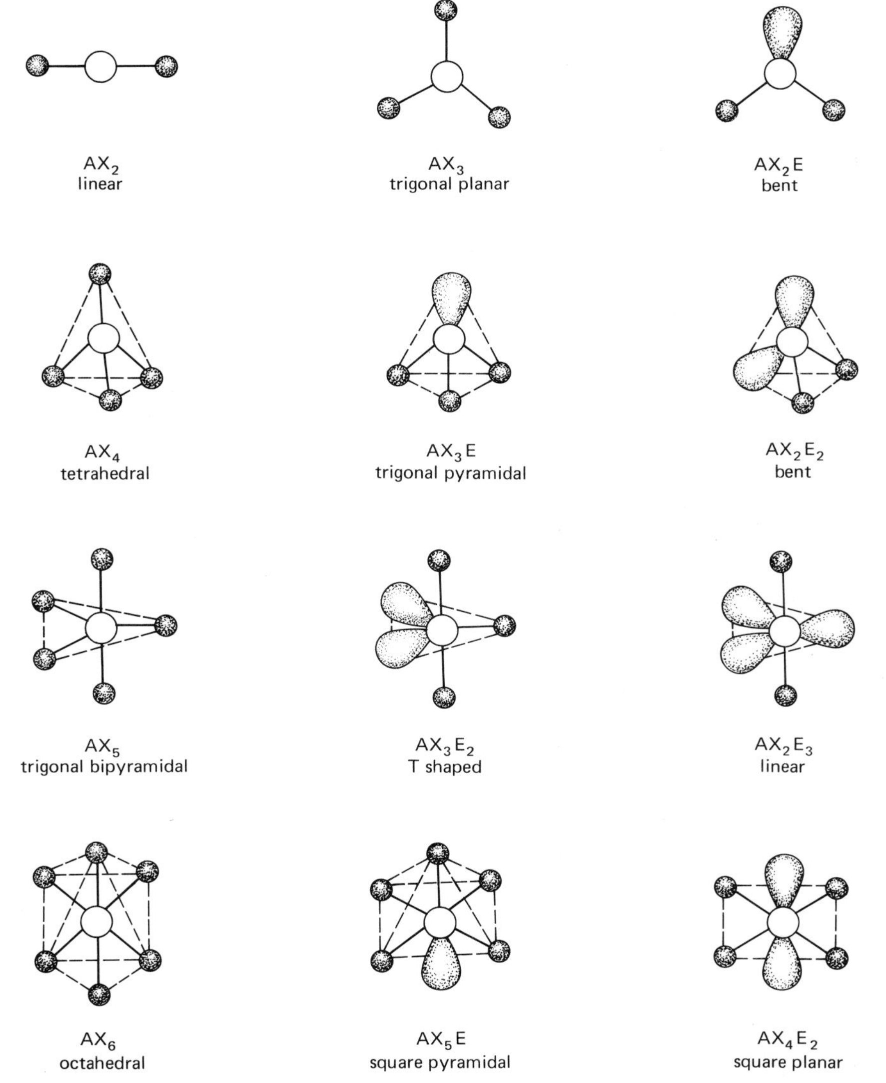

Figure 8.5. Shapes of singly-bonded molecules containing up to six electron pairs in the valence shell. A is the central atom; X is any ligand; E is a lone pair of electrons.

For our illustrative discussion we need two additional postulates. One is that nonbonding or lone electron pairs repel adjacent electron pairs more strongly than do bonding pairs. Bonding electron pairs are constrained between two nuclei and thus occupy a smaller volume of space than a lone pair, which is influenced by only one nucleus. In essence, lone pair orbitals are pictured as "fatter" than bonding pair orbitals. The second postulate is that repulsions exerted by bonding electron pairs decrease as the electronegativity of the ligand increases. A more electronegative ligand will attract the bonding electron pair more strongly and thus reduce repulsions around the central atom.

To show the usefulness of this approach we consider a few examples. Iodine readily dissolves in solutions containing iodide ion to form the triodide ion, I_3^-. Is this ion linear or bent? By writing the electron dot structure we find that the ion is of the form AX_2E_3. Using Figure 8.5, we therefore predict a linear structure (bond angle of 180°), in agreement with experiment.

What is the structure of the interhalogen compound bromine pentafluoride (BrF_5)? From the electron dot structure we find that BrF_5 is of the form AX_5E. It should therefore have a square pyramidal shape with bond angles of 90°. Again this is in accord with experiment.

The VSEPR theory can also be used to explain the bond angles in CH_4, NH_3, and H_2O. The shapes of these molecules were previously discussed in connection with hybrid orbitals. CH_4, NH_3, and H_2O are of the form AX_4, AX_3E, and AX_2E_2 respectively. They should thus all have a basic tetrahedral shape. The presence of one lone pair in NH_3 reduces the bond angle from 109.5°; the presence of two lone pairs in H_2O reduces the angle even further.

Now consider the oxygen difluoride (OF_2) and H_2O molecules. Both have an AX_2E_2 formula, but fluorine is more electronegative than hydrogen. We therefore expect the repulsions between the bonding electron pairs to be less in OF_2 than in H_2O. In agreement with this argument, the OF_2 bond angle is 103.2° compared to the 104.5° angle in H_2O.

Molecular Orbitals

In molecular orbital theory the concept of formation of specific and localized chemical bonds is abandoned. Instead, the nuclei are placed in the positions they occupy in the final molecule. The valence electrons are then distributed throughout the molecule in the force field produced by the nuclei and inner electrons.

We begin our theoretical construction of molecular orbitals with the assumption that these molecular orbitals can be formed from a combination of atomic orbitals of the bonded atoms. We further assume that a molecular orbital that lies close to nucleus A resembles the atomic orbital of an isolated atom A. For a diatomic molecule, A—B, we can view this combination process as either an addition or subtraction of atomic orbitals. Thus two atomic orbitals will produce two molecular orbitals.

In order to minimize repulsions between nuclei, the molecular orbitals involved in forming a stable molecule must have a buildup of electron density in the internuclear region. These orbitals are called **bonding molecular orbitals.** They are formed by addition of one atomic orbital from A and one from B. The energy of this bonding molecular orbital is lower than that of either atomic orbital from which it is formed. A subtraction of these same two atomic orbitals will produce a molecular orbital with the electron density concentrated away from the internuclear region. Electrons in such an orbital do not reduce the repulsions between nuclei and are said to destabilize the molecule. These molecular orbitals are higher in energy than the atomic orbitals which form them and are called **antibonding molecular orbitals.**

We can illustrate these ideas by considering some simple homonuclear molecules, using many of the same concepts that were developed in considering electronic configurations of atoms. We follow an aufbau process in which the lowest energy orbitals are filled first. Each orbital can hold only two electrons, which must be spin-paired (Pauli exclusion principle). Two orbitals of equal energy will be singly occupied before either is doubly occupied in the ground (lowest energy) state (Hund's rule). In this book we will not be concerned with excited (higher energy) states. The "relaxing effect" of an electron in an antibonding orbital is slightly greater than the stabilizing effect of an electron in a bonding orbital. Therefore, molecules that contain equal numbers of bonding and antibonding electrons will be unstable and should not exist.

To construct the molecular orbitals for the H_2 molecule we only need to consider the $1s$ atomic orbitals of each hydrogen. These can be combined as shown in Figure 8.6, where we represent the wave functions for the $1s$ orbitals of H_a and H_b by $1s_a$ and $1s_b$, respectively. Both molecular orbitals are given a σ designation to indicate that they are symmetric to rotation about the molecular axis. This means that, if the orbital is rotated about a line connecting the two nuclei, the orbital will not change in appearance or mathematical sign. Antibonding orbitals are generally marked with an asterisk. Note that there is a zero probability of finding an electron on a nodal plane halfway between the two nuclei in the σ^*_{1s} orbital. The subscript $1s$ in σ_{1s} and σ^*_{1s} denotes the atomic orbitals from which the molecular orbitals were formed. The two valence electrons in H_2 would thus be spin-paired in the σ_{1s} orbital. The energy difference between the atomic orbitals and the σ_{1s} molecular orbital corresponds to the bond energy of the H_2 molecule.

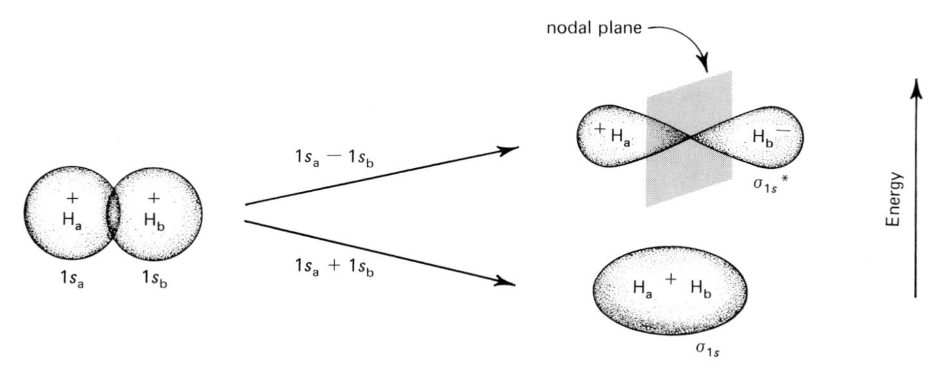

Figure 8.6. Molecular orbital formation by combination of $1s$ atomic orbitals. The plus and minus signs represent the mathematical signs of the wave functions for the orbitals and not electrical charge. The shadings represent electron density, which is related to the square of the molecular wave function.

The molecular orbital diagram in Figure 8.6 can also be used to describe what happens when two helium atoms come together. The energies of the atomic and molecular orbitals of helium would differ from those of hydrogen, but the relative arrangement would be the same. For the two helium atoms there are a total of four valence electrons. Since each molecular orbital can hold only two electrons, both the σ_{1s} and σ^*_{1s} orbitals will be doubly occupied. Consequently, He_2 is energetically unstable with respect to the two separate helium atoms and is not observed.

In elements beyond helium the second principal quantum level involving both *s* and *p* orbitals must be considered. Similar to the 1*s* orbitals, the 2*s* orbitals can combine to produce σ^*_{2s} and σ_{2s} molecular orbitals. We assume that the inner electrons ($n = 1$) are not involved in molecule formation but instead remain in atomic orbitals. When two lithium atoms ($1s^22s^1$) come together, the two valence electrons should occupy the σ_{2s} molecular orbital. Because this is a bonding molecular orbital, the energy of Li_2 should be lower than that of the two lithium atoms. The Li_2 molecule has been observed in the vapor state.

For beryllium ($1s^22s^2$), formation of Be_2 would require a pair of electrons in the σ^*_{2s} orbital as well as a pair in the σ_{2s} orbital. Just as in He_2, the relaxing effect of the antibonding electrons is greater than the stabilizing effect of the bonding electrons so that no stable molecule is formed.

Beginning with boron, the 2*p* orbitals must also be considered. Two different kinds of molecular orbitals can be formed from the p_x, p_y, and p_z atomic orbitals. We arbitrarily designate the *z* axis as the molecular axis, an imaginary line connecting the two nuclei. The $2p_z$ orbitals are parallel to this axis, and the $2p_x$ and $2p_y$ orbitals are perpendicular to it. Overlap of lobes of the $2p_z$ orbitals with the same sign produces a buildup of electron density in the internuclear region. This molecular orbital is therefore a bonding orbital designated σ_z. If lobes of unlike sign overlap there will be a decrease of electron density, and an antibonding molecular orbital, σ^*_z, is formed. These orbitals, shown in Figure 8.7, are both designated σ since they are symmetric to rotation about the molecular axis.

The $2p_x$ atomic orbitals can also be combined as sums or differences. As shown in Figure 8.7, addition of the $2p_x$ orbitals of the two atoms produces a molecular orbital that looks like a spread out version of a p_x atomic orbital. Maximum electron density occurs in two lobes of opposite sign, which are above and below a nodal plane containing the molecular axis. The general appearance of these lobes is that of two small watermelons. The combination involving overlap of p_x lobes of unlike sign produces a four-lobed molecular orbital with a decrease of electron density in the internuclear region. The first orbital is therefore a bonding molecular orbital, designated π_x; the second orbital is antibonding, π^*_x. These π molecular orbitals are not symmetrical to rotation about the molecular axis. Rotation by 180° will produce a similar arrangement, but the lobes will then have the opposite sign.

Combinations of the $2p_y$ atomic orbitals of two atoms will produce the π_y and π^*_y molecular orbitals. The π_y and π^*_y orbitals are identical to the π_x and π^*_x orbitals, except that they are rotated by 90° so as to be perpendicular to the plane of the paper.

To summarize: From the eight atomic orbitals of the $n = 2$ quantum level we

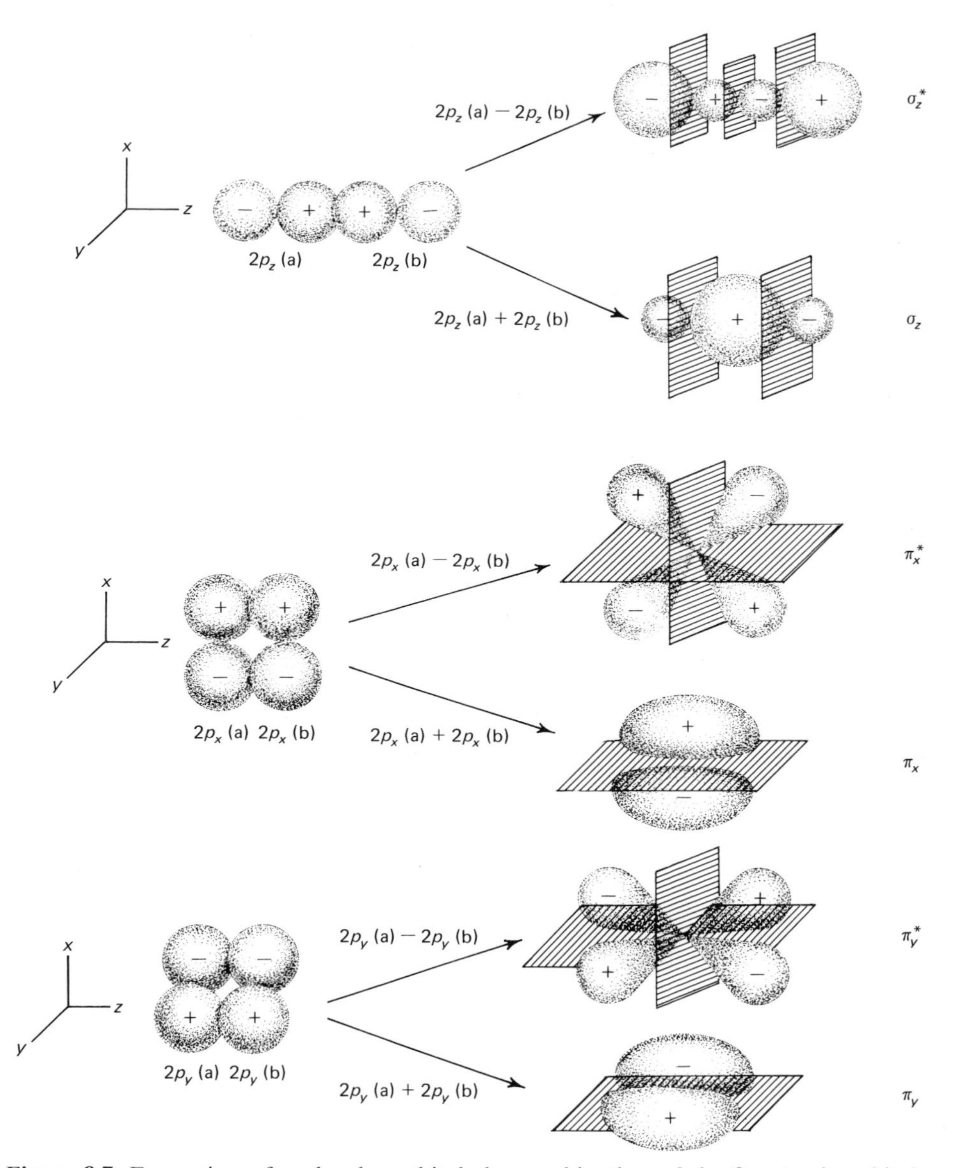

Figure 8.7. Formation of molecular orbitals by combination of the $2p$ atomic orbitals. The z axis is the molecular axis. Plus and minus signs represent the signs of the parts of the wave functions and not electrical charge. The lined nodal planes represent areas of zero electron density.

have obtained eight molecular orbitals. The relative energies of these orbitals are shown in Figure 8.8. Note that the $\pi_{x,y}$ orbitals are shown to be lower in energy than the σ_z orbital. From simple molecular orbital theory we would predict just the opposite arrangement; that is, the σ_z below the $\pi_{x,y}$ orbitals. This relative order was a point of controversy for some time. From careful analysis of spectroscopic

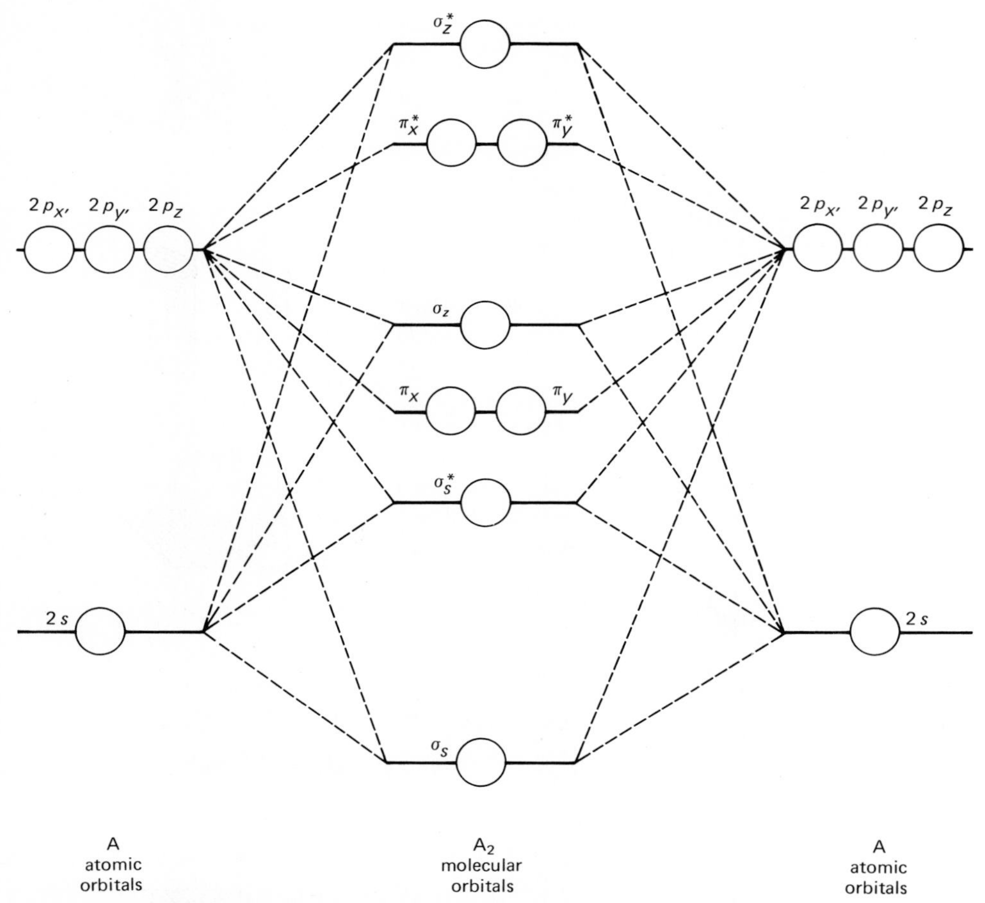

Figure 8.8. Relative order of energy levels for diatomic molecules with mixing of the $2s$ and $2p$ orbitals. If the two atoms were not identical the energy levels would be shifted, but the relative order of molecular orbitals would remain the same. The $1s$ atomic orbitals and corresponding molecular orbitals have been omitted.

data, however, it now appears that in most diatomic molecules the $\pi_{x,y}$ orbitals are lower energy than the σ_z orbitals, as pictured in Figure 8.8. This inversion of the "expected" order can be explained in terms of mixing or hybridization of the $2s$ and $2p_z$ atomic orbitals, but is best discussed in more advanced books in which appropriate mathematics can be used. For our purposes we can simply say that mixing causes a lowering in energy of the σ_s and σ_z orbitals and an increase in energy of the σ_s^* and σ_z^* orbitals, with the most important result being the energy level arrangement pictured in Figure 8.8.

For O_2 and F_2 the spectroscopic evidence indicates that the σ_z orbital is lower in energy than the $\pi_{x,y}$ orbitals, and the concept of mixing does not have to be invoked. In discussing the bond properties of molecules as we are doing, these complications make no important difference because the orbitals under consideration become "inner" filled orbitals.

Following familiar aufbau principles, we can now continue with the discussion of homonuclear diatomic molecules. Our next example is boron ($1s^22s^22p^1$), which has three valence electrons per atom. There are six valence electrons to consider for the diatomic molecule B_2. These six electrons should go into the lowest lying energy levels. We write the electronic configuration as

$$KK(\sigma_s)^2(\sigma_s^*)^2(\pi_x)^1(\pi_y)^1$$

where KK represents the four electrons in the inner shell ($n = 1$) that have practically no effect on bonding. To minimize repulsion between like charges we place one electron in each of the equal energy π orbitals. The B_2 molecule is therefore expected to be stable and paramagnetic. Experiments confirm its existence and the presence of two unpaired electrons.

For C_2, there are eight valence electrons to distribute in molecular orbitals. The two additional electrons (compared to B_2) will complete the π_x and π_y bonding orbitals. Consistent with predictions, the C_2 molecule is diamagnetic. In addition, the bond energy for C_2 is much greater than that of B_2 (150 kcal mole^{-1} and 70 kcal mole^{-1}, respectively). This is consistent with the prediction that C_2 has four bonding electrons, whereas B_2 has only two. Another way of describing the bonding would be to say that C_2 has a double bond and B_2 a single bond.

In N_2 all the bonding molecular orbitals are filled. The molecular orbital designation is

$$KK(\sigma_s)^2(\sigma_s^*)^2(\pi_x)^2(\pi_y)^2(\sigma_z)^2$$

The six net bonding electrons may be considered equivalent to a triple bond. Consistent with predictions, N_2 has the shortest bond (1.10 Å) and grestest bond energy (225 kcal mole^{-1}) of any homonuclear diatomic molecule. It is also diamagnetic.

In O_2, the two additional electrons (compared to N_2) must go into antibonding orbitals. As with B_2, there are two low-lying orbitals of equal energy, in this case π_x^* and π_y^*. We therefore place one electron in each orbital. This simple explanation of the observed paramagnetism of O_2 is an important achievement for molecular orbital theory. There is no straightforward explanation in valence bond theory for this paramagnetism, although the double quartet theory handles it easily (see problem 15). Diatomic oxygen thus has four bonding electrons, equivalent to a double bond. We predict a weaker and longer bond than in N_2, which is confirmed by experiment.

The two additional electrons for F_2 will complete the π_x^* and π_y^* orbitals. This leaves a net of two bonding electrons, equivalent to a single bond. We therefore expect the bond in F_2 to be weaker than that in O_2. In fact, the bond in F_2 is one of the weakest in diatomic molecules, 38 kcal mole^{-1}. This is presumably due to the greater relaxing effect of the antibonding electrons compared to the stabilizing effect of the bonding electrons. As predicted, F_2 is diamagnetic.

Since Ne_2 would have equal numbers of bonding and antibonding electrons, it is unstable compared to the free atoms and is not observed.

Bond lengths and bond energies for the first row diatomic molecules are shown in Table 8.2.

For molecules with more than two atoms, the molecular orbital description

Table 8.2. Properties of Homonuclear Diatomic Molecules

Molecule	*Bond length* (*Å*)	*Bond dissociation energy* (*kcal mole*$^{-1}$)
H_2	0.742	104
Li_2	2.67	25
B_2	1.59	70
C_2	1.31	150
N_2	1.10	225
O_2	1.21	118
F_2	1.42	38

becomes more complicated. For a relatively simple molecule such as methane, CH_4, eight molecular orbitals are required. Each of these molecular orbitals is formed from one of the carbon valence atomic orbitals ($2s$, $2p_x$, $2p_y$, $2p_z$) and part of the $1s$ atomic orbital of each hydrogen. Thus eight atomic orbitals yield eight molecular orbitals, four bonding and four antibonding. The eight valence electrons occupy the four bonding orbitals, leading to four bonds for the molecule. The molecular orbitals must be constructed properly so that there will be a buildup of electron density in regions directed toward the corners of a regular tetrahedron. In addition, the relative electronegativities of the atoms involved must be considered.

Carbon dioxide is also a relatively simple molecule, but with the additional complication of π bonding. We again use the z axis as the molecular axis and designate the atoms as follows:

$$O_a{=}C{=}O_b$$

x, y, z

There must be four σ molecular orbitals, two bonding and two antibonding, composed of the $2s$ and $2p_z$ orbitals of the C atom and the $2p_z$ orbitals of the two oxygen atoms. There will be one π orbital formed from the $2p_x$ orbitals of C and O_a, and another π orbital from the $2p_y$ orbitals of C and O_b. Two antibonding orbitals, designated π^*, will be formed from these same atomic orbitals. Two **nonbonding** molecular orbitals will be formed from the $2p_y$ orbital of O_a and the $2p_x$ orbital of O_b. Nonbonding orbitals have no significant effect on holding the atoms together and are similar to pure atomic orbitals. The $2s$ atomic orbitals of the O atoms are too low in energy to participate in the bonding and will also be nonbonding orbitals. To summarize: twelve atomic orbitals form twelve molecular orbitals—two σ, two σ^*, two π, two π^*, and four nonbonding orbitals. All but the σ^* and π^* orbitals are doubly occupied, leading to two σ and two π bonds for the molecule and accounting for the double bonds between the central carbon atom and each oxygen atom.

Unfortunately, this fairly straightforward explanation of the bonding in CO_2 does not conform entirely with the results of experiments. Calculated bond distances for carbon-oxygen bonds on the basis of this model involving localized bonds are too long. To correct the simple model described above, it is necessary to consider

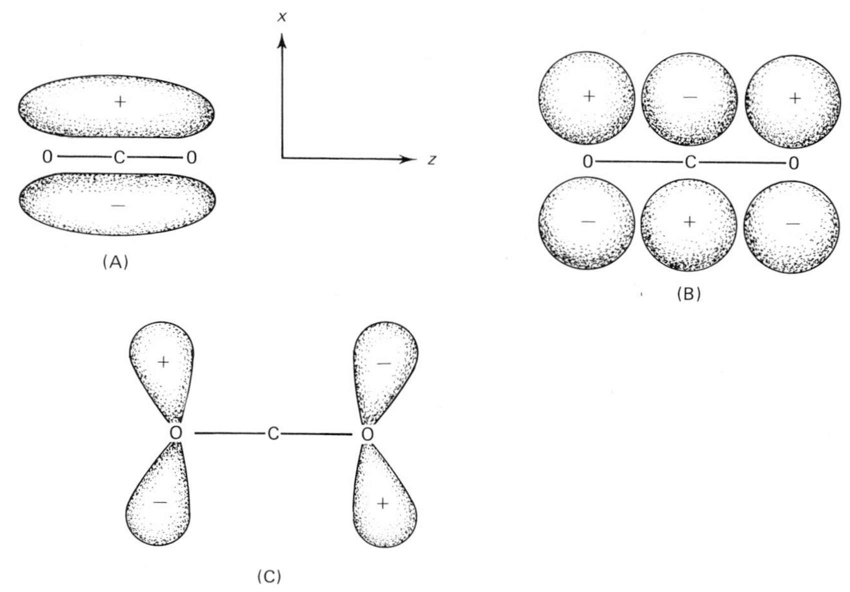

Figure 8.9. Boundary surfaces of the **(A)** bonding, **(B)** antibonding, and **(C)** nonbonding molecular orbitals of CO_2 using a delocalized bond model. The corresponding set of three orbitals from the $2p_y$ atomic orbitals would be rotated by 90°.

the symmetry of the molecule and use delocalized π bonds (spread out over the entire molecule). Three π orbitals, one bonding, one antibonding, and one nonbonding, are formed from the $2p_x$ orbitals of carbon and both oxygens, as illustrated in Figure 8.9. A similar set of three π orbitals is formed from the $2p_y$ orbitals of all three atoms. Electrons are then filled in these orbitals just as in the simpler model already described to yield two σ bonds, two π bonds, and four filled nonbonding orbitals for the molecule.

The concept of delocalization mentioned above is formally analogous to resonance in valence bond theory. It is especially useful in explaining the bonding in organic molecules such as benzene.

References

Gillespie, R. J.: The valence-shell electron-pair repulsion theory of directed valency. *J. Chem. Educ.*, **40:**295 (1963).

Gillespie, R. J.: The electron-pair repulsion model for molecular geometry. *J. Chem. Educ.*, **47:**18 (1970). A summary and extension of the 1963 article.

Gray, H. B.: *Electrons and Chemical Bonding.* W. A. Benjamin, Inc., New York, 1965 (paperback).

Ludeř, W. F.: The electron repulsion theory of the chemical bond. *J. Chem. Educ.*, **44:**206 (1967). The double-quartet approach.

Luder, W. F.: The electron repulsion theory of the chemical bond: II. An alternative to resonance hybrids. *J. Chem. Educ.*, **44:**269 (1967).

Ogryzlo, E. A., and G. B. Porter: Contour surfaces for atomic and molecular orbitals. *J. Chem. Educ.*, **40:**256 (1963).

Pimentel, G. C., and R. D. Spratley: *Chemical Bonding Clarified Through Quantum Mechanics.* Holden-Day, Inc., San Francisco, 1969 (paperback).
Ryschkewitsch, G. E.: *Chemical Bonding and the Geometry of Molecules.* Reinhold Publishing Corp., New York, 1963 (paperback).
Sebera, D. K.: *Electronic Structure and Chemical Bonding,* Blaisdell Publishing Co., Waltham, Mass., 1964 (paperback).

Problems

1. Predict which of the following compounds are ionic and which are covalent:

$CsBr \quad PCl_3 \quad CaO \quad NaH \quad CCl_4 \quad ClF_3 \quad SF_6$

Draw electron dot structures for each. Is it necessary to violate the octet rule in any of these cases?

2. Give three examples for each of the following:
(a) Ions with a noble gas configuration.
(b) Covalent compounds that violate the octet rule.
(c) Molecules with one or two lone pairs of electrons on the central atom.

3. Assuming that the octet rule applies, show that each of the following contains a double or triple bond:

$COCl_2 \quad HN_3 \quad CS_2 \quad HCN \quad P_2 \quad Ca^{2+}C_2^{2-}$

4. The sulfur trioxide (SO_3) molecule is planar with all three S—O bonds the same length. Draw electron dot formulas for SO_3 that do not violate the octet rule, and explain these formulas in terms of resonance.

5. Using electronegativity values (Table 7.5), give a good example for each of the following:
(a) Two atoms that form a more polar bond than C—F.
(b) Two unlike atoms that form a nearly nonpolar bond.
(c) Two atoms that form a more ionic bond than LiI.

6. Sodium nitrate consists of Na^+ and NO_3^- ions. Draw electron dot formulas for NO_3^- that account for the experimental observation that all three N—O bonds are the same length.

7. (a) Draw electron dot formulas for the following organic molecules: ethane, C_2H_6; ethene (also called ethylene), C_2H_4; and acetylene, C_2H_2.
(b) Which of these three molecules should have the longest carbon-carbon bond?
(c) Which molecule should require the most energy to break the carbon-carbon bond?

8. Using VSEPR theory, draw the correct arrangement for the atoms in the following molecules and ions:

$BBr_3 \quad SF_6 \quad NH_4^+ \quad BrF_4^- \quad H_3O^+ \quad Cl_2O$

9. Using electronegativities (Table 7.5) and VSEPR theory, predict which of the following molecules have net dipole moments:

$CF_4 \quad PCl_3 \quad BCl_3 \quad H_2S \quad ClF_3$

10. Explain the relative magnitudes of the dipole moments of the following compounds in terms of electronegativities:

HCl, 1.03 D HBr, 0.79 D HI, 0.30 D

The dipole moments are given in debye (D) units; 1 D is equivalent to 1.0×10^{-18} esu cm.

11. Ozone, O_3, molecules are nonlinear. Both oxygen-oxygen bonds are the same length. Draw electron dot formulas for O_3 and explain the structure in terms of resonance.

12. Nitrogen dioxide, NO_2, is paramagnetic. At room temperature it readily dimerizes to form N_2O_4, dinitrogen tetroxide, which is diamagnetic. Write electron dot formulas that are consistent with the magnetic properties of the monomer and the dimer.

13. A number of noble gas compounds have been reported in the past decade. Use VSEPR theory to predict the general shape of xenon difluoride, XeF_2, and xenon tetrafluoride, XeF_4. Would either of these molecules have a dipole moment?

14. Use the double quartet approach to molecular bonding to describe the N_2 molecule. The structure must account for the facts that the dissociation energy of N_2 is consistent with a triple bond and that N_2 is diamagnetic.

15. Use the double quartet approach to molecular bonding to describe the O_2 molecule. The structure must account for the fact that the dissociation energy of O_2 is consistent with a double bond. It must also show two more electrons of one spin than of the other spin to account for the observed paramagnetism.

16. What is the difference between the concepts of electron sharing in the valence bond and molecular orbital approaches to molecular bonding?

17. Show that the H—C—H bond angle in the tetrahedral CH_4 molecule is 109.5°. It is easiest to consider the tetrahedral molecule inscribed in a cube. The carbon is in the center and the two pairs of hydrogen atoms are at opposite corners of the top and bottom faces.
(*Hint:* It is necessary to use the Pythagorean theorem.)

18. How do you explain the following?
(a) The dipole moment of NH_3 is larger than that of PH_3.
(b) The dipole moment of $SiHCl_3$ is less than that of $SiHF_3$.

19. Which sets of hybrid orbitals account for the following facts?
(a) The $AuCl_2^-$ ion is linear.
(b) The silane molecule, SiH_4, is not planar and does not have a dipole moment.
(c) The C—B—C angle in trimethylborane, $(CH_3)_3B$, is 120°.

20. The BF_3 and NH_3 molecules will react readily in the gas phase to form an addition compound, $F_3B{:}NH_3$.
(a) Describe this reaction in terms of electron dot formulas.
(b) What set of hybrid orbitals should be assigned to boron in the BF_3 molecule and in the $F_3B{:}NH_3$ addition compound?

21. Consider the general molecule A-B with valence electrons in the $n = 3$ level for both atoms. Consider only the $3s$ and $3p$ atomic orbital and use the z axis as the molecular axis.
(a) Which atomic orbitals of A and B will form σ molecular orbitals? Which will form π molecular orbitals?
(b) Draw sketches of all these molecular orbitals. There should be eight of them.

22. (a) Draw the energy level scheme for the molecular orbitals described for the A-B molecule in problem 21.
(b) Assume that atoms A and B have four and six valence electrons, respectively. Show the occupancy of molecular orbitals in the energy level scheme for (a).
(c) Does the bond in A-B correspond to a single, double, or triple bond?

23. On the basis of molecular orbital theory explain the following facts. The bond energy of N_2 is greater than that of N_2^+, whereas the bond energy of O_2 is less than that of O_2^+.

24. Predict whether the following molecules and ions would exist, and whether they would be paramagnetic or diamagnetic:

$$He_2^+ \quad Mg_2 \quad S_2 \quad Li_2^-$$

25. Using molecular orbital theory, explain the following facts. The bond lengths for O_2, O_2^+, and O_2^- are 1.21, 1.12, and 1.26 Å, respectively.
26. Assume that the relative order of energies for molecular orbitals in Figure 8.8 is also valid for molecules of the A-B type. Predict whether nitric oxide, NO, or the nitrosyl ion, NO^+, has the shorter bond length. Is either of these species paramagnetic?
27. Write the molecular orbital designation for CN and CN^-, making the same assumption as in problem 26. Which species should have the highest bond dissociation energy? Is either species paramagnetic?
28. Why is sp^2 hybridization incorrect for the ammonia molecule?
29. For the C_2 molecule, the molecular orbital designation may be written as $KK(\sigma_s)^2(\sigma_s^*)^2(\pi_{y,z})^4$. What does the KK represent? In the same notation, what would be the molecular orbital designation for Si_2?
30. The measured dipole moment of CO is 0.112 D. Which would be the negative end of the molecule? On what basis can this decision be made? Would you expect the dipole moment for the CS molecule to be greater or less than that of CO? Explain.

9

CHEMICAL REACTIONS AND EQUATIONS

Introduction

The various letters H, C, O, H_2O, and CO_2 are often used as abbreviations for the words hydrogen, carbon, oxygen, water, and carbon dioxide. These same letters and the others in a periodic table of the elements are also used as symbols to represent single atoms or molecules. For example, H and O are sometimes used as abbreviations for hydrogen and oxygen and are also sometimes used as symbols to represent single atoms of hydrogen and oxygen. In addition to these uses, the various letters are also used to represent Avogadro's number (1 mole) of atoms or molecules. Thus H and H_2 may represent one mole of hydrogen atoms and one mole of hydrogen molecules. In most cases the way in which we use these letters makes it obvious whether they are abbreviations, symbols that represent one atom or one molecule, or symbols that represent one mole of atoms or molecules.

This chapter is largely concerned with use of the various letters (and numbers) as a means of representing chemical reactions. In particular, we will be concerned with various quantitative relationships between the many substances involved in chemical reactions.

Equations for Chemical Reactions

Properly written chemical reaction equations are of considerable value because they summarize concisely much useful information. In the first place, these equations identify reactants and products. Furthermore, a balanced chemical equation permits us to carry out quantitative calculations relating amounts of various reactants and products.

As our first example we consider the reaction of hydrogen with oxygen to form water. Although this reaction ordinarily involves the combination of molecular hydrogen with molecular oxygen, we first consider the reaction of atomic hydrogen with atomic oxygen for which we write the equation

$$2\,H + O \rightarrow H_2O$$

This equation can mean that two atoms of hydrogen react with one atom of oxygen to form one molecule of water, or it can mean that two moles of hydrogen atoms react with one mole of oxygen atoms to form one mole of water molecules. The atomic weights of hydrogen and oxygen are used with this latter meaning to calculate that 2.0 g of atomic hydrogen react with 16.0 g of atomic oxygen to form 18.0 g of water.

Now we consider the more common reaction of molecular hydrogen with molecular oxygen to form water. We start by writing

$$H_2 + O_2 \rightarrow H_2O$$

This equation indicates the substances that react and the product of the reaction but does not properly indicate the numbers of molecules that react because it is not **balanced.** A properly balanced equation for a chemical reaction must represent the same number of atoms of each element on both sides of the equation. We therefore write a balanced version of the equation above as

$$2\,H_2 + O_2 \rightarrow 2\,H_2O$$

This equation can mean that two molecules of hydrogen (total of four atoms) react with one molecule of oxygen (two atoms) to form two molecules of water (total of four atoms of hydrogen and two atoms of oxygen). Since no atoms of any kind are created or destroyed in chemical reactions, it is necessary that chemical equations be balanced in this sense.

The above equation may also mean that twice Avogadro's number (2 moles) of hydrogen molecules react with Avogadro's number (1 mole) of oxygen molecules to form twice Avogadro's number (2 moles) of water molecules, which is the same as saying that 4.0 g of hydrogen react with 32.0 g of oxygen to form 36.0 g of water. This latter meaning is in agreement with our experience that mass is conserved in chemical reactions.

The reaction of solid sodium with gaseous chlorine to form solid sodium chloride is represented by

$$2\,Na(c) + Cl_2(g) \rightarrow 2\,NaCl(c)$$

where (c) and (g) indicate that the immediately preceding substance is in the crystalline or gaseous state. This equation is balanced and indicates that two moles (2 moles $\times$ 23 g mole^{-1} $= 46$ g) sodium react with one mole (2 moles of Cl $\times$ 35.5 g mole^{-1} $= 71.0$ g) of molecular chlorine to form two moles [$2(23 + 35.5) = 117$ g] of sodium chloride that is made up of two moles of sodium ions and 2 moles of chloride ions.

Equations for chemical reactions must be balanced to account for the conservation of atoms and mass in chemical reactions. Reaction equations must also be balanced so that the shared or transferred electrons are accounted for. The reaction of sodium with chlorine illustrates this point. We know that each atom of sodium easily loses one electron and thereby becomes a positively charged sodium ion represented by Na^+. Similarly, we know that each atom of chlorine may capture one electron and become a negatively charged chloride ion represented by Cl^-. The balanced equation must therefore show that any given number of sodium atoms reacts with the same number of chlorine atoms so that the electrons lost by sodium equal the electrons captured by chlorine. The Na^+ and Cl^- join together in equal numbers to form electrically neutral sodium chloride.

The reaction of aluminum with oxygen can be understood similarly. Each atom of aluminum loses three electrons to form a positively charged ion represented by Al^{3+}. Each oxygen atom gains two electrons to become a negatively charged ion represented by O^{2-}. The simplest formula for aluminum oxide must therefore be Al_2O_3 in order to be consistent with the observed electrical neutrality and to satisfy the requirement that the total number of electrons lost by aluminum must equal the total number of electrons gained by oxygen.

Balanced equations for the reaction of aluminum with molecular oxygen are

$$2\ Al(c) + \tfrac{3}{2}\ O_2(g) \rightarrow Al_2O_3(c)$$

and

$$4\ Al(c) + 3\ O_2(g) \rightarrow 2\ Al_2O_3(c)$$

The first of these equations indicates that two moles of aluminum react with one and one-half moles of oxygen to form one mole of aluminum oxide.

As already explained, one mole of any ionic substance is commonly defined in terms of the simplest formula that can be written for that substance. Thus one mole of aluminum oxide is taken to consist of two moles (twice Avogadro's number) of Al^{3+} ions and three moles of O^{2-} ions. The first equation above also indicates that $2 \text{ moles} \times 27 \text{ g mole}^{-1} = 54 \text{ g}$ of aluminum react with $1.5 \text{ moles} \times 32 \text{ g mole}^{-1} = 48 \text{ g}$ of oxygen to form 102 g of aluminum oxide. The second equation above indicates that four moles (108 g) of aluminum react with three moles (96 g) of oxygen to form two moles (204 g) of aluminum oxide.

In connection with writing formulas for chemical compounds and equations to represent reactions involving these compounds, it is helpful to be familiar with charges, compositions, and names of common ions. A considerable amount of this information is gathered in Table 9.1. Table 9.1 could be memorized in routine fashion, although that would be the hard way to retain all this information. It is simpler and also contributes to understanding to correlate charges and formulas of ions with the electronic structures and with the positions of the elements in the periodic table. For example, all the alkali metals form ions having a positive charge of one electron unit due to the ease with which the single *s* electron in the highest energy level is lost by neutral alkali metal atoms. Similarly, all the alkaline earth metals form positive ions having a positive charge of two electron units.

Now consider the formation of a nitrate ion from one nitrogen atom with five valence electrons and three oxygen atoms, each with six valence electrons. In

Table 9.1. Some Common Ions

Name	*Formula*	*Name*	*Formula*
Acetate	$C_2H_3O_2^-$	Hydroxide	OH^-
Aluminum	Al^{3+}	Hypochlorite	ClO^-
Amide	NH_2^-	Iodate	IO_3^-
Ammonium	NH_4^+	Iodide	I^-
Azide	N_3^-	Lead (plumbous)	Pb^{2+}
Barium	Ba^{2+}	Magnesium	Mg^{2+}
Bicarbonate	HCO_3^-	Manganous	Mn^{2+}
Bromate	BrO_3^-	Mercuric	Hg^{2+}
Bromide	Br^-	Mercurous	Hg_2^{2+}
Cadmium	Cd^{2+}	Nickel	Ni^{2+}
Calcium	Ca^{2+}	Nitrate	NO_3^-
Carbonate	CO_3^{2-}	Nitrite	NO_2^-
Chlorate	ClO_3^-	Oxide	O^{2-}
Chloride	Cl^-	Perchlorate	ClO_4^-
Chlorite	ClO_2^-	Permanganate	MnO_4^-
Chromate	CrO_4^{2-}	Peroxide	O_2^{2-}
Chromic	Cr^{3+}	Peroxydisulfate (persulfate)	$S_2O_8^{2-}$
Chromous	Cr^{2+}	Phosphate	PO_3^{4-}
Cobaltous	Co^{2+}	Potassium	K^+
Cupric	Cu^{2+}	Silver (argentous)	Ag^+
Cuprous	Cu^+	Sodium	Na^+
Cyanide	CN^-	Stannic (tin)	Sn^{4+}
Dichromate	$Cr_2O_7^{2-}$	Stannous (tin)	Sn^{2+}
Ferric (iron)	Fe^{3+}	Strontium	Sr^{2+}
Ferricyanide	$Fe(CN)_6^{3-}$	Sulfate	SO_4^{2-}
Ferrocyanide	$Fe(CN)_6^{4-}$	Sulfide	S^{2-}
Ferrous (iron)	Fe^{2+}	Sulfite	SO_3^{2-}
Fluoride	F^-	Tetrathionate	$S_4O_6^{2-}$
Hydride	H^-	Thiosulfate	$S_2O_3^{2-}$
Hydrogen chromate (bichromate)	$HCrO_4^-$	Triiodide	I_3^-
Hydrogen sulfate (bisulfate)	HSO_4^-	Zinc	Zn^{2+}
Hydrogen sulfide (bisulfide)	HS^-		
Hydrogen	H^+		

addition to these 23 electrons to be shown in our electron dot formula, there is one "extra" electron that makes the whole aggregate carry a net negative charge. An electron dot formula showing 24 electrons is

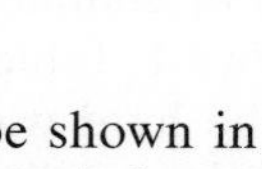

```
      :O:
       ::
     ..N..
   :O:   :O:
```

According to this formula, two of the oxygen atoms are bound to the nitrogen atom by single bonds, whereas the third is bound by a double bond. Because it is an experimental fact that the nitrogen-oxygen bond lengths in NO_3^- are all the same, we know that this simple picture is inadequate. Resonance theory was designed to deal with problems of this sort. We therefore write

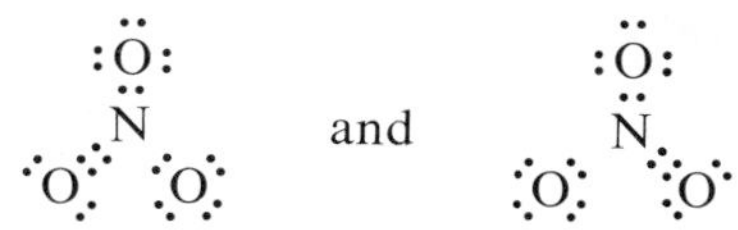

as two more contributing structures to the resonance hybrid structure of the nitrate ion. The true electronic structure of the nitrate ion is a mixture of the three contributing structures pictured above, but it is not easy to describe this hybrid structure in electron dot or atomic orbital language more accurately than to say that each nitrogen-oxygen bond is intermediate between a single and a double bond and that all three bonds are equivalent. Because NO_3^- is diamagnetic, we know that this equivalence of bonds is achieved without any unpaired electrons.

In molecular orbital theory the bonding in the nitrate ion is explained through three localized N—O σ bonds and one delocalized (covering all four atoms) π bond. Thus the concept of resonance may be avoided.

Electronic configurations represented by electron dot formulas can be figured out in much the same way for other polyatomic ions listed in Table 9.1. For some of these ions it is necessary to invoke resonance theory or one of the other bonding theories described in Chapter 8.

It is apparent from Table 9.1 that some common names of ions are derived systematically. For example, if there are two different ions of the same metal, such as Cu^+ and Cu^{2+}, the one with the lower charge is given a name ending in **ous** and the one with the higher charge is given a name ending in **ic.** We use the term **oxidation state** (or sometimes **oxidation number**) to indicate the charge of a monatomic ion. Thus cupr*ous* chloride, CuCl, is said to contain copper in the +1 oxidation state and chlorine in the −1 oxidation state. Cupr*ic* chloride, $CuCl_2$, contains copper in the +2 oxidation state and must have twice as many chloride as cupric ions to make the charges balance. Similarly, Fe^{2+} and Fe^{3+} represent iron in the +2 and +3 oxidation states. These ions and compounds derived from them are called ferr*ous* and ferr*ic*, respectively. The ferr*o*cyanide and ferr*i*cyanide complex ions contain iron in the +2 and +3 oxidation states, respectively, as discussed in more detail in Chapter 20.

Oxidation States

Positive oxidation states or numbers indicate the numbers of electrons that are lost by neutral atoms in forming positively charged ions. Thus the oxidation number also represents the number of electrons that must be added to a positive ion to convert it into a neutral atom. Negative oxidation state numbers indicate how many electrons have been gained by each neutral atom in becoming a negatively charged monatomic ion, or indicate how many electrons each such ion would lose on becoming a neutral atom.

The situation is less simple for substances containing covalent bonds. Although assignment of oxidation states and numbers to atoms in such substances is partly arbitrary, these numbers are still useful in connection with balancing equations, understanding some electrochemical processes, and selecting chemical reagents to effect certain desired reactions. We therefore present a few examples of assignment of oxidation numbers to elements in such compounds.

First consider sodium sulfate, Na_2SO_4. Sodium is present in this compound, as in other sodium containing compounds, in the +1 oxidation state. Now we must consider the more complicated assignment of oxidation numbers to the sulfur and oxygen in SO_4^{2-}. We begin by recalling that oxygen has a considerable tendency to gain two electrons per atom to attain the noble gas electronic configuration of neon. There are many ionic oxides, such as MgO, in which O^{2-} ions are actually present. Oxygen atoms can also attain the stable electronic configuration of neon by sharing electrons with other atoms. In such cases it is useful and customary to say that oxygen is in the −2 oxidation state just as it is in MgO and other simple ionic compounds. We therefore say that sulfur in SO_4^{2-} is in the +6 oxidation state, which accounts for the net charge of −2 electron units on the sulfate ion.

Similarly, the oxidation states of nitrogen in nitrite (NO_2^-) and nitrate (NO_3^-) ions are +3 and +5 respectively. Oxidation states of chromium in dichromate ($Cr_2O_7^{2-}$) and chromate (CrO_4^{2-}) are both +6, whereas the oxidation state of manganese in permanganate (MnO_4^-) is +7.

Our earlier decision to assign an oxidation state of −2 to the oxygen atoms in SO_4^{2-} means that we are, in effect, assigning all of the shared electrons to the oxygen atoms and none to the central sulfur atom. Although this procedure is unrealistic and arbitrary, oxidation numbers or states deduced in this way can be internally consistent and are certainly useful, which provides our justification.

We may summarize the preceding discussion by means of the following conventional rules for assigning oxidation numbers or states:

1. The oxidation number of a monatomic ion is equal to its electric charge expressed in electron units.
2. The oxidation number of each atom in a covalent substance is equal to the charge on each atom when all shared electrons are assigned to the more electronegative of the atoms.
3. The oxidation state of atoms in any elementary substance is zero.

Application of these rules to many substances is straightforward and requires no further comment. On the other hand, there are certain compounds and ions (such as H_2O_2 and $S_2O_3^{2-}$) that are more complicated and will be discussed later in connection with their chemical properties.

Mass Relationships in Chemical Reactions

In order to calculate the amounts of reactants consumed or the amounts of products formed in chemical reactions, it is necessary to have a balanced chemical equation to tell us the relative numbers of atoms, molecules, or moles of various substances involved in the reaction. It is almost always advisable to write down the balanced chemical reaction equation as the first step in working problems concerned with quantities of chemicals in a chemical reaction.

Example Problem 9.1. Water can be decomposed into hydrogen and oxygen by an electric current. How much water must be decomposed to yield 34.5 g of hydrogen?

We begin by writing the balanced equation

$$2\ H_2O(liq) \rightarrow 2\ H_2(g) + O_2(g)$$

from which we see that 2 moles of water yield 2 moles of hydrogen. Next we calculate that 34.5 g of hydrogen are $(34.5\ g)/(2.0\ g\ mole^{-1}) = 17.25$ moles of hydrogen. These 17.25 moles of hydrogen must come from an equal number of moles of water, which is $18.0\ g\ mole^{-1} \times 17.25$ moles $= 310$ g of water. ■

Example Problem 9.2. Benzene, a common organic compound of formula C_6H_6, burns readily in oxygen to form carbon dioxide and water. How much carbon dioxide would be produced by combustion of 35.1 g of benzene?

The balanced equation for the reaction is

$$C_6H_6(liq) + 7.5\ O_2(g) \rightarrow 6\ CO_2(g) + 3\ H_2O(g)$$

The molecular weights of benzene and carbon dioxide are $(6 \times 12.0) + (6 \times 1.0) = 78.0\ g\ mole^{-1}$ and $(12.0) + (2 \times 16.0) = 44.0\ g\ mole^{-1}$.

We first calculate that $(35.1\ g)/(78.0\ g\ mole^{-1}) = 0.45$ mole of benzene is involved in the reaction. The balanced equation shows that 6 moles of CO_2 are formed from every mole of benzene. We therefore know that $6 \times 0.45 = 2.70$ moles of carbon dioxide are formed by combustion of 35.1 g of benzene. The mass of carbon dioxide formed is 2.70 moles $\times\ 44.0\ g\ mole^{-1} = 119$ g. ■

Two examples of mass relationships in chemical reactions have been worked by making use of the balanced reaction equation and atomic weights. The same principles are applicable to the reverse problem; that is, determination of reaction equations and formulas of chemical substances from masses of reactants and products and atomic weights. It is by way of calculations like these that chemists have established formulas of most compounds. Although we now have other means for acquiring such information about new compounds, "old-fashioned" methods based on mass relationships in chemical reactions are still widely used. As a simple example we consider the reaction of zinc with oxygen to form zinc oxide.

Example Problem 9.3. The complete reaction of 2.13 g of zinc with oxygen results in formation of 2.65 g of zinc oxide. Figure out the formula of zinc oxide and write a balanced equation for the reaction of zinc with oxygen.

We have $(2.13\ g)/(65.4\ g\ mole^{-1}) = 0.0326$ mole of zinc involved in the reaction. The zinc oxide formed contains $2.65\ g - 2.13\ g = 0.52$ g of oxygen, which is $(0.52\ g)/(16.0\ g\ mole^{-1}) = 0.0325$ mole of oxygen atoms or oxide ions. Within the accuracy of the experimental data, zinc oxide contains equal numbers of zinc and oxide ions. We therefore write ZnO as the simplest formula of zinc oxide. Because each oxide ion has a net negative charge of two electron units (-2 oxidation state), each zinc ion in zinc oxide must have a net positive charge of two electron units ($+2$ oxidation state).

Balanced equations for the reaction of zinc with oxygen are

$$Zn(c) + \tfrac{1}{2}\ O_2(g) \rightarrow ZnO(c)$$

and

$$2\ Zn(c) + O_2(g) \rightarrow 2\ ZnO(c)$$ ■

Example Problem 9.4. Another problem involving mass relationships commonly arises from a desire to know how much of a given element is contained in a certain amount of some compound. For example, suppose that we want to know how much silver nitrate we must weigh out in order to have 10.0 g of silver available for electroplating purposes.

We first need to know the formula of silver nitrate. From Table 9.1 we see that silver commonly exists in its compounds as Ag^+ ions. Similarly, we see that the formula for nitrate ion is NO_3^-. The formula of silver nitrate is therefore $AgNO_3$ and we calculate its molecular weight to be $107.9 + 14.0 + (3 \times 16.0) = 169.9$ g mole^{-1}.

The 10.0 g of elemental silver that we want amounts to (10.0 g)/(107.9 g mole^{-1}) = 0.0927 mole of silver. Since each mole of $AgNO_3$ contains 1 mole of Ag, we know that we need 0.0927 mole of $AgNO_3$, which amounts to 0.0927 mole $\times$ 169.9 g mole^{-1} = 15.7 g of $AgNO_3$.

Note that similar calculations involving silver sulfate, Ag_2SO_4, must take into account that each mole of silver sulfate contains 2 moles of silver. ■

Example Problem 9.5. An oxide of rhenium has been found to contain 76.9% rhenium and therefore 23.1% oxygen. What is the simplest formula for this compound?

Since we are concerned with *relative* amounts of rhenium and oxygen in this compound, we may do our calculations for any amount of compound that we choose. It is convenient to choose 100 g of compound, which contains 76.9 g of rhenium and 23.1 g of oxygen. This amount of rhenium oxide can also be described as containing (76.9 g)/(186.2 g mole^{-1}) = 0.413 mole of rhenium and (23.1 g)/(16.0 g mole^{-1}) = 1.44 mole of oxide. A formula for this oxide is therefore $Re_{0.413}O_{1.44}$. To express this formula more conveniently, we first calculate $1.44/0.413 \cong 3.5$ and write $Re_1O_{3.5}$. Now it is easy to see that the simplest formula with integers is Re_2O_7. ■

Example Problem 9.6. Some mothballs have been analyzed and found to contain 49.0 mass % carbon, 2.7 mass % hydrogen, and 48.3 mass % chlorine. Measurements of colligative properties (freezing point depression, vapor pressure lowering, boiling point elevation) have shown that the molecular weight is 147 g mole^{-1}. What is the molecular formula of the substance?

In 1 mole of mothball material we have $0.49 \times 147 = 72$ g of carbon, which is also (72 g)/(12 g mole^{-1}) = 6 moles of carbon. We similarly find that 1 mole of this substance contains 4 g of hydrogen and 71 g of chlorine, amounting to 4 moles of hydrogen atoms and 2 moles of chlorine atoms. Thus the formula of the compound is $C_6H_4Cl_2$. ■

Example Problem 9.7. A 2.406 g sample of mixed NaCl and KCl was dissolved in water and treated with excess $AgNO_3$ to precipitate all of the chloride as AgCl, which was collected and found to weigh 5.650 g. What was the mass per cent of NaCl in the original mixture?

We begin by calculating the molecular weights of AgCl, NaCl, and KCl to be 143.32 g mole^{-1}, 58.44 g mole^{-1}, and 74.55 g mole^{-1}, respectively. The total amount of chloride in the sample is found to be (5.650 g)/(143.32 g mole^{-1}) = 0.03942 mole.

We must now combine our knowns and unknowns in suitable algebraic fashion. As our primary unknowns we select the number of moles of NaCl and of KCl in the original mixture, and represent these unknowns by the letters Y and Z, respectively. Our known quantities are various molecular weights, the total number of moles of chloride, and the mass of the original sample that yielded the known number of moles of chloride. We must therefore express what we know in two equations that we can use to solve for our two primary unknowns. Since each mole of both NaCl and KCl yields 1 mole of chloride, we have

$$Y + Z = 0.03942$$

as one of our equations. We also have the mass of NaCl given by (58.44 Y) and the mass of KCl given by (74.55 Z), which we combine to obtain the second equation:

$$58.44\ Y + 74.55\ Z = 2.406$$

One way to handle these equations is to substitute $Z = 0.03942 - Y$ from the first equation into the second equation to obtain

$$58.44\ Y + 74.55(0.03942 - Y) = 2.406$$

and then solve for $Y = 0.03309$ mole of NaCl.

Now we calculate that the original sample contained 0.03309 mole $\times$ 58.44 g mole^{-1} = 1.934 g of NaCl. The sample was therefore $(1.934 \times 100)/2.406 = 80.3$ mass % NaCl.

Interested readers might investigate the effect of a small error in the mass of AgCl on the final calculated composition of sample. ■

Volume Relationships in Chemical Reactions

Volume relationships in reactions involving gases are worked out in the same general way as are mass relationships. We always try to start with a balanced equation for the reaction and then make use of atomic weights, molecular weights, Avogadro's law, and the ideal gas equation as illustrated by the following Example Problems.

Example Problem 9.8. Suppose that 8.26 g of magnesium is placed in a 10.0 liter vessel containing oxygen at 1.12 atm and 27°C. The reaction between magnesium and oxygen is started by heating the bottom of the vessel. After the reaction is complete, the vessel is again maintained at 27°C. What is the final pressure of oxygen in the vessel?

Balanced equations for the reaction are

$$Mg(c) + \tfrac{1}{2}\,O_2(g) \rightarrow MgO(c)$$

and

$$2\ Mg(c) + O_2(g) \rightarrow 2\ MgO(c)$$

Since the amount of oxygen in the vessel at the end of the experiment determines the final pressure, we want to calculate the amount of oxygen initially present. Then we can subtract the amount that is combined with magnesium in the reaction.

From the density of magnesium, we calculate that the volume occupied by the metal was $m/d = (8.26\ \text{g})/(1.74\ \text{g cm}^{-3}) = 4.75\ \text{cm}^3$, which is negligible in comparison with 10.0 liters. We are therefore justified in taking the initial volume of oxygen to be 10.0 liters. A similar calculation for magnesium oxide similarly justifies taking the final volume of oxygen to be 10.0 liters.

Now we use the ideal gas equation $PV = nRT$ in calculating that the number of moles of oxygen originally present was

$$n = \frac{1.12\ \text{atm} \times 10.0\ \text{liters}}{(0.0820\ \text{liter atm deg}^{-1}\ \text{mole}^{-1})(300\ \text{deg})}$$

$$= 0.455\ \text{mole}\ O_2\ \text{originally present}$$

Next we calculate that we had (8.26 g)/(24.3 g $mole^{-1}$) = 0.340 mole of magnesium to start the reaction. Both of the balanced equations show that twice as many moles of magnesium react as do moles of oxygen. We therefore know that 0.340/2 = 0.170 mole of oxygen was converted to oxide in the reaction, leaving 0.455 − 0.170 = 0.285 mole of $O_2(g)$ in the flask.

We now use the ideal gas equation again to calculate the final pressure.

$$P = \frac{(0.285\ \text{mole})(0.0820\ \text{liter atm deg}^{-1}\ \text{mole}^{-1})(300\ \text{deg})}{10.0\ \text{liters}}$$

$$= 0.701\ \text{atm}$$

The information given is also sufficient to permit us to calculate that the experiment yielded 13.7 g of MgO(c). ■

As background information for another Example Problem we consider the reaction of metallic zinc with a solution of hydrogen chloride in water (called hydrochloric acid). The products of this reaction are zinc chloride dissolved in water and gaseous hydrogen. We know from Table 9.1 or from Example Problem 9.3 that the usual oxidation state of zinc in its compounds is +2 and that the formula of zinc chloride is therefore $ZnCl_2$. A balanced equation for the reaction is

$$\text{Zn(c)} + 2\ \text{HCl(aq)} \rightarrow \text{ZnCl}_2\text{(aq)} + \text{H}_2\text{(g)}$$

where (aq) is an abbreviation for aqueous, indicating that the preceding substance is dissolved in water.

It is sometimes useful to write the equation above differently. As shown in Chapter 4, many substances (such as hydrochloric acid and zinc chloride) are present in aqueous solution as ions. Hence it is more realistic to write the reaction equation as

$$\text{Zn(c)} + 2\ \text{H}^+\text{(aq)} + 2\ \text{Cl}^-\text{(aq)} \rightarrow \text{Zn}^{2+}\text{(aq)} + 2\ \text{Cl}^-\text{(aq)} + \text{H}_2\text{(g)}$$

Now this equation can be simplified by crossing out the chloride ions that are unchanged in the reaction and appear on both sides of the equation. We thus obtain

$$\text{Zn(c)} + 2\ \text{H}^+\text{(aq)} \rightarrow \text{Zn}^{2+}\text{(aq)} + \text{H}_2\text{(g)}$$

This last equation is called the net ionic equation for the reaction.

In the following Example Problem it is also necessary to know that copper does not react with dilute or even with moderately concentrated hydrochloric acid at room temperature.

Example Problem 9.9. Suppose that 8.32 g of a mixture of metallic copper and zinc is added to an excess of hydrochloric acid and that the hydrogen evolved is collected over water at 25 °C in an apparatus like that pictured in Figure 9.1. The volume of gas is found to be 223 ml when the atmospheric pressure is 74.2 cm Hg. How much metallic zinc was in the original mixture?

Our first task is to calculate how many moles of hydrogen were liberated by the reaction. For this calculation we need to know the partial pressure of hydrogen in the measuring cylinder. Dalton's law of partial pressures tells us that the total measured pressure is equal

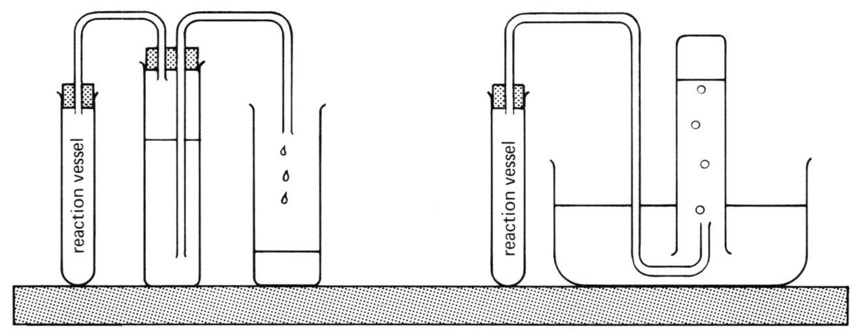

Figure 9.1. Schematic illustrations of two simple arrangements for collecting and measuring the amount of gas evolved by a chemical reaction. For gases that are soluble in water, mercury or a nonvolatile oil may be used as the liquid to be displaced.

to the sum of the partial pressures of all gases present in the cylinder. Since the only gases present are hydrogen and water vapor, we write

$$74.2 \text{ cm Hg} = P_{H_2} + P_{H_2O}$$

which is rearranged to

$$P_{H_2} = 74.2 \text{ cm Hg} - P_{H_2O}$$

The vapor pressure of water at 25°C is given in Appendix VI as 23.8 mm Hg, which we convert to 2.4 cm Hg and use in the above equation to calculate that the partial pressure of hydrogen in the cylinder is 71.8 cm Hg or (71.8 cm Hg)/(76.0 cm Hg atm^{-1}) = 0.945 atm. We now solve $PV = nRT$ for n

$$n = \frac{(0.945 \text{ atm})(0.223 \text{ liter})}{(0.0820 \text{ liter atm deg}^{-1} \text{ mole}^{-1})(298 \text{ deg})}$$
$$= 0.00863 \text{ mole of } H_2$$

We see from any of the balanced equations for reaction of zinc with hydrochloric acid that each mole of zinc results in the formation of 1 mole of hydrogen. There must, therefore, have been 0.00863 mole × 65.4 g mole^{-1} = 0.564 g of zinc in the original mixture. It might also be reported that the mixture contained (0.564 × 100)/8.32 = 6.78 mass % zinc. ■

Example Problem 9.10. Mercuric oxide, HgO, decomposes on heating to yield mercury and oxygen. How many grams of mercuric oxide are required to yield a gas volume of 232 cm^3 measured over water at 751 mm Hg and 298°K?

We begin by writing the balanced reaction equation as

$$HgO(c) \rightarrow Hg(liq) + \tfrac{1}{2} O_2(g)$$

or

$$2\,HgO(c) \rightarrow 2\,Hg(liq) + O_2(g)$$

From Dalton's law of partial pressures we know that the total pressure measured is equal to the sum of the partial pressures of oxygen and water vapor. The partial pressure of oxygen is therefore obtained as

$$P_{O_2} = 751 \text{ mm Hg} - 24 \text{ mm Hg} = 727 \text{ mm Hg}$$

Now we use this pressure in the ideal gas equation to obtain the number of moles of oxygen as

$$n = \frac{(727 \text{ mm Hg})(0.232 \text{ liter})}{(760 \text{ mm Hg atm}^{-1})(0.082 \text{ liter atm deg}^{-1} \text{ mole}^{-1})(298 \text{ deg})}$$
$$= 0.00907 \text{ mole } O_2$$

Both balanced equations show that we must have twice as many moles of HgO as of $O_2(g)$. We therefore require

$$(2 \times 0.00907 \text{ mole})(216.6 \text{ g mole}^{-1}) = 3.93 \text{ g HgO} \quad \blacksquare$$

Example Problem 9.11. A sample of aluminum-magnesium alloy that weighed 0.1030 g was allowed to react with an excess of hydrochloric acid to form aqueous magnesium chloride, aqueous aluminum chloride, and gaseous hydrogen. The hydrogen was collected at 294°K and 0.920 atm over mercury, which has a very low vapor pressure at this temperature, and was found to occupy 120 cm³. What was the composition of the alloy?

We use the ideal gas equation to calculate the number of moles of hydrogen formed as

$$n = \frac{(0.920 \text{ atm})(0.120 \text{ liter})}{(0.082 \text{ liter atm deg}^{-1} \text{ mole}^{-1})(294 \text{ deg})}$$
$$= 0.004576 \text{ mole of } H_2$$

In order to relate this amount of hydrogen to reaction of certain amounts of magnesium and aluminum, we must make use of the following balanced reaction equations:

$$\text{Al(in alloy)} + 3\,H^+(aq) \rightarrow Al^{3+}(aq) + 1.5\,H_2(g)$$
$$\text{Mg(in alloy)} + 2\,H^+(aq) \rightarrow Mg^{2+}(aq) + H_2(g)$$

We now choose letters A and M to represent the unknown numbers of moles of aluminum and magnesium in the original sample. An equation that expresses the mass of sample in terms of atomic weights and unknowns A and M is

$$26.98\,A + 24.31\,M = 0.1030$$

The other equation that we need is obtained from the balanced equations, which tell us that we obtain 1.5 moles of H_2 from each mole of aluminum and 1 mole of H_2 from each mole of magnesium. The equation is

$$1.5A + M = 0.004576$$

We now have two equations and two unknowns. One way to obtain numerical values for A and M is to substitute $M = 0.004576 - 1.5A$ from the second equation into the first equation and then solve for A. Another way is to multiply the second equation through by 24.31 and then subtract the result from the first equation, which yields a single equation with A as the only unknown. Both of these methods lead to $A = 0.00088$ mole of aluminum.

Now we find the mass of aluminum in the original sample to be 0.00088 mole × 26.98 g mole^{-1} and express the composition of the sample as $(0.00088 \times 26.98 \times 100)/0.1030 =$ 23 mass % aluminum.

Interested readers might investigate the effect of a small error in the volume of hydrogen on the final calculated composition and then draw some conclusion about this method of simultaneous analysis for two elements. ■

Another common kind of problem involving volume relationships in chemical reactions is concerned with volumes of solutions. Measurements and calculations of this sort form an important part of quantitative analysis and, indirectly, of many other branches of chemistry.

Suppose that we want to determine the concentration of sodium hydroxide in some solution of this substance. One way to do so is to "titrate" this unknown solution with an acid solution of known concentration. In such a titration experiment one adds acid solution to a known volume of the unknown hydroxide solution until some observable change shows that the hydroxide ions have been completely neutralized. It is then possible to use the measured volumes of solutions with the known concentration of the acid solution to calculate the concentration of the hydroxide in the unknown solution, as shown in the following Example Problem.

Example Problem 9.12. We have a standard solution of hydrochloric acid that is known to be 0.209 M (remember, molarity is represented by $M =$ moles of solute per liter of solution) and have used this for analysis of a solution of sodium hydroxide. In this analysis we have titrated hydrochloric acid solution into a 25.0 ml portion of sodium hydroxide solution to which an indicator had been added. This indicator is a chemical that undergoes a color change at the "end point" of the reaction when exactly enough acid has been added to neutralize the hydroxide. In the analysis under consideration it was observed that 31.4 ml of standard acid solution were required to reach the end point. Calculate the concentration of the unknown sodium hydroxide solution.

We begin by writing a net ionic equation to represent the reaction of acid [H^+(aq)] with base [OH^-(aq)]:

$$H^+(aq) + OH^-(aq) \rightarrow H_2O(liq)$$

Thus we see that at the end point of the reaction we have added exactly the same number of moles of H^+(aq) as there were moles of OH^-(aq) present in the original unknown solution. Recalling the definition of molarity, we write

$$M = 0.209 \text{ moles } H^+(aq)/\text{liter} = n \text{ moles } H^+(aq)/0.0314 \text{ liter}$$

in which we have expressed the volume of hydrochloric acid solution in terms of liters. Now we find that there were

$$n = 0.209 \times 0.0314 \text{ moles } H^+(aq) = 0.00656 \text{ moles } H^+(aq)$$

in the hydrochloric acid solution used in the titration. From the balanced equation above we know that the sample of unknown solution must have contained 0.00656 moles of OH^-(aq) in its original volume of 25.0 ml (0.025 liter). Now we use the defining equation for molarity to calculate

$$M = 0.00656 \text{ mole}/0.025 \text{ liter} = 0.262 \text{ mole liter}^{-1}$$

as the concentration of hydroxide in the original unknown solution. ■

Example Problem 9.13. A balanced net ionic equation for reaction of permanganate ions with ferrous ions in acidic solution is

$$MnO_4^-(aq) + 5\ Fe^{2+}(aq) + 8\ H^+(aq) \rightarrow Mn^{2+}(aq) + 5\ Fe^{3+}(aq) + 4\ H_2O(liq)$$

This reaction has been used in standardization of a potassium permanganate solution as follows.

A 1.202-g sample of very pure "Mohr's salt," formula $FeSO_4 \cdot (NH_4)_2SO_4 \cdot 6H_2O(c)$, was dissolved in about 50 ml of dilute sulfuric acid and then titrated with potassium permanganate solution until persistence of a faint pink color due to $MnO_4^-(aq)$ indicated that reaction according to the equation above was complete. This titration required 37.6 ml of permanganate solution.

The first step in calculating the desired concentration of the permanganate solution is to find out how many moles of iron were contained in the sample of Mohr's salt. We therefore add up the appropriate atomic weights to obtain 392.14 g mole^{-1} as the molecular weight of $FeSO_4 \cdot (NH_4)_2SO_4 \cdot 6H_2O(c)$ and thence calculate that our sample consisted of $(1.202 \text{ g})/(392.14 \text{ g mole}^{-1}) = 3.065 \times 10^{-3}$ mole. Because the balanced equation shows that five $Fe^{2+}(aq)$ react with each $MnO_4^-(aq)$, we know that the complete reaction required $(3.065 \times 10^{-3})/5 = 0.613 \times 10^{-3}$ mole of $MnO_4^-(aq)$. Now we use the defining equation for molarity to calculate

$$M = (0.613 \times 10^{-3} \text{ mole})/(37.6 \times 10^{-3} \text{ liter}) = 0.0163 \text{ mole liter}^{-1}$$

as the concentration of the potassium permanganate solution. ■

Balancing Reaction Equations

Many simple reaction equations are easily balanced by inspection or trial and error methods. On the other hand, trial and error methods can be ridiculously time consuming when applied to some kinds of reaction equations. It is therefore worthwhile to devise systematic methods for balancing chemical reaction equations. The final result of any satisfactory method must be a balanced equation that has the same number of atoms or the same number of moles of atoms of every element involved in the reaction on both sides of the reaction equation. Further, if we are dealing with a net ionic equation that represents a reaction involving ions in solution, the final equation must have the same *net* charge on both sides.

Most complicated equations represent what are called **oxidation-reduction** reactions (sometimes abbreviated as **redox** reactions) in which one or usually more elements in the reaction undergo a change in oxidation state. Our first concern is with being able to recognize oxidation-reduction reactions, as illustrated by the following examples. The reaction of metallic zinc (0 oxidation state) with aqueous cupric ions (+2 oxidation state) to yield metallic copper (0 oxidation state) and aqueous zinc ions (+2 oxidation state) is a typical oxidation-reduction reaction in which two elements undergo changes in oxidation state. The thermal decomposition of potassium chlorate ($KClO_3$ with Cl in +5 oxidation state) to potassium chloride (KCl with Cl in −1 oxidation state) and potassium perchlorate ($KClO_4$ with Cl in +7 oxidation state) is another oxidation-reduction reaction. On the other hand, the reaction of silver nitrate with sodium chloride to yield silver chloride and sodium nitrate is not an oxidation-reduction reaction because none of the elements involved undergoes a change in oxidation state.

The basis for all systematic methods of balancing equations that represent oxidation-reduction reactions is the conservation of electrons. All of the electrons lost by atoms, ions, or molecules must be gained by other atoms, ions, or molecules. This conservation of electrons is illustrated below by the reaction of sodium with chlorine to form solid sodium chloride. First, we note that this is an oxidation-reduction reaction because the oxidation state of sodium is increased from 0 in the elemental state to $+1$ in NaCl and the oxidation state of chlorine is reduced from 0 in the elemental state to -1 in NaCl.

We begin by splitting the reaction of sodium with chlorine into parts. First, we write an equation for the ionization of a sodium atom (or a mole of atoms) to form a sodium ion and an electron as

$$Na \rightarrow Na^+ + e^-$$

Next we write a balanced equation for addition of electrons to molecular chlorine to form chloride ions as

$$2\ e^- + Cl_2 \rightarrow 2\ Cl^-$$

The total reaction equation is given by the sum of the parts, and we must add the parts in such a way that the number of electrons lost by sodium atoms will just equal the number of electrons gained by chlorine. This addition can be done if we first multiply the sodium ionization equation through by two to obtain

$$2\ Na \rightarrow 2\ Na^+ + 2\ e^-$$

Now we add this equation to the equation above for electron capture by chlorine to obtain

$$2\ Na + Cl_2 \rightarrow 2\ Na^+ + 2\ Cl^- = 2\ NaCl$$

in which $2\ e^-$ on each side have been cancelled because exactly the same number of electrons has been gained by chlorine as has been lost by sodium.

We now consider the reaction of aluminum with oxygen to form aluminum oxide. Balanced equations for electron loss by aluminum and electron capture by oxygen are

$$Al \rightarrow Al^{3+} + 3\ e^-$$

and

$$O_2 + 4\ e^- \rightarrow 2\ O^{2-}$$

Now we multiply the first equation through by four ($12\ e^-$ on the right side) and the second equation through by three ($12\ e^-$ on the left side) and add to obtain the desired balanced equation:

$$4\ Al + 3\ O_2 \rightarrow 4\ Al^{3+} + 6\ O^{2-} = 2\ Al_2O_3$$

The reaction above is typical of many reactions involving oxygen in that the oxygen has undergone a change in oxidation state from 0 to -2, corresponding to a capture of electrons by oxygen. It is useful to say that oxygen has acted as an oxidizing agent and has been reduced in the reaction from the 0 to the -2 oxidation state. The aluminum has been oxidized from the 0 oxidation state to the $+3$ oxidation state and has acted as a reducing agent for oxygen. As we shall see in subsequent examples, chemical substances other than oxygen can act as "electron capturers" and are thus appropriately called oxidizing agents.

We now turn to further illustration of the principles of balancing reaction equations and classifying various chemicals as oxidizing or reducing agents. Our first example is the reaction of ferric ions (Fe^{3+}) with stannous ions (Sn^{2+}) in acidic solution to yield ferrous ions (Fe^{2+}) and stannic ions (Sn^{4+}). We begin by writing what we call the **half-reaction** equations, which are

$$Sn^{2+}(aq) \rightarrow Sn^{4+}(aq) + 2\,e^-$$

and

$$Fe^{3+}(aq) + e^- \rightarrow Fe^{2+}(aq)$$

These half-reaction equations are properly balanced with respect to atoms and net charge on each side. Now we multiply the second of these equations through by two and add the result to the first equation to obtain

$$Sn^{2+}(aq) + 2\,Fe^{3+}(aq) \rightarrow Sn^{4+}(aq) + 2\,Fe^{2+}(aq)$$

which is balanced with regard to both number of atoms of each element and net charge on each side of the equation.

The equations above show that tin is oxidized from the +2 oxidation state to the +4 oxidation state, and that iron is reduced from the +3 oxidation state to the +2 oxidation state. Because the ferric ions cause the stannous ions to be oxidized, we call ferric ions the **oxidizing agent** in this reaction. Similarly, we call the stannous ions the **reducing agent** in this reaction. The oxidizing agent is always reduced, and the reducing agent is always oxidized; that is, the oxidizing agent gains electrons from the reducing agent.

Our balanced equation above for reaction of ferric ions with stannous ions is incomplete because it does not show any negatively charged ions, even though we know that there must be enough present to make the net charge in the solution zero. It is entirely proper and often advantageous to leave out the ions (in this case, the negative ions) that are not changed in the reaction. As pointed out before, useful equations of this type are called net ionic equations.

Equations for many reactions that take place in acidic solution cannot be balanced as simply as the preceding equations because hydrogen ions are consumed or produced by the reaction. For example, ferrous ions are oxidized to ferric ions in acidic solution by dissolved oxygen. The oxygen (0 oxidation state) is reduced to oxygen in water (−2 oxidation state).

We begin by writing the half-reaction equations that show gain of electrons by oxygen and loss of electrons by ferrous ions. The iron half-reaction equation is

$$Fe^{2+}(aq) \rightarrow Fe^{3+}(aq) + e^-$$

which is seen to be the reverse of the iron half-reaction equation we wrote earlier. In this reaction ferrous ions are oxidized to ferric ions, whereas in the iron-tin reaction considered earlier ferric ions were reduced to ferrous ions.

Next we must figure out the oxygen-water half-reaction equation. We can write

$$O_2(aq) + ?\,e^- \rightarrow H_2O(liq)$$

on the basis of knowing that oxygen is reduced to water in this reaction, and is therefore the oxidizing agent. (Remember that oxidation corresponds to loss of electrons and reduction corresponds to gain of electrons.) The number of atoms of oxygen on each side can be made the same by changing the right side to read $2\,H_2O(liq)$. Now we must add four hydrogen atoms to the left side, but we cannot do this by writing 2 H_2 or 4 H because neither atoms nor molecules of hydrogen are involved in this reaction. Instead, we must add 4 H^+(aq) to the left side. This choice is reasonable because we have specified that the reaction is carried out in acidic solution where plenty of hydrogen ions are available as reactant. We put all this information together and now write

$$O_2(aq) + 4\ H^+(aq) + ?\ e^- \rightarrow 2\ H_2O(liq)$$

where the ? shows that we must still indicate the number of electrons gained by each molecule of oxygen in being reduced to two molecules of water. To make the half-reaction equation balance electrically, the coefficient in front of the e^- must be 4. Or by figuring that each oxygen atom is reduced from the 0 oxidation state to the -2 oxidation state, we also see that $2 \times 2 = 4$ electrons are required per molecule of oxygen. The completed half-reaction equation is therefore

$$O_2(aq) + 4\ H^+(aq) + 4\ e^- \rightarrow 2\ H_2O(liq)$$

Because the iron half-reaction equation involves only one electron, we must multiply it through by four to obtain a half-reaction equation that can be added to the oxygen-water half-reaction equation to cancel out all the electrons as required when electrons gained and lost are equal. We then obtain

$$\begin{array}{r}
4\ Fe^{2+}(aq) \rightarrow 4\ Fe^{3+}(aq) + 4\ e^- \\
O_2(aq) + 4\ H^+(aq) + 4\ e^- \rightarrow 2\ H_2O(liq) \\
\hline
4\ Fe^{2+}(aq) + O_2(aq) + 4\ H^+(aq) \rightarrow 4\ Fe^{3+}(aq) + 2\ H_2O(liq)
\end{array}$$

This equation is properly balanced with regard to atoms on each side and with regard to net charge on each side.

Now let us consider a balanced equation to represent the oxidation of iron from the $+2$ oxidation state to the $+3$ oxidation state by oxygen in basic or alkaline solution. Because the solution is basic, we will have solid ferrous hydroxide represented by $Fe(OH)_2$(c) in contact with solution rather than Fe^{2+}(aq) ions as a principal reactant. Similarly, we will end up with slightly soluble ferric hydroxide, here represented by $Fe(OH)_3$(c), as a product of the reaction. The oxygen that oxidized iron from the $+2$ to the $+3$ oxidation state is itself reduced from the 0 to the -2 oxidation state in the form of water or hydroxide ions.

We begin by writing as much as we can of the iron half-reaction equation, which is

$$Fe(OH)_2(c) \rightarrow Fe(OH)_3(c) + e^-$$

Here we know that the single electron must be on the right because iron is oxidized (loses an electron) from the $+2$ to the $+3$ oxidation state. This incomplete half-reaction equation must now be balanced by adding water, hydroxide ions, or both, but *not* hydrogen ions. Adding one OH^-(aq) to the left side is sufficient in this case, so we obtain

$$Fe(OH)_2(c) + OH^-(aq) \rightarrow Fe(OH)_3(c) + e^-$$

Now we must write the oxygen half-reaction equation for basic solution. This equation is necessarily different from the oxygen half-reaction equation for acidic solutions because for basic solutions we must write equations in terms of hydroxide ions rather than hydrogen ions. As a beginning we can write

$$O_2(aq) + 4\,e^- \rightarrow 4\,OH^-(aq)$$

We know that the electrons must be on the left side because oxygen gains electrons when it oxidizes anything. Since we know that each oxygen atom can gain two electrons, we write 4 e^-. We also know that there must be 4 $OH^-(aq)$ on the right side to balance the charge of the 4 e^- on the left. Further, we know that we may add $OH^-(aq)$ or $H_2O(liq)$ [but not $H^+(aq)$] as necessary to make the half-reaction equation balance. Now we can make the oxygen and hydrogen atoms balance by adding 2 $H_2O(liq)$ to the left side, giving us

$$O_2(aq) + 2\,H_2O(liq) + 4\,e^- \rightarrow 4\,OH^-(aq)$$

The last step is to add the balanced half-reactions in such a way that the electrons gained and lost are equal and therefore cancel. This step is done by multiplying the iron half-reaction equation by four and adding the result to the oxygen-hydroxide half-reaction equation as follows:

$$\begin{array}{r} 4\,Fe(OH)_2(c) + 4\,OH^-(aq) \rightarrow 4\,Fe(OH)_3(c) + 4\,e^- \\ O_2(aq) + 2\,H_2O(liq) + 4\,e^- \rightarrow 4\,OH^-(aq) \\ \hline 4\,Fe(OH)_2(c) + O_2(aq) + 2\,H_2O(liq) \rightarrow 4\,Fe(OH)_3(c) \end{array}$$

References

Kieffer, W. F.: *The Mole Concept in Chemistry.* Reinhold Publishing Corp., New York, 1963 (paperback).

Nash, L. K.: *Stoichiometry.* Addison-Wesley Publishing Co., Reading, Mass., 1966 (paperback).

Problems

1. How much calcium can be obtained from 3.42 g of calcium chloride?
2. List the oxidation state of each element in the following:
   (a) $KClO_3$ (b) $KClO_4$ (c) H_2SO_4 (d) MnO_2 (e) $KMnO_4$ (f) $MnCl_2$
   (g) Na_2CrO_4 (h) $K_2Cr_2O_7$ (i) Cr_2O_3 (j) UF_6 (k) $NaNO_3$
   (l) $NaNO_2$ (m) NH_3 (n) NH_4^+ (o) NH_2^- (p) IO_3^-
3. The following (unbalanced) equations represent chemical reactions that occur on heating the appropriate reactants. Identify the oxidation-reduction reactions. Identify which substance is oxidized, which is reduced, and label oxidizing and reducing agents in each oxidation-reduction reaction. Balance each equation.
   (a) $Al(c) + Cr_2O_3(c) \rightarrow Al_2O_3(c) + Cr(c)$
   (b) $Ag_2O(c) + H_2(g) \rightarrow Ag(c) + H_2O(g)$
   (c) $CaCO_3(c) \rightarrow CaO(c) + CO_2(g)$
   (d) $KClO_3(c) \rightarrow KCl(c) + O_2(g)$
4. The volume of gas in a cylinder inverted over water is 92 ml when the atmospheric pressure is 745 mm Hg and the temperature is 25°C. How many moles of water vapor are in the cylinder? How many moles of another gas are also in the cylinder?

5. An important compound of uranium has the formula UF_6. How many grams of uranium can be obtained from 100 g of UF_6? How many pounds of uranium can be obtained from 100 lb of UF_6?

6. How many moles of carbon dioxide are produced by complete combustion of 100 g of carbon monoxide?

7. How many moles of carbon dioxide can be obtained by complete combustion of 150 g of ethane, C_2H_6?

8. Aluminum oxide (Al_2O_3) is reduced commercially to metallic aluminum by an electrolytic process. How much aluminum could be obtained from 1000 g of aluminum oxide by a 95% efficient process?

9. What happens to the gas pressure in a constant volume container when acetylene, C_2H_2, reacts completely with hydrogen to form ethane, C_2H_6, at constant temperature? Both ethane and acetylene are gases under the conditions of this experiment.

10. A 500-ml flask is completely filled with oxygen collected over water at 25°C under atmospheric pressure of 742 mm Hg. How many moles of oxygen are in the flask?

11. Magnesium reacts with gaseous hydrogen chloride to yield magnesium chloride and hydrogen. How much magnesium would react with 23 liters of hydrogen chloride measured at 29°C and 1.0 atm?

12. Balance the following oxidation-reduction equations. Since all of the indicated reactions take place in acidic solutions, $H^+(aq)$ and $H_2O(liq)$ may be added where appropriate. List what is oxidized, what is reduced, and what are the oxidizing and reducing agents in each reaction.

(a) $HOCl(aq) + Br_2(aq) \rightarrow BrO_3^-(aq) + Cl^-(aq)$
(b) $MnO_4^-(aq) + HNO_2(aq) \rightarrow Mn^{2+}(aq) + NO_3^-(aq)$
(c) $H_2S(aq) + Fe^{3+}(aq) \rightarrow Fe^{2+}(aq) + S(solid)$
(d) $Cu(c) + NO_3^-(aq) \rightarrow Cu^{2+}(aq) + NO_2(g)$
(e) $Cu(c) + NO_3^-(aq) \rightarrow Cu^{2+}(aq) + NO(g)$
(f) $Zn(c) + NO_3^-(aq) \rightarrow Zn^{2+}(aq) + NH_4^+(aq)$
(g) $MnO_4^-(aq) + H_3AsO_3(aq) \rightarrow Mn^{2+}(aq) + H_3AsO_4(aq)$
(h) $Cu^{2+}(aq) + I^-(aq) \rightarrow CuI(c) + I_3^-(aq)$

When copper reacts with nitric acid, reactions represented by (d) and (e) take place in relative amounts that depend on acid concentration and on temperature.

Reaction (h) occurs when $Cu^{2+}(aq)$ reacts with an excess of $I^-(aq)$. The $I_3^-(aq)$ that is formed is a combination of I^- with I_2 and is called triiodide ion.

All of the reactions given above are useful in analytical chemistry.

13. Balance the following oxidation-reduction equations. Because all of these reactions take place in basic solutions, $OH^-(aq)$ and $H_2O(liq)$ may be added where appropriate. List what is oxidized, what is reduced, and what are the oxidizing and reducing agents in each reaction.

(a) $MnO_4^-(aq) + NO_2^-(aq) \rightarrow MnO_2(c) + NO_3^-(aq)$
(b) $Al(c) + NO_3^-(aq) \rightarrow AlO_2^-(aq) + NH_3(g)$
(c) $Cl_2(aq) \rightarrow Cl^-(aq) + ClO_3^-(aq)$
(d) $Cl_2(aq) \rightarrow Cl^-(aq) + OCl^-(aq)$
(e) $NO_2(aq) \rightarrow NO_2^-(aq) + NO_3^-(aq)$
(f) $Al(c) \rightarrow AlO_2^-(aq) + H_2(g)$

Reaction (c) occurs when chlorine is bubbled into a hot solution of strong base, whereas reaction (d) occurs in cooler solutions.

14. Deduce from comparison of 12(a) and 13(d) whether hypochlorous acid is strong or weak with respect to complete dissociation into ions. Make similar deductions for nitrous acid and nitric acid from 12(b, d, e, f) and 13(b, e).

15. How many $H^+(aq)$ ions are in 10 drops of 0.5 *M* hydrochloric acid solution? The volume of a drop is about 0.05 ml.

16. An acidic solution containing permanganate ions reacts with metallic copper to yield cupric ions and manganous ions. Write a balanced equation for this oxidation-reduction reaction. Identify the oxidizing agent and the substance that is oxidized. What volume of 0.050 *M* permanganate solution is required to react with 1.00 g copper?

17. Reaction of an impure sample of magnesium (0.140 g) with aqueous hydrochloric acid yielded 132 ml of $H_2(g)$ collected over water at 25°C and 750 mm Hg pressure. On the assumption that the impurities did not react to yield any gaseous product, calculate the mass per cent of magnesium in the original sample.

18. Analysis of an oxide of vanadium shows that it is 56 mass % vanadium. What is the simplest formula of this oxide?

19. A certain hydrocarbon has been found to contain 85.8 mass % carbon and 14.2 mass % hydrogen. The density of this gaseous compound is 2.135 g $liter^{-1}$ at 667 mm Hg and 100°C. What is the molecular formula of this compound?

20. Reduction of a 1.00 g sample of tin oxide with hot hydrogen gas yields 0.79 g of metallic tin. What is the simplest formula of the oxide?

21. A sample of metal weighing 1.523 g was treated with acid and was found to liberate 584 ml of hydrogen measured over water at 20°C and 748 mm Hg. What are possible values of the atomic weight of this metal?

22. Calcium carbide reacts with water to yield calcium hydroxide and acetylene as indicated by

$$CaC_2(c) + 2\,H_2O(liq) \rightarrow Ca(OH)_2(c) + C_2H_2(g)$$

How many grams of acetylene will be produced from 250 g of calcium carbide that reacts in this manner?

23. A mixture of $CaCO_3(c)$ and $BaCO_3(c)$ is analyzed by liberating $CO_2(g)$ with strong acid. In this fashion we obtain 30.8 ml of dry $CO_2(g)$ at 24°C and 720 mm Hg from a 0.213 g sample of mixed carbonates. What was the mass per cent of $BaCO_3(c)$ in the original sample?

24. A 1.000 g sample of mixture of $CaCO_3(c)$ and $BaCO_3(c)$ is heated to yield the oxides and 0.00600 mole of $CO_2(g)$. What was the mass per cent of $BaCO_3(c)$ in the sample?

25. A mixture of NaCl(c) and $BaCl_2(c)$ weighing 0.2291 g yielded 0.4138 g of AgCl(c) when treated with aqueous silver nitrate. What was the mass per cent of NaCl(c) in the sample?

26. Find the maximum amount of ammonia that can be synthesized from 12 g of $N_2(g)$ and 12 g of $H_2(g)$ by way of the reaction

$$N_2(g) + 3\,H_2(g) \rightarrow 2\,NH_3(g)$$

27. A sample of mixed NaCl(c) and $BaCl_2(c)$ that weighs 0.394 g is treated with sulfuric acid to precipitate $BaSO_4(c)$, which is found to weigh 0.349 g. What was the mass per cent of $BaCl_2(c)$ in the mixture?

10

HYDROGEN, OXYGEN, AND SULFUR

Introduction

Hydrogen, oxygen, and sulfur are three abundant and widespread elements. Atoms of these elements combine with atoms of most other elements to form a large variety of compounds.

In the ground state, hydrogen has a single electron in its $1s$ energy level. Hydrogen atoms normally attain a lower energy state by pairing to form diatomic H_2 molecules. Reactions of H_2 with most other elements produce compounds containing hydrogen in either a $+1$ or -1 oxidation state. Although it displays some similarities to both the halogens and the alkali metals, hydrogen is best considered as unique in the periodic table.

Oxygen, sulfur, and the other group VI elements are characterized by an outer electronic configuration of ns^2np^4. Atoms of the group VI elements form compounds in which they can be assigned either positive or negative oxidation states. The -2 state is common to all. Oxygen forms compounds in which a $+2$ state is assigned, but this is extremely rare. The other elements of group VI form compounds in which $+4$ and $+6$ states are assigned.

Hydrogen

Three isotopes of hydrogen are known. Protium (^1_1H, or simply H) has a single proton in its nucleus and is by far the most common. Hydrogen atoms that contain one proton and one neutron in the nucleus are called heavy hydrogen or deuterium (^2_1H, or D). Deuterium is much less plentiful; there is about 7000 times as much protium as deuterium. Tritium (^3_1H, or T), with one proton and two neutrons in the nucleus, is radioactive. The short half-life (12.3 years) explains the scarcity of tritium atoms in nature. So little deuterium and tritium are present in natural hydrogen that its properties are substantially those of protium itself.

In general, the chemical and physical properties of isotopes are very similar, but there may be quantitative differences. The uniquely large mass ratios of hydrogen isotopes are responsible for differences far greater than those shown by the isotopes of any other element. For example, the freezing point of H_2O is 0.0 °C whereas that of D_2O is 3.8 °C. The small differences in chemical properties between H and D compounds are often exploited in studies to determine the paths of reactions involving hydrogen compounds.

Molecular hydrogen is very rare in the atmosphere of the earth. From a study of the aurora borealis, traces of it are found to exist in the upper atmosphere. Compounds of hydrogen are, however, very common. Combined hydrogen atoms are among the most abundant atoms in the earth's crust (atmosphere plus 10 miles of material including oceans). The principal hydrogen-containing compounds are water and hydrocarbons, such as petroleum. Analysis of light emitted by stars indicates that most stars are predominantly hydrogen. It appears that hydrogen is the most abundant element in the universe.

In the laboratory, where convenience is the prime consideration, hydrogen is commonly prepared by reaction of zinc with acid, often hydrochloric acid:

$$Zn(c) + 2\,H^+(aq) \rightarrow Zn^{2+}(aq) + H_2(g) \qquad (10.1)$$

In principle any electropositive metal should undergo the above reaction, but overvoltage considerations (Chapter 16) and slow reaction rates limit the possibilities in practice.

Less convenient but more spectacular is the production of hydrogen by reaction of very electropositive metals with water. Sodium and the other alkali metals react vigorously with water to produce hydrogen by reactions such as

$$2\,Na(c) + 2\,H_2O(liq) \rightarrow 2\,Na^+(aq) + 2\,OH^-(aq) + H_2(g) \qquad (10.2)$$

Reaction (10.2) is accompanied by evolution of so much heat that the hydrogen usually becomes hot enough to react explosively with oxygen in the air. This reaction of hydrogen and oxygen to form water causes the light and noise commonly associated with the reaction of sodium or potassium with water.

Other convenient laboratory methods for the production of hydrogen are the reaction of aluminum with aqueous base or reaction of a metal hydride with water.

For commercial preparations, where cost is the prime concern, the raw materials must be cheap and readily available. One industrial route to hydrogen is by passing steam over hot carbon (coke or coal) at temperatures near 1000 °C:

$$H_2O(g) + C(s) \rightarrow CO(g) + H_2(g) \qquad (10.3)$$

A complete separation of carbon monoxide and hydrogen is difficult so that pure H_2 is rarely obtained from this process. Instead the resulting mixture of H_2 and CO may be used as an industrial fuel, called **water gas.**

Very pure (99.9%) hydrogen can be made by the electrolysis of water. Since large amounts of electrical energy are required, electrolytic hydrogen is expensive. Sodium hydroxide or sulfuric acid must be added to the water before it is electrolyzed because pure water is a poor conductor of electricity. If inert electrodes are used in the electrolysis, the net balanced equation for the reaction is

$$2\ H_2O(liq) \xrightarrow{\text{electrolysis}} 2\ H_2(g) + O_2(g) \tag{10.4}$$

Hydrogen is evolved at the cathode and oxygen at the anode. The **cathode** is the electrode that loses electrons, and the **anode** is the electrode that takes electrons from the solution. When sodium hydroxide is the added electrolyte, the electrode reactions are

Cathode: $2\ H_2O(liq) + 2\ e^- \rightarrow H_2(g) + 2\ OH^-(aq)$

Anode: $4\ OH^-(aq) \rightarrow O_2(g) + 2\ H_2O(liq) + 4\ e^-$

Chapter 16 is devoted to a detailed discussion of electrolysis and electrode reactions. We should, however, recognize now the similarity of the half-reaction equations used in balancing oxidation-reduction equations and electrode reactions that take place in batteries or electrolytic cells.

Large amounts of hydrogen are also obtained industrially as a byproduct of the cracking of petroleum hydrocarbons in gasoline refineries. The cracking processes, which are designed to increase the octane rating of the gasoline, are now a principal source of commercial hydrogen.

The low melting (14°K) and boiling (20°K) points of hydrogen indicate that attractive forces between molecules of hydrogen are weak. Liquid hydrogen is used in the laboratory for low temperature studies despite the fact that it is both expensive and hazardous.

Molecular hydrogen has a lower density than any other gas at the same temperature and pressure so that it has been used to fill balloons and dirigibles. However, helium has largely replaced hydrogen for this purpose because of the danger of reaction with atmospheric oxygen. The reaction of hydrogen with oxygen is used to advantage in the oxyhydrogen torch for cutting and welding metals. Large quantities of hydrogen are used in the manufacture of ammonia by direct combination of hydrogen with nitrogen. Hydrogen is also used industrially in the production of methyl alcohol and other organic compounds and is used in the **hydrogenation** of oils to solid fats.

Compounds of Hydrogen

Molecular hydrogen is relatively unreactive at ordinary temperatures. The first step in many reactions of hydrogen involves a dissociation into hydrogen atoms. Since the bond energy is 104 kcal mole^{-1}, large amounts of energy are necessary to initiate these reactions. Consequently, reactions involving hydrogen are often quite slow unless a catalyst is present. The ionization energy of hydrogen is relatively high (13.6 eV), and the electron affinity is quite low (0.75 eV). As a consequence, most compounds contain covalently bound hydrogen, with neither H^+ nor H^- being readily formed.

Binary compounds of hydrogen with most other elements except the noble gases are known. **Saline** (or saltlike) hydrides, containing the hydride ion, H^-, are formed with the alkali and alkaline earth metals. Saline hydrides are characterized by the usual properties of ionic substances, such as high melting points and high electrical conductivity in the fused state. Nonmetals react with hydrogen to form **covalent** hydrides, in which hydrogen may be assigned either a $+1$ or -1 oxidation state depending on the electronegativity of the nonmetal. Covalent hydrides are generally volatile compounds with low melting points and a very low electrical conductivity in the liquid state. **Metallic** hydrides are formed with transition metals and exhibit a wide range of properties. These hydrides generally have metallic structures and are alloylike in most of their characteristics.

Apart from fluorine, which spontaneously explodes with hydrogen in the dark, the other strongly electronegative elements react with hydrogen only on heating or irradiation. The reaction of hydrogen with a halogen may be described by the general equation (10.5)

$$H_2 + X_2 \rightarrow 2\,HX \tag{10.5}$$

where X_2 represents molecular halogen (fluorine, chlorine, bromine, or iodine). The H-X bonds are all polar covalent bonds, and all hydrogen halide molecules have dipole moments. Hydrogen fluoride is a weak acid in water solution, but aqueous solutions of the other hydrogen halides are strong acids.

Some of the other covalent hydrides, the most numerous kind of hydrogen compounds, will be discussed in later chapters. Here we mention the boron hydrides, which have been of considerable interest to chemists because of their unusual bonding properties. The original work on the boron hydrides was done by Alfred Stock and coworkers in Germany between 1912 and 1936. Most of these volatile and very reactive compounds (some being spontaneously inflammable in air) were obtained by the action of aqueous acid on magnesium boride. (There is now some question concerning the exact composition of the magnesium boride that Stock used.)

The simplest boron hydride is BH_3, borane, formed by sharing the three valence electrons of boron with three hydrogen atoms. However, borane is not stable at room temperature and readily dimerizes to diborane, B_2H_6. Part of the interest in this compound is caused by the fact that diborane has only twelve valence electrons although there must be at least seven bonds. Diborane is said to be "electron deficient." In truth, the deficiency is in our ability to describe bonding with simple pictures.

Although it was long a subject of controversy, the molecular structure of diborane now seems well established. As shown in Figure 10.1 the two boron atoms are held together by two **bridging** hydrogen atoms. No simple electron dot structure can be written that will adequately describe these bridge bonds. However, a relatively simple molecular orbital picture can be used.

The four terminal hydrogen atoms are bonded to the two boron atoms through conventional electron pair σ bonds. The bridge bonds are three-center two-electron bonds that extend over the two boron atoms and the bridging hydrogen atom, as shown in Figure 10.1. In essence, there are three atoms, each with an available orbital. From these three atomic orbitals it is possible to form three molecular

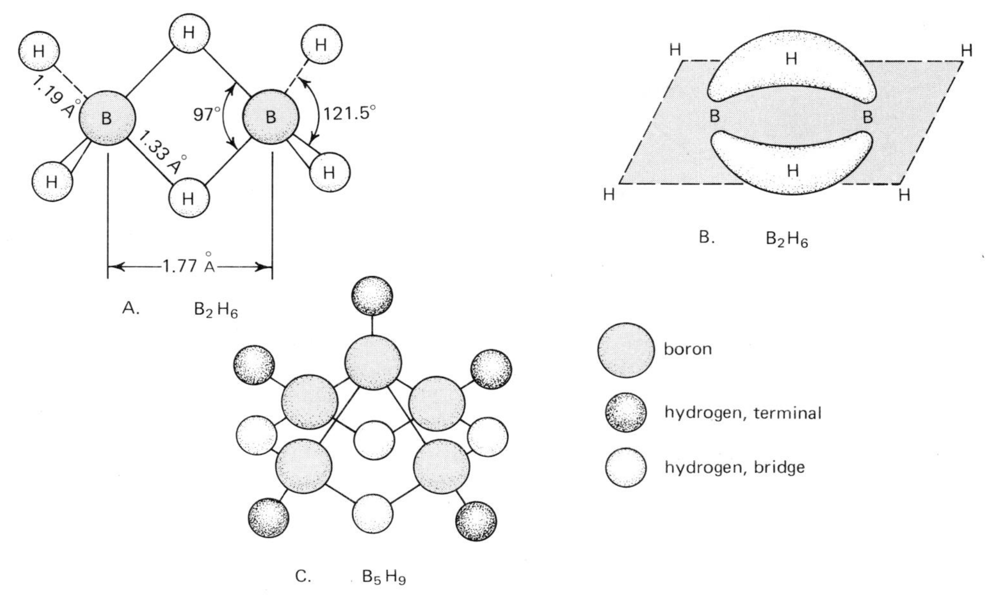

Figure 10.1. Structures of diborane and pentaborane-9.

In diborane (A and B) the boron atoms and bridging hydrogen atoms are in the plane of the paper. The terminal hydrogen atoms are perpendicular to the plane of the paper. The hydrogen bridge bonds may be pictured as "banana bonds."

In pentaborane-9 (C) there are five terminal hydrogen atoms and four bridging hydrogen atoms. In the boron framework there are two two-center, two-electron boron-boron bonds; one three-center, two-electron boron-boron bond; and four three-center, two-electron hydrogen bridge bonds. If the top terminal hydrogen atom is replaced by another B_5H_8 unit the compound decaborane-16, $B_{10}H_{16}$, is obtained.

orbitals, one bonding, one nonbonding, and one antibonding. Because there are only two electrons available, only the lowest energy orbital, the bonding molecular orbital, will be filled. Such a molecular orbital scheme not only gives a proper electron count but is consistent with observed properties, such as a relatively weak bridge bond (compared to normal covalent bonds) and no paramagnetism. A general energy level diagram for the molecular orbital treatment of a three-center bond is shown in Figure 10.2.

There are several other boron hydrides, such as B_4H_{10} (tetraborane-10), $B_{10}H_{14}$ (decaborane-14), and $B_{20}H_{16}$ (icosaborane-16). All exhibit the same type of "electron deficient" bridge bonding found in diborane. In some of the higher boron hydrides three-center bonds between three boron atoms are known. The structure of pentaborane-9, B_5H_9, is also shown in Figure 10.1.

Hydrogen bridge bonds are also found in aluminum hydride, $(AlH_3)_x$, which is a polymeric solid. No simple molecular species, for example Al_2H_6, have been reported.

Finally, a comment on the similarities and differences between hydrogen bridge bonds in the boron hydrides and hydrogen bonding in substances such as water seems in order. Both types of hydrogen bonds can be considered as three-center

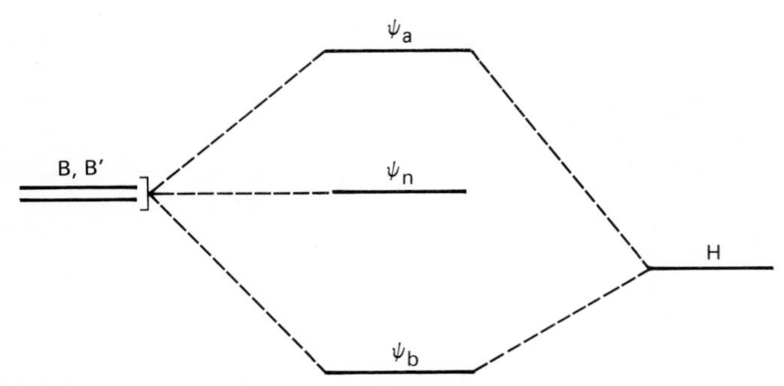

Figure 10.2. Energy level diagram for three-center molecular orbitals.
The B and B′ represent the atomic orbitals of the two borons, and the H represents the atomic orbital of the bridging hydrogen. The molecular orbitals are ψ_b (bonding), ψ_n (nonbonding), and ψ_a (antibonding). For the hydrogen bridge bond in B_2H_6 only the bonding molecular orbital would be filled; there are two hydrogen bridge bonds in diborane. ψ_b is the only molecular orbital that increases the electron density in the internuclear region.

two-electron bonds. The general molecular orbital energy diagram shown in Figure 10.2 is suitable for either. The strengths of the two types of bonds are quite different, however, the hydrogen bridge bonds being much stronger. It requires approximately 30 kcal mole^{-1} to break two hydrogen bridge bonds in diborane and form two BH_3 units. The strength of the hydrogen bond in associated liquids such as water or ammonia is approximately 5 kcal mole^{-1}.

Compounds containing H^-, the hydride ion, are obtained only by reaction of hydrogen with the most electropositive metals. Formation of saline hydrides is not particularly violent and often requires high temperatures. For example, sodium hydride is formed by bubbling hydrogen through molten sodium at 350°C.

$$2\ \mathrm{Na(liq)} + \mathrm{H_2(g)} \rightarrow 2\ \mathrm{NaH(c)} \qquad (10.6)$$

Calcium hydride may be obtained by passing hydrogen over calcium at 150°C.

$$\mathrm{Ca(c)} + \mathrm{H_2(g)} \rightarrow \mathrm{CaH_2(c)} \qquad (10.7)$$

Similar reactions occur between hydrogen and the other alkali and alkaline earth metals.

Metal hydrides react vigorously with water to produce hydrogen gas and hydroxide ions. As a result they are frequently used as drying agents or sources of hydrogen in the laboratory. The hydrolysis of calcium hydride is described by equation (10.8).

$$\mathrm{CaH_2(c)} + 2\ \mathrm{H_2O(liq)} \rightarrow \mathrm{Ca^{2+}(aq)} + 2\ \mathrm{OH^-(aq)} + 2\ \mathrm{H_2(g)} \qquad (10.8)$$

Oxygen

There are three common isotopes of oxygen, the isotope of atomic mass 16 being the most plentiful. The isotopes of atomic mass 17 and 18 together make up about 0.25% of all oxygen. The confusion resulting from the two atomic weight scales based on ^{16}O and naturally occurring oxygen has already been discussed (Chapter 2).

Oxygen is the most abundant element in the earth's crust. About half of the earth's crust by weight is oxygen in the free state or in various forms of chemical combination. Much of the combined oxygen is in the form of water, although it is also found in many minerals, plants, and animals.

Either on a volume or mass basis, approximately 20% of air is oxygen. In the free state, oxygen occurs mainly as O_2 molecules.

The commercial sources of oxygen are air and water. Very pure oxygen is obtained as a byproduct of the manufacture of hydrogen by electrolysis of water. Because electrolysis is expensive, the most important source of industrial oxygen is air.

Besides oxygen and nitrogen, air contains about 1% by volume of the noble gas argon, trace amounts of the other noble gases, variable amounts of carbon dioxide and water, and sometimes various pollutants. As the first step in separation of oxygen from the other components of air, the carbon dioxide and water vapor are removed. Then the remaining gas is liquified by alternate compression, cooling, and expansion. As the liquid air boils, the most volatile components (low boiling points) are vaporized before the least volatile component (relatively high boiling point). Thus nitrogen (normal boiling point = 77°K) and argon (normal boiling point = 87°K) boil away, leaving liquid oxygen (normal boiling point = 90°K). This kind of process, which is called fractional distillation, can lead to oxygen of high purity.

Potassium chlorate, $KClO_3$, is a common laboratory source of small quantities of oxygen. When $KClO_3$ is heated, either of two reactions can occur. One of these is decomposition to potassium chloride (KCl) and potassium perchlorate ($KClO_4$).

$$4\,KClO_3(c) \rightarrow KCl(c) + 3\,KClO_4(c) \qquad (10.9)$$

Note that in this reaction chlorine has gone from a +5 oxidation state to both a lower (−1) and higher (+7) oxidation state. It has been both oxidized and reduced. Such self oxidation-reduction reactions are called **disproportionations** and are a common method of obtaining an element in its highest oxidation state.

The other reaction that occurs when potassium chlorate is heated produces oxygen.

$$2\,KClO_3(c) \rightarrow 2\,KCl(c) + 3\,O_2(g) \qquad (10.10)$$

Reaction (10.10) is favored by addition of a small quantity of manganese dioxide, MnO_2, which is said to **catalyze** the reaction. Any substance that speeds a chemical reaction without itself being consumed is called a **catalyst.** In this case the manganese dioxide is believed to provide a surface from which oxygen gas can evolve. In the second reaction chlorine is reduced to the −1 oxidation state from the +5 oxidation state whereas oxygen is oxidized from the −2 oxidation state to the 0 oxidation state.

All of the common uses of oxygen depend on its oxidizing properties. Because it is both cheap and readily available, oxygen is one of the most widely used industrial oxidizing agents. In purification of many metals, either pure oxygen or air is used to burn off impurities such as carbon or sulfur. Gaseous oxygen and acetylene react in the oxyacetylene torch to produce a hot flame used for cutting and welding metals. Liquid oxygen is used as the oxidizing component of some high energy fuels and explosives. The importance of oxygen for respiration in animals, particularly man, should be familiar.

Oxygen is one of several elements that can exist in more than one form. The existence of an element in two or more forms is termed **allotropy;** it is due either to the existence of molecules containing different numbers of atoms or to the existence of solid phases having more than one possible arrangement of atoms. When gaseous oxygen is passed between two metal plates charged to a potential of several thousand volts, some of the O_2 is converted to ozone, O_3, an allotropic form of oxygen.

$$3\ O_2(g) + \text{energy} \rightarrow 2\ O_3(g) \qquad (10.11)$$

The energy for the conversion is supplied by the electrical discharge. Ozone is also produced in the atmosphere by lightning and by ultraviolet light from the sun.

Ozone is a high energy form of oxygen and is therefore very reactive. The uses of ozone, principally as a bleaching agent and disinfectant, are dependent on the fact that it is a powerful oxidizing agent. In spite of song and story, ozone is harmful to man in more than trace amounts because it attacks (oxidizes) the mucous membranes of the body.

Magnetic studies of molecular oxygen show that it is paramagnetic with two unpaired electrons. The bond strength of O_2, 118 kcal $mole^{-1}$, is roughly consistent with a double bond, being intermediate between that of triply bonded nitrogen (225 kcal) and singly bonded fluorine (38 kcal). There is no way to describe the structure of O_2 adequately using a valence bond approach. The formula showing unpaired electrons

$$:\ddot{\underset{\cdot}{O}}:\ddot{\underset{\cdot}{O}}:$$

is consistent with the observed paramagnetism but not the bond strength. In addition, the above formula violates the octet rule. The formula with all electrons paired

$$:\ddot{O}::\ddot{O}:$$

shows the double bond and is consistent with the octet rule but provides no explanation of the paramagnetism. The molecular orbital description of O_2, already described in Chapter 8, accounts for the observed properties very successfully.

The double quartet approach (Chapter 8) also accounts for the observed paramagnetism and bond strength. The double quartet description shows seven electrons of one spin and five electrons of opposite spin. The tetrahedra may be arranged as

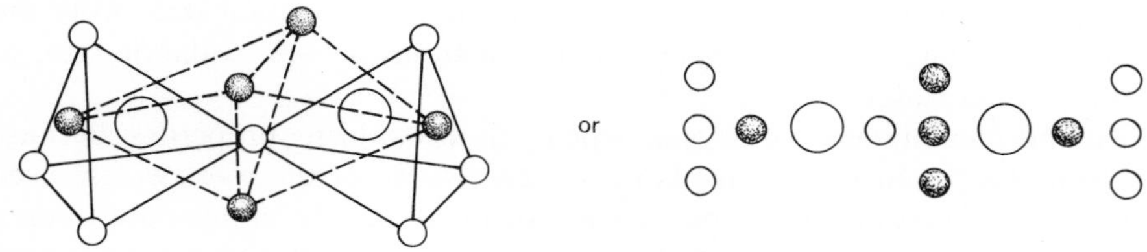

with four bonding electrons equivalent to a double bond.

Table 10.1. Oxygen-Oxygen Bond Distances

Compound	*Bond distance (Å)*
HOOH and F_5SOOSF_5	1.48
O_2^{2-} (peroxide ion)	1.49
O_3	1.28
O_2	1.21

Laboratory measurements show that ozone is diamagnetic and thus has no unpaired electrons. The ozone molecule is bent, and both bonds to the central atom are of equal length. Since the molecule is nonlinear, ozone has a dipole moment. The electron dot formula is written by assuming two contributing resonance forms

$$:\ddot{O}\;\ddot{O}\;\ddot{O}: \;\leftrightarrow\; :\ddot{O}\;\ddot{O}\;\ddot{O}$$

According to this picture, the O—O bond length in ozone should be intermediate between that of a single bond (peroxide) and double bond (O_2). This prediction is verified by the bond length data in Table 10.1. In molecular orbital theory the bonding in ozone is described in terms of strong delocalized π bonds.

Peroxides

The most common oxidation state of oxygen in its compounds is -2. These compounds may be covalent, such as SO_3, or ionic, such as MgO. Compounds that contain the O^{2-} ion are called **oxides,** and those that have complex ions containing oxygen are called **oxy compounds.** Examples of these types of compounds will be discussed in more detail later in this chapter.

There are also compounds in which oxygen is assigned the -1 oxidation state. These compounds are characterized by O—O oxygen single bonds and are called **peroxides.** We write O_2^{2-} for the peroxide ion and H_2O_2 for hydrogen peroxide, which may be considered the parent compound of all peroxides.

Hydrogen peroxide in aqueous solution can be prepared by a two-step synthesis. The first step involves heating barium oxide under about 3 atm pressure of oxygen gas to form barium peroxide.

$$2\ BaO(c) + O_2(g) \rightarrow 2\ BaO_2(c) \tag{10.12}$$

Treatment of barium peroxide with an aqueous solution of sulfuric acid precipitates barium sulfate and leaves aqueous hydrogen peroxide.

$$BaO_2(c) + H^+(aq) + HSO_4^-(aq) \rightarrow BaSO_4(c) + H_2O_2(aq) \tag{10.13}$$

Since H_2O_2 is less volatile than water, the solution may be concentrated by controlled evaporation (fractional distillation).

Note that sulfuric acid has been written as $H^+(aq)$ plus $HSO_4^-(aq)$. Aqueous sulfuric acid is a strong acid with respect to loss of one proton, but HSO_4^-, **hydrogen sulfate** (often called **bisulfate**), is a weak acid. In concentrated solution only a small fraction of the total hydrogen sulfate ions are further dissociated to H^+ and SO_4^{2-} ions.

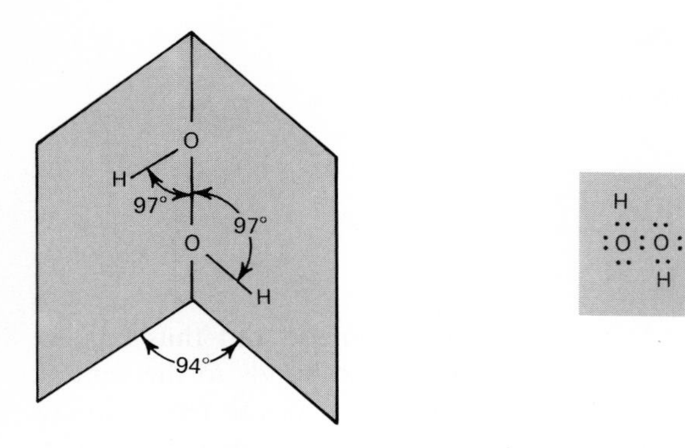

Figure 10.3. Structure of the H_2O_2 molecule. Note that all four atoms are not in the same plane.

Some of the physical properties of pure hydrogen peroxide, such as melting point and high dielectric constant, are quite similar to those of water. The chemical properties differ considerably from water, partly due to the inherent instability of hydrogen peroxide with respect to decomposition into oxygen and water. The equation for this decomposition reaction, which may occur with explosive violence, is

$$2\ H_2O_2(liq) \rightarrow 2\ H_2O(liq) + O_2(g) \qquad (10.14)$$

The decomposition may be slow but is easily catalyzed by light, dust, many metals, and a variety of dissolved compounds. Pure hydrogen peroxide may be stored for reasonable periods of time in dark bottles with various chemicals added to destroy catalysts. Aqueous solutions of hydrogen peroxide also decompose into water and oxygen, but the decomposition proceeds so slowly that dilute solutions can be kept almost indefinitely.

Hydrogen peroxide is a polar covalent substance that, like water, is associated through hydrogen bonding. The O—O bond length corresponds to a single bond. The structure and electron dot formula for hydrogen peroxide are illustrated in Figure 10.3.

The peroxide oxygen is in an oxidation state intermediate between two more stable oxidation states. Peroxides can thus act as oxidizing agents and be reduced to the -2 oxidation state or can act as reducing agents and be oxidized to the 0 oxidation state. In the decomposition reaction discussed above (equation 10.14), the peroxide oxidizes and reduces itself (disproportionates).

Hydrogen peroxide oxidizes ferrous ions to ferric ions in acidic solutions. The half-reaction equations are written

$$H_2O_2(aq) + 2\ H^+(aq) + 2\ e^- \rightarrow 2\ H_2O(liq)$$

and

$$Fe^{2+}(aq) \rightarrow Fe^{3+}(aq) + e^-$$

as the first steps in obtaining a balanced equation. The ferrous-ferric half-reaction is then multiplied through by 2 and added to the hydrogen peroxide-water half-reaction to obtain

$$H_2O_2(aq) + 2\ Fe^{2+}(aq) + 2\ H^+(aq) \rightarrow 2\ H_2O(liq) + 2\ Fe^{3+}(aq) \qquad (10.15)$$

Hydrogen peroxide can also act as a reducing agent and is often used to reduce acidic solutions of potassium permanganate to manganous ions. The half-reaction equations are

$$H_2O_2(aq) \rightarrow O_2(g) + 2\,H^+(aq) + 2\,e^-$$

and

$$MnO_4^-(aq) + 8\,H^+(aq) + 5\,e^- \rightarrow Mn^{2+}(aq) + 4\,H_2O(liq)$$

The complete balanced equation is

$$2\,MnO_4^-(aq) + 5\,H_2O_2(aq) + 6\,H^+(aq) \rightarrow 2\,Mn^{2+}(aq) + 5\,O_2(g) + 8\,H_2O(liq) \qquad (10.16)$$

Students are advised to go through all the steps in obtaining the final equations (10.15) and (10.16).

The heavier alkali metals (potassium, rubidium, and cesium) form colored compounds with general formula MO_2, called **superoxides.** These ionic solids have a paramagnetism consistent with one unpaired electron and the O_2^- anion. In this case oxygen must be assigned a formal oxidation state of $-\frac{1}{2}$. Superoxides exist only in the solid state. They are strong oxidizing agents that react vigorously with water according to equations (10.17) and (10.18).

$$2\,O_2^-(aq) + H_2O(liq) \rightarrow O_2(g) + HO_2^-(aq) + OH^-(aq) \qquad (10.17)$$

$$2\,HO_2^-(aq) \rightarrow 2\,OH^-(aq) + O_2(g) \qquad (10.18)$$

Canisters containing potassium superoxide are used as an emergency source of oxygen by high altitude climbers and underground miners.

Sulfur

Although sulfur constitutes only a small fraction (0.05%) of the earth's crust, it is readily available because it occurs in large beds of the free element. The sulfur obtained from such beds is about 99.5% pure and is used without purification for most industrial purposes. Large quantities of sulfur are also recovered from "sour" natural gas.

Some elemental sulfur is used directly, for example in vulcanization of rubber and in insecticides. Most of it, however, is converted into compounds, principally sulfuric acid. Because of the amount consumed and the wide variety of its uses, sulfur is often considered as one of the five basic raw materials of the chemical industry; the other four are coal, oil, salt, and limestone.

There are several allotropic forms of sulfur, the most important being the **rhombic** and **monoclinic** forms. Rhombic sulfur is the stable form up to 96°C where monoclinic sulfur becomes and remains the stable form up to the melting point at 119°C. Both rhombic and monoclinic sulfur consist of S_8 cyclic units. The eight-membered rings have a crown configuration as shown in Figure 10.4.

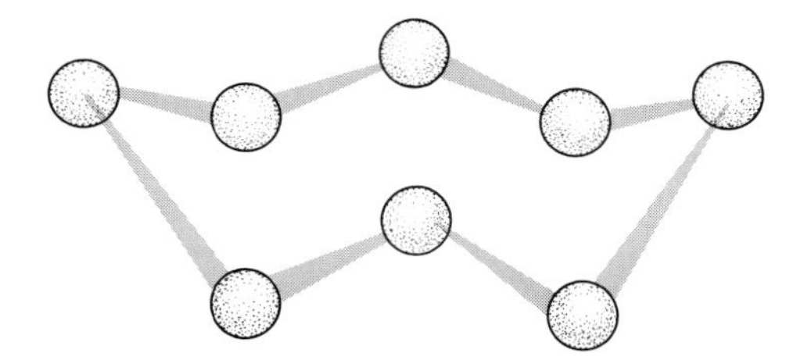

Figure 10.4. Schematic illustration of S_8 molecule, which occurs in rhombic, monoclinic, and gaseous sulfur. The bond lengths and angles are 2.12 Å and 105° in the puckered octagonal ring.

Transitions between the rhombic and monoclinic forms, which differ in their crystal symmetry, are ordinarily slow.

The structural relationships of sulfur in the solid, liquid, and gaseous phases are complex and incompletely understood. Near the melting point it appears that the substance contains three molecular species: S_8, S_6, and S_4. Equilibrium between these and possibly other species is attained slowly, and by rapid heating of rhombic or monoclinic sulfur it is possible to obtain a liquid consisting solely of S_8 molecules. Near the melting point sulfur is a pale yellow liquid of low viscosity. As heating is continued and the temperature is increased, the liquid thickens and reaches its maximum viscosity at about 180°C. It is believed that the increasing viscosity is due to the breakup of S_8 rings into long chains that crosslink. If the thick liquid is poured into water, plastic sulfur results. X-ray studies of plastic sulfur show that it consists of long spiral chains of sulfur atoms. As liquid sulfur is heated closer to the boiling point (444.6°C), the viscosity decreases as the long chains are broken up into smaller fragments.

By vapor density measurements gaseous sulfur has been shown to contain S_8, S_6, S_4, and S_2 molecules, the relative proportions depending on the temperature and pressure. As the temperature is increased the simpler molecules become predominant, and above 2000°C only diatomic and monatomic sulfur are important. Since S_2 is paramagnetic, with two unpaired electrons, the bonding in S_2 is believed to be similar to that in O_2.

Compounds of Sulfur

Although both oxygen and sulfur are characterized by the ns^2np^4 electronic configuration, there are significant differences between the two. Sulfur-sulfur single bonds are much more stable than those of oxygen. The -2 oxidation state is important for both, but sulfur more readily forms compounds in which it has positive oxidation states. The $+4$ and $+6$ states are common for sulfur but unknown for oxygen. In the following paragraphs the discussion of some of the important compounds of sulfur is arranged according to the oxidation state of sulfur.

Sulfides, in which sulfur is in the -2 oxidation state, are generally less stable than the corresponding oxides. Most sulfides can be converted to the corresponding oxide by heating in air. For example, a commercially important process is the roasting of the zinc ore wurtzite, ZnS.

$$2\,ZnS(c) + 3\,O_2(air) \rightarrow 2\,ZnO(c) + 2\,SO_2(g) \qquad (10.19)$$

The differences in physical properties of hydrogen sulfide, H_2S, and H_2O are attributed largely to the extensive hydrogen bonding in water and the relative lack of such bonding in H_2S. Aqueous solutions of H_2S dissociate slightly to form $H^+(aq)$ and $HS^-(aq)$ ions. It is possible to obtain moderate concentrations of $S^{2-}(aq)$ in basic solutions. The corresponding dissociation of $OH^-(aq)$ to $H^+(aq)$ and $O^{2-}(aq)$ does not occur to a detectable extent even in the most basic solution.

Sulfides of many metals are only slightly soluble in water, which explains the occurrence of many sulfide ores in nature. The solubility of sulfides increases with increasing acidity of the solution, eventually leading to evolution of hydrogen sulfide. Selective precipitation of sulfides can be obtained by carefully controlling the acidity of a solution, thereby controlling the sulfide ion (S^{2-}) concentration. This procedure is used as a means of separating and identifying metals in various qualitative analysis schemes.

Sulfur dioxide, SO_2, which contains sulfur in the $+4$ oxidation state, is generally prepared by burning sulfur in air. Because sulfur is readily oxidized to the $+6$ oxidation state, sulfur dioxide is an important industrial reducing agent. Other important uses are bleaching textile fibers and pulp, as a disinfectant in breweries and wineries, and as an intermediate in the production of sulfuric acid. Sulfur dioxide is also a major air pollutant. The difficulties of writing an acceptable electron dot formula for SO_2 were discussed in connection with resonance in Chapter 8.

In acidic solutions sulfur dioxide combines with water to form **sulfurous acid,** H_2SO_3. Sulfurous acid is a weak acid and exists almost entirely as undissociated molecules in acidic solutions. In pure water slight dissociation of H_2SO_3 to HSO_3^- **(hydrogen sulfite** or **bisulfite)** and H^+ ions occurs. A small fraction of HSO_3^- ions is further dissociated to SO_3^{2-} **(sulfite)** and H^+ ions. Sulfur dioxide and sulfurous acid solutions react with strong bases to form salts and water. Sulfite salts are formed if base is present in excess, whereas bisulfites are formed if sulfur dioxide or sulfurous acid is present in excess.

Although sulfur dioxide is primarily used as a reducing agent, it can also function as oxidizing agent and be reduced itself. An example of this is the reaction between sulfur dioxide and hydrogen sulfide.

$$SO_2(g) + 2\,H_2S(g) \rightarrow 4\,S(c) + 2\,H_2O(liq) \tag{10.20}$$

In this case the sulfur in SO_2 is reduced whereas the sulfur in H_2S is oxidized to elemental sulfur.

Sulfur trioxide, SO_3, made industrially by air oxidation of SO_2, contains sulfur in the $+6$ oxidation state. Since the reaction is slow at low temperatures, it is usually carried out at about 500°C in the presence of a catalyst such as finely divided platinum:

$$2\,SO_2(g) + O_2(g) \rightarrow 2\,SO_3(g) \tag{10.21}$$

The sulfur trioxide molecule is planar with 120° angles between the sulfur-oxygen bonds. It may be pictured as a resonance hybrid of the three structures:

```
    ..              ..              
   :O:             :O:             :O:
    ..              ..              ::
    S       <->     S       <->     S
 .O:  :O:        :O:  :O.        :O: :O:
```

We may also describe sulfur trioxide in terms of a strong delocalized π bond.

Although sulfur trioxide is the anhydride of sulfuric acid (H_2SO_4) the direct reaction of SO_3 and water is very slow. In the industrial preparation of sulfuric acid, sulfur trioxide is first added to pure sulfuric acid to form **pyrosulfuric acid,** $H_2S_2O_7$. Then water is added to the pyrosulfuric acid to form more sulfuric acid. These reactions are represented by equations (10.22) and (10.23).

$$SO_3(g) + H_2SO_4(liq) \rightarrow H_2S_2O_7(liq) \tag{10.22}$$

$$H_2S_2O_7(liq) + H_2O(liq) \rightarrow 2\,H_2SO_4(liq) \tag{10.23}$$

In addition to pure H_2SO_4, which can be considered a monohydrate of SO_3, there are other definite compounds formed between water and sulfur trioxide. Two of these are $H_2SO_4 \cdot H_2O$ and $H_2SO_4 \cdot 2H_2O$, which melt at 8.5°C and −39.5°C, respectively. Commercial concentrated sulfuric acid is approximately 93% H_2SO_4 by weight, and may be considered to be a mixture of H_2SO_4 and $H_2SO_4 \cdot H_2O$. Concentrated sulfuric acid solutions are commonly used as dehydrating agents. They may be used in dessicators to dry substances or as a reagent to favor splitting off water in reactions involving organic molecules.

Pure H_2SO_4 melts at 10.4°C and boils at 330°C, a convenient and long liquid range. Since many inorganic and organic substances dissolve in it, sulfuric acid has been extensively used as a nonaqueous solvent.

As already mentioned, in aqueous solution sulfuric acid is a strong acid with respect to dissociation of the first H^+ and a weak acid with respect to the second H^+. There are two common series of salts of sulfuric acid, the **sulfates** containing SO_4^{2-} ions and the **hydrogen sulfates** (or **bisulfates**) containing HSO_4^- ions.

Some compounds and ions of sulfur are not readily classified according to the oxidation state concept. One such species is **thiosulfate** ion, $S_2O_3^{2-}$, formed by reaction of sulfur with sulfite.

$$S(c) + SO_3^{2-}(aq) \rightarrow S_2O_3^{2-}(aq) \tag{10.24}$$

The prefix **thio** indicates substitution of a sulfur atom for an oxygen atom. Experiments with radioactive sulfur, **tracer** experiments, show that there are two different kinds of sulfur atoms in thiosulfate. X-ray studies of thiosulfates show that the ion is tetrahedral with one sulfur atom at the center of the tetrahedron and the other sulfur and three oxygen atoms at the corners.

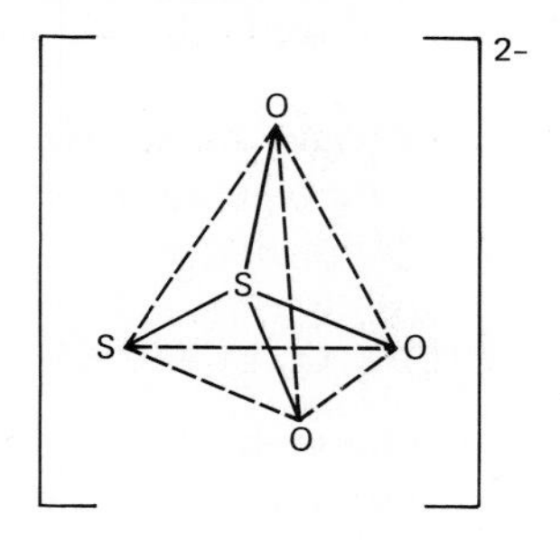

The *average* oxidation state of sulfur in $S_2O_3^{2-}$ is +2. A knowledge of the structure, however, indicates that it is better to assign a +6 state to one sulfur and a −2 state to the other, consistent with the average state. On this basis equation (10.24) represents an oxidation-reduction reaction in which sulfur oxidizes sulfite. The thiosulfate case also indicates the dangers involved in a naive application of the oxidation number concept without additional information.

The most familiar of the thiosulfates is sodium thiosulfate, $Na_2S_2O_3 \cdot 5\,H_2O$, often called **hypo.** Solutions of hypo are used as a "fixer" in developing photographs because they can dissolve the unexposed grains of silver halide on the film. Thiosulfate solutions are also used by analytical chemists as reducing agents, particularly for iodine. In this reaction thiosulfate is oxidized to tetrathionate, $S_4O_6^{2-}$, while iodine is reduced to iodide ions.

Electrolytic oxidation of cold concentrated sulfuric acid gives a solution of **peroxydisulfuric acid,** $H_2S_2O_8$. Similarly, oxidation of potassium hydrogen sulfate, $KHSO_4$, solutions produces potassium peroxydisulfate. The structure of the $S_2O_8^{2-}$ ion is shown in the diagram.

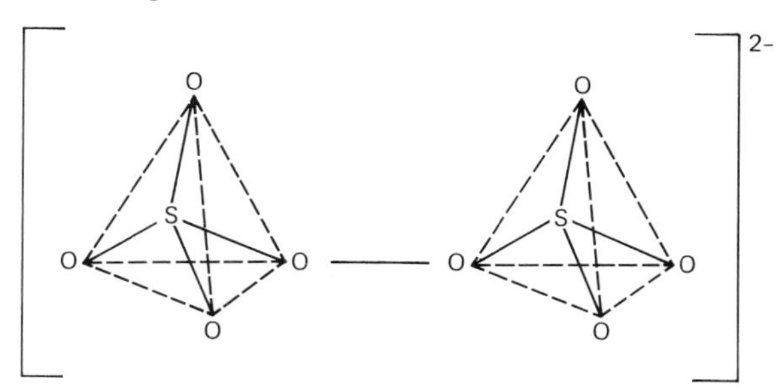

The **peroxy** prefix indicates that the ion has an O—O bond as does hydrogen peroxide. Peroxydisulfuric acid and its salts find use as powerful oxidizing agents.

Acidic and Basic Oxides

We have classified some substances as acids in aqueous solution because they increase the concentration of hydrogen ions and other substances as bases because they increase the concentration of hydroxide ions. Substances such as hydrochloric acid or sodium hydroxide increase the hydrogen or hydroxide ion concentrations directly. Other substances, such as oxides, may increase the hydrogen or hydroxide ion concentrations by reaction with water (hydrolysis).

Sodium oxide and calcium oxide are ionic substances that contain the oxide ion. In aqueous solution oxide ions are hydrolyzed to form hydroxide ions.

$$O^{2-}(aq) + H_2O(liq) \rightarrow 2\ OH^-(aq) \qquad (10.25)$$

As a result of this hydrolysis the concentration of hydroxide ions is greater than that in pure water. Thus ionic oxides or their aqueous solutions neutralize acids and form salts.

Covalent oxides, such as those of sulfur, are acidic because the concentration of hydrogen ions in water is increased as these oxides are dissolved. For example, sulfur trioxide in water produces hydrogen ions by reaction (10.26).

$$SO_3(g) + H_2O(liq) \rightarrow H^+(aq) + HSO_4^-(aq) \qquad (10.26)$$

These oxides or their solutions neutralize bases and form salts.

Solutions of sulfur trioxide and calcium oxide react with each other to precipitate the salt calcium sulfate ($CaSO_4$). The solvent is not necessary for this reaction because sulfur trioxide and calcium oxide can react directly.

$$SO_3(g) + CaO(c) \rightarrow CaSO_4(c) \qquad (10.27)$$

If we use a more general acid-base concept, the direct formation of calcium sulfate can also be classified as an acid-base reaction. According to the Lewis definitions, a base is an electron-pair donor and an acid is an electron-pair acceptor. In reaction (10.27), the oxide ion is a base and sulfur trioxide is an acid.

A less subtle application of the Lewis theory involves the donation of an electron-pair from the Lewis base NH_3 to the Lewis acid H^+ to form NH_4^+ (ammonium ion).

$$\begin{array}{c} \text{H} \\ \text{H}:\ddot{\text{N}}: \\ \text{H} \end{array} + \text{H}^+ \rightarrow \left(\begin{array}{c} \text{H} \\ \text{H}:\ddot{\underset{\cdot\cdot}{\text{N}}}:\text{H} \\ \text{H} \end{array} \right)^+ \qquad (10.28)$$

Other applications of the Lewis acid-base system will be described in later chapters. The principal advantage of the Lewis approach is its generality, which makes it applicable to reactions in the gas phase, in the solid state, and in solution.

Just as some oxides are in between the extremes of ionic and covalent bonding, some oxides are in between the extremes of acidic and basic. Fortunately, there are general rules that help to classify oxides as either acidic or basic and also allow predictions of relative acid or base strength of the resultant solutions. Most of these rules are based on size and electronegativity considerations.

In predicting the acidity or basicity of the oxide of an element that we represent by E, it is convenient to picture the E—O—H bonds formed on hydrolysis. If E is an electropositive element, the covalent bond between oxygen and hydrogen is sufficiently strong that OH^- ions separate as a unit and the solution is basic. On the other hand, if E is a very electronegative element the E—O covalent bond will be strong. We can picture a drift of electron density toward E, thereby weakening the O—H bond. Protons will split off, making the solution acidic.

The ability of element E to pull electrons toward itself (electronegativity) will increase with increasing charge density. Charge density increases with increasing oxidation state, decreasing size of E, or both. In H_2SO_4, sulfur is in a $+6$ oxidation state, whereas it is in a $+4$ state in H_2SO_3. We therefore predict that H_2SO_4 should be a stronger acid (that is, release protons more readily) than H_2SO_3. Experiment confirms this prediction. Between H_2SO_3 and selenous acid, H_2SeO_3, we predict that H_2SO_3 will be stronger. The valence electrons of sulfur are in the $n = 3$ level, whereas those of selenium are in the $n = 4$ level. We therefore expect sulfur to have a smaller radius and greater charge density than selenium. Experimentally, H_2SO_3 does prove to be the stronger of the two.

There are many models like the one just described that allow chemists to predict properties based on limited information. In many cases these models work amazingly well; in others they fail. The "answers" obtained are best considered as plausible "first guesses." There are frequently opposing effects, which are not considered in the simple model, that invalidate these first guesses.

Some oxides, especially those formed by elements near the center of the periodic table, exhibit both acidic and basic properties. Such oxides are called **amphoteric.** For example, zinc oxide undergoes the following two reactions:

$$ZnO(c) + 2\,H^+(aq) \rightarrow Zn^{2+}(aq) + H_2O(liq) \qquad (10.29)$$

$$ZnO(c) + 2\,OH^-(aq) + H_2O(liq) \rightarrow Zn(OH)_4^{2-}(aq) \qquad (10.30)$$

Equation (10.29) shows ZnO acting as a base, whereas equation (10.30) shows it acting as an acid.

Note that we are specifying the nature of the hydration of the anion containing zinc in the last equation. We might also have written the ion as $[ZnO_2 \cdot 2\,H_2O]^{2-}$, indicating two waters of hydration, or simply ZnO_2^{2-}. It is difficult experimentally to distinguish between such formulas as ZnO_2^{2-} and $Zn(OH)_4^{2-}$ for ions in solution. In the absence of reliable detailed information we arbitrarily write such formulas in either the simplest or most convenient way.

Sulfides are also conveniently classified as acidic, basic, or amphoteric on the basis of their solubility in excess acid or in excess sulfide ions. Noting that $S^{2-}(aq)$ and $HS^-(aq)$ are analogous to $O^{2-}(aq)$ and $OH^-(aq)$, this classification is similar to that used for acidic and basic oxides. If we extend the analogy, we would expect that those sulfides in which the central atom is small and in a high oxidation state will be the most covalent and therefore the most acidic. In fact we do find that the solubility of SnS_2 in HS^- or S^{2-} is much greater than that of SnS.

Selenium, Tellurium, and Polonium

Selenium and tellurium are rare, being found mostly as trace contaminants in sulfide ores. Polonium is radioactive and occurs in trace amounts in uranium minerals such as pitchblende.

Selenium exists in several allotropic forms; the most stable at room temperature is "metallic" selenium. This gray form, which has no sulfur analog, contains infinite chains of selenium atoms spiraling around axes parallel to one of the crystal axes. Other modifications contain Se_8 rings.

Unlike other metals, "metallic" selenium is a poor conductor of electricity in the dark, but it is a photoconductor. Its conductivity is increased as much as 200 times by light. This property is utilized in the selenium photocell used in exposure meters for measuring light intensity. Considerable amounts of selenium are used in the glass industry. Small amounts of selenium added to molten glass counteract the green color imparted by iron impurities. Red glass results from larger concentrations of selenium.

Only one crystalline form of tellurium is definitely known. The structure of this silvery white material is similar to that of "metallic" selenium, but its photoconductivity is slight. Because it is a semiconductor, tellurium is sometimes used to make rectifiers. Small amounts of tellurium are also used to improve the properties of various alloys.

The chemical behavior of selenium and tellurium and their compounds is similar to that of sulfur in several respects. For example, aqueous solutions of the oxides are weakly acidic. Investigations of selenium and tellurium chemistry have been hampered by the foul-smelling nature of the compounds. The compounds are taken up by the body and given off in perspiration and breath. Elimination is slow so that the offensive odor remains around the investigator for days or even weeks.

Polonium is constantly being produced by radioactive decay of other elements, but the concentration remains relatively constant because it decays itself. The half-life of the most stable isotope, ^{209}Po, is only 100 years. Polonium isotopes emit highly energetic α particles and are dangerous materials to handle. The chemistry of polonium has not been extensively investigated, but it appears to be the most metallic of the group VI elements. Two allotropic modifications of polonium are known.

References

Cloud, P., and A. Gibor: The oxygen cycle. *Sci. Amer.* **223:**110 (Sept. 1970).
Gregory, D. P.: The hydrogen economy. *Sci. Amer.*, **228:**13 (Jan. 1973).
Heidt, L. J.: The path of oxygen from water to molecular oxygen. *J. Chem. Educ.*, **43:**623 (1966).
Jolly, W. L.: *The Chemistry of the Non-Metals*. Prentice-Hall, Inc., Englewood Cliffs, N.J., 1966 (paperback).
Jurale, B.: Sulfuric acid and the hydrated hydronium ion. *J. Chem. Educ.*, **41:**573 (1964).
Latimer, W. M., and J. H. Hildebrand: *Reference Book of Inorganic Chemistry*. 3rd ed. Macmillan Co., New York, 1951.
Pratt, C. J.: Sulfur. *Sci. Amer.*, **222:**62 (May 1970).
Sanderson, R. T.: Principles of hydrogen chemistry. *J. Chem. Educ.*, **41:**331 (1964).
Sanderson, R. T.: Principles of oxide chemistry. *J. Chem. Educ.*, **41:**415 (1964).
Sanderson, R. T.: Recent improvements in explaining the periodicity of oxygen chemistry. *J. Chem. Educ.*, **46:**635 (1969).
Siegel, B., and J. L. Mack: The boron hydrides. *J. Chem. Educ.*, **34:**314 (1957).

Problems

1. Tritium (^3_1H) undergoes radioactive decay by emission of β^- particles. Write a balanced nuclear reaction equation to represent this decay.
2. A convenient way of making oxygen in the laboratory is by action of water on sodium peroxide, Na_2O_2. Besides O_2, sodium hydroxide is produced. (a) Write a balanced chemical equation for this reaction. (b) What volume of O_2 gas would be obtained at 0.0°C and 1.0 atm if 2.00 g Na_2O_2 were hydrolyzed?
3. The solubility of ozone in water is about 50 times that of O_2. Suggest an explanation.
4. Deuterium compounds are analogous to those of light hydrogen and are often obtained directly from heavy water, D_2O. Using D_2O as the only source of deuterium, write balanced equations to show how the following compounds could be obtained.
   (a) D_2SO_4 (b) NaOD (c) DBr (d) HD (e) LiD
5. Samples weighing 25.0 g of each of the following solids are taken: aluminum metal, lithium metal, sodium hydride, and calcium hydride. Each is treated with excess aqueous acid.
   (a) Write a balanced equation for each reaction.
   (b) Which solid releases the most hydrogen per gram of original solid?
6. Suppose that 1.00 liter of pure liquid hydrogen peroxide at 25°C is decomposed to water and oxygen while the temperature is maintained at 25°C and the pressure at 1.00 atm. The density of H_2O_2 is 1.47 g ml^{-1}. What is the change in volume associated with the decomposition reaction?
7. Although the chemical properties of all isotopes of an element are similar, differences in rates of reactions can be observed. Reactions involving ^1H can occur as much as 15 times more rapidly than those of ^2H, whereas reactions of ^{127}I are only 1.02 times faster than those of ^{129}I. Suggest an explanation for these facts.
8. The bombardment of ^6Li with neutrons in a reactor has been extensively used for production of tritium. Write a balanced nuclear reaction equation to represent this process.
9. The boiling points of H_2 and CO are 20°K and 72°K, respectively, which is far enough apart for an effective separation by fractional distillation. However, it is not considered commercially feasible to separate the two gases when they are obtained by the water gas reaction. Suggest a reason for this.

10. Carbon monoxide present as an impurity may be oxidized to carbon dioxide by cupric oxide, CuO, at 400–600°C. How many moles of carbon dioxide would be obtained from passing 50.0 liters (measured at 0.0°C and 1.0 atm) of water gas over cupric oxide? Assume that water gas is 50% carbon monoxide by volume and that the oxidation is complete.

11. Which gaseous molecule of the following sets would have the largest dipole moment:
(a) HCl, HBr, HI (c) H_2S, H_2Se, H_2Te
(b) CO, CO_2, CS_2 (d) CH_4, NH_3, H_2O
Explain your reasoning in each case.

12. One of the commercial preparations of methyl alcohol, CH_3OH, involves hydrogenation of carbon monoxide over a catalyst.
(a) Write a balanced equation for this reaction.
(b) Starting with 2.75 tons of H_2, how many tons of CH_3OH can be produced?
(c) How many tons of carbon monoxide would be consumed?

13. How much potassium chlorate is needed to produce 11.2 liters of oxygen measured at 0.0°C and 2.0 atm?

14. What volume of air (20% O_2 by volume) at 0.0°C and 1.0 atm pressure contains the same number of oxygen atoms as 1.0 ml of water?

15. What volume of liquid water must be electrolyzed to produce sufficient O_2 to fill a 5.0 liter cylinder at a pressure of 150 atm and a temperature of 25°C? The density of water is 1.0 g ml^{-1}. Although it is not an especially accurate approximation for a gas at such high pressure, use the ideal gas equation of state.

16. Describe the bonding in the ozone molecule in terms of the double quartet approach. Ozone is diamagnetic with a bond length intermediate between that of a double and single bond.

17. Write the molecular orbital designations for each of the following:
(a) O_2^{2-} (b) O_2^+ (c) O_2^- (d) CO (e) S_2
Describe the type of bond that exists between the two atoms in each case. Which would be paramagnetic?

18. An old quantitative procedure for ozone involves the reaction of ozone with excess aqueous iodide ions in a slightly alkaline solution.
(a) If we know that I^- is oxidized to I_2 and that O_3 is reduced to O_2, write a balanced equation for this reaction.
(b) An inert gas containing ozone was passed into an excess of potassium iodide solution. Analysis of the solution showed that 50.8 mg of I_2 were produced. Calculate the mass of O_3 that had been passed into the solution. What volume would this O_3 occupy at 25°C and 1.0 atm?
(The I_2 liberated reacts with excess I^- to form the triiodide ion, I_3^-, but this does not affect the present calculation. In acidic solutions too much iodine is liberated, and high values for ozone are obtained.)

19. Discuss the feasibility of using pure hydrogen peroxide as a chemical solvent.

20. Gaseous sulfur was expanded into a 120 ml bulb at a pressure of 380 mm Hg at 400°C. The weight of the evacuated bulb was 52.4173 g; the weight with sulfur vapor was 52.6553 g. Calculate the average number of sulfur atoms per molecule in the vapor.

21. Metallic copper can be obtained by roasting the ore chalcocite, Cu_2S, in air. A byproduct is SO_2.
(a) Write a balanced equation for this process.
(b) Assuming no effort is made at recovery, how much SO_2 is evolved into the atmosphere when 1.0 ton of Cu_2S is roasted. (It is easiest to think in terms of ton-moles for this problem.)
(c) In reasonably large quantities commercial SO_2 can be purchased for about \$0.10 per lb. How much could be spent to recover the SO_2 evolved from 1.0 ton

Cu_2S? What technological and economic facts of life are being ignored in your answer?

22. Commercially available "concentrated" sulfuric acid is generally about 93% H_2SO_4 by weight with a density of 1.83 g ml^{-1}. Calculate the molarity and molality of this solution.
23. Petroleum and coal contain a number of impurities, principally sulfur and its compounds. When the fuel is burned, (1) sulfur is oxidized to sulfur dioxide. In the vicinity of the flame some of the sulfur dioxide is further oxidized to sulfur trioxide (2). The sulfur trioxide reacts with moisture in the air to form sulfuric acid (3), which is extremely damaging to plant and animal life as well as structural materials such as limestone and marble, $CaCO_3$ (4). Write balanced equations for all of the numbered reactions above.
24. Selenium dissolves in sulfite solutions to give the selenosulfate ion, $SSeO_3^{2-}$. Write the balanced equation for this reaction. What substance is oxidized, and what substance is reduced? Draw the probable structure of $SSeO_3^{2-}$.
25. Using VSEPR theory, predict the structures of the following compounds and ions:

SF_6	sulfur hexafluoride	SO_2F_2	sulfuryl fluoride
$SeCl_4$	selenium tetrachloride	$TeBr_2$	tellurium dibromide
$ClSO_3^-$	chlorosulfonate ion		

26. An excess of hydrochloric acid is added to 3.7 g of sodium sulfite. Calculate the volume of sulfur dioxide evolved if the gas is collected at 23°C and 742 mm Hg.
27. Element A has an electronegativity value of 1.2 and forms an oxide AO. Element B has an electronegativity of 2.4 and forms an oxide BO_2. Write equations for the reactions that occur when these compounds are treated with water.
28. Sulfuryl fluoride (SO_2F_2) is unaffected by water but decomposes in hot basic solution according to the following equation:

$$SO_2F_2(g) + 4\,OH^-(aq) \rightarrow SO_4^{2-}(aq) + 2\,F^-(aq) + 2\,H_2O(liq)$$

A gas sample containing a mixture of SO_2F_2 and an inert gas exerts a pressure of 79.8 cm Hg in a volume of 150 ml at 22°C. The entire sample is bubbled through a hot basic solution, followed by precipitation of sulfate as $BaSO_4$(c). After drying, the $BaSO_4$(c) weighs 3.500 g. Calculate the percentage of SO_2F_2 in the original gas sample.

29. In the puckered octagonal ring of S_8 the bond angles are 105° and the bond lengths are 2.12 Å (see Figure 10.4). Calculate the distance from one sulfur nucleus to the one directly across the ring.
30. Aluminum oxide and aluminum hydroxide are amphoteric. Careful addition of sodium hydroxide to an acidic solution containing Al^{3+}(aq) ions results in precipitation of a white solid that we call aluminum hydroxide. This precipitate is easily dissolved by addition of acid or by further addition of excess base, or the white precipitate may be filtered to separate it from the solution and then converted to the oxide by heating. The oxide may then be dissolved in either acid or excess base. Write balanced equations for all these reactions.
31. Predict which compound of the following sets will be the strongest acid in aqueous solution:

(a) H_2SeO_3, H_2SeO_4
(b) HOCl, HOBr, HOI
(c) CrO, Cr_2O_3, CrO_3
(d) $HOSO_3H$, $HOSO_2Cl$, $HOSO_2F$

32. A route to $^{210}_{84}Po$ involves neutron irradiation of a $^{209}_{83}Bi$ target. ^{210}Po decays by emitting α particles with a half-life of 138 days. (a) Write nuclear reaction equations for the preparation of ^{210}Po and its decay. (b) What percentage of the original ^{210}Po would remain 276 days after it was prepared?

11

CHEMICAL ENERGY AND THE FIRST LAW OF THERMODYNAMICS

Introduction

The accumulation of relationships between energy, heat, work, temperature, and certain related quantities is often called **thermodynamics.** Thermodynamics has proven useful in nearly all fields of scientific and technological endeavor, and some of its most valuable applications are in chemistry. Further, many of the principles of "chemical" thermodynamics, along with their applications, are directly pertinent to interesting and important problems in other areas. Finally, the laws of thermodynamics lead to a variety of insights outside the "traditional" areas of science, some of which we mention briefly later.

This chapter is concerned with the origins of the first law of thermodynamics and with applications of this law to some problems of particular importance in chemistry and related sciences.

The **first law of thermodynamics** may be regarded as a precise statement or generalization of the familiar **law of conservation of energy:** *Energy may be changed from one form to another but is neither created nor destroyed.* Although the idea expressed in this statement may appear obvious today, it was once a matter of considerable dispute.

"Practical" men have long dreamed of getting something for nothing, sometimes by way of what we now call perpetual motion machines. For example, let us consider an imaginary miller who recognized long ago the value that would accrue from owning a mill that could be set up far from a flowing stream. He envisaged a great tank of water above a water wheel. Water would flow from a hole in the bottom of the tank and thereby cause the wheel to turn. The turning wheel would both grind the grain and operate a pump to return the water to the tank, as illustrated in Figure 11.1. Unfortunately, this machine, which could accomplish useful work and run forever without input of fuel or energy, has always failed to operate as described.

Mechanical perpetual motion machines more or less like the one we have just described are intended to accomplish some sort of work and to restore themselves to their original condition so that they can continue doing work forever, all without the aid of a flowing stream or other source of energy. Nowadays rational men refuse to invest in mechanical perpetual motion machines because such machines would have to operate in violation of the law of conservation of energy. Pioneering work by Archimedes, Galileo, Huygens, Newton, Leibniz, d'Alembert, Lagrange, Young, and others provided the basis for the law of conservation of energy as applied to purely mechanical systems. Extension of these principles of mechanics to systems in which heat effects were important, however, presented many difficulties.

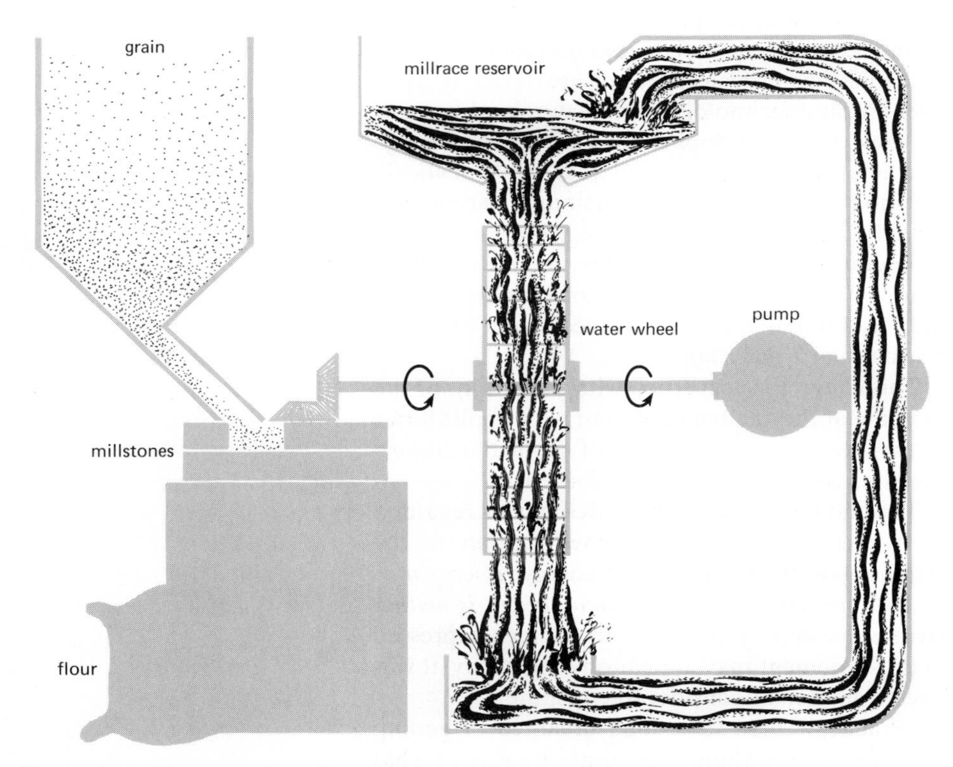

Figure 11.1. The miller's mill without a millstream. His scheme supposes that part of the energy from the water that leaves the tank is used to mill the grain and part is used to pump all of the water back into the tank. This type of device is called a perpetual motion machine of the first kind because it violates the first law of thermodynamics.

The *idea* that heat and motion are intimately connected is an ancient one, having been discussed by Democritus, Heraclitus, and Plato. It was not until many centuries later that some of the early "natural philosophers" began to do experiments to guide them toward some understanding of the nature of heat and the connection between heat, temperature, work, and other quantities. One of the first men who could properly be called an experimental scientist was Robert Boyle, who carried out experiments between 1660 and 1665 that led him to express a connection between heat and motion in nearly the same terms that we use today.

In addition to Boyle, Leibniz and a few others correctly recognized the connection between heat and motion. Most of the ideas of the men who favored a motion theory of heat were soundly based and could account for the observable facts of the day, but these ideas were generally rejected for two reasons. One reason was that Boyle and Leibniz attributed heat to motion of the "tiny particles" (atoms) of matter. It was quite reasonable for skeptical people to reject a theory based upon invisible motions of invisible particles. Probably an even more important reason for the general rejection of the motion theory of heat was the competing **caloric theory,** which was supported by many of the best minds of Europe before the end of the eighteenth century and was not finally laid to rest until the middle of the nineteenth century.

The name "caloric" for the "imponderable matter of heat" was originated in 1789 by Lavoisier, although the theory to which the name applied was already extensively developed. In 1808 John Dalton concisely summarized the conventional wisdom of the day concerning heat as follows: "The most probable opinion concerning the nature of caloric, is, that of its being an elastic fluid of great subtility, the particles of which repel one another, but are attracted by all other bodies."

Friction and vigorous compression of a gas both result in increases in temperature of the system under consideration. Adherents of the caloric theory explained these increases in temperature in terms of squeezing out caloric by friction and release of latent caloric that was made necessary by the decrease in volume of the gas (remember that the particles of caloric were supposed to repel one another). Caloric theorists were similarly able to account for many other experimental observations and in some cases were able to carry out numerical calculations that yielded results that were accurately confirmed by experiments.

Although the caloric theory could account for many experimental observations, it encountered difficulties with the experiments of Count Rumford (Benjamin Thompson). Most of these difficulties centered on the conclusion of Rumford that heat could be produced without limit by means of friction. As a result of his experiments on the heat effects associated with boring of cannons, Rumford reported to the Royal Society as follows: "In reasoning on this subject, we must not forget that most remarkable circumstance, that the source of heat generated by friction in these experiments appeared evidently to be inexhaustible. It is hardly necessary to add that anything which any insulated body or system of bodies can continue to furnish without limitation cannot possibly be a material substance; and it appears to me to be extremely difficult, if not quite impossible, to form any distinct idea of anything capable of being excited and communicated in these experiments, except it be motion."

In the years following Rumford's cannon boring experiments, a number of men were close to recognizing what we now call the first law of thermodynamics. Among those who almost discovered the first law were engineers Sadi Carnot and Marc Seguin, physician William Grove, chemists Justus von Liebig and Karl Mohr, and all-around scientist Michael Faraday. Each of these men had some glimmer of understanding of an appropriate conservation law, and each either carried out or proposed important experiments on heat; none, however, was sufficiently precise in his formulations to develop anything either convincing or useful.

Most of the credit for development of the first law of thermodynamics belongs to Hermann von Helmholtz, James Prescott Joule, and Julius Mayer for work they did from 1840 until about 1850.

As a young physician in Java, Mayer observed in 1840 that the blood of his patients was much brighter red than the blood he had taken from patients in Germany. Since it was known that the red color of venous blood was due to oxygen that had not been used for oxidation of body fuel, Mayer deduced that venous blood in Java was redder than venous blood in Germany because less combustion is needed to supply the needed body heat in hot Java than in relatively cool Germany.

Although Mayer did not know it, others had reached this same kind of conclusion before. But at this point Mayer went one important step further than had others. He reasoned that the heat developed by fuel combustion in the body should be balanced against the loss of heat to the surroundings *and* the work the body performs. That is, Mayer concluded (in present day terminology) that heat and work are merely two different manifestations of a general property called energy, which is conserved. Following his return to Germany, Mayer undertook both experiments and calculations with experimental results of others in an effort to secure supporting evidence for his ideas, but he was largely unsuccessful in convincing the scientists of the day.

James Joule, son of a prosperous brewer, undertook his first serious scientific work in 1837 at age 19 when he built an electric engine powered by a battery. In the course of testing this engine, he discovered what we now call Joule's law, which says that the rate of heat production is proportional to the square of the electric current times the resistance of the wire carrying the current. These measurements led him to investigate the conversion of mechanical work into heat, which he did in a variety of ways that in turn led him to numbers for what we now call the mechanical equivalent of heat. In the years from 1843 to 1850 Joule came to two important conclusions: (1) Heat and work are different manifestations of energy, which is conserved, and (2) the numerical relationship between common units of work (such as foot-pound) and heat (such as calorie) was accurately known from his water-stirring experiments and approximately known from results of various calculations made with earlier experimental results.

In 1847 Hermann Helmholtz, a young (age 26) physician, approached what we now call the first law of thermodynamics in an entirely different way. Nowadays we deny the possibility of mechanical perpetual motion machines because such machines would have to operate contrary to the law of conservation of energy. Helmholtz reversed this argument by pointing out that the assumed impossibility of building such a machine leads to the law of conservation of energy. Helmholtz then showed that heat and work must both be considered as energy and that it is the total that is conserved, rather than either heat or work separately. Finally, Helmholtz used the requirement that energy accounts (including both heat and work) must balance in solving a variety of scientific problems.

Beginning in about 1850, the ideas of Mayer, Joule, and Helmholtz gradually became accepted, initially because of the excellent experimental measurements by Joule and the logical development and application by Helmholtz. It was more than ten years later that Mayer's pioneering work came to be generally appreciated.

The First Law of Thermodynamics

In order to go from a verbal statement of the law of conservation of energy to a useful version of the first law of thermodynamics, it is necessary to enumerate the various forms of energy and then express in an equation the constancy of total energy.

We divide all of what we have called energy into "heat" and "work." Work is then further divided into mechanical work, electrical work, magnetic work, gravitational work, and so on. Of all these, however, only the mechanical work associated with volume changes accomplished against a restraining pressure is of present interest. Later in this chapter we show how to calculate work from experimentally observable quantities.

In a constant volume system (no mechanical work effects), we define heat as the energy that is transferred between system and surroundings due to a temperature difference. The amount of heat transferred can be determined experimentally with certain calorimeters as described later in this chapter.

A common unit of heat is the **calorie,** abbreviated as **cal.** Another common unit is the **kilocalorie,** which is equal to 1000 cal and is abbreviated as **kcal.** The calorie was once defined as the amount of heat required to raise the temperature of 1 g of water by 1 °C. This definition was found to be inadequate because a given amount of heat transferred to water at one temperature does not produce exactly the same change in temperature as the same amount of heat transferred to water at another temperature. This amounts to saying that the specific heat or heat capacity of water is not exactly the same at all temperatures. To get around this difficulty, the so-called 15° cal was defined as the amount of heat required to raise the temperature of 1 g of water from 14.5 °C to 15.5 °C. Still later, because of the accuracy and precision with which some electrical measurements can be carried out, the calorie was redefined in terms of the **joule (J)** (joule = volt coulomb = volt ampere second). Now 1 cal is defined as exactly 4.1840 J and is nearly equal to the old 15° cal.

Although the best and most extensive compilations of chemical data list energies and related quantities in terms of calories and kilocalories, there has been increasing use of joules and kilojoules (1 kJ = 1000 J). As a result, we shall on several occasions have problems in which energies must be converted from one set of units to the other.

Since Einstein has shown by his famous equation $E = mc^2$ that energy and mass are interchangeable, we know that a complete statement of the first law should mention mass. However, because conversion of mass into energy or energy into mass is unimportant in ordinary chemical processes, we shall consider mass and energy to be distinctly different things. Thus we may consider energy separately to be conserved.

We now define a thermodynamic system as that part of the world that is of particular thermodynamic interest in connection with some particular problem. The rest of the world is then taken to be the surroundings of the system. A thermodynamic system may lose heat to its surroundings, may gain heat from its surroundings, may do work on its surroundings, or may have work done on it by its surroundings.

Now let us suppose that some thermodynamic system undergoes a process or series of processes in which it is changed in a variety of ways before returning to its original state. According to the first law, the total energy of the system must be the same at the end of the cyclical process as it was at the beginning. This conservation of energy can be expressed by $E_2 = E_1$ or by $E_2 - E_1 = 0$ in which E_2 represents the energy of the system at the end of the process and E_1 represents the energy of the system at the beginning. We also write

$$\Delta E = 0 \qquad \text{(for cyclical process)} \tag{11.1}$$

in which ΔE represents $E_2 - E_1$. Equation (11.1) is a mathematical statement of the law of conservation of energy and may be regarded as part of a statement of the first law of thermodynamics.

Suppose that we have a thermodynamic system that can exist in two different states that we denote by letters A and B. The energy change in going from state A to state B is ΔE_{AB} and the energy change in going from state B to state A is ΔE_{BA}. The total energy change in going from state A to state B and then back to state A (a cyclical process) is $\Delta E = \Delta E_{\mathrm{AB}} + \Delta E_{\mathrm{BA}}$. If this total $\Delta E \neq 0$, we could construct a machine that would operate around this cycle and yield an energy profit. That is, we could construct a perpetual motion machine that would run forever with no input of fuel, and in addition accomplish some sort of useful work with its energy profit. As stated previously, Helmholtz concluded that this sort of machine is impossible, which requires that $\Delta E = 0$ for a cyclical process. This in turn requires that $\Delta E_{\mathrm{AB}} = -\Delta E_{\mathrm{BA}}$. Since all of these statements are valid for any path from state A to state B or from state B to state A, we can conclude that the value of ΔE for a change from any state to any other state is independent of the path by which the change is carried out.

The reasoning in the paragraph above leads us to describe energy as a state function, meaning that the value of ΔE for transformation of a system from one state to another is dependent only on the particular states and not upon the details of how the transformation is carried out. In mathematical language, we say that dE is an exact differential. As we shall soon see, some other important thermodynamic quantities are state functions, whereas still others are not state functions.

Because we mostly want to apply the first law to noncyclical processes, equation (11.1) is not directly useful. We find it more useful to state the first law in terms of the forms of energy (heat and work) that can be converted into one another or be transferred from one system to another. We say that the energy change undergone by a thermodynamic system is equal to the energy gained in the form of heat minus the energy lost as a result of work done by the system on the surroundings. This statement is expressed concisely by the equation

$$\Delta E = q - w \tag{11.2}$$

in which q represents the heat absorbed by the system and w represents the work done by the system on the surroundings.

Because q represents the heat absorbed by the system, positive q means heat absorbed by the system and negative q means heat evolved by the system (transferred to the surroundings). Because w represents work done by the system on the surroundings, positive w means that the system does work on the surroundings whereas negative w means that work is done on the system by the surroundings. Positive $\Delta E(= E_2 - E_1)$ means that E_2 is greater than E_1 and therefore means that the process under consideration has resulted in a net increase or gain in energy for the system, while the surroundings have lost an equal amount of energy. Negative ΔE means that E_2 is less than E_1, which means that the system has lost energy to its surroundings. Meanings of signs of ΔE, q, and w are illustrated further in example problems later in this chapter.

Work

One of the fundamentals of physics is the definition of mechanical work as force times the displacement or distance through which the force acts. We will now set about using this definition in deriving an equation that we can use for calculating the work associated with expansion or compression of a gas in a cylinder closed by a movable piston as pictured in Figure 11.2.

Because pressure has been defined as force per unit area as in $P = f/A$, we also have

$$f = PA \tag{11.3}$$

Now, remembering that work has been defined as force times distance and recognizing that the relevant distance is the height h through which the piston is moved, we have

$$w = fh = PAh \tag{11.4}$$

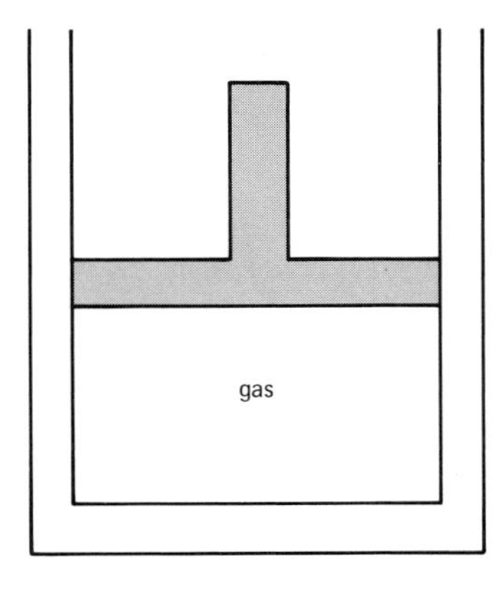

Figure 11.2. The external pressure on the piston is P, and the cross sectional area of the piston is A. The distance through which the piston moves is h, with positive h corresponding to upward movement of the piston. As shown in equations (11.3–11.6), the work done by the gas on the constant pressure surroundings is $w = P\Delta V$.

We now note that the cross-sectional area of the piston times the distance it moves gives the change in volume of the gas in the cylinder, as in

$$V_2 - V_1 = Ah \tag{11.5}$$

Finally, we substitute equation (11.5) into equation (11.4) to obtain

$$w = P(V_2 - V_1) = P\Delta V \tag{11.6}$$

as the desired equation that permits us to calculate the mechanical work of expansion or compression against a constant external pressure represented by P. It is important to keep in mind that equation (11.6) is applicable only in this special, but common, case of constant external pressure; later we shall be concerned with work done against an external pressure that varies in a particular way.

Note that positive ΔV means that the volume of the system is increased and leads by equation (11.6) to positive w, which in turn means that the system has done work on the surroundings. Negative ΔV means that the volume of the system has decreased and leads to negative w, which means that the system has had work done on it by the surroundings.

Because we usually use equation (11.6) to obtain a numerical value for w expressed in some units of pressure times volume (such as atmosphere liters) whereas heat and energy are commonly expressed in terms of calories or joules, it is necessary to convert to a common set of units before we can use equation (11.6) with equation (11.2). This very useful equation is

$$\Delta E = q - P\Delta V \quad \text{(constant pressure)} \tag{11.7}$$

Example Problem 11.1. Suppose that a gas enclosed in a piston and cylinder apparatus like that pictured in Figure 11.2 absorbs 1000 cal from its surroundings and expands against a constant external pressure of 1.5 atm so that the final volume is 6.0 liters larger than the initial volume. Calculate w and ΔE for the gas.

Because the gas expanded, we know that $\Delta V = +6$ liters, which we substitute into equation (11.6) to obtain

$$w = P\Delta V = 1.5 \text{ atm} \times 6.0 \text{ liters} = +9.0 \text{ atm liters}$$

The positive signs for ΔV and w show that the gas has done work on its surroundings by pushing the piston against a restraining pressure.

Because the system absorbed heat from its surroundings, we know that $q = +1000$ cal. Now we insert our values (with units) for q and w into equation (11.2) to obtain

$$\Delta E = q - w = 1000 \text{ cal} - 9.0 \text{ atm liters} \tag{11.8}$$

To obtain the desired ΔE expressed in calories, it is necessary to convert 9.0 atm liters to calories and then carry out the indicated subtraction. We may, if we wish, look up a conversion factor that will enable us to convert atmosphere liters to calories. Instead, we follow the generally useful procedure of remembering or looking up values of the gas constant R expressed in both sets of units and construct our own conversion factor. For our problem the desired values are $R = 1.987$ cal deg^{-1} mole^{-1} and $R = 0.082$ liter atm deg^{-1} mole^{-1}.

We now multiply 9.0 atm liter by R expressed in calories and divide by R expressed in liter atmospheres to obtain

$$\frac{9.0 \text{ atm liter} \times 1.987 \text{ cal deg}^{-1} \text{ mole}^{-1}}{0.082 \text{ liter atm deg}^{-1} \text{ mole}^{-1}} = 218 \text{ cal}$$

Substitution of 218 cal for 9.0 atm liters in equation (11.8) now gives us

$$\Delta E = 1000 \text{ cal} - 218 \text{ cal} = 882 \text{ cal}$$

The positive value of ΔE shows that the system made a net gain of 882 cal from its surroundings. ■

Heat, Energy, and Enthalpy

We are often concerned with the relationship between heat and energy for constant volume processes such as the combustion of carbon in oxygen in a constant volume container called a bomb calorimeter. For a constant volume process, $\Delta V = 0$ and there can be no work of the sort that we are considering so that equation (11.2) leads to

$$\Delta E = q \qquad \text{(constant volume processes)} \qquad (11.9)$$

Processes taking place at constant pressure are even more common in chemistry (also biology and other sciences) than are processes taking place at constant volume. It is therefore useful to consider the first law of thermodynamics in some detail as it applies to constant pressure processes. An example of a common constant pressure process is a chemical reaction that takes place in an open beaker exposed to the (approximately) constant pressure of the atmosphere.

We now set about applying the first law to processes that take place at constant pressure. Our first step is to combine $w = P\Delta V$ for constant pressure processes with the general equation $\Delta E = q - w$ to obtain

$$\Delta E = q - P\Delta V \qquad \text{(constant pressure processes)} \qquad (11.10)$$

Next we write this equation as

$$E_2 - E_1 = q - P(V_2 - V_1)$$

and rearrange to

$$q = (E_2 + PV_2) - (E_1 + PV_1) \qquad (11.11)$$

Because the combination $(E + PV)$ is so closely associated with the heat absorbed or evolved in a process taking place at constant pressure, it is now convenient to define a new thermodynamic quantity H as

$$H = E + PV \qquad (11.12)$$

and substitute in (11.11) to obtain

$$q = H_2 - H_1 = \Delta H \qquad \text{(constant pressure processes)} \qquad (11.13)$$

The quantity represented by H is called the **heat content** or more often the **enthalpy.** The first of these names, at least for constant pressure processes, is indicative of the physical significance of H and can be understood and used as follows. If H_2 is greater than H_1, indicating that the "heat content" of the system in its final state is greater than in its initial state, $\Delta H = q$ is positive. Therefore we know that the system absorbs heat from its surroundings in what is called an **endothermic** process. If H_2 is less than H_1, $\Delta H = q$ is negative and the system loses or transfers energy to its surroundings as heat in what we call an **exothermic** process.

In spite of the descriptive advantage in using heat content as a name for the defined function H, it is probably best to avoid this use because this name implies that heat is a "substance" like "caloric" that can be contained. Further, some more advanced books use the name heat content to represent $H_2 - H_1$ for a system whose temperature is increased or decreased by a specified number of degrees at constant pressure. We shall therefore use the name enthalpy and say that the enthalpy change ($H_2 - H_1 = \Delta H = q$) is a measure of the heat absorbed or evolved by a system undergoing a constant pressure process.

Heat Capacity

A useful quantity in connection with considerations of heat is the **heat capacity** of either a system or a given amount of some substance. The heat capacity of a system may be defined as

$$C = q/\Delta T \tag{11.14}$$

in which q represents the heat absorbed by the system and ΔT is the accompanying increase in temperature of the system.

Heat capacities of most substances are usually expressed in terms of calories per degree per mole (cal deg^{-1} mole^{-1}) although one can use any units that are consistent with

$$\text{heat capacity} = (\text{energy})(\text{degree})^{-1}\ (\text{amount of material})^{-1}$$

At present there is an increasing trend in the direction of expressing heat capacities in terms of joules per degree per mole (J deg^{-1} mole^{-1}). The old term **specific heat** refers to the heat capacity of 1 g of material and is especially useful for substances with unknown molecular weight.

Heat capacities are usually measured under conditions of constant pressure or constant volume. Since these heat capacities are not generally equal, even for the same amount of the same substance at the same temperature, we denote them by different symbols. We follow custom in choosing C_p for the heat capacity at constant pressure and C_v for the heat capacity at constant volume. Remembering that $q = \Delta H$ when the pressure is constant, we have

$$C_p = \Delta H/\Delta T \tag{11.15}$$

For constant volume processes we have $\Delta E = q$ and thence

$$C_v = \Delta E/\Delta T \tag{11.16}$$

The heat capacity of a substance depends on the temperature at which it is measured, and usually increases with increasing temperature. Thus the heat capacities defined by equations (11.15) and (11.16) are average heat capacities over the range of temperatures denoted by ΔT. It is frequently more useful to work with heat capacities that refer to some particular temperature, which can be approached experimentally by making q and therefore ΔT so small that the heat capacity will not change significantly over this small temperature interval. Thus we use the notation of calculus and write equations (11.15) and (11.16) as

$$C_p = \left(\frac{\partial H}{\partial T}\right)_p \tag{11.17}$$

and

$$C_v = \left(\frac{\partial E}{\partial T}\right)_v \tag{11.18}$$

Heats and Enthalpy Changes for Reactions

When 2 moles of gaseous hydrogen react with 1 mole of gaseous oxygen to form 2 moles of liquid water, all at constant pressure of 1.0 atm and at 25 °C, 136.6 kcal are liberated or transferred to the surroundings. One convenient way to express this is by way of the equation

$$2\ H_2(g) + O_2(g) \rightarrow 2\ H_2O(liq) + 136.6\ \text{kcal} \tag{11.19}$$

The reaction is said to be exothermic because heat is liberated or evolved by the system. For the reverse reaction in which 2 moles of liquid water are decomposed into hydrogen and oxygen, all at 1.0 atm and 25 °C, 136.6 kcal are absorbed from the surroundings as indicated by

$$2\ H_2O(liq) + 136.6\ \text{kcal} \rightarrow 2\ H_2(g) + O_2(g) \tag{11.20}$$

This decomposition reaction is said to be endothermic because the chemical system absorbs heat from the surroundings when the reaction is carried out at constant temperature.

From our definition of q as the heat absorbed by the system, we see that $q = -136.6$ kcal for the first reaction, if we regard the thermodynamic system as being the various chemicals. Similarly, $q = +136.6$ kcal for the second reaction. Because we have specified that these reactions take place at constant pressure, equation (11.13) tells us that $q = \Delta H$ for these reactions. It is common and useful practice to specify the heat involved in chemical reactions carried out at constant pressure by giving ΔH values for the reactions as in equations (11.21) and (11.22).

$$2\ H_2(g) + O_2(g) \rightarrow 2\ H_2O(liq) \qquad \Delta H = -136.6\ \text{kcal} \tag{11.21}$$

$$2\ H_2O(liq) \rightarrow 2\ H_2(g) + O_2(g) \qquad \Delta H = +136.6\ \text{kcal} \tag{11.22}$$

It was found experimentally about 1840 by G. H. Hess that quantities of heat associated with chemical reactions can be added just as the balanced equations for the reactions are added. When these q values are measured under constant volume conditions we have $\Delta E = q$, and it follows from our earlier discussion of energy as a state function that this additivity is a necessary consequence of the first law. When these q values are measured under constant pressure conditions, we have $\Delta H = q$. Recalling that enthalpy H is defined as $H = E + PV$ and remembering that E is a state function and that P and V serve to define the state, we see that H is also a state function. Because H is a state function, ΔH is independent of path. As we shall show, this requires the additivity of heats (enthalpies) of reaction as mentioned above.

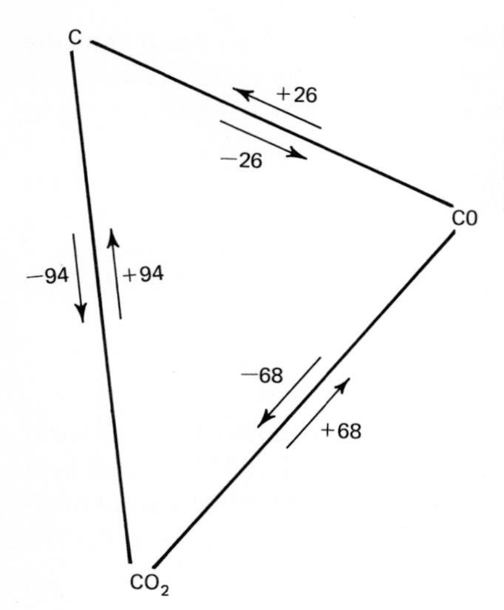

Figure 11.3. The cycle pictured here is consistent with the first law of thermodynamics and the state function nature of enthalpy. Total ΔH for any complete cycle, in either direction, must equal zero. All numbers indicate ΔH values expressed in kilocalories per mole.

Combustion of 1 mole of carbon in the graphite form to yield carbon dioxide is an exothermic reaction as indicated by

$$C(gr) + O_2(g) \rightarrow CO_2(g) \qquad \Delta H = -94 \text{ kcal} \qquad (11.23)$$

Similarly, combustion of 1 mole of carbon monoxide is exothermic, as indicated by

$$CO(g) + \tfrac{1}{2}O_2(g) \rightarrow CO_2(g) \qquad \Delta H = -68 \text{ kcal} \qquad (11.24)$$

Direct measurement of the heat or enthalpy of combustion of carbon in a limited supply of oxygen so that the product will be carbon monoxide is experimentally difficult and is unnecessary. Instead we can make use of the state function properties of enthalpy to get the desired value from equations (11.23) and (11.24).

A diagram representing chemical changes in the C-CO-CO_2 system is shown in Figure 11.3. Because enthalpy H is a state function, we know that the total ΔH for going around this cycle must be zero and that the ΔH for going from carbon to carbon dioxide must be independent of the path. The direct reaction of carbon with oxygen to form carbon dioxide (left-hand path in the figure) has $\Delta H = -94$ kcal mole^{-1}. Consequently, the two-step reaction that goes by way of carbon monoxide (right-hand path in the figure) must also yield a *total* ΔH of -94 kcal mole^{-1}. One of the two steps is already known to have $\Delta H = -68$ kcal mole^{-1}, so we deduce that the other step must have $\Delta H = -26$ kcal mole^{-1} and are able to write

$$C(gr) + \tfrac{1}{2}O_2(g) \rightarrow CO(g) \qquad \Delta H = -26 \text{ kcal} \qquad (11.25)$$

Although it is possible in principle to evaluate enthalpies (constant pressure heats) for a great many reactions by means of other diagrams like that in Figure 11.3, the prospects are not appealing. The actual situation is much worse than working with many-sided figures. Many of the ΔH values that would be used in constructing the many-sided figures of interest would themselves have to be derived from other many-sided figures, and some of the ΔH values in these figures would have come from other figures, and so on. Further, merely tabulating and then indexing the experimental ΔH values to be used in such calculations is not a trivial problem. Happily, the way out of this forest of difficulties is as easy as understanding the heights of mountains and the depths of oceans.

How high is Pike's Peak in Colorado? There is no absolute answer to this question, but a common and useful answer is that the top of the mountain is 14,110 ft above sea level. Similarly, we may say that the bottom of the Mariana Trench in the Pacific Ocean is 36,200 ft below sea level. Or we might say that the height of Pike's Peak is $+14{,}110$ ft and the depth of the Mariana Trench is $-36{,}200$ ft. In both cases we have taken sea level as our zero or reference level for elevation.

Heights of mountains and depths of valleys are illustrated in Figure 11.4 on the basis of sea level as zero elevation. It is easy to see from this figure that simple arithmetic tells us how much higher one mountain is than another. For example, the top of Mount McKinley is $20{,}320 - 14{,}110 = 6{,}210$ ft higher than the top of Pike's Peak. Similarly, the top of Pike's Peak is $14{,}110 - (-280) = 14{,}390$ ft higher than the lowest point in Death Valley.

We can measure enthalpy changes such as $H_2 - H_1 = q$, but we cannot measure

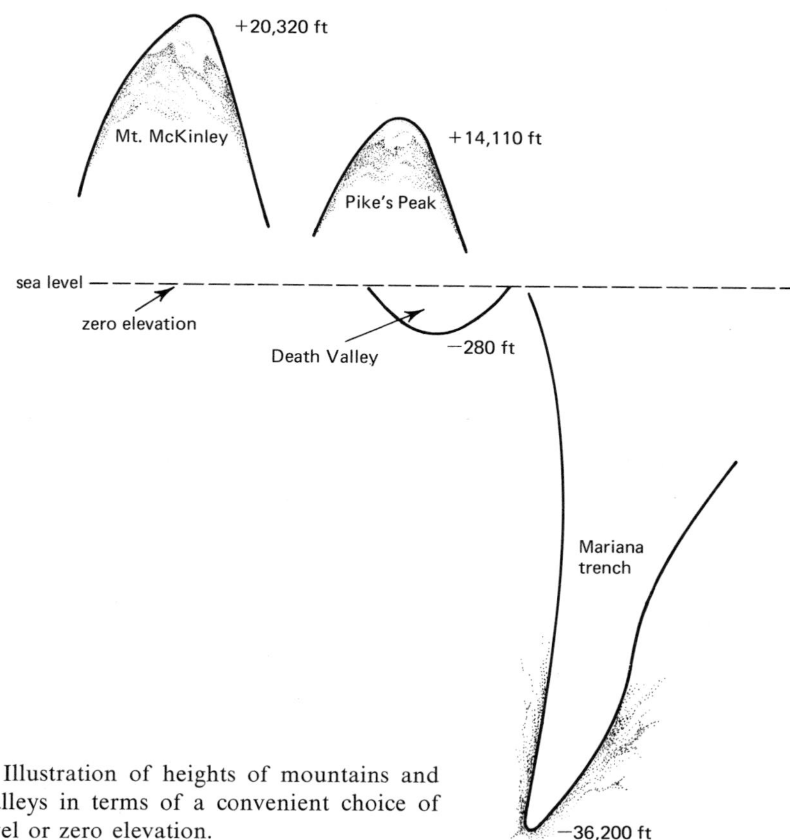

Figure 11.4. Illustration of heights of mountains and depths of valleys in terms of a convenient choice of reference level or zero elevation.

the absolute value of any single enthalpy. We therefore find it convenient and useful to adopt a "sea level" or reference for enthalpies as was done previously for elevations.

By general agreement the zero or reference level for enthalpy is taken to be each of the chemical elements in its ordinary stable state at 298°K and 1.0 atm. Thus, the enthalpies of carbon in the graphite form, gaseous hydrogen, and gaseous oxygen are all taken to be zero and we say that $H^0 = 0$ for C(gr), H_2(g), and O_2(g). The superscript zero is used here to show that we are concerned with the enthalpy of a substance in a standard state—in this case the common one of 298°K and 1.0 atm.

On the basis of our choice of zero levels for enthalpies, we now define the standard enthalpy of formation of any compound as the ΔH^0 for formation of the compound from its elements. An example is

$$H_2(g) + \tfrac{1}{2}O_2(g) \rightarrow H_2O(liq) \qquad \Delta H^0 = -68.315 \text{ kcal mole}^{-1} \qquad (11.26)$$

We now say that $\Delta H_f^0 = -68.315$ kcal mole^{-1} for H_2O(liq). Similarly, on the basis of an accurate value of $\Delta H^0 = -94.051$ kcal mole^{-1} for formation of CO_2(g) from its elements as in equation (11.23), we say that $\Delta H_f^0 = -94.051$ kcal mole^{-1} for CO_2(g).

Now let us consider the ΔH_f^0 for carbon monoxide. As already pointed out, it is not practical to make an accurate measurement of the enthalpy change for the reaction of graphite with oxygen to form carbon monoxide. We must therefore obtain the desired ΔH_f^0 indirectly. One way of doing so has been discussed in connection with Figure 11.3. Another way to do this is to make an "enthalpy elevation" diagram as in Figure 11.5. Still another way is to add two chemical reactions and their ΔH^0 values as follows:

$$C(gr) + O_2(g) \rightarrow CO_2(g) \qquad \Delta H^0 = -94.051 \text{ kcal} \qquad (11.27)$$

$$CO_2(g) \rightarrow CO(g) + \tfrac{1}{2}O_2(g) \qquad \Delta H^0 = +67.634 \text{ kcal} \qquad (11.28)$$

$$C(gr) + \tfrac{1}{2}O_2(g) \rightarrow CO(g) \qquad \Delta H^0 = -94.051 + 67.635 = -26.416 \text{ kcal} \qquad (11.29)$$

On the basis of equation (11.29) and the ΔH^0 value, we now say that $\Delta H_f^0 = -26.416$ kcal mole^{-1} for CO(g). Note that equation (11.28) above is just the reverse of equation (11.24) written earlier and that the enthalpy changes for these two reactions have opposite signs, which is a necessary consequence of the state function property of enthalpy.

We now generalize all of the preceding discussion and calculations by means of the following important equation:

$$\Delta H^0_{\text{reaction}} = \Sigma\Delta H_f^0(\text{products}) - \Sigma\Delta H_f^0(\text{reactants}) \qquad (11.30)$$

Readers may verify that equation (11.30) is in fact consistent with the various $\Delta H^0_{\text{reaction}}$ and ΔH_f^0 values associated with equations (11.20) through (11.29). Here we are using $\Delta H_f^0 = 0$ for elements in their standard states, which is consistent with the earlier choice of $H^0 = 0$.

Equation (11.30) is useful for calculating ΔH_f^0 values from measured $\Delta H^0_{\text{reaction}}$ values, and is also useful for calculating $\Delta H^0_{\text{reaction}}$ values from tabulated ΔH_f^0 values. We illustrate these calculations with some examples.

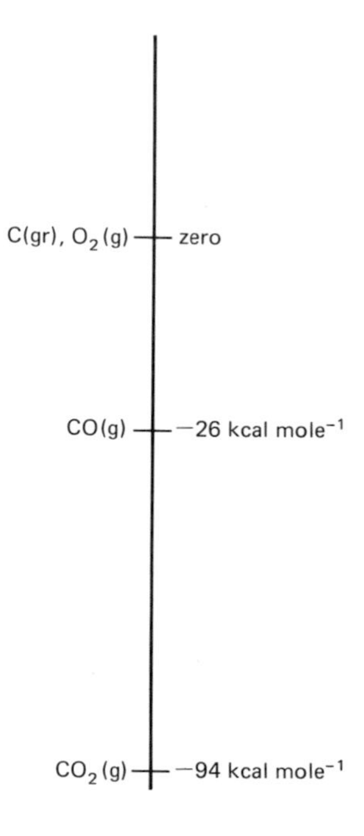

Figure 11.5. "Enthalpy-elevation" map for C(gr), $O_2(g)$, CO(g), and $CO_2(g)$, with the elements in standard state taken to be reference or "sea level" substances with zero enthalpy. Thus we have $\Delta H_f^0 = -26$ kcal mole for CO(g) and $\Delta H_f^0 = -94$ kcal mole^{-1} for $CO_2(g)$.

Example Problem 11.2. Calculate ΔH^0 for the reaction represented by equation (11.31).

$$AgCl(c) + HI(g) \rightarrow AgI(c) + HCl(g) \tag{11.31}$$

PROCEDURE 1. In Appendix II we find ΔH_f^0 values for the substances involved in this reaction.

$$\begin{array}{lll} AgCl(c) & \Delta H_f^0 = & -30.37 \text{ kcal mole}^{-1} \\ AgI(c) & \Delta H_f^0 = & -14.78 \text{ kcal mole}^{-} \\ HI(g) & \Delta H_f^0 = & 6.33 \text{ kcal mole}^{-1} \\ HCl(g) & \Delta H_f^0 = & -22.06 \text{ kcal mole}^{-1} \end{array}$$

Now we substitute these values into equation (11.30) to obtain

$$\begin{aligned} \Delta H_{11.31}^0 &= [(-14.78) + (-22.06)] - [(-30.37) + (6.33)] \\ &= -12.80 \text{ kcal mole}^{-1} \end{aligned} \tag{11.32}$$

PROCEDURE 2. In another approach to a similar use of these ΔH_f^0 values and the equations for the reactions they represent, we begin by writing down the chemical equations for the formation reactions that we refer to when we speak of ΔH_f^0 values. These equations are as follows, with the appropriate ΔH^0 values:

$$\begin{array}{ll} Ag(c) + \tfrac{1}{2} I_2(c) \rightarrow AgI(c) & \Delta H^0 = -14.78 \text{ kcal mole}^{-1} \\ Ag(c) + \tfrac{1}{2} Cl_2(g) \rightarrow AgCl(c) & \Delta H^0 = -30.37 \text{ kcal mole}^{-1} \\ \tfrac{1}{2} H_2(g) + \tfrac{1}{2} Cl_2(g) \rightarrow HCl(g) & \Delta H^0 = -22.06 \text{ kcal mole}^{-1} \\ \tfrac{1}{2} H_2(g) + \tfrac{1}{2} I_2(c) \rightarrow HI(g) & \Delta H^0 = \quad 6.33 \text{ kcal mole}^{-1} \end{array}$$

Now we rewrite the second and fourth equations above in reverse direction, changing the signs of the associated ΔH^0 values. Finally, we add the first and third reaction equations

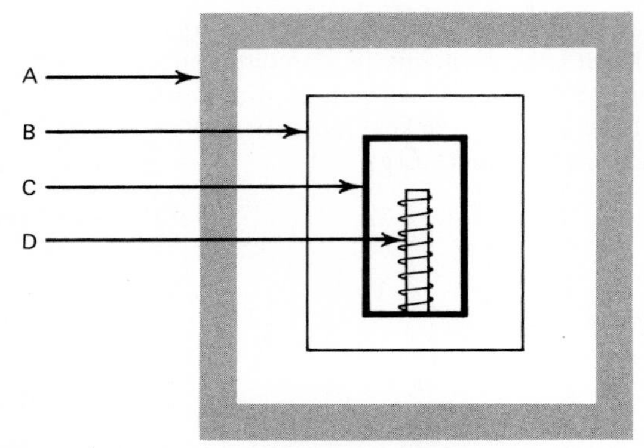

Figure 11.6. Schematic illustration of an apparatus that is used for measurement of heat capacities. A represents the calorimetric environment, which might be a furnace for high temperature investigations or a bath of liquid helium for low temperature investigations. Because the shield B is maintained at the same temperature as the calorimeter C, no heat flows into or out of the calorimeter; B is, therefore, called an adiabatic shield. The calorimeter C contains the material under investigation and also a thermometer and electric heater, both represented by D.

Electrical energy q added to the calorimeter causes an increase in temperature ΔT, and from equation (11.14) the heat capacity is calculated as $C = q/\Delta T$.

to the reversed second and fourth equations and similarly add the ΔH^0 values to obtain the desired ΔH^0 as follows:

$$\begin{array}{ll} Ag(c) + \frac{1}{2}I_2(c) \rightarrow AgI(c) & \Delta H^0 = -14.78 \text{ kcal mole}^{-1} \\ AgCl(c) \rightarrow Ag(c) + \frac{1}{2}Cl_2(g) & \Delta H^0 = +30.37 \text{ kcal mole}^{-1} \\ \frac{1}{2}H_2(g) + \frac{1}{2}Cl_2(g) \rightarrow HCl(g) & \Delta H^0 = -22.06 \text{ kcal mole}^{-1} \\ HI(g) \rightarrow \frac{1}{2}H_2(g) + \frac{1}{2}I_2(g) & \Delta H^0 = -\ \ 6.33 \text{ kcal mole}^{-1} \\ \hline \end{array}$$

$$AgCl(c) + HI(g) \rightarrow AgI(c) + HCl(g) \tag{11.31}$$

$$\Delta H^0_{11.31} = -14.78 + 30.37 - 22.06 - 6.33 \tag{11.32a}$$

$$= -12.80 \text{ kcal mole}^{-1}$$

We see that equations (11.32) and (11.32a) are the same, and thus yield the same value for the desired ΔH^0. It is always possible to work this kind of problem by either Procedure 1 or Procedure 2 illustrated here. The calculation based on equation (11.30) is usually

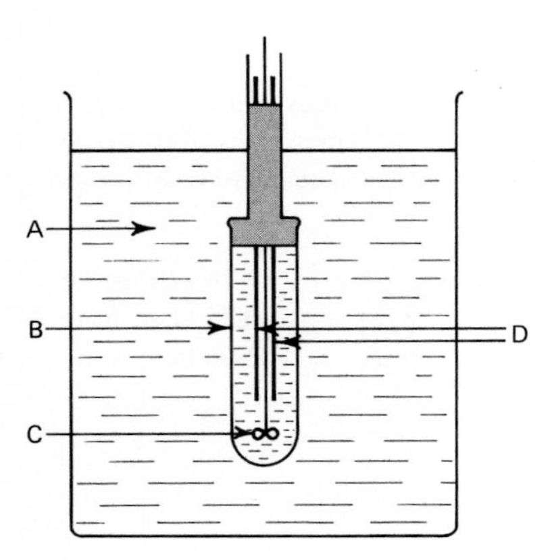

Figure 11.7. Schematic illustration of a calorimeter used for determination of enthalpies of reactions. A, B, and C represent a water bath maintained at constant temperature, a dewar vessel and a stirrer, respectively. D represents an electric heater used for calibration and a thermometer. Means of introducing samples of reacting chemicals are not shown.

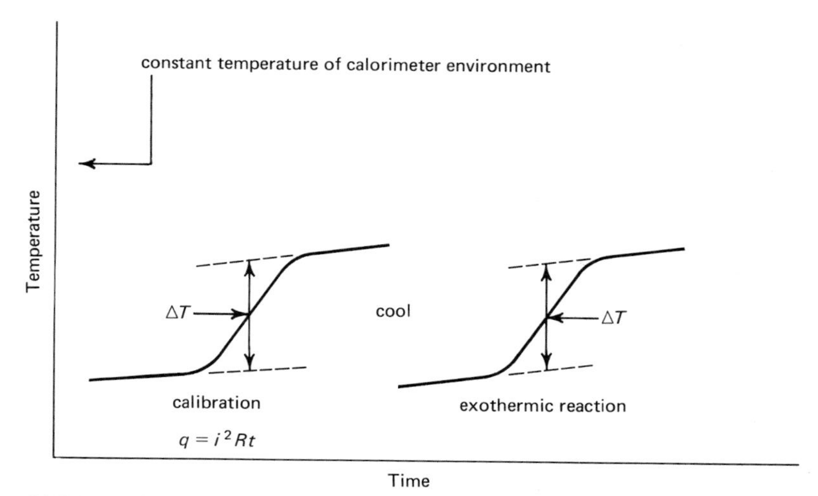

Figure 11.8. Graph of data obtained with a solution calorimeter like that illustrated in Figure 11.7.

The first temperature rise, caused by electrical heat q, permits calculation of the heat capacity of the system as $C = q/\Delta T$. Measurement of the temperature rise associated with the exothermic reaction permits calculation of the reaction heat as $q = C\Delta T$. Because the measurements are carried out at constant pressure, we identify q with ΔH.

the method of choice, simply because it is easier than the method based on explicit addition of reaction equations and enthalpies as done to obtain (11.32a) above.

It should be recognized that Procedures 1 and 2 are based on the same principle; equation (11.30) is simply a convenient generalization of the more cumbersome method shown in Procedure 2. ■

Example Problem 11.3. A result of careful calorimetric measurements is that the ΔH^0 of combustion of propane is -12.034 kcal g^{-1}. Calculate the ΔH_f^0 for propane.

We begin by combining atomic weights to find that the molecular weight of propane (C_3H_8) is 44.10 g mole^{-1}. Thus for combustion of 1 mole of propane we calculate

$$\begin{aligned}\Delta H^0 &= -12.034 \text{ kcal g}^{-1} \times 44.10 \text{ g mole}^{-1}\\ &= -530.6 \text{ kcal mole}^{-1}\end{aligned}$$

Now we write the balanced equation for the combustion reaction as

$$C_3H_8(g) + 5\,O_2(g) \rightarrow 3\,CO_2(g) + 4\,H_2O(liq)$$

Again we make use of equation (11.30). This time we know ΔH^0 for the reaction, and we can look up ΔH_f^0 values for H_2O(liq), CO_2(g), and O_2(g). Substitution of the appropriate values in the general equation (11.30) gives us

$$-530.6 = [3(-94.05) + 4(-68.32)] - [5(0) + (\Delta H_f^0)]$$

This equation is easily solved for

$$\Delta H_f^0(\text{propane}) = -24.8 \text{ kcal mole}^{-1} \quad ■$$

Schematic diagrams of calorimeters used for heat capacity and heat of reaction measurements are shown in Figures 11.6 and 11.7. Experimental data for a heat of reaction experiment are shown in Figure 11.8.

Bond Energies and Enthalpies

In this book we have been and will be further concerned with bond energies in relation to various theories of bonding. It is now appropriate to consider how bond energies (and bond enthalpies) can be obtained from thermodynamic data. We shall also illustrate one use of these quantities.

We begin by considering the bond in a generalized diatomic molecule that we represent by A-B. In order to eliminate complications from intermolecular interactions, we focus our attention on A-B in the ideal gas state at 1.0 atm. Now we represent the dissociation of A-B(g) into separated A and B atoms (also ideal gases at 1.0 atm) by the equation

$$\text{A-B(g)} \rightarrow \text{A(g)} + \text{B(g)} \tag{11.33}$$

for which we have either ΔH^0 or ΔE^0 (often at 298°K). Now we define ΔE^0 as the bond energy of the A-B bond. Although this bond energy may properly be expressed in any units of energy per A-B molecule or energy per mole of A-B molecules, we shall almost always use kilocalories per mole of A-B molecules and thence kilocalories per mole of A-B bonds. Similarly, ΔH^0 for reaction (11.33) is taken to be the bond enthalpy of the A-B bond and is also usually expressed in terms of kilocalories per mole of A-B bonds.

From the definition of enthalpy as $H = E + PV$ we have

$$\Delta H^0 = \Delta E^0 + \Delta(PV) \tag{11.34}$$

in which

$$\Delta(PV) = (PV)_{\text{products}} - (PV)_{\text{reactants}} \tag{11.35}$$

Because products A and B and reactant A-B are all ideal gases for which $PV = nRT$, we can write equation (11.35) as

$$\Delta(PV) = 2RT - RT = RT$$

and substitute RT for $\Delta(PV)$ in equation (11.34) to obtain

$$\Delta H^0 = \Delta E^0 + RT \tag{11.36}$$

for the reaction represented by (11.33). Using $R = 1.987$ cal mole^{-1} deg^{-1} and choosing a typical $T = 298$°K, equation (11.36) becomes

$$\Delta H^0 = \Delta E^0 + 592 \text{ cal mole}^{-1} \tag{11.37}$$

or

$$\Delta H^0 = \Delta E^0 + 0.592 \text{ kcal mole}^{-1} \tag{11.37a}$$

Often we may reasonably ignore this relatively small difference between bond energy and bond enthalpy.

We know from results of reliable measurements that $\Delta H^0 = 104.190$ kcal for dissociation of 1 mole of H_2(g) to 2 moles of H(g), all at 298°K, as represented by

$$H_2(g) \rightarrow 2\,H(g) \qquad \Delta H^0 = 104.190 \text{ kcal}$$

Remembering that $\Delta H_f^0 = 0$ for H_2(g), we see that $\Delta H_f^0 = 52.095$ kcal mole^{-1} for H(g), as given in Appendix II. Because 104.190 kcal mole^{-1} are absorbed when

Avogadro's number of H—H bonds are broken, we take the H—H bond enthalpy to be 104.2 kcal $mole^{-1}$. Note also that $\Delta H^0 = -104.190$ kcal when 2 moles of H(g) unite to form 1 mole of $H_2(g)$, all at 298°K. Because this exothermic process results in formation of 1 mole of H—H bonds, we can again say that the H—H bond enthalpy is 104.2 kcal $mole^{-1}$.

In the paragraph above we have rounded off the assigned value of the H—H bond enthalpy to the nearest tenth of a kilocalorie per mole, and later we shall round other bond enthalpy values off to the nearest kilocalorie per mole. We do so because many of the bond energy and bond enthalpy calculations that are most useful contain approximations of sufficient magnitude that uncertainties and errors often range from a few tenths to several kilocalories per mole.

We have not assigned a sign to the H—H bond enthalpy, but have merely agreed that its numerical value is 104.2 kcal $mole^{-1}$. As illustrated with several following examples, we need no assigned signs for bond enthalpies nor do we need a formula to determine their use if we remember that bond breaking is endothermic (positive ΔH) and bond forming is exothermic (negative ΔH).

We obtain Cl—Cl, Br—Br, N≡N and various other bond enthalpies for diatomic molecules of the A—A type in exactly the same way that we obtained the H—H bond enthalpy above. We can also obtain bond enthalpies of the A—B type in the same way, as illustrated here for H—Cl. Using ΔH_f^0 values from Appendix II we calculate ΔH^0 for dissociation of HCl(g) as follows:

$$\mathrm{HCl(g)} \rightarrow \mathrm{H(g)} + \mathrm{Cl(g)}$$
$$\Delta H^0 = 52.095 + 29.082 - (-22.062)$$
$$= 103.239 \text{ kcal mole}^{-1}$$

Now we assign the value 103.2 kcal $mole^{-1}$ as the H—Cl bond enthalpy. In exactly similar fashion we can obtain bond enthalpy values for H—Br, Br—Cl, and other A—B molecules.

Now let us turn to consideration of bond enthalpy values to be associated with bonds in polyatomic molecules. Our first example is the C—H bond that occurs in many organic molecules, such as CH_4 and C_2H_6.

We begin by *defining* the *average* C—H bond enthalpy as one fourth of the ΔH^0 value associated with breaking all four C—H bonds in $CH_4(g)$, as indicated by the following reaction equation and ΔH^0 calculation with ΔH_f^0 values from Appendix II.

$$\mathrm{H{-}\underset{\substack{|\\H}}{\overset{\substack{H\\|}}{C}}{-}H(g)} \rightarrow \mathrm{C(g)} + 4\,\mathrm{H(g)}$$

$$\Delta H^\circ = 171.3 + 4\,(52.0) - (-17.9)$$
$$= 397.6 \text{ kcal mole}^{-1}$$

$$\text{average C—H bond enthalpy} = 397.6/4 = 99.4 \text{ kcal mole}^{-1}$$

Now we make use of this C—H bond enthalpy in deriving a value for the C—C bond enthalpy. First, we use ΔH_f^0 values from Appendix II in calculating ΔH^0

for dissociation of gaseous ethane into gaseous carbon and hydrogen atoms:

$$\mathrm{H_3C{-}CH_3(g) \rightarrow 2\,C(g) + 6\,H(g)}$$

$$\begin{aligned}\Delta H^\circ &= 2(171.3) + 6(52.1) - (-20.2)\\ &= 675.4 \text{ kcal mole}^{-1}\end{aligned}$$

This total ΔH^0 of reaction is associated with breaking 6 moles of C—H bonds and 1 mole of C—C bonds as expressed in the equation

$$675.4 = 6(99.4) + H_{\mathrm{C-C}}$$

in which $H_{\mathrm{C-C}}$ represents the C—C bond enthalpy. We solve this equation to obtain $H_{\mathrm{C-C}} = 79.0$ kcal mole^{-1}.

Similar calculations can be carried out to yield bond enthalpies for many other bonds. For example, we can calculate the C≡C bond enthalpy from the ΔH^0 of dissociation of acetylene (H—C≡C—H) into gaseous carbon and hydrogen atoms. Similarly, the C=O bond enthalpy can be derived from the ΔH^0 of dissociation of gaseous acetone, $(CH_3)_2C{=}O$.

A list of average bond enthalpies is given in Table 11.1. These bond enthalpies are not *exactly* consistent with many of the ΔH_f^0 values given in Appendix II nor with the bond enthalpies we have just calculated. This inconsistency arises because bond enthalpies are not strictly additive in polyatomic molecules, due to the effects of neighboring atoms on bond strengths. For example, the C—H bond enthalpy in CH_4 is not exactly the same as the C—H bond enthalpy in CH_3OH because of the different effects H and OH have on the bond strengths in the rest of these molecules.

In many molecules these complicating differences are small and it is a reasonable approximation to regard any particular bond enthalpy as nearly independent of the bonds in the rest of the molecule, thus making it practical to compile a table of average bond enthalpy values that we can use for many calculations. Table 11.1 contains such average values, chosen to permit moderately accurate calculations for a wide variety of molecules. We illustrate use of these average bond enthalpies in the following examples.

Example Problem 11.4. Use the bond enthalpies in Table 11.1 for estimation of ΔH^0 for the hydrogenation of propene (also called propylene) to propane, and compare this value with the ΔH^0 calculated from ΔH_f^0 values in Appendix II. The reaction equation is

$$\mathrm{H_2C{=}CH{-}CH_3(g) + H{-}H(g) \rightarrow H_3C{-}CH_2{-}CH_3(g)}$$

The enthalpy change associated with this reaction can be regarded as an endothermic contribution due to breaking 1 mole of H—H bonds, 1 mole of C=C bonds, 1 mole of C—C bonds, and 6 moles of C—H bonds, along with an exothermic contribution due to forming 2 moles of C—C bonds and 8 moles of C—H bonds. The *net* enthalpy change

Table 11.1. Average Bond Enthalpies (in kcal mole^{-1} at 298°K)*

	H	*C*	*N*	*O*	*F*	*Cl*	*Br*	*I*
H—	104	99	93	111	136	103	88	71
C—	99	83	69	82	114	79	66	53
C=		145		174†				
C≡		199	208					
N—	93	69	39		67			
N≡		208	226					
O—	111	82		34				
F—	136	114		67	38			
Cl—	103	79				58		
Br—	88	66					46	
I—	71	53						36
P—	77							
S—	88	66						
Xe—					34			

*See references and problems at the end of this chapter for further discussion of average bond enthalpies. Note also that bond enthalpies defined in a different way are useful in chemical kinetics (Chapter 22).

†For aldehydes and ketones.

is that associated with breaking 1 mole of C=C bonds and 1 mole of H—H bonds while forming 1 mole of C—C bonds and 2 moles of C—H bonds. Numerical calculations follow:

Bonds broken		*Kcal absorbed*	*Bonds formed*		*Kcal evolved*
C=C		145	C—C		83
H—H		104	2(C—H)		198
Total absorbed	=	249	Total evolved	=	281

The net enthalpy effect associated with the reaction is the evolution of 281 − 249 = 32 kcal mole^{-1}. Therefore, our reaction enthalpy estimated from bond enthalpies is $\Delta H^0 \cong -32$ kcal mole^{-1}, with the minus sign chosen to show that the reaction is exothermic (evolves heat).

Now we use ΔH_f^0 values from Appendix II to obtain the reaction ΔH^0 as follows [remember that $\Delta H_f^0 = 0$ for $H_2(g)$]:

$$\Delta H^0 = (-24.82) - (+4.88)$$
$$= -29.70 \text{ kcal mole}^{-1} \quad \blacksquare$$

This value of -29.70 kcal mole^{-1} is truly a reliable "experimental" value in that it is based on ΔH_f^0 values that are themselves based on results of accurate experiments. On the other hand, the ΔH^0 value that is based on average bond enthalpies involves our "picture" of the bonding in the various molecules and also the approximation of average bond enthalpies assumed to be independent of other parts of the molecule. The reasonable agreement between experimental and estimated values shows (in this case, at least) that we have a useful method for estimating ΔH^0 values.

Example Problem 11.5. Use bond enthalpies to estimate the ΔH_f^0 for gaseous ethanol (ethyl alcohol).

We approach this problem by recognizing that bond enthalpies permit us to estimate ΔH^0 for the reaction

$$2\,C(g) + 6\,H(g) + O(g) \rightarrow H{-}\underset{\displaystyle H}{\overset{\displaystyle H}{C}}{-}\underset{\displaystyle H}{\overset{\displaystyle H}{C}}{-}O{-}H(g) \qquad (11.38)$$

In this reaction the isolated unbonded atoms come together to form 1 mole of C—C bonds, 5 moles of C—H bonds, 1 mole of C—O bonds, and 1 mole of O—H bonds. The enthalpy change is calculated as follows:

Bonds formed		*Kcal evolved*
C—C		83
5(C—H)		495
C—O		82
O—H		111
Total kcal evolved	=	771

Thus our estimated enthalpy change for reaction (11.38) is $\Delta H^0 \cong -771$ kcal mole^{-1}, with our choice of minus sign indicating that this is an exothermic reaction.

Now we combine the above estimated ΔH^0 for reaction (11.38) with equation (11.30) and ΔH_f^0 values for C(g), H(g), and O(g) to obtain the desired ΔH_f^0 for gaseous ethanol as follows:

$$-771 \cong \Delta H_f^0\,(\text{ethanol}) - [2(171.3) + 6(52.1) + (59.6)]$$
$$\Delta H_f^0 \cong -56 \text{ kcal mole}^{-1} \text{ for } CH_3CH_2OH(g)$$

In Appendix II we find $\Delta H_f^0 = -56.19$ kcal mole^{-1} for $CH_3CH_2OH(g)$. This accurate "experimental" value is based on the enthalpy of combustion of liquid ethanol and the enthalpy of vaporization of liquid ethanol. The first of these experimental quantities led to the ΔH_f^0 for $CH_3CH_2OH(liq)$, which was then combined with the ΔH^0 of vaporization to yield the "experimental" ΔH_f^0 of $CH_3CH_2OH(g)$ that is listed in Appendix II.

Again we see that there is reasonable agreement between an enthalpy estimated from bond enthalpies and an experimental quantity, thus showing once more that we have a useful means for estimating some ΔH^0 values. ■

A number of additional problems and calculations at the end of this chapter and later in this book further illustrate the utility of bond energy calculations.

Some of these later calculations also illustrate limitations of simple bond enthalpy calculations, and the relationship of these limitations to various theories of bonding. One such important example involves benzene (see Problem 21 at the end of this chapter and the discussion in Chapter 19).

Summary

The first law of thermodynamics has led to definition of a useful function $H = E + PV$ that is called enthalpy and to

$$\Delta H = q \qquad \text{(constant pressure processes)}$$

To facilitate calculation of ΔH^0 values for chemical reactions, we have defined standard enthalpies of formation as being the ΔH associated with formation of 1 mole of the substance of interest from its constituent elements, all in their standard states. All of these values are based on $\Delta H_f^0 = 0$ for all elements in their standard reference states, analogous to the choice of zero elevation for sea level. The ΔH_f^0 quantities are used in the equation

$$\Delta H^0_{\text{reaction}} = \Sigma \Delta H_f^0(\text{products}) - \Sigma \Delta H_f^0(\text{reactants})$$

for convenient calculation of ΔH^0 of reaction from tabluated data.

Average bond enthalpies have been defined and a method for their evaluation from ΔH_f^0 values has been described. Uses of these bond enthalpies for estimation of ΔH^0 of reaction and ΔH_f^0 values have been illustrated with examples.

Some of the uses of ΔH values are fairly obvious. For example, an engineer may want to calculate how much heat can be obtained by burning $1.00 worth of natural gas (mostly methane, CH_4), or a rocket designer may want to calculate how much heat can be obtained from a particular mixture of fuel and oxidizer. Some of the substances under consideration may be sufficiently new or unusual or difficult to handle that experimental thermodynamic data are unavailable; in these cases bond enthalpies can often provide a useful estimate of the desired thermodynamic quantity.

Other uses of ΔH values are less obvious, but just as important as those mentioned above. Our discussion in this chapter has been intended to give a small hint as to the contributions that ΔH considerations can make toward understanding chemical bonding. In the next chapter we begin our study of chemical equilibrium in relation to the laws of thermodynamics, and we shall soon see some of the uses of enthalpies in this very important field.

References

Angrist, S. W.: Perpetual motion machines. *Sci. Amer.*, **218:**115 (Jan. 1968).

Angrist, S. W., and L. G. Hepler: *Order and Chaos: Laws of Energy and Entropy.* Basic Books, Inc., New York, 1967.

Armstrong, G. T.: The calorimeter and its influence on chemistry. *J. Chem. Educ.*, **41:**297 (1964).

Benson, S. W.: Bond energies (resource paper). *J. Chem. Educ.*, **42:**502 (1965).

Bigelow, M. J.: Thermochemistry of hypochlorite oxidations. *J. Chem. Educ.*, **46:**378 (1969).

Cohen, S. R.: The use of energetics in elementary biochemistry and physiology. *J. Chem. Educ.*, **36**:249 (1959).
Cox, J. D., and G. Pilcher: *Thermochemistry of Organic and Organometallic Compounds.* Academic Press, New York, 1970. (This excellent book contains a thorough discussion of more elaborate calculations with bond enthalpies, as well as critically tabulated data for many compounds.)
Goates, J. R., and J. B. Ott: *Chemical Thermodynamics.* Harcourt Brace Jovanovich, Inc., New York, 1971.
Klotz, I. M.: *Introduction to Chemical Thermodynamics.* W. A. Benjamin, Inc., New York, 1964.
Knox, B. E., and H. B. Palmer: The uses and abuses of bond energies. *J. Chem. Educ.*, **38**:292 (1961).
Mahan, B. H.: *Elementary Chemical Thermodynamics.* W. A. Benjamin, Inc., New York, 1963 (paperback).
Nash, L. K.: Elementary chemical thermodynamics (resource paper). *J. Chem. Educ.*, **42**:64 (1965).
Nash, L. K.: *Elements of Chemical Thermodynamics.* Addison Wesley, Reading, Mass., 1970 (paperback).
Neidig, H. A., H. Schneider, and T. G. Teates: Thermochemical investigations for a first-year college chemistry course. *J. Chem. Educ.*, **42**:26 (1965).
O'Hara, W. F., C. H. Wu, and L. G. Hepler: Temperature and power measurements in precision solution calorimetry. *J. Chem. Educ.*, **38**:512 (1961).
Ramsay, J. A.: *A Guide to Thermodynamics.* Chapman and Hall, Ltd., London, 1971 (paperback).
Scheler, V. M.: Thermochemistry and animal metabolism. *J. Chem. Educ.*, **41**:226 (1964).
Ubbelohde, A. R.: *Man and Energy.* Penquin Books, Ltd., Harmondsworth, 1963 (paperback).
Energy and power. *Sci. Amer.*, **225**:1ff (Sept. 1971).

Problems

1. Use ΔH_f^0 values in Appendix II to calculate ΔH^0 for each of the following reactions:
   (a) $C_2H_4(g) + H_2(g) \rightarrow C_2H_6(g)$
   (b) $NH_3(g) + HCl(g) \rightarrow NH_4Cl(c)$
   (c) $SO_2(g) + 2\ H_2S(g) \rightarrow 3\ S(rh) + 2\ H_2O(liq)$
   (d) $CaO(c) + CO_2(g) \rightarrow CaCO_3(c)$
2. Ammonium nitrate is commonly used as an ingredient in fertilizers, and is also a dangerous explosive under some conditions. Calculate ΔH^0 for the reaction

$$NH_4NO_3(c) \rightarrow N_2O(g) + 2\ H_2O(g)$$

   Explain why this reaction might be characterized as leading to an explosion.
3. Calorimetric measurements have led to $\Delta H^0 = -37.32$ kcal mole^{-1} for the reaction

$$MoO_2(c) + \tfrac{1}{2}\,O_2(g) \rightarrow MoO_3(c)$$

   Calculate ΔH_f^0 for $MoO_2(c)$. (See Appendix II.)
4. The average heat capacity of liquid water is 1.0 cal deg^{-1} g^{-1} or 18 cal deg^{-1} mole^{-1} over the range 0–100°C. How many calories are required to heat 30 g of water from 25 to 45°C?
5. Problem 4 is easily solved by means of $C = q/T$. It may also be solved by means of $(\partial H/\partial T)_p = C_p$, which we rearrange to

$$\int_{H \text{ at } T_1}^{H \text{ at } T_2} dH = \int_{T_1}^{T_2} C_p dT$$

In the particular case of heat capacity that does not vary with changing temperature, we take C_p outside the integral sign and obtain

$$\Delta H = C_p(T_2 - T_1) \tag{11.39}$$

Recalling that $\Delta H = q$ at constant pressure, we see that we have obtained a useful expression for solving problem 4.

Now suppose that the heat capacity is not constant, but varies with temperature. One might express the dependence of heat capacity on temperature by means of an equation of the form

$$C_p = a + bT + cT^2 \tag{11.40}$$

(or some other polynomial) in which the lower case letters are constants chosen to make calculated and experimental C_p values agree.

Substitute (11.40) into the integral above and integrate to obtain an equation that will permit calculation of ΔH for changing the temperature of 1 mole of substance from T_1 and T_2. Use this equation to calculate the amount of heat required to heat 1 mole of CO_2(g) at constant pressure from 300 to 400°K. The values for *a*, *b*, and *c* for 1 mole of CO_2 are 6.214, 10.396×10^{-3}, and -35.45×10^{-7} when C_p is expressed in terms of cal deg^{-1} mole^{-1}.

6. Use ΔH_f^0 values from Appendix II to calculate ΔH^0 values for the following reactions:

$$N_2(g) \rightarrow 2\,N(g)$$
$$N_2^+(g) \rightarrow N(g) + N^+(g)$$
$$O_2(g) \rightarrow 2\,O(g)$$
$$O_2^+(g) \rightarrow O(g) + O^+(g)$$

Explain the order of these ΔH^0 values in terms of molecular orbital theory of bonding (See problem 23, Chapter 8).

7. We mix 10 g of water at 20°C with 30 g of water at 70°C (all in a "perfect" Thermos bottle of negligible heat capacity) to obtain 40 g of water at some intermediate temperature. What is this intermediate temperature? (Take C_p = 18 cal deg^{-1} mole^{-1} for water.)

8. Given that a particular system absorbs 900 cal from its surroundings while doing 35 liter atm of work on the surroundings, calculate ΔE.

9. Suppose that a system that consists of 1 mole of ideal gas goes through a complicated process in which the final temperature is equal to the initial temperature of the gas. Because the energy of an ideal gas depends only on temperature, we know that $\Delta E = 0$ for the gas in this process. Evaluate ΔH for the gas in this same process.

10. A current of 0.102 amp passes through a calorimeter heater having resistance of 225 ohms for a period of 90.0 sec. How much heat (expressed in calories and in joules) is developed in the heater?

11. There is an increasing tendency to use the joule (SI units) rather than the calorie as the fundamental unit of chemical energy. Thus, many reports of recent investigations are reported in terms of joules per mole (J mole^{-1}) or kilojoules per mole (kJ mole^{-1}) rather than calories per mole or kilocalories per mole. Many engineering calculations and public discussions of energy are carried out in terms of still other energy units (such as British thermal units, Btu). It is therefore necessary to be able to convert easily from one set of energy units to another. Practice in these and related useful calculations can be obtained by carrying out the following conversions. (See Chapter 1 for useful information.)

(a) Express 2850 cal in terms of joules and kilojoules.
(b) Express 3.62 kJ in terms of calories.
(c) Express 612 liter atm in terms of joules.
(d) Express 109 Btu in terms of both calories and joules.
(e) Express 34 kJ mole^{-1} in terms of ergs molecule^{-1}.
(f) Express 68 kcal in terms of kilowatt hours.

12. Use average bond enthalpies to estimate ΔH values for the following reactions:

$$\text{(a)}\ \mathrm{H{-}\overset{H}{\underset{H}{C}}{-}\overset{H}{\underset{H}{C}}{-}\overset{H}{\underset{H}{C}}{-}\overset{H}{\underset{H}{C}}{-}H(g) \rightarrow H{-}\overset{H}{\underset{H}{C}}{-}\overset{H}{\underset{H}{C}}{-}\overset{H}{C}{=}\overset{H}{C}{-}H(g) + H{-}H(g)}$$

$$\text{(b)}\ \mathrm{H{-}\overset{H}{\underset{H}{C}}{-}\overset{H}{\underset{H}{C}}{-}\overset{H}{\underset{H}{C}}{-}\overset{H}{\underset{H}{C}}{-}H(g) \rightarrow H{-}\overset{H}{\underset{H}{C}}{-}\overset{H}{C}{=}\overset{H}{C}{-}H(g) + H{-}\overset{H}{\underset{H}{C}}{-}H(g)}$$

$$\text{(c)}\ \mathrm{H{-}\overset{H}{\underset{H}{C}}{-}\overset{H}{\underset{H}{C}}{-}\overset{H}{\underset{H}{C}}{-}\overset{H}{\underset{H}{C}}{-}H(g) \rightarrow H{-}\overset{H}{\underset{H}{C}}{-}\overset{H}{\underset{H}{C}}{-}H(g) + H{-}\overset{H}{C}{=}\overset{H}{C}{-}H(g)}$$

$$\text{(d)}\ \mathrm{H{-}\overset{H}{\underset{H}{C}}{-}\overset{H}{\underset{H}{C}}{-}\overset{H}{\underset{H}{C}}{-}\overset{H}{\underset{H}{C}}{-}H(g) \rightarrow H{-}\overset{H}{C}{=}\overset{H}{C}{-}H(g) + H{-}\overset{H}{C}{=}\overset{H}{C}{-}H(g) + H{-}H(g)}$$

Each of these reactions, in which a larger molecule splits into two or more smaller molecules, is of the same type as the "cracking" reactions that are of considerable importance in the conversion of petroleum into useful chemicals.

13. It has been suggested that fat people can lose weight by sucking on ice cubes. Melting the ice requires heat that is obtained by combustion of stored fat. Assuming that the ice cubes are taken from the refrigerator at $-10°C$ and that the cold water produced is warmed to $37°C$ in the body, how much ice must be melted to lead to fat combustion corresponding to 500 kcal? Is this a practical method for losing weight? We have $C_p = 9$ cal deg^{-1} mole^{-1} for ice, $C_p = 18$ cal deg^{-1} mole^{-1} for liquid water, and $\Delta H = 1440$ cal mole^{-1} for melting ice. (Note that the "calories" or "Calories" commonly discussed by weight watchers are really kilocalories.)

14. $\Delta H^0 = -1350$ kcal mole^{-1} for combustion of cane sugar ($C_{12}H_{22}O_{11}$) to form H_2O(liq) and CO_2(g). Write a balanced chemical equation for the combustion reaction and calculate ΔH_f^0 for cane sugar.

15. Use average bond enthalpies to estimate ΔH^0 for the following reaction:

$$\mathrm{H{-}\overset{H}{C}\underset{\underset{H\quad H}{C}}{\diagdown\!\!\!\!-\!\!\!\!-\!\!\!\!\diagup}\overset{H}{C}{-}H(g) + H{-}H(g) \rightarrow H{-}\overset{H}{\underset{H}{C}}{-}\overset{H}{\underset{H}{C}}{-}\overset{H}{\underset{H}{C}}{-}H(g)}$$

The C—C bonds in cyclopropane above are strained so that real cyclopropane molecules are less stable than indicated by average bond enthalpy calculations. State whether the true ΔH^0 of the above reaction should be more negative or more positive than your estimated value.

16. Straight chain hydrocarbons (such as methane, ethane, propane, and so on) can be represented by the general formula C_nH_{2n+2}. Using average bond enthalpies in Table 11.1, derive a general formula of the form $\Delta H_f^0 = a + bn$ for the standard enthalpies of formation of these gaseous straight chain hydrocarbons.

17. As noted in problem 16, straight chain hydrocarbons can be represented by the general formula C_nH_{2n+2}. A related general equation for combustion of compounds of this class is

$$C_nH_{2n+2}(g) + \frac{3n+1}{2} O_2(g) \rightarrow n\ CO_2(g) + (n+1)\ H_2O(g)$$

Using ΔH_f^0 values for $CO_2(g)$ and $H_2O(g)$ from Appendix II and the answer to problem 16, derive a general equation for ΔH^0 of combustion of these gaseous straight chain hydrocarbons.

18. In various equations we have added $E + PV$. Show that the product PV has dimensions of energy.

19. Calorimetric measurements have led to the enthalpy of neutralization of aqueous phenol (C_6H_5OH, represented by HP below) as indicated by

$$HP(aq) + OH^-(aq) \rightarrow H_2O(liq) + P^-(aq) \qquad \Delta H^0 = -7.85 \text{ kcal mole}^{-1}$$

Other calorimetric measurements have led to the enthalpy of ionization of water as

$$H_2O(liq) \rightarrow H^+(aq) + OH^-(aq) \qquad \Delta H^0 = 13.34 \text{ kcal mole}^{-1}$$

Use the information given here to obtain ΔH^0 for ionization of aqueous phenol as represented by

$$HP(aq) \rightarrow H^+(aq) + P^-(aq) \qquad \Delta H^0 = ________$$

20. The electron affinity of Br(g) is 82.6 kcal mole^{-1} and the ionization enthalpy of K(g) is 101.6 kcal mole^{-1}. (You must think about signs.) Calculate ΔH^0 for the reaction

$$K(g) + Br(g) \rightarrow K^+(g) + Br^-(g)$$

21. Use average bond enthalpies to estimate ΔH^0 of decomposition of gaseous benzene as in the reaction represented by

$$C_6H_6(g) \rightarrow 6\ C(g) + 6\ H(g)$$

For this calculation, consider each molecule of benzene to contain three C=C bonds, three C—C bonds, and six C—H bonds as in the structural formula

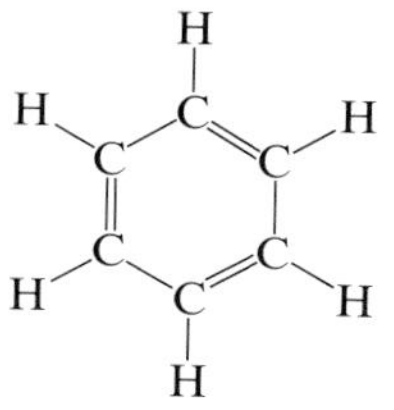

Now use ΔH_f^0 values from Appendix II to calculate a reliable "first law" ΔH^0 for the decomposition reaction above.

The ΔH^0 values obtained in the first part of this problem show that real benzene molecules are considerably more stable than expected on the basis of the bonding pictured above. Explain this difference (commonly called the resonance energy of benzene) in terms of resonance and in terms of delocalized π bonding. In this connection, it may be helpful to review earlier discussions of bonding and to look ahead into Chapter 19.

12

SPONTANEOUS PROCESSES, EQUILIBRIUM, AND THE SECOND LAW OF THERMODYNAMICS

Introduction

We have seen in Chapter 11 how the first law of thermodynamics is applied to the heats (enthalpies) of chemical reactions. As a result, we have useful and convenient methods of dealing with a variety of problems that involve ΔH values associated with chemical reactions and also such processes as melting and vaporization.

The first law can be regarded as a scientific prescription or rule for keeping "accountant's records" for energy (and such related quantities as enthalpy) effects associated with a variety of processes. For example, we know from direct measurement that reaction of CO(g) with O_2(g) to form CO_2(g), all at 298°K and 1.0 atm, yields 67.6 kcal per mole of CO ($\Delta H^0 = -67.6$ kcal mole^{-1}). Now we know from the first law that *if* CO_2(g) decomposes to CO(g) and O_2(g), all at 298°K and 1.0 atm, then $\Delta H^0 = +67.6$ kcal mole^{-1}. However, the first law alone gives us no clue as to whether formation or decomposition of CO_2(g) is spontaneous at 298°K and 1.0 atm; the first law only tells us what ΔH will be *if* the reaction takes place.

Many spontaneous processes are exothermic. For example, when aqueous hydrochloric acid and silver nitrate are mixed, they react spontaneously and exothermically to form a precipitate of silver chloride. Similarly, hydrochloric acid reacts spontaneously and exothermically with sodium hydroxide to form salt and water. Such spontaneous exothermic reactions are so common that it was once thought that ΔH measurements could provide the basis for a systematic analysis of "chemical affinity."

But there are many spontaneous reactions and processes that are endothermic. Crystalline silver nitrate dissolves in water endothermically. Ice, when placed in a warm room, spontaneously melts with absorption of heat. A spontaneous endothermic chemical reaction occurs when hydrochloric acid is added to a solution of sodium bicarbonate:

$$H^+(aq) + HCO_3^-(aq) \rightarrow H_2O(liq) + CO_2(g)$$

Our problem now is to apply thermodynamics to understanding something of the reasons for the spontaneity of the various reactions mentioned above and the nonspontaneity of the reverse reactions under "ordinary" conditions. Our approach to this problem begins with careful consideration of the equilibrium state.

Equilibrium

We know that ice spontaneously melts above 0°C and that water spontaneously freezes below 0°C, at 1.0 atm pressure. Ice and water are said to be in equilibrium at 0°C because they can coexist indefinitely at this temperature in an isolated system. The escaping tendency of ice is just equal to the escaping tendency of water at 0°C. Because vapor pressure is a direct measure of escaping tendency, we expect vapor pressure versus temperature curves for ice and water to intersect at 0°C as shown in Figure 12.1. Below 0°C water has larger escaping tendency

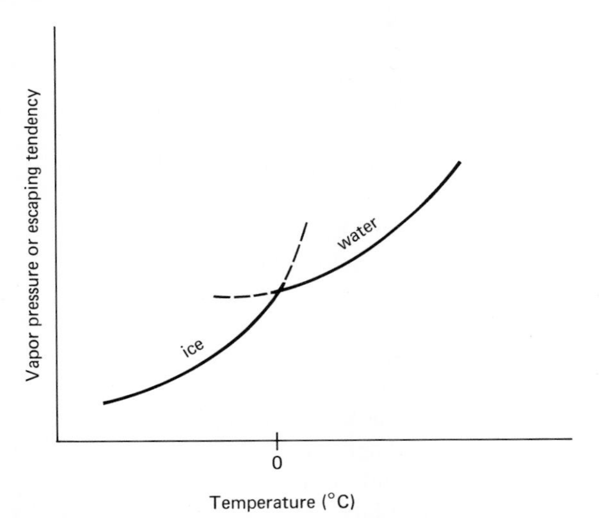

Figure 12.1. We have here a graph of vapor pressures (or escaping tendencies) of ice and water versus temperature. The vapor pressures (escaping tendencies) of ice and water are equal at the melting point.

The dashed lines represent extrapolated vapor pressures of ice above 0°C and of water below 0°C. Because the vapor pressure (escaping tendency) of water at any temperature below 0°C is greater than that of ice at the same temperature, water freezes spontaneously at temperatures below 0°C. Above 0°C, ice melts to water spontaneously because ice has greater escaping tendency than has liquid water.

than has ice, whereas above 0°C ice has larger escaping tendency than has water. (See Chapter 4 for an earlier discussion of escaping tendency.)

We may obtain a saturated salt solution by adding so much salt to a sample of water that solid salt is left in contact with the solution after the dissolving process or reaction is complete. Thus a saturated salt solution is a solution that is in equilibrium with solid salt. In this case we say that the escaping tendency of salt from the solid to the solution is just matched by the tendency of aqueous salt to escape from the solution to the solid phase.

Under some conditions solid calcium carbonate decomposes to calcium oxide and carbon dioxide. Under other conditions, calcium oxide combines with carbon dioxide to form calcium carbonate. Here we have a chemical reaction that can go either way, just as ice can melt or water can freeze. Again, we may describe the equilibrium state in which there is no net change in terms of equal or matching escaping tendencies of carbon dioxide gas and carbon dioxide bound to calcium oxide in calcium carbonate.

One of the purposes of thermodynamics is to give us precise and quantitative means of dealing with escaping tendencies and determining the conditions under which various escaping tendencies are equal so that we have a system in equilibrium. As we have shown before (for example, in connection with freezing point depressions), a variety of problems can be conveniently investigated by considering escaping tendencies. Thus a thermodynamic treatment of escaping tendencies can be very useful. On the other hand, many questions involving spontaneity and equilibrium in chemically reacting systems are often most conveniently approached by way of the equilibrium constant, as discussed later in this book. Fortunately, the same general thermodynamic methods can be used for dealing with escaping tendencies and with chemical equilibrium constants.

In Chapter 22 we shall see that equilibrium can be considered in kinetic terms as being that state in which the rates of competing processes are equal, so that there is no *net* change in the system. Although we can usually obtain more reliable information about the equilibrium state from thermodynamic investigations than from kinetic investigations, the kinetic view is very important because it emphasizes that the equilibrium state is *not* a static state.

A saturated solution of sugar in water (solution in equilibrium with solid sugar) is a system in which some sugar molecules are dissolving and others are precipitating. At equilibrium, each process is taking place at the same rate and the concentration of sugar in the solution does not change with time.

The dynamic nature of the equilibrium state has been demonstrated by a variety of experiments. For example, experiments with radioactive sulfur in lead sulfate have shown that there is exchange between the saturated solution and the solid salt in equilibrium with it.

Probability and Spontaneity

Now let us turn our attention to selected processes in which there is no net energy change. One example is the mixing of red and green balls (identical except for color) in a beaker. More "chemical" examples are the expansion of an ideal gas into a vacuum and the mixing of two different ideal gases. We shall later say that these processes occur spontaneously because they involve the movement of a system from a state of low probability to a state of high probability. The final state is one of greater disorder or randomness than the initial state. A thermodynamic quantity called **entropy** can be used as a measure of the randomness or disorder of such systems. Entropy values can be calculated from statistical considerations or from results of appropriate experiments.

Now let us conduct an imaginary experiment with red and green balls in the bottom of a beaker. If we start with one layer of green balls and one layer of red balls, this can be considered to be a highly ordered arrangement of low probability compared to an arrangement in which the red and green balls are equally divided between the two layers. As a simple example we consider a system that consists of two red balls and two green balls. One arrangement has both red balls in the bottom layer, whereas another arrangement has both red balls in the top layer. These are the only two ways in which we can have segregated or one-color layers. But there are four ways in which we can have mixed color layers, as illustrated in Figure 12.2. Because the balls are identical except for color (same size and mass), each arrangement has the same energy and changing from one arrangement to another has $\Delta E = 0$. Because there are twice as many mixed-color arrangements as one-color arrangements, we expect that we will observe the mixed-color arrangement twice as often as a one-color arrangement in each layer. The mixed-color arrangement is most common simply because it is most probable in a statistical sense.

Now we might carry out the same experiment with a much larger beaker that contains 1000 green balls and 1000 red balls. There are still only two ways in which we can have one layer containing all of the red balls with all of the green balls in the other layer. But there are a large number of ways in which we can have 500 red balls and 500 green balls in each layer. There are also a tremendous number of arrangements that have approximately 500 balls of each color in each layer (say 499 red and 501 green in one layer with 501 green and 499 red in the other layer). We can therefore predict that when we shake a beaker containing

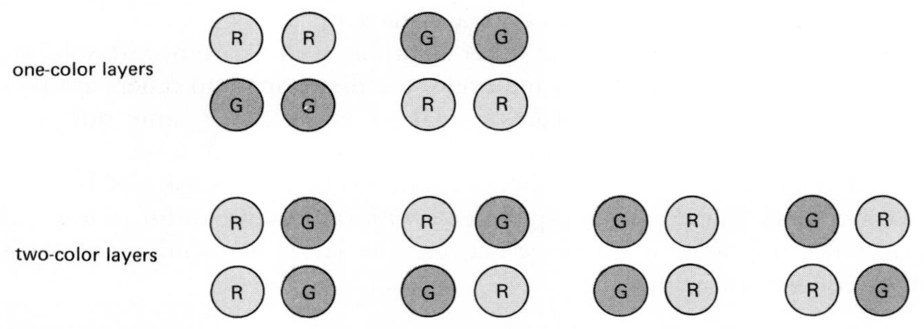

Figure 12.2. Illustration of the greater probability of having two-color layers than one-color layers.

In general, "mixed-up" states are more probable than are "ordered" states, with the difference in the probabilities becoming greater as the number of balls increases. Readers may wish to work out diagrams like those above for a system that consists of some larger number of balls and then derive a general equation that will permit direct calculation of probabilities.

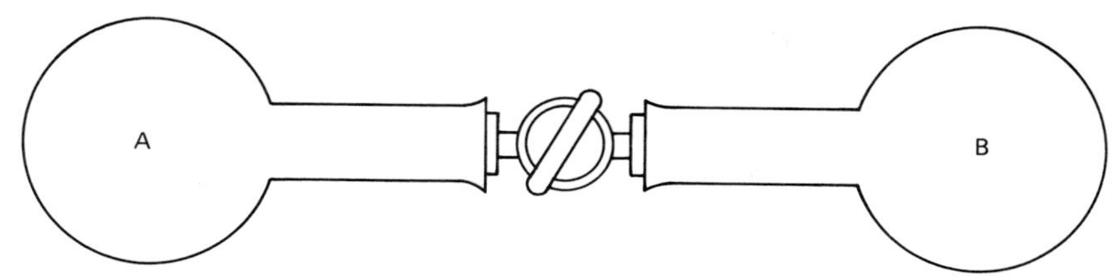

Figure 12.3. Illustration of two flasks connected by a stopcock.

these red and green balls we will end up with approximately 500 balls of each color in each layer. And we should be very much surprised to see further shaking result in separation into one-color layers.

Thus we conclude that a system that consists of two layers of very many red and green balls will at equilibrium be one in which the colors are (nearly) equally distributed between the two layers. Because there are no energy effects ($\Delta E = 0$), this arrangement is determined entirely by probabilities.

Now let us conduct two more imaginary experiments, both with the apparatus pictured in Figure 12.3. In the first of these experiments, flask A contains some ideal gas and flask B is empty (a vacuum). When the stopcock is opened, gas spontaneously flows from flask A into flask B until the gas pressure is the same in both flasks. Because the temperature is constant and the energy of an ideal gas depends only on temperature, we know that $\Delta E = 0$ for this expansion process. Because $\Delta E = 0$ for the expansion, we know from the first law that $\Delta E = 0$ for the reverse process—that is, the process in which the gas in flask B goes back into flask A. Although both processes have $\Delta E = 0$, one occurs spontaneously and the other is so highly improbable that we say that it cannot be spontaneous.

Now let us begin our second imaginary experiment with flask A containing a gas X and flask B containing a gas Y. When the stopcock is opened, the gases mix with each other so that the final state is one in which each gas is equally distributed between the two flasks. Again, $\Delta E = 0$ for this constant temperature process involving ideal gases. We therefore know from the first law that $\Delta E = 0$ for the unmixing of the gases when (if?) the system goes back to its original state. Again we say that the mixing process is spontaneous because it involves change to a highly probable state whereas the reverse process is so improbable that it cannot be spontaneous.

Both the gas expansion and the gas mixing involve the same statistical principle as the experiments with the colored balls, and we will later calculate some probabilities for gas systems using methods similar to those for experiments involving very large numbers of colored balls.

On the basis of the experiments we have just described, it seems reasonable to suggest that disorder or randomness cannot spontaneously decrease by itself. As we shall soon see, this suggestion is equivalent to a crude statement of the second law of thermodynamics.

Now, before we take up the second law in ways that permit us to apply thermodynamics to a variety of chemical problems, let us apply our statistical reasoning to the isotope exchange reaction represented by

$$H_2(g) + D_2(g) \rightleftharpoons 2\ HD(g)$$

Because bond energies of H_2, D_2, and HD are nearly equal, ΔH for this reaction is close to zero. Energy effects are therefore of minimal importance in determining the extent of reaction, and we must consider statistical effects, which we later call entropy effects.

Suppose that we mix 1 mole of H_2 with 1 mole of D_2. Experiment shows that reaction will occur until we have approximately 1 mole of HD with $\frac{1}{2}$ mole each of H_2 and D_2. Or, if we start with 2 moles of HD, rearrangement will occur until we have 1 mole of HD and $\frac{1}{2}$ mole each of H_2 and D_2. In both cases, the reaction proceeds until the system reaches a state of maximum probability (maximum "mixed-upness" or maximum entropy). A graph of entropy versus composition for the H_2-HD-D_2 system is shown in Figure 12.4.

Now let us consider the probabilities of several arrangements of ideal gas molecules in an apparatus like that pictured in Figure 12.5. The total volume of the container is V_2, and the volume of that part of the container to the left of the dotted line is V_1.

Suppose that we have one molecule of gas in the container. The probability of finding the molecule in the entire volume is 1.0, which simply means that the molecule is certainly someplace in the container. The probability of finding the molecule to the left of the dotted line is V_1/V_2, which is necessarily less than unity.

Now suppose that we have two molecules in the entire apparatus. The probability of finding either molecule or both molecules in the entire volume is again unity. The probability of finding either molecule to the left of the dotted line is V_1/V_2, and the probability of finding both molecules simultaneously to the left of the line is $(V_1/V_2)^2$. Similarly, if there are three molecules in the entire volume, the probability of finding all three simultaneously to the left of the dotted line is $(V_1/V_2)^3$. We generalize to the case of N molecules by saying that the probability of finding all N molecules simultaneously to the left of the dotted line is $(V_1/V_2)^N$. N might be Avogadro's number.

Even for V_1/V_2 only very slightly less than unity, $(V_1/V_2)^N$ with large N is exceedingly small, which indicates that it is highly unlikely that all of the molecules in a gas will congregate in any particular part of a container.

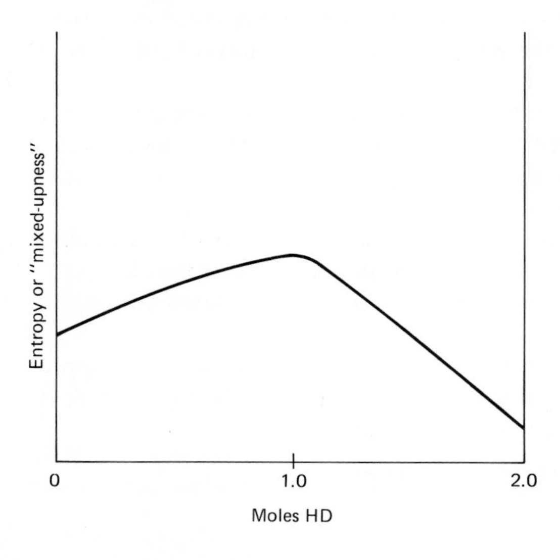

Figure 12.4. Graph of entropy versus composition for a D_2-HD-H_2 system made by mixing 1 mole of D_2 with 1 mole of H_2. The system has maximum entropy when it consists of about $\frac{1}{2}$ mole each of H_2 and D_2 and 1 mole of HD.

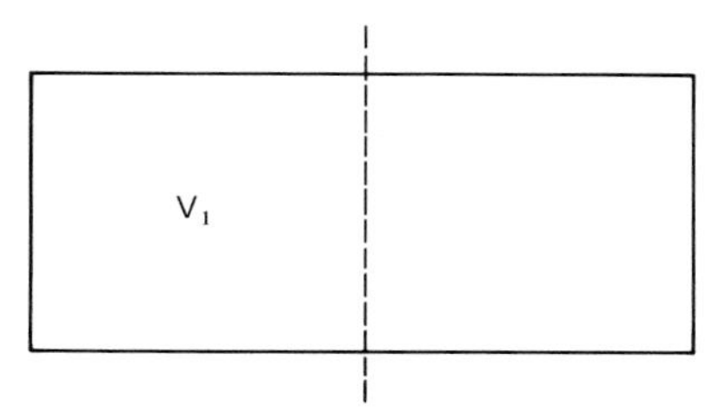

Figure 12.5. Illustration of gas container referred to in the text.

Note the similarity of the statistical calculations described in the text to the statistics of flipping a coin or picking the winners on a perfectly handicapped football parlay card. The probability of flipping a tail with an honest penny is $\frac{1}{2}$. The probability of flipping two consecutive tails is $(\frac{1}{2})^2 = \frac{1}{4}$. The probability of picking a single winner in one choice from a parlay card where the point spreads were picked by an expert is (about) $\frac{1}{2}$. The probability of choosing three winners in three games is $(\frac{1}{2})^3 = \frac{1}{8}$, and the probability of choosing ten winners out of ten games is $(\frac{1}{2})^{10} = \frac{1}{1024}$. Who gets rich when parlay cards offer payoffs of 4 to 1 and 100 to 1 for picking three and ten winners out of three and ten choices?

We are now ready to apply statistics to the expansion of an ideal gas. Suppose that the dotted line in Figure 12.5 represents a movable barrier and that there are Avogadro's number (1 mole) of gas molecules occupying volume V_1 on the left side of the barrier. Now we remove the barrier. The probability that all of the molecules will remain simultaneously in volume V_1 is $(V_1/V_2)^N$, a *very* small number compared to unity. Thus spontaneous expansion of the gas to fill the container corresponds to going from a highly improbable state to a highly probable state. The final equilibrium state, in which gas molecules are spread uniformly throughout the container, is the state of maximum probability for this system.

For the sake of simplicity, we have been considering carefully chosen processes in which there are no energy effects. Statistical or probability considerations have shown why some processes are spontaneous whereas others are so improbable that they are effectively prohibited from occurring spontaneously. Although not yet proven, we can now suggest that for systems in which energy effects are also important (ΔE and ΔH not equal to zero), selection of spontaneous processes and description of the equilibrium state will involve some combination of energy and probability considerations. In order to develop these considerations, we now turn to the relationship between heat and entropy.

Heat and Entropy

We begin by defining the quantity called entropy (represented by S) with the equation

$$dS = \frac{dq_{\text{rev}}}{T} \tag{12.1}$$

In this equation q_{rev} refers to the heat absorbed by the system when the particular process under consideration is carried out reversibly. A reversible process in thermodynamics is a process carried out under conditions that are never more than infinitesimally displaced from equilibrium. Several examples of reversible processes are considered later, along with detailed explanations of the uses of entropy values.

From equation (12.1) we obtain

$$\Delta S = \int_1^2 \frac{dq_{\text{rev}}}{T} \tag{12.2}$$

According to this equation, we can evaluate ΔS for a system undergoing a change from state 1 to state 2 by taking that system from state 1 to state 2 by way of a reversible process. For each infinitesimal step on this path we obtain dq_{rev}/T and add (integrate) all of these dq_{rev}/T values to obtain ΔS. In the special case of a process that is *both* reversible and isothermal (constant temperature), equation (12.2) becomes

$$\Delta S = \frac{1}{T}\int_1^2 dq_{\text{rev}} = \frac{q_{\text{rev}}}{T} \qquad \text{(reversible, isothermal)} \tag{12.3}$$

We are now prepared to state the second law of thermodynamics in three parts as follows:

1. Entropy is a state function, which means that dS is an exact differential and that ΔS for any system undergoing any process depends on the initial and final states but not on the path or way in which the process is carried out. (In this sense, entropy is similar to energy and enthalpy, both of which are also state functions.)
2. For an isolated system, the entropy is a maximum at equilibrium and the entropy of an isolated system undergoing a spontaneous process always increases (ΔS is positive). It is impossible to have a spontaneous process with negative ΔS in an isolated system.
3. In the case of a system that is not isolated, the total ΔS for system plus surroundings is positive for spontaneous processes and is zero for reversible or equilibrium processes. The total ΔS for system plus surroundings cannot be negative.

Neither the general validity nor the important consequences of the second law as stated above are readily obvious. We shall therefore devote considerable attention to these matters. But before we do so, it is appropriate to consider the origins of the second law.

Some perceptive readers may be disappointed that we have not derived the second law of thermodynamics, nor will we derive it. In fact, neither the first law nor the second law has been derived from anything more fundamental. Rather, both laws are concise statements of much human experience. For example, the first law as we have used it (energy is a state function and $\Delta E = q - w$) follows fairly directly from the idea that certain kinds of perpetual motion machines are impossible to construct or even to design.

The first investigation of what we now call the second law of thermodynamics was done by Sadi Carnot in Paris. Carnot's interest in the efficiency of steam engines led him to a general theoretical investigation of efficiencies of heat engines. At age 28, in 1824, Carnot published the results of his studies in a pamphlet titled *Réflexions sur la puissance motrice du feu et sur machines,* which may be translated as *Reflections on the Motive Power of Fire* (*or Heat*). In this pamphlet, Carnot stated a new principle concerning the maximum work obtainable from a heat engine. One form of his new principle, which was the first statement of what we now call the second law of thermodynamics, was as follows: "The motive power of heat is independent of the agents employed to realize it; its quantity is fixed solely by the temperature of the bodies between which is effected, finally, the transfer of caloric."

The second law can be stated in many ways that are superficially different. It is possible, however, to start with any proper statement of the second law and deduce all others, including those we have given and those we will give. For example, either our three-part statement following equation (12.3) or Carnot's original statement given above can be (not necessarily easily) deduced from the other.

It is precisely because the second law of thermodynamics is applicable to so many apparently unrelated problems that it is stated in so many different ways. It is certainly reasonable that engineers, theoretical physicists, cosmologists, chemists, and others should each express the second law in the particular way most directly applicable to problems in each field. We have already stated the second law in one way that will soon be shown to be useful in chemistry. Later in this chapter we will present one more statement that is even more directly useful in chemistry.

Although not of prime importance in the study of chemistry, some alternative statements of the second law are worth repeating because these other statements illustrate some of its important consequences.

1. It is impossible to derive mechanical effect from any portion of matter by cooling it below the temperature of the coolest body of its surroundings. (Early statement by William Thomson, later Lord Kelvin.)
2. It is impossible to construct an engine which will work in a complete cycle and produce no effect except the raising of a weight and the cooling of a reservoir. (Translation of a statement by Max Planck.)
3. Die Energie der Welt ist constant. Die Entropie der Welt strebt einem Maximum zu. (Early statement by Rudolf Clausius, translated as follows. The energy of the world is constant. The entropy of the world increases toward a maximum.)
4. It is impossible to transfer heat from a cold reservoir to a hot reservoir without at the same time converting a certain amount of work into heat. (Translation of another early statement by Clausius.)
5. In any irreversible process the total entropy of all systems concerned is increased. In a reversible process the total increase in entropy of all systems is zero, while the increase in the entropy of any individual system, or part of a system, is equal to the heat which it absorbs divided by its absolute temperature. (Statement by G. N. Lewis and Merle Randall in their pioneering book on application of thermodynamics to chemistry.)

Statements 1, 2, and 4 are clearly pertinent to questions of engineering possibilities and efficiencies. Of these, statement 4 is possibly closest to everyday experience. A specific consequence of this form of the second law is that a refrigerator cannot operate (transfer heat from its cold inside to its relatively warm surroundings) unless it is connected to a source of electrical energy, which is derived from work being done (with entropy increase) in a power plant.

Some quantities with which science is concerned are conserved, such as energy and mass. Others, such as temperature and pressure, can be increased *or* decreased. But entropy and time are different: they always increase. It is true that entropy can be decreased temporarily and in a localized region, but only at the expense of a greater increase in entropy elsewhere. Entropy is therefore a one-way variable that Arthur Eddington has described as "time's arrow." The target of this "arrow" is equilibrium.

Now we return to specific discussion of entropy changes in relation to the defining equation (12.2). Because entropy is a state function, the value of ΔS for a system going from state 1 to state 2 is independent of the path and depends only on the initial and final states. But, to calculate the value of ΔS it is always necessary to consider the transformation of the system from state 1 to state 2 by a reversible path. These statements may appear to be contradictory because in general $dq_{rev}/T \neq dq_{irrev}/T$. The entropy change of the system, ΔS, which is independent of the path, is equal to $\int dq/T$ only when the process is reversible. The quantity $\int dq/T$ depends on how the process is carried out, but ΔS of the system depends only on the initial and final states.

The situation described above is similar to one already encountered in connection with enthalpy. The value of ΔH for a system undergoing any change is independent of the path but is equal to q only when the process is carried out at constant pressure.

Now let us turn to calculation of a specific ΔS to illustrate the general method of such calculations based on equations (12.2) and (12.3). Our particular problem is to calculate ΔS for an ideal gas that undergoes expansion or compression at constant temperature, and our procedure will be to carry out a "thought experiment" in which the gas is expanded or compressed *reversibly* so that we can calculate q_{rev} to use in equation (12.3). Because the expansion or compression is specified as being at constant temperature, we know that $\Delta E = 0$ for the gas and hence that $q = w$. We can therefore obtain the desired q_{rev} by calculating the reversible work denoted by w_{rev}.

The expression for the work of any expansion or compression is

$$w = \int P_{ext}\, dV \tag{12.4}$$

in which P_{ext} represents the external pressure compressing the gas or against which the gas expands. For the special case of a reversible expansion, P_{ext} differs only infinitesimally from the pressure of the gas that we here denote by P_g. Thus we can rewrite equation (12.4) as

$$w_{rev} = \int_{V_1}^{V_2} P_g dV \tag{12.5}$$

in which V_1 and V_2 represent the initial and final volumes of the gas. Because the pressure of an ideal gas is given by $P_g = nRT/V$, we now write equation (12.5) as

$$w_{rev} = \int_{V1_1}^{V_2} \frac{nRT}{V}\, dV = nRT \int_{V_1}^{V_2} \frac{dV}{V}$$

and integrate to obtain

$$w_{rev} = nRT \ln \frac{V_2}{V_1} \tag{12.6}$$

As already pointed out, for expansion or compression of an ideal gas at constant temperature, we have $\Delta E = 0$ and thence $q = w$. Therefore we now have

$$q_{rev} = nRT \ln \frac{V_2}{V_1} \tag{12.7}$$

Finally, substitution of this q_{rev} into equation (12.3) gives us

$$\Delta S = nR \ln \frac{V_2}{V_1} \tag{12.8}$$

For an expansion ($V_2 > V_1$) we have positive ΔS corresponding to an increase in entropy, whereas for a compression ($V_2 < V_1$) we have negative ΔS corresponding to a decrease in entropy. These directions of entropy change are in accord with our earlier discussions of probabilities.

Equation (12.8) gives us the entropy change for an ideal gas undergoing a reversible expansion or compression at constant temperature. Now let us calculate ΔS for the surroundings. When the system (in this case, the ideal gas) absorbs heat denoted by q_{rev} and calculated by way of equation (12.7), the surroundings lose an equal amount. For the surroundings we therefore have

$$\Delta S_{surr} = -nR \ln \frac{V_2}{V_1} \tag{12.9}$$

Now we obtain the total entropy change as the sum of the entropy changes for the gas (equation 12.8) and the surroundings (equation 12.9):

$$\Delta S_{total} = nR \ln (V_2/V_1) - nR \ln (V_2/V_1) = 0$$

Here we find that the total ΔS is zero, as required by the second law of thermodynamics for a reversible (equilibrium) process.

Now let us consider ΔS values to be associated with another expansion of an ideal gas from V_1 to V_2, all at constant temperature. In this case, the gas is *not* expanded reversibly. Instead, the gas is initially in one of the bulbs pictured in Figure 12.3. Then the stopcock is opened to the other (evacuated) bulb so that the gas expands *irreversibly*.

Because entropy is a state function and ΔS depends only on the initial and final states, we know that the entropy change for the gas is again given by equation (12.8). Because the surroundings have not been changed in any way during this process, we also know that $\Delta S_{surr} = 0$. The total entropy change is again given as the sum of ΔS for the gas (equation 12.8) plus the entropy change for the surroundings (zero in this case) so that we have $\Delta S_{total} = nR \ln (V_2/V_1)$. Because $V_2 > V_1$, we have positive total ΔS as required by the second law of thermodynamics for a spontaneous process.

Free Energy and Equilibrium

In spite of the usefulness of the various statements of the second law of thermodynamics given in the preceding section, none conveniently fits many of the applications that are especially important in chemistry and other sciences. We have criteria for spontaneous processes (increasing entropy) and equilibrium or reversible processes (constant entropy) that in principle allow us to ascertain the direction of spontaneous change and the equilibrium conditions in a wide variety of systems. To do so, we must be able to calculate ΔS values for the system and for its surroundings. In one special case, these calculations often become easy—when the system is isolated from the surroundings, in which case we have $\Delta S_{\text{surr}} = 0$. But we are rarely interested in isolated systems. Instead, we are more often interested in chemical systems that are interacting with their surroundings. In particular, we are especially interested in systems that are maintained at constant temperature and pressure by way of interactions with their surroundings.

We say that a system is at equilibrium when it has no tendency to change. For an isolated system (constant V and E), the second law specifies that spontaneous changes occur until the entropy of the isolated system reaches a maximum and equilibrium is attained. On the other hand, for an ordinary mechanical system from which heat effects are excluded (constant S and V), the energy is a minimum at equilibrium. That is, spontaneous processes occur until the energy reaches a minimum and equilibrium is attained.

The push or drive toward equilibrium in a nonisolated system in which there may be heat effects is therefore expected to be related to a combination of two factors. One is the tendency toward a state of maximum entropy, and the other is a tendency toward a state of minimum energy. Our problem is to find and use precise information about how these factors compete with or complement each other in determining the direction of spontaneous change and the final state of equilibrium in nonisolated systems. We presently limit our attention to processes that take place at constant temperature and constant pressure.

For a constant pressure process, we have $q = \Delta H$. Now substitution of ΔH into equation (12.3) gives us

$$\Delta S = \frac{\Delta H}{T} \tag{12.10}$$

with the understanding that this equation is valid only for equilibrium (reversible) processes taking place at constant temperature and constant pressure. Now we rearrange equation (12.10) to

$$\Delta H = T\,\Delta S \qquad \text{or} \qquad \Delta H - T\,\Delta S = 0 \tag{12.11}$$

The form of equation (12.11) might have been anticipated from the discussion in the paragraph preceding equation (12.10). An equilibrium state is attained when the "energy push" represented by ΔH is just balanced by the "entropy or probability push" represented by $T\,\Delta S$.

We now define a new function, which will be a measure of the balance (or lack of balance) of the "energy and entropy pushes" in various processes or reactions. This new function is given the symbol G and is defined as

$$G = H - TS \tag{12.12}$$

The function G is called the Gibbs free energy,* the Gibbs function, or sometimes simply the free energy.

For processes carried out at constant temperature and constant pressure, the free energy change represented by ΔG can be expressed in terms of ΔH and ΔS as

$$\Delta G = \Delta H - T\,\Delta S \tag{12.13}$$

as can be seen from the defining equation (12.12). Now we see from equations (12.11) that $\Delta G = 0$ is a criterion for equilibrium in systems maintained at constant temperature and constant pressure.

We also see from the defining equation (12.12) for G that a decrease in H and an increase in S both correspond to a decrease in G. Because enthalpy (or energy) tends toward a minimum and entropy tends toward a maximum at equilibrium, we anticipate that the free energy tends toward a minimum at equilibrium. The processes that occur as a system (at constant pressure and temperature) approaches equilibrium are spontaneous and are associated with decreasing free energy.

We now state the second law of thermodynamics in a form that is directly and usefully applicable to many processes of chemical interest and importance. This statement is

> *A spontaneous process taking place at constant temperature and constant pressure is always accompanied by a decrease in Gibbs free energy of the system. Hence ΔG is negative for spontaneous processes and is zero at equilibrium.*

As already stated, the Gibbs free energy is a particularly useful function for considerations of systems maintained at constant temperature and constant pressure. However, any particular system can be maintained at a variety of different constant temperatures and constant pressures at different times. We are therefore interested in finding out how free energy depends on both pressure and temperature.

The desired information about the pressure and temperature dependence of free energy can be obtained from the defining equation (12.12). We begin by substituting $E + PV$ for H, which gives us

$$G = E + PV - TS$$

Now differentiation gives us

$$dG = dE + P\,dV + V\,dP - T\,dS - S\,dT$$

Further substitution of $dq - P\,dV$ for dE gives us

$$dG = dq - P\,dV + P\,dV + V\,dP - T\,dS - S\,dT$$

*Another function, called the Helmholtz free energy and represented by A, is defined by

$$A = E - TS$$

The Helmholtz free energy is particularly useful in consideration of systems maintained at constant volume rather than at constant pressure. The letter F has often been used by chemists to represent the Gibbs free energy and by physicists to represent the Helmholtz free energy. To avoid unnecessary confusion, we are following recent practice of most scientists in using G and A for the two free energies. We do not use F at all in this connection.

Finally, we note that $P\,dV$ terms cancel and then substitute $T\,dS$ for dq to obtain

$$dG = V\,dP - S\,dT \tag{12.14}$$

We shall later use this equation (12.14) in several ways to obtain information about how the equilibrium state is influenced by pressure and temperature.

Entropy and Hydrogen Bonding

The normal boiling point of a liquid is that temperature at which the vapor pressure of the liquid is 1.0 atm. We can also say that the liquid is in equilibrium with its vapor at 1.0 atm at the normal boiling point. We therefore know from the second law of thermodynamics that $\Delta G = 0$ for vaporization of a liquid at the normal boiling point when the pressure is 1.0 atm. Hence we have

$$\Delta G = 0 = \Delta H - T\,\Delta S$$

and

$$\Delta S_{\text{vap}} = \Delta H_{\text{vap}}/T_{\text{b}} \tag{12.15}$$

in which ΔS_{vap} and ΔH_{vap} represent the entropy and enthalpy of vaporization and T_{b} is the normal boiling point temperature.

Boiling points of many liquids have been directly observed. Enthalpies of vaporization have been determined both calorimetrically and by calculations with the Clausius-Clapeyron equation as in Example Problem 13.2 in the next chapter. Some of these results are listed in Tables 12.1 and 12.2, along with entropies of vaporization calculated according to equation (12.15). We have listed these values in two groups: Table 12.1 for "normal" liquids and Table 12.2 for hydrogen bonded liquids.

Some of the reasons for describing certain liquids as "normal" and others as hydrogen bonded have already been presented in Chapter 5. The early workers who first recognized the importance of hydrogen bonding were partly led to their conclusions by considerations of entropies of vaporization.

Entropies of vaporization of a wide variety of substances are remarkably close to 21 cal deg^{-1} mole^{-1}, as shown in Table 12.1 for a few substances. On the other hand, entropies of vaporization of liquids that we commonly described as hydrogen bonded are greater than 23 cal deg^{-1} mole^{-1}. Latimer, Rodebush, and others have interpreted this "extra" entropy of vaporization in terms of hydrogen bonding as shown on page 259.

Table 12.1. Boiling Point, ΔH_{vap} and ΔS_{vap} Data for "Normal" Liquids

Substance	*Boiling point* (°*K*)	ΔH_{vap} (*cal mole*$^{-1}$)	ΔS_{vap} (*cal deg*$^{-1}$ *mole*$^{-1}$)
Cl_2	239	4880	20.4
AgBr	1806	37000	20.5
CCl_4	350	7170	20.5
Benzene	353	7350	20.8
$CHCl_3$	334	7020	21.0
Toluene	384	8000	20.9
Hg	630	14160	22.4

Table 12.2. Boiling Point, ΔH_{vap} and ΔS_{vap} Data for Hydrogen Bonded Liquids

Substance	*Boiling point* (°*K*)	ΔH_{vap} (*cal mole*$^{-1}$)	ΔS_{vap} (*cal deg*$^{-1}$ *mole*$^{-1}$)
H_2O	373.0	9720	26.0
C_2H_5OH	351.7	9220	26.2
CH_3OH	337.9	8430	25.0
NH_3	239.8	5580	23.3

We write the general equations

$$\text{liquid} \rightleftharpoons \text{gas} \quad \text{and} \quad \Delta S_{vap} = \Delta H_{vap}/T_b \tag{12.16}$$

for vaporization of any liquid at its normal boiling point. The entropy change for any process is the entropy of the final state minus the entropy of the initial state as shown by

$$\Delta S_{vap} = S_{gas} - S_{liq} \tag{12.17}$$

A "too large" ΔS_{vap} might come about as a result of a "too large" S_{gas} or a "too small" S_{liq}.

The ideal gas equation adequately describes the PVT properties of all the substances listed in Tables 12.1 and 12.2 when these substances are present in the gas state at low or moderate pressures. We therefore tentatively conclude that these substances have "normal" entropies in the gaseous state. In fact, it is possible by methods of statistical thermodynamics to calculate the gas state entropies of all the substances listed in Tables 12.1 and 12.2 in the same way, thereby showing quantitatively that there is nothing unusual about any of these entropies. Therefore we look to the liquids for an explanation of the different ΔS_{vap} values.

Because the entropies of the gases are all "normal," the entropies of the hydrogen bonded liquids must be abnormally low as compared to the entropies of liquids without hydrogen bonds. Remembering that entropy is a measure of molecular disorder or randomness, we conclude that hydrogen bonded liquids are less disordered (more ordered) than are liquids without hydrogen bonds. An explanation of the extra order or unusually small disorder in hydrogen bonded liquids is that hydrogen bond formation requires that the bonded molecules be oriented with respect to one another in certain ways. This orientation corresponds to a loss of directional randomness. Thus a certain amount of molecular order is imposed on liquids by hydrogen bonds.

Trouton's Rule

From the observation that $\Delta S_{vap} \cong 21$ cal deg^{-1} mole^{-1} for "normal" liquids, we write

$$\Delta S_{vap} \cong 21 \text{ cal deg}^{-1} \text{ mole}^{-1} = \Delta H_{vap}/T_b$$

and

$$\Delta H_{vap} \cong 21\ T_b \text{ cal mole}^{-1} \tag{12.18}$$

Equation (12.18) is commonly called Trouton's rule, and is sometimes useful for estimating ΔH_{vap} from a measured boiling point temperature.

Temperature and Entropy

We already have the equations

$$C = dq/dT \qquad \text{(definition of heat capacity)}$$

and

$$dS = dq_{\text{rev}}/T$$

which can be combined to give

$$dS = C\,dT/T$$

Here we shall limit our considerations to systems maintained at constant pressure as indicated by

$$dS = \frac{C_p\,dT}{T} \tag{12.19}$$

Integration of equation (12.19) requires that we know how C_p depends on temperature. For many substances (such as water), it is a very good approximation to take C_p to be constant over moderate ranges of temperature. In this case, integration of (12.19) as definite integrals proceeds as follows:

$$\int_{S \text{ at } T_1}^{S \text{ at } T_2} dS = C_p \int_{T_1}^{T_2} \frac{dT}{T}$$

$$S_2 - S_1 = \Delta S = C_p \ln \frac{T_2}{T_1} = 2.303\ C_p \log \frac{T_2}{T_1} \tag{12.20}$$

Knowing that C_p values are positive (think about what a negative C_p would mean), we see from equation (12.19) or equation (12.20) that increasing temperature leads to increasing entropy. This is precisely what should be expected from our earlier discussion of the relationship between disorder or randomness and entropy.

We now illustrate the use of equation (12.20) with an Example Problem.

Example Problem 12.1. Calculate the entropy increase of 1 mole of water that is heated from $-15\,^\circ$C to $+15\,^\circ$C, all at 1.0 atm. The heat capacity of ice is 0.5 cal deg^{-1} g^{-1} or 9 cal deg^{-1} mole^{-1}, and the heat capacity of liquid water is 1.0 cal deg^{-1} g^{-1} or 18 cal deg^{-1} mole^{-1}. For melting of ice we have $\Delta H = 1440$ cal mole^{-1} at the equilibrium melting point, 273 °K.

We now calculate the desired entropy change in terms of the entropy increase associated with heating the ice to 273 °K, melting the ice at 273 °K, and then heating the water to the final temperature. Entropy changes for the first and third steps are calculated by means of equation (12.20), and the entropy increase in the second step is calculated by means of an equation similar to (12.16). Thus we have

$$\begin{aligned}\Delta S &= 2.303 \times 9\ \text{cal deg}^{-1}\,\text{mole}^{-1} \log \frac{273}{258} + \frac{1440\ \text{cal mole}^{-1}}{273\ \text{deg}} \\ &\quad + 2.303 \times 18\ \text{cal deg}^{-1}\,\text{mole}^{-1} \log \frac{288}{273} \\ &= 6.75\ \text{cal deg}^{-1}\,\text{mole}^{-1} \quad \blacksquare\end{aligned}$$

Here we emphasize that the justification for taking $\Delta S_m = \Delta H_m/T_m$ (subscript m indicates melting) is exactly the same as for taking $\Delta S_{\text{vap}} = \Delta H_{\text{vap}}/T_b$. Melting at the normal melting point is a reversible or equilibrium process, just as vaporization at 1.0 atm and the normal boiling point is a reversible or equilibrium process. Therefore we have $\Delta S = \Delta H/T$ in both cases. We have seen earlier that $\Delta S_{vap} \cong 21$ cal deg^{-1} mole^{-1} for many substances. We have no such generalization for entropies of melting.

References

Angrist, S. W., and L. G. Hepler: *Order and Chaos: Laws of Energy and Entropy*. Basic Books, Inc., New York, 1967.

Bent, H. A.: The second law of thermodynamics. *J. Chem. Educ.*, **39**:491 (1962).

Bent, H. A.: *The Second Law*. Oxford University Press, New York, 1965 (paperback).

Blum, H. F.: *Time's Arrow and Evolution,* 2nd ed. Harper, New York, 1962 (paperback).

Bockhoff, F. J.: A model for introducing the entropy concept. *J. Chem. Educ.*, **39**:340 (1962).

Goates, J. R., and J. B. Ott: *Chemical Thermodynamics*. Harcourt Brace Jovanovich, Inc., New York, 1971 (paperback).

Mahan, B. H.: *Elementary Chemical Thermodynamics*. W. A. Benjamin, Inc., New York, 1963 (paperback).

Mendoza, E. (ed): *Reflections on the Motive Power of Fire by Sadi Carnot and other Papers on the Second Law of Thermodynamics by E. Clapeyron and R. Clausius*. Dover Publications, Inc., New York, 1960 (paperback)

Nash, L. K.: Elementary chemical thermodynamics. *J. Chem. Educ.*, **42**:64 (1965).

Raman, V. V.: Evolution of the second law of thermodynamics. *J. Chem. Educ.*, **47**:331 (1970).

Ramsay, J. A.: *A Guide to Thermodynamics*. Chapman and Hall, Ltd., London, 1971 (paperback).

Ubbelohde, A. R.: *Man and Energy*. Penguin Books, Ltd., Harmondsworth, 1963 (paperback).

Van Ness, H. C.: *Understanding Thermodynamics*. McGraw-Hill Book Co., New York, 1969 (paperback).

Problems

1. Do you expect ΔS for melting a solid to be positive or negative? Explain on the basis of the relative order or disorder of arrangement of molecules in solids and liquids. Also, explain on the basis of the sign of ΔH of melting.
2. Suppose that each flask in Figure 12.3 is filled with a different gas. We intuitively expect and can verify by experiment that the gases will mix when the stopcock is opened. If the gases are ideal, this mixing process does not involve any energy change, yet it is spontaneous. Why?
3. Suppose that both flasks pictured in Figure 12.3 are the same size and that each contains 1 mole of a different ideal gas at the same temperature. Calculate ΔS for the mixing that occurs when the stopcock is opened. (The mixing process under consideration can be regarded as the expansion of the two gases from their initial volumes in one flask to their final volumes in both flasks.)
4. The normal boiling point of cyclohexane is 80.7°C. Estimate ΔH_{vap} for cyclohexane.
5. We know of many processes that really do occur (water freezing to ice is an example) for which we commonly say that ΔS is negative. How can this be consistent with the second law of thermodynamics, which says that there must be an increase in entropy associated with all real processes?

6. Suppose that hot water and cold water are mixed in a "perfect" thermos bottle to yield warm water. What is the sign of the total ΔS for the water?

Because this is an adiabatic process (no heat transferred between the water and the surroundings: $q = 0$), it is tempting to say that $\Delta S = 0$ on the basis of equation (12.3). But equation (12.3) is applicable ONLY to reversible processes, and we are here considering an irreversible process. Have you ever seen warm water in a thermos bottle separate into a hot layer and a cold layer?

Because our system (water) is effectively isolated from its surroundings, we can say that $\Delta S_{\text{surr}} = 0$. Now you predict the sign of ΔS_{water}.

7. When a hot body initially at temperature T_{h} and with heat capacity C_{p}^{h} is brought into contact with a cold body initially at temperature T_{c} and with heat capacity C_{p}^{c}, all in an isolated system such as a "perfect" thermos bottle, equilibrium is finally attained at some final intermediate temperature represented by T_{f}. We can evaluate the final temperature by way of a heat balance equation:

heat lost by hot body = heat gained by cold body

From the definition of heat capacity, we have

$$C_{\text{p}}^{\text{h}}(T_{\text{h}} - T_{\text{f}}) = C_{\text{p}}^{\text{c}}(T_{\text{f}} - T_{\text{c}})$$

After some algebraic manipulation, we obtain

$$T_{\text{f}} = \frac{C_{\text{p}}^{\text{h}}T_{\text{h}} + C_{\text{p}}^{\text{c}}T_{\text{c}}}{C_{\text{p}}^{\text{c}} + C_{\text{p}}^{\text{h}}} \tag{12.21}$$

(a) Simplify equation (12.21) for the special case $C_{\text{p}}^{\text{h}} = C_{\text{p}}^{\text{c}}$, which is appropriate for mixing equal quantities of the same substance.

(b) Use equation (12.21) in calculating the final temperature when 27 g of water initially at 30 °C is mixed with 39 g of water initially at 71 °C. Remember that the heat capacity of water is 1.0 cal deg^{-1} g^{-1}.

8. When equal quantities of hot and cold water are mixed in a "perfect" thermos bottle, we have the final equilibrium temperature given by (see problem 7)

$$T_{\text{f}} = (T_{\text{h}} + T_{\text{c}})/2 \tag{12.22}$$

The hot water is cooled from T_{h} to T_{f} and loses entropy (see equation 12.20) as given by

$$\Delta S_{\text{h}} = C_{\text{p}} \ln (T_{\text{f}}/T_{\text{h}})$$

The cold water is warmed from T_{c} to T_{f} and gains entropy as given by

$$\Delta S_{\text{c}} = C_{\text{p}} \ln (T_{\text{f}}/T_{\text{c}})$$

The total entropy change is

$$\Delta S_{\text{total}} = \Delta S_{\text{h}} + \Delta S_{\text{c}} = C_{\text{p}} \ln (T_{\text{f}}^2/T_{\text{h}}T_{\text{c}}) \tag{12.23}$$

We know from the second law of thermodynamics that ΔS_{total} must be positive, which means that $(T_{\text{f}}^2/T_{\text{h}}T_{\text{c}}) > 1$. Show that this inequality is consistent with equation (12.22).

Hint: take $T_{\text{h}} = T_{\text{c}} + \delta$ and then show that T_{f}^2 is always greater than $T_{\text{h}}T_{\text{c}}$ for positive δ.

9. An ideal gas is compressed adiabatically and reversibly ($q_{\text{rev}} = 0$). What is ΔS for the gas in this process? What happens to the temperature of the gas? Explain your reasoning.

10. Suppose that 1 mole of ice at $-10\,^{\circ}\mathrm{C}$ is added to 5 moles of liquid water initially at $60\,^{\circ}\mathrm{C}$, all in a system with perfect thermal insulation. Calculate the final temperature and ΔS for the process represented by

$$\text{ice} + \text{hot water} \rightarrow \text{cool water}$$

(See problem 13, Chapter 11, for useful data.)

11. We know that the ideal gas equation $PV = nRT$ accurately describes the PVT properties of real gases at low pressures, while more complicated equations of state are required to describe the properties of real gases at higher pressures. One such equation is the van der Waals equation (for 1 mole):

$$\left(P + \frac{a}{V^2}\right)(V - b) = RT$$

Use this equation to obtain an equation analogous to equation (12.6) for an ideal gas.

12. "Instant" cold packs that contain a dry chemical are sometimes used for first aid when refrigeration is unavailable. Addition of water to the dry chemical in a plastic pouch produces "instant cold." The dry chemical (for example, ammonium chloride) must be appreciably soluble in water and must have an endothermic enthalpy of solution. Explain why the temperature drops and "cold is produced."

13. Air conditioners are designed to "pump" heat from a cool room to the warmer outside. Is this a violation of the second law of thermodynamics? Why not?

14. Suppose that you bought a new air conditioner and were too lazy (or ill-informed about thermodynamics) to install it in a window. Instead, you might simply place it on a table and plug it in. If you leave the air conditioner running overnight in your room with all windows and doors closed, will the temperature of the room increase, decrease, or remain unchanged? Why?

15. What do you think about cooling the kitchen by leaving the door of the refrigerator open?

13

FREE ENERGY AND EQUILIBRIUM

Introduction

Thermodynamics is useful in chemistry (and other sciences) because we can use thermodynamic principles in calculating many things that we want to know from a much smaller number of data that we can obtain from results of experiments. In this chapter we apply thermodynamics to the important problem of obtaining information about the equilibrium state in various systems of chemical interest.

Phase Equilibria

We are accustomed to thinking of a chemical reaction as transformation of certain chemical substances into other chemical substances. Thus we might consider a phase change, such as freezing or boiling of a liquid, to be a particular sort of chemical reaction for which a balanced equation can be written and for which we want to know the state of equilibrium under various conditions. We shall first apply thermodynamics to this particular case of equilibrium between pure phases of a single substance and later consider the usual sorts of chemical reactions.

Knowledge of how free energies depend on temperature and pressure can be used in deducing the conditions under which ΔG will be zero for certain processes, and we know from the second law that the conditions for which $\Delta G = 0$ are equilibrium conditions. We can therefore obtain information about how vapor pressures depend on temperature and how melting (or freezing) points depend on pressure.

We now set about deriving the desired thermodynamic equations by considering the equilibrium of some pure substance between two unspecified phases (solid, liquid, gas) as indicated by

$$\text{phase A} \rightleftharpoons \text{phase B} \tag{13.1}$$

When the phases are in equilibrium, we know from the second law of thermodynamics that

$$\Delta G = G_B - G_A = 0$$

from which we see that

$$G_B = G_A$$

and that

$$dG_B = dG_A$$

We have previously found in equation (12.14) a useful expression for dG, which we now write explicitly for dG_A and dG_B:

$$dG_A = V_A\,dP - S_A\,dT \tag{13.2}$$

$$dG_B = V_B\,dP - S_B\,dT \tag{13.3}$$

Because $dG_A = dG_B$, we can equate the right side of (13.2) to the right side of (13.3) to obtain

$$V_A\,dP - S_A\,dT = V_B\,dP - S_B\,dT \tag{13.4}$$

This equation is rearranged to

$$(V_B - V_A)\,dP = (S_B - S_A)\,dT$$

and then to

$$\frac{dP}{dT} = \frac{S_B - S_A}{V_B - V_A} = \frac{\Delta S}{\Delta V} \tag{13.5}$$

Because we have specified that equation (13.1) represents an equilibrium (reversible) process, we are able to substitute q/T for ΔS in equation (13.5) to obtain

$$\frac{dP}{dT} = \frac{q}{T\Delta V} \tag{13.6}$$

In this equation, q represents the heat effect associated with the phase change represented by equation (13.1). Now we identify q with ΔH at any constant pressure and write

$$\frac{dP}{dT} = \frac{\Delta H}{T\Delta V} \tag{13.7}$$

in which ΔH refers to the enthalpy change in the phase transformation represented by equation (13.1). This equation (13.7) is commonly called the Clapeyron equation, after Emile Clapeyron who first derived it in 1834.

We have made no assumptions about the specific nature of the phases represented by A and B other than to state that we are concerned only with pure (single component) systems. Therefore the Clapeyron equation is applicable to equilibria between any two phases of any pure substance. For example, this equation can be used to calculate the changes in equilibrium melting or freezing temperatures caused by changes of pressure.

First, we consider directions of change of melting or freezing temperatures associated with increasing pressure. We know that absolute temperatures represented by T are positive and that melting is an endothermic process with positive ΔH. For melting we also have $\Delta V = V(\text{liquid}) - V(\text{solid})$, which is positive for "typical" substances that expand on melting. Thus all three terms on the right-hand side of equation (13.7) are positive for "typical" substances, which leads to positive dP/dT. The equilibrium melting or freezing temperature for a "typical" substance therefore increases with increasing pressure. On the other hand, water and a few other substances are not "typical." For melting of ice, both T and ΔH are positive, as stated above. But $\Delta V = V(\text{liquid}) - V(\text{solid})$ is *negative* for water. Thus for water (and a few other substances) dP/dT is negative, which means that increasing pressure is associated with a decreasing equilibrium melting or freezing temperature.

Now we turn to numerical calculation of the change of equilibrium melting or freezing temperature associated with a change in pressure.

Example Problem 13.1. Calculate the change in melting point of ice associated with a change in pressure from 1.00 to 2.00 atm. We know that $\Delta H = 1440$ cal mole^{-1} for melting of ice. Densities of ice and water at the melting point are 0.9168 g ml^{-1} and 0.9998 g ml^{-1}, respectively.

We begin by rearranging the Clapeyron equation (13.7) and integrating as follows:

$$\int_{P_1}^{P_2} dP = \frac{\Delta H}{\Delta V}\int_{T_1}^{T_2} \frac{dT}{T}$$

$$P_2 - P_1 = \frac{\Delta H}{\Delta V} \ln \frac{T_2}{T_1}$$

Now we rearrange to obtain

$$\ln \frac{T_2}{T_1} = \frac{\Delta V(P_2 - P_1)}{\Delta H} \tag{13.8}$$

We now use the densities of ice and water with the molecular weight of water to calculate (using $V = M/d$) that $\Delta V = V_{\text{water}} - V_{\text{ice}} = -1.63$ ml mole$^{-1} = -1.63 \times 10^{-3}$ liter mole^{-1}. Substitution of appropriate numerical values in equation (13.8) now gives us

$$\ln \frac{T_2}{T_1} = \frac{-1.63 \times 10^{-2} \text{ liter mole}^{-1} (2 \text{ atm} - 1 \text{ atm})}{1.44 \times 10^{3} \text{ cal mole}^{-1}}$$

$$= -1.13 \times 10^{-6} \frac{\text{liter atm}}{\text{cal}}$$

The quantity on the right-hand side of this equation must ultimately be dimensionless. We therefore multiply by the gas constant R in cal deg^{-1} mole^{-1} and divide by R expressed in liter atm deg^{-1} mole^{-1} to obtain

$$\ln \frac{T_2}{T_1} = \frac{-1.13 \times 10^{-6} \text{ liter atm} \times 1.987 \text{ cal deg}^{-1} \text{ mole}^{-1}}{\text{cal} \times 0.082 \text{ liter atm deg}^{-1} \text{mole}^{-1}}$$

$$= -(2.74 \times 10^{-5}) = 0.9999736 - 1 \tag{13.9}$$

Because $\ln (T_2/T_1)$ is so close to zero, we know that T_2/T_1 is very close to unity—actually a little less than unity so that T_2 is slightly less than T_1. To obtain the change in freezing or melting point represented by $(T_2 - T_1)$, we can calculate T_2/T_1 from seven place logarithm tables. Those who wish to do so may carry out these calculations, but we shall now show how to arrive at the desired end in some easier ways.

One useful approach is to make use of a series expansion of the logarithm in equation (13.9). We begin by noting (see Chapter 1) that

$$\ln x = (x - 1) - 1/2(x - 1)^2 + \cdots$$

For the special case of x only slightly less than unity, we can obtain a very good approximation to $\ln x$ as

$$\ln x \cong (x - 1)$$

Now we let $T_2/T_1 = x$ and obtain

$$\ln \frac{T_2}{T_1} = \frac{T_2}{T_1} - 1 = \frac{T_2 - T_1}{T_1} \tag{13.10}$$

Combination of (13.10) with (13.9) gives us

$$T_2 - T_1 = T_1(-2.74 \times 10^{-5}) = 273°(-2.74 \times 10^{-5}) = -0.0075°$$

Thus we see that the freezing point or melting point of water is lowered by 0.0075° when the pressure is increased from 1.00 to 2.00 atm.

Note that we may also apply (13.10) to (13.8) to obtain the same result:

$$T_2 - T_1 = \frac{T_1\,\Delta V(P_2 - P_1)}{\Delta H} \tag{13.11}$$

Substitution of appropriate values in (13.11) leads to the same result we obtained above.

Another approach to this problem is to replace dP/dT by $\Delta P/\Delta T = (P_2 - P_1)/(T_2 - T_1)$ in the original Clapeyron equation (13.7) to obtain

$$\frac{P_2 - P_1}{T_2 - T_1} = \frac{\Delta H}{T\,\Delta V}$$

Rearrangement of this equation yields (13.11) as obtained above. ■

See problem 17 at the end of this chapter for a brief discussion of LeChatelier's principle, which provides a convenient rule for predicting the direction of change associated with change of an external variable such as pressure.

Now let us turn to the problems of solid-gas and liquid-gas equilibria, which are of considerable importance in chemistry and other sciences. In these problems we shall call the solid or liquid the "condensed phase."

Direct application of the Clapeyron equation (13.7) to equilibria represented by

$$\text{condensed phase} \rightleftharpoons \text{gas} \tag{13.12}$$

is complicated because $\Delta V = V_{\text{gas}} - V_{\text{condensed phase}}$ depends very markedly on *both* the pressure and the temperature. We circumvent the resulting difficulty in integrating equation (13.7) by following a development worked out by Rudolf Clausius.

We begin by noting that the volume of 1 mole of gas is typically very much greater than the volume of 1 mole of either liquid or solid. For example, the volume of 1 mole of water vapor at 1.0 atm and 100°C is about 30.5 liters or

30,500 ml, whereas the volume of 1 mole of liquid water is only about 18 ml. Thus $\Delta V = (30{,}500 - 18)$ ml mole^{-1} under these conditions. Completely ignoring $V_{\text{condensed phase}}$ in this case and taking $\Delta V = V_{\text{gas}} = 30{,}500$ ml mole^{-1} results in an error of only 0.06%. Similar calculations for other substances (at pressures not too high) lead us to the general conclusion that $V_{\text{gas}} \gg V_{\text{condensed phase}}$ so that it is a good approximation to take $\Delta V = V_{\text{gas}}$. We now apply this approximation to equation (13.7) to obtain

$$\frac{dP}{dT} = \frac{\Delta H_{\text{vap}}}{TV_{\text{gas}}} \tag{13.13}$$

Now we note that under conditions (pressure not too high) for which (13.13) is applicable, the behavior of the gas is well represented by the ideal gas equation $PV = nRT$. Thus for 1 mole of gas we have $V = RT/P$, which is substituted into (13.13) to yield

$$\frac{dP}{dT} = \frac{\Delta H_{\text{vap}}}{T(RT/P)} = \frac{P\,\Delta H_{\text{vap}}}{RT^2} \tag{13.14}$$

Remembering that $dP/P = d\ln P$, we rearrange equation (13.13) to

$$\frac{d\ln P}{dT} = \frac{\Delta H_{\text{vap}}}{RT^2} \tag{13.15}$$

Equation (13.15) is known as the Clausius-Clapeyron equation.

The most direct way to make use of equation (13.15) is to integrate it. To do so, we note that enthalpies of vaporization are nearly independent of both temperature and pressure and that R is a constant so that we can write (13.15) as

$$\int d\ln P = \frac{\Delta H_{\text{vap}}}{R}\int\frac{dT}{T^2} \tag{13.16}$$

Integration of (13.16) as an indefinite integral leads to

$$\ln P = -\frac{\Delta H_{\text{vap}}}{RT} + I \tag{13.17}$$

in which I is the constant of integration. Or we can integrate (13.16) as a definite integral with limits P_2 and P_1 on the left and limits T_2 and T_1 on the right to obtain

$$\ln\frac{P_2}{P_1} = -\frac{\Delta H_{\text{vap}}}{R}\left(\frac{1}{T_2} - \frac{1}{T_1}\right) = \frac{\Delta H_{\text{vap}}}{R}\left(\frac{T_2 - T_1}{T_1T_2}\right) \tag{13.18}$$

Note that equation (13.18) can also be obtained by combining two equations of the form of equation (13.17)—one for P_2 and T_2 with another for P_1 and T_1.

Equation (13.17) shows that a graph of $\ln P$ against $1/T$ should give a straight line of slope $-\Delta H_{\text{vap}}/R$. Thus it is possible to calculate the enthalpy of vaporization of a substance from values of the vapor pressure at several temperatures. Enthalpies of vaporization so obtained are in good agreement with results of direct calorimetric measurements. We illustrate this calculation in the following Example Problem.

Table 13.1. Vapor Pressures of Methyl Alcohol at Several Temperatures

Temperature, t(°C)	*Pressure, P(atm)*	*Temperature, T(°K)*	$1/T$	*log P*
10	0.0713	283	0.00353	−1.146
20	0.1251	293	0.00341	−0.903
30	0.2109	303	0.00330	−0.676
40	0.3427	313	0.00320	−0.465
50	0.5288	323	0.00310	−0.269

Example Problem 13.2. Vapor pressure data for methyl alcohol (CH_3OH) are listed in Table 13.1 and are to be used in calculating ΔH of vaporization and the normal boiling point of methyl alcohol. Numbers in the three columns on the right have been calculated from the experimental quantities in the two columns on the left of Table 13.1.

Equation (13.17) can be written (remember that $\ln P = 2.303 \log P$) in the form

$$\log P = -\frac{\Delta H_{\text{vap}}}{2.303R}\left(\frac{1}{T}\right) + \frac{I}{2.303} \tag{13.19}$$

This equation shows that a graph of log P against $1/T$ should give a straight line of slope $-(\Delta H/2.303R)$. In Figure 13.1 we have such a graph constructed from the numbers in Table 13.1. The slope of the straight line is found to be -1985 deg. Thus we have

$$-\frac{\Delta H_{\text{vap}}}{2.303 \times 1.987 \text{ cal deg}^{-1} \text{ mole}^{-1}} = -1.985 \times 10^3 \text{ deg}$$

and thence

$$\Delta H_{\text{vap}} = 9080 \text{ cal mole}^{-1} = 9.08 \text{ kcal mole}^{-1}$$

The normal boiling point is the temperature at which the vapor pressure is 1 atm. We therefore extend the straight line in Figure 13.1 to $\log P = 0$ and see that this corresponds to $1/T = 0.00296$. The normal boiling point is then calculated to be $T_b = 1/0.00296 = 338$°K = 65°C. The experimental value for the normal boiling point is 64.7°C. ■

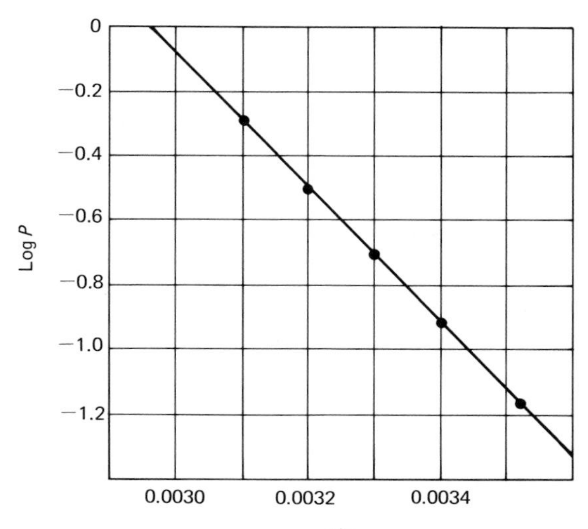

Figure 13.1. Graph of log P versus $1/T$ for methanol, based on vapor pressure data in Table 13.1.

Equilibrium and the Jumping Bean Model

To gain some pictorial understanding of how enthalpy (or energy) and entropy effects combine or compete to determine the state of equilibrium in chemically reacting systems, we first consider an imaginary model system developed by biologist H. F. Blum. After describing this pictorial model and making some nonrigorous deductions based on it, we shall turn to formal derivation of the desired thermodynamic equations; these will give us much useful information about the equilibrium state in systems of reacting chemicals.

Blum's model system is concerned with the distribution of idealized Mexican jumping beans* between two adjoining compartments separated by a wall. Our idealized jumping beans are supposed to be jumping randomly with respect to both height and direction. Further, the average height of jumps is supposed to increase with increasing temperature. We may regard this system as a model for chemical reactions in which beans in the left-hand compartment represent the chemicals on the left side of a reaction equation whereas beans in the right-hand compartment represent the chemicals on the right side of a reaction equation.

Suppose that a number of beans are placed in compartment 1 in Figure 13.2. A bean will occasionally jump from compartment 1 to compartment 2. At first, more beans will jump from compartment 1 to compartment 2 than from 2 to 1 because there are more beans in 1 than 2. When there are more beans in compartment 1 than in 2, net transfer from 1 to 2 is a spontaneous process that brings the system closer to its equilibrium distribution. Because the wall height is the same for beans jumping either direction and because the compartments are the same size, we know that equilibrium will be attained when there are the same number of beans in each compartment. Equal distribution of the beans between the two compartments corresponds to a state of maximum probability or maximum entropy.

The difference in energy for a bean in compartment 1 and in compartment 2 in Figure 13.2 is zero, so we say that $\Delta H = 0$ for transfer of beans from one compartment to the other. When more beans are in compartment 1 than in compartment 2, we say that ΔS is positive for bean transfer from 1 to 2 because this transfer results in a more probable arrangement of beans and therefore corresponds to an increase in entropy. With $\Delta H = 0$ and positive ΔS in $\Delta G = \Delta H - T\,\Delta S$, we have negative ΔG for net transfer of beans from compartment 1 to compartment 2, which is consistent with the second law requirement that ΔG be negative for a spontaneous process. On the other hand, ΔG is positive for the net transfer of beans from compartment 2 to compartment 1, which is a process that does not occur spontaneously under these conditions.

Net transfer of beans from compartment 1 to compartment 2 will continue until a state of equilibrium is reached. In this state of equilibrium, $\Delta G = 0$ for net transfer of beans in either direction. In this case, $\Delta G = 0$ corresponds to $\Delta S = 0$, and ΔS for transfer of beans will be zero when each compartment contains the same number of beans.

*Mexican jumping beans are plant seeds with moth larvae enclosed. When a bean is warmed, the vigorous wiggling of a larva causes the bean to move or "jump"; hence the name.

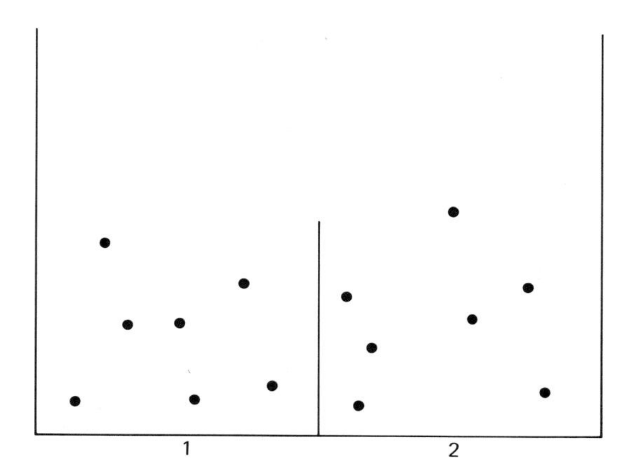

Figure 13.2. Illustration of equilibrium distribution of idealized jumping beans between two compartments. In this case, the "equilibrium constant" $K = 1.0$ at all temperatures.

We now write a "chemical reaction equation" for transfer of beans from left-side compartment (LSC) to right-side compartment (RSC).

$$\text{LSC beans} \rightleftharpoons \text{RSC beans} \tag{13.20}$$

The equilibrium distribution of beans can be expressed as a ratio of number of beans in one compartment to the number in the other compartment. The result is a kind of "equilibrium constant" analogous to those we will write later for chemical reactions.

$$K = \frac{\text{number of beans in RSC}}{\text{number of beans in LSC}} \tag{13.21}$$

We see that $K = 1$ for the system pictured in Figure 13.2.

Now suppose that a number of beans are put in compartments as pictured in Figure 13.3A. Because movement of a bean from compartment 1 to the lower compartment 2 corresponds to a loss of potential energy, we say that ΔH is

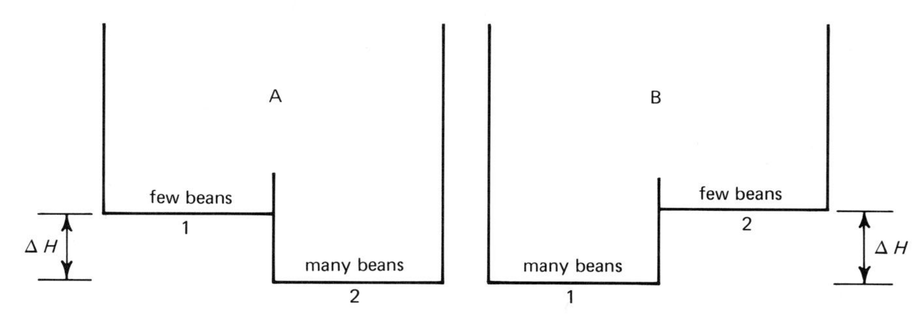

Figure 13.3. Equilibrium distributions of jumping beans between compartments of equal size on different levels.

For the "equilibrium constant" associated with (A) we have $K_A > 1$ at all temperatures. We also have negative (exothermic) ΔH_A and K_A decreasing toward unity as temperature increases. On the other hand, $K_B < 1$ at all temperatures. Here we have positive (endothermic) ΔH_B and K_B increasing toward unity as the temperature increases.

negative (exothermic reaction) for the transfer of beans from compartment 1 to compartment 2. When equal numbers of beans are in both compartments, $\Delta S = 0$ for transfer of beans. With negative ΔH and with $\Delta S = 0$, we see from $\Delta G = \Delta H - T\,\Delta S$ that ΔG is negative for further transfer of beans from compartment 1 to 2.

As this spontaneous transfer occurs, more beans accumulate in compartment 2 than remain in compartment 1 and ΔS for transfer from 1 to 2 gradually becomes more negative. When the number of beans in compartment 2 becomes enough greater than the number in compartment 1, ΔS for transfer from 1 to 2 becomes negative enough that positive $(-T\,\Delta S)$ exactly balances negative ΔH to make $\Delta H - T\,\Delta S = \Delta G = 0$, which corresponds to the final equilibrium distribution indicated in Figure 13.3A. In this case, the "equilibrium constant" K defined by equation (13.21) is some number larger than unity. As temperature increases, the beans jump more vigorously and thus the difference in elevations of the two compartments becomes less important. Therefore at low temperatures we expect to have K much greater than unity, whereas at higher temperatures K is decreased until it is only slightly greater than unity at very high temperature.

We can also apply similar reasoning to the arrangement pictured in Figure 13.3B. In this case, we have K less than unity. At low temperatures, K is very small whereas at high temperature it approaches unity.

Next we consider the equilibrium distributions of beans between large and small compartments on the same level, as pictured in Figure 13.4. In these illustrations we have $\Delta H = 0$, so ΔS determines the states of equilibrium. Because compartment 2 is larger than compartment 1 in A, the most probable (entropy favored) distribution is one in which compartment 2 contains more beans than compartment 1. We therefore have $K_A > 1$. The situation is reversed in B and we have $K_B < 1$. In both cases K is independent of temperature.

In Figure 13.3 the equilibrium distributions are determined by ΔH, whereas in Figures 13.2 and 13.4 the equilibrium distributions are determined by entropy considerations. Now we point out more complicated situations in which both ΔH and ΔS must be considered, as pictured in Figure 13.5, where equilibrium distributions are indicated in the drawings and summarized in the accompanying legend.

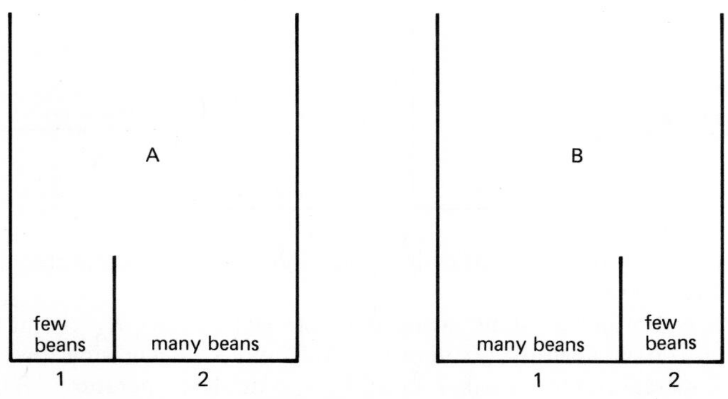

Figure 13.4. The equilibrium distributions are such that we have $K_A > 1$ and $K_B < 1$ at all temperatures. In both (A) and (B) we have $\Delta H = 0$ for transfer of beans.

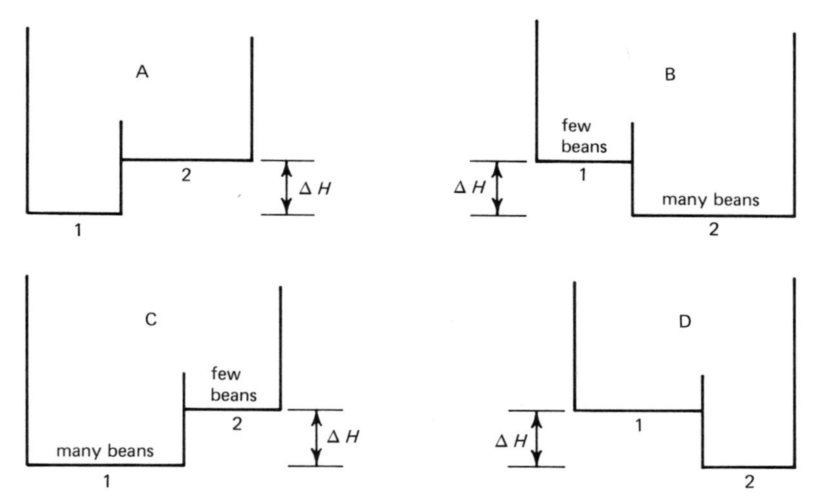

Figure 13.5. Illustration of compartment sizes and elevations leading to equilibrium distributions determined by both ΔH and ΔS.

In (A) we have ΔH (endothermic "reaction") favoring beans in compartment 1, whereas ΔS (probability) favors beans in compartment 2. At low temperatures K_A is relatively small, whereas at higher temperatures K_A becomes larger. In (B) we have both ΔH (exothermic) and ΔS favoring beans in compartment 2 so that $K_B > 1$. As temperature increases, K_B becomes less large because the ΔH "push" toward compartment 2 becomes less important. In (C) we have both ΔH and ΔS favoring beans in compartment 1. The "equilibrium constant" K_C increases as temperature increases but is always less than unity. In (D) we have entropy favoring more beans in 1 than in 2, whereas enthalpy (ΔH is negative) favors more beans in 2. At low temperatures K_D is relatively large, whereas at higher temperatures it becomes smaller.

Now we try to summarize in an equation all of our conclusions about equilibrium distributions of jumping beans. Information that must be included in our summarizing equations follows:

1. When $\Delta H = 0$, K is independent of temperature and is unity if the compartments are the same size.
2. When ΔH is positive, K increases with increasing temperature.
3. When ΔH is negative, K decreases with increasing temperature.
4. When there are no entropy effects (compartments the same size), positive ΔH leads to $K < 1$ and negative ΔH to $K > 1$.

Because we have K passing through unity as the sign of ΔH changes, we first suspect a logarithmic relationship between K and ΔH. The temperature dependence of K is consistent with a relationship of the form $\log K = -k_1 \Delta H/T$, in which k_1 is a positive constant.

In cases where $\Delta H = 0$, we note that K greater or less than unity is consistent with a relationship of form $\log K = k_2 \Delta S^*$, in which ΔS^* represents the entropy change for further transfer when each compartment contains the same number of beans and k_2 is a positive constant.

We put the two relationships above together to obtain

$$\log K = -\frac{k_1 \Delta H}{T} + k_2 \Delta S^* \qquad (13.22)$$

Readers may verify that equation (13.22) is in fact consistent with all of our discussion based on Figures 13.2 through 13.5.

The jumping bean model was originally presented by Blum to give some nonmathematical feeling for how ΔH and ΔS separately and together influence the states of equilibrium in chemical reactions, with the equilibrium states being described in terms of "equilibrium constants." Our nonrigorous analysis of this model has led to an *indication* or suggestion that K should be related to ΔH and ΔS by an equation of the form of (13.22). As we shall soon see, both formal thermodynamics and results of laboratory measurements are in general accord with the informal model analysis.

Free Energy and Equilibrium in Chemical Reactions

To make efficient and convenient use of free energies in relation to chemical reactions, we need to consider standard states and free energies of formation. Our reasons for doing so and the methods of doing so are similar to those already discussed in connection with standard enthalpies of formation represented by ΔH_f^0.

Standard states for free energies of pure substances are the same as for enthalpies and are denoted by superscript 0 just as for ΔH^0 of reaction and ΔH_f^0. These standard states are ordinarily taken to be the solid in its most stable form at 1 atm, the liquid at 1 atm, and the gas in the ideal gas state at 1 atm, all at 298 °K. For some purposes it is convenient to choose the same standard states at some other temperature.

Now we define standard free energies of formation in the same fashion (and for the same reasons) as already done for enthalpies of formation in Chapter 11. Thus the standard free energy of a substance, denoted by ΔG_f^0, is the change in free energy accompanying the formation of 1 mole of the substance in its standard state from its constituent elements in their standard states. According to this convention or choice of "sea level" for free energies, the standard free energy for any element in its standard state is taken to be zero, just as for enthalpies.

For chemical reactions, we are often concerned with the standard free energy change accompanying the transformation of the reactants in their standard states to the products in their standard states. We write an important equation for $\Delta G^0_{\text{reaction}}$ similar to equation (11.30).

$$\Delta G^0_{\text{reaction}} = \Sigma \Delta G_f^0(\text{products}) - \Sigma \Delta G_f^0(\text{reactants}) \qquad (13.23)$$

This equation can be derived in the same fashion as equation (11.30) and has been verified experimentally many times.

Now we derive a thermodynamic equation to relate free energy changes to the state of equilibrium in systems of reacting chemicals. This derivation is first worked out for the special case of a reaction that involves only ideal gases.

We begin with equation (12.14):

$$dG = V\,dP - S\,dT \tag{12.14}$$

At constant temperature ($dT = 0$) this becomes

$$dG = V\,dP \qquad \text{(constant temperature)} \tag{13.24}$$

For an ideal gas we have $V = nRT/P$. We substitute nRT/P for V in (13.24) to obtain

$$dG = \frac{nRT\,dP}{P}$$

which is integrated as a definite integral to yield

$$\Delta G = G_2 - G_1 = nRT \ln \frac{P_2}{P_1} \tag{13.25}$$

(See problem 15 at the end of this chapter for an alternative derivation of this equation.) Equation (13.25) can be used in calculating the free energy change of an ideal gas when it is compressed or expanded at constant temperature. We shall use this equation in obtaining the thermodynamic equation that relates free energy to the state of equilibrium in systems of reacting chemicals.

Now that we are ready to apply thermodynamics to consideration of the state of equilibrium in systems of reacting ideal gases, let us consider the general system represented by

$$a\,\text{A (gas at } P_\text{A}) + b\,\text{B (gas at } P_\text{B}) \rightleftharpoons c\,\text{C (gas at } P_\text{C}) + d\,\text{D (gas at } P_\text{D}) \tag{13.26}$$

The lower case letters (a, b, c, d) specify the number of moles of substances in the reaction represented by the balanced equation (13.26).

Reaction (13.26) as described in terms of specified pressures (P_A, P_B, and so on) will be accompanied by some free energy change ΔG. Our problem now is to calculate the standard state free energy change ΔG^0 (note the superscript 0) for reaction of the same substances with all partial pressures equal to 1.0 atm. We make this calculation by adding several equations based on equation (13.25). These equations and the means of combining them are chosen so that the equation finally obtained will indicate substances A and B in their standard states being converted to substances C and D in their standard states, with an accompanying standard free energy change represented by ΔG^0.

The equations to be added, with accompanying ΔG values, are as follows:

$$a\,\text{A}(P_\text{A}) + b\,\text{B}(P_\text{B}) \rightleftharpoons c\,\text{C}(P_\text{C}) + d\,\text{D}(P_\text{D}) \qquad \Delta G \tag{13.26}$$

$$a\,\text{A}(P = 1\text{ atm}) \rightleftharpoons a\,\text{A}(P_\text{A}) \qquad \Delta G = aRT \ln (P_\text{A}/1) \tag{13.27}$$

$$b\,\text{B}(P = 1\text{ atm}) \rightleftharpoons b\,\text{B}(P_\text{B}) \qquad \Delta G = bRT \ln (P_\text{B}/1) \tag{13.28}$$

$$c\,\text{C}(P_\text{C}) \rightleftharpoons c\,\text{C}(P = 1\text{ atm}) \qquad \Delta G = cRT \ln (1/P_\text{C}) \tag{13.29}$$

$$d\,\text{D}(P_\text{D}) \rightleftharpoons d\,\text{D}(P = 1\text{ atm}) \qquad \Delta G = dRT \ln (1/P_\text{D}) \tag{13.30}$$

$$a\,\text{A}(P = 1\text{ atm}) + b\,\text{B}(P = 1\text{ atm}) \rightleftharpoons c\,\text{C}(P = 1\text{ atm}) + d\,\text{D}(P = 1\text{ atm}) \tag{13.31}$$

$$\Delta G^0 = \Delta G + aRT \ln P_\text{A} + bRT \ln P_\text{B} + cRT \ln (1/P_\text{C}) + dRT \ln (1/P_\text{D}) \tag{13.32}$$

We simplify equation (13.32) by combining all of the logarithm terms to obtain

$$\Delta G^0 = \Delta G + RT \ln \frac{(P_A)^a(P_B)^b}{(P_C)^c(P_D)^d} \tag{13.33}$$

Inversion of the ratio in the last term on the right side of (13.33) with change of sign now gives us

$$\Delta G^0 = \Delta G - RT \ln \frac{(P_C)^c(P_D)^d}{(P_A)^a(P_B)^b} \tag{13.34}$$

Because our principal purpose in deriving equation (13.34) has been to obtain information about the equilibrium state, we now apply this equation to the specific state or set of conditions that correspond to A and B in equilibrium with C and D. It has already been shown in our earlier discussion of the second law of thermodynamics that the relevant criterion for equilibrium is $\Delta G = 0$. Thus, *for the equilibrium state,* we set $\Delta G = 0$ and write equation (13.34) as

$$\Delta G^0 = -RT \ln \frac{(P_C)^c(P_D)^d}{(P_A)^a(P_B)^b} \tag{13.35}$$

Now we use the symbol K to represent the ratio of *equilibrium* pressures in (13.35).

$$K = \frac{(P_C)^c(P_D)^d}{(P_A)^a(P_B)^b} \tag{13.36}$$

This K is called the **equilibrium constant** for the reaction under consideration. We combine equations (13.35) and (13.36) to obtain the very important equation

$$\Delta G^0 = -RT \ln K \tag{13.37}$$

The form of the equilibrium constant K in (13.36) is perfectly general for all reactions involving ideal gases. The numerator of the equilibrium constant for a reaction involving ideal gases is the product of all the *equilibrium* partial pressures of product gases, each raised to the power of the coefficient of that gas in the balanced equation for the reaction. The denominator is a similar product of *equilibrium* partial pressures of all reactant gases, each raised to the power of the coefficient of that gas in the balanced equation for the reaction.

For reactions involving gases at such high pressures that the ideal gas equation is unsatisfactory, a more realistic equation of state can be used with equation (13.25) to obtain appropriate modifications of the equations we have derived here.

Now let us consider the meaning and uses of the equilibrium constant. The standard free energy change ΔG^0 for a reaction is a constant at any particular temperature because it is the change in free energy accompanying a particular definitely specified process. Because K and ΔG^0 are directly related by equation (13.37), K is also truly a constant for any particular reaction at any particular temperature. The name **equilibrium constant** is therefore appropriate.

The knowledge that equilibrium constants exist for all reactions and are of the same general form for each reaction is of considerable value. For example, some A and B might be mixed and allowed to react until a state of equilibrium is reached. Then the amounts or partial pressures of A, B, C, and D could be determined experimentally. These equilibrium pressures then permit calculation of a numerical value for K by way of equation (13.36). The known value of K can then be used to calculate the equilibrium partial pressures in any chemical system involving A, B, C, and D at the specified temperature.

We can see from the defining equation (13.36) that a large value for the equilibrium constant K means that the system is mostly C and D at equilibrium, with only a little A and B present. If we start with only A and B, they will react until they are mostly converted into C and D. Or, if we start with C and D, they will react only a little to form A and B because the equilibrium condition for this system is one in which there is more C and D than there is A and B.

Similarly, a small value for K means that the system at equilibrium contains mostly A and B. In such systems C and D will react to a considerable extent to form A and B, but A and B will react to form only a little C and D.

Before turning to a number of illustrative calculations involving equilibrium constants, we combine certain equations already given to obtain information about how equilibrium constants depend on temperature.

Equilibrium Constants at Various Temperatures

To see how equilibrium constants depend on temperature, we combine equation (13.37) with equation (12.13) as follows:

$$\Delta G^0 = -RT \ln K \tag{13.37}$$

$$\Delta G^0 = \Delta H^0 - T\Delta S^0 \tag{12.13}$$

$$-RT \ln K = \Delta H^0 - T\Delta S^0$$

$$\ln K = -\frac{\Delta H^0}{RT} + \frac{\Delta S^0}{R} \tag{13.38}$$

$$\log K = -\frac{\Delta H^0}{2.303RT} + \frac{\Delta S^0}{2.303R} \tag{13.39}$$

Equation (13.39), which we have derived by way of traditional thermodynamics, is seen to be of the same form as equation (13.22), which we deduced from consideration of Blum's jumping bean model system.

According to equation (13.39), we expect $\log K$ to vary linearly with $1/T$, with slope equal to $-\Delta H^0/2.203R$. We may therefore make use of equation (13.39) for calculation of ΔH^0 values from K values determined experimentally at several temperatures. We may also use a known K value at one temperature with a known ΔH^0 value for calculation of K values at other temperatures.

Before we turn to illustration of a variety of calculations involving equilibrium constants, we should derive one more useful equation. To do so, we write equation (13.39) in terms of temperature T_2 and corresponding equilibrium constant K_2. Then we write the same equation in terms of T_1 and K_1. Subtraction of the second equation from the first gives

$$\log K_2 - \log K_1 = -\frac{\Delta H^0}{2.303R}\left(\frac{1}{T_2} - \frac{1}{T_1}\right) \tag{13.40}$$

Some calculations are more conveniently done with a rearranged form of equation (13.40).

$$\log\left(\frac{K_2}{K_1}\right) = \frac{\Delta H^0}{2.303R}\left(\frac{T_2 - T_1}{T_2 T_1}\right) \tag{13.41}$$

When two of the three quantities represented by K_1, K_2, and ΔH^0 are known, the other may be calculated. The choice between equations (13.39), (13.40), and (13.41) is usually made on the basis of convenience.

Illustrative Calculations

We now illustrate a number of useful calculations involving equilibrium constants.

Example Problem 13.3. Experiments have been carried out in which HI(g) or mixtures of HI(g), H_2(g), and I_2(g) are maintained at constant temperature until equilibrium is established in the reaction system represented by the chemical equation

$$H_2(g) + I_2(g) \rightleftharpoons 2\ HI(g) \tag{13.42}$$

Results of one such experiment led to the following equilibrium partial pressures at 731°K: $P_{H_2} = 0.297$ atm, $P_{I_2} = 0.0314$ atm, and $P_{HI} = 0.672$ atm. Use these results to calculate a numerical value for the equilibrium constant for the reaction represented by equation (13.42).

Comparison of the general reaction equation (13.26) and the corresponding general equilibrium constant expression (13.36) with reaction equation (13.42) leads us to

$$K = \frac{(P_{HI})^2}{(P_{H_2})(P_{I_2})}$$

Insertion of the experimental partial pressures into this expression leads to $K = 48.4$. Results of a number of similar experiments at the same temperature have led to an average $K = 48.5$, which we will use in some subsequent calculations. ■

Example Problem 13.4. In a typical student's experiment, a 500 ml vessel was connected to a vacuum pump, evacuated, and weighed. It was then filled with gas (NO_2 and N_2O_4 as it came out of a tank) to a total pressure of 1.00 atm at 25°C and weighed again. The vessel with gas in it weighed 1.586 g more than it did when empty.

Use the information given above to calculate the equilibrium constant for the reaction

$$N_2O_4(g) \rightleftharpoons 2\ NO_2(g) \tag{13.43}$$

for which

$$K = \frac{(P_{NO_2})^2}{(P_{N_2O_4})}$$

We see that our first task is to calculate the partial pressures of NO_2 and N_2O_4, which we can obtain from knowledge of the mole fractions. One way to obtain the mole fractions of the gases is to obtain the average molecular weight of gas under the specified experimental conditions. From the ideal gas equation $PV = nRT$ we can derive (see Chapter 3) $M = dRT/P$, which we use with the experimental density $d = 1.586$ g/0.500 liter = 3.172 g liter^{-1} as follows:

$$M_{av} = \frac{3.172 \text{ g liter}^{-1} \times 0.08205 \text{ liter atm deg}^{-1} \text{ mole}^{-1} \times 298 \text{ deg}}{1.00 \text{ atm}}$$

$$= 77.5 \text{ g mole}^{-1}$$

We can express this average molecular weight (M_{av}) in terms of the numbers of moles of each gas and the individual molecular weights. Using n_1 for the number of moles of NO_2, M_1 for the molecular weight of NO_2, n_2 for the number of moles of N_2O_4, and M_2 for the molecular weight of N_2O_4, we have

$$M_{av} = \frac{n_1 M_1 + n_2 M_2}{n_1 + n_2} \qquad (13.44)$$

Now we calculate the total number of moles of gas from $n = PV/RT$ as

$$n = \frac{1.00 \text{ atm} \times 0.500 \text{ liter}}{0.08205 \text{ liter atm deg}^{-1} \text{ mole}^{-1} \times 298 \text{ deg}}$$

$$= 0.02045 \text{ mole} = n_1 + n_2$$

Substitution of 0.02045 in the denominator of (13.44) and $(0.02045 - n_1)$ for n_2 in the numerator, along with appropriate molecular weights, gives us

$$77.5 = \frac{46.0 n_1 + 92.0(0.02045 - n_1)}{0.02045}$$

This equation is solved for $n_1 = 0.00645$, which leads to $n_2 = 0.02045 - n_1 = 0.0140$. Now the mole fraction of NO_2 is $X_1 = 0.00645/0.02045 = 0.315$ and that of N_2O_4 is $X_2 = 0.0140/0.02045 = 0.685$. (Notice that $X_1 + X_2 = 1.0$, as they must.)

Partial pressures are calculated as $P_i = X_i P_{tot}$, so that in this case we have $P_i = X_i$. Appropriate substitutions in the equilibrium constant expression lead to

$$K = (0.315)^2/(0.685) = 0.145$$

It is tempting to write the equilibrium constant as $K = 0.145$ atm above, and this is often done to emphasize that the partial pressures are to be expressed in terms of atmospheres. But this equilibrium constant is actually dimensionless because equation (13.32), which led us to the general expression for the equilibrium constant in terms of partial pressures, is derived in terms of (P/P^0) in which the standard state pressure represented by P^0 is taken to be 1.0 atm. Thus the pressures that appear in equilibrium constant expressions are really ratios of equilibrium pressures to a standard pressure, which is ordinarily taken to be 1.0 atm. ■

Example Problem 13.5. We now consider another way of working the problem already done in Example Problem 13.4. This alternative approach may be either easier or more convenient for solving some problems, and certainly suits the tastes of some problem solvers.

Let us suppose that we begin with n moles of pure N_2O_4 and that a certain fraction α of this N_2O_4 dissociates to NO_2, so that $(1 - \alpha)$ is the fraction of N_2O_4 remaining at equilibrium. Again letting n_1 represent the number of moles of NO_2 and n_2 represent the number of moles of N_2O_4, we have

$$n_2 = n(1 - \alpha)$$
$$n_1 = 2n\alpha$$

Now the total number of moles at equilibrium is given by

$$n_{tot} = n(1 - \alpha) + 2n\alpha = n(1 + \alpha)$$

Making use of Dalton's law of partial pressures in the form $P_i = X_i P$, we have the partial pressures of NO_2 and N_2O_4, represented by P_1 and P_2:

$$P_1 = \frac{n_1}{n_{tot}} P = \frac{2n\alpha P}{n(1 + \alpha)} = \frac{2\alpha P}{(1 + \alpha)}$$

$$P_2 = \frac{n_2}{n_{tot}} P = \frac{n(1 - \alpha)P}{n(1 + \alpha)} = \frac{(1 - \alpha)P}{(1 + \alpha)}$$

Now we substitute these partial pressures in

$$K = (P_1)^2/(P_2)$$

to obtain

$$K = \frac{[2\alpha P/(1 + \alpha)]^2}{[(1 - \alpha)P/(1 + \alpha)]}$$

and then

$$K = \frac{4\alpha^2 P}{1 - \alpha^2}$$

This equation provides a direct relationship between K, P, and α that is useful in solving some problems. For now, we must obtain a relationship between the average molecular weight (M_{av}) and α. The desired equation is

$$M_{av} = \frac{2\alpha n M_1 + n(1 - \alpha)M_2}{n(1 + \alpha)}$$

in which we have represented molecular weights of NO_2 and N_2O_4 by M_1 and M_2. Now we note that n cancels from this equation and substitute appropriate values already cited in Example Problem 13.4 for M_{av}, M_1, and M_2 to obtain

$$77.5 = \frac{2\alpha(46) + (1 - \alpha)92}{1 + \alpha}$$

This equation is then solved for

$$\alpha = 0.187$$

We can therefore conclude that N_2O_4 is 18.7% dissociated to NO_2 under the conditions cited in Example Problem 13.4.

Now insertion of $\alpha = 0.187$ and $P = 1.0$ into the expression for K gives us

$$K = \frac{4(0.187)^2}{1 - (0.187)^2} = 0.145 \quad \blacksquare$$

Example Problem 13.6. Use ΔG_f^0 values from Appendix II to obtain the equilibrium constant for the reaction

$$N_2O_4(g) \rightleftharpoons 2\ NO_2(g)$$

at 298°K. Then use ΔH_f^0 values to predict how this equilibrium constant changes with temperature and for calculation of the value of K at 313°K.

In Appendix II we find that $\Delta G_f^0 = 12.26$ kcal mole^{-1} for $NO_2(g)$ and that $\Delta G_f^0 =$ 23.38 kcal mole^{-1} for $N_2O_4(g)$. Combining these values as required by equation (13.23) gives us

$$\Delta G^0 = 2(12.26) - (23.38) = 1.14 \text{ kcal mole}^{-1} = 1140 \text{ cal mole}^{-1}$$

We use this value in equation (13.37) as follows:

$$\Delta G^0 = -RT \ln K = -2.303RT \log K$$

$$\log K = -\frac{1140 \text{ cal mole}^{-1}}{2.303 \times 1.987 \text{ cal deg}^{-1} \text{ mole}^{-1} \times 298 \text{ deg}}$$

$$\log K = -0.835 = 0.165 - 1$$

$$K = 0.146$$

ΔH_f^0 values from Appendix II are 7.93 kcal mole^{-1} for NO_2(g) and 2.19 kcal mole^{-1} for N_2O_4(g). These values lead us to

$$\Delta H^0 = 2(7.93) - (2.19) = 13.67 \text{ kcal mole}^{-1}$$

for the enthalpy of dissociation of N_2O_4(g) to yield 2 NO_2(g).

Equation (13.41) now tells us that for $T_2 > T_1$ with positive ΔH^0, $\log (K_2/K_1)$ is positive and therefore K_2 is larger than K_1. This means that the equilibrium constant for the reaction under consideration increases with increasing temperature. Insertion of appropriate numerical values in equation (13.41) gives us

$$\log \frac{K_2}{K_1} = \frac{13{,}670 \text{ cal mole}^{-1}}{2.303 \times 1.987 \text{ cal deg}^{-1} \text{ mole}} \left(\frac{313 - 298}{313 \times 298}\right)$$

$$= 0.482$$

$$\frac{K_2}{K_1} = 3.03$$

$$K_2 = 3.03 \times 0.146 = 0.442 \quad \blacksquare$$

Example Problem 13.7. It is often convenient to do calculations with equilibrium constants that are expressed in terms of concentrations rather than in terms of partial pressures. We should therefore derive a general equation to relate equilibrium constants expressed in terms of concentrations. Then we apply this general expression to calculation of a numerical value for the "concentration equilibrium constant" for the reaction considered in Example Problems 13.4, 13.5, and 13.6.

We rearrange the ideal gas equation to $P = (n/V)RT$ and note that n/V is a concentration. Thus we have the pressure of a gas i given by

$$P_i = [i]RT \tag{13.45}$$

in which P_i represents the partial pressure of the gas under consideration and $[i]$ represents the concentration of the same gas. Square brackets [ ] or parentheses ( ) are commonly used to denote concentration, usually expressed in terms of moles liter^{-1}.

We now substitute (13.45) into the general expression (13.36) to obtain

$$K = \frac{[C]^c(RT)^c[D]^d(RT)^d}{[A]^a(RT)^a[B]^b(RT)^b}$$

Rearrangement gives

$$K = \frac{[C]^c[D]^d}{[A]^a[B]^b}(RT)^{c+d-a-b}$$

and then

$$\frac{K}{(RT)^{\Delta n}} = \frac{[C]^c[D]^d}{[A]^a[B]^b}$$

where we have written Δn for $c + d - a - b$, which is the change in number of moles for the complete reaction according to the generalized balanced equation (13.26).

At constant temperature the equilibrium constant (in terms of partial pressures expressed in atmospheres) denoted by K is truly a constant for a given reaction, and Δn is also a constant for a given reaction. Therefore $K/(RT)^{\Delta n}$ is a constant that we will represent by K_c. Thus the equation above becomes

$$\frac{K}{(RT)^{\Delta n}} = \frac{[C]^c[D]^d}{[A]^a[B]^b} = K_c \qquad (13.46)$$

Now we apply equation (13.46) to the reaction represented by

$$N_2O_4(g) \rightleftharpoons 2\ NO_2(g)$$

for which $\Delta n = 2 - 1 = 1$. At 298°K we therefore have

$$K_c = \frac{0.146}{0.082 \times 298} = 0.0597 = \frac{[NO_2]^2}{[N_2O_4]}$$

with concentrations expressed in terms of moles per liter. ■

Example Problem 13.8. Suppose that 0.01350 mole of HI(g) is put in a 1.00 liter container maintained at 731°K. How much of the HI remains and how much is decomposed into $H_2(g)$ and $I_2(g)$ at equilibrium?

We nearly always begin equilibrium constant problems by writing down the reaction equation (or its reverse):

$$H_2(g) + I_2(g) \rightleftharpoons 2\ HI(g)$$

Next, we write the equilibrium constant expression for this reaction. Here we have a choice between the equilibrium constant expressed in terms of partial pressures (denoted by K or sometimes K_p) and the equilibrium constant expressed in terms of concentrations (denoted by K_c). Because it appears that this problem is most easily solved by way of K_c, we write

$$K_c = \frac{[HI]^2}{[H_2][I_2]}$$

in which square brackets indicate concentrations. In Example Problem 13.3 we found that $K = 48.5$ for this reaction and in Example Problem 13.7 we found that $K_c = K/(RT)^{\Delta n}$. Because $\Delta n = 0$ for this reaction, we have $K_c = K = 48.5$ and now write

$$48.5 = \frac{[HI]^2}{[H_2][I_2]}$$

We see from the balanced equation for the reaction that the concentrations of $H_2(g)$ and $I_2(g)$ will be equal, and therefore let X represent this as yet unknown quantity. The concentration of HI(g) at equilibrium will be the initial concentration minus that which disappears due to decomposition. Thus we have $[HI] = 0.01350 - 2X$. Substitution into the equilibrium constant expression above now gives us

$$48.5 = \frac{(0.01350 - 2X)^2}{X^2}$$

This equation can be solved for X by multiplying both sides by X^2, expanding the squared term in the numerator, rearranging, and applying the quadratic formula. It is easier, however, to take the square root of both sides to obtain

$$6.97 = \frac{0.01350 - 2X}{X}$$

This equation is easily solved, and yields $X = 0.00151$. Now we know that the concentrations of $I_2(g)$ and $H_2(g)$ are each 0.00151 mole liter^{-1} and the concentration of HI(g) is $0.01350 - 0.00302 = 0.01048$ mole liter^{-1}.

Readers may wish to work this problem in terms of partial pressures. ■

Example Problem 13.9. It has been found that $K = 0.12$ at 500°K for the reaction

$$N_2(g) + 3\,H_2(g) \rightleftharpoons 2\,NH_3(g)$$

Suppose that 1 mole each of N_2, H_2, and NH_3 are placed in a container at 500°K of such size that the total pressure is 3.0 atm and the partial pressure of each gas is therefore 1.0 atm. As reaction proceeds, the total pressure is maintained constant at 3.0 atm by changing the volume of the container. How much of each gas will be present at equilibrium?

The equilibrium constant (expressed in terms of partial pressures) for the reaction is

$$0.12 = \frac{(P_{NH_3})^2}{(P_{N_2})(P_{H_2})^3}$$

If the system is already at equilibrium so that no reaction will take place, the initial partial pressures substituted into the above expression will work out to be equal to the value of K. We see that they actually give

$$\frac{(1.0)^2}{(1.0)(1.0)^3} = 1.0$$

rather than 0.12 as they must at equilibrium. To make this ratio of partial pressures come out equal to 0.12, as it must at equilibrium, we must have a smaller partial pressure of NH_3 and larger partial pressures of H_2 and N_2, which corresponds to decomposition of NH_3 to give N_2 and H_2.

We use the letter n with appropriate subscript to indicate the number of moles of a particular gas and use X to represent the unknown number of moles of $NH_3(g)$ that decompose. Thus at equilibrium we have

$$n_{NH_3} = 1.0 - X$$

$$n_{N_2} = 1.0 + \frac{X}{2}$$

$$n_{H_2} = 1.0 + \frac{3X}{2}$$

and total number of moles

$$\Sigma n = (1.0 - X) + \left(1.0 + \frac{X}{2}\right) + \left(1.0 + \frac{3X}{2}\right) = 3.0 + X$$

From Dalton's law of partial pressures we know that the partial pressure of each gas is given by the mole fraction of that gas times the total pressure. We therefore have the following equilibrium partial pressures

$$P_{NH_3} = \left(\frac{1 - X}{3 + X}\right)3$$

$$P_{N_2} = \left(\frac{1 + \frac{X}{2}}{3 + X}\right)3$$

$$P_{H_2} = \left(\frac{1 + \frac{3X}{2}}{3 + X}\right)3$$

Substituting these partial pressures into the equilibrium constant expression gives

$$0.12 = \frac{\left(\frac{1 - X}{3 + X}\right)^2 3^2}{\left(\frac{1 + \frac{X}{2}}{3 + X}\right) 3 \left(\frac{1 + \frac{3X}{2}}{3 + X}\right)^3 3^3}$$

This complicated equation is fairly easy to simplify to

$$1.08 = \frac{(1 - X)^2(3 + X)^2}{\left(1 + \frac{X}{2}\right)\left(1 + \frac{3X}{2}\right)^3} \tag{13.47}$$

Equation (13.47) contains only one unknown but is quartic (fourth power) in the unknown. Because we have no simple way of solving such equations by algebraic manipulation, we proceed by a method of approximation that is designed to lead to the desired solution to any desired degree of accuracy.

We know that X is greater than zero because we have shown that some ammonia must decompose in the system under consideration. We also know that X is less than unity because not all of the ammonia can decompose. (If all of the ammonia decomposed so that P_{NH_3} became zero, the ratio of partial pressures would also equal zero, rather than 0.12 as required at equilibrium.)

We now guess a value of X between zero and unity and use this value to evaluate the right side of equation (13.47). If we have guessed the correct value of X, the right side will come out equal to 1.08. It is much more likely that we will not guess X exactly right on the first try, in which case the right side will not be equal to 1.08. We therefore guess another value of X and obtain the corresponding value for the right side of equation (13.47). Figure 13.6 shows a graph of values of X. On this graph, we look for the value of X that makes the right side of equation (13.47) equal 1.08 and choose $X = 0.36$. Substitution of 0.36 for X makes the right side of (13.47) equal 1.07, which may be regarded

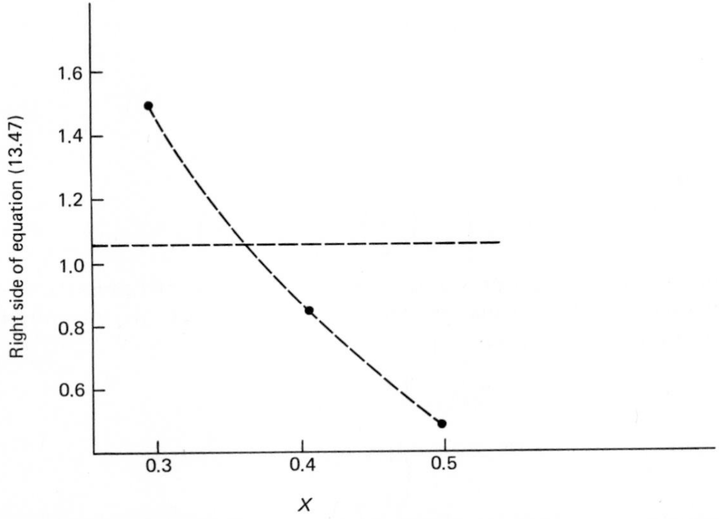

Figure 13.6. Graph of right side of equation (13.47) against various values of X. The value of X that makes the right side of equation (13.47) equal 1.08 is the desired solution.

Table 13.2. Equilibrium Constants for the Reaction $H_2(g) + I_2(g) \rightleftharpoons 2\ HI(g)$

Equilibrium constant, K	*Temperature, T(°K)*
61.0	667
55.2	699
48.5	731
45.8	764

as close enough. Or we can try $X = 0.35$ and find that the right side of equation (13.47) now becomes 1.14. We therefore conclude that X is slightly smaller than 0.36 and can obtain the desired value more precisely by further trials, possibly making use of an expanded scale graph like that shown in Figure 13.6.

Using $X = 0.36$, we find

$$n_{NH_3} = 1.00 - X = 0.64$$

$$n_{N_2} = 1.00 + \frac{X}{2} = 1.18$$

$$n_{H_2} = 1.00 + \frac{3X}{2} = 1.54$$

We can also insert $X = 0.36$ in expressions given earlier and thereby calculate the equilibrium partial pressures of the various gases. ■

Example Problem 13.10. Calculate ΔH^0 for the reaction

$$H_2(g) + I_2(g) \rightleftharpoons 2\ HI(g)$$

from the equilibrium constants listed in Table 13.2.

One way to work this problem is by way of equation (13.41). We can insert any two of the K values and corresponding T values from Table 13.2 in equation (14.41) and then calculate the desired ΔH^0. Then, because we have used only part of the information given in Table 13.2, we should carry out the same calculation with all other possible combinations of K and T values.

We actually work this problem with a graphical method suggested by equations (13.38) and (13.39). According to equation (13.39), a graph of $\log K$ versus $1/T$ should yield a straight line with slope equal to $-\Delta H^0/2.303R$. A graph of $\log K$ versus $1/T$, based on K and T values in Table 13.2, is shown in Figure 13.7. The slope of the straight line is found to be 660 deg. Therefore we write

$$660 \text{ deg} = -\frac{\Delta H^0}{2.303 \times 1.987 \text{ cal deg}^{-1} \text{ mole}^{-1}}$$

and solve for

$$\Delta H^0 = -3020 \text{ cal mole}^{-1} \quad ■$$

Example Problem 13.11. The equilibrium constant for the reaction

$$C_2H_6(g) \rightleftharpoons C_2H_4(g) + H_2(g)$$

is 5.77×10^{-2} at 900°K, and $\Delta H^0 = 34.4$ kcal mole^{-1} for this reaction. Choose that temperature either 100° above or below 900° at which K is larger and then calculate K at that temperature.

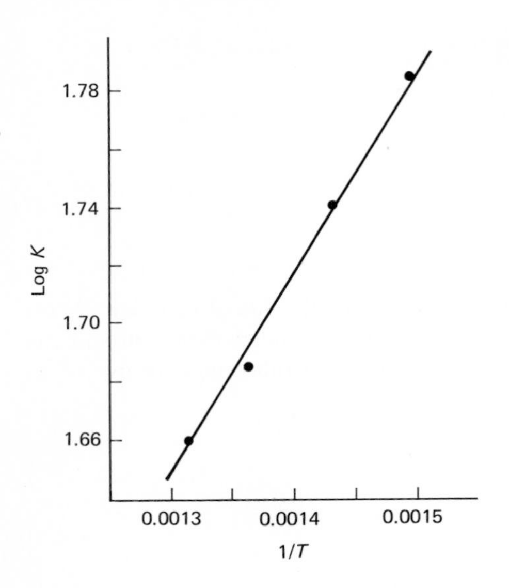

Figure 13.7. A graph of log K [for the reaction $H_2(g) + I_2(g) \rightleftharpoons 2\ HI(g)$] against $1/T$ as suggested by equation (13.39).

Because ΔH^0 is positive (endothermic reaction), we know from our discussion of Blum's jumping bean model or from our thermodynamic equations that the equilibrium constant for this reaction increases with increasing temperature. (Also see problem 17 at the end of this chapter.) We therefore want to calculate K at 1000°K, which we denote by K_{1000}.

Following equation (13.41) we now write

$$\log \frac{K_{1000}}{5.77 \times 10^{-2}} = \left(\frac{34.4 \times 10^3}{1.987 \times 2.303}\right)\left(\frac{1000 - 900}{1000 \times 900}\right) = 0.836$$

(Note that we have expressed both ΔH^0 and R in terms of calories.) Then we find

$$\frac{K_{1000}}{5.77 \times 10^{-2}} = 6.85$$

and

$$K_{1000} = 0.395$$

References

Angrist, S. W., and L. G. Hepler: *Order and Chaos: Laws of Energy and Entropy*. Basic Books, Inc., New York, 1967.

Bent, H. A.: *The Second Law*. Oxford University Press, New York, 1965 (paperback).

Blum, H. F.: *Time's Arrow and Evolution*. Harper and Brothers, New York, 1962 (paperback).

Boggs, J. E.: The logarithm of ten apples. *J. Chem. Educ.*, **35**:30 (1958).

Mahan, B. H.: *Elementary Chemical Thermodynamics*. W. A. Benjamin, Inc., New York, 1963 (paperback).

Nash, L. K.: *Elements of Chemical Thermodynamics*, 2nd ed. Addison-Wesley Publishing Co., Reading, Mass., 1970 (paperback).

Problems

1. Write out the equilibrium constant expression for each of the following reactions:

(a) $NO(g) + \frac{1}{2}O_2(g) \rightleftharpoons NO_2(g)$

(b) $Cl_2(g) + 2\ HBr(g) \rightleftharpoons Br_2(g) + 2\ HCl(g)$

(c) $N_2(g) + 3\,H_2(g) \rightleftharpoons 2\,NH_3(g)$
(d) $SO_2(g) + \frac{1}{2}O_2(g) \rightleftharpoons SO_3(g)$

2. Use ΔG_f^0 values from Appendix II for calculating K values (at 298°K) for reactions in problem 1.
3. Use ΔH_f^0 values from Appendix II for calculating ΔH^0 values for reactions in problem 1. Predict whether equilibrium constants for the reactions increase or decrease with increasing temperature.
4. The barometric pressure at the top of Mount Everest is about 250 mm Hg. For vaporization of water we have $\Delta H = 10.7$ kcal mole^{-1}. Estimate the boiling point of water on top of Mount Everest. Remember that the normal boiling point is that temperature at which the vapor pressure is 1.0 atm.
5. Explain why pressure cookers are commonly used at high altitudes.
6. The normal boiling point of benzene is 80.1°C. There is no evidence of hydrogen bonding between molecules of benzene in the liquid phase nor is there any other evidence to indicate that Trouton's rule should not describe the vaporization of benzene. Estimate the vapor pressure of benzene at 50°C on the assumption that Trouton's rule is applicable to benzene.
7. Vapor pressures of carbon tetrachloride at three temperatures are as follows:

Pressure, P(mm Hg)	*Temperature, t(°C)*
91	20
216	40
451	60

Calculate ΔH of vaporization and the normal boiling point of carbon tetrachloride.

8. Vapor pressures are commonly represented by equations of the form

$$\log P = \frac{A}{T} + B$$

Relate A to the ΔH of vaporization.

9. Derive a general equation for the vapor pressure of a liquid for which $\Delta H_{vap}/T_b = 21$ cal deg^{-1} mole^{-1} (Trouton's rule). Your final equation should involve only T_b that is characteristic of the particular liquid under consideration.
10. The normal melting point of mercury is -38.87°C. The densities of the solid and liquid at this temperature are 14.19 and 13.69 g ml^{-1}, respectively. For the melting of mercury we have $\Delta H = 2.33$ cal g^{-1} (note units). What is the melting point of mercury at $P = 100$ atm?
11. Phosphorus pentachloride dissociates as indicated by the equation

$$PCl_5(g) \rightleftharpoons PCl_3(g) + Cl_2(g)$$

The density of the equilibrium mixture of PCl_5, PCl_3, and Cl_2 obtained on heating PCl_5 is 2.695 g liter^{-1} at 523°K and total pressure 1.0 atm. Calculate the equilibrium constant for the dissociation reaction. (See Example Problem 13.4 for guidance.)

12. Using the density given in problem 11, calculate the equilibrium constant for the dissociation of PCl_5 by the method illustrated in Example Problem 13.5. Here it should be noted $K \neq 4\alpha^2 P/(1 - \alpha^2)$ for dissociation of PCl_5. Rather, we have $K = \alpha^2 P/(1 - \alpha^2)$, which readers should derive for themselves.
13. We have $K = 0.145$ for the dissociation of N_2O_4 at 25°C as shown in Example Problems 13.4 and 13.5. Calculate the fraction of N_2O_4 dissociated when the total pressure is

0.50 atm at 25°C. What is the average molecular weight of the mixture of N_2O_4 and NO_2 under these conditions?

14. At 2150°K and 1.0 atm steam is 1.2% decomposed into hydrogen and oxygen. Calculate K for the reaction

$$2\ H_2O(g) \rightleftharpoons 2\ H_2(g) + O_2(g)$$

15. In the text we have derived equation (13.25) for the free energy change associated with expansion or compression of an ideal gas at constant pressure. Show that this same equation can be obtained in another way as follows. First, what is ΔH for isothermal compression or expansion of an ideal gas? Second, what is ΔS for isothermal expansion or compression of an ideal gas? (See Chapter 12.) Combine ΔH with $T\,\Delta S$ to obtain the desired equation for ΔG.

16. Consider the chemical system described in Example Problem 13.9. What would be the equilibrium composition of this system if the size of the containing vessel were changed so that the total pressure is maintained at 1.0 atm?

17. The effects of variables such as temperature and pressure on chemical equilibrium were summarized qualitatively in 1888 by LeChatelier who wrote, "Any change in one of the variables that determine the state of a system in equilibrium causes a shift in the position of equilibrium in a direction that tends to counteract the change in the variable under consideration." This statement is called LeChatelier's principle and may be regarded as a verbal statement of thermodynamic conclusions based on equation (12.14) and related equations.

(a) Show that LeChatelier's principle is consistent with the Clapeyron equation (13.7) applied to effects of pressure on melting/freezing points.

(b) Show that LeChatelier's principle is consistent with equations (13.38) through (13.41) for the temperature dependence of the equilibrium constant.

18. One approach to the relationship of K, ΔH^0, and T is given by equations (13.38) through (13.41). Here is another approach.

Restricting equation (12.41) to constant pressure processes and to substances in their standard states gives us

$$dG^0 = -S^0\,dT$$

Our problem is concerned with the standard change in free energy for a reaction rather than for a single substance, so we write the equation above as

$$d(\Delta G^0) = -(\Delta S^0)dT \qquad \text{or} \qquad \left(\frac{\partial \Delta G^0}{\partial T}\right)_P = -\Delta S^0$$

Now we differentiate $\Delta G^0 = -RT \ln K$ with respect to temperature to obtain

$$\left(\frac{\partial \Delta G^0}{\partial T}\right)_P = -R\left[T\left(\frac{d \ln K}{dT}\right) + \ln K\right] = -\Delta S^0$$

This equation is easily rearranged to

$$RT^2\left(\frac{d \ln K}{dT}\right) = -RT \ln K + T\,\Delta S^0$$

Because $-RT \ln K = \Delta G^0$ and $\Delta G^0 + T\,\Delta S^0 = \Delta H^0$, the right side of this equation is simply ΔH^0. Division of both sides by RT^2 gives

$$\frac{d \ln K}{dT} = \frac{\Delta H^0}{RT^2} \qquad (13.48)$$

The important equation (13.48) is called the van't Hoff equation, which can be integrated as follows:

$$\int d \ln K = \frac{1}{R} \int \frac{\Delta H^0}{T^2}\, dT \qquad (13.49)$$

(a) Taking ΔH^0 to be a constant (independent of temperature), show that integration of (13.49) as a definite integral leads to equations (13.40) and (13.41) obtained in the text in a different way.

(b) Integrate equation (13.49) as an indefinite integral and show that the result is of the same form as equation (13.38). What is the relationship between the constant of integration (commonly given the symbol I) and the standard entropy change represented by ΔS^0?

19. The reaction of $N_2(g)$ with $H_2(g)$ to form ammonia as represented by

$$N_2(g) + 3\, H_2(g) \rightleftharpoons 2\, NH_3(g)$$

is important because it is the principal source of the ammonia essential to the highly productive agriculture necessary to feed the huge population of the earth. It is desirable to carry out this reaction under such conditions that the yield of ammonia is as large as possible. On the basis of LeChatelier's principle (problem 17), predict the conditions of pressure (high or low) and temperature (high or low) that will lead to the largest yield of ammonia. You will need to know the sign of ΔH^0 for the reaction.

14

EQUILIBRIA INVOLVING SOLIDS AND SOLUTIONS

Introduction

In Chapter 13 we saw how equilibrium constants for reactions involving only gases are defined and used. In this chapter we will be concerned with equilibrium constant definitions and calculations for reactions involving solids, liquids, and solutions as well as gases.

Reactions of Solids and Gases

When solid calcium carbonate ($CaCO_3$) is heated, it decomposes to yield solid calcium oxide (CaO) and gaseous carbon dioxide (CO_2) as shown by

$$CaCO_3(c) \rightleftharpoons CaO(c) + CO_2(g) \qquad (14.1)$$

If this reaction involved only gases, we could immediately write the equilibrium constant expression in terms of partial pressures as in the defining equation (13.13). This reaction, however, actually involves two solids along with one gas, and we must consider carefully how the equilibrium constant expression should be written and used.

We know from our previous discussions of equilibrium constants that the equilibrium constant for reaction (14.1) should eventually be expressed in terms of thermodynamic activities or escaping tendencies of the various substances involved in the reaction. Following our earlier practice, we express the thermodynamic activity of $CO_2(g)$ as the pressure (or partial pressure) of $CO_2(g)$ denoted by P_{CO_2}. We also know that the vapor pressure of a solid is a convenient measure of its escaping tendency and therefore conclude that we can use equilibrium vapor pressures represented by P_{CaCO_3} and P_{CaO} as measures of escaping tendencies or thermodynamic activities of $CaCO_3$(c) and CaO(c). On this basis we begin by writing an equilibrium constant for reaction (14.1) as

$$K' = \frac{P_{CO_2} P_{CaO}}{P_{CaCO_3}} \tag{14.2}$$

Unfortunately, the vapor pressures of these (and many other) solids are so low at most temperatures of interest that they are very difficult to measure. We can get along without knowing particular values for these vapor pressures, however, if we recognize that the equilibrium vapor pressure of any pure solid depends only on the temperature. Thus at any specified temperature, P_{CaO} and P_{CaCO_3} in equation (14.2) are constants that can be represented by C_1 and C_2 so that (14.2) becomes

$$K' = \frac{C_1 P_{CO_2}}{C_2}$$

We rearrange this equation to

$$\frac{K'C_2}{C_1} = P_{CO_2}$$

and recognize that the left side of this equation is a constant at any particular temperature so that we can also write

$$K = P_{CO_2} \tag{14.3}$$

The K in (14.3) is the most commonly used equilibrium constant for the reaction represented by equation (14.1)

The same approach may be used in working out the form of equilibrium constants for other reactions involving solids and gases. For example, consider the decomposition of solid ammonium chloride to gaseous ammonia and hydrogen chloride.

$$NH_4Cl(c) \rightleftharpoons NH_3(g) + HCl(g) \tag{14.4}$$

In this case, we first express the equilibrium constant as

$$K' = \frac{P_{NH_3} P_{HCl}}{P_{NH_4Cl}} = \frac{P_{NH_3} P_{HCl}}{C} \tag{14.5}$$

Rearrangement gives

$$K'C = P_{NH_3} P_{HCl}$$

Because the left hand side of this equation is a constant at any specified temperature, we write

$$K = P_{NH_3} P_{HCl} \tag{14.6}$$

The K in equation (14.6) is the most commonly used equilibrium constant for the reaction represented by equation (14.4)

We might also have taken another approach to the problem of formulating equilibrium constants for reactions involving solids and gases. Because equilibrium constants for gas reactions may be written in terms of concentrations, as shown earlier, we may reasonably assume that the same thing may be done for other reactions such as (14.1) and (14.4). To do so we must recognize that the *concentration* of a pure solid is a constant at any particular temperature and pressure. The *amount* of that solid may change considerably as the reaction proceeds, but the amount of solid that is present in a given volume of itself (amount/volume = concentration) is independent of the total amount present. The concentration of a pure solid is therefore determined by its density and molecular weight and not by the amount that happens to be present.

Because concentrations of solids are constant when temperature and total pressure are constant, we need not include these values in equilibrium constant expressions. Even large changes in total pressure have little effect on densities of most solids, so it is usually unnecessary to hold rigorously to the requirement of constant pressure.

More extensive thermodynamic treatment of standard states than is appropriate here shows that taking thermodynamic activities or concentrations of pure solids to be unity in equilibrium constant expressions (such as 14.3 and 14.6) permits straightforward application of $\Delta G^0 = -RT \ln K$ and convenient use of K values obtained on this basis.

We now illustrate some calculations involving equilibrium constants for reactions involving solids and gases.

Example Problem 14.1. The pressure of CO_2(g) in equilibrium with $CaCO_3$(c) and CaO(c) at 800°C has been found to be 0.220 atm. Hence the equilibrium constant (equation 14.3) for reaction (14.1) has the value 0.220 at this temperature. Now suppose that 50 cm^3 of $CaCO_3$(c) is placed in an otherwise empty container of volume 800 cm^3 and maintained at 800°C. How much $CaCO_3$(c) will be decomposed and how much CaO(c) will be formed?

The first step is to calculate now many moles of CO_2(g) will be produced in order to bring P_{CO_2} from its initial value of zero up to the 0.220 atm that is required at equilibrium. Taking the volume of gas to be 800 cm^3 – 50 cm^3 = 750 cm^3 = 0.75 liter, we use $n = PV/RT$ as follows:

$$n = \frac{0.220 \text{ atm} \times 0.75 \text{ liter}}{0.082 \text{ liter atm mole}^{-1} \text{ deg}^{-1} \times 1073 \text{ deg}} = 1.87 \times 10^{-3} \text{ mole}$$

The balanced equation (14.1) shows that each mole of CO_2(g) must have been formed from 1 mole of $CaCO_3$(c). We therefore know that 1.87×10^{-3} mole of $CaCO_3$(c) is converted into 1.87×10^{-3} mole of CaO(c). ■

Example Problem 14.2. What would happen as CO_2(g) is pumped into a container maintained at 800°C in which there is some CaO(c)? We know the value for the equilibrium constant for reaction (14.1) from Example Problem 14.1.

As CO_2(g) is pumped in, its pressure (P_{CO_2}) will steadily increase until it reaches 0.220 atm. As long as P_{CO_2} is less than 0.220 atm at this temperature, none of the CaO(c) will react to form $CaCO_3$(c). Once a pressure of 0.220 atm is reached, further addition of CO_2(g) results in reaction of CO_2(g) with CaO(c) to form $CaCO_3$(c) in just sufficient amounts to maintain the pressure of $CO_2(g)$ at 0.220 atm. When enough $CO_2(g)$ has been added to convert all the CaO(c) to $CaCO_3(c)$, so that we are no longer concerned with the equilibrium represented by equation (14.1), the pressure of $CO_2(g)$ will again increase as more $CO_2(g)$ is pumped into the container.

If we want to convert $CaCO_3(c)$ into CaO(c) by heating at 800°C, we must provide means for the $CO_2(g)$ produced by reaction (14.1) to escape so that P_{CO_2} cannot build up to 0.220 atm. Then the decomposition reaction can continue until all the $CaCO_3(c)$ is converted to CaO(c). This decomposition reaction is important in preparation of lime (CaO) and thence calcium hydroxide, $Ca(OH)_2$, from limestone ($CaCO_3$). ■

Example Problem 14.3. Ferrous sulfate ($FeSO_4$) undergoes a thermal decomposition reaction that is described by the reaction equation

$$2\ FeSO_4(c) \rightleftharpoons Fe_2O_3(c) + SO_2(g) + SO_3(g) \tag{14.7}$$

It has been found experimentally that the total gas pressure (SO_2 and SO_3) is 0.90 atm at 650°C. Write the equilibrium constant expression for this reaction and calculate a value of K for the reaction at 650°C.

Following the procedure discussed earlier in this chapter, we set the thermodynamic activities of the pure solids equal to unity and thence obtain the thermodynamic equilibrium constant expression

$$K = P_{SO_2}P_{SO_3}$$

The balanced reaction equation shows us that equal quantities of SO_2 and SO_3 are produced in the decomposition reaction. We therefore know that half of the observed total pressure is due to $SO_2(g)$ and half to $SO_3(g)$. Thus we obtain

$$K = 0.45 \times 0.45 = 0.20 \quad ■$$

Example Problem 14.4. Suppose that an excess of $FeSO_4(c)$ is placed in a flask maintained at 650°C in which the initial pressure of $SO_2(g)$ is 0.60 atm, with no $SO_3(g)$ present initially. Calculate the final gas pressure in the system and the partial pressure of each gas on the basis of the reaction and equilibrium constant discussed in Example Problem 14.3.

We now let x represent the pressure (in atmospheres) of $SO_3(g)$ at equilibrium that results from reaction (14.7). Because the reaction equation shows that $SO_2(g)$ and $SO_3(g)$ are produced in equal amounts, we see that

$$P_{SO_2} = 0.60 + x$$

Substitution into the equilibrium constant expression $K = P_{SO_2}P_{SO_3}$ gives

$$0.20 = (0.60 + x)(x) = 0.60x + x^2 \tag{14.8}$$

Now we rearrange this equation to the standard quadratic form

$$x^2 + 0.60x - 0.20 = 0$$

Application of the quadratic formula leads to

$$x = \frac{-0.60 \pm \sqrt{(0.60)^2 + 4(0.20)}}{2} = \frac{-0.60 \pm 1.08}{2}$$

The two solutions for the quadratic equation are $x = -0.84$ and $x = +0.24$. Because negative x is not chemically significant in this problem, we know that the solution we want is $x = 0.24$ and can verify by substitution in (14.8) that we do indeed have a correct solution.

From our original definition of x, we have $P_{SO_3} = 0.24$ atm and then $P_{SO_2} = 0.60 + 0.24 = 0.84$ atm. The total pressure is $0.24 + 0.84 = 1.08$ atm. ■

Reactions in Solutions

Detailed and rigorous thermodynamic treatment of the equilibrium concept and its application to reactions in solution leads to the same conclusions that we will now develop by simple analogy with reactions involving only gases or reactions involving gases and solids.

Suppose that we are concerned with a reaction that we represent by the general equation

$$a\,A + b\,B \rightleftharpoons c\,C + d\,D \tag{14.9}$$

in which the capital letters represent chemical substances in solution. Following the procedure already developed for reactions involving gases, we write the equilibrium constant expression in terms of concentrations:

$$K = \frac{[C]^c[D]^d}{[A]^a[B]^b} \tag{14.10}$$

In this equilibrium constant expression we have used square brackets to indicate concentrations of the various solute species. Because we are usually concerned with aqueous solutions, we ordinarily express concentrations in terms of molality or molarity.

Before we go on with evaluating and using equilibrium constants for reactions in solution, we should consider the limitations imposed on use of equilibrium constants expressed in terms of concentrations. Again, we proceed by analogy with reactions that involve only gases.

The thermodynamic treatment in Chapter 13 of equilibrium in systems of reacting gases was based on the ideal gas equation $PV = nRT$. Thus the treatment given there is exact for gases at very low pressure but only approximate for gases at other pressures. Similar treatment of the same problem for real gases (equation of state more complicated than $PV = nRT$) leads to formulation of equilibrium constants in terms of concentrations or partial pressures, each multiplied by a numerical factor that can be evaluated from PVT data or from a realistic equation of state. These numerical factors, sometimes called **activity coefficients**, approach unity as the pressure approaches zero.

Complete thermodynamic treatment of equilibrium constants for reactions in solution shows that equilibrium constant expression (14.10) really ought to be written in terms of **activities** rather than concentrations. Just as for gases, these solute activities are obtained as products of concentrations and numerical factors called activity coefficients. These activity coefficients can be obtained from such measured quantities as freezing point depressions, vapor pressures, cell potentials, or conductivities of solutions. Because these activity coefficients approach unity for very dilute solutions, the exact formulation of the equilibrium constant in terms of activities and our formulation (14.10) in terms of concentrations become identical in the limit of infinitely dilute solutions. For moderately dilute solutions, expression of equilibrium constants in terms of concentrations is a good approximation to exact formulation in terms of activities.

Because we are primarily interested in the principles of equilibrium constant calculations rather than the ultimate in numerical accuracy, we will express equilibrium constants in terms of concentrations. This is equivalent to setting all activity coefficients equal to unity, which is exactly right for infinitely dilute solutions and a good approximation for reasonably dilute solutions.

The usual standard state for aqueous solutes is based on concentration expressed in terms of molality. Equilibrium constants calculated from ΔG_f^0 values and $\Delta G^0 = -RT \ln K$ should therefore be expressed in terms of molalities. Because molality is approximately equal to ($\cong$) molarity for dilute aqueous solutions, we commonly substitute molarity for molality when this makes problem solving easier.

Ionization of Water

Pure water is a poor conductor of electricity, but its small conductivity does indicate the presence of low concentrations of ions. Because this small conductivity is present in the purest water samples, we conclude that the ions are derived from water itself rather than from some dissolved salt or other substance. A variety of evidence suggested long ago that these ions result from ionization of a small fraction of the water molecules as represented by

$$H_2O(\text{liq}) \rightleftharpoons H^+(\text{aq}) + OH^-(\text{aq}) \tag{14.11}$$

Subsequent investigations have confirmed that this reaction does occur and is responsible for the presence of ions in pure water.

Following the scheme indicated by equations (14.9) and (14.10), we write the equilibrium constant for the ionization of water in (14.11) as

$$K' = \frac{[H^+][OH^-]}{[H_2O]} \tag{14.12}$$

As usual, the square brackets are taken to represent concentrations or activities.

Conductivity measurements at 25°C have shown that the concentrations of $H^+(\text{aq})$ and $OH^-(\text{aq})$ are both 1.0×10^{-7} molal. The concentration of water itself in pure water is

$$\frac{\text{moles}}{\text{kg}} = \frac{1000 \text{ g}/18.0 \text{ g mole}^{-1}}{1.00 \text{ kg}} = 55.6\ m \text{ (molal)}$$

Now we use these concentrations in (14.12) to obtain

$$K' = \frac{(1.0 \times 10^{-7})(1.0 \times 10^{-7})}{(55.6)} = 1.8 \times 10^{-16} \tag{14.13}$$

It is more convenient and hence more common to work with another formulation of the equilibrium constant for the ionization of water. Because the concentration of water in any aqueous solution is 55.6 m, we can choose to incorporate this value into the equilibrium constant expression as follows:

$$K'(H_2O) = (1.8 \times 10^{-16})(55.6) = 1.0 \times 10^{-14} = [H^+][OH^-] \qquad (14.14)$$

We now use K_w as a special symbol to represent the product $K'(H_2O)$ and on the basis of (14.14) write

$$K_w = 1.0 \times 10^{-14} = [H^+][OH^-] \qquad \text{(at 25°C)} \qquad (14.15)$$

as the most commonly used equilibrium constant expression and value for the ionization of water at 25°C.

Before going on with illustration of the use of (14.15), we point out that there is another way of obtaining this important relationship. First, we write the equilibrium constant expression for the ionization reaction in terms of activities as

$$K = \frac{a_{H^+}a_{OH^-}}{a_{H_2O}} \qquad (14.16)$$

Because we are here concerned with very dilute solutions, we can immediately substitute concentrations of H^+(aq) and OH^-(aq) for activities denoted by a_{H^+} and a_{OH^-}. We are free to choose any standard state that we like for water and elect to choose pure water. Thus the activity of pure water is unity and (14.16) becomes identical with (14.15), which we obtained previously.

Application of (14.16) to aqueous solutions of various solutes requires that we consider the effect of solutes on the activity of water. Raoult's law shows that the activity of any liquid is decreased when any solute is dissolved in it. But because the amount of decrease is small for dilute solutions, we can still take the activity of water to be close to unity. Again, we obtain (14.15) from (14.16).

Example Problem 14.5. Given that the concentration of H^+(aq) in a particular solution at 25°C is 0.05 m, calculate the concentration of OH^-(aq) in this same solution.

From (14.15) we have

$$K_w = 1 \times 10^{-14} = [H^+][OH^-] = (5 \times 10^{-2})[OH^-]$$

that is solved for

$$[OH^-] = (1 \times 10^{-14})/(5 \times 10^{-2}) = 0.2 \times 10^{-12} \text{ molal}$$

Because the solution is very dilute, it would have been a good approximation to take $[H^+] = 0.05\ M$ (molar) and thence $[OH^-] = 0.2 \times 10^{-12}\ M$.

Example Problem 14.6. It has been reported that the ionization constant of water (based on molalities of H^+ and OH^- and unit activity for water) is 9.6×10^{-14} at 60°C. Thus we have

$$K_w = 9.6 \times 10^{-14} = [H^+][OH^-] \qquad \text{(at 60°C)}$$

Calculate the concentrations of H^+(aq) and OH^-(aq) in pure water at this temperature and deduce the sign of ΔH^0 for the reaction

$$H_2O(liq) \rightleftharpoons H^+(aq) + OH^-(aq)$$

We see from the balanced equation for the ionization reaction that H^+(aq) and OH^-(aq) are produced in equal numbers by the ionization reaction so that

$$[H^+] = [OH^-]$$

Substitution into the equilibrium constant expression gives

$$9.6 \times 10^{-14} = [H^+][H^+] = [H^+]^2$$

We solve for

$$[H^+] = 3.1 \times 10^{-7}\ m$$

and thence also

$$[OH^-] = 3.1 \times 10^{-7}\ m$$

in pure water at 60 °C.

Because K_w is larger at high temperature than at low temperature, we know that ΔH^0 is positive (endothermic) for ionization of water. ■

The pH Scale

Chemists and others commonly work with acidic solutions that have $H^+(aq)$ concentrations as high as 10 *M*. Even reasonably dilute acidic solutions that are presently of most interest to us may have $H^+(aq)$ concentrations as large as about 1 *M* (or *m*). We see from a calculation like that in Example Problem 14.5 that a solution with $[H^+] \cong 1\ m$ (or *M*) has $[OH^-] \cong 10^{-14}\ m$ (or *M*). Similarly, an alkaline solution with $[OH^-] \cong 1\ m$ has $[H^+] \cong 10^{-14}\ m$. Thus we see that concentrations of $H^+(aq)$ in various commonly used solutions may range from about 1 *m* to about $10^{-14}\ m$. (Remember that we use *m* for molality and *M* for molarity. For dilute aqueous solutions, $m \cong M$.)

Partly because of the convenience of working with numbers that range from ~0 to ~14 rather than with numbers that range from ~1 to ~10^{-14}, the logarithmic pH scale has been defined and widely used. The definition that we use is

$$\text{pH} = \log \frac{1}{[H^+]} = -\log [H^+] \tag{14.17}$$

Concentrations of H^+ and OH^- (25 °C) and corresponding pH values are listed in Table 14.1. Calculations involving pH values are illustrated in the following Example Problems.

Table 14.1. Corresponding Values of pH, $[H^+]$, and $[OH^-]$ in Water at 25 °C

pH	$[H^+]$	$[OH^-]$
0.0	$1 \times 10^0 = 1.0$	1×10^{-14}
1.0	$1 \times 10^{-1} = 0.1$	1×10^{-13}
7.0	1×10^{-7}	1×10^{-7}
13.0	1×10^{-13}	$1 \times 10^{-1} = 0.1$
14.0	1×10^{-14}	$1 \times 10^0 = 1.0$

Example Problem 14.7. The pH of a solution is 4.22 at 25°C. What are the concentrations of hydrogen and hydroxide ions in this solution?

From equation (14.17) we write

$$4.22 = \log \frac{1}{[H^+]}$$

and find in a log table or on a slide rule that 0.22 is the log of 1.66 so that we have

$$1.66 \times 10^4 = \frac{1}{[H^+]}$$

This equation leads to

$$[H^+] = 0.60 \times 10^{-4} = 6.0 \times 10^{-5}$$

We might also have calculated $[H^+]$ from (14.17) in the form

$$\log [H^+] = -\text{pH} = -4.22 = 0.78 - 5.00$$

With a slide rule or log table we now find that 0.78 is the log of 6.0 and thence obtain

$$[H^+] = 6.0 \times 10^{-5}$$

We use (14.15) to calculate the concentration of hydroxide ions as follows:

$$K_w = 1.0 \times 10^{-14} = [H^+][OH^-] = (6.0 \times 10^{-5})(OH^-)$$
$$[OH^-] = (1.0 \times 10^{-14})/(6.0 \times 10^{-5}) = 0.17 \times 10^{-9} = 1.7 \times 10^{-10} \quad \blacksquare$$

Example Problem 14.8. Calculate the pH of water (or of a neutral aqueous solution) at 60°C.

From Example Problem 14.6 we have $[H^+] = 3.1 \times 10^{-7}\ m$ in pure water (or neutral solution) at 60°C. In order to use

$$\text{pH} = -\log [H^+] = -\log (3.1 \times 10^{-7})$$

we find $\log 3.1 = 0.49$ with slide rule or log table and know that $\log 10^{-7} = -7$. Therefore,

$$\log (3.1 \times 10^{-7}) = 0.49 - 7 = -6.51$$

and

$$\text{pH} = 6.51.$$

We might also have used

$$\text{pH} = \log \frac{1}{[H^+]} = \log \frac{1}{(3.1 \times 10^{-7})} = \log (3.23 \times 10^6)$$

Taking $\log 3.23 = 0.51$ from slide rule or log table now leads to pH = 6.51 as above. ■

Weak Acids and Bases

Hydrogen chloride, sulfur trioxide, acetic acid, and many other substances dissolve in water to yield acidic solutions. Still other substances, such as sodium hydroxide and ammonia, dissolve in water to yield alkaline solutions. Use of equilibrium constants permits us to understand many of the properties of these acidic and alkaline solutions and also permits us to use a relatively small amount of input information to obtain a great deal of other useful information.

Weak Acids

Suppose that we make up separate 1.0 molal solutions of HCl (hydrogen chloride) and CH_3CO_2H (acetic acid) and measure $[H^+]$ in both of these solutions. Such measurements can be made with electrochemical cells discussed in Chapter 16, with a pH meter and glass electrode, or with various indicator substances whose colors are related to $[H^+]$.

These measurements lead to $[H^+] = 1.0\ m$ in $1.0\ m$ HCl solution. We therefore conclude that hydrogen chloride, which exists as HCl molecules in the gas phase and in some "inert" solvents, is entirely ionized to yield $H^+(aq)$ and $Cl^-(aq)$ ions in aqueous solution. This conclusion is consistent with results of freezing point depression measurements discussed earlier in Chapter 4. Here we have no need to consider an equilibrium constant.

Similar measurements lead to $[H^+] \cong 4 \times 10^{-3}\ m$ in the $1.0\ m$ acetic acid solution. Because this $[H^+]$ is considerably larger than the value in pure water ($1 \times 10^{-7}\ m$), we know that the acetic acid has produced hydrogen ions. And because this $[H^+]$ is considerably less than $1.0\ m$, we know that only a tiny fraction of the acetic acid is ionized in this aqueous solution. We therefore consider the equilibrium constant for the ionization of acetic acid.

Acetic acid molecules are represented by the structural formula

```
   H  O
   |  ‖
H—C—C—O—H
   |
   H
```

and the acetate ion formed by loss of one H^+ from an acetic acid molecule is represented as a resonance hybrid of the structures

```
 /  H  O  \−          /  H  O  \−
(   |  ‖   )    and  (   |  |   )
(H—C—C—O   )         (H—C—C═O   )
 \  |     /           \  |     /
    H                    H
```

It is convenient and common to represent acetic acid by HAc and acetate ion by Ac^-. We therefore write

$$HAc(aq) \rightleftharpoons H^+(aq) + Ac^-(aq) \tag{14.18}$$

for ionization of acetic acid in aqueous solution. The equilibrium constant for this acid ionization reaction is

$$K = \frac{[H^+][Ac^-]}{[HAc]} \tag{14.19}$$

in which the square brackets indicate activities or concentrations as usual. Sometimes it is convenient to use subscript letters (as in K_w) to indicate that we are concerned with an equilibrium constant for a particular reaction or class of reactions. According to this practice, K_a is sometimes used as symbol for equilibrium constants such as (14.19) for acid ionization reactions such as (14.18).

Now we show in several Example Problems how equilibrium constants (K or

K_a) are calculated from experimental data and how these equilibrium constants are then used in a number of useful calculations.

Example Problem 14.9. An acetic acid (HAc) solution has been made by dissolving 0.100 mole HAc in 1.00 kg of water. This solution is commonly described as being 0.100 *m* HAc. Measurements at 25°C have shown that $[H^+] = 1.33 \times 10^{-3}$ *m* and also that $[Ac^-] = 1.33 \times 10^{-3}$ *m*. Use these experimental results in calculating a numerical value for the equilibrium constant for the reaction represented by equation (14.18).

The equilibrium constant expression in (14.19) is written in terms of $[H^+]$, $[Ac^-]$, and [HAc]. Both $[H^+]$ and $[Ac^-]$ are known as a result of measurements. Measurements might be made to yield the concentration of un-ionized HAc but are unnecessary because we can deduce this quantity from the information we already have. That is, the concentration of HAc at equilibrium is obtained as the initial concentration minus the concentration of Ac^- that have appeared as a result of the ionization reaction. We therefore know that the equilibrium concentration of HAc(aq) is $0.100 - 0.001 = 0.099$ *m*.

Now we insert the various equilibrium concentrations cited above into the equilibrium constant expression (14.19) as follows:

$$K = \frac{(1.33 \times 10^{-3})(1.33 \times 10^{-3})}{0.099} = 1.8 \times 10^{-5}$$

Similar measurements made on other acetic acid solutions lead to the same value of *K*, which we will use in solving several other problems. ■

Example Problem 14.10. Using the equilibrium constant for ionization of aqueous acetic acid from Example Problem 14.9, calculate concentrations of all species at 25°C in a solution made by dissolving 0.300 mole acetic acid in 1.00 kg of water. We are particularly interested in $[H^+]$ and the pH of the solution.

PROCEDURE 1. Our first and simplest approach to this problem is to write down a balanced equation for the ionization of acetic acid with the corresponding equilibrium constant expression:

$$HAc(aq) \rightleftharpoons H^+(aq) + Ac^-(aq)$$

$$K = 1.8 \times 10^{-5} = \frac{[H^+][Ac^-]}{[HAc]}$$

To solve this equilibrium constant expression for the desired $[H^+]$, we must eliminate $[Ac^-]$ and [HAc] as unknowns. Although we have no immediate way of knowing $[Ac^-]$, we can see from the balanced equation for the ionization reaction that equal numbers of $H^+(aq)$ and $Ac^-(aq)$ are produced by the ionization of HAc. We therefore set $[Ac^-] = [H^+]$. The concentration of un-ionized HAc represented by [HAc] is also unknown, but we do know that the total of HAc(aq) and $Ac^-(aq)$ in the solution at equilibrium must correspond to the amount of HAc originally put into the solution. We therefore have $[HAc] + [Ac^-] = 0.300$ or $[HAc] = 0.300 - [Ac^-]$ and thence $(HAc) = 0.300 - (H^+)$ to substitute in the equilibrium constant expression.

Carrying out the substitutions described above gives us

$$1.8 \times 10^{-5} = \frac{[H^+][H^+]}{0.300 - [H^+]} \tag{14.20}$$

This equation, which contains only one unknown, can be solved by algebraic methods.

One approach is to multiply through by $0.300 - [H^+]$, rearrange to the form

$$a[H^+]^2 + b[H^+] + c = 0$$

and then apply the quadratic formula to obtain

$$[H^+] = 2.32 \times 10^{-3}\ \text{m}$$

Readers should go through this calculation.

PROCEDURE 2. We might have solved equation (14.20) more easily by using chemical reasoning to simplify the algebra. Many experimental results show that HAc is a weak electrolyte in aqueous solution, and the small value of the equilibrium constant for ionization is a quantitative expression of this knowledge. Because the value for K is small, we infer that only a tiny fraction of the HAc molecules is ionized. It is therefore reasonable to proceed by *assuming* that $[Ac^-] \ll [HAc]$ or that $[H^+] \ll 0.300$ so that the denominator of equation (14.20) can be approximated by 0.300. We will make this approximation, obtain a tentative value for $[H^+]$, and then check the accuracy of our approximation.
The approximation described above leads to

$$1.8 \times 10^{-5} \cong \frac{[H^+]^2}{0.300} \tag{14.20a}$$

This approximation to equation (14.20) is easily solved for

$$[H^+]^2 \cong 0.54 \times 10^{-5}$$
$$\cong 5.40 \times 10^{-6}$$

and then

$$[H^+] = 2.32 \times 10^{-3}\ m \quad (?)$$

We have written (?) to emphasize that this is a tentative answer based on a reasonable but as yet unverified approximation. First, we see that this $[H^+]$ is indeed considerably smaller than 0.300 m and then substitute this value in the right side of equation (14.20) and verify that it is indeed a satisfactory solution to our problem.
Now we calculate that

$$\text{pH} = -\log[H^+] = -\log(2.32 \times 10^{-3})$$
$$= (0.37 - 3) = -(-2.63) = 2.63$$

Because we originally set $[H^+] = [Ac^-]$, we now have $[Ac^-] = 2.32 \times 10^{-3}\ m$. We also have $[HAc] = 0.300 - 0.00232 = 0.298\ m$.
Finally, we calculate the concentration of OH^- from our $[H^+]$ and K_w as follows:

$$K_w = 1.0 \times 10^{-14} = [H^+][OH^-] = (2.32 \times 10^{-3})[OH^-]$$
$$[OH^-] = (1.0 \times 10^{-14})/(2.32 \times 10^{-3}) = 0.43 \times 10^{-11}\ m \quad \blacksquare$$

Now readers should stop for a moment to think about another approximation (in this case justified) that we have made without saying so. The approximation referred to consisted in setting $[H^+] = [Ac^-]$, which is exactly correct if acetic acid is the only source of both $H^+(aq)$ and $Ac^-(aq)$ in the solution of interest. Actually, acetic acid *is* the only source of acetate ions in this solution. But hydrogen ions can be derived both from the ionization of acetic acid and from the ionization of water. Because water is so slightly ionized as compared to acetic acid ($K_w = 1.0 \times 10^{-14}$ and $K_a = 1.8 \times 10^{-5}$), it is actually a very good approximation to ignore water as a source of hydrogen ions in the solution of present interest. In the following Example Problem we both verify this statement about the unimportance of ionization of water in this solution and show how to solve this problem in a more general way that can be extended to many other problems.

Example Problem 14.11. Set up equations that will permit evaluation of concentrations of $H^+(aq)$, $Ac^-(aq)$, $HAc(aq)$, and $OH^-(aq)$ in 0.300 molal acetic acid solution at 25 °C. Do not initially make any approximations such as $[H^+] \cong [Ac^-]$.

We begin by writing down equations for the pertinent equilibria and the corresponding equilibrium constants as follows:

$$HAc(aq) \rightleftharpoons H^+(aq) + Ac^-(aq)$$

$$H_2O(liq) \rightleftharpoons H^+(aq) + OH^-(aq)$$

$$K_a = 1.8 \times 10^{-5} = \frac{[H^+][Ac^-]}{[HAc]} \tag{14.21}$$

$$K_w = 1.0 \times 10^{-14} = [H^+][OH^-] \tag{14.22}$$

Because we have four unknowns ($[H^+]$, $[Ac^-]$, $[HAc]$, and $[OH^-]$), we must have a total of four equations. Therefore, we need two more equations to go with (14.21) and (14.22). One of these is the so-called mass balance or chemical content equation that expresses the chemical composition of the solution:

$$[HAc] + [Ac^-] = 0.300 \tag{14.23}$$

The other equation expresses the fact that every solution is electrically neutral. That is, positive charges balance negative charges as shown by

$$[H^+] = [Ac^-] + [OH^-] \tag{14.24}$$

This system of four equations is sufficient to permit evaluation of all four unknowns. Here are several procedures for doing so.

PROCEDURE 1. We know that a solution of acetic acid in water is acidic, which means that $[H^+] > 1 \times 10^{-7}$ and therefore that $[OH^-] < 1.0 \times 10^{-7}$. This observation *suggests* that we can neglect $[OH^-]$ in comparison with $[H^+]$ and $[Ac^-]$ in equation (14.24) and thereby obtain

$$[H^+] \cong [Ac^-] \tag{14.24a}$$

Combination of (14.24a) with (14.23) leads to

$$[HAc] + [H^+] \cong 0.300$$

and thence

$$[HAc] = 0.300 - [H^+] \tag{14.25}$$

Substitution of (14.24a) and (14.25) in the equilibrium constant expression (14.21) gives

$$1.8 \times 10^{-5} \cong \frac{[H^+][H^+]}{0.300 - [H^+]} \tag{14.26}$$

This expression, which is based on the clearly specified assumption that $[OH^-]$ is negligible compared to $[H^+]$ and $[Ac^-]$, is seen to be identical with equation (14.20) obtained in Example Problem 14.10. We have already shown in Example Problem 14.10 that $[H^+] = 2.32 \times 10^{-3}\ m$ is a mathematically satisfactory solution to this expression.

Now it remains to find out if this value of $[H^+]$ is consistent with our system of four equations (14.21–14.24) and the approximation that we have made. We begin by using

$[H^+] = 2.32 \times 10^{-3}$ with equation (14.22) to find that $[OH^-] = 0.43 \times 10^{-11}\ m$. Now we use (14.24) to obtain

$$[Ac^-] = [H^+] - [OH^-] = (2.32 \times 10^{-3}) - (0.43 \times 10^{-11})$$

It is easy to see that (0.43×10^{-11}) is truly negligible compared to (2.32×10^{-3}) so that

$$[Ac^-] = 2.32 \times 10^{-3} = [H^+]$$

is an excellent approximation and our whole calculation is justified.

PROCEDURE 2. We will proceed by algebraic methods (no approximations) to obtain a single (probably complicated) equation in terms of one unknown. We might do this in terms of any of the four unknowns, but choose to do so in terms of $[H^+]$ by eliminating other unknowns as follows:

We rearrange (14.22) to obtain

$$[OH^-] = K_w/[H^+]$$

and substitute into (14.24) to obtain

$$[H^+] = [Ac^-] + \frac{K_w}{[H^+]}$$

and thence

$$[Ac^-] = [H^+] - \frac{K_w}{[H^+]} \qquad (14.27)$$

Substitution of (14.27) in (14.23) and rearrangement gives

$$[HAc] = 0.300 - [H^+] + \frac{K_w}{[H^+]} \qquad (14.28)$$

Finally, we substitute both (14.27) and (14.28) in the equilibrium constant expression (14.21) to obtain

$$1.8 \times 10^{-5} = \frac{[H^+]\left([H^+] - \dfrac{K_w}{[H^+]}\right)}{0.300 - [H^+] + \dfrac{K_w}{[H^+]}} \qquad (14.29)$$

in which $K_w = 1.0 \times 10^{-14}$.

There are several ways to deal with equation (14.29) and similar equations for other problems. Because this is a cubic equation in $[H^+]$, as may be shown by suitable algebraic rearrangement, no simple algebraic solution is practical.

One somewhat tedious but entirely correct procedure is to find by trial and error a value for $[H^+]$ that makes the right hand side of equation (14.29) come out 1.8×10^{-5}. Another approach is to anticipate that $[H^+]$ for this solution is of order of $\sim 10^{-3}$ (say between 10^{-4} and 10^{-2}). This $[H^+]$ is small enough to be negligible compared to 0.300 and large enough that $K_w/[H^+]$ is negligible compared to $[H^+]$ in the numerator and 0.300 in the denominator of (14.29). Thus we can simplify equation (14.29) to the approximate relationship

$$1.8 \times 10^{-5} \cong \frac{[H^+]^2}{0.300} \qquad (14.29a)$$

Now we see that (14.29a) is the same as (14.20a) obtained previously in Example Problem 14.10, and is easily solved for $[H^+] = 2.32 \times 10^{-3}\ m$. Finally, as before, it is necessary to show that the mathematical approximations made in going from (14.29) to (14.29a) are justified. ■

Before we proceed with further Example Problems that involve equilibrium constants, it is useful to summarize briefly two different methods of approaching such problems.

One approach, which relies heavily on chemical reasoning, was illustrated in Example Problem 14.10. That approach was based on our recognizing that ionization of aqueous acetic acid was so much more important than ionization of water that we could properly ignore the latter and set the concentration of hydrogen ions equal to the concentration of acetate ions. Then, in order to avoid bothering with the quadratic formula, a mathematical approximation was made, based on our chemical experience that only a small fraction of the aqueous acetic acid is ionized.

The other approach, illustrated in Example Problem 14.11, is one in which we write down equations (equilibrium constants, chemical content, and charge balance) equal in number to the number of unknowns. The resulting system of equations is likely to be so complicated that simple algebraic evaluation of unknowns is impossible. One procedure for this evaluation is to introduce mathematical simplifications based on our chemical knowledge. Another procedure is to combine the system of equations into one new equation that contains only one unknown. Then this complicated equation can be solved by means of various approximations or by a more or less efficient trial and error procedure.

Readers may well wonder which of the approaches described in the two preceding paragraphs is "best." Unfortunately, there is no single answer to this question. We can, however, offer some generalizations as follows.

The "chemical reasoning" approach illustrated in Example Problem 14.10 is generally easiest IF one is able to identify in advance the most important chemical reaction and IF certain simplifying mathematical assumptions are both recognizable and valid. But beginners and experienced chemists are sometimes alike in being unable to recognize in advance the most important reaction. Furthermore, it may well be that more than one reaction really must be considered in some problems. And finally, mathematical approximations may either be overlooked or may not be justified. In such cases it is necessary to use the method of Example Problem 14.11 in which we work with several equations that must all be satisfied. Actual solution of the system of simultaneous equations may be carried out in a variety of ways.

No matter what approach is adopted, it is always necessary to check all answers to be sure that they are consistent with appropriate chemical equilibrium constants and with the chemical composition of the system. If any mathematical approximations have been made, it is also necessary to check to be sure that these approximations are justified.

Now we turn to some more Example Problems to illustrate further calculations with equilibrium constants for ionization of weak acids.

Example Problem 14.12. It is known from results of experiments cited in a later Example Problem that $K_a = 6.3 \times 10^{-5}$ for ionization of aqueous benzoic acid at 25°C. The structure of benzoic acid is represented by

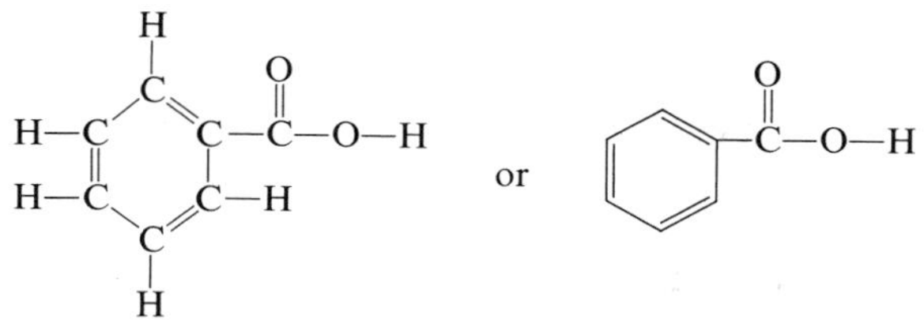

or

and that of benzoate ion is represented as a resonance hybrid of

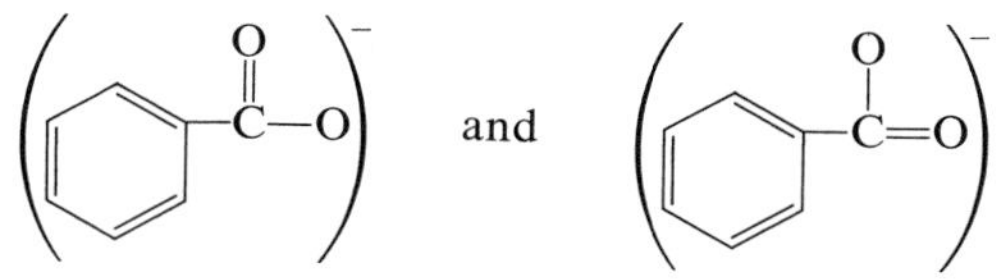

and

It is convenient and common to represent benzoic acid by HBz and benzoate ion by Bz^- so that we have

$$HBz(aq) \rightleftharpoons H^+(aq) + Bz^-(aq)$$

and

$$K_a = 6.3 \times 10^{-5} = \frac{[H^+][Bz^-]}{[HBz]}$$

Calculate the concentration of $H^+(aq)$ in a solution made by dissolving 0.00100 mole of benzoic acid in 100 g of water at 25°C.

Let us begin by *assuming* that the ionization of water is negligible compared to that of the HBz in this solution, which means that we can set $[H^+] = [Bz^-]$. The total concentration of HBz and Bz^- is 0.00100 mole/0.100 kg = 0.0100 *m*. Therefore the concentration of un-ionized HBz(aq) at equilibrium is $[HBz] = 0.0100 - [Bz^-] = 0.0100 - [H^+]$. Substitution into the equilibrium constant expression above gives

$$6.3 \times 10^{-5} = \frac{[H^+]^2}{0.0100 - [H^+]} \tag{14.30}$$

Equation (14.30) is a quadratic and may be solved by rearranging to the standard form that permits application of the usual quadratic formula. But the same reasoning that we used in Example Problem 14.10 suggests a simplifying approximation. It is reasonable to *assume* that $[H^+]$ is small enough that $0.0100 - [H^+] \cong 0.0100$ will be a satisfactory approximation to make in the denominator of (14.30). On this basis we write

$$6.3 \times 10^{-5} \cong \frac{[H^+]^2}{0.0100} \tag{14.30a}$$

and solve for

$$[H^+]^2 = 6.3 \times 10^{-7} = 63 \times 10^{-8}$$
$$[H^+] \cong 7.9 \times 10^{-4} = 0.00079 \quad (?)$$

We have written (?) to emphasize that this value of $[H^+]$ is a satisfactory solution to (14.30) only if $0.0100 - [H^+] \cong 0.0100$ is a satisfactory approximation. We see that $0.0100 - 0.0008 = 0.0092$ rather than 0.0100 used in (14.30a) above. If we are content with an accuracy of a few percent, we can stop here and take $[H^+] \cong 8 \times 10^{-4}$ as our answer. But if we want a more accurate answer, we must do more calculations.

One approach is to start all over and apply the quadratic formula to (14.30). Another approach is to use 0.0092 from the paragraph above as a better approximation to the denominator of (14.30) to obtain

$$6.3 \times 10^{-5} \cong \frac{[H^+]^2}{0.0092} \qquad (14.30b)$$

(14.30b) is a second and presumably better approximation to (14.30) than (14.30a). From (14.30b) we obtain

$$[H^+] \cong 7.6 \times 10^{-4} = 0.00076 \quad (?)$$

Now we have $0.0100 - [H^+] = 0.00924$, which is seen to be in good agreement with 0.0092 used in (14.30b). Thus we expect that $[H^+] = 7.6 \times 10^{-4}$ is an accurate solution of (14.30) and confirm that this is true by substituting this value in the right side of (14.30) to find that this right side really does equal 6.3×10^{-5} as required.

We might have arrived at this same result by rearranging (14.30) to the standard quadratic form

$$[H^+]^2 + 6.3 \times 10^{-5}[H^+] - 6.3 \times 10^{-7} = 0$$

Then application of the quadratic formula gives

$$[H^+] = \frac{-(6.3 \times 10^{-5}) \pm \sqrt{(6.3 \times 10^{-5})^2 + 4(6.3 \times 10^{-7})}}{2}$$

$$= \frac{-(6.3 \times 10^{-5}) \pm (1.59 \times 10^{-3})}{2}$$

Because the negative sign in the middle of the numerator above leads to an unacceptable negative concentration of H^+ we use the positive sign and calculate

$$[H^+] = 7.6 \times 10^{-4}$$

Readers should decide for themselves whether it is easier to solve (14.30) by successive approximations or by the quadratic formula.

One more check is necessary. Remember that we began by assuming that the ionization of water is negligible in 0.0100 *m* HBz solution. Now we use $[H^+] = 7.6 \times 10^{-4}$ with K_w to calculate

$$[OH^-] = (1.0 \times 10^{-14})/(7.6 \times 10^{-4}) = 0.13 \times 10^{-10}$$

The charge balance equation

$$[H^+] = [Bz^-] + [OH^-] = [Bz^-] + (0.13 \times 10^{-10})$$

now shows that taking $[H^+] = [Bz^-]$ to obtain (14.30) was justified.

Readers may wish to set up and solve this problem by the method illustrated in Example Problem 14.11. ■

Bronsted-Lowry Acids and Bases

We have seen that solutions of hydrogen chloride, acetic acid, and benzoic acid in water are acidic. Each of these acids increases the $[H^+]$ in water as a result of ionization. Hydrogen chloride is 100% ionized in dilute aqueous solution, and we have no need for equilibrium constant calculations. Acetic acid and benzoic acid, however, are both only partly ionized in water, and we have shown how to carry out some equilibrium constant calculations to account for properties of these weak electrolytes.

Now we note that compounds of some metals (for example, $AlCl_3$ and $FeCl_3$) also dissolve in water to yield acidic solutions, although direct addition of hydrogen ions to water by these compounds is impossible. After a brief discussion of aqueous bases, we shall return to an explanation of these acidities and to their quantitative description in terms of equilibrium constants.

The alkali metal hydroxides (LiOH, NaOH, KOH, and so on) have been called bases partly because the concentration of hydroxide ions in water is increased when these substances are dissolved. Freezing point lowering and electrical conductivity measurements have shown that aqueous solutions of sodium hydroxide and other alkali metal hydroxides are completely dissociated in dilute solution and are completely or nearly completely dissociated in concentrated solutions. The alkali metal hydroxides are therefore strong electrolytes in aqueous solution and we have no need for equilibrium constants in connection with their basic properties.

Hydroxides of the alkaline earth metals [$Mg(OH)_2$, $Ca(OH)_2$, and so on] are generally considered to be strong bases in aqueous solution. But association of hydroxide ions with the metal ions to form what are called ion pairs occurs to an appreciable extent in all but the most dilute solutions. This formation of ion pairs occurs to a greater extent with the alkaline earth ions than with the alkali metal ions because the +2 charges of the alkaline earth ions attract the negatively charged hydroxide ions more powerfully than do the +1 charges of the alkali metal ions.

The action of the alkali metal and alkaline earth hydroxides as bases in water is easy to understand in terms of the hydroxide ions that they put into aqueous solution. Other substances, such as ammonia and sodium acetate, which cannot directly add hydroxide solutions to water, also dissolve to yield basic or alkaline solutions.

The acid-base behavior of such compounds as ammonia, sodium acetate, aluminum chloride, and ferric chloride in water can be understood in terms of the Bronsted-Lowry concept of acids and bases already discussed in Chapter 5. Bronsted and Lowry considered an acid to be a proton donor and a base to be a proton acceptor, as summarized by the equation

$$\text{acid} \rightleftharpoons \text{base} + \text{proton} \qquad (14.31)$$

Hydrogen chloride and acetic acid in aqueous solution are acids in the Bronsted-Lowry system because they donate protons to water, which acts as a base in such solutions.

Ammonia is a base in the Bronsted-Lowry system because it is a proton acceptor. When dissolved in water, ammonia molecules accept or take protons from water molecules so that ammonium ions and hydroxide ions are produced. Sodium acetate is also a base in aqueous solution because acetate ions accept or take protons from the water to form hydroxide ions and molecules of acetic acid.

Water can act as a proton acceptor (base) or as a proton donor (acid). The actual behavior of water in a particular solution depends on what solutes are present to donate or accept protons and also on the concentrations of these solutes.

Acetic acid is an acid in aqueous solution because some of the acetic acid molecules donate protons to the water to yield acetate ions and hydrated protons, which we may call hydronium ions or represent concisely by $H^+(aq)$. But acetate ions are bases in aqueous solution because some of them accept protons from water, resulting in the formation of acetic acid molecules and hydroxide ions. It might appear that the acidic behavior of acetic acid and the basic behavior of acetate ions described here are contradictory; in one case we have acetic acid molecules losing protons to water and in the other we have acetate ions taking protons from water. The explanation is that there is an equilibrium between the various species. We can account for all of the experimental observations by means of the acetic acid ionization equilibrium constant K_a and the ionization constant for water K_w.

The discussion above has been partly designed to show that acidic and basic behavior are not separate, but go together. This combination of acid-base behavior in solution is illustrated by the following general equations:

$$\text{acid (1)} \rightleftharpoons \text{base (1)} + \text{proton} \qquad (14.32)$$

$$\text{proton} + \text{base (2)} \rightleftharpoons \text{acid (2)} \qquad (14.33)$$

$$\text{acid (1)} + \text{base (2)} \rightleftharpoons \text{acid (2)} + \text{base (1)} \qquad (14.34)$$

We have previously written

$$HAc(aq) \rightleftharpoons H^+(aq) + Ac^-(aq) \qquad (14.35)$$

and

$$K_a = \frac{[H^+][Ac^-]}{[HAc]} \qquad (14.36)$$

for the acid ionization of aqueous acetic acid. Now, in order to show that water and acetate ions are bases in this system we write

$$HAc(aq) + H_2O(liq) \rightleftharpoons H_3O^+(aq) + Ac^-(aq) \qquad (14.37)$$

and

$$K'_a = \frac{[H_3O^+][Ac^-]}{[HAc](a_{H_2O})} \qquad (14.38)$$

Comparison of equations (14.34) and (14.37) now leads to the following description of the aqueous acetic acid system. We identify $HAc(aq)$ as acid (1) and $H_3O^+(aq)$ as acid (2). Both $HAc(aq)$ and $H_3O^+(aq)$ are proton donors and are therefore properly called acids in the Bronsted-Lowry scheme. We also identify $H_2O(liq)$ as base (2) and $Ac^-(aq)$ as base (1). Both $Ac^-(aq)$ and $H_2O(liq)$ are proton acceptors and are therefore properly called bases in the Bronsted-Lowry scheme. The acetic acid–acetate ion pair is sometimes called a **conjugate pair.** Similarly, H_3O^+–H_2O can be called a conjugate pair.

We now compare the equilibrium constants K_a and K'_a defined in equations (14.36) and (14.38). In K_a we have $[H^+]$ whereas in K'_a we have $[H_3O^+]$. These symbols are merely two different ways of representing the same quantity—the concentration of aqueous hydrogen ions. This quantity is the same whether we describe it as $[H^+]$ or as $[H_3O^+]$ or as $[H(H_2O)_n^+]$ or in any other way. In K'_a as in (14.38) we have a_{H_2O}, which does not appear explicitly in K_a in (14.36). But we have previously shown that $a_{H_2O} \cong 1.0$ in dilute solution. Therefore $K_a = K'_a$ so that we have only one ionization constant to consider in this case.

Weak Bases

We are now ready to consider the observed basicity of sodium acetate solutions in terms of conjugate pairs of acids and bases and appropriate equilibrium constants. To begin, we write an equation to represent reaction of proton-accepting base (acetate ion) with proton-donating acid (water) to form a conjugate acid (acetic acid) and a conjugate base (hydroxide ion).

$$Ac^-(aq) + H_2O(liq) \rightleftharpoons HAc(aq) + OH^-(aq) \tag{14.39}$$

Remembering that $a_{H_2O} = 1.0$, we have

$$K(\text{often designated } K_b \text{ or } K_h) = \frac{[HAc][OH^-]}{[Ac^-]} \tag{14.40}$$

Here we have used subscript b to indicate that we are concerned with an equilibrium constant for reaction of $Ac^-(aq)$ as a base. Or the subscript h may be used to indicate that this is a so-called hydrolysis reaction in which acetate ion reacts with the solvent water. We shall use the symbol K_b for this equilibrium constant and for similar equilibrium constants for other reactions.

The numerical value for the equilibrium constant K_b for reaction (14.39) can be evaluated directly from results of appropriate measurements on solutions of sodium acetate. But we may also calculate the value from information already at hand as follows.

We have

$$H_2O(liq) \rightleftharpoons H^+(aq) + OH^-(aq) \qquad K_w = [H^+][OH^-]$$

and

$$HAc(aq) \rightleftharpoons H^+(aq) + Ac^-(aq) \qquad K_a = [H^+][Ac^-]/[HAc]$$

Dividing K_w by K_a gives us

$$\frac{K_w}{K_a} = \frac{[H^+][OH^-]}{1} \cdot \frac{[HAc]}{[H^+][Ac^-]} = \frac{[OH^-][HAc]}{[Ac^-]} \tag{14.41}$$

Comparison of (14.40) with (14.41) shows that

$$K_b = \frac{K_w}{K_a} \tag{14.42}$$

Thus the numerical value for the equilibrium constant for reaction (14.39) is

$$K_b = \frac{1.0 \times 10^{-14}}{1.8 \times 10^{-5}} = 0.56 \times 10^{-9} \tag{14.43}$$

We show how to use this equilibrium constant in the following Example Problem.

Example Problem 14.13. Enough water is added to 0.32 mole of sodium acetate to form exactly 1 liter of solution at 25°C. The resultant solution is said to be 0.32 *M* sodium acetate. Calculate the pH of this solution.

We know that sodium acetate, like many other salts, is a strong electrolyte. The reaction of interest and the corresponding equilibrium constant have been given in the preceding discussion as

$$Ac^-(aq) + H_2O(liq) \rightleftharpoons HAc(aq) + OH^-(aq) \tag{14.39}$$

and

$$K_b = 0.56 \times 10^{-9} = \frac{[HAc][OH^-]}{[Ac^-]} \qquad (14.40\text{–}14.43)$$

On the basis of the balanced equation for the principal reaction (14.39), we have $[HAc] = [OH^-]$ and $[HAc] + [Ac^-] = 0.32$, which lead to $[Ac^-] = 0.32 - [OH^-]$. Substitution into the equilibrium constant expression gives

$$0.56 \times 10^{-9} = \frac{[OH^-]^2}{0.32 - [OH^-]} \qquad (14.44)$$

Quadratic equation (14.44) can be solved for the unknown $[OH^-]$ by rearranging and applying the standard quadratic formula. It is easier, however, to recognize that $[OH^-]$ is very small so that $0.32 - [OH^-] \cong 0.32$ in the denominator of (14.44) is likely to be a very good approximation. In this way we obtain

$$0.56 \times 10^{-9} \cong \frac{[OH^-]^2}{0.32} \qquad (14.44a)$$

We solve (14.44a) for

$$[OH^-]^2 \cong 0.179 \times 10^{-9} = 1.79 \times 10^{-10}$$

and thence

$$[OH^-] \cong 1.34 \times 10^{-5} \quad (?)$$

The (?) indicates that this value can only be regarded as tentative until we check to make sure that all approximations that we made are justified.

First, we write $0.32 - (1.34 \times 10^{-5}) = 0.32 - 0.0000134$ and see that writing the denominator of (14.44) as 0.32 was in fact a very good approximation. We might also test the above value for $[OH^-]$ by inserting it in (14.44) and finding that the right side does indeed come out equal to 0.56×10^{-9}.

Finally, we use $[OH^-] = 1.34 \times 10^{-5}$ in $K_w = 1.0 \times 10^{-14} = [H^+][OH^-]$ to find

$$[H^+] = (1.0 \times 10^{-14})/(1.34 \times 10^{-5}) = 0.75 \times 10^{-9} = 7.5 \times 10^{-10}$$

Now we use log table or slide rule to find that $\log 7.5 = 0.88$ and combine with $\log 10^{-10} = -10.00$ to obtain $\log(7.5 \times 10^{-10}) = 0.88 - 10.00 = -9.12$. Insertion of this value in the defining equation $pH = -\log[H^+]$ gives the answer: pH = 9.12 for this solution of sodium acetate.

See problem 19 at the end of this chapter for another approach to this calculation. ■

We now consider the basicity of aqueous ammonia, following the general scheme of reaction (14.34) just illustrated by equation (14.39) for the basicity of acetate ion. For ammonia the reaction is

$$NH_3(aq) + H_2O(liq) \rightleftharpoons NH_4^+(aq) + OH^-(aq) \qquad (14.45)$$

and the equilibrium constant is

$$K_b = \frac{[NH_4^+][OH^-]}{[NH_3]} \qquad (14.46)$$

This equilibrium constant K_b can be evaluated from results of measurements (as in the next Example Problem) or from prior knowledge of the acidity constant for aqueous ammonium ion:

$$NH_4^+(aq) \rightleftharpoons H^+(aq) + NH_3(aq) \qquad (14.47)$$

$$K_a = \frac{[H^+][NH_3]}{[NH_4^+]} \tag{14.48}$$

Here we see that

$$K_a \times K_b = K_w \tag{14.49}$$

which is another way of writing (14.42) that was previously derived for the acetic acid–acetate ion conjugate pair in aqueous solution.

Example Problem 14.14. A solution containing 1.70 g of ammonia in 200 g of water is found to have pH = 11.48 at 25°C. Evaluate the equilibrium constant K_b for the reaction represented by equation (14.45).

There is

$$\frac{1.70 \text{ g}}{17.0 \text{ g mole}^{-1}} = 0.100 \text{ mole}$$

of ammonia in 200 g (0.200 kg) of water. The concentration of ammonia [temporarily neglecting reaction (14.45)] is therefore

$$0.100 \text{ mole}/0.200 \text{ kg} = 0.50\ m$$

The concentration of hydrogen ions is calculated from the measured pH, and then the concentration of hydroxide ions is obtained as follows:

$$\begin{aligned} -\text{pH} &= -11.48 = 0.52 - 12 = \log [H^+] \\ [H^+] &= 3.3 \times 10^{-12} \\ K_w &= 1.0 \times 10^{-14} = [H^+][OH^-] = (3.3 \times 10^{-12})[OH^-] \\ [OH^-] &= 3.0 \times 10^{-3} \end{aligned}$$

The balanced reaction equation (14.45) leads to $[OH^-] = [NH_4^+]$. Chemical material balance leads to $0.50 = [NH_3] + [NH_4^+]$ and thence to $[NH_3] = 0.50 - [OH^-]$. Substitution into the equilibrium constant expression (14.46) now gives

$$\begin{aligned} K_b &= \frac{(3.0 \times 10^{-3})^2}{0.50 - 0.003} = \frac{9.0 \times 10^{-6}}{0.497} \\ &= 1.8 \times 10^{-5} \end{aligned}$$

Readers may remember that $K_a = 1.8 \times 10^{-5}$ for ionization of aqueous acetic acid. It is only a coincidence that these two equilibrium constants have the same value at 25°C. Because the two reactions have different ΔH^0 values, their equilibrium constants change differently with changing temperature and are not equal to each other at other temperatures. ■

Example Problem 14.15. What is the concentration of H^+(aq) in a solution made by dissolving 0.40 mole of NH_4Cl(c) in enough water to make 0.50 liter of solution at 25°C?

PROCEDURE 1. The acid ionization reaction equation for aqueous ammonium ion (a proton donor) is

$$NH_4^+(aq) \rightleftharpoons H^+(aq) + NH_3(aq) \tag{14.50}$$

with equilibrium constant

$$K_a = \frac{[H^+][NH_3]}{[NH_4^+]} \tag{14.51}$$

We have already shown [equation (14.42)] that the acid ionization constant of acetic acid and the basic constant for acetate ion are related by $K_b = K_w/K_a$. Readers should now use the same method to show that this relationship also holds between the acid constant of ammonium ion and the basic constant for ammonia. A convenient way to remember this relationship is

$$K_a K_b = K_w \tag{14.49}$$

Using $K_b = 1.8 \times 10^{-5}$ from Example Problem 14.14 and rearranging (14.49), we obtain

$$K_a = (1.0 \times 10^{-14})/(1.8 \times 10^{-5}) = 0.56 \times 10^{-9}.$$

The concentration of ammonium chloride in our solution is 0.40 mole/0.50 liter = 0.80 M. Ammonium chloride is a strong electrolyte in aqueous solution. We therefore know that $[Cl^-] = 0.80\ M$ and that, neglecting reaction (14.50), $[NH_4^+] = 0.80\ M$.

On the basis of reaction (14.50) we set $[NH_3] = [H^+]$ and see from the material balance equation that $0.80 = [NH_3] + [NH_4^+]$ or $[NH_4^+] = 0.80 - [H^+]$. Substitution into the equilibrium constant expression (14.51) now gives

$$0.56 \times 10^{-9} = \frac{[H^+]^2}{0.80 - [H^+]} \tag{14.52}$$

Equation (14.53) is recognizable as a quadratic that can be rearranged to the standard form, which will permit straightforward application of the usual quadratic formula. But we can also see from the familiar form of this equation and the small magnitude of the equilibrium constant that $[H^+]$ is likely to be so much smaller than 0.80 that $0.80 - [H^+] \cong 0.80$ is probably an excellent approximation. On this basis we obtain

$$0.56 \times 10^{-9} \cong \frac{[H^+]^2}{0.80} \tag{14.52a}$$

and thence

$$[H^+]^2 \cong 0.448 \times 10^{-9} = 4.48 \times 10^{-10}$$
$$[H^+] \cong 2.1 \times 10^{-5} \quad (?)$$

As before, we have written (?) to emphasize that this value must be regarded as tentative until we test the validity of the approximation $0.80 - [H^+] \cong 0.80$. In this case it is easily seen that the approximation made was a very good one, so we can accept $[H^+] = 2.1 \times 10^{-5}\ M$ as our answer.

PROCEDURE 2. Another way to work this problem is to list all unknowns: $[H^+]$, $[OH^-]$, $[NH_4^+]$, and $[NH_3]$. Next we must have an equal number of independent equations. One is the chemical material balance equation:

$$0.80 = [NH_4^+] + [NH_3]$$

Another is the charge balance equation:

$$[H^+] + [NH_4^+] = [OH^-] + 0.80.$$

[Remember that $[Cl^-] = 0.80$.] For our other two equations we may choose any two from the set of K_a, K_b, and K_w.

Readers should either use $[H^+] = 2.1 \times 10^{-5}$ in this set of four equations to show that we have indeed found the right value for this unknown, or start from the beginning and solve the set of four equations for all four unknowns. ■

Buffer Solutions

Various mixtures of weak acids, bases, and salts are called **buffer solutions.** These buffer solutions are commonly used to maintain approximately constant pH, usually in the range between pH $\cong$ 3 and pH $\cong$ 11. It is also common to make use of the properties of buffer solutions in the determination of K_a and K_b values.

Several calculations involving equilibrium constants and the acid-base properties of buffer solutions are illustrated in the following Example Problems.

Example Problem 14.16. Calculate $[H^+]$ in a solution made by dissolving 0.20 mole of acetic acid (HAc) and 0.20 mole of sodium acetate (NaAc) in 1.0 kg of water at 25°C. (Remember that sodium acetate, like many other salts, is a strong electrolyte in dilute aqueous solution.)

PROCEDURE 1. The principal reaction in this solution is represented by

$$HAc(aq) \rightleftharpoons H^+(aq) + Ac^-(aq)$$

for which we have

$$K_a = 1.8 \times 10^{-5} = \frac{[H^+][Ac^-]}{[HAc]}$$

We see from the balanced equation for the principal reaction that the number of $H^+(aq)$ from ionization of HAc(aq) equals the number of $Ac^-(aq)$ derived from ionization of HAc(aq). The total concentration of $Ac^-(aq)$ is therefore 0.20 (from NaAc) plus $[H^+]$. Un-ionized acetic acid concentration is obtained as the original concentration minus the amount that disappears due to the ionization reaction. Appropriate substitution into the equilibrium constant expression above now gives us

$$1.8 \times 10^{-5} = \frac{[H^+](0.20 + [H^+])}{0.20 - [H^+]} \tag{14.53}$$

Once again we have a quadratic equation that can be solved for the unknown $[H^+]$ by rearrangement and application of the usual quadratic formula. But we might also recognize that $[H^+]$ is surely small in this solution. Therefore it is reasonable to try $0.20 + [H^+] \cong 0.20$ and $0.20 - [H^+] \cong 0.20$ as approximations in (14.53), which gives us

$$1.8 \times 10^{-5} \cong \frac{[H^+](0.20)}{(0.20)} \tag{14.53a}$$

and thence

$$[H^+] = 1.8 \times 10^{-5} \quad (?)$$

To test whether the tentative value indicated by (?) is valid, we substitute this value in (14.53) and find that it does lead to the right side equal to 1.8×10^{-5} as it should if our approximations were justified.

PROCEDURE 2. We might also have solved this problem by listing all unknowns, which in this case are $[H^+]$, $[OH^-]$, [HAc], and $[Ac^-]$. Two of the four equations needed for evaluation of these unknowns are given by $K_a = [H^+][Ac^-]/[HAc]$ and $K_w = [H^+][OH^-]$. Another equation that expresses the total acetic acid and acetate ion in the solution is

$$0.20 + 0.20 = [HAc] + [Ac^-]$$

Remembering that $[Na^+] = 0.20$, we can also write the electrical charge balance equation

$$0.20 + [H^+] = [Ac^-] + [OH^-]$$

Readers should go through this system of four equations to find values for all four unknowns. One way is to start with $[H^+] = 1.8 \times 10^{-5}$ already found and verify that it is a satisfactory solution while evaluating the other three unknowns. Another way is to recognize that $[OH^-]$ in this acidic solution is likely to be small enough that it can be neglected in the charge balance equation, thereby eventually leading to (14.53). ■

Example Problem 14.17. We have a solution that consists of 0.15 mole of NH_3 dissolved in 500 g of water. How much NH_4Cl must be added to this solution to give us a buffer solution with pH = 9.40 at 25°C?

The principal reaction of interest is

$$NH_3(aq) + H_2O(liq) \rightleftharpoons NH_4^+(aq) + OH^-(aq)$$

for which we have

$$K_b = 1.8 \times 10^{-5} = \frac{[NH_4^+][OH^-]}{[NH_3]}$$

Our task is to calculate $[OH^-]$ from the above given pH and combine this with the concentration of ammonia and the equilibrium constant expression to obtain information about the concentration of ammonium ion and thence how much ammonium chloride is needed.

We have $\log [H^+] = -pH = -9.40 = 0.60 - 10$ and use log tables or slide rule to find that $[H^+] = 3.98 \times 10^{-10}$. Then combination of this value with K_w gives

$$[OH^-] = (1.0 \times 10^{-14})/(3.98 \times 10^{-10}) = 0.25 \times 10^{-4}$$

We must also calculate the concentration of ammonia as 0.15 mole/0.500 kg = 0.30 m. The final equilibrium concentration of ammonia is less than this figure because of the principal reaction in the solution so that we have $[NH_3] = 0.30 - [OH^-] = 0.30 - 0.000025$, which is adequately approximated by 0.30.

Substitution into the equilibrium constant expression gives us

$$1.8 \times 10^{-5} = \frac{[NH_4^+](0.25 \times 10^{-4})}{(0.30)}$$

This equation is easily solved for $[NH_4^+] = 0.216\ m$. The final equilibrium concentration of $NH_4^+(aq)$ that we have found to be 0.216 m consists of a contribution from the added ammonium chloride and a contribution from the principal reaction. We see from the balanced equation for this reaction that each hydroxide ion formed is accompanied by formation of an ammonium ion and therefore we are able to write (using subscripts a and f to indicate contributions from *added* and from *formed* ammonium ions)

$$0.216 = [NH_4^+]_a + [NH_4^+]_f = [NH_4^+]_a + [OH^-]$$

and thence

$$[NH_4^+]_a = 0.216 - 0.000025 \cong 0.216\ m$$

So we finally know that we must add enough NH_4Cl to correspond to 0.216 mole kg^{-1} of water. Because we have 0.500 kg, the amount of ammonium chloride needed is

$$0.216 \text{ mole kg}^{-1} \times 0.500 \text{ kg} = 0.108 \text{ mole}$$

Finally, we add the atomic weights to find that the molecular weight of NH_4Cl is 53.5 g $mole^{-1}$ and calculate that we need

$$0.108 \text{ mole} \times 53.5 \text{ g mole}^{-1} = 5.78 \text{ g } NH_4Cl(c) \quad ■$$

Example Problem 14.18. In Example Problem 14.12 we made use of $K_a = 6.3 \times 10^{-5}$ for aqueous benzoic acid, HBz(aq), at 25 °C. Here we shall show how this value has been obtained from pH measurements on buffer solutions.

A benzoic acid–sodium benzoate (HBz–NaBz) buffer solution has been made by dissolving 9.0×10^{-4} mole of benzoic acid and 9.0×10^{-4} mole of sodium benzoate in enough water to yield 100 ml of solution. Measurements with a pH meter have led to pH = 4.21 for this solution at 25 °C. Calculate K_a for HBz(aq).

First, we have $\log [H^+] = -pH = -4.21 = 0.79 - 5$ and use log table or slide rule to find that $[H^+] = 6.2 \times 10^{-5}$.

The equation to represent ionization of aqueous benzoic acid is

$$HBz(aq) \rightleftharpoons H^+(aq) + Bz^-(aq)$$

and the equilibrium constant expression is

$$K_a = \frac{[H^+][Bz^-]}{[HBz]}$$

We already have $[H^+] = 6.2 \times 10^{-5}$ for substitution into the equilibrium constant expression. Now we calculate that the concentrations of HBz(aq) and $Bz^-(aq)$ are both (neglecting the ionization reaction for now) equal to $(9.0 \times 10^{-4} \text{ mole})/(0.10 \text{ liter}) = 9.0 \times 10^{-3}\ M$. The balanced equation for the ionization reaction shows that the final equilibrium concentration of $Bz^-(aq)$ is given by $9.0 \times 10^{-3} + [H^+] = 0.00900 + 0.00006 = 9.06 \times 10^{-3}\ M$ and that the equilibrium concentration of HBz(aq) is $9.0 \times 10^{-3} - [H^+] = 8.94 \times 10^{-3}\ M$. Substitution of these quantities into the equilibrium constant expression gives

$$K_a = \frac{[H^+][Bz^-]}{[HBz]} = \frac{(6.2 \times 10^{-5})(9.06 \times 10^{-3})}{(8.94 \times 10^{-3})}$$
$$= 6.3 \times 10^{-5}$$

Readers should note that the solution described here ($9.0 \times 10^{-3}\ M$ HBz and $9.0 \times 10^{-3}\ M$ NaBz) can be prepared in several ways. For example, one might start with 18×10^{-4} mole of HBz in 100 ml of solution and add 9.0×10^{-4} mole of NaOH to neutralize half of the HBz, leaving 9.0×10^{-4} mole of HBz(aq) and forming 9.0×10^{-4} mole of dissolved NaBz. Or the solution might have been formed by starting with 18×10^{-4} mole of NaBz dissolved in 100 ml of solution and adding 9.0×10^{-4} mole of HCl to convert half of the $Bz^-(aq)$ to HBz(aq). Naturally, such experiments can be done with various other ratios of HBz(aq) to $Bz^-(aq)$. ■

Solubility

The solubility of a given solid substance in a particular solvent is given by the concentration of solute in saturated solution. A saturated solution is one in equilibrium with the solid solute substance.

First, consider a hypothetical salt MX that dissolves in water to yield $M^+(aq)$ and $X^-(aq)$ ions. The equilibrium between solid (crystalline) MX and its aqueous ions in saturated solution is represented by

$$MX(c) \rightleftharpoons M^+(aq) + X^-(aq) \tag{14.54}$$

for which we write the equilibrium constant expression

$$K_{sp} = [M^+]_{ss}[X^-]_{ss} \tag{14.55}$$

The activity of pure solid MX(c) has been taken to be unity in accord with our previous considerations of equilibria involving pure solids. Subscripts ss have been written to emphasize that the activities or concentrations to be used in (14.55) must be activities or concentrations of the various species in **saturated solution** in equilibrium with the solid solute substance. The subscript sp has been applied to K because equilibrium constants for solubility equilibria like (14.54) are commonly called **solubility products.**

Suppose that the solubility of MX(c) in water is small enough that the saturated solution in equilibrium with solid salt is quite dilute. Then activity coefficients can properly be taken to be unity, and we can use concentrations in the solubility product expression. We represent the solubility expressed as molality or molarity of the saturated solution by S. Both $[M^+]_{ss}$ and $[X^-]_{ss}$ in (14.55) are equal to S so that we have

$$K_{sp} = S^2 \tag{14.56}$$

We might use free energy data (ΔG_f^0 values) to obtain ΔG^0 for reaction (14.54) and then calculate K_{sp} by way of the general equation $\Delta G^0 = -RT \ln K$. Then the solubility of MX(c) in water could be calculated from

$$S = (K_{sp})^{\frac{1}{2}} \tag{14.57}$$

It is also possible to use the solubility product expression (14.55) to calculate the solubility of MX(c) in presence of either $M^+(aq)$ or $X^-(aq)$ from some other salt.

Example Problem 14.19. The solubility product of MX(c) is 4.0×10^{-10} at 25°C. Calculate the solubility of MX(c) in water and in 0.01 M NaX, which is a strong electrolyte.

We can immediately calculate the solubility of MX(c) in water from equation (14.57) as

$$S = (4.0 \times 10^{-10})^{\frac{1}{2}} = 2.0 \times 10^{-5}$$

This means that the concentrations of $M^+(aq)$ and $X^-(aq)$ are 2.0×10^{-5} M in saturated solution at 25°C.

We already have the reaction equation (14.54) and now write the solubility product expression as

$$4.0 \times 10^{-10} = [M^+][X^-] \tag{14.58}$$

in which we have omitted the ss subscripts for convenience. Because all of the $M^+(aq)$ in the 0.01 M NaX solution must come from the MX(c) that dissolves, we have $[M^+] = S$. The initial concentration of $X^-(aq)$ due to NaX is increased at equilibrium by an amount corresponding to the amount of MX(c) that dissolves. Therefore we have $[X^-] = 0.01 + S$. Substitution in the solubility product expression (14.58) gives

$$4.0 \times 10^{-10} = (S)(0.01 + S) \tag{14.59}$$

Equation (14.59) is a quadratic that can be solved by rearrangement and application of the usual quadratic formula. Because K_{sp} is small, we expect that S will be small. In fact, from LeChatelier's principle we expect that S in this solution will be even smaller than S for pure water; that is, we expect that S is smaller than 2×10^{-5}. Thus it should be a good approximation to take $(0.01 + S) \cong 0.01$, which leads to

$$4.0 \times 10^{-10} \cong (S)(0.01) \tag{14.59a}$$

Now it is easy to solve (14.59a) for

$$S \cong 4.0 \times 10^{-8} \quad (?)$$

As usual, we have used (?) to indicate that this value is only tentative until our approximation has been checked. In this case it is clear that our approximation was justified and we can take $S = 4.0 \times 10^{-8}$ *M* as the solubility of MX(c) in 0.01 *M* NaX solution. ■

The decreased solubility of a salt, MX, in NaX solution as compared to its solubility in pure water is sometimes called the **common ion effect.**

There are several slightly soluble salts of the $+1:-1$ type represented by MX(c) in our preceding discussion. Among these are AgCl, AgBr, AgI, CuCl, CuBr, CuI, $AgIO_3$, TlBr, and several others. There are also many $+2:-2$ salts for which the same calculations can be carried out in the same way, including many carbonates, sulfates, and sulfides.

In addition to the symmetrical $+1:-1$ and $+2:-2$ salt types discussed above, there are various unsymmetrical types such as $+2:-1$, $+1:-2$, and so on. Examples of slightly soluble $+2:-1$ salts are several hydroxides [such as $Fe(OH)_2$ and $Mg(OH)_2$], several fluorides [such as CaF_2 and PbF_2], and a few other classes of compounds. An example of a common slightly soluble $+1:-2$ compound is Ag_2CrO_4.

To illustrate calculations with solubility products for unsymmetrical salts, we choose the hypothetical salt represented by M_2Y, which dissolves to yield $M^+(aq)$ and $Y^{2-}(aq)$ ions as shown by

$$M_2Y(c) \rightleftharpoons 2\ M^+(aq) + Y^{2-}(aq) \tag{14.60}$$

The equilibrium constant for solid M_2Y in equilibrium with its saturated solution is again called the solubility product:

$$K_{sp} = [M^+]^2[Y^{2-}] \tag{14.61}$$

The activity of solid M_2Y has been set equal to unity as usual, and the exponent of $[M^+]$ is in accord with our definition of the equilibrium constant in Chapter 13. For convenience we have omitted the subscripts ss, but it is important to remember that the solubility product equation (14.61) is applicable *only* to saturated solutions.

We first consider how the solubility product in (14.61) might be evaluated from solubility data. First, excess $M_2Y(c)$ may be placed in contact with water for a long enough time to insure that the solution formed is saturated. Then measured samples (often called aliquots) of the solution could be analyzed for both M^+ and Y^{2-}. If the saturated solution is quite dilute, as it would be if $M_2Y(c)$ were slightly soluble, these analyses might present serious experimental difficulties. Such analyses are often carried out by use of radioactive tracers, optical instruments, or electrochemical methods. It is partly because solubility measurements on slightly soluble compounds are generally difficult that solubility product calculations are so useful.

Suppose that the analyses mentioned above have been carried out on a saturated solution of M_2Y and that the concentrations of $M^+(aq)$ and $Y^{2-}(aq)$ in saturated solution have been found to be 8×10^{-5} and 4×10^{-5} M, respectively. We substitute these values in the solubility product expression (14.61) to obtain

$$K_{sp} = (8 \times 10^{-5})^2(4 \times 10^{-5}) = 2.56 \times 10^{-13}$$

If the analyses leading to solubility data are difficult, we might choose to analyze for either $M^+(aq)$ or for $Y^{2-}(aq)$, rather than for both of these. Then, from the balanced equation (14.60) we deduce that the concentration of $M^+(aq)$ in saturated solution is twice that of $Y^{2-}(aq)$, or that the concentration of $Y^{2-}(aq)$ is half that of $M^+(aq)$.

Once we know a value for K_{sp}, we can use this value in calculating solubilities of M_2Y in many solutions.

Example Problem 14.20. How much M_2Y will dissolve in 1 liter of 0.10 M solution of Na_2Y?

Equation (14.60) has already been written to represent the solution reaction and we have

$$K_{sp} = 2.56 \times 10^{-13} = [M^+]^2[Y^{2-}]$$

from the illustrative calculation that preceded this Example Problem.

We let S represent the number of moles of M_2Y that dissolve in 1 liter of 0.10 M Na_2Y. The equilibrium concentrations in saturated solution are $[M^+] = 2S$ and $[Y^{2-}] = 0.10 + S$. We substitute these concentrations in the solubility product expression to obtain

$$2.56 \times 10^{-13} = (2S)^2(0.10 + S) \qquad (14.62)$$

This cubic equation (containing S^3 and S^2) is difficult to solve by purely algebraic methods, but it is easy to solve after we simplify by making an approximation based on chemical reasoning.

Both the small value of K_{sp} and the solubility data that led to K_{sp} show that M_2Y is only slightly soluble in water. Because we expect M_2Y to be even less soluble in Na_2Y solution than in pure water (common ion effect or LeChatelier's principle), we anticipate that $0.10 + S \cong 0.10$ should be an excellent approximation. Thus (14.62) becomes

$$2.56 \times 10^{-13} \cong (2S)^2(0.10) = 0.4S^2 \qquad (14.62a)$$

Now it is easy to obtain

$$S^2 = 6.4 \times 10^{-13} = 64 \times 10^{-14}$$

and then

$$S \cong 8 \times 10^{-7} \quad (?)$$

This calculated value of S is small enough to justify the approximation and we can accept it as the solubility of M_2Y in 0.10 M Na_2Y solution. ■

Now let us turn from solubility calculations for hypothetical salts to similar calculations for real salts that are slightly soluble. In many cases we can apply exactly the same principles that we have already illustrated for the hypothetical

Table 14.2. Silver Chloride Solubility Data

(All concentrations are expressed in molarities and refer to saturated solutions at 25°C.)

$[Ag^+]$	$[Cl^-]$	$K_{sp} = [Ag^+][Cl^-]$
*1.33×10^{-5}	1.33×10^{-5}	1.77×10^{-10}
†1.97×10^{-5}	0.90×10^{-5}	1.77×10^{-10}
†2.87×10^{-5}	0.62×10^{-5}	1.78×10^{-10}

*Solubility in pure water.
†Solubility in the presence of *very* dilute $AgNO_3$.

salts represented by MX and M_2Y. For some compounds, however, we must introduce complications in our calculations in order to account for complications in chemical behavior.

As a real example of a slightly soluble salt, we now consider silver chloride. Because many investigations have yielded solubility data for silver chloride in water and in aqueous solutions containing various added solutes, we are able to check results of our calculations.

First, we consider the solubility data given in Table 14.2, along with the calculated values of K_{sp}. Because the data in Table 14.2 lead to consistent values of K_{sp}, we conclude that the chemical behavior of the systems under consideration is adequately represented by the reaction

$$AgCl(c) \rightleftharpoons Ag^+(aq) + Cl^-(aq) \tag{14.63}$$

Because only a small range of $Ag^+(aq)$ and $Cl^-(aq)$ concentrations is covered in Table 14.2, we should now compare results of calculations based on reaction (14.63) with results of experiments on considerably different solutions. For this comparison we calculate the solubility of AgCl in 0.0055 M NaCl solution and compare with the measured value of $S = 57 \times 10^{-8}$ M. Using $K_{sp} = 1.77 \times 10^{-10}$ from Table 14.2 and letting S represent the solubility expressed in moles liter^{-1}, we find the following:

$$1.77 \times 10^{-10} = [Ag^+][Cl^-] = (S)(0.0055 + S)$$
$$S = 3.2 \times 10^{-8}$$

Because the measured $S = 57 \times 10^{-8}$ M is more than 16 times as large as this calculated S, we know that something is wrong with our calculation.

One possible explanation for the discrepancy between measured and calculated solubilities above might be our neglect of activity coefficients. But activity coefficients that we have implicitly taken to be unity are ordinarily about 0.93 for 0.005 M solutions of $+1:-1$ electrolytes. Because these values are too close to unity to account for the observed error in the calculations, we must look elsewhere for an explanation.

Results of several investigations of related chemical systems *suggest* that reaction (14.63) is an adequate representation of the properties of AgCl(aq) only when *both* $Ag^+(aq)$ and $Cl^-(aq)$ are *very* dilute. In other solutions in which there are larger concentrations of either $Ag^+(aq)$ or $Cl^-(aq)$, it may be necessary to allow for association of Ag^+ and Cl^- to form a combined AgCl associated species that is sometimes called an **ion pair.** Or we can say that AgCl(aq) is a weak electrolyte that may be only partly dissociated into separate ions.

We now want to test the suggestion that Ag^+ and Cl^- may be associated in solution. This means that we want to try to account for solubility data in terms of the solubility reaction represented by equation (14.63) *and* an association (or dissociation) reaction.

We now set about interpreting experimental solubility data in terms of the solubility equilibrium represented by equation (14.63) and the new equilibrium represented by

$$AgCl(aq) \rightleftharpoons Ag^+(aq) + Cl^-(aq) \tag{14.64}$$

Equilibrium constants for these two reactions are

$$K_{sp} = [Ag^+][Cl^-] \tag{14.65}$$

and

$$K_d = \frac{[Ag^+][Cl^-]}{[AgCl]} \tag{14.66}$$

In (14.64) we have used AgCl(aq) to represent dissolved but undissociated silver chloride, while [AgCl] in (14.66) represents the activity or concentration of these AgCl(aq) molecules or ion pairs in the solution. We have used a subscript to indicate the class of equilibrium constant, in this case d to indicate a dissociation equilibrium.

Our experimental data consist of solubilities of silver chloride at different values of $Cl^-(aq)$ concentration. Because these solubilities are a measure of the total amount of silver chloride dissolved in various dilute HCl and NaCl solutions, we write

$$S = [Ag^+] + [AgCl] \tag{14.67}$$

In order to evaluate K_{sp} and K_d from the S and $[Cl^-]$ data, we must start with equation (14.67) and obtain a new equation that expresses S in terms of K_{sp}, K_d, and $[Cl^-]$.

First, we use (14.66) to obtain $[AgCl] = [Ag^+][Cl^-]/K_d$ and substitute this in (14.67) to obtain

$$S = [Ag^+] + \frac{[Ag^+][Cl^-]}{K_d} \tag{14.68}$$

From (14.65) we obtain $[Ag^+] = K_{sp}/[Cl^-]$ to substitute in (14.68) to obtain

$$S = \frac{K_{sp}}{[Cl^-]} + \frac{K_{sp}}{K_d} \tag{14.69}$$

Equation (14.69) is of the form: $y = mx + b$. A graph of solubility S against $1/[Cl^-]$, based on data in Table 14.3, is shown in Figure 14.1. The slope of the straight line in the Figure is 1.72×10^{-10}, which gives us $K_{sp} = 1.72 \times 10^{-10}$. The intercept of this line is seen to be 0.05×10^{-5} so that we have

$$0.05 \times 10^{-5} = \frac{K_{sp}}{K_d} = \frac{1.72 \times 10^{-10}}{K_d}$$

and thence $K_d = 3.4 \times 10^{-4}$.

Table 14.3. Solubility of Silver Chloride in Aqueous Chloride Solutions

(All concentrations are expressed in molarities and refer to 25°C.)

S	$[Cl^-]$	$1/[Cl^-]$
0.33×10^{-5}	5.92×10^{-5}	16,900
0.20×10^{-5}	1.12×10^{-4}	8,930
0.10×10^{-5}	3.44×10^{-4}	2,910
0.069×10^{-5}	9.66×10^{-4}	1,040
0.060×10^{-5}	1.27×10^{-3}	788
0.056×10^{-5}	1.59×10^{-3}	629
0.049×10^{-5}	2.75×10^{-3}	364
0.057×10^{-5}	5.50×10^{-3}	182

Because all of the points based on experimental data in Table 14.3 fall on or very close to the straight line in Figure 14.1, we know that our picture of this chemical system in terms of equilibria represented by equations (14.63) and (14.64) is consistent with all data that we have so far considered. Further, the value of K_{sp} obtained above (1.72×10^{-10}) is in excellent agreement with the value (1.77×10^{-10}) given in Table 14.2. Also, electrochemical measurements cited in Chapter 16 lead to a closely similar value. And finally, the agreement can be made even better by taking into account the activity coefficients that are close to but not exactly equal to unity.

Let us test our interpretation and derived equilibrium constant values by calculating some solubilities to compare with measured solubilities. The graph in Figure 14.1 shows that such calculations and experimental solubilities will be in good agreement for $Cl^-(aq)$ concentrations ranging from 0.6×10^{-5} to 5×10^{-3} M, so we choose to make this comparison for some larger $[Cl^-]$ values. For these calculations we insert values of K_{sp} and K_d in equation (14.69) to obtain

$$S = \frac{1.72 \times 10^{-10}}{[Cl^-]} + 0.05 \times 10^{-5} \tag{14.70}$$

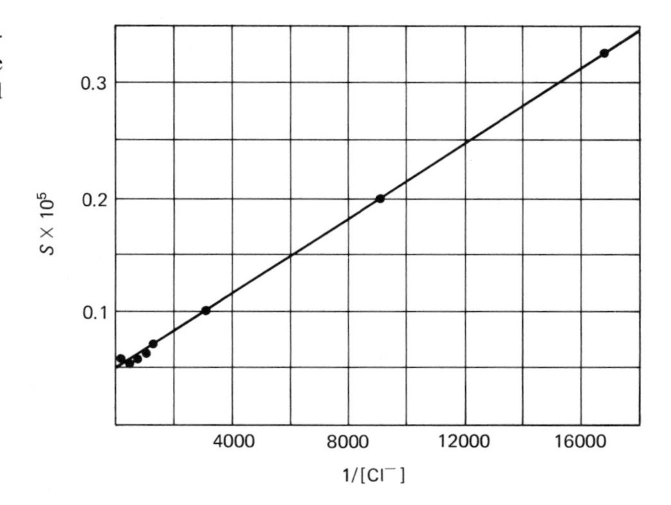

Figure 14.1. Graph of solubility (S) of silver chloride against $1/[Cl^-]$ as suggested by equation (14.69).

Table 14.4. Solubility of Silver Chloride in Aqueous Chloride Solutions

(All concentrations are expressed in molarities and refer to 25°C.)

S	$[Cl^-]$
0.0575×10^{-5}	5.50×10^{-3}
0.0660×10^{-5}	1.10×10^{-2}
0.110×10^{-5}	2.75×10^{-2}
0.195×10^{-5}	5.50×10^{-2}
0.380×10^{-5}	1.10×10^{-1}

First, we calculate the solubility of silver chloride in 0.011 M $Cl^-(aq)$ and find $S = 0.052 \times 10^{-5}$ M. The experimental value (0.066×10^{-5} M from Table 14.4) is 1.3 times as large as our calculated value. This discrepancy is only a little greater than can be accounted for by our neglect of activity coefficients, but it does indicate that our picture in terms of reactions (14.63) and (14.64) may be slightly incomplete for solutions with $[Cl^-]$ as large as ~0.01 M.

Again using equation (14.70), we calculate the solubility of silver chloride in 0.11 M $Cl^-(aq)$ and find $S = 0.05 \times 10^{-5}$. In this case the experimental value (0.38×10^{-5} from Table 14.4) is more than 7 times as large as the calculated value. Thus we see that the discrepancy between calculated and experimental solubility becomes larger as we consider more concentrated solutions. We also see that calculated solubilities based on equation (14.70) will remain approximately constant at 0.05×10^{-5} for all $[Cl^-]$ larger than about 10^{-2}, although the experimental solubilities (Table 14.4) are clearly increasing with increasing $[Cl^-]$.

Electrochemical measurements on solutions of silver chloride in moderately concentrated $Cl^-(aq)$ solutions show that some of the dissolved silver is in the form of negatively charged ions. One possible negatively charged ion containing silver could be formed by addition of a chloride ion to an undissociated silver chloride molecule or ion pair as represented by

$$AgCl(aq) + Cl^-(aq) \rightleftharpoons AgCl_2^-(aq) \tag{14.71}$$

The equilibrium constant for this complex ion forming reaction is

$$K_c = \frac{[AgCl_2^-]}{[AgCl][Cl^-]} \tag{14.72}$$

Because we are now postulating $Ag^+(aq)$, $AgCl(aq)$, and $AgCl_2^-(aq)$ as possible dissolved species in saturated silver chloride solutions, we extend (14.67) to

$$S = [Ag^+] + [AgCl] + [AgCl_2^-] \tag{14.73}$$

We use (14.72) to obtain $[AgCl_2^-] = K_c[AgCl][Cl^-]$ and substitute this in (14.73) to obtain

$$S = [Ag^+] + [AgCl] + K_c[AgCl][Cl^-]$$

Further substitution for $[Ag^+]$ and $[AgCl]$ as in obtaining (14.69) from (14.67) gives

$$S = \frac{K_{sp}}{[Cl^-]} + \frac{K_{sp}}{K_d} + \frac{K_c K_{sp}[Cl^-]}{K_d} \tag{14.74}$$

Finally, we rearrange to

$$\frac{K_d\left(S - \dfrac{K_{sp}}{[Cl^-]} - \dfrac{K_{sp}}{K_d}\right)}{K_{sp}} = K_c[Cl^-] \tag{14.75}$$

The left side of equation (14.75) involves only experimental values of S and $[Cl^-]$ in combination with the now known values of K_{sp} and K_d. We use data in Table 14.4 to evaluate the left side of (14.75) and then construct a graph of this quantity against $[Cl^-]$ as shown in Figure 14.2. The points fall on a fair straight line with slope of about 50. We therefore take $K_c \cong 50$.

If we extend the considerations of the last few paragraphs to still more concentrated chloride solutions, we again find discrepancies between calculated and observed solubilities. These discrepancies suggest that it may be necessary to allow for presence of small concentrations of such species as $AgCl_3^{2-}(aq)$ in moderately concentrated chloride solutions. Because activity coefficients are distinctly less than unity for these solutions, it is not appropriate for us to pursue the matter further in this book.

Using values of K_{sp}, K_d, and K_c found in our preceding discussion and calculations, we can write equation (14.74) as

$$S = \frac{4.72 \times 10^{-10}}{[Cl^-]} + 5.0 \times 10^{-7} + 2.50 \times 10^{-5}[Cl^-] \tag{14.76}$$

This equation accounts accurately for the solubility of silver chloride in solutions containing $Cl^-(aq)$ at concentrations up to a bit more than 0.1 M. At very low $Cl^-(aq)$ concentrations, only the first term on the right contributes appreciably

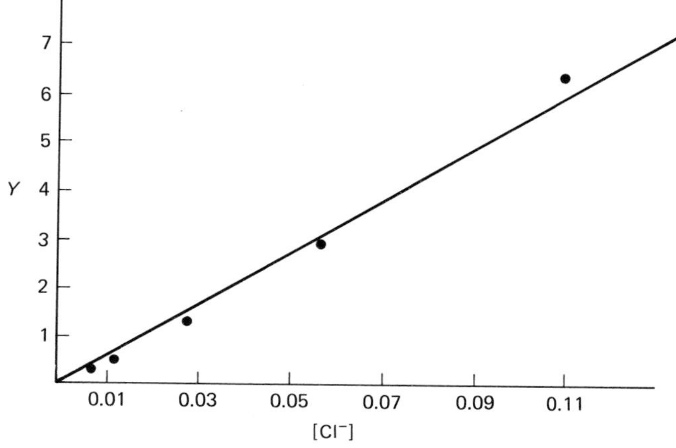

Figure 14.2. Graph of left side of equation (14.75), here represented by Y, against $[Cl^-]$.

to S, which is equivalent to considering only $Ag^+(aq)$ as an important solute species. At slightly higher $Cl^-(aq)$ concentrations, it is necessary to consider the second term as well as the first, which means that we are now considering both $Ag^+(aq)$ and AgCl(aq) to be important solute species. At still larger $Cl^-(aq)$ concentrations, we must also include the third term on the right, which means that $AgCl_2^-(aq)$ is now an important silver-containing solute species.

For some slightly soluble salts the simple calculations illustrated earlier for hypothetical salts MX and M_2Y are entirely appropriate. But for some other salts we can achieve agreement between calculations and experiments only by bringing in complications as we have done for silver chloride. Here it should be understood that these complicated calculations are not done for their own sake, but are necessary because certain real chemical systems are themselves complicated and can be understood only in terms of more than one chemical reaction. When we have more than one important chemical reaction in a solution, we almost always have a complicated calculation.

References

Blackburn, T. R.: *Equilibrium: A Chemistry of Solutions.* Holt, Rinehart and Winston, Inc., New York, 1969 (paperback).

Bulter, J. N.: *Solubility and pH Calculations.* Addison-Wesley Publishing Co., Inc., Reading, Mass., 1964 (paperback).

Cooper, J. N.: Solubility of lead bromide in nitrate media. *J. Chem. Educ.*, **49**:282 (1972).

Davidson, D.: Amphoteric molecules, ions, and salts. *J. Chem. Educ.*, **32**:550 (1955).

Fischer, R. B., and D. G. Peters: *Chemical Equilibrium.* W. B. Saunders Co., Philadelphia, 1970 (paperback).

Fleck, G. M.: *Equilibria in Solution.* Holt, Rinehart and Winston, Inc., New York, 1966.

Guyon, J. C., and B. E. Jones: *Introduction to Solution Equilibrium.* Allyn and Bacon, Boston, 1969 (paperback).

King, E. L.: Calculating the concentrations of the species present in complex buffers. *J. Chem. Educ.*, **31**:183 (1954).

Morris, K.: *Principles of Chemical Equilibrium,* 2nd. ed. Van Nostrand and Reinhold, New York, 1967 (paperback).

Nightingale, E. R., Jr.: The use of exact expressions in calculating hydrogen ion concentrations. *J. Chem. Educ.*, **34**:277 (1957).

Nyman, C. J., and R. E. Hamm: *Chemical Equilibrium.* Raytheon Education Co., 1968 (paperback).

Ramette, R. W.: Solubility and equilibria of silver chloride. *J. Chem. Educ.*, **37**:348 (1960).

Ramette, R. W.: The nature of dissolved silver acetate. *J. Chem. Educ.*, **43**:299 (1966).

Robbins, O., Jr.: *Ionic Reactions and Equilibria.* Macmillan, New York, 1967 (paperback).

VanderWerf, C. A.: *Acids, Bases, and the Chemistry of the Covalent Bond.* Reinhold Publishing Corp., New York, 1961 (paperback).

West, A. C.: The effects of chloride ion and temperature on lead chloride solubility. *J. Chem. Educ.*, **46**:773 (1969).

Problems

1. Calculate the concentration of $H^+(aq)$ in a solution made by dissolving 0.21 mole of HAc in 750 g of water at 25 °C.
2. A particular solution has pH = 6.12 at 25 °C. Calculate the concentrations of $H^+(aq)$ and $OH^-(aq)$ ions in this solution.

3. Hypochlorous acid ionizes in aqueous solution as shown by

$$HOCl(aq) \rightleftharpoons H^+(aq) + OCl^-(aq)$$

The equilibrium constant for this reaction at 25°C is $K_a = 2.8 \times 10^{-8}$. What is the concentration of $H^+(aq)$ in a solution that is 0.08 m in HOCl?

4. What is the pH of a solution made by dissolving 0.10 mole of sodium hypochlorite (NaOCl, a strong electrolyte) in enough water to make 0.50 liter of solution at 25°C? (See problem 3.)

5. When 0.10 mole of HAc is neutralized with 0.10 mole of NaOH in 500 ml of water at 25°C, the resulting solution is slightly alkaline. Account for this slight alkalinity by writing the relevant reaction equation, and then calculate the pH of the solution. This calculation shows that the "end point" or "equivalence point" in a titration of HAc by NaOH is not at pH = 7.0. This is a fact that must be considered in careful analytical work.

6. It is easy to see that the solution made by dissolving 0.10 mole of HCl (a strong electrolyte) in 1.0 liter of water at 25°C has pH = 1.0. Similarly, a solution made by dissolving 10^{-4} mole of HCl in 1.0 liter of water has pH = 4.0. Now it is tempting to say that the solution made by dissolving 10^{-7} mole of HCl in 1 liter of water has pH = 7.0 and also to say that the solution made by dissolving 10^{-8} mole of HCl in 1 liter of water has pH = 8.0.

   A little thought shows that the stated pH values in the last sentence above are ridiculous. How can we add even tiny quantities of acid to water to obtain a neutral or alkaline solution? We can't.

   Resolve this question by doing some careful calculations in which you consider the ionization of water.

7. Calculate the concentrations of all species in a solution made by dissolving 1.0×10^{-7} mole of acetic acid in 1.0 liter of water at 25°C. Compare your $[H^+]$ with that for 10^{-7} M hydrochloric acid found in problem 6.

8. Hydrogen cyanide ionizes in aqueous solution as shown by

$$HCN(aq) \rightleftharpoons H^+(aq) + CN^-(aq)$$

for which we have $K_a = 6 \times 10^{-10}$ at 25°C.

   (a) Calculate $[H^+]$ in a solution made by dissolving 0.0010 mole of HCN in 100 ml of water.

   (b) Calculate the pH of a solution made by adding 0.0005 mole of NaOH to the solution mentioned above.

   (c) Calculate the pH of a solution made by dissolving 0.01 mole of sodium cyanide (NaCN, a strong electrolyte) in 500 ml of water.

9. A weak acid that we represent by HQ ionizes in water as

$$HQ(aq) \rightleftharpoons H^+(aq) + Q^-(aq)$$

The equilibrium constant for this reaction is $K_a = 1.0 \times 10^{-13}$. What are the concentrations of all species in a solution made by dissolving 1.0 mole of HQ in 1.0 liter of water at 25°C. At what pH will the ratio $[HQ]/[Q^-]$ be 2.0?

10. A buffer solution is prepared by dissolving 6.0 g of acetic acid and 4.0 g of sodium acetate in 750 g of water at 25°C. What is the pH of the solution?

11. Fifty ml of 0.10 M hydrochloric acid and 100 ml of 0.08 M ammonia solution are mixed at 25°C. What are the concentrations of all species in the resulting solution?

12. Calculate the concentrations of all species (Na^+, H^+, OH^-, HAc, and Ac^-) in the solution that results from adding 30.0 g of pure acetic acid (CH_3CO_2H) to 500 ml of 0.50 M sodium hydroxide solution at 25°C.

13. Aniline, $C_6H_5NH_2$, is a base that is in many ways similar to ammonia. One mole of aniline dissolved in 1.0 liter of water at 25°C gives a solution with $[OH^-] = 2.04 \times 10^{-5}$ *M*. Calculate the equilibrium constant (K_b) for the reaction

$$C_6H_5NH_2(aq) + H_2O(liq) \rightleftharpoons C_6H_5NH_3^+(aq) + OH^-(aq)$$

Also calculate the equilibrium constant (K_a) for the reaction

$$C_6H_5NH_3^+(aq) \rightleftharpoons C_6H_5NH_2(aq) + H^+(aq)$$

It has been reported in a recent critical review that the "best" available results lead to $\Delta G^0 = 6.27$ kcal mole^{-1} for the acid ionization of anilinium ion as in the equation above. Use this ΔG^0 to obtain a "best" K_a value. Can you think of any explanation for the modest difference between this "best" value and the value calculated from the $[OH^-]$ value given above?

14. Calculate the concentration of hydrogen ions and the pH of a solution made by dissolving 0.20 mole of anilinium chloride (a strong electrolyte having formula $C_6H_5NH_3Cl$) in 400 ml of water 25°C. (See problem 13.)

15. Explain why a solution made by adding 0.01 mole of NaAc and 0.01 mole of HCl to 1 liter of water has the same pH as a solution made by adding 0.01 mole of HAc to 1 liter of water.

16. Using the definition of pH with $K_w = 1.0 \times 10^{-14}$ and $\Delta H^0 = 13.34$ kcal mole^{-1} for ionization of water, derive an equation for the pH of water as a function of temperature.

17. It has been reported that $K = 5 \times 10^{-3}$ at 25°C for

$$Fe^{3+}(aq) + H_2O(liq) \rightleftharpoons FeOH^{2+}(aq) + H^+(aq)$$

Calculate the pH of a solution that is 0.01 *M* $Fe^{3+}(aq)$.

18. Adding acid to a solution containing $Fe^{3+}(aq)$ and $FeOH^{2+}(aq)$ causes the equilibrium to shift in the direction of $Fe^{3+}(aq)$. At what pH will a solution that contains 0.02 mole liter^{-1} of iron in the +3 oxidation state be 99% $Fe^{3+}(aq)$ and 1.0% $FeOH^{2+}(aq)$?

19. In working Example Problem 14.13 we implicitly assumed that equation (14.39) represents the only important reaction in this solution, which amounts to saying that $OH^-(aq)$ ions from ionization of water need not be considered. Although this assumption is justified by the result we obtained, the problem can be worked in another way as follows.

We have four unknowns: $[H^+]$, $[OH^-]$, [HAc], and $[Ac^-]$. We need four equations, which are provided by two equilibrium constant expressions (K_a and K_w), a chemical content equation, and a charge balance equation. Set up these four equations.

Using the $[H^+]$ and $[OH^-]$ values found in Example Problem 14.13 with the set of equations, evaluate [HAc] and $[Ac^-]$ and show that all four values are consistent with the set of four equations.

Pretend that you do not know the answers to this problem and start with the set of four equations. You may proceed either by trial and error or by making the reasonable approximation that $[H^+]$ is small enough to be neglected in the charge balance equation.

20. Calculate the $[H^+]$ in a solution that is 0.100 *M* HAc. Then calculate the $[H^+]$ in a solution that is 0.100 *M* HAc and also contains 0.0002 *M* HCl as an impurity.

Suppose that we were using the second $[H^+]$ above for evaluation of K_a for HAc, without knowing that the solution contained HCl(aq) as an impurity. What value would you find for K_a?

In the text it was stated that buffer solutions are often used in the determination of K_a values. We illustrate this use with the following calculations.

Suppose that 0.0500 mole of NaOH is added to 1 liter of the 0.100 M HAc that contains 0.0002 M HCl as an impurity. Calculate the $[H^+]$ in the resulting solution. Finally, pretend that you did not know about the presence of the impurity and use this calculated $[H^+]$ in calculation of K_a for HAc.

What conclusion can you draw about using buffer solutions as protection against substantial error in K_a evaluation due to acidic impurities in chemicals?

21. In the text it was stated that aqueous solutions of certain ions (such as Al^{3+} and Fe^{3+}) are slightly acidic. These hydrated aqueous ions are proton donors and are therefore properly called Bronsted acids. A general reaction equation to illustrate the Bronsted acid properties of certain metal ions is

$$M(H_2O)_n^{+x} + H_2O \rightleftharpoons M(OH)(H_2O)_{n-1}^{+x-1} + H_3O^+ \qquad (14.77)$$

Write an equilibrium constant for reaction (14.77) and call this equilibrium constant K_a.

Some people choose to write the reaction that accounts for acidity of metal ions as

$$M^{+x}(aq) + H_2O \rightleftharpoons MOH^{+x-1}(aq) + H^+(aq) \qquad (14.78)$$

Write an equilibrium constant expression for reaction (14.78) and label it K_h.

Explain why K_a (for reaction 14.77) $\equiv K_h$ (for reaction 14.78).

22. Following our earlier definition of $pH = -\log[H^+]$, we now define

$$pOH = -\log[OH^-]$$

and

$$pK = -\log K$$

Derive an equation that relates pH and pOH of a solution to pK_w.

23. A solution is saturated with both AgCl and AgI so that both AgCl(c) and AgI(c) lie at the bottom of the solution. Calculate the ratio $[Cl^-]/[I^-]$ in this solution at 25°C. Use K_{sp} values from Appendix III.

24. Calculate the solubility of silver chromate (Ag_2CrO_4) in water at 25°C.

Assuming that dissolved Ag_2CrO_4 is present as $Ag^+(aq)$ and $CrO_4^{2-}(aq)$ ions, how much Ag_2CrO_4 will dissolve in 100 ml of 0.01 M sodium chromate (Na_2CrO_4, a strong electrolyte)?

If some dissolved Ag_2CrO_4 is present as $AgCrO_4^-(aq)$, how will the real solubility compare with that calculated above? (K_{sp} for Ag_2CrO_4 is given in Appendix III.)

25. Leussing and Kolthoff have carefully investigated the solubility of solid $Fe(OH)_2$ in several alkaline solutions of known $[OH^-]$. One way of treating their data is to plot $S[OH^-]$ against $1/[OH^-]$. The slope and intercept of the resulting straight line are 1×10^{-15} and 5×10^{-10}, respectively.

We anticipate that important species in ferrous hydroxide solutions might be $Fe^{2+}(aq)$ and $FeOH^+(aq)$ and therefore write

$$S = [Fe^{2+}] + [FeOH^+]$$

Evaluate equilibrium constants for all of the following reactions:

$$Fe(OH)_2(c) \rightleftharpoons Fe^{2+}(aq) + 2\,OH^-(aq) \qquad K_{sp}$$
$$Fe^{2+}(aq) + OH^-(aq) \rightleftharpoons FeOH^+(aq) \qquad K_c$$
$$Fe^{2+}(aq) + H_2O(liq) \rightleftharpoons FeOH^+(aq) + H^+(aq) \qquad K_a$$

26. Using data from Tables 14.2–14.4, construct a graph of S versus $[Cl^-]$ and find the value of $[Cl^-]$ that corresponds to minimum solubility of silver chloride.

Differentiate equation (14.74) to obtain $dS/d[Cl^-]$ and set this derivative equal to zero to find a relationship between equilibrium constants and $[Cl^-]$ at minimum S.

Compare the results of the two paragraphs above.

27. Show that the solubility product for a slightly soluble salt MX_3 is related to the molar solubility S by $K_{sp} = 27S^4$.

28. We have concluded from study of the solubility of AgCl in chloride solutions that AgCl(aq) is not completely dissociated in all solutions and that $AgCl_2^-$(aq) ions are formed in presence of enough Cl^-(aq) ions. On this basis we made use of equations (14.63–14.76).

Various workers have also investigated the solubility of AgCl in $AgNO_3$ solutions. It is possible that AgCl(aq) and Ag_2Cl^+(aq) are important species in these solutions. Show how you could make use of such solubility data for determining the equilibrium constant for the reaction

$$AgCl(aq) + Ag^+(aq) \rightleftharpoons Ag_2Cl^+(aq)$$

15

CHEMISTRY OF ALKALI METALS AND ALKALINE EARTH METALS

Introduction

Atoms of the alkali metals (lithium, sodium, potassium, rubidium, cesium, and francium) have similar electronic configurations. Lithium atoms have the configuration $1s^22s^1$, which is the same as the configuration of the noble gas helium plus one electron in the next lowest available energy level. The configuration of sodium atoms is $1s^22s^22p^63s^1$, which is the same as the configuration of the noble gas neon plus one electron in the next lowest available energy level. Similarly, each atom of potassium, rubidium, cesium, and francium has the same electronic configuration as the preceding noble gas plus one electron in the next lowest available energy level. Atoms of all the alkali metals lose one electron each to form M^+ ions with the same electronic configurations as the preceding noble gas. The $+1$ oxidation state is the only important one for compounds of the alkali metals.

The alkaline earth metals are beryllium, magnesium, calcium, strontium, barium, and radium. Atoms of these elements have the same electronic configurations as the preceding noble gas plus two *s* electrons. Only the $+2$ oxidation state is important in the chemistry of compounds of the alkaline earths.

Natural Occurrence

Sea water contains about 2.8% sodium chloride, 0.8% potassium chloride, and lesser amounts of the

Table 15.1. Some Slightly Soluble Compounds of Alkali Metals

Name	*Formula*	*Comments*
Spodumene	$LiAl(SiO_3)_2$	An important mineral source of lithium
Potassium feldspar	$KAlSi_3O_8$	A constituent of granite
Pollucite	$H_2Cs_4Al_4(SiO_3)_9$	An important mineral source of cesium
Cryolite	Na_3AlF_6	Used in electrolytic production of aluminum

other alkali metals. Because all simple salts of the alkali metals are quite soluble in water, these compounds are commonly found only in solution in the ocean or in solid deposits in places that are now dry. Evaporation of water from ancient inland seas has resulted in large deposits of various alkali metal salts in several places. Many complex compounds of the alkali metals are only slightly soluble in water and are found in the earth's crust. Some of these are listed in Table 15.1.

The naturally occurring isotope of francium, ^{223}Fr, which has a half-life of 23 min for decay by β emission, was discovered in 1939 by Mlle. Perey in France. Other isotopes of francium have been made by nuclear bombardment reactions, and all are radioactive with shorter half-lives.

Many alkaline earth compounds are only slightly soluble in water and are therefore found in the earth's crust. A few of these compounds are listed in Table 15.2.

Production and Reactions of Alkali and Alkaline Earth Metals

All the alkali metals are easily oxidized, largely because each atom of these elements loses one electron so easily. Conversely, it is difficult to reduce the alkali metal M^+ ions to the corresponding metals. The alkali metals are said to be very electropositive, which means that the metals are easily oxidized (are strong reducing agents) and that the M^+ ions are difficult to reduce.

In 1807 Humphrey Davy first prepared metallic sodium and potassium by electrolysis of their molten hydroxides. Electrolysis of molten hydroxides and molten halides has continued to be the usual method for preparing the alkali metals.

Some reactions of the alkali metals are summarized in Table 15.3.

Table 15.2. Some Important Alkaline Earth Minerals

Name	*Formula*	*Comments*
Beryl	$Be_3Al_2(SiO_3)_6$	Beryl crystals that are colored by small amounts of chromium are called emeralds
Asbestos	$CaMg_3(SiO_3)_4$	Used as thermal insulation
Limestone	$CaCO_3$	Source of calcium in many calcium-containing compounds
Gypsum	$CaSO_4 \cdot 2\,H_2O$	Used in cement; partly dehydrated to form plaster of paris
Fluorspar	CaF_2	A primary source of fluorine
Barite	$BaSO_4$	Used as a pigment

Table 15.3. Some Reactions of the Alkali Metals

Reaction equations	*Comments*
$4\,Li + O_2 \rightarrow 2\,Li_2O$	Similar reaction occurs with sodium in limited supply of dry oxygen
$10\,K + 2\,KNO_3 \rightarrow 6\,K_2O + N_2$	Similar reactions occur with other alkali metals
$K + O_2 \rightarrow KO_2$	Similar superoxides of heavier alkali metals are made in the same way
$2\,M + 2\,H_2O \rightarrow 2\,M^+ + 2\,OH^- + H_2$	Occurs vigorously with all alkali metals
$2\,M + X_2 \rightarrow 2\,MX$	All alkali metals with all halogens
$2\,M + 2\,NH_3 \rightarrow 2\,MNH_2 + H_2$	Occurs with all alkali metals if catalyst present
$2\,M + H_2 \rightarrow 2\,MH$	These hydrides contain M^+ and H^- ions

Large quantities of sodium are used in a lead-sodium alloy for production of tetraethyl lead, $Pb(C_2H_5)_4$, which is an antiknock additive in gasoline. Large scale production and use of tetraethyl lead for this purpose some 50 years ago was an important step in widespread application of the internal combustion engine in automobiles. For many years there was little appreciation of the dangers to be associated with increased concentrations of lead in our environment.

Now that the ill effects that may be caused by ingestion of lead are becoming better known, there has been considerable interest in alternatives for tetraethyl lead in gasoline. Among other additives that have been used are ethyl alcohol and various aromatic organic compounds such as benzene and toluene. There have also been changes in design and operating characteristics of the internal combustion engine.

Problems associated with lead (and other pollutants) from automobiles illustrate the technological, economic, and human complexities that must be resolved to avoid a general environmental catastrophe. Here we cite only a few of these complications.

1. Some effective antiknock additives apparently substitute one pollutant for another, with the "advantage" that the second pollutant *may* be less harmful than the first.
2. Natural gas (mostly methane, CH_4) may be used as fuel to replace gasoline, with considerable decrease in several pollutants. Among the complications are human attitudes toward change and possible increased "costs." Considerations of cost should include direct dollar costs, harm to environment, and consumption of nonrenewable resources. Costs of these latter sorts are too often ignored.
3. Most pollution from automobiles could be eliminated by eliminating most automobile use. This drastic step would have far-reaching political, economic, and social consequences that everyone should consider.

Metallic lithium, sodium, and potassium are widely used in chemical research and in industrial production of various chemicals. These uses are largely based on the strong reducing powers of the alkali metals.

Visible light readily ejects electrons from metallic cesium, which is often used in photocells ("electric eyes") that convert a light signal to an electrical signal.

Table 15.4. Some Reactions of Alkaline Earth Metals

Reaction equations	*Comments*
$2\,M + O_2 \rightarrow 2\,MO$	Barium also forms the peroxide, BaO_2
$M + 2\,H_2O \rightarrow M(OH)_2 + H_2$	Slow with beryllium and magnesium
$Ca + H_2 \rightarrow CaH_2$	Strontium and barium also form hydrides
$M + X_2 \rightarrow MX_2$	All alkaline earth metals with all halogens

The photoelectric effect was discussed in Chapter 6 in connection with the origins of quantum theory.

Lithium and sodium both have high specific heats and long liquid ranges. These properties make them attractive as heat exchangers in certain atomic reactors, particularly fast breeder reactors. The heat exchange liquid absorbs energy from the reactor and releases it to the water used to run the steam turbines. Unfortunately, the corrosiveness and chemical reactivity of the liquid metals present some practical disadvantages.

Like the alkali metals, the alkaline earth metals are quite electropositive. As shown later, the electropositive character of the alkaline earth metals is largely due to their relatively low ionization energies (as compared to other elements that commonly form M^{2+} ions). Some reactions of alkaline earth metals are summarized in Table 15.4.

Thermodynamic Considerations in Formation of Positive Ions

We now consider the particular properties that account for the alkali metals and alkaline earth metals being more electropositive than various other metals. Our previous discussions of thermodynamics have shown that ΔG provides a useful measure of the driving force of a reaction. Because $T\,\Delta S^0$ values for the reactions that we want to consider are all about the same and are considerably smaller than corresponding ΔH^0 values, we are presently justified in regarding ΔH^0 values as providing good approximate measures of the chemical driving forces for certain similar reactions.

To compare the tendency of an alkali metal to form $+1$ aqueous ions with the tendency of silver to form $+1$ aqueous ions, we carry out some thermodynamic calculations for the reactions represented by

$$Na(c) \rightarrow Na^+(aq) + e^- \tag{15.1}$$

and

$$Ag(c) \rightarrow Ag^+(aq) + e^- \tag{15.2}$$

From Appendix II we have $\Delta H_f^0 = 0$ for both Na(c) and Ag(c), $\Delta H_f^0 = -57.39$ kcal mole^{-1} for $Na^+(aq)$, and $\Delta H_f^0 = +25.23$ kcal mole^{-1} for $Ag^+(aq)$. Because e^- in (15.1) is the same as e^- in (15.2), the ΔH_f^0 value assigned to the electron is of no consequence in our comparative calculations. We can therefore follow the usual custom of taking $\Delta H_f^0 = 0$ for e^-. The ΔH_f^0 values above lead to $\Delta H^0 = -57.39$ kcal mole^{-1} for reaction (15.1) and $\Delta H^0 = +25.23$ kcal mole^{-1} for reaction (15.2). Remember that we presently take $\Delta G^0 \cong \Delta H^0$ and that negative ΔG^0 values correspond to larger driving forces for spontaneous reaction than positive ΔG^0 values. We then see that the standard state driving force for reaction (15.1) is about 83 kcal mole^{-1} greater than the standard state driving force for reaction (15.2). We thus have a relative measure of electropositive characters of sodium and silver, but as yet no clear explanation for the considerable difference between these elements.

A useful way to approach understanding of ΔH^0 values for reactions (15.1) and (15.2) is to consider these reactions as occurring in steps, which we may compare separately. This procedure is permitted because the ΔH^0 value for any process or reaction depends only on initial and final states and is therefore independent of the particular path or choice of steps by which the final state is reached. The steps that we choose to consider are the following:

$$\mathrm{M(c)} \rightarrow \mathrm{M(g)} \qquad \Delta H^0_{\mathrm{subl}} \qquad (15.3)$$
$$\mathrm{M(g)} \rightarrow \mathrm{M^+(g)} + e^- \qquad \Delta H^0_{\mathrm{ioniz}} \qquad (15.4)$$
$$\mathrm{M^+(g)} \rightarrow \mathrm{M^+(aq)} \qquad \Delta H^0_{\mathrm{hyd}} \qquad (15.5)$$
$$\mathrm{M(c)} \rightarrow \mathrm{M^+(aq)} + e^- \qquad \Delta H^0_{\mathrm{f}} = \Delta H^0_{\mathrm{subl}} + \Delta H^0_{\mathrm{ioniz}} + \Delta H^0_{\mathrm{hyd}} \qquad (15.6)$$

Enthalpies of sublimation ($\Delta H^0_{\mathrm{subl}}$) of metals have been calculated on the basis of the Clausius-Clapeyron equation from slopes of the log of vapor pressure versus $1/T$. Enthalpies of ionization ($\Delta H^0_{\mathrm{ioniz}}$) are obtained from ionization energies (often called ionization potentials). This calculation usually involves conversion of units of energy from electron volts per atom to kilocalories per mole. The results of all these calculations for sodium and silver (and other elements) have led to ΔH^0_{f} values listed in Appendix II for M(g) and M^+(g) species.

Enthalpies of hydration ($\Delta H^0_{\mathrm{hyd}}$) cannot be obtained directly from results of laboratory measurements. It is possible, however, to obtain total $\Delta H^0_{\mathrm{hyd}}$ values for any pair of oppositely charged ions (such as Na^+ and Cl^-) or to obtain the difference between $\Delta H^0_{\mathrm{hyd}}$ values for ions having the same charge (such as Na^+ and Ag^+). If some theory or new measurement could give a thoroughly reliable value for $\Delta H^0_{\mathrm{hyd}}$ for any one ion, we could then obtain $\Delta H^0_{\mathrm{hyd}}$ values for many other ions. Because we currently have no such reliable assignment of absolute $\Delta H^0_{\mathrm{hyd}}$ values, we will restrict our attention to relative $\Delta H^0_{\mathrm{hyd}}$ values for ions having the same charge.

The easiest way to deal with these relative hydration enthalpies is to make use of tabulated ΔH^0_{f} values for aqueous ions. All of these values in Appendix II and similar tabulations elsewhere are based on the convention or choice of "sea level" that $\Delta H^0_{\mathrm{f}} = 0$ for H^+(aq). This is equivalent to an arbitrary assignment of $\Delta H^0_{\mathrm{hyd}} = -367.161$ kcal mole^{-1} for H^+. We repeat that this procedure is satisfactory as long as we limit ourselves to comparing ions having the same charge (such as Na^+ and Ag^+ of present interest).

Using ΔH^0_{f} values from Appendix II, we calculate ΔH^0 values for reactions (15.3, 15.4, and 15.5) for sodium and silver to obtain the results shown in Table 15.5.

Highly electropositive metals have large negative ΔH^0_{f} and ΔG^0_{f} values for formation of their aqueous ions. As shown by equations (15.3–15.6), we may express ΔH^0_{f} in terms of the sum

$$\Delta H^0_{\mathrm{f}} = \Delta H^0_{\mathrm{subl}} + \Delta H^0_{\mathrm{ioniz}} + \Delta H^0_{\mathrm{hyd}} \qquad (15.7)$$

Table 15.5. Enthalpies of Formation of Aqueous Ions of Sodium and Silver

	ΔH^0 (*kcal mole*$^{-1}$)		ΔH^0 (*kcal mole*$^{-1}$)
$Na(c) \rightarrow Na(g)$	+25.98	$Ag(c) \rightarrow Ag(g)$	+68.01
$Na(g) \rightarrow Na^+(g) + e^-$	+120.04	$Ag(g) \rightarrow Ag^+(g) + e^-$	+175.58
$Na^+(g) \rightarrow Na^+(aq)$	−203.41	$Ag^+(g) \rightarrow Ag^+(aq)$	−218.36
$Na(c) \rightarrow Na^+(aq) + e^-$	−57.39	$Ag(c) \rightarrow Ag^+(aq) + e^-$	+25.23

In general, ΔH^0_{subl} and ΔH^0_{ioniz} values are positive, corresponding to endothermic processes, whereas ΔH^0_{hyd} values are negative, corresponding to exothermic processes. We can therefore expect that the low ionization energies of the alkali metals as compared to other metals will be an important factor in making the alkali metals highly electropositive. The ΔH^0 values in Table 15.5 are in accord with this generalization.

We also see that a low ΔH^0_{subl} value contributes to the final large negative ΔH^0_f for $Na^+(aq)$. Because ΔH^0_{subl} values are roughly proportional to melting and boiling points of metals, we conclude that, other things being equal, low melting metals will be more electropositive than high melting metals.

To summarize, we can say that it is a combination of relatively low ΔH^0_{subl} and ΔH^0_{ioniz} values that accounts for the highly electropositive properties of the alkali metals as compared to other elements.

The alkaline earth metals are also highly electropositive. On the basis of our discussion above, it is reasonable to expect that the dominant factor in making the alkaline earth metals more electropositive than other metals that form M^{2+} ions will be the relatively low ionization energies of the alkaline earth metals. The results of calculations like those already done for sodium and silver, which are shown in Table 15.6, are in accord with this expectation that relatively low ionization energies are an important factor in making the alkaline earths more electropositive than many other elements.

The ΔH^0 values in Table 15.6 show clearly that it is a low ΔH^0_{ioniz} value for calcium that makes calcium much more electropositive (negative ΔH^0_f, more powerful reducing agent) than zinc, lead, and mercury. The ΔH^0 values in Table 15.6 also show that it is primarily a relatively large (negative) ΔH^0_{hyd} value that makes zinc more electropositive than lead, whereas the very large (positive) ΔH^0_{ioniz} for mercury is mainly responsible for its chemical inertness (resistance to oxidation).

Table 15.6. ΔH^0 Data for Formation of Aqueous Ions of Calcium, Zinc, Lead, and Mercury

	ΔH^0 (*kcal mole*$^{-1}$)			
	Ca	*Zn*	*Pb*	*Hg*
$M(c) \rightarrow M(g)$	+42.6	+31.24	+46.6	+14.7
$M(g) \rightarrow M^{2+}(g) + 2e^-$	+417.7	+633.85	+520.6	+676.1
$M^{2+}(g) \rightarrow M^{2+}(aq)$	−590.0	−701.87	−567.6	−649.9
$M(c) \rightarrow M^{2+}(aq) + 2e^-$	−129.7	−36.78	−0.4	+40.9

Compounds of the Alkali Metals

Some important compounds of the alkali metals are listed in Table 15.7, along with their common names and some uses. Other compounds of the alkali metals are discussed later in connection with their anions.

Sodium carbonate is manufactured in large quantities from salt (NaCl) and limestone ($CaCO_3$) by way of the Solvay process. In this process carbon dioxide is obtained from thermal decomposition of limestone as shown by

$$CaCO_3(c) \xrightarrow{\text{heat}} CaO(c) + CO_2(g)$$

Then the carbon dioxide is reacted with an aqueous solution of ammonia and sodium chloride to yield a precipitate of sodium bicarbonate.

$$CO_2(g) + NH_3(aq) + Na^+(aq) + H_2O(liq) \rightarrow NaHCO_3(c) + NH_4^+(aq)$$

Heating the sodium bicarbonate yields sodium carbonate as indicated by

$$2\ NaHCO_3(c) \xrightarrow{\text{heat}} Na_2CO_3(c) + H_2O(g) + CO_2(g)$$

Ammonia is recovered from the solution from which sodium bicarbonate was precipitated by the reaction

$$2\ NH_4^+(aq) + CaO(c) \xrightarrow{\text{heat}} 2\ NH_3(g) + Ca^{2+}(aq) + H_2O(liq)$$

Finally, the recovered ammonia is used with salt and carbon dioxide to continue the process.

The major by-product of the Solvay process is calcium chloride, $CaCl_2$, which presents a large disposal problem. Although some calcium chloride is used to "melt" ice on roads, the supply far exceeds the annual demand. As a result most of the calcium chloride from Solvay plants is either dumped, creating water pollution problems, or stored in holding ponds.

Despite the apparent efficiency of the Solvay process, it is gradually becoming less important in the production of sodium carbonate. More rigid pollution standards, which are difficult for Solvay plants to meet, are one factor. Another

Table 15.7. Some Compounds of the Alkali Metals

Formula and common name	*Uses*
$LiClO_4$	Oxidizing component in high energy solid fuels
Li_2CO_3	In ceramics, glass, medicine, as a chemical intermediate
Na_2CO_3, soda ash	In glass and soap, also in manufacture of many chemicals
$Na_2CO_3 \cdot 10\ H_2O$	Washing soda
NaOH, caustic soda, lye	In chemical manufacture, pulp and paper treatment, synthetic fiber production
Na_2SO_4	In glass manufacture and wood pulp treatment
NaCN	In electroplating and metallurgical processes
NaBr and KBr	In photography and medicine
$NaNO_3$ and KNO_3, saltpeter	Fertilizer
NaCl	Original source of sodium in most compounds produced industrially

important reason is the presence of vast deposits of trona ($Na_2CO_3 \cdot NaHCO_3 \cdot 2H_2O$) in California and Wyoming. These deposits of natural soda ash are easily worked, and the wastes present no ecological problems.

Sodium ions play an important role in regulating the flow of body fluids through cell walls by osmosis. It is for this reason that "low sodium" diets are part of the treatment of certain ailments. On the other hand, some of the unpleasant symptoms of diarrhea are related to excessive loss of sodium ions from the body.

Potassium is the principal positive ion in the cells, just as sodium is the principal positive ion in the body fluids. Excessive concentrations of potassium ions in the body fluids are harmful and can lead to paralysis and even death. Healthy kidneys, however, ordinarily eliminate potassium ions from the body fluids.

Compounds of the Alkaline Earth Metals

Some important compounds of the alkaline earth metals are listed in Table 15.8, along with their common names and some uses. Other compounds of the alkaline earth elements are discussed later in connection with their anions.

Magnesium sulfate in the form of the heptahydrate, $MgSO_4 \cdot 7\,H_2O$, commonly called Epsom salts, is used in medicine as a purgative. This effect is achieved because $Mg^{2+}(aq)$ ions in the alimentary canal promote passage of water into the bowel from surrounding body fluids. (See Chapter 21 for a discussion of osmosis.)

Deficiency of calcium (normally obtained from milk and certain other foods) is associated with rickets and unsatisfactory bone development. Because both Ca^{2+} and chemically similar Sr^{2+} ions are deposited in bones, ^{90}Sr fallout from nuclear bombs is more harmful than many other sources of radioactivity.

An important industrial reaction is the production of calcium carbide from lime (made by heating limestone, $CaCO_3$) and carbon as shown by

$$2\,CaO(c) + 5\,C(s) \xrightarrow{\text{heat}} 2\,CaC_2(c) + CO_2(g)$$

The equation for reaction of calcium carbide with water to yield acetylene is

$$CaC_2(c) + 2\,H_2O(liq) \rightarrow Ca^{2+}(aq) + 2\,OH^-(aq) + C_2H_2(g)$$

If there is a relatively small amount of water present, calcium hydroxide will precipitate.

Acetylene, produced as shown above, was once fairly widely used to produce

Table 15.8. Some Compounds of the Alkaline Earth Metals

Formula and common name	*Uses*
MgO, magnesia	In fire bricks and as thermal insulation
$MgCO_3$	In tooth powder and metal polishes
$MgSO_4 \cdot 7\,H_2O$, Epsom salts	In medicine
CaC_2, calcium carbide	In manufacture of acetylene
$CaCl_2$	As a drying agent and to "melt" snow and ice
CaO, lime, quicklime	In cement, plaster, agriculture, and chemical manufacture
$Ca(OH)_2$, slaked lime	In mortar and hard water treatment
$BaSO_4$	As a component of lithopone pigment and in medicine

light by burning in so-called "carbide lamps." A principal use now is in welding and cutting metals. It is possible that acetylene produced in this way will someday be a primary material used in manufacture of organic compounds now derived from petroleum.

Calcium carbide also reacts with nitrogen at high temperatures to form calcium cyanamide, as shown by

$$CaC_2(c) + N_2(g) \rightarrow CaCN_2(c) + C(s)$$

Calcium cyanamide reacts with water to yield calcium carbonate and ammonia:

$$CaCN_2(c) + 3\ H_2O(liq) \rightarrow CaCO_3(c) + 2\ NH_3(aq)$$

Barium ions, like the ions of most heavy elements, are poisonous. But the solubility of barium sulfate is so low that $BaSO_4$ is safely taken into the stomach to aid in taking x-ray pictures of the digestive system. The scattering of x rays is proportional to the density of the electrons in the scattering medium. Barium ions, Ba^{2+}, have $56 - 2 = 54$ electrons in a small volume and therefore scatter x rays more than do ions of most other elements.

Thermodynamic Considerations for Stabilities and Solubilities

Only the +2 oxidation state is important in the chemistry of alkaline earth elements. We will carry out some approximate calculations of thermodynamic properties of the hypothetical compound CaCl(c) to gain some understanding of the instability of compounds containing alkaline earth metals in the +1 oxidation state and the consequent stability of the +2 oxidation state.

The thermodynamic property that we most want to know for CaCl(c) is its ΔG_f^0. Next best will be its ΔH_f^0. But because this compound has never been isolated and subjected to thermodynamic investigation, we are unable to look up these quantities in the usual tabulations. Instead, we must calculate them by appropriate combinations of experimental results with quantities that we estimate.

First, we estimate the size of a Ca^+ ion. Pauling has deduced from results of x-ray investigations of many crystals that the radius of a Ca^{2+} ion is about 0.99 Å (that is, 0.99×10^{-8} cm). We know that a Ca^+ ion would be larger than a Ca^{2+} ion because Ca^+ contains one more electron than Ca^{2+}. Pauling has also found that the radius of the K^+ ion is about 1.33 Å. From comparisons of known radii of ions of elements that are adjacent to each other in the periodic table, it is expected that the radius (r) of the Ca^+ ion is less than that of the K^+ ion. Pauling's estimate of $r = 1.18$ Å for Ca^+ is consistent with all of this discussion. Noting that $r = 0.95$ Å for Na^+ and $r = 1.33$ Å for K^+, we are confident that the Ca^+ ion ($r \cong 1.18$ Å) is intermediate in size between Na^+ and K^+.

Now we estimate what is sometimes called the **lattice energy** or the **lattice enthalpy** of CaCl(c). By lattice energy or enthalpy we mean the change in energy or enthalpy associated with combination of one mole of $Ca^+(g)$ ions with one mole of $Cl^-(g)$ ions to form one mole of CaCl(c). Most of the lattice energy or enthalpy comes from electrostatic interactions of the ionic charges with each other. We therefore expect that the lattice enthalpy of CaCl(c) should be intermediate between the lattice enthalpies of NaCl(c) and KCl(c). (Remember that the radius of Ca^+ is intermediate between radii of Na^+ and K^+.) We therefore use ΔH_f^0 values from Appendix II to calculate the following:

$$Na^+(g) + Cl^-(g) \rightarrow NaCl(c) \qquad \Delta H^0 = -186 \text{ kcal mole}^{-1}$$
$$K^+(g) + Cl^-(g) \rightarrow KCl(c) \qquad \Delta H^0 = -169 \text{ kcal mole}^{-1}$$

Then we estimate the lattice enthalpy of CaCl(c) as

$$Ca^+(g) + Cl^-(g) \rightarrow CaCl(c) \qquad \Delta H^0 \cong -180 \text{ kcal mole}^{-1}$$

Finally, we use this ΔH^0 with reliable ΔH_f^0 values for $Ca^+(g)$ and $Cl^-(g)$ to obtain

$$-180 = \Delta H_f^0(CaCl) - \Delta H_f^0(Ca^+) - \Delta H_f^0(Cl^-)$$
$$-180 = \Delta H_f^0(CaCl) - (185) - (-59)$$
$$\Delta H_f^0(CaCl) \cong -54 \text{ kcal mole}^{-1}$$

Now that we have an approximate ΔH_f^0 for CaCl(c), let us calculate ΔH^0 for the reaction

$$2\,CaCl(c) \rightleftharpoons Ca(c) + CaCl_2(c) \qquad (15.8)$$

For this calculation we have $\Delta H_f^0 = 0$ for Ca(c) and $\Delta H_f^0 = -190$ kcal mole^{-1} for $CaCl_2(c)$ to use with our value for CaCl(c) as follows:

$$\Delta H^0 = -190 - 2(-54) = -82 \text{ kcal mole}^{-1}$$

On the basis of the reasonable approximation that $T\Delta S^0$ may be neglected in comparison to ΔH^0 for this reaction, we also have $\Delta G^0 \cong -82$ kcal mole^{-1}. This large negative ΔG^0 value tells us that the equilibrium constant for reaction (15.8) is very large, which means that CaCl(c) is unstable with respect to decomposition to Ca(c) and $CaCl_2(c)$. We also have $\Delta G^0 \cong +82$ kcal mole^{-1} for the reverse of reaction (15.8), which tells us that Ca(c) and $CaCl_2(c)$ do not react spontaneously to form CaCl(c).

Even if the error in our estimated lattice enthalpy of CaCl(c) is as large as 10 kcal mole^{-1} and the $T\Delta S^0$ term that we have ignored is also as large as 10 kcal mole^{-1}, our qualitative conclusions are still sound. On the other hand, there are some compounds for which it is necessary to have more accurate estimates. In such cases it is not acceptable to ignore $T\Delta S^0$ and more elaborate methods of estimating such quantities as lattice enthalpies are used.

Readers who care to do some more thermodynamic calculations with ΔH_f^0 values in Appendix II may verify that the large lattice enthalpy of $CaCl_2(c)$ is the dominant factor in making this compound more stable than CaCl(c), in spite of the second ionization energy of calcium being larger than the first.

Solubilities of the alkaline earth sulfates in water are shown in Figure 15.1 and solubilities of alkaline earth hydroxides are shown in Figure 15.2. The solubilities of the carbonates are roughly parallel to those of the sulfates, and the solubilities of the fluorides are roughly parallel to those of the hydroxides. Solubility products can be calculated from these measured solubilities and then used for other calculations.

The various trends in solubilities can be understood in terms of lattice enthalpies and enthalpies of hydration. We consider the sulfates first and then the hydroxides.

Figure 15.1. Comparison of solubilities (S is expressed in moles per liter) of alkaline earth sulfates in water at 25°C.

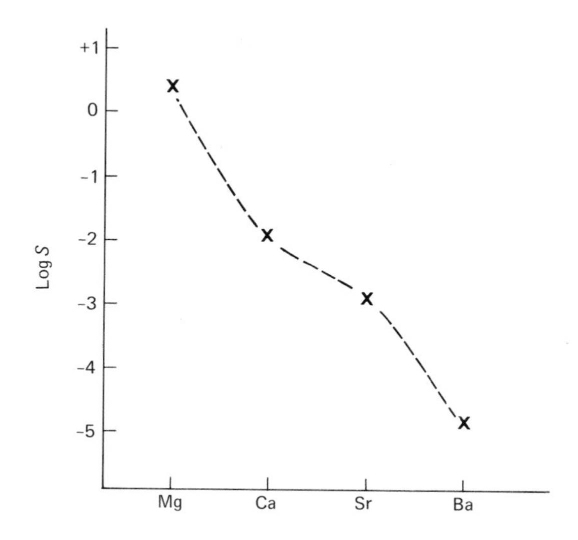

For the alkaline earth sulfates, general equations for reactions to be considered are the following:

$$MSO_4(c) \rightarrow M^{2+}(g) + SO_4^{2-}(g) \qquad \Delta H_9^0 \qquad (15.9)$$
$$M^{2+}(g) \rightarrow M^{2+}(aq) \qquad \Delta H_{10}^0 \qquad (15.10)$$
$$SO_4^{2-}(g) \rightarrow SO_4^{2-}(aq) \qquad \Delta H_{11}^0 \qquad (15.11)$$
$$MSO_4(c) \rightarrow M^{2+}(aq) + SO_4^{2-}(aq) \qquad \Delta H_{soln}^0 \qquad (15.12)$$

Trends in the heat or enthalpy of solution represented by ΔH_{soln}^0 largely determine trends in ΔG^0 of solution and thence trends in solubility. The magnitude of ΔH_{11}^0 is independent of which positive ion is considered; thus, ΔH_{11}^0 has nothing to do with determining trends in ΔH_{soln}^0 and need not be considered now.

All the enthalpies of hydration of M^{2+} ions represented by ΔH_{10}^0 are negative. These enthalpies of hydration become more negative as we go from the relatively

Figure 15.2. Comparison of solubilities (S is expressed in moles per liter) of alkaline earth hydroxides.

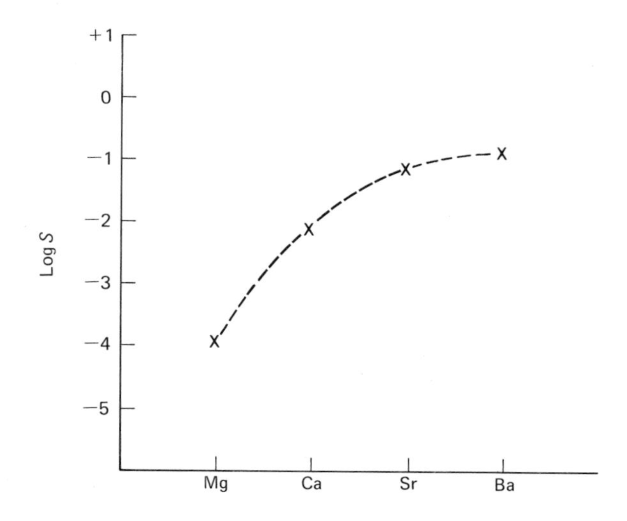

large Ra^{2+} ion to the smaller Mg^{2+} ion. Thus these enthalpies of hydration tend to make ΔH^0_{soln} more exothermic for the sulfates containing small M^{2+} ions than for those containing larger M^{2+} ions. Because SO_4^{2-} ions are considerably larger than M^{2+} ions, the lattice enthalpies are largely determined by the packing of SO_4^{2-} ions and changing from Mg^{2+} to Ra^{2+} makes little difference in the lattice enthalpy represented by ΔH^0_9. Thus we conclude that the trend in ΔH^0_{soln} values for the alkaline earth sulfates is largely due to the trend in hydration enthalpies of the M^{2+} alkaline earth ions. Consistent with these arguments, $MgSO_4$ is the most soluble alkaline earth sulfate.

The trend in the solubilities of the hydroxides of the alkaline earths may also be understood in terms of hydration and lattice enthalpies. General equations to be considered are the following:

$$M(OH)_2(c) \rightarrow M^{2+}(g) + 2\,OH^-(g) \qquad \Delta H^0_{13} \qquad (15.13)$$

$$M^{2+}(g) \rightarrow M^{2+}(aq) \qquad \Delta H^0_{14} \qquad (15.14)$$

$$2\,OH^-(g) \rightarrow 2\,OH^-(aq) \qquad \Delta H^0_{15} \qquad (15.15)$$

$$M(OH)_2(c) \rightarrow M^{2+}(aq) + 2\,OH^-(aq) \qquad \Delta H^0_{soln} \qquad (15.16)$$

Again, trends in ΔH^0_{soln} largely determine trends in ΔG^0 of solution and thence trends in solubility. The enthalpy represented by ΔH^0_{15} is constant and need not be considered now.

The crystal lattice decomposition reactions represented by (15.13) are endothermic, corresponding to positive ΔH^0_{13}. This lattice enthalpy is largest for the smallest M^{2+} ion and smallest for the largest M^{2+} ion. (Think about Coulomb's law.) The hydration reactions represented by (15.14) are exothermic, corresponding to negative ΔH^0_{14}. The enthalpies of hydration are more exothermic for the smaller M^{2+} ions and less exothermic for the larger M^{2+} ions. Thus the trends in ΔH^0_{13} and ΔH^0_{14} values are opposite, and we have not yet accounted for the trend in ΔH^0_{soln} values.

The interaction between the M^{2+} ion and the dipoles of nearby water molecules in aqueous solution is weaker than the interaction between the M^{2+} ion and nearby SO_4^{-2} ions in the crystal. (Again, think about Coulomb's law.) Therefore, the variation in hydration enthalpies is not as large as the variation in lattice enthalpies. We conclude that the trend in ΔH^0_{soln} values should follow the trend in lattice enthalpies, and $Mg(OH)_2$ should be less soluble than $Ba(OH)_2$, as is observed.

Hard Water

It has been known for several thousand years that natural waters leave a residue of mineral matter when they are boiled or allowed to evaporate. It was noted more than 2000 years ago that some of this residue was usually a hard rocklike scale that adhered to the vessel in which the water was boiled. Ancient natural philosophers explained this in terms of the four "elements"—earth, air, fire, and water. It was postulated that, when water and fire came in contact, they unite to form earth (precipitated mineral matter) and air (water vapor). A "chemical equation" for this "reaction" is

$$\text{fire} + \text{water} \rightarrow \text{earth} + \text{air}$$

Although there are few pertinent records from antiquity, it seems certain that scale in cooking pots was a minor but persistent nuisance for many centuries before anyone used the term "hard water."

The term "hard water" originated because such waters were hard to wash in. It was known long ago that it required more soap to wash with "hard water" than with "soft water."

We now know that the "hard" properties of water result from presence of $Ca^{2+}(aq)$ ions, and sometimes also from $Mg^{2+}(aq)$, $Fe^{2+}(aq)$, and other metal ions. Hardness due to $Ca^{2+}(aq)$ ions is common because limestone is so widespread that most ground water contains small but still important concentrations of $Ca^{2+}(aq)$ ions. Because our atmosphere contains about 0.04% $CO_2(g)$, ground water often contains dissolved CO_2. This ground water that contains $CO_2(aq)$ can dissolve limestone by the reaction represented by

$$CaCO_3(c) + CO_2(aq) + H_2O(liq) \rightleftharpoons Ca^{2+}(aq) + 2\,HCO_3^-(aq) \quad (15.17)$$

A similar reaction of $CO_2(aq)$ with dolomite ($MgCO_3 \cdot CaCO_3$) leads to $Mg^{2+}(aq)$ in some ground water.

When soap, often largely sodium stearate, is added to hard water, calcium ions react with stearate ions to form a precipitate of slightly soluble calcium stearate. Calcium stearate has no cleansing powers so that this precipitate wastes soap and also appears as an undesirable scum, commonly known as "ring around the collar" or "tattle tale gray." The equation for this reaction is

$$Ca^{2+}(aq) + 2\,C_{18}H_{35}O_2^-(aq) \rightleftharpoons Ca(C_{18}H_{35}O_2)_2(\downarrow)$$

where ($\downarrow$) indicates a precipitate.

Water that contains bicarbonate ions as well as calcium ions is often called **temporary** or **carbonate hard water.** Such water can be softened by boiling and then separating the precipitated calcium carbonate from the water. The reaction equation is

$$Ca^{2+}(aq) + 2\,HCO_3^-(aq) \rightleftharpoons CaCO_3(c) + H_2O(liq) + CO_2(g) \quad (15.18)$$

Note that reaction (15.18) is nearly the same as the reverse of reaction (15.17) by which $Ca^{2+}(aq)$ ions get into water from limestone in the first place. At high temperature the solubility of CO_2 in water is small, and the reaction represented by (15.18) proceeds readily. At lower temperatures the equilibrium concentration of dissolved CO_2 is larger, and calcium carbonate dissolves as in (15.17).

Although boiling and the resultant reaction (15.18) does soften temporary or carbonate hard water, the process is too expensive (requires too much energy) to apply to large quantities of water.

We also note that reaction (15.18) illustrates another reason why hard water is objectionable. Calcium carbonate is precipitated out of hard water that is boiled or even heated in various industrial boilers, home hot water heaters, tea kettles, and so on. The calcium carbonate that is precipitated in this way is called boiler scale.

It has been known since 1730 or maybe longer that various alkaline substances could soften water. The development and widespread utilization of the steam engine made the need for water softening ever greater and led Thomas Clark,

Professor of Chemistry in Aberdeen University in Scotland, to develop in 1841 a practical method for softening hard water. Clark's method was based on adding lime water or slaked lime to the hard water. The hydroxide ions lead to precipitation of $CaCO_3(c)$ as shown by

$$Ca^{2+}(aq) + HCO_3^-(aq) + OH^-(aq) \rightleftharpoons CaCO_3(c) + H_2O(liq) \qquad (15.19)$$

Although the added limewater contains $Ca^{2+}(aq)$ ions, its addition to temporary hard water results in net removal of calcium because each mole of $Ca(OH)_2$ furnishes two moles of hydroxide ions to neutralize two moles of bicarbonate ions and precipitate two moles of $CaCO_3$.

Following Clark's work it was discovered that adding soda ash, Na_2CO_3, or washing soda, $Na_2CO_3 \cdot 10H_2O$, also softens hard water. Again $Ca^{2+}(aq)$ ions are removed by precipitation of $CaCO_3$.

Another way of softening water is to replace the $Ca^{2+}(aq)$ ions by other ions, such as Na^+ (aq), which do not form precipitates with carbonate or stearate ions. This exchange of ions of one metal for ions of another is called **ion exchange** and is done with various **zeolites** and **ion exchange resins.** The chief advantage is that no precipitate results to clog pipes or washing machines.

Zeolites, as originally identified by Swedish geologist Cronstedt in 1756, are hydrated alkali or alkaline earth oxides in combination with various aluminum silicates. Both natural and synthetic zeolites now play an important role in water treatment and also in such other areas as soil chemistry.

Among the various synthetic ion exchange resins are polymers consisting of a hydrocarbon framework with $—SO_3^-$ groups bonded to some of the carbon atoms. Positive ions are held in the resin by these negative charges so that the resin as a whole is electrically neutral.

We can illustrate the use of ion exchange in water softening by considering hard water passed through an ion exchange resin that is loaded with Na^+ ions. Because the resin holds Ca^{2+} ions more strongly then Na^+ ions, it removes Ca^{2+} ions from solution and releases Na^+ ions to the solution. An equation for this process or reaction is

$$Ca^{2+}(aq) + 2\ Na^+(\text{in resin}) \rightleftharpoons Ca^{2+}(\text{in resin}) + 2\ Na^+(aq) \qquad (15.20)$$

The equilibrium constant for reaction (15.20) is large, so that the resin removes nearly all $Ca^{2+}(aq)$ ions from hard water.

After a certain quantity of hard water has been treated as described above, the resin becomes saturated with bound calcium ions and is able to release no more sodium ions. Then the resin can be regenerated by flowing an excess of concentrated sodium chloride solution through it. By maintaining a large concentration of $Na^+(aq)$ ions, and a small concentration of $Ca^{2+}(aq)$ ions in the regenerating solution, we reverse reaction (15.20) so that the resin again becomes loaded with bound sodium ions and can be used to soften more water.

Ion exchange resins in which $—N(CH_3)_3^+$ groups are bound to some of the carbon atoms have been synthesized and are used as anion or negative ion exchangers. Combining an anion (negative ion) exchanger with a cation (positive ion) exchanger makes it possible to remove all the ions from water to obtain what is called deionized water. The cation exchanger puts $H^+(aq)$ ions into the water and removes all other positively charged ions; the anion exchanger puts $OH^-(aq)$ ions into the water and removes all other negatively charged ions. The $H^+(aq)$ ions and $OH^-(aq)$ ions combine to form water. The cation exchange resin is regenerated by treating it with a flow of moderately concentrated hydrochloric acid solution, and the anion exchanger is regenerated by similar treatment with sodium hydroxide solution.

Another convenient method of removing Ca^{2+} ions from hard water is by formation of complexes with sequestering agents such as the polyphosphates discussed in Chapter 18. Because these complexes are very stable, Ca^{2+} ions are effectively isolated but without the disadvantages of precipitate formation.

References

Bassow, H., D. Hamilton, B. Schneeberg, and B. Stad: A study of the physical and chemical rates of $CaCO_3$ dissolution in HCl. *J. Chem. Educ.*, **48**:327 (1971).

Dyrssen, D., E. Ivanova, and K. Aren: The solubility curves for calcium and strontium sulfates. *J. Chem. Educ.*, **46**:252 (1969).

Glanville, J. and E. Rau: Soda ash manufacture—an example of what?, *J. Chem. Educ.*, **50**:64 (1973).

Hecht, C. E.: Desalination of water by reverse osmosis. *J. Chem. Educ.*, **44**:53 (1967).

Morton S. D., and G. F. Lee: Calcium carbonate equilibria in lakes. *J. Chem. Educ.*, **45**:511 (1968).

Morton, S. D., and G. F. Lee: Calcium carbonate equilibria in the oceans. *J. Chem. Educ.*, **45**:513 (1968).

Moynihan, C. T., A low temperature fused salt experiment: the conductivity, viscosity, and density of molten calcium nitrate tetrahydrate. *J. Chem. Educ.*, **44**:53 (1967).

Problems

1. Write an equation to represent the radioactive disintegration of ^{223}Fr by β particle emission.

2. How much calcium carbide is needed for preparation of enough calcium cyanamide to yield 1.0 mole of ammonia?

3. The hydrides of sodium and calcium react with water to yield hydrogen gas and the corresponding hydroxide as indicated by the following (unbalanced) equations:

$$NaH(c) + H_2O(liq) \rightarrow Na^+(aq) + OH^-(aq) + H_2(g)$$
$$CaH_2(c) + 2\,H_2O(liq) \rightarrow Ca^{2+}(aq) + 2\,OH^-(aq) + 2\,H_2(g)$$

(a) How much hydrogen can be obtained from 1.0 lb of NaH?
(b) How much hydrogen can be obtained from 1.0 lb of CaH_2?
(c) Suppose that we want to obtain the largest possible amount of $H_2(g)$ from the smallest possible *total* mass of metal hydride plus water. Would you use NaH or CaH_2? What ratio of (mass of hydride)/(mass of water) would you use?

4. Many photographic flash bulbs contain magnesium wire in an oxygen atmosphere. When the bulb is fired, an electric current heats the wire to start the reaction with oxygen to yield MgO(c).
(a) How many cal g^{-1} of Mg are evolved by this reaction?
(b) How much magnesium is required to react with all of the oxygen in a 25 cm^3 bulb filled with oxygen at 1.0 atm and 25°C?
(c) One way to gain some understanding of why so much light is obtained from this

reaction is to make a simple calculation of a highly idealized maximum temperature achieved as a result of the combustion. For this calculation consider reaction of 1.0 mole of Mg(c) with exactly one half mole of O_2(g) so that all reactants are converted to MgO. From the ΔH_f^0 of MgO(c) we know that 144 kcal are evolved. Assuming that the reaction takes place in a perfectly adiabatic enclosure (with negligible heat capacity) so that no energy can be transferred to the surroundings, and that the average heat capacity of MgO is 12 cal deg^{-1} $mole^{-1}$, calculate the temperature reached by the MgO. Of course the MgO never reaches this very high temperature, in part because it radiates away a significant fraction of the 144 kcal produced by the reaction.

5. Solutions of sodium carbonate have pH > 7. Why?

6. Write a balanced equation to describe what happens when a pellet of sodium is dropped into water. It often happens that an explosion accompanies the reaction of sodium (or other alkali metal) with water. What is the reaction that is directly responsible for the explosion?

7. Write balanced equations for the reactions that occur when limestone is strongly heated and then the solid product (after cooling) is added to water.

8. Some consideration has been given to the production of ammonia by the following sequence of reactions:
(i) Magnesium plus nitrogen (from air) to yield magnesium nitride (Mg_3N_2).
(ii) Magnesium nitride plus water to yield ammonia plus magnesium hydroxide.
(iii) Conversion of magnesium hydroxide to magnesium chloride with hydrochloric acid.
(iv) Electrolytic reduction of magnesium chloride to metallic magnesium to be used again for reaction with nitrogen.
(a) Write balanced equations for all of these reactions.
(b) Why do you think this process is too expensive to make it industrially attractive?

9. How much calcium bicarbonate is removed from temporary hard water and how much calcium carbonate is precipitated when 1.0 mole of calcium hydroxide is added to an excess of hard water?

10. Reactions involved in production of metallic magnesium from sea water are summarized by the following equations:

$$Mg^{2+}(\text{sea water}) + Ca(OH)_2(c) \rightarrow Mg(OH)_2(c) + Ca^{2+}(aq)$$
$$Mg(OH)_2(c) + 2\,H^+(aq) + 2\,Cl^-(aq) \rightarrow Mg^{2+}(aq) + 2\,Cl^-(aq) + 2\,H_2O(liq)$$

Concentration of the magnesium chloride solution yields a precipitate of $MgCl_2 \cdot H_2O$(c), which is dehydrated and melted to permit electrolysis as indicated by

$$MgCl_2(liq) \xrightarrow{\text{elect.}} Mg + Cl_2$$

(a) How much hydrogen chloride is used in the production of 1.0 mole of magnesium?
(b) How might the calcium hydroxide used in the first step be made from limestone? How much chlorine is produced as a by-product?

11. Use ΔG_f^0 values in Appendix II to calculate ΔG^0 for the reaction

$$BaSO_4(c) \rightleftharpoons Ba^{2+}(aq) + SO_4^{2-}(aq)$$

and then calculate K_{sp} for barium sulfate at 298°K.

12. Use ΔH_f^0 values in Appendix II to calculate ΔH^0 for the reaction in problem 11. Then calculate K_{sp} for barium sulfate at 40°C.

13. Two crystal forms of calcium carbonate are called calcite and aragonite.

(a) Without doing any calculations, predict from the ΔG_f^0 values in Appendix II which of these forms is least soluble in water at 298°K. Explain your reasoning.
(b) Calculate solubility products for both calcite and aragonite at 298°K.
(c) The ΔG_f^0 values show that calcite is more stable than aragonite at 298°K and 1.0 atm. Is there a temperature at which calcite and aragonite are in equilibrium at 1.0 atm? If so, what is this temperature?
(d) The density of aragonite is greater than that of calcite. Can calcite be converted to aragonite by high pressure applied at 298°K?

14. In the text of this chapter we have stated that it is, for some reactions, a satisfactory approximation to take $\Delta G^0 \cong \Delta H^0$. By analogy with such compounds as NaCl(c), KCl(c), AgCl(c), and CuCl(c), estimate the entropy of CaCl(c) at 298°K and use this value with known entropies of Ca(c) and $CaCl_2$(c) to obtain ΔS^0 for reaction (15.8). Then calculate $T\,\Delta S^0$ at 298°K and compare this value with our value for ΔH^0 of reaction (15.8). How does the error associated with neglect of $T\,\Delta S^0$ compare with the likely uncertainty in our estimated lattice enthalpy of CaCl(c)? Thermodynamic data are tabulated in Appendix II.

15. Taking the lattice enthalpy of the hypothetical KCl_2(c) to be the same as that of $MgCl_2$(c), calculate ΔH^0 values for the reactions

$$KCl_2(c) + K(c) \rightleftharpoons 2\ KCl(c)$$
$$2\ KCl_2(c) \rightleftharpoons 2\ KCl(c) + Cl_2(g)$$

What can you conclude about the possibilities for synthesizing KCl_2(c) and the chances for keeping it if someone should give you a sample.

16. Why are alkali metals and alkaline earth metals not found in nature as free metals?

17. Crushed limestone is widely used by farmers, partly to change the pH of soil. Is it used to increase or to decrease the pH? Explain your reasoning.

18. Lithium is a very low density metal. Why it it not used in airplane construction?

16

ELECTROCHEMISTRY

Introduction

Much of electrochemistry is concerned with chemical reactions that cause or are caused by passage of an electrical current. For example, a battery produces an electric current as a result of a chemical reaction, whereas passage of a current through a silver plating cell causes chemical reactions to occur.

Electrochemistry is also concerned with electrical conductivities of various substances and their solutions. In the past much of this interest has centered on aqueous solutions of both weak and strong electrolytes. More recently this interest in aqueous solutions has been complemented by increasing emphasis on investigations of conductivities of nonaqueous solutions, of solids, and of molten salts.

Electrochemistry is important in chemistry and other sciences for several reasons. Intelligent design and use of batteries and of electroplating procedures obviously depend on the principles of electrochemistry. It may be less obvious but it is equally true that many of our thermodynamic data for inorganic compounds and aqueous ions are derived from results of electrochemical measurements. Most of our equilibrium constants for ionization of weak acids are also derived from results of electrochemical measurements. Further, substantial parts of our growing knowledge about such diverse subjects as corrosion of metals, nerve impulse transmission, and interactions of aqueous ions with each other and with surrounding water molecules is based on information obtained by way of electrochemical measurements and principles.

Electrochemical Cells

When an electric current is passed through molten sodium chloride, as schematically indicated in Figure 16.1, chemical changes take place where the current enters and leaves the molten salt. Electrons flow out of the terminal on the current source that is marked ⊖ and to the cell electrode that is labelled **cathode.** An equal number of electrons leave the cell electrode labelled **anode** and go to the terminal marked ⊕ on the current source.

To minimize confusion that often arises in discussions of electrochemistry, we shall use the symbols ⊕ and ⊖ to indicate positive and negative in an electrical sense. The uncircled symbols + and − will be used as usual to indicate positive and negative in an algebraic sense.

The electrons that go into the cathode of the cell pictured in Figure 16.1 are captured in reaction with the sodium ions as shown by

$$Na^+ + e^- \rightarrow Na \qquad \text{(cathode reaction)} \tag{16.1}$$

The electrons that leave the anode are released by the reaction

$$Cl^- \rightarrow \tfrac{1}{2}Cl_2 + e^- \qquad \text{(anode reaction)} \tag{16.2}$$

Cations (positively charged ions) are attracted to the cathode in this cell while anions (negatively charged ions) are attracted to the anode. As stated in Figure 16.1 and shown by equation (16.1), reduction takes place at the cathode. The oxidation represented by equation (16.2) takes place at the anode.

Combination of the anode and cathode reactions above yields

$$Na^+ + Cl^- \rightarrow Na + \tfrac{1}{2}Cl_2$$

for the total electrochemical reaction. Or we may double both the anode and cathode reaction equations and add them to obtain

$$2\ Na^+ + 2\ Cl^- \rightarrow 2\ Na + Cl_2$$

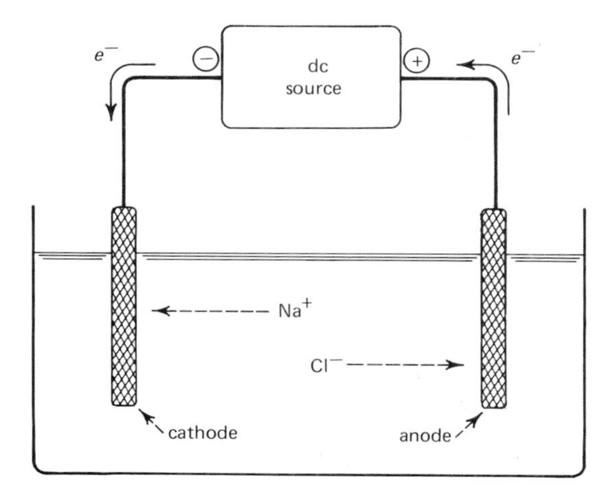

Figure 16.1. Schematic illustration of electrolysis of molten NaCl. The cathode is defined as the electrode at which reduction takes place, while the anode is the electrode at which oxidation takes place.

for the total cell reaction equation. Equations representing the total cell reaction must be balanced because equal numbers of electrons pass through each electrode.

Passage of one mole of electrons through the cell results in production of one mole of sodium as shown by equation (16.1), and one half mole of chlorine, as shown by equation (16.2). One mole of electrons is Avogadro's number of electrons and has a total charge of 96,487 coulombs. Because a coulomb is an ampere second, 96,487 coulombs is the total charge passed when a conductor carries a current of 1 ampere (amp) for 96,487 sec, or carries a current of 10 amp for 9648.7 sec.

The total charge of Avogadro's number of electrons (one mole) is called one Faraday in honor of Michael Faraday (1791–1867), who first carried out quantitative investigations of the chemical effects of electric currents. The Faraday is commonly given the symbol $\mathscr{F}$. We often express the Faraday as 96,500 coulombs per mole of electrons.

Example Problem 16.1. How much sodium can be produced per minute by passing a current of 10 amperes through a cell like that pictured in Figure 16.1?

To find the number of moles of sodium produced per minute, we must calculate the number of moles of electrons passing through the cell per minute, which requires that we express the given information in terms of coulombs per minute.

Because there are 60 sec min^{-1}, we have

$$\begin{aligned} 60 \text{ sec min}^{-1} \times 10 \text{ amp} &= 600 \text{ amp-sec min}^{-1} \\ &= 600 \text{ coulomb min}^{-1} \end{aligned}$$

passing through the cell. Now we combine this quantity with the Faraday to obtain

$$\frac{600 \text{ coulomb min}^{-1}}{96{,}500 \text{ coulomb (mole } e^-)^{-1}} = 6.22 \times 10^{-3} \frac{\text{mole } e^-}{\text{min}}$$

Because equation (16.1) shows that each mole of electrons corresponds to one mole of sodium formed, we know that 6.22×10^{-3} mole of sodium can be produced per minute. Production of sodium by a current of 10 amp in this cell is

$$6.22 \times 10^{-3} \text{ mole min}^{-1} \times 23.0 \text{ g mole}^{-1} = 0.143 \text{ g min}^{-1}. \quad \blacksquare$$

Aluminum is produced electrochemically on a large scale. The process now used was discovered in 1886 by Paul Heroult in France and by Charles Hall in the United States, each working without knowledge of the other. The process involves electrochemical decomposition of aluminum oxide to aluminum and oxygen. The aluminum oxide to be electrolyzed is dissolved in molten cryolite, Na_3AlF_6, at about 1000°C.

Electrolytic production of aluminum is illustrated schematically in Figure 16.2. Although the electrode reactions are complicated and not yet entirely understood, the total cell reaction may be represented by

$$2\ Al_2O_3(\text{in cryolite}) \rightarrow 4\ Al(\text{liq}) + 3\ O_2(g)$$

Some of the oxygen produced by this reaction reacts with the carbon anodes to form carbon dioxide, thus corroding the anodes.

Aluminum oxide to be used in the production of aluminum is ordinarily obtained from bauxite, a hydrated aluminum oxide, that is commonly contami-

Figure 16.2. Schematic illustration of electrolytic production of aluminum by the Hall-Heroult process.

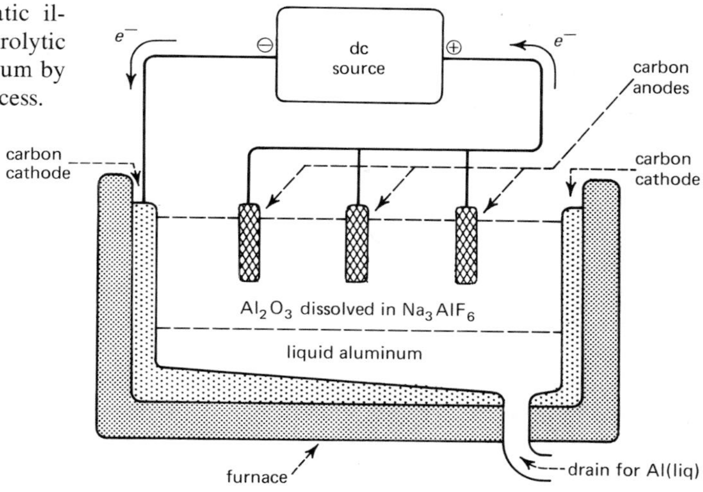

nated with ferric oxide and sometimes with other substances. Because aluminum oxide is amphoteric and ferric oxide is not, bauxite is purified by treatment with concentrated sodium hydroxide solution. This solution dissolves the aluminum oxide, as shown by

$$Al_2O_3(c) + 2\ OH^-(aq) + 3\ H_2O(liq) \rightarrow 2\ Al(OH)_4^-(aq)$$

and leaves the non-amphoteric ferric oxide undissolved. After the ferric oxide and solution are separated, the solution is made less basic by treatment with carbon dioxide from the air so that aluminum hydroxide is precipitated. The wet (hydrated) aluminum oxide is converted to the desired pure oxide by heating to drive off water.

Example Problem 16.2. Calculate the electric current that must pass through a cell like that pictured in Figure 16.2 to yield aluminum at the rate of 1.0 lb min^{-1}.

We know that there are 454 g lb^{-1} and calculate that 1.0 lb aluminum is (454 g)/(27 g $mole^{-1}$) = 16.8 moles of aluminum. Each mole of elemental aluminum that is produced from Al_2O_3 (containing aluminum in the +3 oxidation state) requires 3 moles of electrons. The required electric current must therefore supply 3 × 16.8 = 50.4 moles of electrons per minute or 50.4/60 = 0.84 moles of electrons per second. The required current is therefore

$$0.84\frac{\text{mole } e^-}{\text{sec}} \times 96{,}500\frac{\text{coulomb}}{\text{mole } e^-} = 81 \times 10^3\frac{\text{coulomb}}{\text{sec}}$$

$$= 81 \times 10^3 \text{ amp}$$

The huge currents needed for industrial production of aluminum account for location of these plants near sites of plentiful and inexpensive hydroelectric power. ■

Now consider reactions that occur when an electric current is passed through water that contains dissolved ions. Electrons flow from the external power source to the cell electrode that we call the cathode, where a reduction reaction takes place. This cathode reaction may be either reduction of ions in the solution or reduction of the solvent water to gaseous hydrogen. At the anode there is an oxidation reaction and flow of electrons toward the external power source. The anode reaction may be either oxidation of ions in the solution or oxidation of the solvent water to gaseous oxygen.

Our efforts to understand and predict the results of electrolysis experiments must be based on considerations of the relative *ease* and relative *rate* of each possible electrode reaction. Thermodynamics provides answers to questions of relative ease (or difficulty) of various electrode reactions. Conversely, results of electrochemical experiments can yield thermodynamic data. Rates of electrode reactions usually must be considered separately from thermodynamics.

Consideration of the ΔG values for the various possible electrode reactions usually permits us to select the reactions that will actually take place in a particular electrolytic cell or battery. Certain electrode reactions, however, often proceed quite slowly unless extra free energy (called **overvoltage**) is supplied. The overvoltage is the difference between the voltage required to make the cell reaction proceed at the desired rate and reversible ("zero rate") voltage.

Several reactions might take place when an aqueous solution of sodium flouride is electrolyzed in a cell containing inert electrodes that are neither oxidized nor reduced. At the cathode where reduction occurs, the following reations are possible:

$$Na^+(aq) + e^- \rightarrow Na(c) \tag{16.3}$$
$$2\,H_2O(liq) + 2\,e^- \rightarrow H_2(g) + 2\,OH^-(aq) \tag{16.4}$$
$$2\,H^+(aq) + 2\,e^- \rightarrow H_2(g) \tag{16.5}$$

Oxidation reactions that might occur at the anode are:

$$2\,F^-(aq) \rightarrow F_2(g) + 2\,e^- \tag{16.6}$$
$$2\,H_2O(liq) \rightarrow O_2(g) + 4\,H^+(aq) + 4\,e^- \tag{16.7}$$
$$4\,OH^-(aq) \rightarrow 2\,H_2O(liq) + O_2(g) + 4\,e^- \tag{16.8}$$

Before going on to further consideration of electrode reactions, it should be noted that electrode reaction equations (16.3–16.8) are balanced with respect to atoms and with respect to net charge.

Because electrolysis of aqueous sodium fluoride solution actually yields hydrogen at the cathode, we choose either (16.4) or (16.5) to represent the cathode reaction. Similarly, the observed production of oxygen at the anode leads us to choose either (16.7) or (16.8) to represent the anode reaction. Finally, because concentrations of $H^+(aq)$ and $OH^-(aq)$ are both very small in aqueous sodium fluoride solutions, we choose (16.4) and (16.7) as the most realistic representations of the electrode reactions under consideration.

The total cell reaction equation is obtained by adding the electrode reactions. As written above, the cathode reaction equation (16.4) involves only half as many electrons as does the anode reaction equation (16.7). Because the same number of electrons must go through both electrodes, we double (16.4) before adding it to (16.7) as follows:

$$4\,H_2O(liq) + 4\,e^- \rightarrow 2\,H_2(g) + 4\,OH^-(aq)$$
$$2\,H_2O(liq) \rightarrow O_2(g) + 4\,H^+(aq) + 4\,e^-$$
$$\overline{6\,H_2O(liq) \rightarrow 2\,H_2(g) + O_2(g) + 4\,OH^-(aq) + 4\,H^+(aq)} \tag{16.9}$$

Hydrogen ions and hydroxide ions appear in the total cell reaction equation (16.9) as products of the electrolysis, which is consistent with the observation that the solution near the cathode becomes basic while the solution near the anode becomes acidic. Figure 16.3 presents a schematic illustration and summary of the electrolysis of aqueous sodium fluoride.

If one waits a long time or removes the porous divider pictured in Figure 16.3, the $OH^-(aq)$ ions in the cathode compartment and the $H^+(aq)$ ions in the anode compartment will diffuse together and react to form water as indicated by

$$4\,H^+(aq) + 4\,OH^-(aq) \rightarrow 4\,H_2O(liq) \qquad (16.10)$$

Combination of total reaction equation (16.9) with this neutralization equation (16.10) gives

$$2\,H_2O(liq) \rightarrow 2\,H_2(g) + O_2(g) \qquad (16.11)$$

as the *net* cell reaction for the electrolysis of an aqueous solution of sodium fluoride.

Now consider the electrolysis of an aqueous solution containing $Cu^{2+}(aq)$ and $SO_4^{2-}(aq)$ ions, again with inert electrodes. Possible reactions at the cathode are (16.4) as in our earlier example and

$$Cu^{2+}(aq) + 2\,e^- \rightarrow Cu(c, \text{cathode}) \qquad (16.12)$$

It is reaction (16.12) that actually occurs because it is the "easier" of the two possible reactions. The anode reaction is (16.7) because this is the "easiest" oxidation reaction that can take place at an inert electrode in this system.

If the electrolysis of aqueous copper sulfate is carried out with a copper anode instead of an inert platinum or carbon anode as specified above, we have to consider the possibility that the anode reaction will now be

$$Cu(c, \text{anode}) \rightarrow Cu^{2+}(aq) + 2\,e^- \qquad (16.13)$$

As found in practice and predicted from thermodynamic data, reaction (16.13) actually occurs at the anode rather than (16.7) as before with an inert electrode.

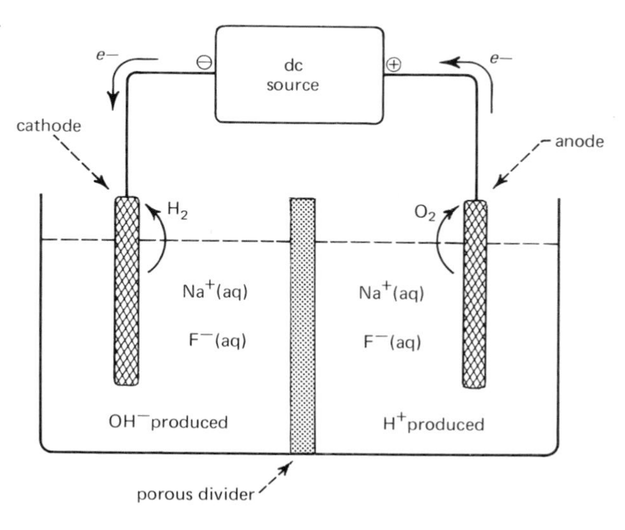

Figure 16.3. Illustration of electrolysis of aqueous sodium fluoride with inert electrodes. Equations (16.4), (16.7), and (16.9) represent the cathode reaction, the anode reaction, and the total cell reaction, respectively.

In this case, the net effect of electrolysis is transfer of metallic copper from anode to cathode. This is a useful procedure if the cathode consisted of something that one wanted electroplated with copper, and is also useful in the large scale purification of copper.

Large quantities of copper are produced from various sulfide ores. The metal first obtained from such ores is ordinarily only about 98% copper and requires further refining and purification. In the commonly used electrolytic purification process, the impure copper is used as a large anode while a smaller quantity of pure copper is used for the cathode. Careful control of the electrolysis voltage results in transporting the copper in the anode to the cathode. Some impurities in the original anode are left undissolved and settle out as "anode sludge" while others are oxidized to ions that remain in the electrolyte solution.

Electroplating silver on other metals is another example of a useful electrolytic process. Pure silver is used as anode and the metallic object to be plated is used as cathode. The electrolyte might be any soluble salt of silver, such as silver nitrate. In this case, the anode reaction is the oxidation of Ag(c) to Ag^+(aq) and the cathode reaction is the reduction of Ag^+(aq) ions to metallic silver on the object being plated.

Electrolysis of solutions of simple salts such as silver nitrate usually results in deposition of coarse crystals that do not adhere well to the cathode surface. On the other hand, slow deposition from solutions containing negatively charged complex ions often results in the formation of smooth and fine grained metallic deposits that adhere well to the surface of the cathode. Thus silver plating is often done with solutions that contain most of the dissolved silver in the form of $Ag(CN)_2^-$(aq) complex ions.

We have several times been faced with questions as to which of several possible electrode reactions is the "easiest." Following our discussion of batteries, we will have the background needed to obtain answers to these questions.

Batteries

We have been discussing electrochemical cells in which an external current source forces electrons to flow through the cell to make an oxidation reaction take place at an anode and a reduction take place at a cathode. Conversely, batteries consist of chemicals arranged in such fashion that an oxidation-reduction reaction can force electrons to flow through an external circuit.

Batteries are sometimes called **galvanic** or **voltaic** cells in honor of Luigi Galvani (1737–1798) and Alessandro Volta (1745–1827) who made the basic discovery of transforming chemical energy directly into electrical energy.

When metallic zinc is dipped into a solution containing Cu^{2+}(aq) ions, the zinc is oxidized to Zn^{2+}(aq) ions and the Cu^{2+}(aq) ions are reduced to metallic copper as shown by

$$\mathrm{Zn(c)} + \mathrm{Cu^{2+}(aq)} \rightarrow \mathrm{Zn^{2+}(aq)} + \mathrm{Cu(c)} \qquad (16.14)$$

The half-reactions that make up this total reaction are

$$\mathrm{Zn(c)} \rightarrow \mathrm{Zn^{2+}(aq)} + 2\,e^- \qquad \text{(anode)} \qquad (16.15)$$

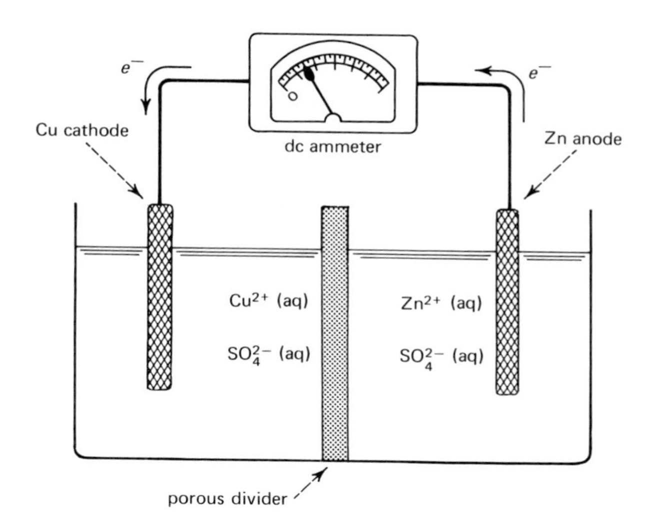

Figure 16.4. Schematic illustration of a battery driving an electric current through an external circuit. The chemical reaction that drives the current is represented by equation (16.14). The battery illustrated here is sometimes called a Daniell cell.

and

$$Cu^{2+}(aq) + 2\,e^- \rightarrow Cu(c) \qquad \text{(cathode)} \tag{16.16}$$

By physically separating the two half-reactions so that electrons cannot be transferred directly from Zn(c) to Cu^{2+}(aq), we can use the "driving force" or free energy of the spontaneous reaction (16.14) to cause electrons to flow from Zn(c) to Cu^{2+}(aq) through an external circuit. The useful result is a battery, as illustrated in Figure 16.4. Because the terms anode and cathode are used to denote the electrodes at which oxidation and reduction take place, the zinc electrode is called the anode and the copper electrode is called the cathode.

As the battery illustrated in Figure 16.4 causes current to flow in the external circuit, the spontaneous reaction represented by equation (16.14) results in a decrease in concentration of Cu^{2+}(aq) ions and an increase in concentration of Zn^{2+}(aq) ions. These concentration changes lead to a decrease in "driving force" or output voltage of the battery. The battery is said to be discharging as it delivers current to the external circuit. As the spontaneous reaction represented by equation (16.14) proceeds toward equilibrium, the free energy of the chemical system decreases and ΔG for the chemical reaction approaches zero. When the system finally reaches chemical equilibrium, it can no longer cause current to flow and the battery is said to be discharged. In practice, the battery will cease to function usefully before it reaches the final equilibrium state.

Connection of the copper–zinc cell to an external current source (battery charger) as shown in Figure 16.5 results in flow of electrons in the direction opposite to that shown in Figure 16.4. The half-reactions taking place at the electrodes shown in Figure 16.5 are

$$Zn^{2+}(aq) + 2\,e^- \rightarrow Zn(c) \qquad \text{(cathode)}$$

and

$$Cu(c) \rightarrow Cu^{2+}(aq) + 2\,e^- \qquad \text{(anode)}$$

Because copper is being oxidized and Zn^{2+}(aq) ions are being reduced, we now say that the copper electrode is the anode and the zinc electrode is the cathode.

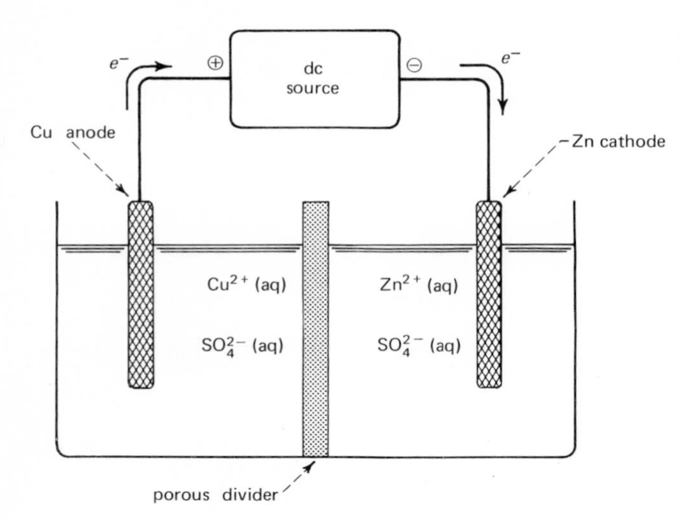

Figure 16.5. Schematic illustration of the Daniell cell already shown in Figure 16.4 being "charged." Note that designation of electrodes as anode and cathode is here reversed from Figure 16.4.

The total cell reaction is

$$Cu(c) + Zn^{2+}(aq) \rightarrow Cu^{2+}(aq) + Zn(c) \qquad (16.17)$$

Because reaction (16.17) is the reverse of spontaneous reaction (16.14), we now say that the battery is being "charged" so that it can later function as shown in Figure 16.4.

The equilibrium constant for reaction (16.14) is very large. Metallic zinc spontaneously reduces cupric ions to metallic copper. The equilibrium constant for the reverse reaction (16.17) is the reciprocal of the very large equilibrium constant for reaction (16.14) and is thus very small, in accord with the inability of metallic copper to reduce aqueous zinc ions to metallic zinc. The spontaneous reaction (16.14) is associated with a battery or cell from which we may obtain useful electrical energy. Reaction (16.17), with a very small equilibrium constant, is associated with a cell into which we must put electrical energy.

One of the most common batteries is the lead storage cell. In its simplest form the lead cell consists of a sheet of lead and a sheet of lead dioxide, both immersed in sulfuric acid, with terminals on both sheets for connection to an external circuit. In actual practice, the lead dioxide is usually supported on a grid of metallic lead while the lead electrode consists of "spongy" lead supported on a grid of ordinary lead. In most commercially available lead cells, several lead and lead dioxide electrodes, which are kept apart by insulating separators, are connected in series.

During discharge of a lead cell, the lead electrode (anode) is oxidized as shown by

$$Pb(c) + HSO_4^-(aq) \rightarrow PbSO_4(c) + H^+(aq) + 2\,e^- \qquad (16.18)$$

Reduction occurs at the lead dioxide cathode where the discharge reaction is

$$PbO_2(c) + HSO_4^-(aq) + 3\,H^+(aq) + 2\,e^- \rightarrow PbSO_4(c) + 2\,H_2O(liq) \qquad (16.19)$$

The total reaction for discharge of the cell is obtained by adding the two electrode reactions:

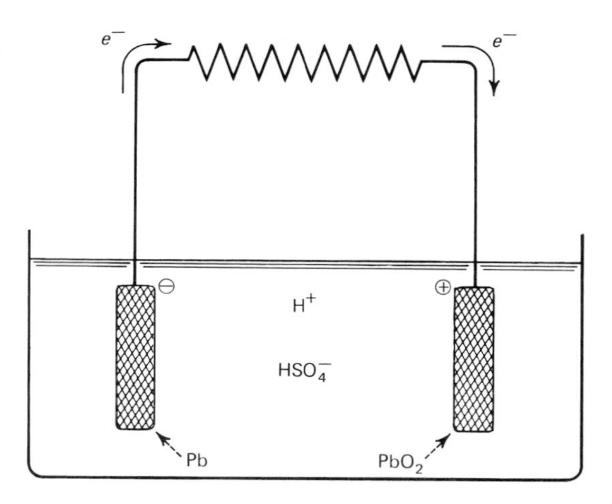

Figure 16.6. Schematic illustration of a lead cell.

The lead electrode is the anode and the lead dioxide electrode is the cathode when the cell is discharging (note direction of electron flow indicated by arrow). Electrode reactions and total cell reaction for discharge are represented by equations (16.18–16.20). When the cell is being charged as a result of electron flow forced by an external power source, labels anode and cathode are switched and electrode reactions and total cell reaction become the reverse of those represented by equations (16.18–16.20).

$$Pb(c) + 2\,HSO_4^-(aq) + 2\,H^+(aq) + PbO_2(c) \rightleftharpoons 2\,PbSO_4(c) + 2\,H_2O(liq) \qquad (16.20)$$

Equation (16.20) shows that lead and lead dioxide are both converted to lead sulfate when a lead cell is used to supply electrical power. Because sulfuric acid ($H^+ + HSO_4^-$) is consumed and water is formed in the discharge reaction, the acid in the cell becomes more dilute as the cell is discharged. Density of the concentrated acid in an adequately charged cell is about 1.20 g ml^{-1} and is about 1.05 g ml^{-1} when the cell has been discharged and needs to be recharged.

A lead cell is pictured schematically in Figure 16.6. The lead electrode (anode when the cell is discharging) is labelled ⊖ and the lead dioxide electrode (cathode when the cell is discharging) is labelled ⊕, in accord with usual practice.

Example Problem 16.3. Lead cells to be used as batteries in automobiles are often advertized in terms of ampere-hour rating. Calculate the number of moles of $PbSO_4(c)$ that will be formed when a 100 amp-hr battery is completely discharged.

Our task is to carry out the following sequence of calculations:

$$\text{amp-hr} \rightarrow \text{amp-sec} = \text{coulombs} \rightarrow \text{moles } e^- \rightarrow \text{moles } PbSO_4$$

This calculation offers a convenient opportunity to review and illustrate conversion from one set of units to another.

We know that 1 hr = 60 min or 60 min/1 hr = 1 and that 1 min = 60 sec or 60 sec/1 min = 1. Thus we can convert the given 100 amp-hr to amp-sec as follows:

$$100 \text{ amp-hr} \times 60 \text{ min hr}^{-1} \times 60 \text{ sec min}^{-1} = 36 \times 10^4 \text{ amp-sec}$$

From fundamental definitions we have coulomb = ampere-seconds so we know that the 100 amp-hr battery can deliver 36×10^4 coulomb.

As previously noted, the quantity called the Faraday ($\mathscr{F}$) is 96,500 coulomb (mole e^-)$^{-1}$, from which we calculate that discharge of the 100 amp-hr battery corresponds to "circulation" of

$$\frac{36 \times 10^4 \text{ coulomb}}{96.5 \times 10^3 \text{ coulomb (mole } e^-)^{-1}} = 3.7 \text{ moles } e^-$$

Equation (16.18) shows that passage of two moles of electrons leads to formation of one mole of $PbSO_4$ at the anode. The same two moles of electrons lead (equation 16.19) to formation of another mole of $PbSO_4$ at the cathode. We therefore conclude that two moles of electrons correspond to two moles $PbSO_4$ in this battery. Thus discharge of the advertised 100 amp-hr battery results in formation of a total of 3.7 moles of $PbSO_4$(c)—half of this at the anode and half at the cathode. ■

Before proceeding with application of thermodynamics to electrochemistry, a brief reveiw to emphasize meanings of some words is in order. The words **anode** and **cathode** are always associated with particular kinds of reactions that occur at electrodes. Oxidation is associated with an anode and reduction with a cathode. Electrons flow from the anode through the external circuit to the cathode. A particular electrode may be either an anode or a cathode, depending on whether oxidation or reduction takes place at that electrode, which in turn depends on whether the cell is being used to force electrons to flow through an external circuit or whether a power source in the external circuit is forcing electrons to flow through the cell.

Electrochemical Thermodynamics

We have written the first law of thermodynamics as

$$\Delta E = q - w$$

and have taken $w = P\,\Delta V$ when we have been concerned only with PV work at constant pressure. Now we are also interested in electrical work so that the total work done by the system (at constant pressure) is represented by

$$w = P\,\Delta V + w_{el}$$

Now the first law is written as

$$\Delta E = q - P\,\Delta V - w_{el} \tag{16.21}$$

We have earlier seen that the enthalpy is defined by $H = E + PV$ so that we also have

$$\Delta H = \Delta E + P\,\Delta V \tag{16.22}$$

Substitution of (16.21) in (16.22) yields

$$\Delta H = q - w_{el} \tag{16.23}$$

Now, because we are mostly interested in free energy changes, we substitute (16.23) in $\Delta G = \Delta H - T\,\Delta S$ to obtain

$$\Delta G = q - T\,\Delta S - w_{el} \tag{16.24}$$

For a reversible process, $q = T\,\Delta S$, which is combined with (16.24) to yield

$$\Delta G = -w_{rev\ el} \tag{16.25}$$

in which ΔG represents the free energy change to be associated with the cell reaction and $w_{rev\ el}$ is the reversible electric work done by the system (cell) on the surroundings (external circuit).

The electrical work done by a cell is given by the quantity of electricity made to flow multiplied by the potential difference $\mathscr{E}$ that causes the current to flow. We are ordinarily interested in the electrical work (and thence ΔG) associated with reaction of one mole of one of the cell components. In this case the quantity of electricity that flows is obtained as the product of number of moles of electrons involved in the specified cell reaction times the number of coulombs per mole of electrons. We therefore have

$$w_{el} = n\mathscr{F}\mathscr{E} \tag{16.26}$$

in which n represents the number of moles of electrons per mole of reaction and $\mathscr{F}$ represents the Faraday.

Now we consider the particular case of reversible electrical work and write (16.26) as

$$w_{rev\ el} = n\mathscr{F}\mathscr{E}_{rev} \tag{16.27}$$

and combine (16.27) with (16.25) to obtain

$$\Delta G = -n\mathscr{F}\mathscr{E}_{rev} \tag{16.28}$$

in which $\mathscr{E}_{rev}$ represents the "reversible potential" of the cell. This reversible potential is the cell potential corresponding to infinitesimal current flow. Reversible potentials are ordinarily measured with a potentiometer (illustrated schematically in Figure 16.7) but can be estimated from voltage measurements with a very high resistance volt meter.

It is customary to write the important equation (16.28) without the subscript rev. We henceforth follow this practice and work with this equation in the form

$$\Delta G = -n\mathscr{F}\mathscr{E} \tag{16.29}$$

with the understanding that $\mathscr{E}$ represents the reversible potential of the cell.

Following our usual practice, we use superscript ^{0} to indicate thermodynamic quantities referring to reactions or processes in which all substances are in their standard states. Thus $\mathscr{E}^0$ represents the standard cell potential and ΔG^0 represents the standard free energy change associated with the cell reaction. We therefore have

$$\Delta G^0 = -n\mathscr{F}\mathscr{E}^0 \tag{16.30}$$

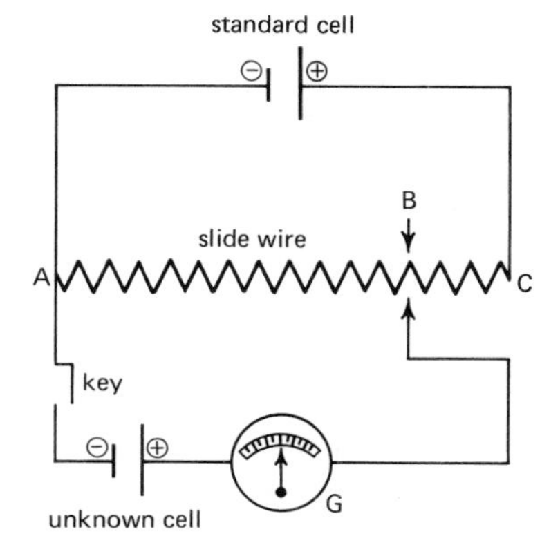

Figure 16.7. Schematic illustration of a simple potentiometer in which the slide wire resistance is adjusted until zero current flows through the galvanometer G. When no current flows through G, the unknown cell potential $\mathscr{E}$ and the standard or reference cell potential $\mathscr{E}_s$ are related by $\mathscr{E} = \mathscr{E}_s R_{AB}/R_{AC}$.

from (16.29). Before taking up the procedure for obtaining the *standard* cell potential ($\mathscr{E}^0$) from *measured* cell potentials ($\mathscr{E}$), we must discuss and illustrate a very important use of $\mathscr{E}^0$ in relation to ΔG^0 and the equilibrium constant K.

We combine (16.30) with

$$\Delta G^0 = -RT \ln K \tag{13.37}$$

to obtain

$$-RT \ln K = -n\mathscr{F}\mathscr{E}^0$$

and then rearrange to

$$\log K = \frac{n\mathscr{F}}{2.303\ RT}\mathscr{E}^0 \tag{16.31}$$

Equation (16.31) is a very useful relationship between $\mathscr{E}^0$ of a cell and the equilibrium constant for the cell reaction. This equation permits easy calculation of equilibrium constants from tabulated potentials as shown later in this chapter.

To obtain a direct numerical relationship between equilibrium constant K and standard potential $\mathscr{E}^0$, we must insert appropriate numerical values for $\mathscr{F}$, R, and T in equation (16.31). Because many cell measurements and most tabulated potentials refer to 298°K, we use this value along with $R = 1.987$ cal deg^{-1} mole^{-1} and $\mathscr{F} = 23{,}060$ cal volt^{-1} (mole e^-)$^{-1}$ to obtain

$$\log K = 16.9n\mathscr{E}^0 \tag{16.32}$$

or

$$\log K = \frac{n\mathscr{E}^0}{0.05916} \tag{16.33}$$

These equations show that positive $\mathscr{E}^0$ corresponds to K larger than unity, while negative $\mathscr{E}^0$ corresponds to K less than unity. Also, when $\mathscr{E}^0 = 0$, $K = 1$. We later return to further discussion of signs of various potentials.

Example Problem 16.4. Measurements on cells constructed with a silver electrode in contact with Ag^+(aq) ions and a copper electrode in contact with Cu^{2+}(aq) ions lead to $\mathscr{E}^0 = 0.46$ volt at 298°K. The total cell reaction is

$$2\ Ag^+(aq) + Cu(c) \rightleftharpoons 2\ Ag(c) + Cu^{2+}(aq)$$

Calculate the equilibrium constant for this reaction.

Our first problem is to decide on the value of n to use in equation (16.32) or (16.33). To do so we note that reduction of two moles of Ag^+(aq) and oxidation of one mole of Cu(c) requires transfer of two moles of electrons. Therefore, $n = 2$. Or we might write out the electrode reactions ($2\ Ag^+ + 2e^- \rightleftharpoons 2\ Ag$ and $Cu \rightleftharpoons Cu^{2+} + 2e^-$) and reach the same conclusion.

Now we insert $n = 2$ and $\mathscr{E}^0 = 0.46$ in equation (16.32) to obtain

$$\log K = 16.9 \times 2 \times 0.46 = 15.5$$

and then

$$K = 3.2 \times 10^{15}$$

This large value of K shows that we can ordinarily expect Ag^+(aq) ions to oxidize metallic copper to Cu^{2+}(aq) ions. Or we can say that metallic copper reduces Ag^+(aq) ions to metallic silver. The net result is a reaction in which Ag^+(aq) ions are reduced to Ag(c) and Cu(c) is oxidized to Cu^{2+}(aq) ions. Metallic copper is a stronger reducing agent than metallic silver, while Ag^{+}(aq) is a stronger oxidizing agent than Cu^{2+}(aq). ■

Now that we have briefly illustrated the connection between $\mathscr{E}^0$ and K, we consider how one obtains $\mathscr{E}^0$ values from measured $\mathscr{E}$ values and the reverse problem of calculating actual cell potentials from $\mathscr{E}^0$ values that are readily available in tabular form.

In deducing the general form of the equilibrium constant expression and deriving the thermodynamic equation relating ΔG^0 to K, we obtained equation (13.34), which we now rewrite as

$$\Delta G^0 = \Delta G - RT \ln Q \tag{16.34}$$

In equation (16.34), Q has been used as a convenient symbol for an expression of the form of the appropriate equilibrium constant *but with the actual rather than the equilibrium concentrations or activities.* Combining (16.29) and (16.30) with (16.34) now gives

$$-n\mathscr{F}\mathscr{E}^0 = -n\mathscr{F}\mathscr{E} - RT \ln Q$$

and thence

$$\mathscr{E}^0 = \mathscr{E} + \frac{RT}{n\mathscr{F}} \ln Q \tag{16.35}$$

This useful equation is commonly called the Nernst equation, in honor of German chemist W. H. Nernst (1864–1941). At 298°K, the Nernst equation becomes

$$\mathscr{E}^0 = \mathscr{E} + \frac{0.05916}{n} \log Q \tag{16.36}$$

In so far as it is an acceptable approximation to use concentrations rather than activities in Q in the Nernst equation, we can use (16.36) in direct fashion as follows. (1) We may set up a cell with particular known concentrations, which permits straightforward calculation of Q. Then a measurement of the reversible potential $\mathscr{E}$ of this cell permits direct calculation of $\mathscr{E}^0$. (2) We may also obtain $\mathscr{E}^0$ from standard state thermodynamic data (as illustrated later). Then we may choose the particular concentrations of interest to us and calculate the corresponding value of Q. These values of $\mathscr{E}^0$ and Q lead directly to the reversible voltage $\mathscr{E}$ of this cell.

Example Problem 16.5. In Example Problem 16.4 we have quoted $\mathscr{E}^0 = 0.46$ volt at 25°C for the cell in which the reaction is

$$2\,Ag^+(aq) + Cu(c) \rightleftharpoons 2\,Ag(c) + Cu^{2+}(aq)$$

Calculate the reversible potential of this cell when concentrations of Ag^+(aq) and of Cu^{2+}(aq) are each 0.01 M, making the approximation that activity is approximately equal to concentration.

Here we have

$$Q = \frac{[Cu^{2+}]}{[Ag^+]^2}$$

in which square brackets indicate concentrations. Insertion of the stated values gives

$$Q = (1 \times 10^{-2})/(1 \times 10^{-2})^2 = 1 \times 10^2$$

Now we solve the Nernst equation (16.36) for $\mathscr{E}$ and have

$$\begin{aligned}\mathscr{E} &= \mathscr{E}^0 - \frac{0.05916}{n} \log Q \\ &= \mathscr{E}^0 - \frac{0.05916}{2} \log (1 \times 10^2) \\ &= 0.46 - 0.059 = 0.40 \text{ volt}\end{aligned}$$

as the reversible potential of the specified cell. ■

The calculation described above may also be reversed. That is, we combine a measured $\mathscr{E}$ value with a corresponding Q value calculated from the known concentrations and thence obtain the $\mathscr{E}^0$ value. It is in this way that many useful $\mathscr{E}^0$ values have been obtained.

Example Problem 16.6. Use ΔG_f^0 values listed in Appendix II for calculation of $\mathscr{E}^0$ at 298°K for the cell (illustrated in Figure 16.8) in which the reaction is

$$\tfrac{1}{2} H_2(g) + AgCl(c) \rightleftharpoons H^+(aq) + Cl^-(aq) + Ag(c)$$

Then calculate the potential $\mathscr{E}$ for this cell when $P_{H_2} = 0.940$ atm and the hydrochloric acid is 0.02 *m*. As first approximation, take the activities of aqueous ions to be equal to their molalities.

In Appendix II we find $\Delta G_f^0 = -31.372$ kcal mole^{-1} for $Cl^-(aq)$ and $\Delta G_f^0 = -26.244$ kcal mole^{-1} for AgCl(c). It is unnecessary to look up ΔG_f^0 values for $H_2(g)$, $H^+(aq)$, and Ag(c) because our choice of reference state or "sea level" for free energy has made all of these equal to zero. Now we calculate the standard free energy change for the cell reaction to be

$$\Delta G_f^0 = -31.372 - (-26.244) = -5.128 \text{ kcal mole}^{-1}$$

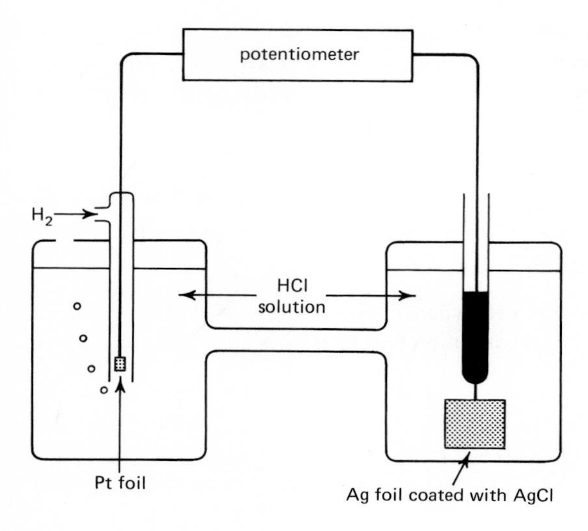

Figure 16.8. Schematic illustration of the cell discussed in Example Problem 16.6.

Insertion of this value (converted to cal mole^{-1}) in $\Delta G^0 = -n\mathscr{F}\mathscr{E}^0$ leads to

$$\mathscr{E}^0 = 5128/23060 = 0.2224 \text{ volt}$$

(Readers should work problem 15 at the end of this chapter to make sure of understanding the units and choice of value for $\mathscr{F}$ in this calculation.)

The Nernst equation (16.36) is written as

$$\mathscr{E} = \mathscr{E}^0 - \frac{0.05916}{n} \log Q$$

in which

$$Q = \frac{[H^+][Cl^-]}{(P_{H_2})^{\frac{1}{2}}} = \frac{(2 \times 10^{-2})^2}{(0.94)^{\frac{1}{2}}} = 4.13 \times 10^{-4}$$

Then we find $\log Q = -3.384$ and

$$\mathscr{E} = 0.2224 - (0.05916)(-3.384) = 0.4226 \text{ volt}$$

The small difference between the actually measured $\mathscr{E}$ for this cell and the value calculated here is due to the small difference between concentration and activity of the aqueous ions, and can therefore be used for evaluation of activity coefficients. ■

Cell Potentials, Electrode Potentials, and Oxidation-Reduction Potentials

A large number of electrochemical investigations have led to reliable values for many standard cell potentials. Also, many investigations of chemical equilibria and thermodynamic properties of chemical substances have led by way of $\Delta G^0 = -n\mathscr{F}\mathscr{E}^0$ to many more standard cell potentials. In this section we are concerned with efficient and convenient tabulation and use of all of these potentials.

In general, every cell reaction can be regarded as a combination of an anode reaction and a cathode reaction. Similarly, we can regard each cell potential as a corresponding combination of an anode electrode potential and a cathode electrode potential. Unfortunately, no one has ever succeeded in measuring the potential of a *single* electrode, and some argue that a single electrode potential is fundamentally impossible to measure. Because our experimental knowledge leads us only to total cell potentials, we obtain useful single electrode potentials only with respect to some arbitrarily chosen reference electrode or "sea level." The conventional choice is to take the standard potential of the hydrogen electrode to be zero. Before turning to detailed consideration of consequences and application of this choice, we carefully consider some terminology and "sign conventions."

The reactions that occur at particular electrodes are called electrode reactions. These same reactions can also occur in non-electrochemical systems, in which case they are ordinarily called half-reactions. Potentials may be associated with electrodes, with electrode reactions, and with half-reactions, as discussed below.

Cell potentials measured in the laboratory are positive or negative in an electrical sense, which we may denote by ⊕ and ⊖. These electrical cell potentials lead to electrode potentials (based on an arbitrary value for a reference electrode) that are also positive or negative in an electrical sense. The electrical signs of these electrode potentials are independent of how we humans choose to describe the electrodes or write the electrode reactions. For example, the standard potential of the electrode that consists of metallic copper in a solution of $Cu^{2+}(aq)$ is ⊕0.34 volt. Thus we might write

$$Cu^{2+}|Cu \quad \text{or} \quad Cu|Cu^{2+} \qquad \mathscr{E}^0 = \oplus 0.34 \text{ volt} \quad (16.37)$$

Although not recommended, we might choose to represent this electrode by an electrode reaction, still with standard electrode potential $\mathscr{E}^0 = \oplus 0.34$ volt, no matter how we write the reaction. Similarly, the potential of the electrode that consists of metallic zinc in a solution of $Zn^{2+}(aq)$ has standard potential $\mathscr{E}^0 = \ominus 0.76$ volt as shown by

$$Zn^{2+}|Zn \quad \text{or} \quad Zn|Zn^{2+} \qquad \mathscr{E}^0 = \ominus 0.76 \text{ volt} \quad (16.38)$$

The electrode potentials discussed above are sign invariant. That is, the sign of the standard copper electrode is always ⊕ with respect to the standard hydrogen electrode, independent of how we choose to represent the electrodes or the cell reaction or the electrode reactions. This means that the copper electrode is (electrically) positive with respect to the hydrogen electrode, as shown in Figure 16.9. Similarly, the standard zinc electrode is always ⊖ with respect to the standard hydrogen electrode, as shown in Figure 16.10.

If we now consider a cell consisting of a (standard) copper electrode and a (standard) zinc electrode as pictured in Figure 16.4, we can see that the difference between the potentials of the two electrodes amounts to 1.10 volt. Because the zinc electrode is ⊖ with respect to the hydrogen electrode and the copper electrode is ⊕ with respect to the hydrogen electrode, we know that the zinc electrode must be electrically negative with respect to the copper electrode. Thus the spontaneous flow of electrons must occur from the ⊖ zinc electrode to the ⊕ copper electrode, as already pictured in Figure 16.4.

We may also associate potentials with the reactions that occur at various electrodes or with the half-reactions that go to make up a total reaction that occurs

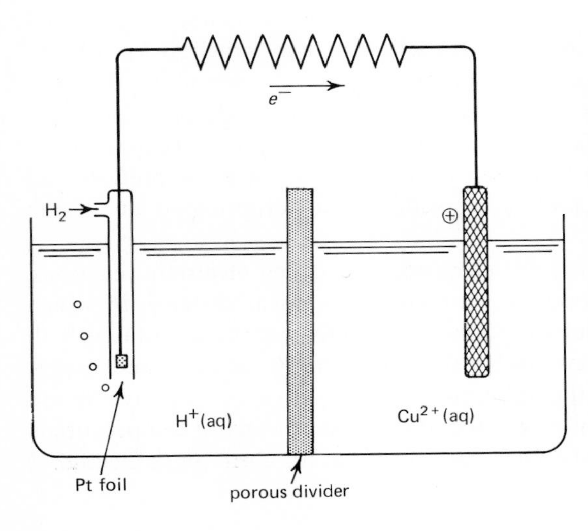

Figure 16.9. Schematic illustration of a (standard state) cell consisting of copper and hydrogen electrodes. The arrow indicates the direction of spontaneous electron flow through the external circuit.

Because electrons (negatively charged) spontaneously flow toward the positive electrode, we say that the standard copper electrode is ⊕ with respect to the standard hydrogen electrode. The chemical reaction associated with delivery of current through the external circuit is

$$Cu^{2+}(aq) + H_2(g) \rightarrow Cu(c) + 2\,H^+(aq)$$

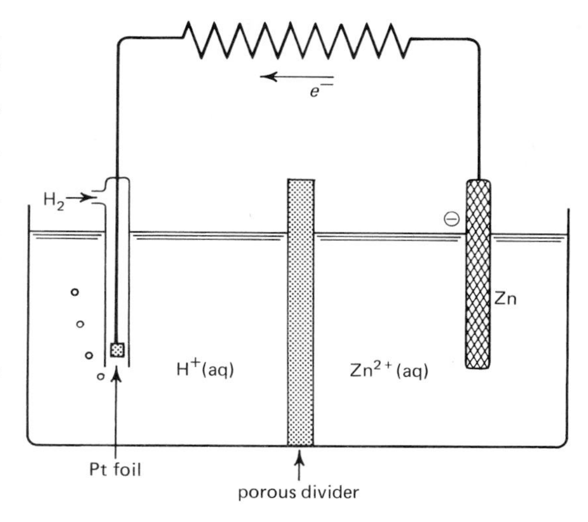

Figure 16.10. Schematic illustration of a (standard state) cell consisting of zinc and hydrogen electrodes. The arrow indicates the direction of spontaneous electron flow through the external circuit.

Because electrons (negatively charged) spontaneously flow away from the negative electrode, we say that the standard zinc electrode is $\ominus$ with respect to the standard hydrogen electrode. The chemical reaction associated with delivery of current through the external circuit is

$$Zn(c) + 2\ H^+(aq) \rightarrow Zn^{2+}(aq) + H_2(g)$$

in some non-electrochemical system. One important and useful reason for associating potentials with half-reactions is that these potentials can be used in combination with $\Delta G^0 = -n\mathscr{F}\mathscr{E}^0$ and $\Delta G^0 = -RT \ln K$ [as in equations (16.32) and (16.33)] to obtain information about chemical equilibria and the direction of spontaneous reactions.

Under different conditions, a particular half-reaction or electrode reaction may be spontaneous in either direction. For example, the reaction associated with the hydrogen electrode in the cell pictured in Figure 16.9 is indicated by

$$H_2(g) + 2\,e^- \rightarrow 2\ H^+(aq)$$

On the other hand, the spontaneous reaction associated with the hydrogen electrode in the cell pictured in Figure 16.10 is

$$2\ H^+(aq) \rightarrow H_2(g) + 2\,e^-$$

Here it should be clearly understood that the hydrogen electrode is the same in both cells: it is the reaction that occurs at the electrode that is reversed.

Because we are often interested in using potentials in various algebraic equations (such as 16.32), these potentials must have *algebraic* rather than *electrical* signs. Further, because we are interested in chemical equilibria and reversible reactions, we must recognize that the *algebraic* sign depends upon the direction in which we write the half-reaction, in contrast to *electrode* potentials with *electrical* signs that are independent of how we represent the electrode. Thus the standard electrode potential of the copper electrode is $\oplus$0.34 volt (with respect to the hydrogen electrode), no matter how we choose to represent the electrode. On the other hand, the algebraic sign of the standard potential associated with the half-reaction involving metallic copper, Cu^{2+}(aq) ions, and electrons depends on the direction in which we write the half-reaction. As shown later, these directions of reaction and algebraic signs of half-reaction potentials are as follows:

$$Cu(c) \rightleftharpoons Cu^{2+} + 2\,e^- \qquad \mathscr{E}^0 = -0.34 \text{ volt} \quad (16.39)$$

$$Cu^{2+} + 2\,e^- \rightleftharpoons Cu(c) \qquad \mathscr{E}^0 = +0.34 \text{ volt} \quad (16.40)$$

The potential associated with half-reaction (16.39), which is an oxidation, is often called an **oxidation potential.** Similarly, the potential associated with half-reaction (16.40), which is a reduction, is often called a **reduction potential.**

We are now faced with a choice or choices about how we are to list electrodes or electrode reactions and how we are to tabulate corresponding electrode potentials or half-reaction potentials. Some scientists, particularly those who are directly concerned with electrical measurements on various cells, have quite reasonably chosen to focus on electrodes and corresponding electrode potentials, and have therefore listed electrodes and potentials as in (16.37) and (16.38). Other scientists, particularly those who are interested in thermodynamic calculations in relation to chemical equilibrium, have more commonly tabulated and used half-reactions and oxidation potentials as in (16.39). Still others, particularly analytical chemists who are interested in electrical reduction of metal ions to the metals, have tabulated and used reduction half-reactions and reduction potentials as in (16.40). Furthermore, some scientists have used different symbols (such as E^0, $\mathscr{V}^0$, etc.) to distinguish between electrode potentials and half-reaction potentials.

In the remainder of this book we shall list and use reduction half-reactions and reduction potentials with algebraic signs as in (16.40). Each reduction half-reaction may be reversed to give an oxidation half-reaction, which requires that the sign of each reduction potential be changed to give the sign of the corresponding oxidation potential. One reason that this particular choice of "sign convention" is becoming increasingly widely adopted is that the algebraic signs associated with reduction half-reaction potentials are interchangeable with the corresponding electrical signs associated with the corresponding electrodes. For example, the standard reduction potential of the $Cu^{2+}(aq) + 2\,e^- \rightleftharpoons Cu(c)$ half-reaction is (algebraic) + 0.34 volt, and the standard electrode potential of the copper electrode is (electrical) $\oplus$0.34 volt.

Now we turn to some examples of evaluation and use of various potentials.

One obvious way in which to obtain electrode potentials and thence half-reaction potentials (both relative to the H_2–H^+ combination) is to determine the standard potential of a cell consisting of the electrode of interest and the hydrogen electrode. This approach is illustrated in Figures 16.8, 16.9, and 16.10. Standard potentials of these cells lead to the standard potential for the copper electrode (16.37) and the standard potentials for the copper–cupric ion half-reactions (16.39) and (16.40). Similarly, we have the standard electrode potential of the zinc electrode as in (16.38), from which we deduce the following half-reaction potentials:

$$Zn^{2+}(aq) + 2\,e^- \rightleftharpoons Zn(c) \qquad \mathscr{E}^0 = -0.76 \text{ volt} \qquad (16.41)$$
$$Zn(c) \rightleftharpoons Zn^{2+}(aq) + 2\,e^- \qquad \mathscr{E}^0 = +0.76 \text{ volt} \qquad (16.42)$$

The first of these half-reaction potentials is the reduction potential that we shall later tabulate and use. In the same fashion, the cell discussed in Example Problem 16.6 and illustrated in Figure 16.8 has a standard potential that leads to the following standard electrode potential for the silver–silver chloride electrode

$$Ag|AgCl \quad \text{or} \quad AgCl|Ag \qquad \mathscr{E}^0 = \oplus 0.22 \text{ volt} \qquad (16.43)$$

From this standard electrode potential we deduce the following standard half-reaction potentials

$$AgCl(c) + e^- \rightleftharpoons Ag(c) + Cl^-(aq) \qquad \mathscr{E}^0 = +0.22 \text{ volt} \qquad (16.44)$$
$$Ag(c) + Cl^-(aq) \rightleftharpoons AgCl(c)\, e^- \qquad \mathscr{E}^0 = -0.22 \text{ volt} \qquad (16.45)$$

Again, it is the reduction half-reaction and reduction potential (16.44) that we shall tabulate.

It is not always convenient, practical, or even safe (think about the reaction of hydrogen with atmospheric oxygen) to work with the hydrogen electrode as a means of determining standard potentials of other electrodes and thence other half-reaction potentials. But this presents no special difficulty once we know a few potentials from previous work with the hydrogen electrode. For example, we can combine the silver–silver chloride electrode with the mercury–mercurous chloride electrode as illustrated in Figure 16.11. Voltage measurements lead to $\mathscr{E}^0 = 0.05$ volt as the standard potential for this cell, in which the spontaneous reaction is

$$Hg_2Cl_2(c) + 2\,Ag(c) \rightleftharpoons 2\,Hg(liq) + 2\,AgCl(c) \qquad (16.46)$$

From the direction of electron flow indicated in Figure 16.11, we know that the mercury–mercurous chloride electrode is $\oplus$ with respect to the silver–silver chloride electrode. Now combination of the standard potential of the cell (0.05 volt) with the already known standard potential for the silver–silver chloride electrode [$\oplus$0.22 volt as in (16.43)] gives us the following standard electrode potential for the mercury–mercurous chloride electrode:

$$Hg\,|\,Hg_2Cl_2 \quad \text{or} \quad Hg_2Cl_2\,|\,Hg \qquad \mathscr{E}^0 = \oplus 0.27 \text{ volt} \qquad (16.47)$$

Now we can write the half-reaction and the corresponding reduction and oxidation potentials:

$$Hg_2Cl_2(c) + 2\,e^- \rightleftharpoons 2\,Hg(liq) + 2\,Cl^-(aq) \qquad \mathscr{E}^0 = +0.27 \text{ volt} \qquad (16.48)$$
$$2\,Hg(liq) + 2\,Cl^-(aq) \rightleftharpoons Hg_2Cl_2(c) + 2\,e^- \qquad \mathscr{E}^0 = -0.27 \text{ volt} \qquad (16.49)$$

Again, it is the reduction half-reaction and reduction potential (16.48) that we shall tabulate.

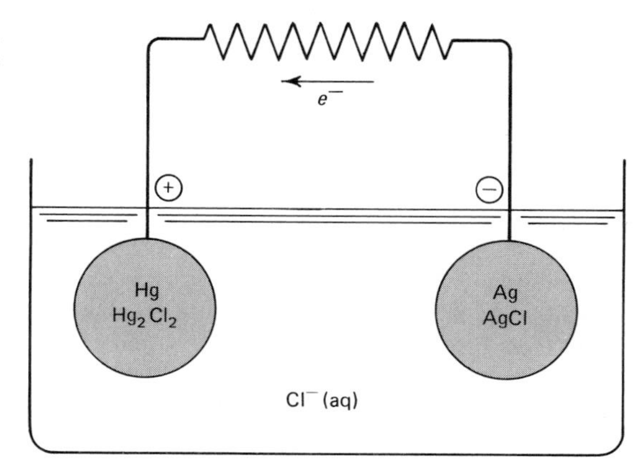

Figure 16.11. Schematic illustration of the cell discussed in the text, for which equation (16.46) represents the spontaneous cell reaction.

The difference between the two electrode potentials given in (16.43) and (16.47) gives the standard potential for this cell. Half-reactions for the two electrodes are given by equations (16.45) and (16.48).

Many reduction half-reactions and corresponding reduction potentials have been derived from electrochemical cells, while many others have been derived from ΔG_f^0 values. Various calculations involving these potentials are illustrated in the following Example Problems and in later chapters that are concerned with the chemical properties of specific elements.

Example Problem 16.7 Use reduction half-reaction potentials listed in Appendix IV to evaluate the standard potential of a cell consisting of a metallic zinc–zinc ion electrode and a metallic cadmium–cadmium ion electrode. Then deduce the direction of spontaneous electron flow through an external circuit and the equilibrium constant for the cell reaction.

We begin by finding the following half-reactions and reduction potentials:

$$Zn^{2+}(aq) + 2\,e^- \rightleftharpoons Zn(c) \qquad \mathscr{E}^0 = -0.762 \text{ volt}$$
$$Cd^{2+}(aq) + 2\,e^- \rightleftharpoons Cd(c) \qquad \mathscr{E}^0 = -0.402 \text{ volt}$$

Now these reduction half-reactions must be combined to yield a complete reaction that consists of an oxidation and a reduction, which requires that we reverse one half-reaction. Because the potentials above indicate that $Cd^{2+}(aq)$ is easier to reduce than $Zn^{2+}(aq)$ and that Zn(c) is easier to oxidize than Cd(c), we choose to reverse the zinc half-reaction and combine it with the cadmium half-reaction as follows:

$$Zn(c) \rightleftharpoons Zn^{2+}(aq) + 2\,e^- \qquad \mathscr{E}^0 = +0.762 \text{ volt}$$
$$Cd^{2+}(aq) + 2\,e^- \rightleftharpoons Cd(c) \qquad \mathscr{E}^0 = -0.402 \text{ volt}$$
$$Zn(c) + Cd^{2+}(aq) \rightleftharpoons Zn^{2+}(aq) + Cd(c) \qquad \mathscr{E}^0 = +0.360 \text{ volt} \quad (16.50)$$

We see from (16.50) that the standard potential of the cell under consideration is 0.360 volt, and we use this potential in equation (16.32) to obtain the equilibrium constant for the cell reaction (16.50) as

$$\log K = 16.9n\mathscr{E}^0 = 16.9 \times 2 \times 0.36 = 12.2$$
$$K = 1.6 \times 10^{12}$$

In this calculation we have used $n = 2$ because two moles of electrons are transferred per mole of reaction, as shown by the half-reactions that go to make up the complete cell reaction (16.50).

We see from the reduction potentials (signs equivalent to signs of electrode potentials) that the zinc electrode is more negative (with respect to the hydrogen electrode) than is the cadmium electrode. Thus the zinc electrode is negative with respect to the cadmium electrode and electrons flow from the zinc electrode through the external circuit to the cadmium electrode, as indicated in Figure 16.12. We can also deduce the direction of electron flow from the half-reactions. As shown by the half-reactions above equation (16.50), metallic zinc loses electrons when $Zn^{2+}(aq)$ ions are formed and metallic cadmium is formed when $Cd^{2+}(aq)$ ions capture electrons from the external circuit.

We may also deduce some chemical reaction information from the results of our calculations in this problem. Because the equilibrium constant for reaction (16.50) is large, we know that metallic zinc will react spontaneously with $Cd^{2+}(aq)$ ions to yield metallic cadmium and $Zn^{2+}(aq)$ ions. Thus we may say that zinc is a stronger reducing agent than is cadmium, and that $Cd^{2+}(aq)$ ions are stronger oxidizing agents than $Zn^{2+}(aq)$ ions.

The equilibrium constant for the reaction that is the reverse of (16.50) is simply the reciprocal of the equilibrium constant for (16.50), and the sign of the reaction potential is reversed as may be seen from the half-reaction potentials:

$$Zn^{2+}(aq) + Cd(c) \rightleftharpoons Zn(c) + Cd^{2+}(aq) \qquad \mathscr{E}^0 = -0.360 \text{ volt} \quad (16.51)$$
$$K_{16.51} = 1/K_{16.50} = 6.3 \times 10^{-13}$$

Figure 16.12. Illustration of the cell discussed in Example Problem 16.7.

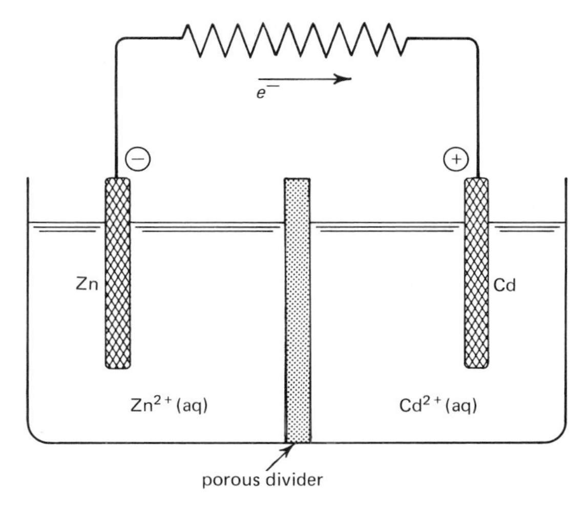

We may also obtain $K_{16.51}$ from the potential as follows:

$$\log K_{16.51} = 16.9 \times 2 \times (-0.360) = -12.2 = 0.8 - 13$$
$$K_{16.51} = 6.3 \times 10^{-13} \quad \blacksquare$$

Example Problem 16.8. Use reduction potentials from Appendix IV to deduce the standard potential of a cell with silver and copper electrodes in solutions of their aqueous ions. Then decide whether $Ag^{+}(aq)$ ions will react with Cu(c) or whether $Cu^{2+}(aq)$ ions will react with Ag(c).

The half-reactions and reduction potentials that we need are

$$Ag^{+}(aq) + e^{-} \rightleftharpoons Ag(c) \qquad \mathcal{E}^{0} = +0.799 \text{ volt} \quad (16.52)$$
$$Cu^{2+}(aq) + 2\,e^{-} \rightleftharpoons Cu(c) \qquad \mathcal{E}^{0} = +0.340 \text{ volt} \quad (16.53)$$

First, we must change one of these reduction half-reactions to an oxidation half-reaction, with accompanying change in sign of potential. Because $Ag^{+}(aq)$ ions are more easily reduced (stronger oxidizing agent) than $Cu^{2+}(aq)$ ions, we reverse the copper half-reaction. In order to have our final equation balanced with respect to charge as well as atoms and to show that electrons gained are equivalent to electrons lost, we must also double the silver half-reaction equation:

$$2\,Ag^{+}(aq) + 2\,e^{-} \rightleftharpoons 2\,Ag(c) \qquad \mathcal{E}^{0} = +0.799 \text{ volt} \quad (16.52a)$$
$$Cu(c) \rightleftharpoons Cu^{2+}(aq) + 2e^{-} \qquad \mathcal{E}^{0} = -0.340 \text{ volt} \quad (16.54)$$
$$2\,Ag^{+}(aq) + Cu(c) \rightleftharpoons 2\,Ag(c) + Cu^{2+}(aq) \qquad \mathcal{E}^{0} = +0.459 \text{ volt} \quad (16.55)$$

Here it should be noted that the $\mathcal{E}^{0}$ for the silver half-reaction has NOT been doubled. See problem 16 at the end of this chapter for formal proof that this potential should not be doubled. Here we will only state informally that a big battery and a small battery have the same potential, although the big battery can last longer and deliver more current.

Now we substitute the potential in equation (16.32) to obtain

$$\log K = 16.9 \times 2 \times 0.459 = 15.51$$
$$K = 3.24 \times 10^{15}$$

We have used $n = 2$ because the balanced equation (16.55) to which K applies is based on two moles of electrons transferred.

Because the equilibrium constant for reaction (16.55) is large, we know that $Ag^+(aq)$ ions react with Cu(c) until the ratio of $[Cu^{2+}]/[Ag^+]^2$ is very large. We can also conclude that spontaneous reaction of Ag(c) with Cu^{2+} [from right to left of equation (16.55)] is negligible. ■

Example Problem 16.9. Use half-reactions and reduction potentials from Appendix IV for calculation of the solubility product of silver chloride in water at 25°C.

The desired solubility product is the equilibrium constant for the reaction

$$AgCl(c) \rightleftharpoons Ag^+(aq) + Cl^-(aq)$$

and can be calculated from $\mathscr{E}^0$ for this reaction. The standard potential for the Ag^+/Ag half-reaction is +0.799 volt and that for the $AgCl/Ag,Cl^-$ half-reaction is +0.222 volt. These half-reactions and their potentials are combined as follows:

$$\begin{array}{lr} Ag(c) \rightleftharpoons Ag^+(aq) + e^- & \mathscr{E}^0 = -0.799 \text{ volt} \\ AgCl(c) + e^- \rightleftharpoons Ag(c) + Cl^-(aq) & \mathscr{E}^0 = +0.222 \text{ volt} \\ \hline AgCl(c) \rightleftharpoons Ag^+(aq) + Cl^-(aq) & \mathscr{E}^0 = -0.577 \text{ volt} \end{array}$$

Insertion of this potential in equation (16.32) gives

$$\log K = 16.9 \times 1 \times (-0.577) = -9.75 = 0.25 - 10$$

and

$$K = 1.78 \times 10^{-10}$$

This equilibrium constant (solubility product) is in good agreement with the value calculated in Chapter 14 from solubility data.

Example Problem 16.10. When a cell containing 1.0 m $Ag^+(aq)$ and 1.0 m $Cu^{2+}(aq)$ is electrolyzed, metal will be plated out on the cathode. Which metal is plated out first? Can the metals be separated by electrolysis?

Reduction potentials pertinent to this problem are the following:

$$\begin{array}{lr} Ag^+(aq) + e^- \rightleftharpoons Ag(c) & \mathscr{E}^0 = +0.799 \text{ volt} \\ Cu^{2+}(aq) + 2\,e^- \rightleftharpoons Cu(c) & \mathscr{E}^0 = +0.340 \text{ volt} \end{array}$$

We see from these potentials that $Ag^+(aq)$ ions are easier to reduce than are $Cu^{2+}(aq)$ ions. Or we can say that $Ag^+(aq)$ ions are stronger oxidizing agents than $Cu^{2+}(aq)$ ions. Either mode of expression leads to the conclusion that silver will be plated out before copper. Electroplating silver on the cathode will continue until the concentration of $Ag^+(aq)$ remaining in solution is so small that the potential required to plate out more silver is equal to the potential required to begin plating out copper.

In order to decide whether electrolysis will lead to satisfactory separation of silver from copper, it is now necessary for us to calculate the residual concentration of $Ag^+(aq)$ ions when copper begins to plate out. That is, we want to know the concentration of $Ag^+(aq)$ ions that will lower the Ag^+/Ag potential from 0.799 volt to 0.340 volt. We therefore write the Nernst equation as

$$\mathscr{E} = \mathscr{E}^0 - \frac{0.05916}{n} \log \frac{1}{[Ag^+]}$$

$$0.340 = 0.799 + 0.05916 \log [Ag^+]$$

$$\log [Ag^+] = -7.76 = 0.24 - 8$$

$$[Ag^+] = 1.7 \times 10^{-8}$$

This calculation shows that the concentration of $Ag^+(aq)$ is *very small* when copper begins to plate out from 1.0 m solution. We therefore conclude that almost all of the silver is plated out before copper begins to plate out, which means that silver can be effectively separated from copper by electrolysis. ■

Effect of Temperature on Cell and Reaction Potentials

It should seem nearly obvious now that reaction and cell potentials (related to free energies) depend on the temperature and that the relationship between potential and temperature can be deduced from proper application of thermodynamics. To obtain this relationship we combine $\Delta G^0 = -n\mathscr{F}\mathscr{E}^0$ with $\Delta G^0 = \Delta H^0 - T\Delta S^0$, which yields

$$-n\mathscr{F}\mathscr{E}^0 = \Delta H^0 - T\Delta S^0$$

Division of both sides by $-n\mathscr{F}$ yields

$$\mathscr{E}^0 = -\frac{\Delta H^0}{n\mathscr{F}} + \frac{\Delta S^0}{n\mathscr{F}}T \tag{16.56}$$

According to this equation, a graph of $\mathscr{E}^0$ versus T should yield a straight line with slope equal to $\Delta S^0/n\mathscr{F}$ and intercept equal to $-\Delta H^0/n\mathscr{F}$. Thus it is possible to deduce both ΔH^0 and ΔS^0 for a cell reaction from measurements of the cell potential at several temperatures. It is also possible to calculate the cell potential at various temperatures if one already has thermodynamic data for the cell reaction.

Equation (16.56), in a slightly different form as in problem 17 at the end of this chapter, is commonly called the Gibbs-Helmholtz equation.

Electrical Conductivity

The electrical conductivity of pure water is low and practically unaffected by adding such nonelectrolytes as methyl alcohol, ethyl alcohol, sugar, or acetone. Solutions of acids, bases, and salts are better conductors than pure water because these solutes furnish ions that act as current carriers.

The conductivity of a solution depends on the concentration of ions largely because the electrolyte concentration determines the number of ions in a given volume of solution between the electrodes. To a lesser extent, the conductivity also depends on concentration because the ions interact with each other, which influences their mobility in the solution. This interaction depends on the average distance of the ions from each other, and the distance depends on the concentration. Conductivity also depends on the nature of the electrolyte because different ions have different mobilities and are therefore able to carry the electrical current differently.

Electrical interactions between ions are the principal cause of nonideality in electrolyte solutions. This nonideality of electrical origin is responsible for deviations of observed freezing point depression from the ideal freezing point equation (4.18). Expressed differently, the electrically caused nonideality of solutions of electrolytes is largely responsible for the deviations of activity coefficients of ionic solutes from unity. Activity coefficients calculated from conductivities are in excellent agreement with those obtained from freezing point depressions and from cell potentials.

Because the conductivity of a solution depends on the number of ions present in a given volume of that solution, conductivity measurements can be looked on as a means of "ion counting." It is therefore possible to use conductivities for evaluation of equilibrium constants for various reactions that involve ions.

References

Anson, F. C.: Electrode sign conventions. *J. Chem. Educ.*, **36**:394 (1959).

Hill, D. L., S. J. Moss, and R. L. Strong: Heat of reaction in aqueous solution by potentiometry and calorimetry (metal replacement reaction). *J. Chem. Educ.*, **42**:541 (1965).

Jolly, W. L.: The use of oxidation potentials in inorganic chemistry. *J. Chem. Educ.*, **43**:198 (1966).

Licht, T. S., and A. J. deBethune: Recent developments concerning the signs of electrode potentials. *J. Chem. Educ.*, **34**:433 (1957).

MacInnes, D. A.: *The Principles of Electrochemistry*. Dover Publications, Inc., New York (paperback).

Miller, S. I.: Combining half-reactions and their standard electrode potentials. *J. Chem. Educ.*, **29**:140 (1952).

Moss, S. J., and D. L. Hill: Heat of reaction in aqueous solution by potentiometry and calorimetry (redox reaction). *J. Chem. Educ.*, **42**:544 (1965).

Ramette, R. W.: Silver equilibria via cell measurements. *J. Chem. Educ.*, **49**:423 (1972).

Ramsay, J. B.: The basis for a new conception of an electrode potential. *J. Chem. Educ.*, **38**:353 (1961).

Robbins, O., Jr.: *Ionic Reactions and Equilibria*. The Macmillan Co., New York, 1967 (paperback).

Tackett, S. L.: Potentiometric determination of solubility product constants. *J. Chem. Educ.*, **46**:857 (1969).

Vincent, C. A.: Thermodynamic parameters from an electrochemical cell. *J. Chem. Educ.*, **47**:365 (1970).

Problems

1. How much silver can be plated out in 10 minutes by a steady current of 3.0 amperes?
2. How much copper can be plated out in 10 minutes by a steady current of 3.0 amperes?
3. How much aluminum can be produced by passage of 100 amperes through a Hall-Heroult cell for one hour?
4. What volume of hydrogen, measured at 298°K and total pressure 726 mm Hg over water, is produced by passage of 2.10 amperes for 10 minutes through an aqueous solution of sodium fluoride?
5. Write equations for the anode reaction and the cathode reaction when an aqueous solution of sodium hydroxide is electrolyzed.
6. Suppose that copper that is contaminated with silver and zinc is to be purified electrolytically. Use half-reaction potentials to predict which metal will remain in the anode (or fall to the bottom of the cell as anode sludge) and which will dissolve but not be reduced at the cathode.

7. Use half-reaction potentials to calculate the equilibrium constant at 298°K for the reaction

$$Ag(CN)_2^-(aq) \rightleftharpoons Ag^+(aq) + 2\ CN^-(aq)$$

8. Use half-reaction potentials to obtain the equilibrium constant at 25°C for the reaction

$$Fe^{2+}(aq) + Ag^+(aq) \rightleftharpoons Fe^{3+}(aq) + Ag(c)$$

Using this equilibrium constant and taking all activity coefficients to be unity, calculate the concentrations of $Ag^+(aq)$, $Fe^{2+}(aq)$, and $Fe^{3+}(aq)$ at equilibrium after an excess of metallic silver has been added to a solution that originally contained 0.05 molar $Fe^{3+}(aq)$.

9. Using half-reaction potentials, calculate the equilibrium constants at 25°C for the following reactions:

$$Cu^{2+}(aq) + 2\ Ag(c) \rightleftharpoons Cu(c) + 2\ Ag^+(aq)$$
$$\tfrac{1}{2}Cu(c) + Ag^+(aq) \rightleftharpoons \tfrac{1}{2}Cu^{2+}(aq) + Ag(c)$$

How could you calculate either of these equilibrium constants directly from the other?

10. (a) Calculate the standard potential of a cell (25°C) consisting of a Zn/Zn^{2+} electrode and a Fe/Fe^{2+} electrode. (b) Write a balanced equation for the reaction that occurs when the cell delivers current to an external circuit and calculate the equilibrium constant for this reaction. (c) Sketch a diagram of the cell and label the electrodes as anode and cathode. Indicate the direction of electron flow through the external circuit. (d) What would be the potential of this cell when the concentration of $Zn^{2+}(aq)$ ions is 10 times the concentration of $Fe^{2+}(aq)$ ions?

11. An excess of Cu(c) is added to a solution containing 0.10 M $Ag^+(aq)$ at 25°C. What is the final equilibrium concentration of $Ag^+(aq)$?

12. One electrode of a cell consists of a mixture of mercury and mercurous chloride (Hg_2Cl_2) in contact with aqueous hydrochloric acid. This electrode is commonly called a calomel electrode or mercury/mercurous chloride electrode. The other electrode is a Cl_2/Cl^- electrode that consists of a platinum gauze dipping in hydrochloric acid with Cl_2 being bubbled over the gauze. (a) Use half-reaction potentials for calculation of the standard potential (25°C) of this cell and to decide the spontaneous direction of the cell reaction. (b) Sketch a schematic illustration of this cell with indication of the direction of electron flow through the external circuit when the cell is delivering current.

13. A cell may be constructed by dipping Ag/AgBr and Hg/Hg_2Br_2 electrodes into the same sodium bromide solution. One way to represent this cell is as follows:

$$Ag(c),AgBr(c)\,|\,NaBr(aq)\,|\,Hg_2Br_2(c),Hg(liq)$$

(a) Use half-reaction potentials to calculate the standard potential for this cell at 25°C.
(b) Write an equation to represent the spontaneous cell reaction.
(c) Does the cell potential depend on the concentration of sodium bromide?

14. Using half-reaction potentials, calculate the equilibrium constant (solubility product) at 25°C for the reaction

$$Ag_2CrO_4(c) \rightleftharpoons 2\ Ag^+(aq) + CrO_4^{2-}(aq)$$

15. In Example Problem 16.6 we have used $\mathscr{F} = 23{,}060$ (no units given). Look up the units for this quantity (Chapter 1) and combine with units for ΔG^0, n, and $\mathscr{E}^0$ to show that the calculation of standard potential in Example Problem 16.6 has in fact been done with an appropriate value for $\mathscr{F}$.

16. Use ΔG_f^0 values (Appendix II) to calculate ΔG^0 for the following reactions:
(a) $2\ Ag^+(aq) + H_2(g) \rightleftharpoons 2\ Ag(c) + 2\ H^+(aq)$
(b) $Ag^+(aq) + \frac{1}{2} H_2(g) \rightleftharpoons Ag(c) + H^+(aq)$
It is entirely reasonable that $\Delta G_a^0 = 2\ \Delta G_b^0$ because ΔG^0 is an extensive quantity (like mass or heat capacity) that is directly proportional to the amount of material.

Now consider application of $\Delta G^0 = -n\mathcal{F}\mathcal{E}^0$ to reactions (a) and (b) above. First, explain why $n_a = 2\ n_b$. Second, show that $\mathcal{E}_a^0 = \mathcal{E}_b^0$. This last equality means that the cell potential (an intensive quantity, like density) is not proportional to the amount of material.

If our selection of "sea level" or reference states for free energies includes specification of $\Delta G_f^0 = 0$ for e^-, show that considerations above are justification for our earlier calculation of $\mathcal{E}^0$ for reaction (16.55) in Example Problem 16.8.

17. Equation (16.56) presents one approach to the relationship between cell potential, temperature, and thermodynamic quantities. Here we present another approach—readers should fill in missing steps in the following derivation.

Differentiation of $\Delta G^0 = -n\mathcal{F}\mathcal{E}^0$ with respect to temperature gives $d\Delta G^0/dT = -n\mathcal{F}(d\mathcal{E}^0/dT)$. Combination of this equation with $d\Delta G^0/dT = -\Delta S^0$ (look up the source of this equation) gives

$$-\Delta S^0 = -n\mathcal{F}(d\mathcal{E}^0/dT) \tag{16.57}$$

Now substitute $-n\mathcal{F}\ \mathcal{E}^0$ for ΔG^0 and $-n\mathcal{F}\ (d\mathcal{E}^0/dT)$ for $-\Delta S^0$ in $\Delta G^0 = \Delta H^0 - T\,\Delta S^0$ to obtain

$$-n\mathcal{F}\mathcal{E}^0 = \Delta H^0 - Tn\mathcal{F}(d\mathcal{E}^0/dT)$$

This last equation can be rearranged to

$$\Delta H^0 = n\mathcal{F}\left(T\frac{d\mathcal{E}^0}{dT} - \mathcal{E}^0\right) \tag{16.58}$$

Equation (16.58) is commonly called the Gibbs-Helmholtz equation.
(a) Integrate equation (16.57) as an indefinite integral and compare the result with equation (16.56).
(b) Use entropy data from Appendix II to calculate ΔS^0 for the reaction for the cell discussed in Example Problem 16.6. Then, using equation (16.57), predict the sign of $(d\mathcal{E}^0/dT)$ and state whether the standard potential of this cell increases or decreases with increasing temperature.
(c) Using a numerical value for $(d\mathcal{E}^0/dT)$ obtained from equation (16.57) with $\mathcal{E}^0 = 0.2224$ volt at 298°K from Example Problem 16.6, calculate ΔH^0 for the cell reaction by way of equation (16.58).
(d) Compare the ΔH^0 calculated above in (c) with values calculated in each of the following ways:
(i) Use ΔH_f^0 values from Appendix II, and (ii) combine ΔG^0 already calculated in Example Problem 16.6 with ΔS^0 obtained in (b) above.
(e) Some student measurements led to $(d\mathcal{E}^0/dT) = -0.000645$ volt deg^{-1} for this cell. How does this value compare with the value of $(d\mathcal{E}^0/dT)$ obtained from thermodynamic data?

18. The Nernst equation for the cell discussed in Example Problem 16.6 and illustrated in Figure 16.8 is

$$\mathcal{E} = \mathcal{E}^0 - 0.05916 \log [H^+][Cl^-]$$

when $P_{H_2} = 1.00$ atm. This equation is easily rearranged to

$$-\log[H^+] = pH = \frac{\mathscr{E} - \mathscr{E}^0}{0.05916} + \log[Cl^-]$$

Suppose that the electrolyte in this cell is $5.00 \times 10^{-3}\ m$ NaCl, $5.00 \times 10^{-3}\ m$ HZ, and $5.00 \times 10^{-3}\ m$ NaZ. The measured potential is $\mathscr{E} = 0.6084$ volt.

(a) Calculate the pH and $[H^+]$ for this solution.

(b) Calculate the equilibrium constant for ionization of weak acid HZ as in

$$HZ(aq) \rightleftharpoons H^+(aq) + Z^-(aq)$$

19. Explain why corrosion of an iron pipe is accelerated if it is attached to a copper wire. Strips of "active" metals such as magnesium or zinc are frequently attached to underground iron (or steel) pipes to retard corrosion. Justify the term "sacrificial anode" for the active metals and explain how they protect the iron pipes. You may wish to refer to potentials in Appendix IV.

20. Compare the potentials listed in Appendix IV for iron and aluminum. Which metal should be most susceptible to oxidation by air? Explain why aluminum pans ordinarily "rust" less than do iron pans.

21. One alternative to lead storage batteries is the Edison cell, which involves reaction between iron and nickel(III) in alkaline solution. The composition of the compound that contains nickel in the +3 oxidation state is uncertain, but is commonly represented as Ni_2O_3. On this basis, the spontaneous electrode reactions may be represented by

$$Ni_2O_3(c) + 3\,H_2O(liq) + 2\,e^- \rightleftharpoons 2\,Ni(OH)_2(c) + 2\,OH^-(aq)$$

and

$$Fe(c) + 2\,OH^-(aq) \rightleftharpoons Fe(OH)_2(c) + 2\,e^-$$

Identify the cathode and anode reactions and write the balanced cell reaction equation. Explain why there is no change in output voltage as the battery is discharged, which is a distinct advantage over the lead storage battery.

22. A significant contribution to pollution of natural water systems by mercury has come from cells used in electrolytic production of chlorine from brine (NaCl) solutions. Mercury is used as an electrode material in these cells.

(a) Write the cathode, anode, and total cell reaction equations for electrolysis of aqueous sodium chloride solution.

(b) If we regard Cl_2 as the principal product, what is the principal by-product of the electrolysis?

(c) When electrolysis has proceeded for a while and the concentration of NaCl has become too low, the cells are flushed out and a new brine solution is added. How does this account for some mercury pollution?

(d) Look up a few of the chemical properties of mercury and suggest ways in which it might be possible to eliminate or at least greatly diminish the pollution from mercury cells. Might your "cure" be worse than the original problem? What kind of electrode material might be substituted for mercury in these cells?

23. One method of cleaning partially oxidized copper coins at the site of archaeological "digs" is to attach the coin to the negative terminal of a battery with a copper wire and suspend the coin in a 2.5% NaOH solution. A graphite electrode is attached to the positive terminal of the battery. What reaction occurs at the surface of the coin to clean it?

17

HALOGENS

Introduction

The elements of group VII (fluorine, chlorine, bromine, iodine, and astatine) are called halogens, meaning salt producers. With an outer electronic configuration of ns^2np^5, a halogen atom can achieve a noble gas configuration by gaining one electron to form a uninegative halide ion or by forming a single covalent bond. With the exception of astatine, the halogens are found extensively in nature in the form of halide salts.

Our knowledge of astatine is rather meager, largely due to the rapid radioactive decay of all its isotopes. About 20 isotopes of astatine are now known, the longest lived being ^{210}At with a half-life of only 8.3 hours. Because of the short half-lives, macroscopic quantities of astatine cannot be accumulated. Evidence from tracer studies, however, indicates that astatine is a "well-behaved" halogen.

A recent and exciting branch of halide chemistry is that of noble gas compounds. Before 1962 no true compounds of the noble gases were known, and it was generally accepted that none could exist. After Neil Bartlett's initial report in 1962, the field of noble gas chemistry developed very rapidly. Most of the noble gas compounds are fluorides, oxides, or oxyfluorides, although some xenon compounds containing chlorine or other very electronegative groups have also been prepared.

Group Properties

Fluorine, chlorine, bromine, and iodine normally exist as diatomic molecules. At high temperature dissociation takes place, as indicated by

$$X_2(g) \rightleftharpoons 2\,X(g) \tag{17.1}$$

where X_2 and X represent any halogen molecule and the corresponding atom. From the results of spectroscopic and PVT measurements, equilibrium constants for reactions represented by (17.1) can be calculated. Values of ΔH^0 of dissociation are obtained from the slopes of the lines in graphs of $\log K$ against $1/T$. This enthalpy is sometimes called a dissociation energy or bond energy. Some properties of the halogens are summarized in Table 17.1.

The dissociation enthalpy of fluorine is unexpectedly low compared to those of the other halogens, which can be explained by considering repulsions between nonbonding electrons in the small fluorine atoms. Similar repulsions in the heavier halogens are less effective due to larger size and the availability of low energy *d* orbitals; these low energy orbitals permit a spreading of electron density, either by hybridization of orbitals or partial multiple bonding. Support for this general argument comes from the fact that single bonds of other first row elements, for example O—O in peroxides or N—N in hydrazines, are also relatively weak.

Although the bonds between halogen atoms in X_2 molecules are fairly strong, attractions between neighboring X_2 molecules are quite weak. These van der Waals intermolecular attractive forces increase with increasing numbers of electrons per molecule. As a result, the melting points and boiling points of the halogens increase in going from F_2 to I_2. In addition, the outer part of the electron cloud is less tightly held in the heavier atoms, as indicated by decreasing ionization energies. The resultant increased polarizability also contributes to increasing van der Waals intermolecular forces.

Electron affinities of halogen atoms are ΔH^0 values for processes of the type

$$X(g) + e^- \rightleftharpoons X^-(g)$$

and were discussed in Chapter 7.

Table 17.1. Some Properties of the Halogens

	Fluorine	*Chlorine*	*Bromine*	*Iodine*
Melting point (°C)	−223	−102	−7	114
Boiling point (°C)	−188	−35	59	184
Ionic radius, X^- (Å)	1.36	1.81	1.95	2.16
Covalent radius (Å)	0.64	0.99	1.14	1.33
First ionization enthalpy (kcal mole^{-1})	403	302	275	243
Electron affinity (kcal mole^{-1})	82.4	84.8	79.1	72.1
Dissociation enthalpy (kcal mole^{-1})	37.8	58.2	46.1	36.1
Color of gas	pale yellow	yellow-green	red-brown	violet

When light strikes a substance, some is absorbed, some is transmitted, and some is reflected. It is the color of transmitted light that corresponds to the characteristic color of a substance. The colors of the halogens arise from absorption of visible light that excites one or more electrons to higher energy states. Energies necessary for excitation of electrons in halogen molecules (electronic absorption energies) roughly parallel energies necessary for removal of electrons from corresponding halogen atoms (ionization energies). Less energy is required to ionize an iodine atom than any other halogen atom, and less energy is required to excite an electron to a higher energy state in an iodine molecule than in any other halogen molecule. In general as the size of an atom or molecule increases the energy states are closer together. Iodine absorbs relatively low energy (low frequency, long wavelength) yellow-green light and appears violet. Fluorine absorbs higher energy (higher frequency, shorter wavelength) violet light and appears yellow.

Due to their high reactivities, halogens do not occur in the elemental state in nature. The free halogens are ordinarily prepared by oxidation of the corresponding halides, electrolytically in the case of fluorine and chemically or electrolytically in the case of the others. Iodine is also produced by reduction of compounds containing iodine in a positive oxidation state, for example sodium iodate, $NaIO_3$.

The free halogens are commonly used oxidizing agents in the laboratory because they are readily reduced to the X^- state. Fluorine oxidizes water so it is not much used in aqueous solutions, but it is an excellent oxidizing agent for gas phase and some nonaqueous solvent work. The standard reduction potentials for the halogens are given in Table 17.2.

The reduction potentials show that fluorine is the strongest and iodine the weakest oxidizing agent of the halogens. Conversely, fluoride ions are most difficult to oxidize and are the weakest reducing agents of the halide ions. Thus fluorine is able to oxidize aqueous chloride, bromide, and iodide ions. Chlorine can oxidize aqueous bromide and iodide ions but cannot oxidize fluoride ions. Bromine can oxidize iodide ions but cannot oxidize fluoride or chloride ions.

The trends in reduction potentials of the halogens can be understood in terms of thermodynamic cycles such as those used in Chapter 15. The ΔH^0 values in Table 17.3 (bottom line) show that the F_2/F^- half-reaction is the most exothermic and the I_2/I^- half-reaction is the least exothermic. Remembering that $\Delta G^0 = -n\mathcal{F}\mathcal{E}^0$ and that ΔG^0 may be approximated by ΔH^0, we anticipate that the thermochemical data in Table 17.3 will be useful in providing some understanding of the trend in reduction potentials of the halogens.

All of the ΔH^0 values listed in Table 17.3 have been calculated from ΔH_f^0 values in Appendix II. These ΔH_f^0 values have themselves been calculated from results

Table 17.2. Standard Reduction Potentials for the Halogens (25°)

$\frac{1}{2}F_2(g) + e^- \rightleftharpoons F^-(aq)$	$\mathcal{E}^0 = 2.87$ volt
$\frac{1}{2}Cl_2(g) + e^- \rightleftharpoons Cl^-(aq)$	$\mathcal{E}^0 = 1.36$ volt
$\frac{1}{2}Br_2(liq) + e^- \rightleftharpoons Br^-(aq)$	$\mathcal{E}^0 = 1.08$ volt
$\frac{1}{2}I_2(c) + e^- \rightleftharpoons I^-(aq)$	$\mathcal{E}^0 = 0.54$ volt

Table 17.3. ΔH^0 Data for $X_2 + 2\,e^- \rightleftharpoons 2\,X^-(aq)$

(All values are in kcal mole^{-1} at 25°C.)

	F	*Cl*	*Br*	*I*
$X_2(c) \rightleftharpoons X_2(g)$	—	—	—	+14.9
$X_2(liq) \rightleftharpoons X_2(g)$	—	—	+7.4	—
$X_2(g) \rightleftharpoons 2\,X(g)$	+37.8	+58.2	+46.0	+36.2
$2\,X(g) + 2\,e^- \rightleftharpoons 2\,X^-(g)$	−167.2	−175.8	−165.3	−145.1
$2\,X^-(g) \rightleftharpoons 2\,X^-(aq)$	−29.6	+37.7	+53.7	+67.6
$X_2(\text{std state}) + 2\,e^- \rightleftharpoons 2\,X^-(aq)$	−159.0	−79.9	−58.2	−26.4

of various measurements in combination with the usual choices of "sea level" for enthalpies: $\Delta H_f^0 = 0$ for all pure elements in their standard states and also for $H^+(aq)$.

As pointed out in Chapter 15, calculations based on these ΔH_f^0 values lead to the correct total ΔH_{hyd}^0 for any pair of oppositely charged ions (such as Na^+ and Cl^-) and to correct differences between ΔH_{hyd}^0 values for like charged ions (such as Cl^- and Br^- of present interest). We therefore know that the ΔH_{hyd}^0 values for $2\,X^-(g) \rightleftharpoons 2\,X^-(aq)$ in Table 17.3 correctly indicate *relative* enthalpies of hydration of these ions, but all values are displaced from the unknown true enthalpies by some constant amount. We shall therefore proceed to discuss and make use of these values, with emphasis on comparisons of similar ions rather than on the value for any single ion. (See problem 9 at the end of this chapter for some calculations with "absolute" hydration enthalpies.)

The ΔH^0 values in Table 17.3 show that it is largely the trend in enthalpies of hydration that accounts for the trend in ΔH^0 values (and thence in ΔG^0 and $\mathscr{E}^0$ values) for the X_2/X^- half-reactions. The observed trend in ΔH_{hyd}^0 values is what we should expect on the basis of ion size considerations. Attractive interactions between the ion charge and surrounding water dipoles are larger for small ions (high charge density) than for large ions (low charge density). We therefore expect to find that ΔH_{hyd}^0 values become increasingly exothermic in going from large I^- to small F^-, or increasingly endothermic in going from small F^- to large I^-. The ΔH_{hyd}^0 values in Table 17.3 are in accord with our expectations.

Halides

Metals with low ionization energies generally form ionic halides with high melting points, high solubility in water, and high conductivities in aqueous solution or in the fused state. The degree of ionic character in MX compounds is roughly proportional to the electronegativity difference between M and X. We therefore expect MF compounds to be more ionic than corresponding MI compounds.

Melting points of the largely ionic halides of the alkali metals and the alkaline earth metals are high, mostly in the range 500–800°C, whereas the melting points of the covalent halogen compounds of carbon, silicon, nitrogen, and phosphorus are mostly well below room temperature.

For most of the alkali and alkaline earth halides, solubilities in water increase in going from the fluorides to the iodides. Because of the low solubilities of fluoride salts, very little fluoride is found in ground water or sea water. In contrast, chlorine, bromine, and iodine are all produced industrially from halides in sea water or underground brines.

Most of the nonmetal halides are volatile covalent compounds that react vigorously with water. Consider the following molecular reactions:

$$BF_3 + 3\,H_2O \rightarrow 3\,HF + B(OH)_3 \tag{17.2}$$

$$SiCl_4 + 4\,H_2O \rightarrow 4\,HCl + Si(OH)_4 \tag{17.3}$$

$$PBr_3 + 3\,H_2O \rightarrow 3\,HBr + P(OH)_3 \tag{17.4}$$

In each case the hydrolysis products of the covalent halide are hydrogen halide and the hydroxy compound of the nonmetal. In (17.2) the boron containing product is boric acid, more commonly written as H_3BO_3. The silicon product in (17.3) is better written as hydrated silica, $SiO_2 \cdot 2H_2O$. The phosphorus product in (17.4) is phosphorous acid, H_3PO_3. The guiding principle for remembering the products of such reactions is the polarity of bonds as indicated by

$$\begin{array}{ccc} \delta+ & & \delta- \\ M & / & X \\ HO & / & H \\ \delta- & & \delta+ \end{array} \rightarrow MOH + HX$$

This same principle is useful for predicting the products of many other reactions.

The transition metal halides show a wide gradation of properties between ionic and covalent substances. Some of these compounds will be considered in more detail in Chapter 20.

Because of the great oxidizing strength of fluorine, many nonmetals exhibit their highest oxidation state in combination with fluorine. Some examples are sulfur hexafluoride, SF_6, and iodine heptafluoride, IF_7. In addition, the small size of the fluoride ion permits relatively large numbers of them to be attached to a central atom in a complex ion. Some examples of complex fluoride ions in which the central atom has a high coordination number (meaning number of nearest neighbors) are AlF_6^{3-}, PF_6^-, and TaF_8^{3-}.

Hydrogen Halides

Hydrogen halides dissolve in water to give acidic solutions. But in contrast to the other hydrogen halides, aqueous solutions of hydrofluoric acid are weakly acidic. For the equilibrium

$$HF(aq) \rightleftharpoons H^+(aq) + F^-(aq)$$

$K = 7 \times 10^{-4}$ at 25°C. The weakness of HF(aq) as an acid may be attributed to the great strength of the H—F bond (see Table 11.1). Anhydrous hydrofluoric acid is, however, a very strong acid.

Aqueous hydrogen chloride is so strong an acid that we have no quantitative knowledge of the magnitude of K for the equilibrium

$$HCl(aq) \rightleftharpoons H^+(aq) + Cl^-(aq) \tag{17.5}$$

Two simple calculations illustrate the difficulty in determining ionization constants for strong acids. Consider the equilibrium constant expression for ionization of 0.1000 *M* HCl(aq).

$$K = \frac{[H^+][Cl^-]}{[HCl]} = \frac{[H^+][H^+]}{0.1000 - [H^+]}$$

If the acid is 99.9% dissociated, $[H^+] = 0.0999$ *M*, and $K \cong 1000$. If the acid is 99.0% dissociated, $[H^+] = 0.0990$, and $K \cong 10$. Calculations such as these show that measurements of hydrogen ion concentration cannot lead to accurate evaluation of the exceedingly small concentrations of un-ionized acid present in dilute solutions of acids with K greater than 1. For some moderately strong acids, spectroscopic measurements can be used to determine the concentration of un-ionized acid. By combining these values with determinations of hydrogen ion concentrations, reasonably accurate values for K can be obtained.

Because hydrogen chloride, hydrogen bromide, and hydrogen iodide are all so nearly completely ionized in aqueous solution, we cannot determine which is the stronger acid. This tendency of a solvent to make it impossible to differentiate between the strengths of several acids is called the **leveling effect** of that solvent. It is possible, however, to study the extent of ionization of the HX acids in other solvents. One approach is to use a solvent with a dielectric constant lower than that of water, which results in increased attractive forces between oppositely charged ions and therefore decreases the extent of ionization.

Another approach is based on consideration of the equilibrium

$$HCl + S \rightleftharpoons HS^+ + Cl^- \qquad (17.6)$$

in which S represents a molecule of solvent that accepts a proton from HCl. When the solvent is water, equilibrium (17.6) is so far to the right that accurate evaluation of the equilibrium constant is impossible. But by using a solvent that is more reluctant than water to accept a proton, it is possible to make meaningful equilibrium measurements. Remembering the Bronsted-Lowry definition of a base as a proton acceptor, we see that the desired solvent should be less basic (more acidic) than water.

Pure acetic acid (called glacial acetic acid), which has a lower dielectric constant and is more acidic than water, permits us to distinguish between the acid strengths of the hydrogen halides. Measurements on solutions of HCl, HBr, and HI in acetic acid show that equilibrium constants are in the order $K_{HCl} < K_{HBr} < K_{HI}$ for the equilibria

$$HX + HAc \rightleftharpoons H_2Ac^+ + X^-$$

where HX and X^- represent hydrogen halide and the corresponding halide ion and HAc represents acetic acid.

Solubilities of Metal Halides

Silver halides are among the few metal halides that have low solubility in water, the other common ones being those of mercury(I) and lead. Aqueous ammonia readily dissolves AgCl, forming the complex ion $Ag(NH_3)_2^+(aq)$. Silver bromide

Table 17.4. Solubility Data for Silver Halides (25°)

Silver Halide	K_{sp}	$\Delta G^0_{17.8}$ (*kcal mole*$^{-1}$)	$\Delta H^0_{17.8}$ (*kcal mole*$^{-1}$)
AgCl	1.8×10^{-10}	+13.3	+15.7
AgBr	5.0×10^{-13}	+16.7	+20.2
AgI	8.7×10^{-17}	+21.9	+26.8

is less soluble in aqueous ammonia than is AgCl, and AgI is still less soluble. Treatment of AgCl with Br^-(aq) yields AgBr as shown by equation (17.7).

$$AgCl(c) + Br^-(aq) \rightleftharpoons AgBr(c) + Cl^-(aq) \qquad K = 350 \quad (17.7)$$

Addition of I^-(aq) to AgCl(c) or AgBr(c) yields AgI(c). These observations from competition experiments show that the solubilities and solubility products are in the order AgCl > AgBr > AgI.

Quantitative values for solubility products of the silver halides have been obtained from solubility data and also from potentials for the AgX/Ag half-reactions. Enthalpies of solution have been calculated from $(d \ln K)/dT = \Delta H^0/RT^2$ and have also been evaluated from calorimetrically determined enthalpies of precipitation. Solubility products of AgCl, AgBr, and AgI are given in Table 17.4, along with values of ΔG^0 and ΔH^0 for the solution reaction represented by

$$AgX(c) \rightleftharpoons Ag^+(aq) + X^-(aq) \qquad (17.8)$$

The K_{sp} data in Table 17.4 are in accord with the results of competition experiments as expected. The free energies of solution, calculated from $\Delta G^0 = -RT \ln K_{sp}$, are naturally in the same order as the solubilities and solubility products. Values of $\Delta H^0{}_{17.8}$ are also in the same order and help account for the observed trends.

Lattice enthalpies and enthalpies of hydration calculated from enthalpies of formation are given in Table 17.5. Using data from the appendices, readers should carry out appropriate calculations for an alkali halide (say NaCl) to compare with the data given in Table 17.5.

The data in Table 17.5 show that lattice enthalpies and enthalpies of hydration vary oppositely in going from AgCl to AgI. Since chloride ions have a smaller radius, they have a higher charge density than iodide ions. We therefore observe, as expected, that more energy is required to convert AgCl(c) into gaseous ions than is required for AgI(c). This effect favors a low solubility for AgCl as compared

Table 17.5. Thermochemical Data for Silver Halides

	ΔH^0 (*kcal mole*$^{-1}$)		
Reaction	*AgCl*	*AgBr*	*AgI*
$AgX(c) \rightleftharpoons Ag^+(g) + X^-(g)$	+215.2	+211.7	+211.4
$Ag^+(g) \rightleftharpoons Ag^+(aq)$	−218.4	−218.4	−218.4
$X^-(g) \rightleftharpoons X^-(aq)$	+18.9	+26.9	+33.8
$AgX(c) \rightleftharpoons Ag^+(aq) + X^-(aq)$	+15.7	+20.2	+26.8

to AgBr and AgI. On the other hand, hydration of chloride ions is more exothermic (see problem 9 at the end of this chapter) than hydration of larger bromide and iodide ions, which favors greater solubility for AgCl than for AgBr and AgI. But the difference in lattice enthalpies is only 4 kcal $mole^{-1}$, as compared to a difference of 15 kcal $mole^{-1}$ in enthalpies of hydration of chloride and iodide ions. Thus the trend in enthalpies of hydration determines the trend in enthalpies of solution and solubilities of these silver halides.

The relatively small difference between lattice enthalpies of AgCl and AgI can be explained in terms of some covalent bonding in these silver halides. The decreasing electrostatic attraction in going from AgCl to AgI is partly compensated by increasing attraction due to electron sharing, which tends to make the difference in lattice enthalpies small. The increasing tendency to take part in electron sharing in going from chloride ion to iodide ion is consistent with the increasing polarizabilities of the ions in going from chloride to iodide.

Comparison of lattice enthalpies of silver halides with potassium halides as in Table 17.6 helps illustrate the importance of covalent bonding in silver halides. The silver ion is only slightly smaller than the potassium ion. If only ionic forces were involved, we would expect the lattice enthalpies of the silver halides to be only slightly greater than those of the corresponding potassium halides. Actually, the lattice enthalpies of the silver halides are about 55 kcal $mole^{-1}$ larger than those of the corresponding potassium halides, which suggests that there are additional forces binding the ions together in the silver halides.

It might also be noted that the silver halides have lower melting points than the corresponding potassium halides, which is again consistent with the idea that there is some covalent bonding in the silver halides.

On the basis of the preceeding discussions of silver halides, we predict that AgF should be more soluble than AgCl. Although the prediction is correct, it is not possible to make detailed calculations as in Table 17.5 because the dihydrate $AgF \cdot 2H_2O(c)$ rather than AgF(c) is the solid phase in equilibrium with a saturated aqueous solution of silver fluoride. All of the information needed for carrying out these calculations for solid hydrates is not available.

Although AgCl, AgBr, and AgI are only very slightly soluble in water, they are more soluble in water that contains various solutes such as $NH_3(aq)$ or $CN^-(aq)$. This enhanced solubility is due to formation of complex ions such as

Table 17.6. Comparison of Potassium and Silver Halides

Halide	*Ionic radius of M^+ (Å)*	*Lattice enthalpy (kcal $mole^{-1}$)*	*Melting point (°C)*
KCl	1.33	169	776
KBr	1.33	161	730
KI	1.33	155	686
AgCl	1.26	215	455
AgBr	1.26	212	432
AgI	1.26	211	558*

*At 146°C the low temperature form α-AgI undergoes a solid state transition to β-AgI, which melts at 558°C.

$Ag(NH_3)_2^+(aq)$ and $Ag(CN)_2^-(aq)$. In the following Example Problem we show quantitatively with equilibrium constant caculations how the presence of $NH_3(aq)$ increases the solubility of AgBr.

Example Problem 17.1. How much silver bromide will dissolve in one liter of 0.50 *M* aqueous ammonia at 25°C?

For the reaction

$$Ag^+(aq) + 2\,NH_3(aq) \rightleftharpoons Ag(NH_3)_2^+(aq) \qquad (17.9)$$

we have

$$K = 1.7 \times 10^7 = \frac{[Ag(NH_3)_2^+]}{[Ag^+][NH_3]^2} \qquad (17.10)$$

and from Table 17.4 we have

$$K_{sp} = 5.0 \times 10^{-13} = [Ag^+][Br^-] \qquad (17.11)$$

for the reaction

$$AgBr(c) \rightleftharpoons Ag^+(aq) + Br^-(aq) \qquad (17.12)$$

As discussed in Chapter 14, a general method of solving equilibrium constant problems is to set up a group of simultaneous equations equal in number to the number of chemical unknowns. Then the mathematics frequently can be simplified by making approximations suggested by chemical knowledge.

In this problem the unknowns are the concentrations of Ag^+, $Ag(NH_3)_2^+$, Br^-, and NH_3. Two of the four equations needed are (17.10) and (17.11) above. Accounting for all the ammonia in the solution gives

$$0.50 = 2\,[Ag(NH_3)_2^+] + [NH_3] \qquad (17.13)$$

The concentration of silver-ammonia complex is multiplied by two because each complex ion contains two molecules of ammonia. Expressing the electroneutrality of the solution gives

$$[Ag(NH_3)_2^+] + [Ag^+] = [Br^-] \qquad (17.14)$$

as the last equation needed. The four equations (17.10), (17.11), (17.13), and (17.14) can now be solved by purely mathematical methods for the four unknowns, and the sum of $[Ag(NH_3)_2^+]$ plus $[Ag^+]$ is then the desired solubility. It is much easier, however, to make an approximation suggested by the knowledge that silver bromide is slightly soluble in water.

The silver-ammonia complex is quite stable [as shown by the large value of K for reaction (17.9)] and silver bromide is only slightly soluble in water. We therefore expect that most of the silver in the saturated solution should be in the form of $Ag(NH_3)_2^+$ complex ions. We then *assume* that $[Ag(NH_3)_2^+]$ is enough larger than $[Ag^+]$ that equation (17.14) can be simplified to

$$[Ag(NH_3)_2^+] \cong [Br^-] \qquad (17.15)$$

Next we rearrange (17.11) to

$$[Ag^+] = (5.0 \times 10^{-13})/[Br^-]$$

and substitute into equation (17.10) to obtain

$$1.7 \times 10^{7} = \frac{[Ag(NH_3)_2^+][Br^-]}{(5.0 \times 10^{-13})[NH_3]^2}$$

and then

$$8.5 \times 10^{-6} = \frac{[Ag(NH_3)_2^+][Br^-]}{[NH_3]^2} \tag{17.16}$$

We let S represent the concentration of $Ag(NH_3)_2^+$, and see by (17.15) that $[Br^-] = S$. The concentration of ammonia is given by $0.5 - 2S$; that is, the final concentration of ammonia is the initial concentration minus the concentration of ammonia molecules held in the form of silver-ammonia complex ions. Substituting in (17.16) gives

$$8.5 \times 10^{-6} = \frac{(S)(S)}{(0.50 - 2S)^2} \tag{17.17}$$

Taking the square root of both sides of the equation (17.17) leads to

$$2.9 \times 10^{-3} = \frac{S}{0.5 - 2S}$$

which is solved for $S = 0.0014$ mole of $Ag(NH_3)_2^+$ per liter.

It is now necessary to check the assumption leading to the approximate equation (17.15). According to this equation and the value of S given above, $[Br^-] = 0.0014\ M$. Substitution of this value in (17.11) leads to $[Ag^+] = 3.6 \times 10^{-10}\ M$. This concentration of Ag^+ ions is indeed negligible compared to the concentration of $Ag(NH_3)_2^+$ ions. The approximation leading to equation (17.15) is thereby justified.

Readers may find it instructive to evaluate all four unknowns and show that the values found are consistent with the four equations (17.10), (17.11), (17.13), and (17.14).

We might also have solved this problem by beginning with the equation for the *principal* reaction

$$AgBr(c) + 2\ NH_3(aq) \rightleftharpoons Ag(NH_3)_2^+(aq) + Br^-(aq) \tag{17.18}$$

The equilibrium constant expression for this reaction is

$$K = 8.5 \times 10^{-6} = \frac{[Ag(NH_3)_2^+][Br^-]}{[NH_3]^2} \tag{17.19}$$

where the value for $K_{17.19}$ has been calculated as the product of $K_{17.10} \times K_{17.11}$. It may be seen in either of the two following ways that $K_{17.19}$ can be obtained as the above product. (a) The reaction (17.18) is the sum of reactions (17.9) and (17.12), so ΔG^0 for (17.18) is the sum of the ΔG^0 values for reactions (17.9) and (17.12). Since $\Delta G^0 = -RT \ln K$, adding free energies corresponds to multiplying equilibrium constants. (b) The equilibrium constant expression given in (17.19) for reaction (17.18) is obtained as the product of the equilibrium constant expressions (17.10) and (17.11). Hence the value of $K_{17.19}$ must be given by $K_{17.10} \times K_{17.11}$.

Because equations (17.19) and (17.16) are identical, we may solve (17.19) for the concentration of $Ag(NH_3)_2^+$ just as we solved (17.16).

After solving (17.19), it is necessary to show that equation (17.18) really does represent the principal reaction relevant to this problem. Using $[Br^-] = 0.0014\ M$, we calculate from (17.11) that $[Ag^+] = 3.6 \times 10^{-10}\ M$ and see that the extent of reaction (17.12) is truly negligible compared to reaction (17.18). ■

Halogen-Oxygen Compounds and Ions

Since fluorine is the most electronegative element, it is questionable whether it should be assigned positive oxidation states. Compounds of oxygen with fluorine are called fluorides whereas those of the remaining halogens are properly called halogen oxides. Conventional practice places the most electronegative element last in descriptive names of compounds.

When fluorine is passed through dilute aqueous sodium hydroxide at room temperature, a reactive compound called oxygen difluoride is formed:

$$2\ F_2(g) + 2\ OH^-(aq) \rightarrow 2\ F^-(aq) + OF_2(g) + H_2O(liq) \qquad (17.20)$$

With metals and nonmetals OF_2 is both a strong oxidizing and fluorinating agent.

In 1971 E. Appelman reported isolation of the first oxyacid of fluorine. By passing gaseous F_2 over water at 0°C for short contact times, small amounts of hypofluorous acid, HOF, were obtained. Hypofluorous acid is unstable and a very strong oxidizing agent. Although fluorine can be assigned a +1 oxidation state in HOF, the validity of this assignment is debatable.

Passing O_2 and F_2 through an electric discharge at low temperatures and pressures produces compounds such as O_2F_2, O_3F_2, and O_4F_2. All of these oxygen fluorides are very reactive and have been considered as oxidizers for rocket fuels.

Some important compounds and ions containing chlorine, bromine, or iodine in combination with oxygen are listed and named in Table 17.7. Electron dot formulas for these species frequently prove unsatisfactory. It is apparent that for an "odd electron" molecule such as ClO_2 the octet rule must be violated. A less obvious problem is presented with the perchlorate ion, for which an electron dot

Table 17.7. Some Halogen-Oxygen Compounds and Ions

Oxidation state of halogen	*Formula*	*Name*
+1	Cl_2O	Chlorine monoxide
	HOCl	Hypochlorous acid
	ClO^-	Hypochlorite ion
	HOBr	Hypobromous acid
	BrO^-	Hypobromite ion
+3	$HClO_2$	Chlorous acid
	ClO_2^-	Chlorite ion
+4	ClO_2	Chlorine dioxide
+5	ClO_3^-	Chlorate ion
	BrO_3^-	Bromate ion
	I_2O_5	Iodine pentoxide
	HIO_3	Iodic acid
	IO_3^-	Iodate ion
+6	Cl_2O_6	Chlorine hexoxide
+7	Cl_2O_7	Chlorine heptoxide
	$HClO_4$	Perchloric acid
	ClO_4^-	Perchlorate ion
	BrO_4^-	Perbromate ion
	HIO_4	Metaperiodic acid
	H_5IO_6	Paraperiodic acid

structure with single bonds can easily be drawn. The experimentally observed Cl—O bond distances in ClO_4^- are 1.44Å, considerably shorter than the value of 1.66 Å calculated from the ionic radii for Cl^{7+} and O^{2-}. Apparently there is a considerable amount of double bond character in these Cl—O bonds.

Chlorine monoxide (Cl_2O) is prepared by the reaction

$$2\ Cl_2(g) + HgO(c) \rightleftharpoons Cl_2O(g) + HgCl_2(c)$$

in which chlorine is both oxidized to the +1 state and reduced to the −1 state. Gaseous Cl_2O is unstable and explodes rather easily on heating to yield Cl_2 and O_2. Dissolving Cl_2O in water yields aqueous hypochlorous acid (HOCl), as shown by

$$Cl_2O(g) + H_2O(liq) \rightleftharpoons 2\ HOCl(aq)$$

Small concentrations of HOCl(aq) can also be obtained by dissolving chlorine in water:

$$Cl_2(g) + H_2O(liq) \rightleftharpoons HOCl(aq) + H^+(aq) + Cl^-(aq) \qquad (17.21)$$

Since hypochlorous acid is a weak acid with ionization constant $K = 3 \times 10^{-8}$ at 25°, we have written undissociated HOCl(aq) rather than $H^+(aq)$ plus $ClO^-(aq)$. Despite the electronegativity rule concerning order of listing elements in compounds, hypochlorous acid is usually written HOCl to emphasize that oxygen and not chlorine is the central element.

Because the equilibrium constant for reaction (17.21) is small ($K = 2.8 \times 10^{-5}$ at 25°), not much chlorine undergoes self oxidation-reduction (disproportionation) in neutral or acidic solution. However, from either LeChatelier's principle or equilibrium constant calculations, we can see that the equilibrium in reaction (17.21) can be shifted towards the right by decreasing the concentration of $H^+(aq)$. Chlorine readily disproportionates in aqueous hydroxide solutions as shown by

$$Cl_2(aq) + 2\ OH^-(aq) \rightleftharpoons ClO^-(aq) + Cl^-(aq) + H_2O(liq) \qquad (17.22)$$

Equation (17.22) is written in terms of hypochlorite ion (ClO^-) because in alkaline solution HOCl is neutralized, making $ClO^-(aq)$ the principal species present.

As indicated by the half-reaction and its potential,

$$ClO^-(aq) + H_2O(liq) + 2\ e^- \rightleftharpoons Cl^-(aq) + 2\ OH^-(aq) \qquad \mathscr{E}^0 = +0.89 \text{ volt}$$

alkaline solutions of hypochlorite are strong oxidizing agents. Such solutions are commonly used as laundry bleaches and household disinfectants.

Hypochlorite ions are unstable in aqueous alkaline solution with respect to disproportionation to the +5 and −1 oxidation states. The reaction equation is

$$3\ ClO^-(aq) \rightleftharpoons ClO_3^-(aq) + 2\ Cl^-(aq) \qquad (17.23)$$

Although the equilibrium constant for reaction (17.23) is large, the reaction is slow unless the solution is heated. We refer to cold alkaline hypochlorite solutions as being thermodynamically unstable but kinetically stable. Combining equations (17.22) and (17.23) yields

$$3\ Cl_2(aq) + 6\ OH^-(aq) \rightleftharpoons ClO_3^-(aq) + 5\ Cl^-(aq) + 3\ H_2O(liq) \quad (17.24)$$

as the net equation for the reaction that occurs when chlorine is dissolved in hot aqueous base.

Half-reaction potential diagrams as developed by Latimer offer a convenient and useful way of concisely representing a lot of chemical information. We illustrate this approach for chlorine in alkaline solution, with emphasis on using a potential diagram to decide which species are stable or unstable with respect to disproportionation (self oxidation-reduction).

We begin by writing down all of the aqueous species ranging from the highest oxidation state on the left (in this case, ClO_4^-) to the lowest oxidation state on the right (in this case, Cl^-). Then we look up in Appendix IV the potentials relating these species and fill in the diagram as follows:

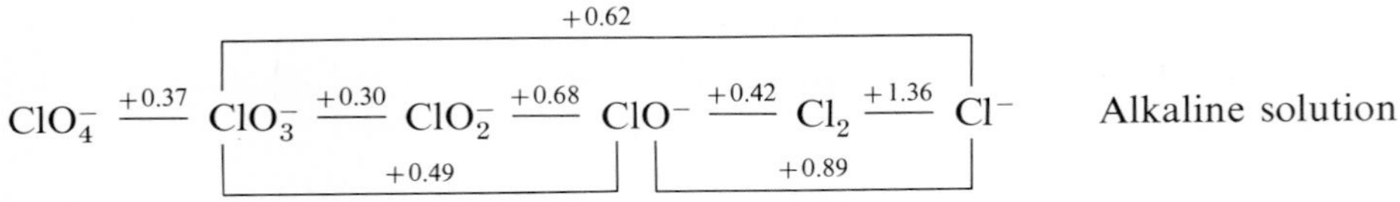

$$\overbrace{}^{+0.62\ (ClO_3^- \to Cl^-)}$$

$$ClO_4^- \xrightarrow{+0.37} ClO_3^- \xrightarrow{+0.30} ClO_2^- \xrightarrow{+0.68} ClO^- \xrightarrow{+0.42} Cl_2 \xrightarrow{+1.36} Cl^- \qquad \text{Alkaline solution}$$

$$\underbrace{}_{+0.49\ (ClO_3^- \to ClO^-)} \qquad \underbrace{}_{+0.89\ (ClO^- \to Cl^-)}$$

We know from $\Delta G^0 = -n\mathscr{F}\mathscr{E}^0 = -RT \ln K$ that a positive $\mathscr{E}^0$ for a reaction corresponds to $K > 1$ and therefore to a reaction that will be thermodynamically spontaneous under most conditions of interest. To illustrate application of the diagram above, we consider some specific reactions.

First, let us deduce from the potentials whether Cl_2 spontaneously disproportionates to Cl^- and ClO^-. The half-reactions and their potentials are given below, where they are also combined to give the overall reaction and corresponding potential:

$$\begin{array}{lr}
\tfrac{1}{2}Cl_2(g) + e^- \rightleftharpoons Cl^-(aq) & \mathscr{E}^0 = 1.36 \text{ volt} \\
\tfrac{1}{2}Cl_2(g) + 2\,OH^-(aq) \rightleftharpoons ClO^-(aq) + H_2O(liq) + e^- & \mathscr{E}^0 = -0.42 \text{ volt} \\
\hline
Cl_2(g) + 2\,OH^-(aq) \rightleftharpoons Cl^-(aq) + ClO^-(aq) + H_2O(liq) & \mathscr{E}^0 = 0.94 \text{ volt}
\end{array}$$

Now we see from the reaction potential $\mathscr{E}^0 = 0.94$ volt that the equilibrium constant for this reaction is very large [see equations (16.32) and (16.33)] and conclude that Cl_2 will spontaneously disproportionate to $Cl^-(aq)$ and $ClO^-(aq)$ in alkaline solution.

On the basis of the calculation above and other similar calculations for other possible disproportionation reactions, we can state the following important generalization:

If the potential to the right of a particular species in a potential diagram is algebraically larger than the potential to the left, the species is thermodynamically unstable with respect to disproportionation.

The generalization above leads us to state the $ClO^-(aq)$ is unstable with respect to disproportionation to $ClO_3^-(aq)$ and $Cl^-(aq)$ in alkaline solution, but is stable with respect to $ClO_2^-(aq)$ and $Cl_2(aq)$.

We can conclude from the potential diagram that Cl_2 bubbled into aqueous base leads to $ClO_4^-(aq)$ and $Cl^-(aq)$ *at equilibrium*. Although the disproportionation of $ClO_3^-(aq)$ is thermodynamically favorable, it is kinetically slow so that the principal products of reaction are $ClO_3^-(aq)$ and $Cl^-(aq)$.

The chemical properties of HOBr and BrO^- are similar to those of HOCl and

ClO^-. Solutions of hypobromite disproportionate to bromide and bromate much more readily than do those of hypochlorite, however. Disproportionation is moderately fast even at room temperature, and solutions of hypobromite can only be kept at around 0°C. Hypoiodous acid and hypoiodite ion, HOI and IO^-, are so unstable with respect to disproportionation that they are of little chemical importance.

The only definitely known halous acid and halites are those of chlorine. The equilibrium constant for the ionization

$$HClO_2(aq) \rightleftharpoons H^+(aq) + ClO_2^-(aq)$$

is 1×10^{-2} at 25°C. Chlorous acid exists only in solution and attempts to isolate it lead to rapid decomposition.

Chlorine dioxide (ClO_2) is a strong and rapid oxidizing agent that is much used for oxidizing (or bleaching) flour, paper, and textiles. Because of its explosiveness, chlorine dioxide is prepared as needed rather than shipped. Some preparative methods for chlorine dioxide are indicated by the following equations:

$$2\ NaClO_2(c) + Cl_2(g) \rightarrow 2\ NaCl(c) + 2\ ClO_2(g) \qquad (17.25)$$
$$2\ ClO_3^-(aq) + SO_2(g) + H^+(aq) \rightarrow 2\ ClO_2(g) + HSO_4^-(aq) \qquad (17.26)$$
$$2\ ClO_3^-(aq) + H_2C_2O_4(aq) + 2\ H^+(aq) \rightarrow 2\ ClO_2(g) + 2\ CO_2(g) + 2\ H_2O(liq) \qquad (17.27)$$

Equation (17.27) is written in terms of undissociated $H_2C_2O_4$ because oxalic acid is a weak acid. Chlorine dioxide disproportionates in alkaline solution to form chlorite and chlorate ions, as shown by

$$2\ ClO_2(g) + 2\ OH^-(aq) \rightleftharpoons ClO_2^-(aq) + ClO_3^-(aq) + H_2O(liq) \qquad (17.28)$$

Halates may be obtained by reaction of a halogen with hot aqueous base, as shown for chlorine in equation (17.24). In addition, bromates and iodates are frequently prepared by reaction of bromine or iodine with powerful oxidizing agents such as hydrogen peroxide or concentrated nitric acid. They are also prepared industrially by electrolytic oxidation.

Probably the most commercially important halate salt is potassium chlorate, which is used as an oxidizing agent in matches, fireworks, and some explosives. Both bromate, $BrO_3^-(aq)$, and iodate, $IO_3^-(aq)$, are strong and rapid oxidizing agents in aqueous solution. Many useful procedures in quantitative analysis depend on the use of bromate or iodate as oxidizing agents. The pure salts potassium bromate, $KBrO_3$, and potassium iodate, KIO_3, are moderately soluble in water and easily obtained.

Of the halic acids only iodic acid, HIO_3, is known in the free state. Iodic acid is isolated as a white solid, which when heated is dehydrated to form iodine pentoxide, I_2O_5. The equilibrium constant for

$$HIO_3(aq) \rightleftharpoons H^+(aq) + IO_3^-(aq)$$

is 0.16 at 25°C, making $HIO_3(aq)$ a slightly stronger acid than $HSO_4^-(aq)$.

Chloric and bromic acids are best obtained in solution by treating the corresponding barium halates with sulfuric acid. Both acids appear to be completely dissociated in dilute solutions, but neither can be isolated as a pure compound.

Efforts to isolate $HClO_3$ from concentrated solutions containing $ClO_3^-(aq)$ and $H^+(aq)$ ions have been unsuccessful because of violent explosions.

Until recently the only perhalates known were those of chlorine and iodine. There were even several theoretical papers suggesting reasons why perbromates could not be obtained. It was not until 1968 that E. Appelman first prepared a perbromate by a hot-atom process involving β decay of radioactive ^{83}Se incorporated into a selenate:

$$^{83}SeO_4^{2-} \rightarrow {}^{83}BrO_4^- + \beta^-$$

Since that time more conventional chemical routes to perbromates have been devised. The most convenient synthesis now involves oxidation of bromate in alkaline solution by molecular fluorine. In general, the properties of the perbromates are intermediate between those of perchlorates and periodates, as might be expected.

As mentioned previously, disproportionation of ClO_3^- to ClO_4^- and Cl^- is thermodynamically favorable but occurs too slowly in solution to be a useful synthetic procedure. When solid potassium chlorate is carefully heated in the absence of catalysts, it disproportionates according to the equation

$$4\,KClO_3(c) \rightarrow 3\,KClO_4(c) + KCl(c) \qquad (17.29)$$

In the presence of a catalyst, such as MnO_2, molecular oxygen is evolved (see Chapter 10). Perchlorates are most commonly prepared by electrolytic oxidation of aqueous chlorate solutions. Only a few chemical oxidizing agents are capable of oxidizing ClO_3^- to ClO_4^-; one of these is peroxydisulfate:

$$S_2O_8^{2-}(aq) + ClO_3^-(aq) + H_2O \rightarrow 2\,SO_4^{2-}(aq) + ClO_4^-(aq) + 2\,H^+(aq) \qquad (17.30)$$

Perchlorates are powerful oxidizing agents. Both lithium perchlorate and ammonium perchlorate have been used as components of rocket fuels. Aqueous perchlorate ion is a powerful oxidizing agent in acidic solution, but it generally reacts very slowly with reducing agents unless the solution is hot. Another important property of the perchlorate ion is its slight tendency to form ion pairs or serve as a ligand in complex ions. Thus perchlorates are widely used in studies of transition metal complexes (Chapter 20), the assumption being made that no correction for the concentration of perchlorate complexes need be made.

Perchloric acid ($HClO_4$) is the only oxy acid of chorine that can be prepared in the free state. Pure colorless liquid perchloric acid may be obtained by distilling a mixture of a perchlorate salt and concentrated sulfuric acid. Perchloric acid is both a strong oxidizing agent and a strong acid. Contact with organic substances such as wood, paper, or rubber leads to violent explosions. Several crystalline hydrates of perchloric acid are known of which the monohydrate is particularly interesting. In $HClO_4 \cdot H_2O$ the lattice positions in the crystal are occupied by H_3O^+ and ClO_4^- ions, providing direct experimental evidence for the existence of the hydronium ion.

Oxidation of iodates electrolytically or with chlorine in alkaline solution yields species containing iodine in the +7 oxidation state. From such solutions solid paraperiodic acid, H_5IO_6, may be obtained. At 100°C this form loses water and is converted into metaperiodic acid, HIO_4. The equilibrium constant for

$$H_5IO_6(aq) \rightleftharpoons H^+(aq) + H_4IO_6^-(aq)$$

is about 5×10^{-4} at 25°C, showing that paraperiodic acid is a much weaker acid than the other perhalic acids. The $H_4IO_6^-(aq)$ ion is in equilibrium with $H_3IO_6^{2-}(aq)$ and also with metaperiodate ion $IO_4^-(aq)$ as shown by

$$H_4IO_6^-(aq) \rightleftharpoons H^+(aq) + H_3IO_6^{2-}(aq) \qquad K = 10^{-7}$$

and

$$H_4IO_6^-(aq) \rightleftharpoons IO_4^-(aq) + 2\,H_2O(liq) \qquad K = 40$$

The periodic acids and periodates are powerful oxidizing agents that usually react smoothly and rapidly. They are particularly useful in quantitative analysis.

Much of the aqueous chemistry of the halogen-oxygen species discussed in the preceding pages is concisely summarized by the reduction potentials listed in Appendix IV. The use of these potentials is a great aid in systematizing and understanding chemical reactions. Many of the problems at the end of the chapter are designed to develop skill in applying this powerful tool.

The concept of electron drift (inductive effect) can be helpful in remembering relative acid strengths of the oxyhalogen acids. For example, consider the hypohalous acids. Because chlorine is more electronegative than iodine or bromine, there should be a greater drift of electron density away from the O—H bond in HOCl than in HOI or HOBr. The net result will be to make the proton more available and HOCl the strongest acid in the HOX series. We can use similar reasoning for the oxychlorine acids, noting that oxygen is a very electronegative element. Increasing the number of oxygens attached to the central chlorine atom should also lead to a weakening of the O—H bond. Thus, consistent with experiment, the predicted order of acid strength is $HClO_4 > HClO_3 > HClO_2 > HOCl$.

The shapes of the oxyhalogen compounds and ions are readily predicted by VSEPR theory. Hypohalite ions are of course linear. Halite ions are AX_2E_2 cases and are therefore bent. Halate ions are trigonal pyramidal (AX_3E), and perhalate ions are tetrahedral (AX_4). In the paraperiodates six oxygen atoms are bound octahedrally to a central iodine atom, an AX_6 case.

Interhalogen Compounds

Since all the halogens readily form diatomic molecules, it is not surprising that many interhalogen compounds can be obtained. Most of these compounds are prepared by direct combination of the elements.

There are several interhalogen compounds of the XY type, with all combinations except IF known. The physical properties of XY compounds are generally intermediate between those of the parent halogens, X_2 and Y_2. As an example, the boiling point of chlorine monofluoride, ClF, is −100°C, while that of Cl_2 is −35°C and that of F_2 is −188°C. The XY compounds are somewhat more reactive than the parent halogens. They behave as oxidizing agents and attack most other elements to yield mixtures of the halides.

In addition to the XY type of interhalogen compounds, there are also XY_3, XY_5, and XY_7 compounds. With the exception of iodine trichloride, all of these higher interhalogen compounds are fluorides. There are three examples known of both the XY_3 type (ClF_3, BrF_3, and ICl_3) and XY_5 type (ClF_5, BrF_5, and IF_5). To date the only XY_7 type reported is iodine heptafluoride, IF_7.

All of the XY_3, XY_5, and XY_7 compounds are of interest as examples of failure of the octet rule. In the cases where the structures are known, they are consistent with the predictions of VSEPR theory. Some of the higher interhalogens, particularly BrF_3, have been used extensively as solvents for reactions involving reactive fluorides. Some of them have also been investigated as components of high energy fuels.

Noble Gas Compounds

Most of the group 0 elements (helium, neon, argon, krypton, xenon, and radon) were discovered in the late nineteenth century. Because of their general lack of reactivity they were commonly termed "inert gases," and bonding theories of the early twentieth century centered around atoms attaining noble gas configurations. The absence of noble gas compounds eventually led to a "closed shell–closed mind" attitude. Even after discovery of compounds such as IF_7 and SF_6, which clearly violate the octet concept, there was no rush to prepare noble gas compounds. Although science is an evolutionary process, individual scientists often resist change of well established concepts.

Before 1962 no true compounds of the noble gases were known. Several **clathrates,** in which the noble gas atom is trapped in cage-like holes in crystalline lattices, were reported for the heavier elements. Some diatomic ions, such as HeH^+, were also obtained under high energy conditions in discharge tubes. These, however, were considered exceptions and not compounds in the true sense of the word. In 1933 Linus Pauling predicted the existence of hexafluorides of xenon and krypton, but experimental attempts by others to prepare them were unsuccessful.

The first step toward achieving noble gas compounds was the discovery of the remarkable oxidizing power of platinum hexafluoride, PtF_6, by Neil Bartlett and D. H. Lohmann in 1962. Molecular oxygen reacts with PtF_6 to produce a red solid, which was shown by x-ray diffraction studies to consist of O_2^+ and PtF_6^- ions:

$$O_2(g) + PtF_6(g) \rightarrow O_2^+PtF_6^-(c) \qquad (17.31)$$

Noting that the ionization energy of Xe (12.1 eV) is very close to that of O_2 (12.2 eV), Bartlett correctly surmised that PtF_6 should also be capable of oxidizing xenon. The room temperature reaction between dark red PtF_6 vapor and colorless xenon to produce a yellow crystalline solid was both visually dramatic and chemically significant:

$$Xe(g) + PtF_6(g) \rightarrow Xe^+PtF_6^-(c) \qquad (17.32)$$

Xenon hexafluoroplatinate (V) was the first real compound of a noble gas element.

Bartlett's discovery aroused much interest and activity in the chemical world, particularly at Argonne National Laboratory near Chicago where PtF_6 was first prepared. Workers there (including H. H. Claassen, H. Selig, and J. G. Malm) soon reported several new compounds of xenon with other heavy metal fluorides. Various evidence suggested that the heavy metal fluorides acted as fluorinating agents as well as electron acceptors.

The first binary noble gas compound obtained by Claassen, Selig, and Malm was xenon tetrafluoride. Heating a 5:1 mixture of fluorine and xenon at 400°C in a nickel can, followed by rapid cooling, produced a colorless solid, XeF_4:

$$Xe(g) + 2\ F_2(g) \rightarrow XeF_4(c) \tag{17.33}$$

By varying the reactant ratio and experimental conditions, XeF_2 and XeF_6 were also obtained.

Chemists all over the world were soon contributing to the rapidly expanding literature of noble gas compounds. Compounds in which xenon is bonded to fluorine, chlorine, or oxygen are now known, with the fluorides being the most stable. Several oxyfluorides, such as XeO_2F_2 and $XeOF_4$, have been isolated. Less stable fluorides of radon and krypton are also known. Considering the relative ease of obtaining noble gas compounds and their fairly "normal" properties, it may seem surprising that these compounds were not isolated earlier. This illustrates two points: we rarely find what we do not seek, and chemistry is still an experimental science in which surprises are possible.

Hydrolysis of xenon fluorides produces several possible reactions, for example:

$$XeF_6 + H_2O \rightarrow XeOF_4 + 2\ HF$$
$$XeF_6 + 2\ H_2O \rightarrow XeO_2F_2 + 4\ HF$$
$$XeF_6 + 3\ H_2O \rightarrow XeO_3 + 6\ HF$$

The reactions may be accomplished in the vapor phase or by freezing the reactants with liquid nitrogen followed by slow warming, but separation of products is difficult. Xenon oxide tetrafluoride is a relatively stable colorless liquid. Xenon dioxide difluoride forms colorless crystals at room temperature but is thermally unstable, as indicated by its positive enthalpy of formation. Solid xenon trioxide is a dangerous explosive (comparable to TNT), consistent with its ΔH_f^0 of $+96$ kcal mole^{-1}. Since XeO_3 is a common end product of xenon chemistry, investigations often prove exciting in more than a scientific sense.

The aqueous chemistry of xenon has also proved interesting, although the actual species are not well characterized. In aqueous solution, XeF_6 yields various products depending on the pH of the solution. At low pH, Xe(VI) species sometimes referred to as "xenic acid" predominate. Xenic acid is probably best considered as hydrated xenon trioxide. At high pH, Xe(VI) disproportionates to Xe(VIII) species and xenon gas. The strong oxidizing powers of the aqueous xenon species are indicated by the following approximate reduction potentials:

Acidic solution: $H_4XeO_6 \xrightarrow{\sim 2.3} XeO_3 \xrightarrow{\sim 1.8} Xe$

Basic solution: $HXeO_6^{3-} \xrightarrow{\sim 0.9} HXeO_4^- \xrightarrow{\sim 0.9} Xe$

In dilute acid solution perxenic acid (H_4XeO_6) rapidly oxidizes manganous ion to permanganate ion.

The shapes of noble gas compounds are consistent with predictions of VSEPR theory. Although the structure of XeF_6 has not been unambigously determined,

Table 17.8. Properties of Some Xenon Compounds

Compound	*Structure*	*Melting point (°C)*	ΔH_f^0 *(kcal mole^{-1})*	*Average bond enthalpy (kcal mole^{-1})*
XeF_2	Linear	129	-39	31
XeF_4	Square planar	117	-66	31
XeF_6	Distorted octahedron (?)	49	-96	30
XeO_3	Trigonal pyramid	Decomposes	$+96$	~ 15
$XeOF_4$	Square pyramid	~ -41	$\sim +35$	—

a distorted octahedral structure is predicted by VSEPR. The shapes and physical properties of some simple xenon compounds are listed in Table 17.8.

It is important to realize that no new or exotic bonding schemes are necessary to explain formation of noble gas compounds. Bond strengths and bond lengths are consistent with extrapolations from analogous chemical species such as interhalogen ions. Although valence bond and molecular orbital theories generally lead to identical answers, the molecular orbital approach for noble gas compounds is more popular. The use of relatively high energy *d* orbitals is avoided in molecular orbital theory.

The molecular orbital description for XeF_2 is illustrated in Figure 17.1. From the $5p_z$ orbitals of xenon and the $2p_z$ orbitals of fluorine, three molecular orbitals can be formed. The bonding and nonbonding molecular orbitals are occupied, but the antibonding orbital is vacant. The result is a three-center, four-electron bond for XeF_2. This bond is analogous to the bridge bonds in B_2H_6 (Chapter 10), but is an "electron-excess" rather than "electron-deficient" bond. This electron population of orbitals is consistent with the low xenon-fluorine bond energy.

The above bonding approach can be extended to XeF_4 and XeF_6 by using the $5p_x$ and $5p_y$ orbitals of xenon to form additional molecular orbitals. Since these molecular orbitals are mutually perpendicular, square planar and octahedral structures are predicted for XeF_4 and XeF_6, respectively. Because valence bond theory (and VSEPR) predict a distorted octahedron for XeF_6, there is a conflict. Although the exact structure of XeF_6 is uncertain, most evidence favors a distorted octahedron. It is probable that successively combining the *p* orbitals is too simplistic an approach.

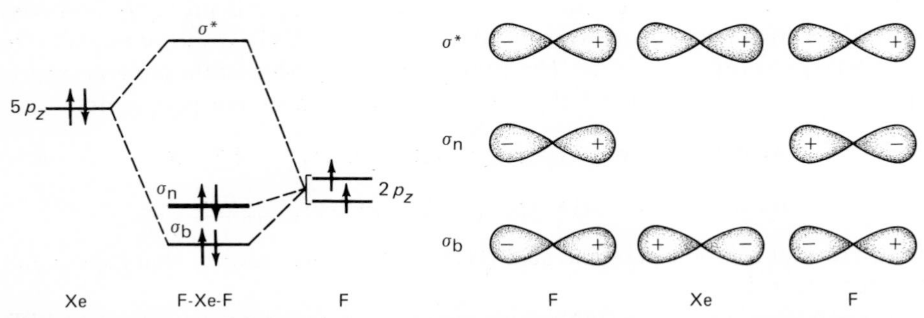

Figure 17.1. Molecular orbital description for XeF_2. The *z* axis is taken as the molecular axis. The remaining Xe ($5s$, $5p_x$, $5p_y$) and F ($2s$, $2p_x$, $2p_y$) valence orbitals would be occupied nonbonding orbitals in XeF_2.

The eventual solution to understanding bonding in noble gas compounds may involve an approach quite different from either present valence bond or molecular orbital theory. The discovery of noble gas compounds may well provide the impetus for improved bonding theories, just as previous failures to react noble gases led to earlier theories.

References

Bartlett, N.: The chemistry of the noble gases. *Endeavour,* **23**:3 (1964).
Bartlett, N.: Noble-gas compounds. *Endeavor,* **31**:107 (Sept. 1972).
Herrell, A. Y., and K. H. Gayer: The elusive perbromates. *J. Chem. Educ.,* **49**:583 (1972).
Hyman, H. H.: The chemistry of noble gas compounds. *Science,* **145**:773 (1964).
Jolly, W. L.: *The Chemistry of the Non-Metals.* Prentice-Hall, Inc., Englewood Cliffs, N.J., 1966 (paperback).
Jolly, W. L.: The use of oxidation potentials in inorganic chemistry. *J. Chem. Educ.,* **43**:198 (1966).
Pauling, L.: Why is HF a weak acid. *J. Chem. Educ.,* **33**:16 (1956).
Sanderson, R. T.: Principles of halogen chemistry. *J. Chem. Educ.,* **41**:361 (1964).
Teddy, J. J.: Salt—a pillar of the chemical industry. *J. Chem. Educ.,* **47**:386 (1970).
Turner, J. J.: Oxygen fluorides. *Endeavour,* **27**:42 (1968).
Selig, H., J. G. Malm, and H. H. Claassen: The chemistry of the noble gases, *Sci. Amer.,* **210**:66 (May, 1964).

Problems

1. Predict the order of boiling points for the carbon tetrahalides: CF_4, CCl_4, CBr_4, CI_4. Suggest an explanation for the trend.

2. An important source of iodine has been sodium iodate from Chile. Iodine is obtained by reduction of the iodate with sodium bisulfite, as shown by

$$2\,IO_3^-(aq) + 5\,HSO_3^-(aq) \rightarrow I_2(c) + 5\,SO_4^{2-}(aq) + 3\,H^+(aq) + H_2O(liq)$$

Sodium bisulfite for this reaction is formed by oxidation of sulfur to sulfur dioxide, which is then dissolved in sodium hydroxide. How much iodine can be produced by use of 32 g of sulfur?

3. Use reduction potentials in Appendix IV to aid in selecting three reagents that are capable of oxidizing $Cl^-(aq)$ to Cl_2 in acidic solutions.

4. Use reduction potentials in Appendix IV to predict the extent of reactions of the type

$$2\,Fe^{3+}(aq) + 2\,X^-(aq) \rightleftharpoons 2\,Fe^{2+}(aq) + X_2$$

where X^- and X_2 represent the halide ions and corresponding halogens.

5. The concentration of $F^-(aq)$ in sea water is very low. Many bodies of water contain appreciable concentrations of $Ca^{2+}(aq)$ due to the weathering of limestone ($CaCO_3$). Explain how these two statements are interrelated.

6. The bond lengths in chlorine dioxide, ClO_2, are consistent with chlorine-oxygen double bonds. Draw an electron dot structure for ClO_2, predict the approximate O—Cl—O bond angle, and state whether the compound is diamagnetic or paramagnetic. Because the energies of 3*d* orbitals are closer to those of the valence orbitals in second row elements, it is more reasonable to assume that chlorine "expands" its octet than oxygen.

7. How much fluorine can be obtained by electrolysis of molten potassium bifluoride, KHF_2, for 10 hr with a constant current of 10 amp?

8. Most of the world production of bromine is from sea water. The first step in the process involves chlorination of sea water after adjusting the pH to between 1 and 4:

$$Cl_2(g) + 2\,Br^-(aq) \rightleftharpoons Br_2(liq) + 2\,Cl^-(aq)$$

Using reduction potentials, calculate the standard free energy change for the above reaction.

9. As pointed out earlier in this chapter and in Chapter 15, enthalpies of hydration have been calculated on the basis of arbitrary selection of $\Delta H_f^0 = 0$ for $H^+(aq)$ and thence of $\Delta H^0 = -367$ kcal mole^{-1} for hydration of H^+. We may summarize by writing

$$H^+(g) \rightleftharpoons H^+(aq) \qquad \Delta H^0_{conv} = -367 \text{ kcal mole}^{-1}$$

in which the subscript emphasizes that this is based on a conventional choice of ΔH_f^0 for $H^+(aq)$. Various theoretical investigations suggest quite convincingly that the enthalpy of hydration of H^+ is really about -269 kcal mole^{-1}, which we indicate by

$$H^+(g) \rightleftharpoons H^+(aq) \qquad \Delta H^0_{theo} \cong -269 \text{ kcal mole}^{-1}$$

Now we see that ΔH^0_{hyd} values for all M^+ ions should be about 98 kcal mole^{-1} less exothermic than calculated from conventional ΔH_f^0 values. Thus we obtain the following (in kcal mole^{-1}):

$$Na^+(g) \rightleftharpoons Na^+(aq) \qquad \Delta H^0_{conv} = -203 \quad \Delta H^0_{theo} \cong -105$$
$$Ag^+(g) \rightleftharpoons Ag^+(aq) \qquad \Delta H^0_{conv} = -218 \quad \Delta H^0_{theo} \cong -120$$

(a) Use ΔH_f^0 values (Appendix II) to calculate ΔH^0 for the reaction

$$Na^+(g) + Cl^-(g) \rightleftharpoons Na^+(aq) + Cl^-(aq) \qquad (17.34)$$

The ΔH^0 value so obtained is free of any arbitrary or conventional factor based on $\Delta H_f^0 = 0$ for $H^+(aq)$ because of compensating effects on ΔH_f^0 values for oppositely charged $Na^+(aq)$ and $Cl^-(aq)$.

(b) Calculate ΔH^0_{conv} for

$$Cl^-(g) \rightleftharpoons Cl^-(aq)$$

(Note that your value should be half that listed in Table 17.3 for twice this reaction.) Now combine your ΔH^0_{conv} for Cl^- with that above for Na^+ to obtain ΔH^0 for reaction (17.34) to confirm the last sentence of part (a).

(c) It should be apparent that ΔH^0_{theo} for hydration of Cl^- is about 98 kcal mole^{-1} more exothermic than ΔH^0_{conv}. Obtain this value and combine it with ΔH^0_{theo} above for Na^+ to obtain ΔH^0 for reaction (17.34).

(d) We expect that ΔH^0_{hyd} values for all X^- halide ions should be exothermic, but the (conventional) values given in Table 17.3 are endothermic for Cl^-, Br^-, and I^-. Because we were concerned only with relative values (comparing various X^- ions with each other), we did not need to deal with this problem earlier. Now show that there really is no problem on this score by calculating ΔH^0_{theo} values for hydration of Cl^-, Br^-, and I^-.

10. Predict the products and write balanced equations for the following reactions:

(a) $AsCl_3 + H_2O \rightarrow$

(b) $Al_2Br_6 + H_2O \rightarrow$

(c) $BBr_3 + D_2O \rightarrow$

11. Gaseous HF and HCl are usually prepared in the laboratory by heating a metal halide with sulfuric acid.

(a) Write equations for the reactions of $CaCl_2$ and BaF_2 with hot H_2SO_4 solution.

(b) Assuming an 85% yield, how many moles of HCl(g) could be obtained from 1.2 g $CaCl_2$?

12. Hot concentrated H_2SO_4 solutions are powerful oxidizing agents and can not be used for preparation of HBr and HI because Br^- and I^- ions are too easily oxidized. One way to avoid the problem is to react the metal halide with a nonoxidizing acid such as orthophosphoric acid, H_3PO_4. An alternative procedure involves hydrolysis of a covalent halide such as PBr_3. What volume of HBr, measured at 65°C and 0.50 atm, could be obtained by complete hydrolysis of all the PBr_3 that could be made from 100 g of phosphorus?

13. Calculate ΔH^0 values for the following reactions and explain their signs in terms of lattice enthalpies, bond enthalpies, and hydration enthalpies.

$$AgCl(c) + I^-(aq) \rightleftharpoons AgI(c) + Cl^-(aq)$$
$$AgCl(c) + HI(g) \rightleftharpoons AgI(c) + HCl(g)$$

14. How much ammonia must be added to 1.0 liter of water to enable the resulting solution to dissolve 0.0020 mole of AgBr at 25°C?

15. How much AgI will dissolve in 1.0 liter of 1.0 M aqueous ammonia at 25°C?

16. How much AgCl will dissolve in 1.0 liter of 0.20 M aqueous ammonia at 25°C?

17. Suppose that silver nitrate is slowly added to a solution containing 0.00060 M $Cl^-(aq)$ and 0.0010 M $Br^-(aq)$.

(a) What precipitate will appear first?

(b) What will be the concentration of the first halide to precipitate when the second halide begins to precipitate?

18. Using data from Appendix II calculate ΔH^0 for

$$KBr(c) \rightleftharpoons K^+(aq) + Br^-(aq)$$

and compare with the value for AgBr in Table 17.5.

19. Using VSEPR theory, predict the structure of the gaseous HOF molecule. If fluorine were assigned an oxidation state of -1, what would the oxidation state of oxygen be in this molecule?

20. Write electron dot structures and predict the general shapes of the species: ClO_2^-, Cl_2O, BrO_3^-, HIO_4, I_3^-, IO_6^{5-}.

21. Use $\mathscr{E}^0$ values to calculate the equilibrium constant at 25°C for the hydrolysis of Cl_2, equation (17.21).

22. The yield of HOCl(aq) in equation (17.21) can be increased by adding silver oxide, Ag_2O. Suggest a reason for this fact.

23. Given that $K = 2.8 \times 10^{-5}$ for reaction (17.21) and that $K = 2.8 \times 10^{-8}$ for ionization of aqueous hypochlorous acid, calculate K for reaction (17.22).

24. How much potassium chlorate might be obtained by solution of 1.0 mole of chlorine in an excess of hot aqueous potassium hydroxide?

25. What volume of oxygen, measured at 25°C and 740 mm Hg, can be obtained by thermal decomposition of 25 g of $KClO_3$?

26. A common household disinfectant is made by dissolving chlorine in sodium hydroxide solution to form hypochlorite and chloride ions as shown in equation (17.22). If some of this disinfectant spills, would you suggest neutralizing the spill with vinegar (dilute acetic acid)? Why?

27. Using the reduction potential diagram in Appendix IV, predict which bromine species are stable with respect to disproportionation in aqueous acid solution.

28. Although barium salts are more expensive than sodium salts, chloric and bromic acids are obtained by treating the corresponding barium halate with sulfuric acid. Why not use the sodium halate? Write a balanced equation for the preparation of aqueous bromic acid.

29. Predict the relative acid strength of
(a) $HBrO_4$ compared to the other perhalic acids.
(b) HOF compared to the other hypohalous acids.
(c) HAt compared to the other hydrogen halides.

30. Predict the hydrolysis products of iodine monobromide, IBr, and chlorine trifluoride, ClF_3. (*Hint:* Consider the polarity of the bonds.)

31. Using VSEPR theory, predict the general shapes of the following interhalogen species: ICl_2^-, BrF_3, ClF_5, BrF_6^-.

32. Why is calcium fluoride (CaF_2) more soluble in acidic solution than in neutral solution? Calculate the solubility of CaF_2 in water and show how to calculate the solubility of CaF_2 in 0.10 M $H^+(aq)$ by setting up the appropriate equations.

33. Relative to the other halogens, predict for astatine: (a) ionization energy, (b) electronegativity, (c) covalent radius, (d) melting point of At_2, (e) stability of the +7 oxidation state, and (f) acid strength of HOAt.

34. Using S^0 values from Appendix II, calculate ΔS^0 values and then $T\Delta S^0$ values (at 298°K) for the solution reactions of AgCl, AgBr, and AgI. On the basis of these values, was it reasonable to concentrate our interpretation of solubilities of these compounds on their enthalpies of solution as in Table 17.5?
(Note that our ΔH^0 and ΔG^0 values are given in terms of kcal $mole^{-1}$, while entropies are given in terms of cal deg^{-1} $mole^{-1}$.)

35. Given that the ionization enthalpy of $O_2(g)$ is 281 kcal $mole^{-1}$, the calculated lattice enthalpy for $O_2^+PtF_6^-$ is -125 kcal $mole^{-1}$, and that reaction (17.31) is exothermic, calculate a minimum value for the electron affinity of PtF_6. Compare this value with the electron affinity of elemental fluorine.

36. The vapor pressure of solid XeF_4 is 4 mm Hg at 25°C and the enthalpy of vaporization is 15.3 kcal $mole^{-1}$. Estimate the normal boiling point of XeF_4 using the Clausius-Clapeyron equation. Would this be a reliable estimate?

37. Predict the geometry of perxenate ion, XeO_6^{4-}, and xenon tetroxide, XeO_4. Both species have been observed and characterized.

38. Comment on the probability of obtaining compounds of argon or neon.

39. Write balanced equations for each of the following:
(a) formation of $XeCl_4$ from $^{129}ICl_4^-$ by β emission
(b) formation of XeO_2F_2 from XeF_6 and XeO_3
(c) hydrolysis of XeF_2 to yield elemental xenon
(d) reaction of H_2 with $XeOF_4$, one of the products being water
(e) disproportionation of $HXeO_4^-$ in alkaline solution.

40. The melting points of the alkali metal halides, MX, and alkaline earth halides, MX_2, are given in the following table:

	Melting points (°C)			
	F	Cl	Br	I
K	846	776	730	686
Rb	775	715	682	642
Cs	682	646	636	621
Ca	1360	772	730*	740
Sr	1450	873	643	515
Ba	1280	963	847	740

*Decomposes slowly.

On a graph of melting points versus halogen, plot the melting points for each of the metal halides.

(a) Are the trends more regular for alkali halides or alkaline earth halides?

(b) Suggest an explanation for the change in melting point in going from KF to KI or from SrF_2 to SrI_2.

(c) Suggest an explanation for the change in melting point in going from $CaCl_2$ to $BaCl_2$. Is this same explanation valid for the change in melting point from KF to CsF?

(d) Estimate the melting points of the halides of francium.

18

NITROGEN AND PHOSPHORUS

Introduction

Group V elements exhibit a wide range of properties from nonmetallic to metallic in going down the group. The lighter elements, nitrogen and phosphorus, are typical nonmetals and form acidic oxides. The intermediate members, arsenic and antimony, are metalloid in character and form amphoteric oxides. Bismuth is a metal and forms a basic oxide.

The outer electronic configuration of group V elements is ns^2np^3. Although some metal nitrides and phosphides with considerable ionic character are known, most compounds containing these elements are covalent. The most common oxidation states are -3, $+3$, and $+5$, the stability of the -3 state decreasing down the group. Nitrogen and phosphorus are unusual in that they exhibit all integral oxidation states from -3 to $+5$ inclusive.

Nitrogen

Nitrogen (N_2) constitutes about 78 volume per cent of dry air. It is found in compounds to a lesser extent, principally as Chile saltpeter ($NaNO_3$) and in proteins, which average about 15% nitrogen. Nitrogen is obtained commercially by liquefaction and fractional distillation of air. Since N_2 has a lower boiling point (77°K) than O_2 (90°K), it is more volatile and comes off in the first fractions.

Nitrogen obtained from air often contains 2–3% argon (boiling point, 87°K). In 1894 Lord Rayleigh first noted that nitrogen from air was about 0.1% more dense than nitrogen obtained from decomposition of ammonium nitrite. In conjunction with Sir William Ramsay, Rayleigh reacted nitrogen isolated from air with hot magnesium to form magnesium nitride, Mg_3N_2. There remained a gas that was unreactive, unlike any other gas known at the time. The name argon was chosen from the Greek word meaning lazy. Later spectroscopic investigations, mostly by Ramsay, showed that all the noble gases exist in the earth's atmosphere. Argon, with an abundance of 0.94%, is by far the most common.

Very pure nitrogen may be prepared by a variety of methods, most of which involve the oxidation of either ammonia or ammonium ions. The thermal decomposition of ammonium nitrite

$$NH_4NO_2(c) \rightarrow N_2(g) + 2\ H_2O(g) \tag{18.1}$$

is an oxidation-reduction reaction in which nitrogen is oxidized from a -3 oxidation state in NH_4^+ and reduced from a $+3$ oxidation state in NO_2^- to the zero oxidation state in N_2. Passage of ammonia over hot copper oxide is another route to pure nitrogen:

$$2\ NH_3(g) + 3\ CuO(c) \rightarrow N_2(g) + 3\ H_2O(g) + 3\ Cu(c)$$

Very pure nitrogen may be obtained by thermal decomposition of sodium azide (NaN_3) to its elements.

Due to its high bonding energy, nitrogen is relatively unreactive and is commonly used as an inert atmosphere. Incandescent bulbs contain nitrogen to prevent air oxidation of the hot filament, and preparations of oxygen sensitive compounds are frequently done under a nitrogen "blanket." The molecular orbital description (Chapter 8) of nitrogen indicates a triple bond, which is consistent with the large bonding energy.

$$N_2(g) \rightleftharpoons 2\ N(g) \qquad \Delta H^0 = 226 \text{ kcal mole}^{-1} \tag{18.2}$$

Because dissociation into atoms is endothermic, the equilibrium constant and extent of dissociation increase with increasing temperature. The highly positive ΔH^0 for reaction (18.2) makes ΔG^0 positive and K small, even at high temperatures. For example $K_{18.2}$ is $\sim 10^{-15}$ at 2000°C, indicating negligible dissociation of N_2 even at this high temperature.

Nitrogen Fixation

Despite its great stability with respect to dissociation into atoms, nitrogen does undergo some reactions at room temperature. The combination of atmospheric nitrogen into compounds, called **nitrogen fixation,** is of considerable scientific interest and industrial value because of the importance of nitrogen compounds in agriculture and in chemical industries.

Every crop harvested removes essential nutrients from the soil, the primary nutrients being nitrogen, phosphorus, and potassium. To keep land productive, these nutrients must be replenished. Since natural sources such as manure are not sufficiently abundant or concentrated, chemical fertilizers are necessary.

Ammonia, containing 82.4% nitrogen, is the most practical concentrated source of nitrogen and is the basis of the world's nitrogen fertilizer industry. The dollar value of ammonia produced annually exceeds that of any other inorganic chemical.

Nitrogen-fixing bacteria, which live on the roots of leguminous plants such as beans and clover, convert atmospheric nitrogen into nitrogen compounds that the plants convert into proteins. Some blue-green algae also have the ability to fix nitrogen. Unfortunately, the mechanism by which these reactions occur is not well understood.

There has been considerable interest in inorganic nitrogen fixation under mild conditions since 1965 when two Canadians, Allen and Senoff, reported the first transition metal complex containing N_2 as a neutral ligand. Some of the metals that form metal-N_2 bonds under mild conditions are ruthenium(II), osmium(II), and cobalt(I). In some of these compounds the combined nitrogen can be reduced, at least partly, to ammonia.

Lithium is the only metal that combines directly with nitrogen at moderate temperatures; the product is lithium nitride, Li_3N. The alkaline earth metals, boron, and aluminum must be heated to red heat before they form nitrides. The nitrides of alkali metals other than lithium cannot be made by direct combination of the elements.

Another form of nitrogen fixation that is becoming increasingly important involves the internal combustion engine. Under the oxidizing conditions of an automobile engine, nitrogen and oxygen in the air react to form nitric oxide, NO, which is further oxidized to nitrogen dioxide, NO_2. Nitrogen dioxide is one of the prime components of photochemical smog (Los Angeles type) and can combine with water to form nitric acid, a very corrosive material to man and buildings. Although the same series of reactions can be initiated in the atmosphere by lightning bolts, man has less control over thunderstorms than automobiles.

The only industrially important method of nitrogen fixation is production of ammonia by the Haber process, for which the reaction equation is

$$\tfrac{1}{2}\,N_2(g) + \tfrac{3}{2}\,H_2(g) \rightleftharpoons NH_3(g) \qquad \Delta H^0 = -11 \text{ kcal mole}^{-1} \qquad (18.3)$$

According to LeChatelier's principle, the equilibrium formation of ammonia from nitrogen and hydrogen is favored by low temperatures and high pressures. Unfortunately, in the absence of catalysts the reaction is too slow at low temperatures to be useful. At very high temperatures the equilibrium constant is too small to permit a satisfactory yield of ammonia. Fritz Haber, a German chemist, first found that catalysts consisting of iron oxides with traces of other metals made the rate of reaction fast enough at pressures around 300 atm and temperatures around 500°C to make the production of ammonia from the elements economically feasible. Contrast these conditions to those required by nitrogen-fixing bacteria and algae, so-called simple organisms.

The Haber process for ammonia is an excellent example of the compromise that must frequently be made between kinetics and thermodynamics to obtain reasonable yields in reasonable times. A graph showing the *equilibrium* yield of ammonia at different pressures and temperatures is given in Figure 18.1.

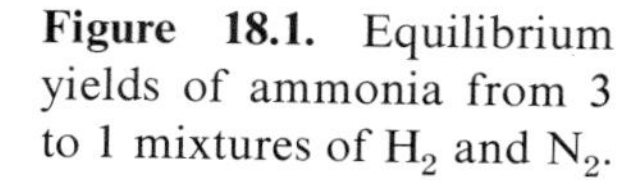

Figure 18.1. Equilibrium yields of ammonia from 3 to 1 mixtures of H_2 and N_2.

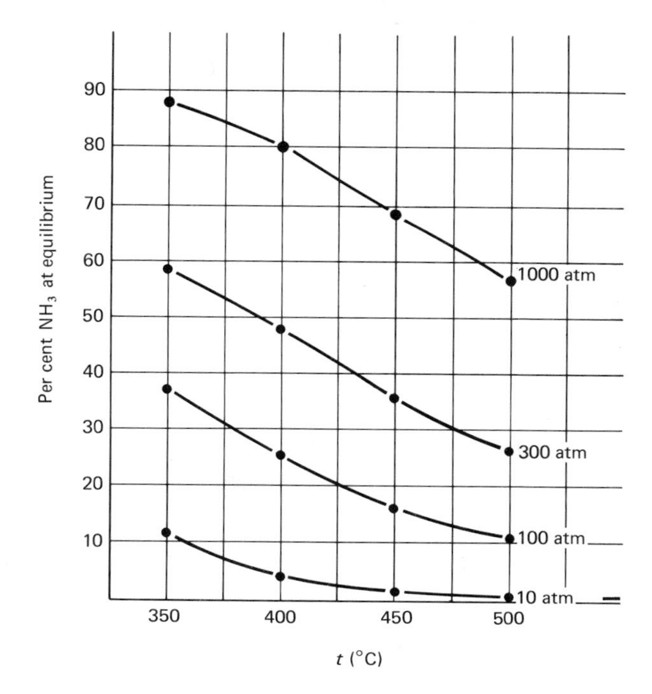

Compounds of Nitrogen

Although not as numerous as those of carbon, the compounds of nitrogen are just as diverse. Their chemical reactions are perhaps more complicated, because it is often more difficult to predict products and establish reaction patterns in nitrogen chemistry. The use of half-reaction potentials is very helpful in this regard. Potentials summarizing some of the aqueous chemistry of nitrogen compounds and ions are given in Table 18.1 and in the diagrams of Figure 18.2.

Some compounds and ions containing nitrogen in formal oxidation states ranging from +5 to −3 are listed and named in Table 18.2. Only a few of these nitrogen compounds and their reactions can be considered here. Our discussion will start with the highest oxidation state and proceed toward the lowest.

Compounds with nitrogen in oxidation states greater than +5 have been reported but are not well characterized. Both the peroxide, NO_3, and peroxynitric acid, HNO_4, are best considered as unstable reaction intermediates.

Nitrate ions (NO_3^-) are easily reduced (positive reduction potentials) in acidic solutions and are thus powerful oxidizing agents. Aqueous nitric acid is a common oxidizing agent in industry and in the laboratory. Solid nitrates have long been used as the oxidizing agent in gunpowder.

Reduction of nitrate in acid solution can lead to species with nitrogen in any oxidation state from +4 to −3. A mixture of products is usually obtained, depending on the reducing agent, concentration of nitric acid, and temperature. For instance, reduction of *dilute* nitric acid with copper gives mostly colorless nitric oxide (NO) as shown by

$$3\ Cu(c) + 8\ H^+(aq) + 2\ NO_3^-(aq) \rightarrow 3\ Cu^{2+}(aq) + 2\ NO(g) + 4\ H_2O(liq) \quad (18.4)$$

Table 18.1. Some Standard Reduction Potentials For Nitrogen Species in Aqueous Solutions at 25°C

Reaction	$\mathcal{E}^0$
Acidic Solution	
$NO_3^-(aq) + 2\,H^+(aq) + e^- \rightleftharpoons NO_2(g) + H_2O(liq)$	$\mathcal{E}^0 = +0.77$ volt
$NO_3^-(aq) + 3\,H^+(aq) + 2\,e^- \rightleftharpoons HNO_2(aq) + H_2O(liq)$	$\mathcal{E}^0 = +0.94$ volt
$NO_3^-(aq) + 4\,H^+(aq) + 3\,e^- \rightleftharpoons NO(g) + 2\,H_2O(liq)$	$\mathcal{E}^0 = +0.96$ volt
$HNO_2(aq) + H^+(aq) + e^- \rightleftharpoons NO(g) + H_2O(liq)$	$\mathcal{E}^0 = +0.99$ volt
$2\,HNO_2(aq) + 4\,H^+(aq) + 4\,e^- \rightleftharpoons N_2O(g) + 3\,H_2O(liq)$	$\mathcal{E}^0 = +1.29$ volt
$NH_3OH^+(aq) + 2\,H^+(aq) + 2\,e^- \rightleftharpoons NH_4^+(aq) + H_2O(liq)$	$\mathcal{E}^0 = +1.35$ volt
$N_2H_5^+(aq) + 3\,H^+(aq) + 2\,e^- \rightleftharpoons 2\,NH_4^+(aq)$	$\mathcal{E}^0 = +1.25$ volt
$HN_3(aq) + 11\,H^+(aq) + 8\,e^- \rightleftharpoons 3\,NH_4^+(aq)$	$\mathcal{E}^0 = +0.7$ volt
$HN_3(aq) + 3\,H^+(aq) + 2\,e^- \rightleftharpoons N_2(g) + NH_4^+(aq)$	$\mathcal{E}^0 = +2.1$ volt
$3\,N_2(g) + 2\,H^+(aq) + 2\,e^- \rightleftharpoons 2\,HN_3(aq)$	$\mathcal{E}^0 = -3.3$ volt
Basic Solution	
$NO_3^-(aq) + H_2O(liq) + 2\,e^- \rightleftharpoons NO_2^-(aq) + 2\,OH^-(aq)$	$\mathcal{E}^0 = +0.02$ volt
$NO_3^-(aq) + 6\,H_2O(liq) + 8\,e^- \rightleftharpoons NH_3(aq) + 9\,OH^-(aq)$	$\mathcal{E}^0 = -0.12$ volt
$2\,NH_2OH(aq) + 2\,e^- \rightleftharpoons N_2H_4(aq) + 2\,OH^-(aq)$	$\mathcal{E}^0 = +0.7$ volt
$N_3^-(aq) + 9\,H_2O(liq) + 8\,e^- \rightleftharpoons 3\,NH_3(aq) + 9\,OH^-(aq)$	$\mathcal{E}^0 = -0.4$ volt
$N_3^-(aq) + 3\,H_2O(liq) + 2\,e^- \rightleftharpoons N_2(g) + NH_3(aq) + 3\,OH^-(aq)$	$\mathcal{E}^0 = -3.0$ volt
$3\,N_2(g) + 2\,e^- \rightleftharpoons 2\,N_3^-(aq)$	$\mathcal{E}^0 = -3.6$ volt

Acidic Solution

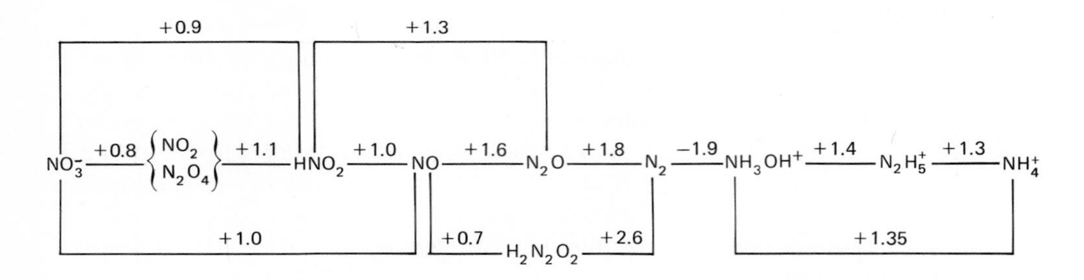

Basic Solution

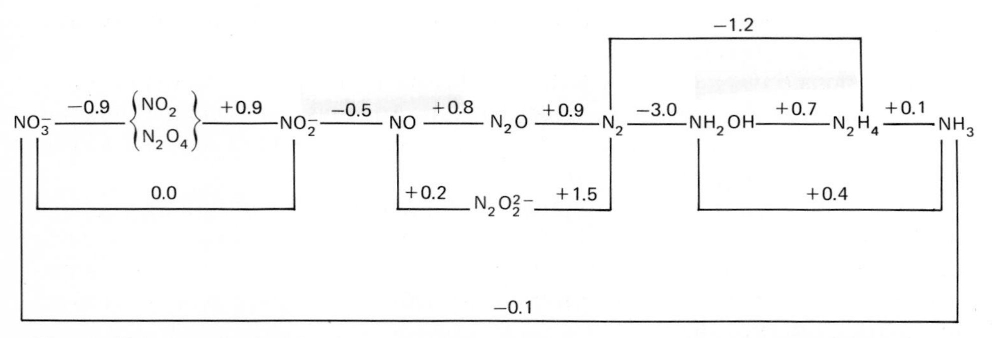

Figure 18.2. Standard reduction potential diagrams for nitrogen in aqueous solution at 25°C.

Table 18.2. Some Compounds and Ions Containing Nitrogen

Oxidation state of nitrogen	*Formula*	*Name*
+5	HNO_3	Nitric acid
	NO_3^-	Nitrate ion
	NO_2^+	Nitronium ion
	N_2O_5	Dinitrogen pentoxide
	NO_2Cl	Nitryl chloride
+4	NO_2	Nitrogen dioxide
	N_2O_4	Dinitrogen tetroxide
+3	HNO_2	Nitrous acid
	NO_2^-	Nitrite ion
	$NOCl$	Nitrosyl chloride
+2	NO	Nitric oxide
+1	$H_2N_2O_2$	Hyponitrous acid
	N_2O	Nitrous oxide
−1	NH_2OH	Hydroxylamine
−2	N_2H_4	Hydrazine
	$N_2H_5^+$	Hydrazinium ion
	NH_3	Ammonia
	NH_4^+	Ammonium ion
−3	N^{3-}	Nitride ion
	NH_2^-	Amide ion
	H_2NCl	Chloramine

Similar reaction of copper with *concentrated* nitric acid yields mostly brown nitrogen dioxide (NO_2). Since nitric oxide is easily oxidized by air to nitrogen dioxide, brown fumes may also appear when dilute nitric acid is used.

Reduction of nitrate ions with aluminum in alkaline solution gives ammonia. In dilute acid solution, weaker reducing agents such as zinc and tin reduce nitrate to hydroxylamine (NH_2OH). Electrolytic reduction of nitrate ion at most cathodes yields ammonia, but hydroxylamine or hyponitrous acid ($H_2N_2O_2$) is obtained at a mercury cathode.

The ability of nitric acid to dissolve the less active metals is due to the oxidizing power of nitrate ion. Some metals such as copper and silver will not dissolve in dilute hydrochloric acid; the $H^+(aq)$ ion is not a sufficiently strong oxidizing agent. Nitric acid, which contains the additional oxidizing agent NO_3^-, dissolves them readily. Gold and platinum are insoluble in either hydrochloric or nitric acids alone but do dissolve in a mixture of the two. The mixture, called **aqua regia,** usually consists of one part concentrated nitric acid to three parts concentrated hydrochloric acid. The dissolving power of aqua regia is due to the oxidizing ability of nitrate ions in acid solution coupled with the complexing ability of chloride ions. Complexation will be considered in more detail in Chapter 20.

Nitric acid is now usually manufactured by converting atmospheric nitrogen to ammonia, oxidizing ammonia over a platinum-rhodium catalyst to nitric oxide, and absorbing the nitric oxide in presence of oxygen in water. The procedure involved in converting ammonia to nitric acid, equations (18.5) to (18.7), is called the Ostwald process.

$$4\ NH_3(g) + 5\ O_2(g) \xrightarrow[\text{heat}]{\text{Pt/Rh}} 4\ NO(g) + 6\ H_2O(g) \quad (18.5)$$

$$2\ NO(g) + O_2(g) \rightarrow 2\ NO_2(g) \quad (18.6)$$

$$3\ NO_2(g) + H_2O(liq) \rightarrow 2\ H^+(aq) + 2\ NO_3^-(aq) + NO(g) \quad (18.7)$$

The nitric oxide produced in (18.7) is recycled.

Anhydrous nitric acid (100% HNO_3) is prepared by distillation of a mixture of aqueous nitric acid and concentrated sulfuric acid. Commercial "concentrated" aqueous nitric acid (70% HNO_3 by weight) is colorless, but often becomes yellow as a result of photochemical decomposition to give nitrogen dioxide:

$$4\ HNO_3 \xrightarrow{\text{light}} 4\ NO_2 + O_2 + 2\ H_2O$$

Nitric acid is completely dissociated into $H^+(aq)$ and $NO_3^-(aq)$ ions in dilute aqueous solutions.

The nitrate ion is planar and symmetrical, with all three nitrogen–oxygen bonds equivalent. The structure may be explained in valence bond theory as a resonance hybrid of the three contributing structures shown in Figure 18.3. In molecular orbital theory nitrogen is considered to form three σ bonds using sp^2 hybrid orbitals, and the p_z orbitals of nitrogen and three oxygen atoms combine to form a delocalized π bonding molecular orbital containing two electrons. Gaseous HNO_3 has the structure shown in Figure 18.3. This planar molecule may also be represented by resonance structures in valence bond theory or with a delocalized π bond in molecular orbital theory.

Gentle heating (~250°C) of molten ammonium nitrate yields nitrous oxide by the reaction

$$NH_4NO_3(liq) \rightarrow N_2O(g) + 2\ H_2O(g) \quad (18.8)$$

while more vigorous heating also yields nitrogen and oxygen as shown by

$$NH_4NO_3(liq) \rightarrow N_2(g) + \tfrac{1}{2}\ O_2(g) + 2\ H_2O(g) \quad (18.9)$$

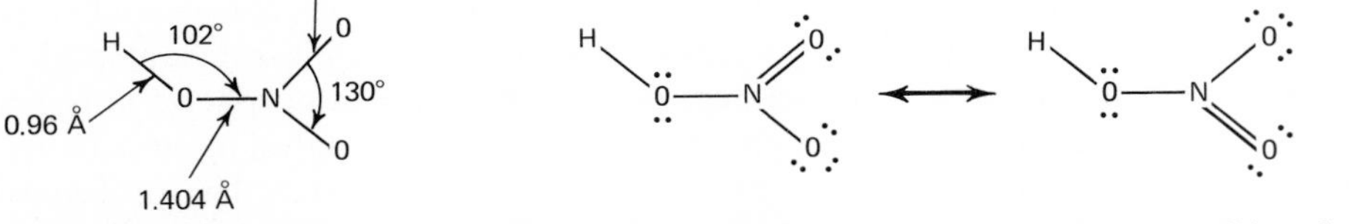

Figure 18.3. Structures and resonance representations of nitrate ion and nitric acid molecule.

In each reaction, nitrogen in nitrate is reduced and nitrogen in ammonium ion is oxidized. In (18.9) oxygen is also oxidized. If heating is too rapid, ammonium nitrate may detonate, and several disastrous explosions have been caused by ammonium nitrate. Ammonium nitrate is used as fertilizer and as an explosive, particularly in admixture with easily oxidized materials such as aluminum or trinitrotoluene (TNT).

Nitrogen dioxide is a brown paramagnetic gas that exists in a strongly temperature- and pressure-dependent equilibrium with its colorless diamagnetic dimer dinitrogen tetroxide, N_2O_4:

$$2\ NO_2(g) \rightleftharpoons N_2O_4(g) \qquad \Delta H^0 = -13.7\ \text{kcal mole}^{-1} \qquad (18.10)$$

In the solid state the oxide is entirely dinitrogen tetroxide. Partial dissociation to nitrogen dioxide occurs in the liquid state. At the melting point (−11.2°C) the liquid is pale yellow, becoming red-brown near the boiling point (21.2°C).

The paramagnetism and ready dimerization of nitrogen dioxide are consistent with the fact that each molecule has an odd number of valence electrons. Nitrogen dioxide molecules are nonlinear and the bonding may be represented by resonance structures (valence bond theory) or with a delocalized π bond (molecular orbital theory). Nitrogen dioxide loses its odd electron to give NO_2^+, the nitronium ion. The most stable form of dinitrogen tetroxide is a planar $O_2N—NO_2$ arrangement.

In alkaline solution dinitrogen tetroxide is unstable with respect to nitrite and nitrate as shown by the potentials in Figure 18.2. The equation for the disproportionation of N_2O_4 in alkaline solution is

$$N_2O_4(g) + 2\ OH^-(aq) \rightleftharpoons NO_2^-(aq) + NO_3^-(aq) + H_2O(liq)$$

In neutral or acidic solutions N_2O_4 or NO_2 reacts to form HNO_3 and NO, as shown in equation (18.7).

Heating molten sodium nitrate causes evolution of oxygen and formation of sodium nitrite in another self oxidation-reduction reaction:

$$2\ NaNO_3(liq) \rightarrow 2\ NaNO_2(liq) + O_2(g) \qquad (18.12)$$

The principal use of nitrites in industry is as a source of the $—NO_2$ group for making synthetic dyes.

In acid solution, nitrites form nitrous acid, HNO_2. The dissociation constant for $HNO_2(aq)$ to its aqueous ions is 5.9×10^{-4}, showing that it is a weak acid. Since nitrogen in HNO_2 is in an intermediate oxidation state, HNO_2 may function as either an oxidizing or reducing agent. It is unstable when heated, decomposing by several reactions to yield mostly nitric acid, nitric oxide, and nitrogen dioxide.

Nitric oxide, NO, can be obtained by a variety of reactions involving reduction of nitric acid or solutions of nitrates or nitrites. The reaction of copper with dilute nitric acid has already been mentioned. Nitric oxide is obtained industrially by catalytic oxidation of ammonia (equation 18.5). Direct combination of nitrogen with oxygen occurs only at high temperatures, and the yields are so low that it has never been a practical process. In air, nitric oxide is rapidly oxidized to brown nitrogen dioxide (equation 18.6).

Nitric oxide is a paramagnetic "odd electron molecule" with an uneven number of valence electrons. But unlike NO_2, NO is colorless and does not dimerize in the gaseous state. There is evidence of association in the liquid and solid states, however. The electronic structure of NO may be treated by valence bond theory using resonance structures, with the "odd" electron on either nitrogen or oxygen, but it is advantageous to use molecular orbital theory. NO contains one more electron than does N_2; according to a simple molecular orbital description (Chapter 8), this electron should go in an antibonding π^* orbital. Consistent with this picture, the bond properties of NO are intermediate between those of N_2 (triple bond) and O_2 (double bond). For example, the bond length in NO is 1.14 Å compared to 1.10 Å for N_2 and 1.21 Å for O_2. In addition, the first ionization energy of NO is lower than for similar molecules. The resultant nitrosonium ion, NO^+, is a well-characterized species.

Nitrous oxide, N_2O, is prepared industrially by careful thermal decomposition of ammonium nitrate (NH_4NO_3) (equation 18.8). N_2O is a colorless, relatively unreactive gas, decomposing to the elements only at high temperatures (around 500 °C). It is far less toxic than the other oxides of nitrogen. Small doses of N_2O produce a mild hysteria, and it has been used in dentistry as a general anesthetic, called laughing gas. It has a moderate solubility in cream and finds extensive use as the propellant gas in "whipped" cream aerosol cans.

The N_2O molecule is linear with a nitrogen as the central atom. In valence bond theory it is considered as a resonance hybrid:

$$\underset{\delta-}{:\ddot{N}}=\underset{\delta+}{N}=\ddot{O}: \leftrightarrow :N\equiv\underset{\delta+}{N}-\underset{\delta-}{:\ddot{\underset{..}{O}}:}$$

The molecular orbital treatment is similar to that of carbon dioxide, with which it is **isosteric** (same number of atoms and same number of valence electrons). In the molecular orbital description of N_2O there are two σ bonds and two delocalized π bonds.

Sodium azide, NaN_3, was mentioned previously as a source of very pure nitrogen. In the linear azide ion, N_3^-, nitrogen has a formal oxidation state of $-\frac{1}{3}$, which must represent an average value. Because of many similarities to halide ions, azides are commonly called **pseudohalides.** For example, HN_3 and ClN_3 are known compounds, and AgN_3 is only slightly soluble in water. The analog to the diatomic halogen molecules, namely $(N_3)_2$, is not known, however. Hydrazoic acid, HN_3, is both a weak acid and very explosive. Azides of lead and mercury are also extremely explosive, which accounts for their use as detonators.

Hydroxylamine, NH_2OH, may be considered a derivative of ammonia, but its usual preparation involves reduction of nitrates or nitrites either electrolytically or chemically. Because pure hydroxylamine is an unstable solid at room temperature, it is generally used in aqueous solution. Like ammonia, it has a lone pair of electrons and can function as a base. For the equilibrium

$$NH_2OH(aq) + H_2O(liq) \rightleftharpoons NH_3OH^+(aq) + OH^-(aq)$$

we have $K = 1 \times 10^{-8}$ at 25 °C, showing that hydroxylamine is a weaker base than ammonia. This can be rationalized by noting that the OH group of H_2NOH is more electron withdrawing than the corresponding H group of H_2NH. As a consequence, the lone pair in hydroxylamine is "less available." Since nitrogen is in an intermediate oxidation state, hydroxylamine can act as either an oxidizing or reducing agent.

Hydrazine, N_2H_4, may be pictured as a derivative of ammonia in which a hydrogen is replaced by an $—NH_2$ group. It is commonly obtained by the Raschig process in which chlorine is bubbled through aqueous ammonia containing glue or gelatin. Hypochlorite is formed initially and reacts with ammonia to produce chloramine, NH_2Cl, which then slowly reacts with more ammonia to form hydrazine. The net reaction is

$$Cl_2(g) + 4\,NH_3(aq) \rightarrow N_2H_4(aq) + 2\,NH_4^+(aq) + 2\,Cl^-(aq) \qquad (18.13)$$

A competing reaction that reduces the overall yield of hydrazine is

$$N_2H_4(aq) + 2\,NH_2Cl(aq) \rightarrow N_2(g) + 2\,NH_4^+(aq) + 2\,Cl^-(aq)$$

This latter reaction is catalyzed by traces of metal ions (especially copper). Glue or gelatin in the solution forms complex ions with the metal ions and thus inhibits the undesired side reaction.

Pure hydrazine is a colorless liquid and may be obtained by distilling the aqueous material in the presence of a dehydrating agent such as sodium hydroxide. Like ammonia, hydrazine is a good solvent for many salts. It burns in air with considerable evolution of heat,

$$N_2H_4(liq) + O_2(g) \rightarrow N_2(g) + 2H_2O(liq) \qquad \Delta H^0 = -149 \text{ kcal mole}^{-1} \quad (18.14)$$

which accounts for the interest in hydrazine and its derivatives as rocket fuels.

As expected, hydrazine is a weaker base then ammonia. It has two basic sites and both $N_2H_5^+$ and $N_2H_6^{2+}$ are known. Hydrazine is a strong reducing agent in alkaline solution, often being oxidized to N_2. In acid solution hydrazine is a good oxidizing agent, but reactions are frequently slow.

In addition to ammonia and ammonium salts containing nitrogen in the -3 oxidation state, various nitrides containing N^{3-} ions are known. These nitrides are generally made by direct combination of nitrogen with the hot metal. Ionic nitrides such as Na_3N and Mg_3N_2 are quite reactive and combine with water to form ammonia:

$$Mg_3N_2(c) + 6\,H_2O(liq) \rightarrow 2\,NH_3(g) + 3\,Mg(OH)_2(c) \qquad (18.15)$$

Nitrides of the transition metals, such as TiN and TaN, are inert hard solids that are used in making containers for high temperature reactions.

The above introduction to some nitrogen compounds should give some idea of the complexity and variety of nitrogen chemistry.

Liquid Ammonia Chemistry

Beginning around 1900 with the work of Cady, Franklin, and Kraus, liquid ammonia has been the most studied of all nonaqueous solvents. Most studies have been conducted near the normal boiling point ($-33.3\,°C$) of ammonia, and various specialized techniques have been developed to permit investigations at this low temperature.

Ammonia has many waterlike properties and some notable differences. Liquid ammonia undergoes self-ionization to ammonium and amide ions, just as water ionizes to form hydronium and hydroxide ions:

$$2\,NH_3(liq) \rightleftharpoons NH_4^+(am) + NH_2^-(am) \qquad (18.16)$$
$$2\,H_2O(liq) \rightleftharpoons H_3O^+(aq) + OH^-(aq) \qquad (18.17)$$

The notation (am) indicates that the ions are in liquid ammonia solution just as (aq) indicates aqueous solutions. The extent of self-ionization in ammonia is much less than in water; at $-33\,°C$, $K \cong 10^{-30}$ for the reaction represented by equation (18.16).

The reverse of both (18.16) and (18.17) are equations for neutralization reactions; acid plus base produces solvent. Compounds that increase the ammonium ion concentration are acids in liquid ammonia, whereas compounds that increase the amide ion concentration are bases. Solutions of NH_4Cl(am) may be titrated with $NaNH_2$(am) using phenolphthalein as an indicator for the end point, just as HCl(aq) may be titrated with NaOH(aq).

Since ammonia has a greater affinity for protons than does water, ammonia is a more basic solvent than water. As a result, moderately weak acids in water such as hydrofluoric acid or acetic acid become strong acids in liquid ammonia. This is another example of the "leveling effect" of a solvent.

The dielectric constant of liquid ammonia (23 at $-33\,°C$ and 17 at $25\,°C$) is low relative to that of water (78.5 at $25\,°C$) but still high enough to make ammonia a fairly good ionizing solvent. Solubilities in the two solvents are quite different, however. Many organic compounds are soluble in liquid ammonia, whereas most salts containing doubly charged ions are nearly insoluble.

Perhaps the most notable difference between the two solvents is their behavior toward alkali metals. By analogy to alkali metal reactions with water, the following reaction might be expected in liquid ammonia:

$$2\,Na(c) + 2\,NH_3(liq) \rightarrow H_2(g) + 2\,Na^+(am) + 2\,NH_2^-(am) \qquad (18.18)$$

Indeed this reaction does occur very slowly, or it may occur rapidly if a catalyst such as a transition metal salt is present. In the absence of a catalyst, alkali metals (and alkaline earth metals except beryllium) dissolve in liquid ammonia without hydrogen evolution. The dilute solutions have a beautiful blue color (due to the short-wavelength tail of an absorption band at ~15,000 Å), and the concentrated solutions have the appearance of molten bronze. Electrical conductivities of these solutions are intermediate between those of aqueous solutions and free metals. Evaporation of ammonia from these highly paramagnetic solutions yields the original metal.

The conductivities and other properties of these metal-ammonia solutions can be explained by assuming the metal "dissociates" to give ammoniated cations and electrons:

$$M(c) \xrightleftharpoons{NH_3(liq)} M^+(am) + e^-(am) \qquad (18.19)$$

In dilute solutions the electrons occupy cavities surrounded by ammonia molecules. As the concentration of the solution is increased, the M^+ and e^- species associate. As might be expected, these solutions containing "ammoniated electrons" are strong reducing agents. Reactions involving metal-ammonia solutions may be followed either by color or conductivity changes.

Carbon-Nitrogen Chemistry

"Organic" carbon-nitrogen compounds, including amino acids and proteins, will be considered in Chapter 19. The most important "inorganic" carbon-nitrogen species are cyanide, cyanate, and thiocyanate ions, and their derivatives.

Cyanogen, $(CN)_2$, is a colorless, very poisonous gas with the odor of bitter almonds. It may be obtained by heating various heavy metal cyanides, for example

$$Hg(CN)_2(c) \xrightarrow{400\,°C} Hg(liq) + C_2N_2(g) \tag{18.20}$$

Cyanogen tends to polymerize but is remarkably stable considering its $\Delta G_f^0 = +71$ kcal mole^{-1}. Because of similarities to halogen chemistry, cyanogen is also called a pseudohalogen. In aqueous base cyanogen disproportionates to cyanide and cyanate ions:

$$(CN)_2(g) + 2\,OH^-(aq) \rightleftharpoons CN^-(aq) + OCN^-(aq) + H_2O(liq) \tag{18.21}$$

Compare this reaction to that of chlorine with aqueous base as in equation (17.22).

Hydrogen cyanide, HCN, is a poisonous gas evolved when cyanides are treated with acids. The boiling point (26 °C) and dielectric constant (107 at 25 °C) of hydrogen cyanide are relatively high, suggesting intermolecular association through hydrogen bonding. In aqueous solution hydrogen cyanide is a very weak acid ($K_a = 6 \times 10^{-10}$ at 25 °C), even weaker than hydrogen fluoride.

Cyanide ion, CN^-, is **isoelectronic** (same number of electrons) with N_2. Its short bond length of 1.14 Å is also consistent with a triple bond. Like the halides, silver and mercurous cyanides are only slightly soluble in water. The cyanide ion forms a large number of complexes with transition metals; some, such as $Ag(CN)_2^-$ and $Au(CN)_2^-$, are of technical importance.

The cyanate ion, OCN^-, and thiocyanate ion, SCN^-, are also considered pseudohalides, although the analogies are not as strong as they are with CN^-. There is little evidence for $(OCN)_2$ although $(SCN)_2$ exists in certain solvents. Thiocyanates are obtained by fusing alkali metal cyanides with sulfur.

Passing nitrogen gas over calcium carbide at 1000 °C produces calcium cyanamide, $CaCN_2$:

$$CaC_2 + N_2 \rightarrow CaCN_2 + C \tag{18.22}$$

Before the Haber process was developed, this was the only important industrial method of nitrogen fixation. Although its use has decreased, calcium cyanamide is still the preferred fertilizer for certain crops and soils.

Alkali metal cyanides are made by fusing calcium cyanamide with carbon and the appropriate carbonate, for example

$$CaCN_2 + C + Na_2CO_3 \rightarrow CaCO_3 + 2\,NaCN \tag{18.23}$$

Heating sodium amide with carbon is another route to sodium cyanide.

Phosphorus

Phosphorus is the most abundant of the group V elements (~0.12% of the earth's crust) and is the only one that does not exist as the free element in nature. It occurs principally as phosphates, and is also found in many biologically important molecules such as proteins. Calcium phosphate, $Ca_3(PO_4)_2$, is the major constituent of bones and teeth. Since phosphorus is an essential nutrient for plant life, much of the phosphate mined is converted into fertilizers.

Elemental phosphorus is generally made by heating phosphate minerals with coke and silica in an electric furnace. The reaction equation may be written as

$$2\ Ca_3(PO_4)_2 + 6\ SiO_2 + 10\ C \rightarrow P_4(g) + 6\ CaSiO_3 + 10\ CO(g)$$

The product gases are passed through water, which condenses the phosphorus and protects it from oxidation by air. There are several allotropic forms of phosphorus, the most important being white, red, and black. White phosphorus melts at 44.1 °C and boils at 280.5 °C. It is very reactive and is spontaneously inflammable in air.

In solid and liquid white phosphorus there are tetrahedral P_4 molecules, as illustrated in Figure 18.4. Gaseous phosphorus consists mainly of P_4 molecules below ~800 °C, but above that temperature appreciable dissociation into P_2 molecules occurs. The bonding in P_2 is believed to be similar to that of N_2, that is, one σ and two π bonds, but the bond dissociation enthalpy (116 kcal $mole^{-1}$) is considerably less.

Red phosphorus, the stable form at room temperature, is made by heating the white form at 400 °C for several hours in the absence of air. The exact structure of red phosphorus is uncertain; at least six modifications of it are believed to exist. The evidence suggests that it consists of P_4 tetrahedra linked together. Because of its polymeric nature the red form is less volatile, less soluble in nonpolar solvents, and less reactive than white phosphorus.

Crystalline black phosphorus is obtained by heating the white form under pressure for several days. It has a layer structure in which adjacent sheets of phosphorus atoms are held together by weak van der Waals forces. Black phosphorus is relatively unreactive.

Almost all metals form binary compounds with phosphorus in the −3 state. The compositions and structures of many of these compounds, called phosphides, are quite complex. When alkaline earth metals are heated with red phosphorus, compounds of the general formula M_3P_2 are formed. Hydrolysis of alkaline earth phosphides produces gaseous phosphine, PH_3, for example

$$Ca_3P_2(c) + 6\ H_2O(liq) \rightarrow 2\ PH_3(g) + 3\ Ca(OH)_2(c) \qquad (18.24)$$

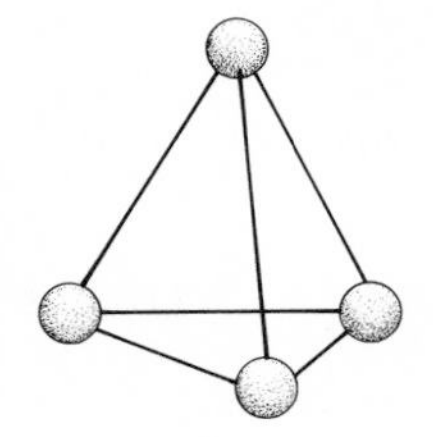

Figure 18.4. Structure of P_4 molecule. The P—P—P bond angles are 60°. The P—P bond distance is 2.21 Å.

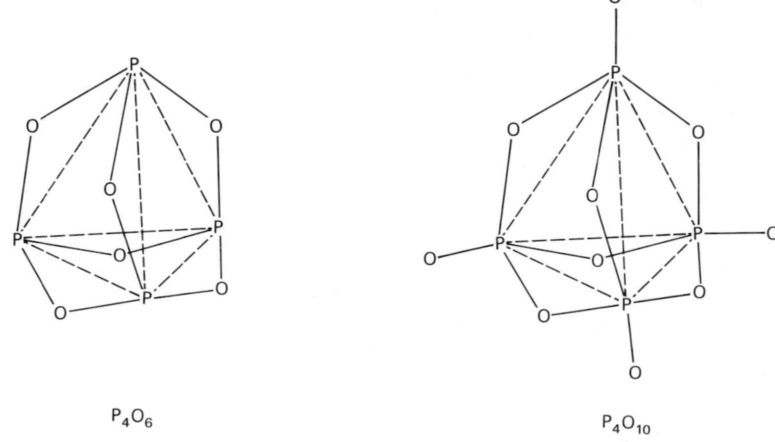

Figure 18.5. Structures of P_4O_6 and P_4O_{10} molecules in the vapor state.

Molecular phosphine is pyramidal, like ammonia. The H—P—H bond angle (93.5°) suggests that nearly pure *p* orbitals are used in bonding. Due to absence of hydrogen bonding, phosphine is more volatile (boiling point, −88°C) than ammonia. Phosphine is also less basic than NH_3, which may be rationalized by considering that the lone pair in PH_3 is in a larger, more diffuse orbital. As a result, phosphonium salts such as PH_4Cl decompose far more readily than the corresponding ammonium salts. On exposure to air, PH_3 usually bursts into flame, presumably due to spontaneous oxidation of the impurity diphosphine, P_2H_4.

The binary halides of phosphorus are of two types: the trihalides and pentahalides. With the exception of a pentaiodide, all possible binary trihalides and pentahalides are known. Most of these compounds are obtained by direct combination of the elements, the relative amount of halogen determining whether PX_3 or PX_5 is obtained. Phosphorus halides are quite reactive and are commonly used as halogenating agents.

Unlike the nitrogen oxides, none of the oxy compounds or ions of phosphorus are particularly strong oxidizing agents. There are two well-characterized oxides of phosphorus: P_4O_6 and P_4O_{10}. When phosphorus is burned in a limited supply of oxygen, phosphorus trioxide, P_4O_6, is formed. If an excess of oxygen is available, phosphorus pentoxide, P_4O_{10}, is obtained. The oxides are named according to their simplest formulas for historical reasons. Both of the phosphorus oxides are white solids, P_4O_6 melting at 23.8°C and P_4O_{10} subliming at 360°C (1.0 atm).

The structures of P_4O_6 and P_4O_{10} gaseous molecules may be pictured as being derived from the tetrahedral P_4 molecule. In P_4O_6 an oxygen atom is inserted between each pair of phosphorus atoms; in P_4O_{10} each phosphorus has, in addition, one terminal oxygen atom. The structures are shown in Figure 18.5. Discrete molecules of P_4O_{10} also exist in the solid, but the structure of crystalline P_4O_6 is not known.

When P_4O_6 is added to cold water, phosphorous acid, H_3PO_3, is formed. Although phosphorous acid contains three hydrogens, only two will dissociate in aqueous solution. The equilibrium constants at 25°C are:

$$H_3PO_3(aq) \rightleftharpoons H^+(aq) + H_2PO_3^-(aq) \qquad K_1 = 1.6 \times 10^{-2}$$
$$H_2PO_3^-(aq) \rightleftharpoons H^+(aq) + HPO_3^{2-}(aq) \qquad K_2 = 7 \times 10^{-7}$$

For this reason phosphorous acid is better formulated as $HPO(OH)_2$ than as $P(OH)_3$. The existence of a P—H bond has been demonstrated by a variety of structural studies. With two ionizable hydrogens, two series of salts from phosphorous acid are known: the dihydrogen phosphites, (for example, NaH_2PO_3), and the monohydrogen phosphites, (for example, Na_2HPO_3). Salts such as Na_3PO_3 are unknown.

Phosphorous pentoxide shows a great affinity for water and is one of the most effective drying agents at room temperature. On exposure to moisture, the solid becomes syrupy. With an excess of water, the principal product is orthophosphoric acid, H_3PO_4. This is a triprotic acid with the following stepwise dissociation constants at 25°C:

$$H_3PO_4(aq) \rightleftharpoons H^+(aq) + H_2PO_4^-(aq) \qquad K_1 = 7.1 \times 10^{-3}$$
$$H_2PO_4^-(aq) \rightleftharpoons H^+(aq) + HPO_4^{2-}(aq) \qquad K_2 = 6.3 \times 10^{-8}$$
$$HPO_4^{2-}(aq) \rightleftharpoons H^+(aq) + PO_4^{3-}(aq) \qquad K_3 = 5 \times 10^{-13}$$

The pure acid and its crystalline hydrates have tetrahedral PO_4 groups connected by hydrogen bonds. Hydrogen bonding persists in the concentrated solutions, which accounts for the syrupy nature. There are three series of salts derived from H_3PO_4 depending on whether one, two, or three hydrogens have been replaced by metal ions.

Orthophosphoric acid described above is only one of a series of phosphoric acids that can be formed by hydration of P_4O_{10}. With smaller amounts of water, solutions are obtained that contain various polyphosphoric acids. Two examples are $H_4P_2O_7$, diphosphoric or pyrophosphoric acid, and $H_5P_3O_{10}$, triphosphoric acid. These polyphosphoric acids and their salts have chain structures containing PO_4 tetrahedra linked together by sharing an oxygen atom. They can also be obtained by careful heating of orthophosphates.

The metaphosphoric acids can also be obtained by controlled addition of water to P_4O_{10}. These acids are all polymeric with the general formula $(HPO_3)_n$. They can be pictured as built up from PO_3 units in which each phosphorus has four oxygens tetrahedrally arranged around it. This means that two of the oxygens must be bridge oxygens, as shown in Figure 18.6. It is frequently more convenient to obtain a metaphosphate by thermal dehydration of an orthophosphate rather than neutralization of a metaphosphoric acid. For example, sodium trimetaphosphate is generally prepared by heating NaH_2PO_4:

$$3\ NaH_2PO_4(c) \xrightarrow{\text{heat}} Na_3P_3O_9(c) + 3\ H_2O(g) \qquad (18.25)$$

Metaphosphates and linear polyphosphates, such as sodium triphosphate ($Na_5P_3O_{10}$), have been much used as water softeners and for "building" soaps. They form very stable complexes with calcium and magnesium ions, and thus make these ions unavailable for precipitation as soap scums. Phosphates are also strongly absorbed on dirt particles, making it easier to remove the dirt during washing.

The use of phosphates in soaps and detergents has come under attack and much effort has been expended toward finding substitutes. Phosphorus is a plant nutrient

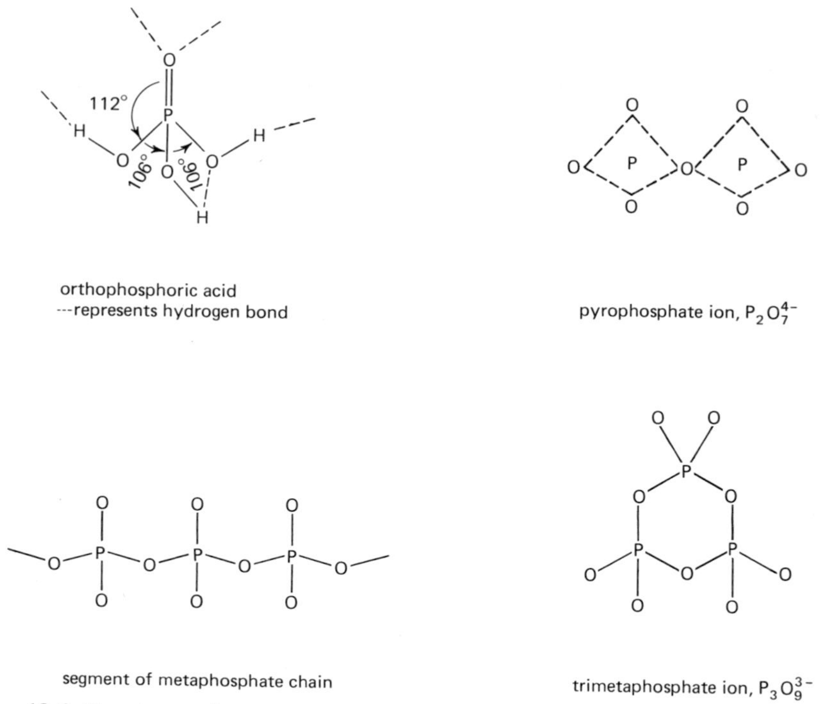

orthophosphoric acid
---represents hydrogen bond

pyrophosphate ion, $P_2O_7^{4-}$

segment of metaphosphate chain

trimetaphosphate ion, $P_3O_9^{3-}$

Figure 18.6. Structures of some common phosphate species.

that is frequently in short supply in natural waters. Discharge of wash water containing phosphates can thus promote algal growth and accelerate eutrophication or "aging" of lakes. Many factors contribute to eutrophication, and the controlling factors are not yet clear. Regardless of the outcome of eutrophication arguments, it is likely that phosphates will be removed from soaps anyway. Known phosphate reserves will last approximately 60 years at current usage rates. This diminishing supply is better used as fertilizer to increase food production than to make cleaning easier.

Organic phosphates are of major importance in biological processes. For example, conversion of adenosine triphosphate (ATP) to adenosine diphosphate (ADP) involves splitting off a phosphate group. This exothermic reaction, which can be described as the hydrolysis of a C—O—P bond, releases approximately 10 kcal mole^{-1}, which can be used for the mechanical work of muscle contraction. Various enzymes catalyze the hydrolysis and formation of C—O—P bonds.

Arsenic, Antimony, and Bismuth

The remaining elements of group V are not very abundant (10^{-4}% or less in the earth's crust); they are commonly found as sulfides. These elements illustrate the usual trend toward increasing metallic character in going down a group. Both arsenic and antimony have unstable yellow allotropes, presumably containing As_4 and Sb_4 molecules, which are soluble in nonpolar solvents. Bismuth does not have a nonmetallic yellow form. The normal black forms of As, Sb, and Bi are bright and metallic in appearance with crystal structures similar to that of black phosphorus; they all have low electrical and thermal conductivities.

All three elements exhibit -3 oxidation states in binary compounds with alkali metals. The stability of these *-ide* salts decreases in going from arsenic to bismuth. Hydrolysis of the -ide salt yields hydrides of the general formula MH_3, in which the formal oxidation state of the group V element is $+3$. For example, arsine (AsH_3) may be produced by the following reaction:

$$Na_3As(c) + 3\ H_2O(liq) \rightarrow AsH_3(g) + 3\ NaOH(c) \tag{18.26}$$

Stibine (SbH_3) and bismuthine (BiH_3) are also known but are not thermally stable.

Binary halides of arsenic, antimony, and bismuth are of two main types: MX_3 and MX_5. All the group V halides are rapidly attacked by water. Most of the halides, particularly the pentafluorides, are strong Lewis acids or halide ion acceptors. In the gas phase or in nonaqueous solvents, reactions such as (18.27) readily occur:

$$BrF_3 + SbF_5 \rightleftharpoons BrF_2^+SbF_6^- \tag{18.27}$$

Many of the trihalides, particularly those of arsenic, have been studied as ionizing solvents.

Oxides of the type M_4O_6 and M_2O_5 are known for arsenic and antimony, but the pentaoxides are not well characterized. The only well-established oxide of bismuth is Bi_2O_3. Unlike the arsenic and antimony oxides, Bi_2O_3 shows no acidic character. The antimony oxides are amphoteric.

Aqueous solutions of As_4O_6 are not well characterized, but orthoarsenic acid, H_3AsO_4, and arsenates are readily obtained from the pentaoxide. H_3AsO_4 is a triprotic acid; it is a slightly weaker acid and considerably stronger oxidizing agent than H_3PO_4. Aqueous antimony and bismuth species are not as well defined.

Compounds of arsenic are cumulative systemic poisons and are almost tasteless. Their well-known use for homicidal purposes is now less attractive than formerly due to sensitive tests for trace amounts of arsenic compounds. Arsenates, such as $Ca_3(AsO_4)_2$, are commonly used as insectides.

Aqueous solutions containing bismuth in the $+5$ oxidation state are powerful oxidizing agents. For example, in aqueous perchloric acid Bi(V) oxidizes halide ions (except fluoride) to yield the free halogen. The potential for the half-reaction

$$Bi^{5+}(aq) + 2\ e^- \rightleftharpoons Bi^{3+}(aq)$$

has an approximate value of $+2.0$ volts.

References

Alexander, M. D.: Gas laws, equilibrium, and the commercial synthesis of nitric acid (a simple demonstration). *J. Chem. Educ.*, **48:**839 (1971).

Bent, H. A.: Isoelectronic systems. *J. Chem. Educ.*, **43:**170 (1966).

Delwiche, C. C.: The nitrogen cycle. *Sci. Amer.* **223:**136 (Sept. 1970).

Haber, L. F.: Fritz Haber and the nitrogen problem. *Endeavour,* **27:**150 (1968).

Hammond, A. L.: Phosphate replacements: problems with the washday miracle. *Science,* **172:**361 (23 April 1971)

Holmes, R. R.: Ionic and molecular halides of the phosphorus family. *J. Chem. Educ.*, **40:**125 (1963).

Huheey, J. E.: Chemistry of disphosphorus compounds. *J. Chem. Educ.*, **40:**153 (1963).

Jolly, W. L.: *The Inorganic Chemistry of Nitrogen.* W. A. Benjamin, Inc., New York, 1964.

Jolly, W. L.: *The Chemistry of the Non-Metals.* Prentice-Hall, Inc., Englewood Cliffs, N.J., 1966 (paperback).

Jolly, W. L.: The use of oxidation potentials in inorganic chemistry. *J. Chem. Educ.*, **43:**198 (1966).

Kriz, G. S., Jr., and K. D. Kriz: Analysis of phosphates in detergents. *J. Chem. Educ.*, **48:**551 (1971).

Pratt, C. J.: Chemical fertilizers. *Sci. Amer.*, **212:**62 (June 1965).

Schneller, S. W.: Nitrogen fixation: an interdisciplinary frontier. *J. Chem. Educ.*, **49:**786 (1972).

Watt, G. W.: Reactions in liquid ammonia. *J. Chem. Educ.*, **34:**538 (1957).

Problems

1. (a) Assume that it behaves as an ideal gas and calculate the density of pure nitrogen gas at 0.0°C and 1.0 atm.
(b) The density of "nitrogen" obtained from air after removal of oxygen, carbon dioxide, and water is 1.2572 g liter^{-1} at 0.0°C and 1.0 atm pressure. Assuming that the only impurity is argon, calculate the number of moles of argon obtainable from 3.0 liters (at 0°C and 1.0 atm) of "nitrogen" from air.

2. (a) Write balanced equations for hydrolysis of lithium nitride and magnesium nitride.
(b) Starting with 100 g each of magnesium and lithium, what is the maximum amount of ammonia that can be obtained by converting the metals to nitrides and hydrolyzing the resulting nitrides?

3. Write a balanced equation for the principal reaction of concentrated nitric acid with copper.

4. Write a balanced equation for reduction of dilute nitric acid to hydroxylamine (existing as NH_3OH^+ in acidic solution) by zinc. Calculate ΔG^0 for this reaction at 25°C (using reduction potentials).

5. Use reduction potentials to predict which nitrogen containing species would result from bubbling N_2O_4 through an acidic solution.

6. Explain why vigorous heating of ammonium nitrate generally results in explosions. Would ΔS^0 for this reaction be large or small, positive or negative?

7. Assuming that anhydrous nitric acid undergoes autodissociation, what ions are most likely formed?

8. Write balanced equations that account for formation of NO, NO_2, N_2, NH_4^+, and sulfur when H_2S is bubbled into aqueous nitric acid.

9. Explain why concentrated sulfuric acid (rather than concentrated hydrochloric acid) is used to prepare anhydrous nitric acid by distillation.

10. Heating ammonium chloride causes it to decompose into ammonia and hydrogen chloride. If ammonium chloride is heated in one end of a long tube open only at the other end, a piece of moist litmus paper at the open end first turns blue and then turns red. Explain.

11. Calculate the pH of a solution made by dissolving 0.50 mole of ammonia in 750 ml of water at 25°C.

12. Is reduction of HNO_2 to N_2O or to $H_2N_2O_2$ more favored thermodynamically as indicated by standard reduction potentials? Is it possible to prepare $H_2N_2O_2$ by hydrating N_2O?

13. Draw resonance structures for NO_2. In the nitrogen dioxide-dinitrogen tetroxide equilibrium, equation (18.10), what will be the effect on K of increasing temperature?

14. Using standard reduction potentials, show that disproportionation of N_2O_4 to nitrite and nitrate is spontaneous in aqueous base at 25°C.

15. Suppose that the pseudohalogen N_3—N_3 could be prepared. Predict the products of bubbling N_3—N_3 through alkaline solution.

16. How much ammonium chloride is required to neutralize a solution made by dissolving 10.0 g of sodium amide in liquid ammonia?

17. In the "nitrogen system of compounds" the amide ion is considered analogous to hydroxide in the oxygen system of compounds. To what compound would silver imide, Ag_2NH, be formally analogous? Would Ag_2NH function as an acid or base in liquid ammonia?

18. (a) Calculate the percentage of nitrogen in ammonium nitrate, a common ingredient in fertilizers, and compare to the per cent nitrogen in NH_3.
(b) Write a series of balanced equations that represent feasible industrial processes for preparing solid ammonium nitrate.

19. Hydrolysis of calcium cyanamide produces ammonia. Write a balanced equation for this reaction and calculate the weight of ammonia available from a 100-lb bag of calcium cyanamide. Explain how this fertilizer helps to neutralize acidic soils.

20. (a) Draw an electron dot structure for the thiocyanate ion, SCN^-. Do you expect this ion to be linear or bent?
(b) Write a balanced equation to represent disproportionation of $(SCN)_2$ in alkaline solution.

21. Describe the bonding in the P_2 molecule using electron dot pictures and molecular orbital theory.

22. Write balanced equations for each of the following reactions:
(a) Red phosphorus is burned in an excess of oxygen.
(b) The product from (a) is dissolved in a large quantity of water.
(c) One mole of the product from (b) is reacted with one mole of NaOH.
(d) The product from (c) is heated at temperatures above 100°C. Give a commercial use for the product from (d).

23. Write electron dot structures and predict the general shapes of the following molecules and ions: PBr_3, PF_5, $POCl_3$, PF_6^-, PCl_4^+.

24. Phosphorus oxychloride, $POCl_3$, is an important intermediate in industrial chemistry. It may be prepared by reacting phosphorus trichloride with oxygen. Starting with 1.0 ton of apatite, what is the maximum amount of $POCl_3$ that can be obtained? Use $Ca_3(PO_4)_2$ as the formula of apatite.

25. Calculate the pH at 25°C of an aqueous solution that is labeled 0.020 M H_3PO_4.

26. The radioactive isotope ^{90}Sr is especially dangerous to man because it concentrates in bones and teeth. Suggest a reason for this behavior.

27. Predict the following properties of the as yet unknown element 115 in comparison with those of the other group V elements:
(a) Electrical conductivity
(b) Stability of -3 oxidation state
(c) Acidity of its oxides
(d) Thermal stability of its hydride

28. Assuming a 60% yield, how many grams of Ca_3P_2 must be hydrolyzed to yield 1.0 mole of PH_3?

29. Given an aqueous solution that is labeled 1.0 M H_3PO_4, calculate the concentration of H^+, $H_2PO_4^-$, HPO_4^{2-}, and PO_4^{3-} at 25°C.

30. What concentration of HCl is required to make the HPO_3^{2-} concentration 2.5×10^{-3} *M* in a 0.20 *M* H_3PO_3 aqueous solution at 25°C?

31. Given the following reduction potential in acidic solution:

$$H_3AsO_4(aq) + 2\,H^+ + 2\,e^- \rightleftharpoons HAsO_2(aq) + 2\,H_2O(liq) \qquad \mathcal{E}^0 = 0.58 \text{ volt}$$

Decide whether arsenic acid is a strong enough oxidizing agent to oxidize any of the halide ions to the free halogen. Would the conditions for oxidation be more favorable at high or low pH?

32. How are the standard potentials in Table 18.1 consistent with the observation that aqueous nitric acid becomes a stronger oxidizing agent as the pH is lowered?

33. Any oxidizing agent that will oxidize $NH_4^+(aq)$ to $NH_3OH^+(aq)$ in acid solution is strong enough to carry the oxidation on to $NO_3^-(aq)$. Explain how this observation is consistent with the potentials in Figure 18.2.

34. Use the potentials in Table 18.1 to show that hydrazoic acid (HN_3) is unstable with respect to oxidizing and reducing itself in acidic solution. Similarly, show that azide ion (N_3^-) is unstable in alkaline solution.

19

ORGANIC CHEMISTRY

Introduction

The term organic chemistry originally referred to the chemistry of substances derived from living organisms. Organic substances were believed to possess a "vital force" that accounted for many of their reactions and sharply differentiated them from "inorganic" materials, which did not possess this mysterious vital force. Although the concept of vital force is no longer important, the term organic chemistry has been retained; it now refers to the study of compounds that contain carbon together with hydrogen and a few other elements. As with most general definitions, the borderline between organic and inorganic chemistry is indistinct. Both inorganic and organic chemists are apt to study the same compounds, for example, organometallic compounds such as tetraethyllead, $(C_2H_5)_4Pb$.

The demise of the vitalistic theory began in 1828 when Friedrich Wohler, a German chemist, obtained urea by heating ammonium cyanate, NH_4OCN, a familiar inorganic substance:

$$NH_4OCN \xrightarrow{\text{heat}} H_2NCONH_2$$

Urea, $CO(NH_2)_2$, had already been isolated from many animal sources and therefore should possess the vital force. Wohler's experiment demonstrated the continuity between organic and inorganic substances, but old theories die hard. Theories generally succumb slowly to accumulated evidence rather than to a single brilliant experiment; in this case another 20 years of experimentation was necessary. By 1850 other chemists, such as Hermann Kolbe, had demonstrated that many organic substances could be converted into others in the laboratory. According to the vitalistic theory, only living things could accomplish such conversions.

A striking feature of organic chemistry is the great variety of organic compounds. Well over a million are now known with thousands more added each year, either by discovery in nature or preparation in the laboratory. There are several reasons why so many carbon compounds are known: (1) carbon atoms bond to each other to form chains of varying length **(catenation);** (2) adjacent carbon atoms can be joined by single, double, or triple bonds; (3) since each carbon atom can form four bonds, different arrangements of atoms can lead to compounds with the same general formula but different structures and properties **(isomers);** and (4) substitution of other elements for hydrogen leads to a large number of derivatives.

The primary sources of organic compounds are coal, petroleum, wood, and agricultural products. Each of these primary sources was originally produced by photosynthesis, and each is therefore a reservoir of solar energy. This energy may be liberated as heat on combustion. A simplified energy cycle for many processes may be represented by the following equation:

$$CO_2 + H_2O + \text{energy} \underset{\text{combustion}}{\overset{\text{photosynthesis}}{\rightleftharpoons}} \text{organic substance} + O_2$$

The only major exceptions to this cycle are hydroelectric, tidal, and nuclear energy.

Organic chemicals are put to many uses, which accounts for the large size of the organic chemical industry. Among the common uses of organic compounds are as fuels, detergents, insecticides, dyes, explosives, food additives, drugs, and plastics.

In this chapter we will discuss some of the simpler organic compounds, their structures, and reactions. We shall also briefly consider some compounds of biological interest, such as proteins, carbohydrates, and enzymes. Biochemistry, the chemistry of living systems, deals with these topics in detail and is one of the fastest growing branches of chemistry.

Alkanes

Carbon has four valence electrons and an intermediate electronegativity (2.5); we therefore expect carbon to form four covalent bonds. From VSEPR theory (Chapter 8) we predict that these bonds should be directed towards the corners of a regular tetrahedron.

Alkanes contain only carbon-carbon single bonds and carbon-hydrogen bonds, and are frequently called **saturated** or **aliphatic** hydrocarbons. Alkanes may be represented by the general formula C_nH_{2n+2}, where n is an integer. Some of the lighter members of the series are CH_4 (methane), C_2H_6 (ethane), C_3H_8 (propane), and C_4H_{10} (butane). As shown in Figure 19.1, there are several ways of drawing structural formulas for these compounds. Although alkanes are frequently called

CH_4 CH_3CH_3 $CH_3CH_2CH_3$ $CH_3CH_2CH_2CH_3$

methane ethane propane butane

Figure 19.1. Formulas of the lighter alkanes. The top graphic formulas show bonds but give no indication of geometry. The condensed structural formulas in the middle, in which individual bonds are omitted, are versatile and commonly used. The ball and stick formulas at the bottom show geometry and bonds but are tedious to draw.

straight chain compounds, it should be realized that the carbon chains are bent and that axial rotation about C—C bonds is possible.

The experimental bond angles in alkanes (H—C—H, H—C—C, or C—C—C) are all very close to the tetrahedral angle of 109.5°. In methane all four C—H bonds are equivalent. Valence bond theory provides the simplest explanation for the observed bond properties. The valence orbitals of carbon are the 2*s* and the three 2*p* orbitals. Mathematically mixing the wave expressions for these four orbitals produces a set of four equivalent sp^3 hybrid orbitals that have large positive lobes directed towards the corners of a regular tetrahedron. A pictorial representation to illustrate formation of sp^3 hybrid orbitals is shown in Figure 19.2. Overlap of the sp^3 orbitals of one carbon with the 1*s* orbital of hydrogen or sp^3 orbitals of another carbon produces the observed bond angles. (Methane is also represented in Figure 8.2.)

Compounds that have the same molecular formula but different structures and

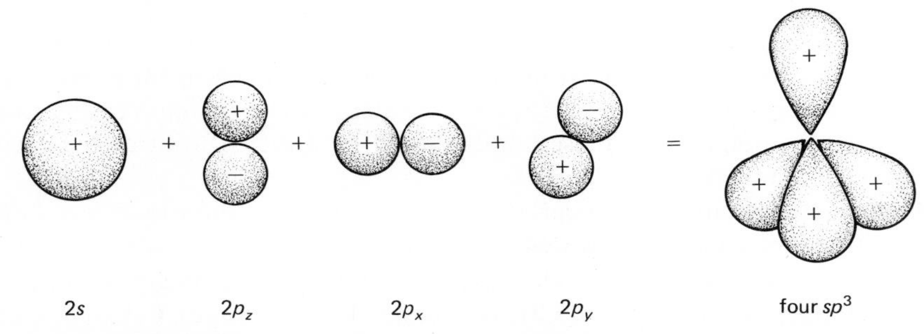

$2s$ $2p_z$ $2p_x$ $2p_y$ four sp^3

Figure 19.2. Pictorial representation of formation of the four equivalent sp^3 hybrid orbitals for carbon. The small negative lobes of the sp^3 orbitals are not shown.

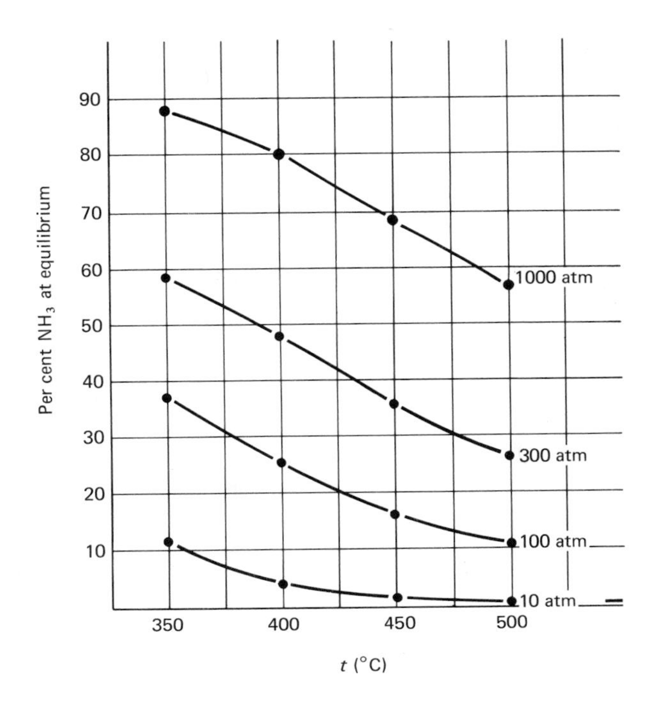

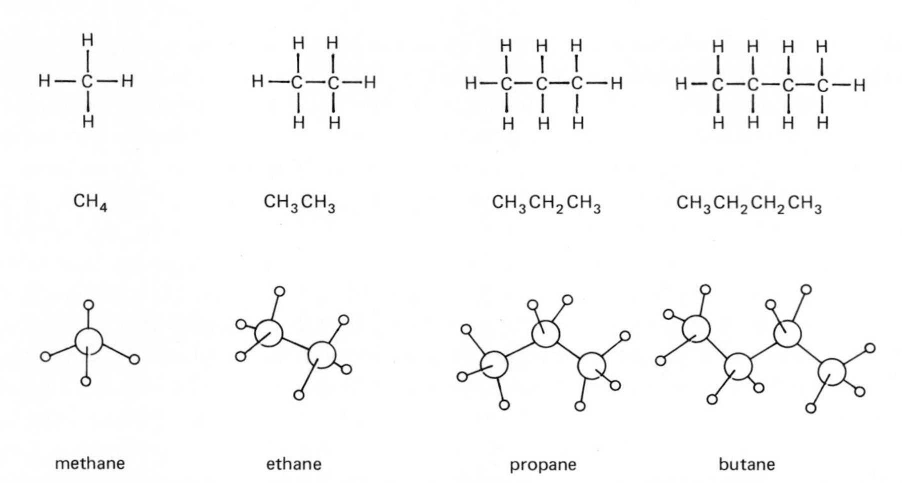

properties are called **isomers.** In the alkane series, butane is the first member to exhibit **structural isomerism.** The formulas

$$\underset{\text{butane}}{CH_3CH_2CH_2CH_3} \quad \text{and} \quad \underset{\text{isobutane}}{CH_3\overset{\overset{\displaystyle CH_3}{|}}{C}HCH_3}$$

represent two different compounds although both have the formula C_4H_{10}. As an example of their different properties, the normal boiling point of butane is $-0.5\,°C$ whereas that of isobutane is $-12\,°C$. It is possible to draw other two-dimensional structures of butane, but these will be equivalent to one of the two above. It is important to realize that the actual molecules are three dimensional; the true spatial relationships can best be visualized by using ball and stick or other three-dimensional models.

As the length of the carbon chain increases the number of structural isomers increases greatly. There are three isomers of pentane (C_5H_{12}), five isomers of hexane (C_6H_{14}), 75 possible isomers of decane ($C_{10}H_{22}$), and 336,319 possible isomers of eicosane ($C_{20}H_{42}$). Needless to say, not all of these possible isomers have been isolated and characterized. Although isomers always have different properties, sometimes these differences are so slight that separation is almost impossible.

Ring formation is another complicating feature of hydrocarbon chemistry that increases the number of isomers. **Cycloalkanes** are ring compounds containing carbon-carbon single bonds and have the general formula C_nH_{2n}. The most stable members of the cycloalkane series are

```
          H2
          C
        /   \
    H2C       CH2              H2                     
      \       /                C
      H2C---CH2          H2C/     \CH2
                   and     |       |
                         H2C\     /CH2
                                C
                                H2
    cyclopentane             cyclohexane
```

The rings in these compounds are puckered so that bond angles can closely approach the 109.5° tetrahedral angle. In three- and four-membered rings the bond angles are much less than the tetrahedral angle, and the bonds are said to be "strained." For example, cyclopropane (C_3H_6) is a planar molecule with C—C—C bond angles of 60°. Consistent with the strain theory, cyclopropane is readily converted to derivatives; in the process the ring is broken.

As a class, saturated hydrocarbons are not very reactive and show little affinity for most chemical reagents. Alkanes are nearly insoluble in water and do not react with aqueous solutions of acids, bases, or oxidizing agents at ordinary temperatures. A partial explanation of the low reactivity of alkanes can be based on the relatively large bond energies for C—C (83 kcal $mole^{-1}$) and C—H (99 kcal $mole^{-1}$) bonds.

Chlorine and bromine react with alkanes to form alkyl halides, but the reaction is not a simple substitution. For example, the overall reaction of methane with bromine to form methyl bromide is

$$CH_4 + Br_2 \rightarrow CH_3Br + HBr \qquad (19.1)$$

The actual mechanism of this reaction, however, is far more complicated than indicated by equation (19.1). The first step of the reaction involves dissociation of Br_2 into reactive bromine atoms, $\cdot Br$. The bromine atoms attack methane to produce HBr and $\cdot CH_3$, a methyl radical. Methyl radicals react with Br_2 to form CH_3Br and more $\cdot Br$, thus forming a chain reaction. The chain reaction continues until all the free radicals have been consumed.

Unlike the other halogen-hydrocarbon reactions, fluorine substitution reactions proceed rapidly, are very exothermic, and generally lead to complete fluorination of the hydrocarbon. For the reaction,

$$R_3CH + F_2 \rightarrow R_3CF + HF \qquad (19.2)$$

where R is used to represent any hydrocarbon group, the enthalpy of reaction using bond energy data is -113 kcal mole^{-1} compared to -25 kcal mole^{-1} for the corresponding chlorination reaction. To avoid breakdown of the hydrocarbon skeleton, some means must be provided to dissipate the excess heat. This is generally accomplished by carrying out fluorinations in a flow reactor containing a silver-plated copper screen as catalyst–heat exchanger and by diluting the reactants with nitrogen gas.

Another important route to fluorinated compounds is electrochemical fluorination. This involves electrolysis between nickel electrodes of a solution of the organic compound in anhydrous hydrogen fluoride.

Fluorocarbons find a variety of uses because of their high thermal stability, chemical inertness, and resistance to electrical breakdown. The reasons for the exceptional inertness of fluorocarbons are not yet clear, but one possible explanation is that the fluorine atoms pack closely together, thus shielding the carbon-carbon bonds from chemical attack.

One important carbon-fluorine substance is Teflon, a trade name for the polymer of tetrafluoroethylene, $F_2C{=}CF_2$. The self-lubricating properties of Teflon make it very useful in valves and sealed bearings. Another commercially important compound is dichlorodifluoromethane, CCl_2F_2, one of a number of compounds called Freons. Dichlorodifluoromethane is widely used as the "working" liquid in refrigerators because of its low boiling point, noninflammability, and physiological inertness.

As far as man is concerned the most important reaction that hydrocarbons undergo is air oxidation, which is accompanied by evolution of heat and is the basis for their use as fuels. For example, the combustion of octane

$$C_8H_{18}(liq) + \tfrac{25}{2}\,O_2(g) \rightarrow 8\,CO_2(g) + 9\,H_2O(g) \qquad \Delta H^0 = -1213 \text{ kcal mole}^{-1} \qquad (19.3)$$

releases large amounts of energy. The final products of combustion of hydrocarbons when oxygen is plentiful are carbon dioxide and water. When oxygen is limited, which frequently happens in the internal combustion engine, carbon monoxide is produced.

Although human beings can tolerate carbon dioxide concentrations up to 10% before suffocation, carbon monoxide is far more toxic. Concentrations of 1000 ppm (parts per million by volume) of carbon monoxide can produce unconsciousness in one hour and death in four hours. The explanation of the extreme sensitivity of man to carbon monoxide is related to the ability of this substance to form very stable coordination compounds. This subject is discussed further in Chapter 20 in connection with hemoglobin, a protein that contains iron and is the oxygen carrying component of the blood.

The principal source of hydrocarbons is petroleum. Crude petroleum contains a wide range of hydrocarbons, which are separated into fractions by distillation. The economics of the petroleum industry is centered on the production of gasoline, a mixture of hydrocarbons in the C_5 to C_{12} range. To increase the yield of gasoline from petroleum, higher alkanes are heated to high temperatures in the absence of air to "crack" them (break them into smaller molecules). In addition, hydrocarbons below the C_5 range can be alkylated to form larger molecules. Hydrocarbons obtained from petroleum are major starting materials in industrial synthetic chemistry. Including rubber, plastics, detergents, and textiles, more than 2000 products are obtained from petroleum and natural gas.

Alkenes and Alkynes

Alkenes, or **olefins,** are hydrocarbons that contain carbon-carbon double bonds and have the general formula C_nH_{2n}. **Alkynes,** or **acetylenes,** contain carbon-carbon triple bonds and have the general formula C_nH_{2n-2}. Because they contain fewer than the maximum number of hydrogen atoms, alkenes and alkynes are collectively referred to as **unsaturated** hydrocarbons. In general, unsaturated compounds are more reactive than saturated hydrocarbons. The multiple bonds are reactive sites and readily undergo addition reactions with small molecules such as hydrogen, halogens, water, or hydrogen halides.

It is a general rule (although not inviolate) that multiple bonds are both shorter and stronger, with respect to dissociation, than single bonds. This is well illustrated in comparisons of alkanes, alkenes, and alkynes. The average bond dissociation energies and average bond distances for these three classes of compounds are given below:

Compound	*Average carbon-carbon bond energy (kcal mole^{-1})*	*Bond distance (Å)*
$RCH_2—CH_2R$	83	1.54
$RCH=CHR$	145	1.35
$RC\equiv CR$	199	1.20

The symbol R is commonly used by organic chemists to designate H or an alkyl group, such as methyl (CH_3), ethyl (C_2H_5), propyl (C_3H_8), or butyl (C_4H_9). In discussions about general reactions or classes of compounds, it is a convenient abbreviation.

It is common practice to describe the σ bonds in unsaturated molecules in terms of hybrid orbitals (valence bond theory) and the π bonds using molecular orbital theory. The justification for this amalgamation of theories is simply that it proves convenient. Molecular geometry is readily explained using hybrid orbitals; reactivity and spectra are easily interpreted using molecular orbital terminology. Since the σ bonding orbitals are generally much lower in energy than π orbitals, the highest energy electrons, which are the chemically and spectroscopically "active" electrons, are in π orbitals. This bonding approach will be illustrated for ethene, which is the simplest member of the alkene series.

Ethene, also called ethylene, has the formula $H_2C{=}CH_2$. All six atoms lie in the same plane, and the bond angles are approximately 120°. From Table 8.1 we see that the appropriate hybrid orbitals for carbon are sp^2. If we place the C_2H_4 molecule in the yz plane, the $2p_x$ orbitals are available to form bonding and antibonding π molecular orbitals, as shown in Figure 19.3.

Ethene contains twelve valence electrons, four from each carbon, one from each hydrogen. From Figure 19.3 we see that there are five σ bonds in the molecule, meaning that ten electrons will be in relatively low energy σ orbitals. The remaining two electrons will thus be in the π bonding orbital, leading to the double bond (one σ and one π) between the two carbon atoms.

If one of the hydrogen atoms in ethene is replaced by a methyl group, propene (also called propylene), $CH_3CH{=}CH_2$, is obtained. The bonding picture is similar to that for ethene with the chemically active site again being the π electron system. Due to the perturbing effect of the methyl group, the absolute values for the energies of the π and π^* orbitals are different, but the relative positions remain the same.

With butene, C_4H_8, two new types of isomerism arise, leading to a total of four butene isomers. The isomers are obtained because the double bond can be located between either the first two carbons or the middle two carbons:

$$\overset{1}{C}H_2{=}\overset{2}{C}H\overset{3}{C}H_2\overset{4}{C}H_3 \qquad \text{and} \qquad \overset{1}{C}H_3\overset{2}{C}H{=}\overset{3}{C}H\overset{4}{C}H_3$$

1-butene   2-butene

Placing the double bond between C_3 and C_4 does not produce a new isomer but is simply another way of writing 1-butene. Since 1- and 2-butene differ only in the position of the double bond, they are called **position isomers.** The carbon atoms in the chain are numbered, and the number before the name indicates the position of the double bond or other functional group.

A third butene isomer is obtained owing to branching of the alkyl chain:

$$CH_2{=}\overset{\overset{\displaystyle CH_3}{|}}{C}CH_3$$

2-methylpropene

Compounds that contain branched chains are properly named as derivatives of the longest alkyl chain. No chain branching is possible in 2-butene.

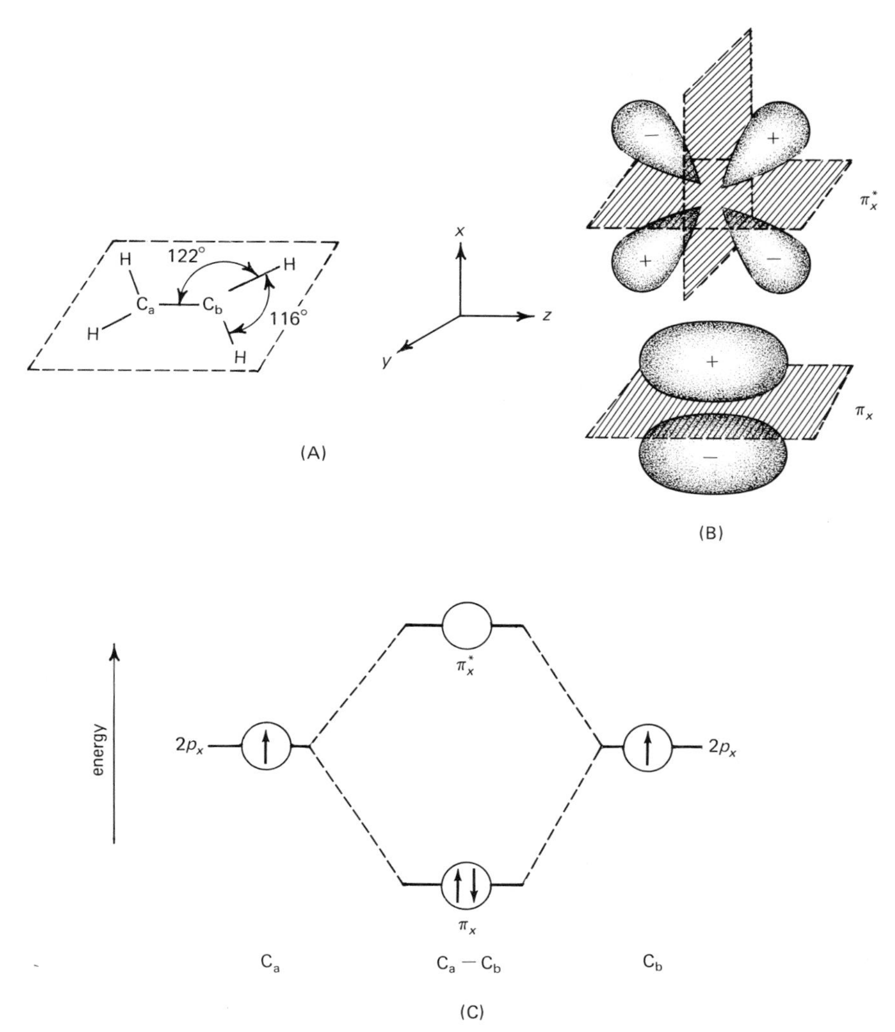

Figure 19.3. Bonding in the ethene molecule. In (A) the planar ethene molecule is in the *yz* plane. Each line represents a σ bond formed by overlap of sp^2 hybrid orbitals of carbon with each other or with a l*s* orbital of hydrogen. The π_x and π_x^* molecular orbitals are shown schematically in (B), and the molecular orbital energy level diagram for the π system in (C). The two electrons for the π system go into the low energy π bonding molecular orbital.

Unlike carbon-carbon single bonds, which have a very small energy barrier to rotation (about 3 kcal $mole^{-1}$), rotation about carbon-carbon double bonds requires considerable energy (about 60 kcal $mole^{-1}$). Rotation about a σ bond does not greatly affect the net overlap of orbitals and thus requires little energy. But rotation about a π bond seriously reduces the orbital overlap, which provides a partial explanation for the high energy barrier to rotation about multiple bonds.

Because of the restricted rotation about a double bond, two different spatial arrangements for 2-butene are possible. All four carbon atoms are in the same plane. The two methyl groups may be on the same side of the double bond (*cis*), or on opposite sides (*trans*):

$$\begin{array}{ccc} CH_3 & & CH_3 \\ & C{=}C & \\ H & & H \end{array} \qquad \begin{array}{ccc} CH_3 & & H \\ & C{=}C & \\ H & & CH_3 \end{array}$$

cis-2-butene *trans*-2-butene

This is one example of **geometric isomerism,** which is a type of stereoisomerism. Steroisomers have the same structural formulas but differ in the arrangement of atoms in space.

Finally, it should be noted that the four isomers of C_4H_8 have slightly different physical properties and can be separated. For example, the normal boiling points are: 2-methylpropene, −6.6 °C; 1-butene, −6.3 °C; *cis*-2-butene, 3.7 °C; and *trans*-2-butene, 0.9 °C.

Other stereoisomers exhibit even wider variations in their physical properties than those found with the butenes. For example, the normal boiling point of *cis*-1, 2-dichloroethene is 60.3 °C, while that of the *trans* isomer is 47.5 °C.

$$\begin{array}{ccc} Cl & & Cl \\ & C{=}C & \\ H & & H \end{array} \qquad \begin{array}{ccc} Cl & & H \\ & C{=}C & \\ H & & Cl \end{array}$$

cis-1,2-dichloroethene *trans*-1,2-dichloroethene

The simplest member of the alkyne series is ethyne, HC≡CH, more commonly called acetylene. It is a linear molecule. Thus the appropriate orbitals of carbon for the σ framework are *sp* hybrid orbitals. Taking the *z* axis as the molecular axis, the $2p_x$ and $2p_y$ orbitals of carbon can form π bonds. Since acetylene is isoelectronic with molecular nitrogen, we expect similar bonding in the two molecules. The bonding in the acetylene molecule is illustrated in Figure 19.4. The π_x and π_y bonding molecular orbitals are of equal energy **(degenerate),** as are the π_x^* and π_y^* antibonding orbitals. There are ten valence electrons in acetylene, four from each carbon and one from each hydrogen. Six electrons are used in the σ framework to form three σ bonds, one C—C and two C—H. The remaining four electrons occupy the π_x and π_y bonding orbitals, the next lowest energy orbitals, forming two π bonds. The triple bond characteristics of acetylene are thus readily explained.

By substitution of alkyl groups for hydrogen in acetylene, higher members of the alkyne series are obtained. Because the parent molecule is linear, the isomerism possibilities are not as great as in the alkene series. For example, position isomers of butyne are known, but there are no stereoisomers of 2-butyne.

$$HC{\equiv}CCH_2CH_3 \qquad CH_3C{\equiv}CCH_3$$

1-butyne 2-butyne

In 2-butyne all four carbon atoms are in a straight line, and no *cis-trans* isomerism is possible.

Alkenes are obtained in industrial quantities by the cracking of petroleum. The

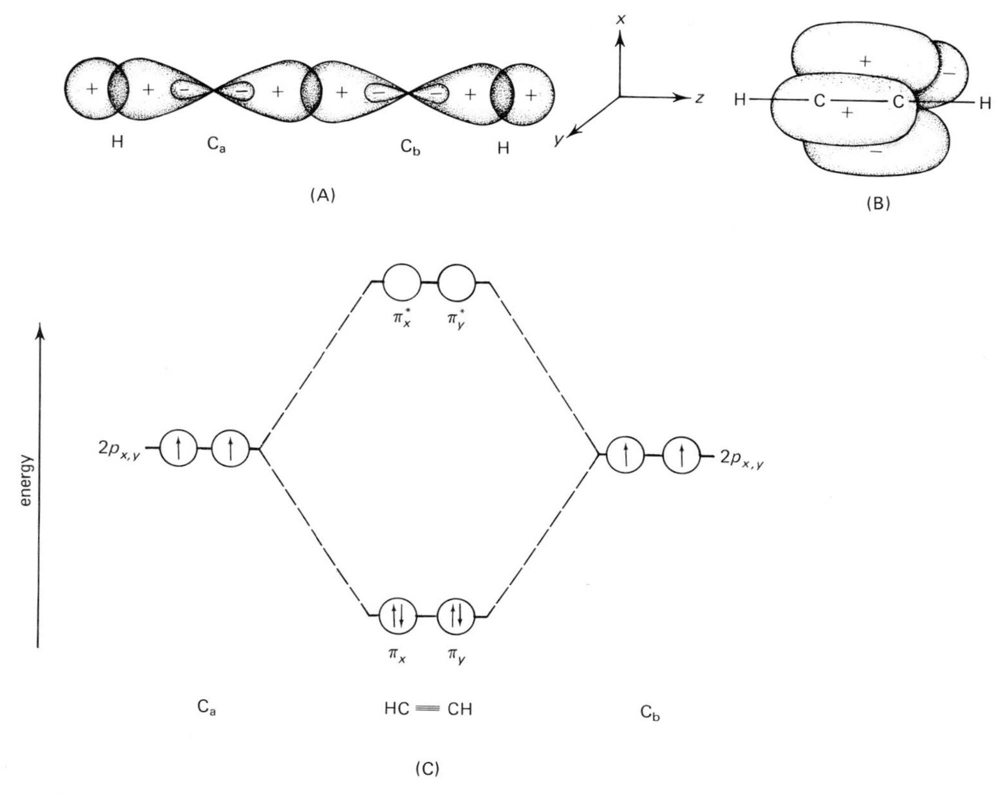

Figure 19.4. Bonding in the acetylene molecule. Only the σ framework is shown in (A), the π orbitals in (B). Since overlap of the $2p_x$ orbitals is comparable to that of the $2p_y$ orbitals, the resultant π orbitals are degenerate in energy, as shown in the molecular orbital energy diagram in (C).

smaller alkenes are obtained in pure form by fractional distillation. Higher alkenes, which are difficult to separate, remain as valuable components of gasoline.

In the laboratory, alkenes are frequently obtained by removal of a small molecule from a larger molecule, commonly referred to as **elimination reactions.** One of the best procedures is dehydration of an alcohol using sulfuric acid. For example, heating 2-butanol with 65% H_2SO_4 yields both 1- and 2-butene, with 2-butene being the major product:

$$CH_3CH_2\underset{}{\overset{OH}{\overset{|}{C}}}HCH_3 \xrightarrow[\text{heat}]{H_2SO_4} CH_3CH_2CH{=}CH_2 + CH_3CH{=}CHCH_3 \qquad (19.4)$$

In organic chemistry it is common to obtain more than one product. By careful selection of reaction conditions and catalysts, it is often possible to maximize the yields of a desired product.

The only alkyne of industrial importance is acetylene. It is prepared either by partial oxidation of methane or hydrolysis of calcium carbide, CaC_2. In the latter process, acetylene is obtained in a few steps from three abundant, cheap raw materials: water, coal, and limestone.

$$CaCO_3(c) \xrightarrow{\text{heat}} CaO(c) + CO_2(g) \tag{19.5}$$

$$CaO(c) + 3\ C(c) \rightarrow CaC_2(c) + CO(g) \tag{19.6}$$

$$CaC_2(c) + 2\ H_2O(liq) \rightarrow HC{\equiv}CH(g) + Ca(OH)_2(c) \tag{19.7}$$

Acetylene is an important starting material for synthesis of many organic compounds, including acetic acid (CH_3COOH) and several unsaturated compounds used for polymerization to plastics and rubber. Many of the synthetic uses of acetylene resulted from work in Germany during World War II, when supplies of coal were more abundant than petroleum. If petroleum usage continues at present rates, this same situation will soon face the world in general. Large quantities of acetylene are also used as a fuel for the oxyacetylene torch, which produces a very hot flame (about 2800°C).

The most significant distinction between the reactions of saturated compounds and unsaturated compounds is that alkanes undergo substitution reactions (equation 19.1), whereas alkenes and alkynes undergo addition reactions. Less rigorous reaction conditions are required for additions than substitutions, and unsaturated compounds are said to be more chemically reactive. Some examples of addition reactions are (1) hydrogenation, in which the choice of catalyst is very important:

$$RC{\equiv}CH \xrightarrow[\text{Pb/Pd}]{H_2} RCH{=}CH_2 \xrightarrow[\text{Ni}]{H_2} RCH_2CH_3 \tag{19.8}$$

(2) halogenation,

$$RCH{=}CH_2 + Br_2 \rightarrow \underset{\substack{|\\ Br}}{}\!\!RCH(Br){-}CH_2(Br) \tag{19.9}$$

and (3) hydration

$$CH_2{=}CH_2 + HOH \rightarrow \underset{\substack{\text{ethanol}\\ \text{(ethyl alcohol)}}}{CH_3CH_2OH} \tag{19.10}$$

Another important property of alkenes is their tendency to link end-to-end to form polymers. The polymerization of tetrafluoroethylene to Teflon, $\left(CF_2CF_2\right)_n$, has already been mentioned. Another example is polypropylene, which is formed by combining individual propene (also called propylene) molecules to produce a long chain polymer:

$$n\ HC(CH_3){=}CH_2 \xrightarrow{\text{catalyst}} \left(HC(CH_3){-}CH_2\right)_n$$

Again, careful selection of catalysts and conditions is necessary to produce the desired type of polymer. Polymerization reactions and polymers will be considered further in Chapter 21.

A distinctive feature of acetylene chemistry is the acidity (in the Lewis sense) of the proton attached to a carbon-carbon triple bond. For example, sodium reacts with acetylene to liberate hydrogen gas and form sodium acetylide:

$$HC{\equiv}CH + Na \rightarrow HC{\equiv}C^{-}Na^{+} + \tfrac{1}{2}H_2 \qquad (19.11)$$

Sodium acetylide reacts with water to regenerate acetylene:

$$HC{\equiv}C^{-}Na^{+} + HOH \rightarrow HC{\equiv}CH + NaOH \qquad (19.12)$$

We thus conclude that acetylene is a weaker acid than water, or conversely, that acetylide ion is a stronger base than hydroxide.

If sodium acetylide is made to react with an alkyl halide, such as bromoethane (also called ethyl bromide), a substituted acetylene results:

$$HC{\equiv}C^{-}Na^{+} + BrCH_2CH_3 \rightarrow \underset{\text{1-butyne}}{HC{\equiv}CCH_2CH_3} + NaBr \qquad (19.13)$$

Sodium acetylide is used in the synthesis of most higher alkynes. By continuing the reaction, the triple bond can be positioned at any desired place in the chain.

Aromatic Compounds

The term "aromatic compounds" arose originally because many of these substances have strong aromas. It was later discovered that these aromatic compounds have some structural and chemical properties in common and the distinction of odor was lost. Today, chemists generally consider aromatic compounds as being derivatives of benzene, C_6H_6, the simplest member of the series; however, the general concept of aromatic behavior can be extended to many other ring compounds.

The structural properties of aromatic compounds can be illustrated with benzene. As shown in Figure 19.5A, benzene has a planar ring structure with 120° bond angles. The carbon-carbon bond distances are all equal and intermediate between carbon-carbon single (1.54 Å) and double (1.35 Å) bond distances. In valence bond theory, with emphasis on hypothetical structures, benzene can be represented by the two resonance structures shown in Figure 19.5B. The appropriate hybrid orbitals for carbon are sp^2. If the yz plane is chosen as the molecular plane, the π bonds are formed by utilizing the $2p_x$ orbitals of each carbon. The concept of alternate single and double bonds in resonance structures qualitatively explains the observed bond distances.

In the molecular orbital picture it is again convenient to consider the σ and π orbitals independently. Benzene has 30 valence electrons, four from each carbon and one from each hydrogen. To form the six C—C σ bonds and six C—H σ bonds, 24 electrons are needed. The remaining six electrons comprise the π system of the entire molecule. Construction of the π orbitals for benzene involves symmetry and mathematical arguments that will not be presented here. The conclusions, however, are straightforward: six $2p_x$ atomic orbitals form six π molecular orbitals, three bonding and three antibonding. The six remaining electrons in benzene occupy the three lower energy π molecular orbitals. Because the electrons are viewed as belonging to the molecule as a whole, they are said to be "delocalized" and not restricted between any two atoms. The symbol in Figure 19.5D designates

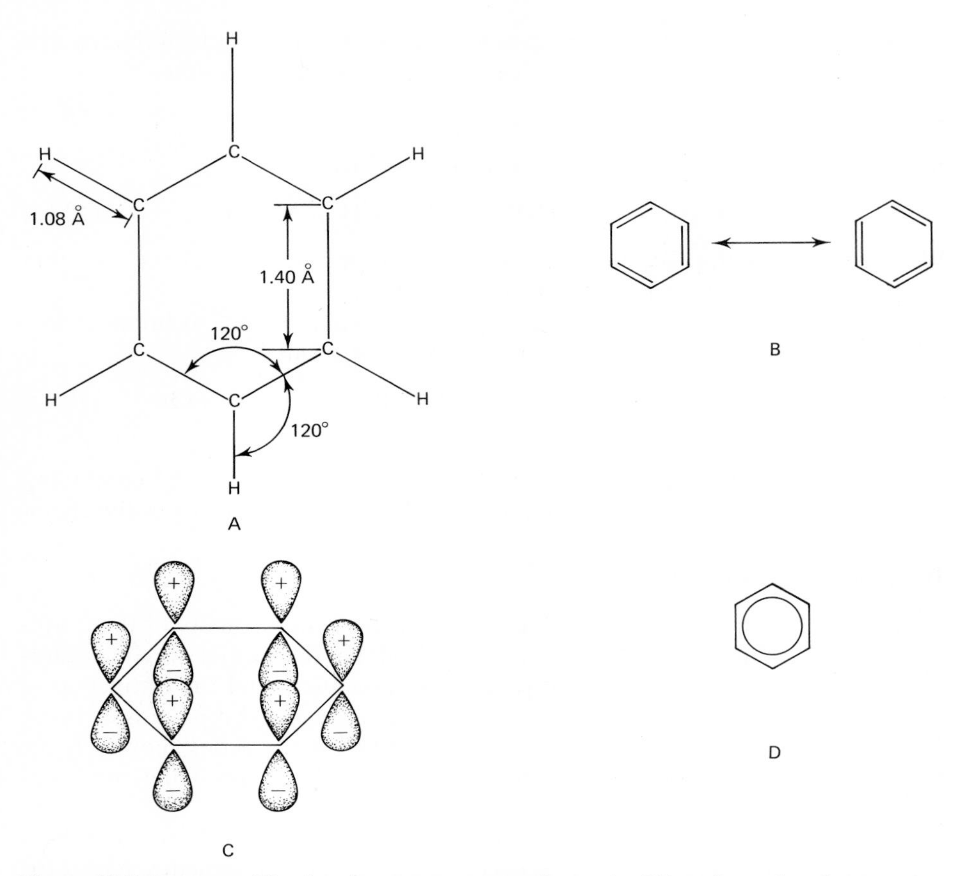

Figure 19.5. Benzene. The bond parameters are shown in (A), valence bond resonance structures in (B), the atomic $2p_x$ orbitals available to form the π system in (C), and the molecular orbital representation suggesting delocalized π electrons in (D). Each point of the hexagons in (B) and (D) represents a carbon atom with a hydrogen bonded to it. The hydrogen atoms are not drawn but are understood to be there.

this delocalization. Finally, a word of caution about symbols: no matter how perfect, they must never be confused with the things symbolized; in the words of Professor Ingold, they serve only as a sort of "intellectual scaffolding."

Consistent with the delocalized π bond structure, aromatic compounds do not readily undergo addition reactions similar to those of alkenes and alkynes. The ring structure is remarkably stable towards strong oxidizing and reducing agents as well as heat. Most reactions of benzene involve substitution of other groups for one or more hydrogen atoms. Obviously, the number of possible derivatives is huge. Only a few of the more important types of substitution reactions can be considered here.

Halogenation of the benzene ring is believed to proceed via an ionic mechanism. Metallic iron is a common catalyst, but it is probably first oxidized to Fe(III) as indicated for the following chlorination reaction:

$$Fe + \tfrac{3}{2}Cl_2 \rightarrow FeCl_3 \qquad (19.14)$$

$$FeCl_3 + Cl_2 \rightarrow FeCl_4^- + Cl^+ \qquad (19.15)$$

$$+ Cl^+ \rightarrow \quad \text{Cl} \quad + H^+ \qquad (19.16)$$

$$H^+ + FeCl_4^- \rightarrow HCl + FeCl_3 \qquad (19.17)$$

The net reaction may be simply described by

$$+ Cl_2 \xrightarrow{FeCl_3} \quad \text{Cl} \quad + HCl \qquad (19.18)$$

chlorobenzene

Free Cl^+ ions as indicated in equation (19.15) are probably not actually formed. It seems more likely that the ferric chloride only polarizes the chlorine molecule, $Cl^{\delta-}{-}Cl^{\delta+}$, and the positive end of the molecule then attacks the π electron cloud. In any event, the reaction is properly described as an acid-base reaction. Ferric chloride functions as an acidic catalyst and benzene as a Lewis base.

Substitutions of aliphatic groups on the benzene ring are called alkylation reactions. Like the halogenation reaction, alkylation reactions are believed to proceed through an ionic acid-base mechanism. As an example consider formation of ethylbenzene using chloroethane as a reagent and aluminum chloride as an acid catalyst:

$$CH_3CH_2Cl + AlCl_3 \rightarrow AlCl_4^- + CH_3CH_2^+ \qquad (19.19)$$

$$+ CH_3CH_2^+ + AlCl_4^- \rightarrow \quad {-}CH_2CH_3 + HCl + AlCl_3 \qquad (19.20)$$

Clearly the net reaction is

$$+ CH_3CH_2Cl \xrightarrow{AlCl_3} \quad {-}CH_2CH_3 + HCl \qquad (19.21)$$

Whether the free ethyl ion (called a **carbonium** ion) actually exists is also questionable, but it does provide a relatively simple means of visualizing the process.

Nitration reactions are also pictured as proceeding through an ionic substitution mechanism. In this case sulfuric acid is used as the acid catalyst and nitric acid as the nitrating agent. The nitronium ion, NO_2^+, is considered to be the reactive intermediate:

$$HNO_3 + H_2SO_4 \rightarrow NO_2^+ + HSO_4^- + H_2O \qquad (19.22)$$

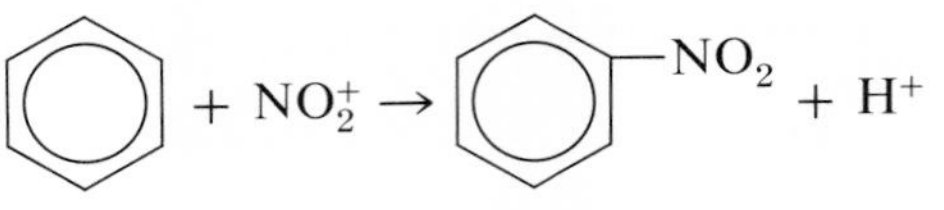

$$C_6H_6 + NO_2^+ \rightarrow C_6H_5NO_2 + H^+ \qquad (19.23)$$

nitrobenzene

The formation of the nitronium ion in (19.22) seems likely because it is known to exist in compounds such as solid nitronium perchlorate, $NO_2^+ClO_4^-$. In addition, sulfuric acid has a strong affinity for water, which should tend to shift reaction (19.22) to the right (recall LeChatelier's principle).

All three substitution reactions considered above are believed to involve attack by an electrophilic ("electron loving") reagent, E^+, on the π cloud of benzene. The resultant intermediate, itself a carbonium ion, is less stable than the substituted benzene and undergoes a hydrogen abstraction reaction by some base, B^-:

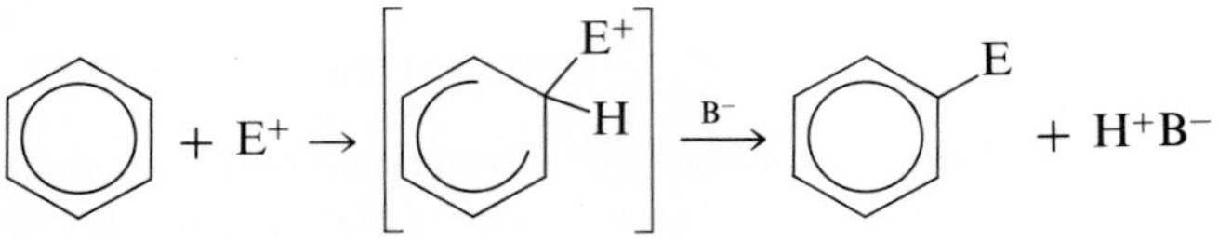

$$C_6H_6 + E^+ \rightarrow \left[C_6H_6E^+\right] \xrightarrow{B^-} C_6H_5E + H^+B^- \qquad (19.24)$$

The base, B^-, must be a stronger base than benzene, a condition easily met since benzene is a very weak base.

In the above reactions the substituent has always been placed at the same position on the ring. This was merely for convenience, since all carbons of the benzene ring are equivalent.

A question now arises: At which of the five remaining carbons does the second substituent go? If all of these sites were equivalent, the distribution would be statistical as follows:

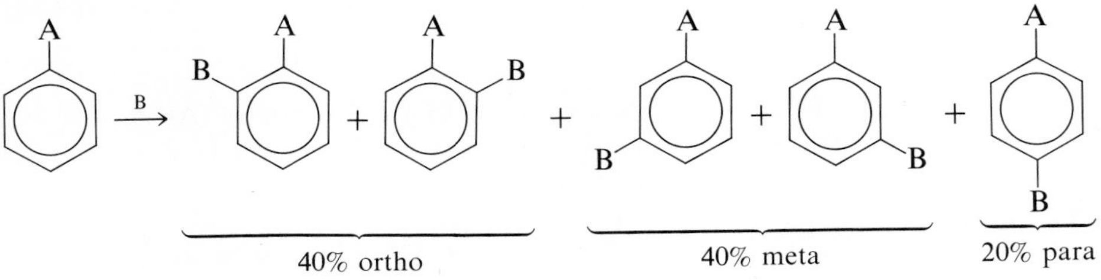

$$C_6H_5A \xrightarrow{B} \underbrace{o\text{-}C_6H_4AB + o\text{-}C_6H_4AB}_{40\%\ \text{ortho}} + \underbrace{m\text{-}C_6H_4AB + m\text{-}C_6H_4AB}_{40\%\ \text{meta}} + \underbrace{p\text{-}C_6H_4AB}_{20\%\ \text{para}}$$

Ortho refers to substituents on adjacent carbons, **meta** to substituents with one carbon atom between them, and **para** to substituents diagonally across the ring.

Substitution, however, is not random. Some substituents cause primarily ortho-para substitution, whereas other substituents give mostly meta derivatives. How a substituent alters the electron density distribution in the ring is believed to be responsible for the directing effect by making attack at certain carbons more likely than at others. The explanation of just how electron density in the ring is shifted is best left for organic chemistry texts. We can, however, consider some of the practical applications for synthetic organic chemists.

Assume that we are interested in preparing some trinitrotoluene (TNT). Noting that nitro groups are meta directors and methyl groups are ortho-para directors, it is clear that we should do an alkylation reaction first to form toluene, $C_6H_5CH_3$, and then nitrate the toluene:

CH_3Cl / $AlCl_3$ → CH_3 (toluene) → HNO_3 / H_2SO_4 → CH_3, O_2N, NO_2, NO_2 (trinitrotoluene)

toluene trinitrotoluene

Nitrating first would not give us the desired product. (Please note that we do *not* suggest that students attempt this reaction. TNT is exceedingly dangerous.)

We should also note that several **condensed ring** aromatic compounds are known. Naphthalene ($C_{10}H_8$), anthracene ($C_{14}H_{10}$), and phenanthrene ($C_{14}H_{10}$) are three of the simpler members of the series.

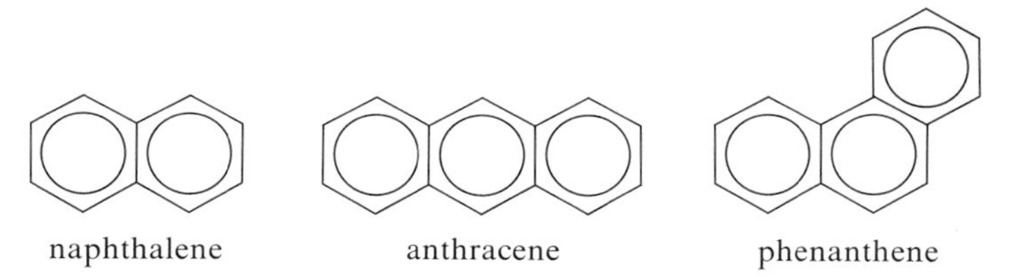

naphthalene anthracene phenanthene

The relationship between these compounds and benzene should be obvious. The bonding for condensed ring compounds is considered to be similar to that in benzene but with greater delocalization effects. In molecular orbital terminology the electrons can move throughout the entire molecule; the larger the molecule, the greater is the stabilization achieved through delocalization. Condensed ring compounds undergo substitution reactions similar to those of benzene.

The chief sources of aromatic compounds today are petroleum and coal tar, which is the nonvolatile liquid that remains after coal is converted to coke by heating in the absence of air. Aromatic compounds find wide use in the chemical industry as starting materials for dyes, pharmaceuticals, and plastics. Another use, which is questionable considering their carcinogenic (cancer-producing) properties, is as an additive to gasoline to improve octane ratings.

Alcohols and Ethers

Alcohols and ethers are two general classes of organic compounds that contain carbon-oxygen single bonds.

The atom or group of atoms that defines the structure of a family of organic compounds, and also determines their properties, is called a **functional group.** Using R to represent a hydrocarbon group, alcohols have the general formula ROH, whereas ethers are ROR. Thus the functional group of alcohols is —OH and that of ethers is —O—. Although the alkyl portion of the molecule can also undergo reactions, the functional group is generally the most reactive part of any molecule.

Table 19.1. Names, Formulas, and Properties of Alcohols

Formula	*Name*	*Melting point (°C)*	*Boiling point (°C)*
CH_3OH	Methanol (methyl alcohol)	−98	65
CH_3CH_2OH	Ethanol (ethyl alcohol)	−117	79
$CH_3(CH_2)_2OH$	1-Propanol (*n*-propyl alcohol)	−127	97
$CH_3CH(OH)CH_3$	2-Propanol (*iso*-propyl alcohol)	−89	82
$CH_3(CH_2)_3OH$	1-Butanol (*n*-butyl alcohol)	−90	118
$CH_3(CH_2)_4OH$	1-Pentanol (*n*-pentyl alcohol)	−79	137
$(CH_3)_3COH$	2-Methyl-2-propanol (*tert*-butyl alcohol, *t*-butyl alcohol)	25	82

A few of the simpler alcohols are listed in Table 19.1 along with their normal melting and boiling points. Since both systematic and common names (in parentheses) are often used, students should be familiar with both.

Recall that a number preceding the name of an organic compound indicates placement of the functional group. Thus in 2-propanol (isopropyl alcohol), the OH group is on the second carbon of a three-carbon chain.

$$\begin{array}{c} OH \\ | \\ CH_3—CH—CH_3 \end{array} \qquad \begin{array}{c} OH \\ | \\ CH_3—C—CH_3 \\ | \\ CH_3 \end{array}$$

2-propanol   2-methyl-2-propanol

In 2-methyl-2-propanol (tertiary butyl alcohol), two groups are attached to the second carbon of the three-carbon chain. It should be obvious that moving the OH group around the carbon chain leads to position isomers.

Note that the primary alcohols, in which the OH group is attached to the end carbon, have large liquid ranges. The high boiling points of alcohols compared to the parent hydrocarbons can be attributed to hydrogen bonding. We therefore expect, and observe, that alcohols have larger than "normal" values for the entropy of vaporization (Trouton's constant). The "structure" in liquid alcohols is less extensive than in water however, since only one hydrogen per molecule is capable of hydrogen bonding.

Alcohols may be regarded as derivatives of water in which one hydrogen is replaced by a hydrocarbon group. Due to the similarity in structure and hydrogen bonding, the first three alcohols listed in Table 19.1 are completely miscible in water. As the hydrocarbon part of the molecule becomes larger, it exerts greater influence; therefore the solubility in water of alcohols with large R is slight. For example, the solubility of 1-butanol in water is approximately 1.1 mole liter^{-1} whereas that of 1-pentanol is 0.3 mole liter^{-1}, both at 20°C. Higher alcohols are even less soluble.

Like water, alcohols can function either as acids or bases. Alcohols will accept a proton from a strong acid such as sulfuric acid

$$H_2SO_4 + ROH \rightleftharpoons ROH_2^+ + HSO_4^- \tag{19.25}$$

or donate protons to strong bases such as hydride

$$ROH + NaH \rightleftharpoons NaOR + H_2 \qquad (19.26)$$

If ethanol is reacted with sodium hydride and excess ethanol evaporated after evolution of hydrogen has ceased, sodium ethoxide ($NaOC_2H_5$) is recovered as a white solid. The RO^- ion has the general name of alkoxide. Since alcohols are more weakly acidic than water, reactions such as (19.26) are generally less violent in alcohols than in water.

The large liquid ranges plus the solubility of many organic and inorganic compounds in alcohols make them useful solvents for a variety of reactions.

Methanol, commonly called wood alcohol because it was formerly obtained by destructive distillation of wood, is obtained industrially now by reacting carbon monoxide with hydrogen:

$$CO(g) + 2\,H_2(g) \rightarrow CH_3OH(g) \qquad \Delta H^0 = -21.5 \text{ kcal mole}^{-1} \qquad (19.27)$$
$$\Delta G^0 = -5.9 \text{ kcal mole}^{-1}$$

High temperatures (300–400°C), high pressures (200–300 atm), and catalysts such as $ZnO\text{-}Cr_2O_3$ mixtures are necessary for good yields in short times. Methanol is widely used as a solvent and chemical intermediate.

Ethanol (grain alcohol) is traditionally and extensively prepared by fermentation of sugars by yeast. Yeast contains a mixture of enzymes, commonly referred to as "zymase," which catalyze breakdown of sugar molecules, as indicated in the oversimplified reaction equation:

$$C_6H_{12}O_6 \xrightarrow{\text{zymase}} 2\,CO_2 + 2\,C_2H_5OH \qquad (19.28)$$

This is probably the oldest synthetic chemical process known to man; an ancient Babylonian tablet gives a recipe for making beer. Beer is made by fermenting the sugar from sprouted barley (malt) and adding hops to give the characteristic flavor. It is thus a dilute solution of ethanol and carbohydrates in water (3–5% C_2H_5OH by volume). If grape sugar is used, wine results. Since yeast cannot survive concentrations more than about 14% alcohol by volume, there is a natural limitation on the concentration of alcohol in wines. The higher alcohol concentrations in whiskeys are obtained by distillation.

Ethanol is also obtained industrially by addition of water to ethene. The hydration reaction proceeds in two steps using sulfuric acid as an intermediate:

$$CH_2{=}CH_2 + HOSO_3H \rightarrow CH_3CH_2OSO_3H$$
$$CH_3CH_2OSO_3H + H_2O \rightarrow CH_3CH_2OH + H_2SO_4$$

Hydration of the appropriate alkene and other specialized processes are used to obtain the higher alcohols.

Many organic compounds contain more than one functional group. One example is 1,2-ethanediol, commonly known as ethylene glycol:

$$\begin{array}{cc} OH & OH \\ | & | \\ H_2C & \!\!-\!\!- CH_2 \end{array}$$

ethylene glycol

Because of its low melting point (−16 °C), high boiling point (197 °C), and great miscibility with water, aqueous solutions of ethylene glycol are extensively used as "permanent antifreeze" in car radiators.

If the OH group is attached to a benzene ring, the properties of the compound are unlike those of typical aliphatic alcohols. For example, phenol

$$C_6H_5OH(aq) \rightarrow C_6H_5O^-(aq) + H^+(aq) \qquad (19.29)$$

phenol — phenoxide ion

is quite acidic, having an acid ionization constant of 1.1×10^{-10}. By comparison, the acid ionization constant of ethanol is about 10^{-16}. The greater ionization of phenol has been attributed to the stability of the phenoxide ion in which the charge is delocalized in the π system of the benzene ring.

The existence of intermolecular hydrogen bonding in phenol is indicated by its high boiling point, 182 °C. The solubility of phenol in water is about 1 mole liter^{-1} at 20 °C, comparable to that of aliphatic alcohols of similar molecular weight. Aqueous solutions of phenol, commonly called carbolic acid, are used as general disinfectants.

Concentrated sulfuric acid can function not only as a protonating agent towards alcohols but also as an acid catalyst and dehydrating agent. The product obtained is dependent upon conditions. For example, heating ethanol with concentrated sulfuric acid to 180 °C produces ethene:

$$CH_3CH_2OH + H_2SO_4 \xrightarrow{180\,^\circ C} CH_2{=}CH_2 + H_2SO_4 \cdot H_2O \qquad (19.30)$$

Heating a similar mixture to 140 °C yields diethyl ether, $C_2H_5{-}O{-}C_2H_5$. In effect, one molecule of water is removed from two molecules of ethanol:

$$2\,CH_3CH_2OH + H_2SO_4 \xrightarrow{140\,^\circ C} (C_2H_5)_2O + H_2SO_4 \cdot H_2O$$

Ethers, with general formula ROR, may also be considered as formally related to water. Like water, ether molecules are bent with C—O—C angles of ~110°, as predicted from VSEPR theory. Ether molecules thus have a small net dipole moment (1.18 D for diethyl ether), with the oxygen end of the molecule being negative. This polarity does not greatly affect the boiling points of ethers, which are comparable to alkanes of similar molecular weights. Since all hydrogens are attached to carbons, ethers do not exhibit hydrogen bonding. Solubility of ethers in water is also considerably less than that of alcohols.

The contrasts in physical properties between ethers and alcohols are well illustrated with dimethyl ether, CH_3OCH_3, and ethanol. Because the two compounds are isomeric, they have the same molecular weight. In contrast to the high boiling point (78°) and complete miscibility of ethanol with water, dimethyl ether boils at −24 °C and has a solubility in water of about 1.7 moles liter^{-1} at 20 °C. We must conclude that the existence of hydrogen bonding is very important in creating waterlike properties in compounds.

The most distinctive property of ethers is their relative lack of reactivity. The ether linkage is not affected by strong oxidizing or reducing agents nor by dilute acids or bases. This chemical stability, plus the polarity of ether molecules which enhances solvent properties, make ethers the solvent of choice for many chemical reactions. Hazards involved in the use of ethers include their high volatility, flammability, and tendency to be air-oxidized to explosive peroxides on long storage.

Dimethyl ether is a gas at room temperature and is used as a propellant for aerosol sprays. Diethyl ether was once used extensively as a general anesthetic, but has been replaced by halogenated hydrocarbons (such as $CHBrClCF_3$), which have less serious side effects. Higher ethers are commonly used as solvents for varnishes and lacquers.

Carbonyl Compounds

Aldehydes are compounds of the general formula RCHO; ketones have the general formula RR′CO. The groups R and R′ may be aliphatic or aromatic. The C=O group is called the carbonyl group, and aldehydes and ketones are collectively called carbonyl compounds.

The carbonyl carbon is bonded to three other atoms with bond angles of approximately 120°. We therefore say that this carbon uses sp^2 hybrid orbitals to form σ bonds. The remaining $2p$ orbital of carbon overlaps with the corresponding $2p$ orbital of oxygen to form bonding and antibonding π orbitals. Only the π bonding orbital is occupied, leading to one σ and one π bond for the carbonyl group.

Although the carbonyl double bond is a very strong bond with respect to dissociation (174 kcal $mole^{-1}$), it is relatively reactive. The reactivity is undoubtedly due to the polarity of the bond, as indicated by the large dipole moments (2.3–2.8 D) for aldehydes and ketones. Oxygen is not only more electronegative than carbon, causing an unequal sharing of the bonding electron cloud, but also has two lone pairs of electrons.

Some aldehydes and most ketones are prepared by oxidation or dehydrogenation of the appropriate alcohol. Dehydrogenation, removing a molecule of hydrogen, is generally accomplished in the absence of air to avoid secondary air-oxidation of the desired product. For example, dehydrogenation of methanol yields formaldehyde, HCHO, the simplest aldehyde:

$$CH_3OH \xrightarrow[550\text{–}600\,^\circ C]{Cu} HCHO + H_2 \tag{19.31}$$

Formaldehyde, also called methanal, is a gas at room temperature with a normal boiling point of −21°C. Since both hydrogens are attached to carbon, hydrogen bonding is not observed. Formaldehyde readily polymerizes to a solid, however. Aqueous solutions of formaldehyde, called formalin, are widely used as disinfectants and for preservation of biological specimens.

The most important ketone is acetone, also called dimethyl ketone or propanone. It is also obtained by dehydrogenation of an alcohol, in this case 2-propanol:

$$CH_3\overset{\overset{\displaystyle OH}{|}}{C}HCH_3 \xrightarrow[200\text{–}300\,^\circ C]{Cu} CH_3\overset{\overset{\displaystyle O}{\|}}{C}CH_3 + H_2 \tag{19.32}$$

Acetone is a liquid at room temperature with a normal boiling point of 56°C. Its high volatility and ability to dissolve a variety of organic compounds make it a common solvent for varnishes, lacquers, and plastics. Acetone is miscible with water in all proportions and is commonly used for drying glassware in the laboratory.

The lighter aliphatic aldehydes have sharp unpleasant odors, but those with 8–12 carbon atoms have flowery odors that make them useful in perfumery. Many ketones also have pleasing odors. Aldehydes and ketones occur widely in nature, often contributing flavor and odor to plants. Some naturally occurring compounds and their sources are shown below.

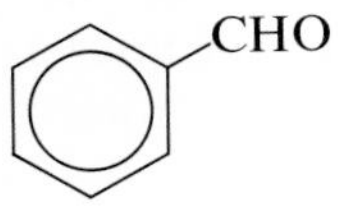

benzaldehyde (almonds)	vanillin (vanilla)	testosterone (male sex hormone)

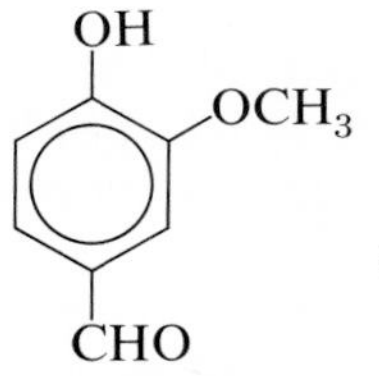

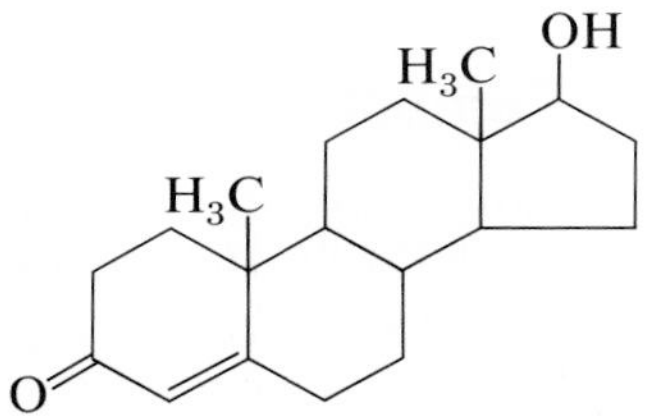

Aldehydes are readily oxidized to carboxylic acids, which contain the —COOH functional group and are discussed in the next section. Even mild oxidizing agents such as $Ag(NH_3)_2^+$ ions can oxidize aldehydes. On reduction of $Ag(NH_3)_2^+$ a silver mirror

$$RCHO + 2\,Ag(NH_3)_2^+ + 3\,OH^- \rightarrow 2\,Ag + RCOO^- + 4\,NH_3 + 2\,H_2O \quad (19.33)$$

forms on the surface of the container. The appearance of a silver mirror on treatment of an organic compound with ammoniacal silver solution is commonly used as a qualitative test for the aldehyde group.

Ketones are less easily oxidized than aldehydes; for example, they do not react with ammoniacal silver ions. But stronger oxidizing agents, such as acidic permanganate solutions, cause cleavage of the carbon chain to produce two carboxylic acids:

$$R{-}\overset{\overset{\displaystyle O}{\|}}{C}{-}R' \xrightarrow[H^+(aq)]{MnO_4^-} R\overset{\overset{\displaystyle O}{\|}}{C}{-}OH + R'\overset{\overset{\displaystyle O}{\|}}{C}{-}OH \quad (19.34)$$

Cleavage may occur on either side of the carbonyl group to produce a variety of products.

Aldehydes and ketones can be reduced to the corresponding alcohols by catalytic hydrogenation, or with strong reducing agents such as metal hydrides. Since carbonyl compounds can be both oxidized and reduced, they represent an intermediate oxidation state. However, disproportionation reactions are not characteristic and occur only in very special cases (when there is no hydrogen on a carbon adjacent to the carbonyl group).

The carbonyl group undergoes addition reactions with many polar molecules. For example, addition of hydrogen cyanide to aldehydes and the more reactive ketones is catalyzed by traces of base:

$$\underset{\delta+}{CH_3\overset{\overset{O^{\delta-}}{\|}}{C}CH_3} + \overset{\delta+}{H}—\overset{\delta-}{CN} \xrightarrow{OH^-(aq)} CH_3\underset{\underset{CN}{|}}{\overset{\overset{OH}{|}}{C}}CH_3 \qquad (19.35)$$

acetone cyanohydrin

Acetone cyanohydrin is an intermediate in the preparation of methyl methacrylate, $CH_2{=}C(CH_3)CO_2CH_3$, which polymerizes to form transparent plastics known as Plexiglas and Lucite.

Acids and Esters

Organic compounds that contain the carboxyl group,

$$—\overset{\overset{O}{\|}}{C}—OH$$

are called **carboxylic acids.** Some of the simpler carboxylic acids and their physical properties are listed in Table 19.2. Many of these compounds may be found in nature. For example, formic acid may be isolated from ants; acetic acid is found in vinegar and is the end product of fermentation of many agricultural products.

The high boiling points of carboxylic acids are indicative of hydrogen bonding. Except for benzoic acid, the acids listed are very soluble in water, which may also be attributed to hydrogen bonding. As is true for alcohols and carbonyl compounds, the solubility in water decreases as the size of R attached to the functional group increases.

The acid ionization constants in Table 19.2 are for the process

$$R\overset{\overset{O}{\|}}{C}—OH + H_2O \rightleftharpoons R\overset{\overset{O}{\|}}{C}—O^- + H_3O^+ \qquad (19.36)$$

Note that, except for formic acid, the dissociation constants are all nearly the same; the nature of R has little effect on the functional group in these cases. [See problem 34 for further discussion of equilibria represented by equation (19.36)]

Table 19.2. Physical Properties of Carboxylic Acids

Formula	*Name*	*Boiling point (°C)*	K_a *(25°C)*
HCOOH	Formic	101	1.8×10^{-4}
CH_3COOH	Acetic	118	1.8×10^{-5}
CH_3CH_2COOH	Propionic	141	1.4×10^{-5}
$CH_3(CH_2)_2COOH$	Butyric	164	1.5×10^{-5}
C_6H_5COOH	Benzoic	249	6.5×10^{-5}

When the hydrogens on the carbon adjacent to the carboxyl group are replaced with electronegative atoms, however, a pronounced effect on the dissociation constant can be obtained. The K_a of monochloroacetic acid, $ClCH_2COOH$, is 1.4×10^{-3} and that of monofluoroacetic acid, FCH_2COOH, is 2.6×10^{-3}. The extent of ionization of these acids is almost 100 times greater than that of acetic acid itself. This effect may be explained in terms of a shift of electron density toward the electronegative substituent, thus weakening the oxygen-hydrogen bond in the carboxyl group.

The most important sources of aliphatic carboxylic acids are animal and vegetable fats. Acetic acid is obtained by air oxidation of acetaldehyde, either industrially or through natural fermentation processes.

Reaction of organic acids with inorganic bases produces salts. Two important ones are the sodium salt of acetylsalicyclic acid and stearic acid.

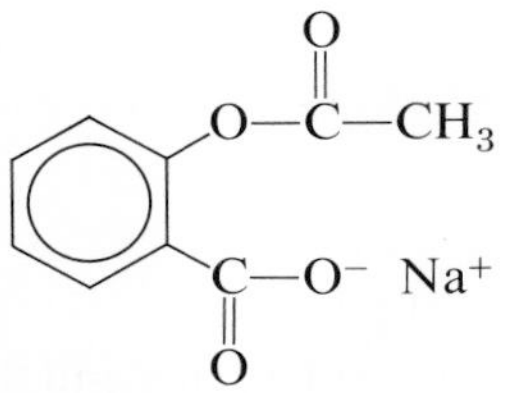

$$C_6H_4(O{-}\overset{O}{\overset{\|}{C}}{-}CH_3)(\underset{\underset{O}{\|}}{C}{-}O^-\ Na^+)$$

sodium acetylsalicylate
(aspirin)

$$CH_3(CH_2)_{16}\overset{O}{\overset{\|}{C}}{-}O^-\ Na^+$$

sodium stearate
(soap)

The cleansing action of soap is believed to arise from the solubility of the hydrocarbon "tail" of soap in grease and solubility of the polar "head" in water. The greasy "dirt" of soiled clothes can thus be rinsed away.

Organic acids react with alcohols to produce esters with the general formula, RCOOR′. Esterification reactions, described by the general equation

$$R\overset{O}{\overset{\|}{C}}{-}O{-}H + R'{-}O{-}H \rightleftharpoons R\overset{O}{\overset{\|}{C}}{-}O{-}R' + H_2O \qquad (19.37)$$

involve splitting off water. Tracer studies using ^{18}O have shown that the oxygen in water produced in (19.37) comes from the acid rather than the alcohol. In esterification it is the carbon-oxygen bond that breaks in carboxylic acids rather than the oxygen-hydrogen bond.

Equilibrium constants for esterification reactions are close to unity. Therefore yields of esters are poor unless conditions are adjusted to favor the forward reaction. Two common ways to increase yields are by use of a large excess of one reactant, usually alcohol, or by removing water by distillation or with a drying agent. Both methods are practical applications of LeChatelier's principle.

Many esters are found in nature. Unlike the pungent odors of the parent carboxylic acids, odors of esters are very pleasant. The characteristic fragrances of many flowers and fruits are due to esters. Consequently, esters are much used in perfumes and as flavoring agents for foods. The essence of bananas is amyl acetate, $CH_3COO(CH_2)_4CH_3$, and that of oranges is octyl acetate, $CH_3COO(CH_2)_7CH_3$.

Most esters are colorless liquids that are nearly insoluble in water. Boiling points are generally lower than alcohols or acids of comparable carbon content because of the absence of hydrogen bonding.

Amino Acids and Proteins

Carboxylic acid molecules that contain an amino group, $—NH_2$, on the carbon adjacent to the carboxyl group are called **amino acids.** The amino group is a proton acceptor and the carboxyl group is a proton donor; thus amino acids can function as either acids or bases.

In aqueous solution amino acids can exist in two forms, as illustrated for glycine, the simplest member of the series:

$$H_2NCH_2\overset{\overset{\large O}{\|}}{C}—OH \rightleftharpoons {}^{+}H_3NCH_2\overset{\overset{\large O}{\|}}{C}—O^- \tag{19.38}$$

Equation (19.38) shows that intramolecular proton transfer from the carboxyl to the amino group can occur in amino acids. The form of the amino acid on the right is called a **zwitterion.** The net charge on a zwitterion is zero, but it is highly polar because of the two oppositely charged groups.

At a pH of 6.0 glycine exists primarily as a zwitterion and shows no net migration in an electric field. This pH, called the **isoelectric point,** is frequently used to characterize different amino acids. At very low pH glycine exists as a positive ion with both ends protonated, $H_3^+NCH_2COOH$. At high pH both protons are removed, and glycine is present as a negative ion, $H_2NCH_2COO^-$.

Amino acids constitute the fundamental units of animal and plant proteins. Although many amino acids have been found in nature, only about 20 commonly occur in proteins. Some of the simpler amino acids are listed in Table 19.3. Note that they all have the general formula $RCH(NH_2)COOH$. They are sometimes called α-amino acids, since the carbon adjacent to the carboxyl group is given the α designation. Amino acids are crystalline solids that melt at high temperatures, usually with decomposition. The amino acids are moderately soluble in water, being least soluble at the isoelectric point.

Our bodies are able to synthesize only about half the amino acids used to make proteins. The other amino acids, called essential amino acids, must be present in our food. In a dietician's terms, a "complete" protein contains all the essential

Table 19.3. Some Common Amino Acids

Name	*Symbol*	*R in $RCH(NH_2)COOH$*
Glycine	Gly	H
Alanine	Ala	CH_3
Valine	Val	$(CH_3)_2CH$
Leucine	Leu	$(CH_3)_2CHCH_2$
Phenylalanine	Phe	$C_6H_5CH_2$
Cysteine	Cys	$HSCH_2$
Lysine	Lys	$H_2N(CH_2)_4$
Glutamic Acid	Glu	$HOOCCH_2CH_2$

amino acids, whereas an "incomplete" protein lacks one or more of them. Casein, a protein found in milk and cheese, is one example of a complete protein. Most proteins are low or deficient in one or more amino acids. A proper diet consists of a mixture of proteins that are not all deficient in the same way.

By elimination of water between two amino acid molecules a **peptide bond**

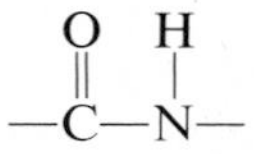

$$-\overset{\overset{\displaystyle O}{\|}}{C}-\overset{\overset{\displaystyle H}{|}}{N}-$$

is formed, as illustrated in equation (19.39):

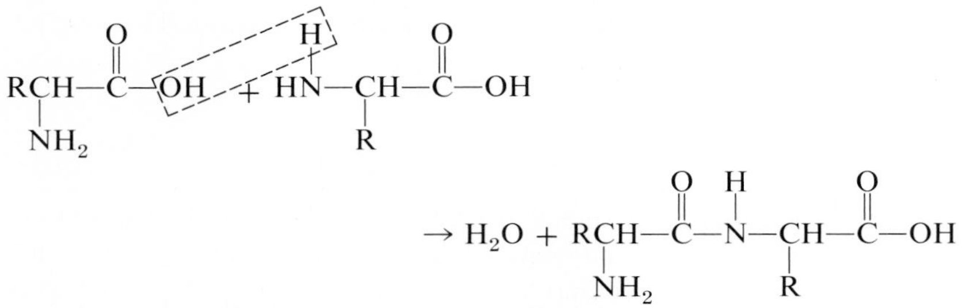

$$\underset{\underset{\displaystyle NH_2}{|}}{RCH}-\overset{\overset{\displaystyle O}{\|}}{C}-OH + \overset{\overset{\displaystyle H}{|}}{HN}-\underset{\underset{\displaystyle R}{|}}{CH}-\overset{\overset{\displaystyle O}{\|}}{C}-OH$$

$$\rightarrow H_2O + \underset{\underset{\displaystyle NH_2}{|}}{RCH}-\overset{\overset{\displaystyle O}{\|}}{C}-\overset{\overset{\displaystyle H}{|}}{N}-\underset{\underset{\displaystyle R}{|}}{CH}-\overset{\overset{\displaystyle O}{\|}}{C}-OH \quad (19.39)$$

The two linked amino acids form a **dipeptide.** Since the dipeptide also contains amino and carboxyl functional groups, more peptide links can be formed to produce **polypeptides.**

It is common practice to group long sequences of amino acids linked by peptide bonds into two classes: **proteins** or **peptides.** Proteins are naturally occurring substances with molecular weights greater than about 10,000 g $mole^{-1}$. The term peptide is ordinarily used for substances with molecular weights less than about 10,000 g $mole^{-1}$. Although the distinction is arbitrary, it is not trivial. Proteins have many more intricate three-dimensional arrangements than peptides have.

Many biologically important substances are peptides, particularly hormones. One example is **oxytocin,** a pituitary hormone that controls lactation and strong uterine contractions. Oxytocin consists of a sequence of nine amino acids. Vincent Du Vigneaud, an American biochemist, first identified the sequence and then synthesized the hormone. For this achievement Du Vigneaud received the Nobel Prize in chemistry in 1955. Other peptide hormones have since been fully characterized. Many antibiotics are also peptides, some of which contain amino acids not found in nature.

Proteins are found in almost every part of every living thing. **Fibrous proteins** are structural proteins that occur in hair, nails, bone, and connective tissue (collagen). **Globular proteins** are more complex and are the edible proteins of meat, milk, eggs, and similar foods. Another type of protein, called **conjugated proteins,** contains nonpeptide parts. An example is hemoglobin, which consists of 96% protein (globin) and 4% of a complex iron-containing molecule (hemin). Another important class of proteins is **enzymes,** the catalysts for the many chemical reactions necessary for life.

Investigation of detailed structures of proteins is one of the most active and important areas of current biochemical research. The identity, number, and sequence of amino acids as well as their spatial configuration all contribute to

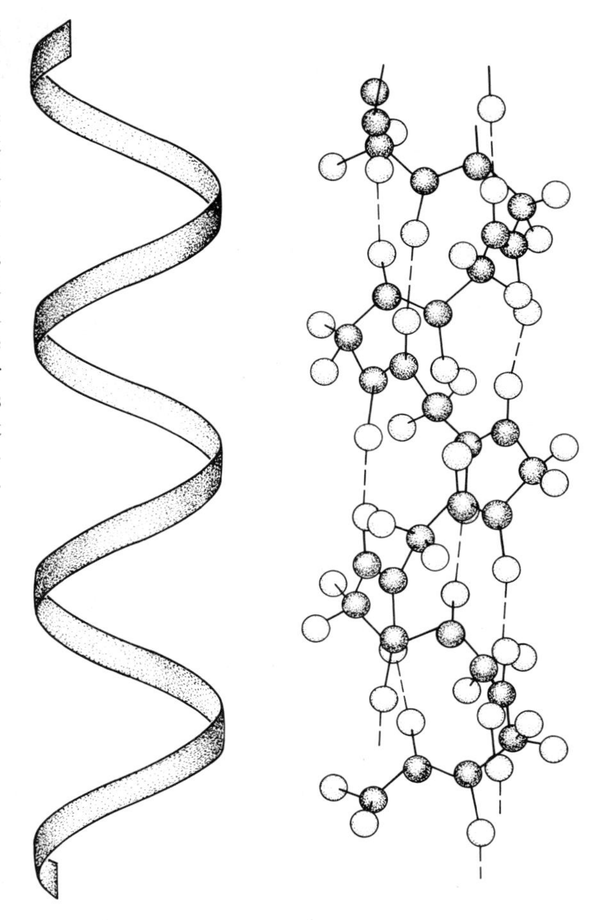

Figure 19.6. The alpha helix is the single most stable spatial configuration of a protein chain.

An important feature of the alpha helix is that it is right handed; that is, the amino acid chain forms the pattern of the thread on a right hand screw. There are 3.7 amino acid residues per turn in the alpha helix, and each NH group forms a hydrogen bond to the carboxyl group of the third amino acid residue further along the chain. Hydrogen bonds can also form with water so that the configurations of proteins frequently change in aqueous solution.

the physical and biological properties of any given protein. X-ray diffraction studies have helped elucidate the structures of many of these very complex molecules. Linus Pauling and Robert Corey were the first to suggest that protein chains could be coiled in a helix. Hydrogen bonding interactions are very important in giving some proteins their exact helical arrangement, as illustrated in Figure 19.6. Of course, not all proteins are helical.

One example of the effect of sequence of amino acids on biological properties of a protein may be found in sickle cell anemia. This hereditary disease, particularly prevalent among American Blacks, is often fatal. Sufferers of this malady synthesize hemoglobin in which two amino acids (both glutamic acid) in the hemoglobin molecule of nearly six hundred amino acids are replaced by another one (valine). This seemingly minor chemical spelling error has very profound effects. In glutamic acid the R group (see Table 19.3) is polar and water soluble, whereas the nonpolar hydrocarbon group in valine is insoluble in water. The insoluble hemoglobin distorts the cells into a sickle shape, clogs capillaries with blood clots, and deprives tissue of oxygen.

Carbohydrates

Carbohydrates, which include sugars, starches, and cellulose, constitute the most abundant single class of organic compounds in nature. Their abundance is due to the fact that they are regularly formed in green plants by photosynthesis.

In essence, green plants utilize energy from the sun to perform the following reaction:

$$n\,CO_2 + n\,H_2O \xrightarrow{h\nu} (CH_2O)_n + n\,O_2$$

The formula $(CH_2O)_n$ indicates a general carbohydrate in that plants are able to synthesize a variety ranging from the simple sugar glucose to the polymer cellulose. The exact mechanism of photosynthesis is not yet clear, but it is established that the green pigment chlorophyll, a complicated coordination compound of magnesium, is important in conversion of sunlight to chemical energy. Photosynthesis is a highly endothermic process, requiring more than 100 kcal of energy per mole of carbon dioxide consumed. Combustion, the reverse of the above process, is thus highly exothermic, which accounts for the utilization of photosynthetic products as fuels.

The name carbohydrate comes from the empirical formula of these substances, $C_x(H_2O)_y$. Although correct, this representation is an oversimplification and misleading. Carbohydrates contain both hydroxyl and carbonyl functional groups that determine their chemical properties. Since each carbohydrate molecule contains several functional groups, their chemistry becomes involved and challenging.

Carbohydrates may be divided into three classes: **monosaccharides** or simple sugars; **oligosaccharides** containing a few sugars; and **polysaccharides,** complex polymers including starches and cellulose.

Most of the naturally occurring monosaccharides are **hexoses.** This means that they contain six carbon atoms and have the general formula $C_6H_{12}O_6$. Two of the most important hexoses are glucose and fructose.

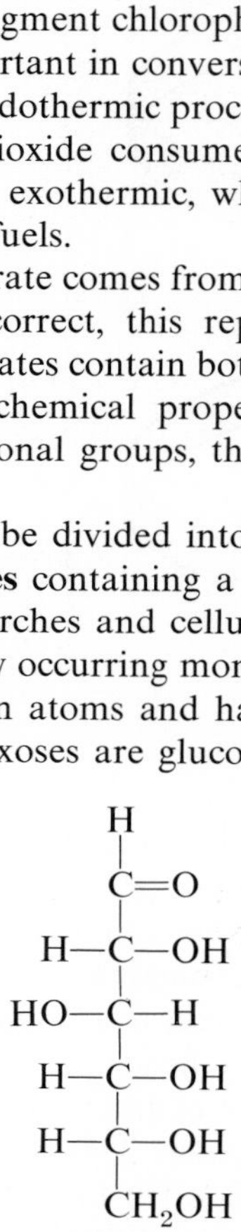

$$\begin{array}{rcl} & H & \\ & | & \\ & C{=}O & \\ & | & \\ H- & C & -OH \\ & | & \\ HO- & C & -H \\ & | & \\ H- & C & -OH \\ & | & \\ H- & C & -OH \\ & | & \\ & CH_2OH & \end{array} \qquad \begin{array}{rcl} & CH_2OH & \\ & | & \\ & C{=}O & \\ & | & \\ HO- & C & -H \\ & | & \\ H- & C & -OH \\ & | & \\ H- & C & -OH \\ & | & \\ & CH_2OH & \end{array}$$

D-glucose 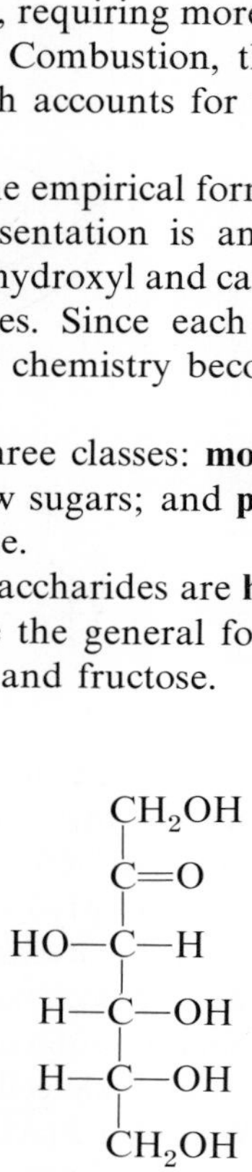 D-fructose

Note that glucose is an aldehyde and fructose is a ketone.

The D notation signifies one of the two optical isomers of glucose. The properties of compounds that exist in D or L forms are identical in almost all respects except their behavior toward plane-polarized light. When polarized light vibrating in a given plane is passed through an optically active substance, it emerges vibrating in a different plane.

Optical isomerism is observed when a molecule does not have a superimposable mirror image. It is common among organic compounds, particularly naturally occurring substances such as sugars and amino acids. Frequently only one of the two optical isomers is found in nature. The simplest optical isomers are compounds that contain an asymmetric carbon atom—one to which four different groups are attached. Glucose contains four asymmetric carbon atoms and fructose has three. Some transition metal complexes also exhibit optical isomerism.

A further problem in the structure of monosaccharides is their tendency to form cyclic molecules. D-Glucose forms a six-membered puckered ring, whereas D-fructose can form either five- or six-membered rings. Each of the ring structures is also optically active, and only one of the two optical isomers is found in nature.

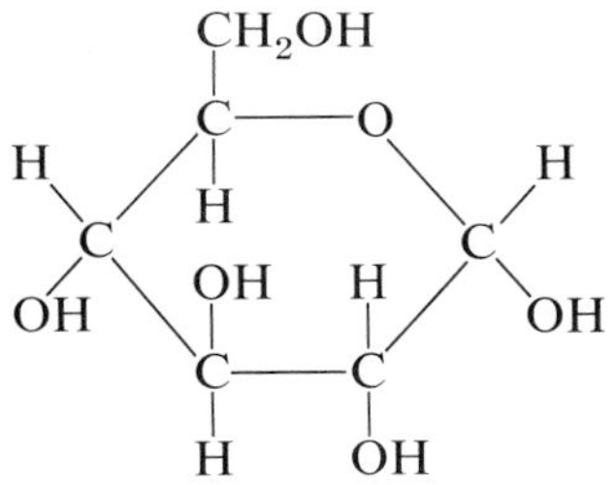

D-glucose 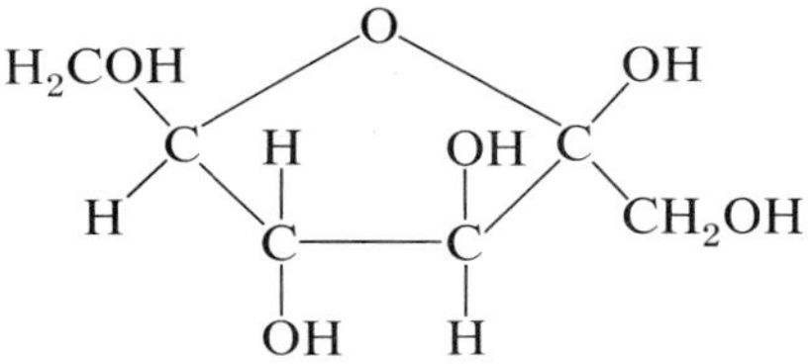 D-fructose

Glucose, also called dextrose or grape sugar, occurs widely in fruits such as grapes. Glucose is the principal sugar in blood, all other sugars ingested being first converted into glucose. When oral nutrition is not possible, saline solutions containing glucose are administered intravenously. Fructose also occurs widely in fruits and is the principal sugar in honey.

Oligosaccharides are obtained by linking together a few (two to ten) simple sugars. The **glycoside** link is obtained by splitting out water between two alcohol groups. Sucrose, ordinary table sugar, is a disaccharide of D-glucose and D-fructose.

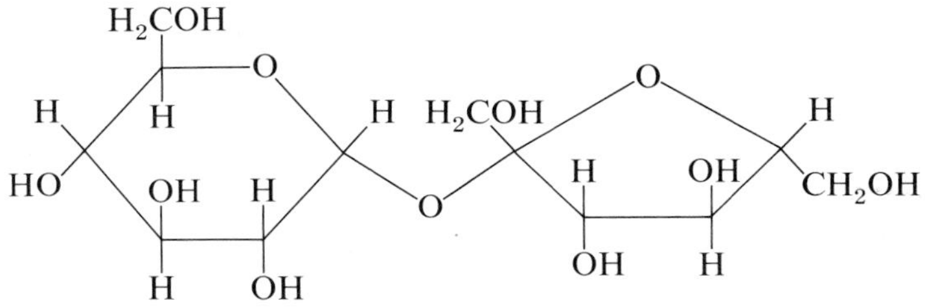

sucrose

Acid hydrolysis of sucrose breaks the glycoside link to yield the two monosaccharides.

Important natural sources of sucrose are sugar cane and sugar beets. Both contain about 15% sucrose, but the yield of sugar per acre is greater from cane. Sugar cane grows only in tropical climates, and today about half of the world's supply of sucrose comes from sugar beets.

Lactose, or milk sugar, makes up about 5% of milk. It is not very sweet and is not fermented by yeast. Lactose is the disaccharide of D-glucose and D-galactose, another hexose. **Maltose,** which does not occur free in nature, is a disaccharide containing two glucose units.

Polysaccharides are polymers composed of repeating units of a monosaccharide. Both starch and cellulose are polymers of D-glucose; cellulose is the larger polymer with molecular weights ranging up to 2 million.

There is an important distinction in the stereochemistry of the glycoside linkage between starch and cellulose, however. In starch, the glycoside link is below the plane of the two rings, which is called an α linkage.

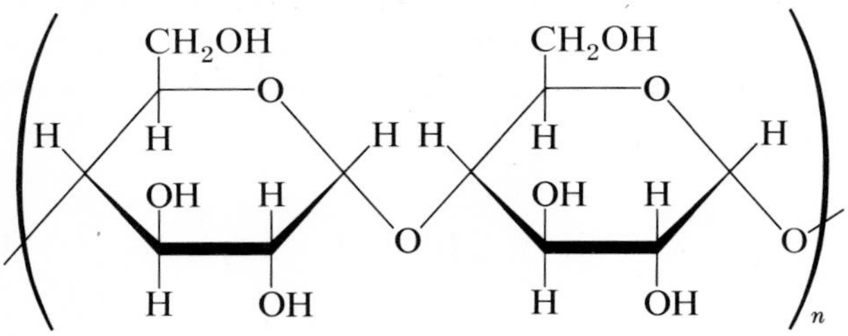

starch

The D-glucose units in cellulose are joined with β linkages, in which the glycoside linkage is above the plane of the two rings.

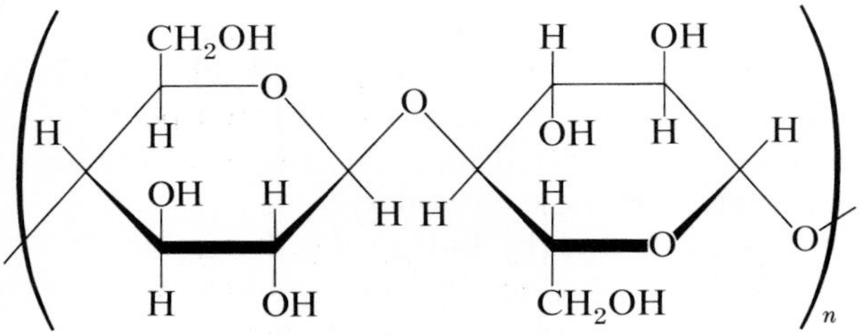

cellulose

Starch is an important food material, but cellulose, the principal structural material of wood and cotton, can not be digested by man. Man and other carnivorous animals are unable to hydrolyze cellulose because they do not have suitable enzymes to catalyze breaking of the β-glycoside linkage. Termites, notorious for eating wood, host microorganisms in their digestive tracts that do have this ability.

Fats

Of the food we eat, fats have the highest energy content. On complete combustion, fats furnish 9 kcal g^{-1}; proteins provide 5.6 kcal g^{-1}; and carbohydrates, 4.2 kcal g^{-1}. Whereas proteins have mainly structural functions, carbohydrates are sources of ready energy, and fats function as energy reserves. On oxidation, fats produce not only energy but also water. The camel's hump is one example of water storage through the use of fatty tissues.

Fats are esters of long chain carboxylic acids with the polyalcohol glycerol (also called glycerin). The general formula for fats, which belong to the class of compounds called **lipids,** is

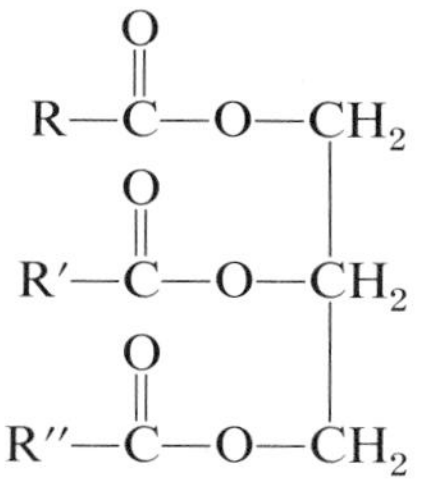

$$\begin{array}{l} R-\overset{\displaystyle O}{\overset{\|}{C}}-O-CH_2 \\ \qquad\qquad\quad | \\ R'-\overset{\displaystyle O}{\overset{\|}{C}}-O-CH_2 \\ \qquad\qquad\quad | \\ R''-\overset{\displaystyle O}{\overset{\|}{C}}-O-CH_2 \end{array}$$

Some common fatty acids obtained on hydrolysis of fats are

palmitic acid	$CH_3(CH_2)_{14}COOH$
stearic acid	$CH_3(CH_2)_{16}COOH$
oleic acid	$CH_3(CH_2)_7CH{=}CH(CH_2)_7COOH$
linoleic acid	$CH_3(CH_2)_4CH{=}CHCH_2CH{=}CH(CH_2)_7COOH$

Animal fats contain a high percentage of saturated fatty acids and are solids at room temperature. Vegetable fats, or oils, have predominantly unsaturated fatty acids and are liquids at room temperature.

By catalytic hydrogenation it is possible to saturate the double bonds in vegetable oils to produce solids. In this way margarine is made from corn oil, in which the fatty acid is 46% oleic and 42% linoleic. The unsaturated oils are subject to air oxidation and often give products having disagreeable odors or tastes. There is statistical evidence that suggests some health advantages in using unsaturated fats for cooking. People who regularly eat solid fats seem particularly subject to arterial clotting. The statistical correlation is not perfect but has led to increased use of polyunsaturated oils in foods and cooking.

Basic hydrolysis of fats to produce soap is a very old chemical process, although not as old as production of ethanol. (Apparently man's desire for intoxication predates his desire for cleanliness.) The saponification (soap making) process may be described by the general equation

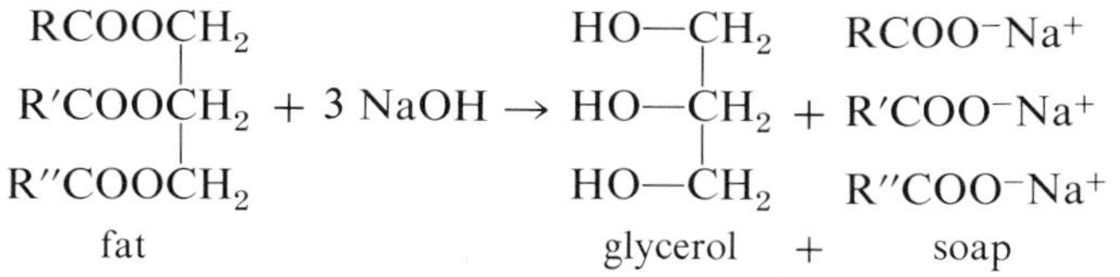

$$\begin{array}{ccccc} RCOOCH_2 & & HO-CH_2 & & RCOO^-Na^+ \\ | & & | & & \\ R'COOCH_2 & + 3\,NaOH \rightarrow & HO-CH_2 & + & R'COO^-Na^+ \\ | & & | & & \\ R''COOCH_2 & & HO-CH_2 & & R''COO^-Na^+ \\ \text{fat} & & \text{glycerol} & + & \text{soap} \end{array}$$

Ordinary soap is thus a mixture of sodium (or potassium) salts of long chain fatty acids. Different fats may be used, perfumes or germicides added, but chemically all soaps are pretty much the same.

Hard water contains calcium and magnesium ions, which form insoluble precipitates with soap, leading to "bathtub ring" and "ring around the collar." To avoid these problems and because the supply of natural fats is limited, synthetic cleansing agents called **detergents** were developed.

Although the synthetic detergents vary considerably in their chemical structures, they still have a feature in common with soap: a large nonpolar hydrocarbon "tail" that is oil soluble and a polar end that is water soluble. One big advantage of the synthetic detergents is that their calcium and magnesium salts are water soluble. Thus synthetic detergents retain their efficiency in hard water.

Perhaps the most widely used detergents are sodium salts of alkylbenzene sulfonic acids with the general formula:

SO_3^- Na^+

R

R, which may be attached at various places on the ring, is commonly a C_{12} straight hydrocarbon chain. Initially, highly branched side chains were attached to the benzene ring, but these detergents were not biodegradable. The result was sudsy streams and contaminated drinking water.

Making detergents biodegradable does not solve all our problems, however. Degradation of detergents by bacterial action uses oxygen. If oxygen depletion of natural waters proceeds far enough, some fish can no longer survive. The difficulties associated with phosphate additives in detergents as plant nutrients have already been mentioned in Chapter 18. There are no simple answers in the clean water versus clean clothes dilemma, but it does seem clear that the ultimate answer will involve a compromise between cost and convenience.

References

Beishline, R. R.: Directive effects in electrophilic aromatic substitution. *J. Chem. Educ.*, **49:**128 (1972).

Breslow, R.: The nature of aromatic molecules. *Sci. Amer.*, **227:**32 (Aug. 1972).

Caserio, M. C.: Reaction mechanisms in organic chemistry. *J. Chem. Educ.*, **42:**570 (1965).

Caserio, M. C.: Reaction mechanisms in organic chemistry: concerted reactions. *J. Chem. Educ.*, **48:**782 (1971).

Church, L. B.: The chemistry of winemaking. *J. Chem. Educ.*, **49:**174 (1972).

Dawkins, M. J. R., and D. Hull: The production of heat by fat. *Sci. Amer.*, **213:**62 (Aug. 1965).

Elias, W. E.: The natural origin of optically active compounds. *J. Chem. Educ.*, **49:**448 (1972).

Frieden, E.: The chemical elements of life. *Sci. Amer.*, **227:**52 (July 1972).

Herz, W.: *The Shape of Carbon Compounds.* W. A. Benjamin Co., New York, 1964 (paperback).

Lambert, J. B.: The shapes of organic molecules. *Sci. Amer.*, **222:**58 (Jan. 1970).

Light, R. L.: *A Brief Introduction to Biochemistry,* W. A. Benjamin Co., New York, 1966 (paperback).

Rice, R.: Drug receptors. *J. Chem. Educ.*, **44:**565 (1967).

Roberts, J. D.: Organic chemical reactions. *Sci. Amer.*, **197:**117 (Nov. 1957).

Sebastian, J. F.: The electronic effects of alkyl groups. *J. Chem. Educ.*, **48:**97 (1971).

Stein, W. H., and S. Moore: The chemical structure of proteins. *Sci. Amer.*, **204:**81 (Feb. 1961).

Problems

1. Use ΔH_f^0 values from Appendix II to calculate the ΔH^0 for the conversion of ammonium cyanate to urea. Predict whether the equilibrium constant would increase or decrease with increasing temperature.
2. Explain why chemists use hybrid orbitals of carbon and not simply 2*s* and 2*p* orbitals to explain the structure of CH_4.
3. (a) Draw the three structural isomers of pentane, C_5H_{12}.
   (b) Draw the five structural isomers of hexane, C_6H_{14}.
4. Free radicals, such as those described for reaction (19.1), are very reactive species. They readily react with other free radicals and also abstract atoms from stable molecules to produce a variety of products. Predict two probable byproducts for the following reaction:
$$CH_3CH_3(g) + Cl_2(g) \rightarrow CH_3CH_2Cl(g) + HCl(g)$$
5. In rush hour traffic on busy streets, concentrations of CO often approach 50 parts per million (ppm by volume), which is a safety limit for exposure of a healthy industrial worker over an 8-hr period. Assuming that CO behaves as an ideal gas, calculate the number of molecules and number of moles of CO in 1.0 liter of air at a pressure of 1.0 atm and 25°C when the CO concentration is 50 ppm. Is 50 ppm a large or small number?
6. Organic compounds that contain multiple bonds have larger bond enthalpies but are more reactive than compounds containing carbon-carbon single bonds. Explain why this is not a contradiction.
7. Draw formulas for at least six isomers with the formula C_5H_{10}.
8. Use ΔH_f^0 values from Appendix II to calculate the average C-F bond enthalpy in CF_4. Suggest an explanation for the difference between this value and that given in Table 11.1. How does the C-F bond enthalpy compare with other single bond enthalpies containing carbon?
9. (a) Complete combustion of 0.1050 g of compound A yielded 0.3247 g of CO_2 and 0.1490 g of H_2O. If chemical tests indicate that A is an alkane and the error in combustion analysis is negligible, what is the molecular formula for A?
   (b) Calculate the percentage composition of an alkane with one more carbon atom. With an error of 0.5% in the analysis, could it be used to distinguish between the two compounds?
10. A student suggested that it would be possible to distinguish between 1-propanol and 2-propanol by dehydrating the alcohol with sulfuric acid and identifying the alkene produced. Explain why this would or would not be a suitable approach.
11. Write equations for a four-step procedure by which 2-pentyne, $CH_3C{\equiv}CCH_2CH_3$, could be obtained from acetylene and the appropriate bromoalkånes.
12. The ΔH_f^0 value for gaseous benzene is 19.8 kcal mole^{-1}. Use bond enthalpy values from Table 11.1 and data from Appendix II to calculate ΔH^0 for the reaction:
$$6\ C(g) + 6\ H(g) \rightarrow C_6H_6(g)$$
    For the calculation assume that benzene has alternate single and double carbon-carbon bonds. The difference between ΔH^0 and ΔH_f^0 is called the resonance stabilization enthalpy and indicates that benzene is more stable than predicted from an alternate single bond, double bond structure.
13. (a) Draw structures for *ortho*-, *meta*-, and *para*-dichlorobenzene. *Para*-Dichlorobenzene is commonly used as a moth repellent.
    (b) If a third chlorine is added to each of the dichlorobenzenes, how many isomers can be produced in each case? Note that Cl is an ortho-para director.

14. Chrysene, $C_{18}H_{12}$, and 1,2,5,6-dibenzanthracene, $C_{22}H_{14}$, are two more condensed ring aromatic compounds. Is it possible to devise a general formula for these two compounds as well as for naphthalene, anthracene, and phenanthrene?

15. Use data in Table 19.1 to graph boiling points of primary alcohols versus the number of carbons in the alcohol. Predict the boiling points of 1-hexanol, $C_6H_{13}OH$, and 1-heptanol, $C_7H_{15}OH$, and compare with the values found in some reference work such as a handbook of chemistry and physics.

16. (a) Given the following spontaneous reactions for the generalized case:

$$RO^- \, Na^+ + HOH \rightarrow Na^+ + OH^- + ROH$$
$$HC{\equiv}C^- \, Na^+ + ROH \rightarrow RO^- \, Na^+ + HC{\equiv}CH$$

Arrange OH^-, OR^-, and $C{\equiv}CH^-$ in order of decreasing basicity.

(b) Would CH_3OH or HOH be the stronger acid? Would it be feasible to attempt a preparation of sodium methoxide by reacting methanol with aqueous sodium hydroxide?

17. (a) Consider the preparation of methanol from carbon monoxide and hydrogen, equation (19.27). Explain why high temperature, high pressure, and catalysts are used and relate your explanation to LeChatelier's principle.

(b) Use data in Appendix II to calculate ΔS^0 and ΔG^0 for this reaction.

18. Write formulas for at least five of the six ethers with the formula $C_5H_{12}O$.

19. Suggest an explanation for the fact that diethyl ether is not very soluble in water but is highly miscible with 36% aqueous hydrochloric acid.

20. Laboratory preparations of aldehydes are simplified by the fact that aldehydes always have a lower boiling point than the alcohol from which they are formed. For example, acetaldehyde (CH_3CHO) has a normal boiling point of 20°C, whereas that of ethanol (CH_3CH_2OH) is 78°C. Suggest a reason for this difference in boiling points.

21. Tollens' reagent contains $Ag(NH_3)_2^+$ in an alkaline solution and is used as a means of detecting aldehydes.

(a) On oxidation of the aldehyde to $RCOO^-$, the metallic silver produced forms a mirror under the proper conditions. Write a balanced equation for the Tollens' test on acetaldehyde, CH_3CHO.

(b) Draw structures for at least three compounds of formula $C_5H_{10}O$ that would give a positive Tollens' test and four compounds of the same formula that would give a negative Tollens' test.

22. In the text we have written the carboxylate ion as $R{-}\overset{\overset{\displaystyle O}{\|}}{C}{-}O^-$. This representation indicates that there should be one long and one short carbon-oxygen bond. Experimentally, both bonds are found to be equivalent. In general terms explain how this ion would be described in both valence bond and molecular orbital theory.

23. Acetic acid forms a dimer with two hydrogen bonds:

```
       O---H—O
      //      \
CH3—C          C—CH3
      \       //
       O—H---O
```

At 110°C and 454 mm pressure, 0.110 g acetic acid vapor occupies 63.7 cm^3; at 156°C and 458 mm, 0.081 g of vapor occupies 66.4 cm^3. Assume that the vapor obeys the ideal gas law and calculate the molecular weight of gaseous acetic acid at each temperature. How do you interpret these results?

24. After looking at equation (19.27), predict the sign of ΔS^0 for this reaction. Then calculate ΔS^0 in two ways: (a) From the ΔH^0 and ΔG^0 values given with the reaction equation, and (b) from S^0 values in Appendix II.

25. Arrange the following acids in order of increasing values of K_a: $BrCH_2COOH$, CH_3COOH, CF_3COOH, $ClCH_2COOH$, CCl_3COOH. Give a brief reason for your order.

26. An early and important experiment using $H_2^{18}O$ helped answer the question: "In the hydrolysis of an organic ester, where does the molecule break?" The two possibilities for methyl acetate are indicated by the dashed lines a and b:

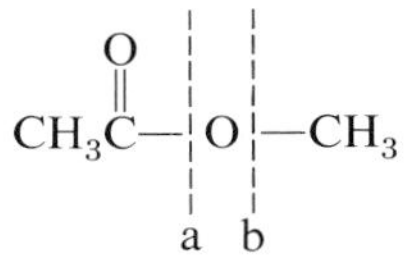

$$CH_3\overset{\overset{\displaystyle O}{\|}}{C}\text{—}\,\vdots\,O\,\vdots\,\text{—}CH_3 \qquad (\text{a: C—O bond; b: O—CH}_3\text{ bond})$$

a b

Knowing that $CH_3\overset{\overset{\displaystyle O}{\|}}{C}\text{—}^{18}O\text{—}H$ was the observed product containing ^{18}O, answer the question.

27. (a) Write equations to show the amino acid alanine functioning as an acid and as a base.

(b) Suggest a reason for the fact that amino acids are least soluble in water at the isoelectric point.

28. Consider that a peptide consists of four different amino acids: A, B, C, and D. How many different peptides can be formed from all possible combinations of these four amino acids?

29. One means of distinguishing between hexoses is the Tollens' test (see problem 21). Sugars that give positive Tollens' tests (form a silver mirror) are called reducing sugars. Which of the two hexoses, glucose and fructose, would give a positive Tollens' test?

30. If a sample of starch has an average molecular weight of 400,000 g $mole^{-1}$, how many glucose units does it contain?

31. Of the three principal classes of foods (fats, proteins, and carbohydrates), carbohydrates have the lowest energy content per gram. Yet carbohydrates are commonly used by athletes and others as a source of quick energy. Suggest a reason for this.

32. The widely used but controversial pesticide DDT has the proper name 2,2-bis-(4-chlorophenyl)-1,1,1-trichloroethane. Draw its structure.

33. The preparation of the amino acid cystine is a common laboratory exercise in biochemistry. Student preparations of cystine are often contaminated with NaCl. When treated with aqueous $AgNO_3$, 9.82 g of one such preparation gave 2.15 g AgCl. Find the percentage of cystine in the sample.

34. In earlier chapters we have written equations to represent ionization of carboxylic acids as

$$R\overset{\overset{\displaystyle O}{\|}}{C}\text{—}OH(aq) \rightleftharpoons H^+(aq) + R\overset{\overset{\displaystyle O}{\|}}{C}\text{—}O^-(aq)$$

Explain why the equilibrium constant for the acid ionization is independent of whether we represent the ionization by the equation above or by equation (19.36). Explain the consequences of the fact that it is relatively easy to determine concentrations of dissolved species but very difficult to obtain detailed and reliable information about their extent of hydration.

20

TRANSITION METAL CHEMISTRY

Introduction

In between the alkaline earth metals and the boron family in the long form of the periodic table are elements that are called **transition elements** or **transition metals.** As we shall see later, there are ambiguities in classifying some of these elements. In this book, we take elements 21 (scandium) through 30 (zinc) to be the first row transition elements. Elements 39 (yttrium) through 48 (cadmium) are the second row transition elements. Elements 57 (lanthanum) through 71 (lutetium) are sometimes called inner transition elements, but are more often labeled **lanthanides** or **rare earths.** Elements 72 (hafnium) through 80 (mercury) are the third row transition elements. Elements 89 (actinum) through 103 (lawrencium) are a second inner transition element series and are called **actinides.** Elements 104 and 105, as yet unnamed, resume the fourth row transition element series.

The *d* Orbitals

The shapes of *s* and *p* orbitals were considered in Chapter 6. Now we turn our attention to the *d* orbitals, which are of particular importance in the chemistry of the transition elements.

The 3*d* orbitals have $n = 3$, $l = 2$. Since there are five possible values of m_l, there must be five *d* orbitals, each able to accommodate two electrons. These *d* orbitals have directional properties that

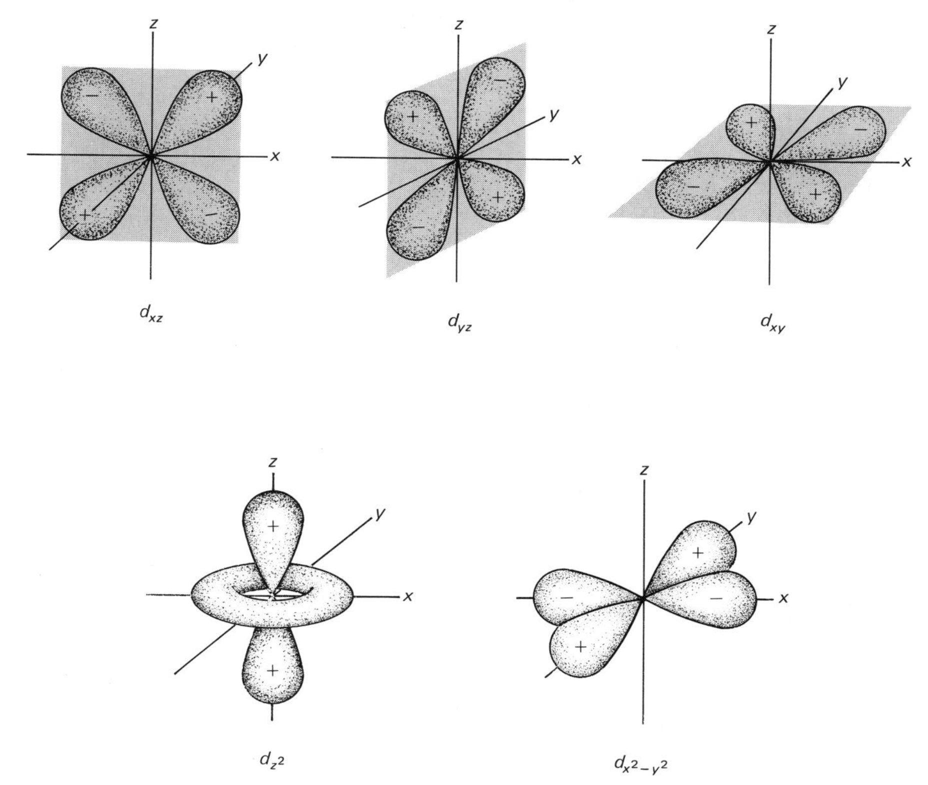

Figure 20.1. Boundary contours for the five 3*d* orbitals.
The 4*d*, 5*d*, and 6*d* orbitals can be considered to be essentially the same except for an increase in size. Remember that the boundary contours are related to ψ^2, whereas the plus and minus signs are related to ψ. Note how the sign of the wave function changes from one lobe to the next in a given orbital.

are illustrated in Figure 20.1 and discussed later in this chapter in connection with bonding in various compounds of the transition elements. Our first concern will be with the number of *d* orbitals and their energies in relation to energies of other orbitals and electronic configurations of single atoms of the transition elements.

Electronic Configurations of Transition Elements

Electronic configurations of the first eighteen elements of the periodic table were discussed in Chapter 7. As shown in Table 7.2, the electronic configuration of argon is $1s^2 2s^2 2p^6 3s^2 3p^6$. Since there are only three *p* orbitals and each orbital can contain no more than two electrons, we cannot assign any more electrons to the 3*p* orbitals in our imaginary aufbau process. Therefore, higher energy orbitals must be used for elements with atomic numbers greater than 18. The problem then becomes one of knowing the relative energies of the various orbitals.

The order of orbital energies for the lighter elements has been shown in Figure

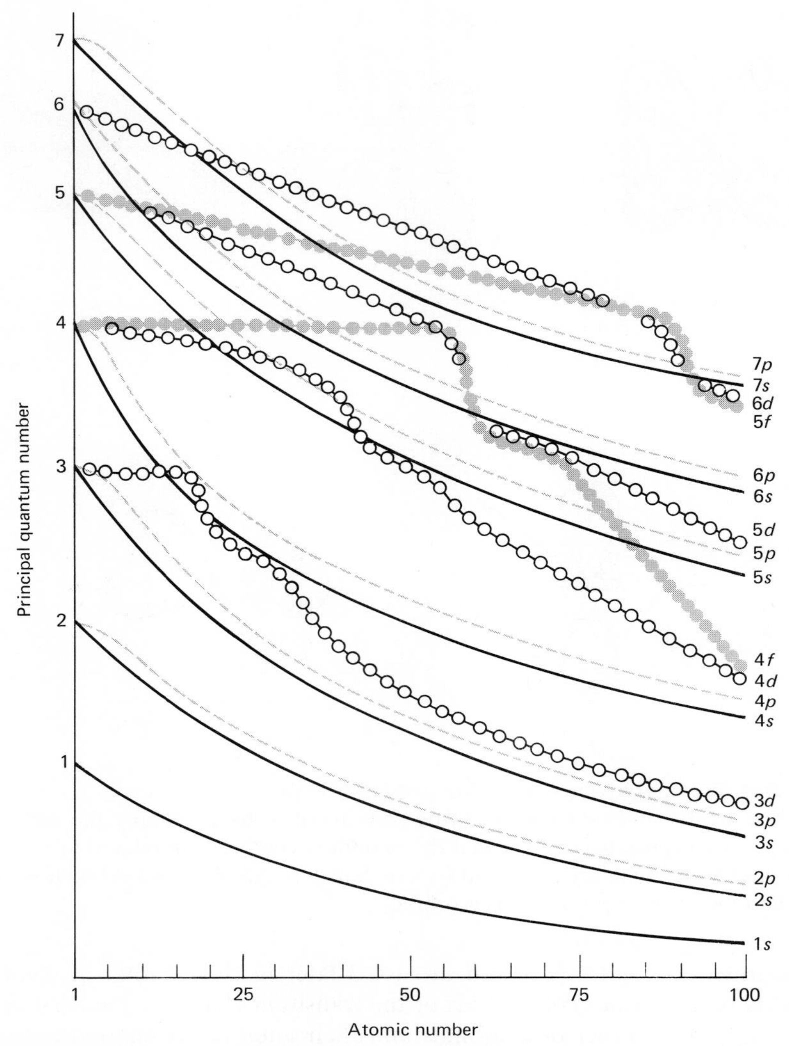

Figure 20.2. Dependence of orbital energies on atomic number.

The curves are based on analyses of optical and x-ray spectra. Note that there are several crossovers at higher values of n; we are especially interested in the $3d$-$4s$ crossover. Also note that all orbital energies become closer together as atomic number increases.

7.1. But analyses of atomic spectra have shown that relative energies of orbitals are dependent on atomic number, as indicated in Figure 20.2. (Also see problem 3 at the end of this chapter for a mnemonic device.)

For potassium (atomic number $Z = 19$) we conclude that the nineteenth electron will go into a $4s$ orbital in our imaginary aufbau process. At this atomic number, $4s$ orbital energy is lower than $3d$ orbital energy, as shown in Figure 20.2. Further, the properties of potassium are very similar to those of sodium, which provides additional support for assigning the (Ar)$4s^1$ configuration to potassium.

Atoms of calcium have 20 electrons each, with the highest energy electron being the second one in the 4*s* orbital. The electronic configuration of calcium is thus (Ar)$4s^2$. This configuration is consistent with spectroscopic evidence, with the fact that calcium atoms are diamagnetic with all electron spins paired, and is similar to the electronic configuration of magnesium, which has many properties that are similar to those of calcium.

With scandium ($Z = 21$) we must again decide on the next lowest energy orbital. Spectroscopic evidence indicates that the 3*d* ($n = 3$, $l = 2$) orbitals are lower in energy than the 4*p* orbitals. This is also consistent with the properties of the elements between calcium and krypton ($Z = 36$), the next noble gas. If the 4*p* orbitals were at lower energy than the 3*d* orbitals, we would expect to find a noble gas with $Z = 26$, which is actually iron. But with the 3*d* orbitals at lower energy than the 4*p* orbitals, we would expect to find a noble gas at $Z = 20 + 10 + 6 = 36$, which is krypton. We therefore express the electronic configuration of scandium as (Ar)$3d^14s^2$. It is customary to write the 3*d* before the 4*s* in such representations, despite the lower energy of the 4*s* orbital as compared to the 3*d* orbitals.

The element with atomic number 22 is titanium, with electronic configuration (Ar)$3d^24s^2$. From Hund's rule we expect the configuration with two electrons in different *d* orbitals (parallel spins) to be lower in energy than the configuration with both electrons in the same *d* orbital (spins paired). Consistent with this expectation, gaseous titanium atoms have paramagnetism corresponding to two unpaired electrons per atom.

The electronic configuration of the next element, vanadium, is (Ar)$3d^34s^2$. Vanadium atoms have paramagnetism corresponding to three unpaired electrons, as predicted from Hund's rule.

The electronic configuration of chromium ($Z = 24$) is not what we might at first expect. Spectroscopic evidence shows that the chromium configuration is (Ar)$3d^54s^1$, which corresponds to a lower energy state than does the (Ar)$3d^44s^2$ configuration that might have been expected. Paramagnetism of chromium gaseous atoms corresponds to six unpaired electrons per atom as in (Ar)$3d^54s^1$. We have previously seen that species containing half-filled subshells are more stable than similar configurations without such half-filled subshells. Thus the first ionization energy of nitrogen ($1s^22s^22p^3$) is larger than that of the immediately preceding or succeeding elements (carbon and oxygen). We find this same special stability for half-filled *d* and *f* orbitals, corresponding to d^5 and f^7 configurations.

The configuration of manganese is (Ar)$3d^54s^2$; that of iron is (Ar)$3d^64s^2$; that of cobalt is (Ar)$3d^74s^2$; and that of nickel is (Ar)$3d^84s^2$. As predicted by Hund's rule, manganese has five unpaired electrons per atom, iron has four, cobalt has three, and nickel has two.

Just as noble gas configurations with filled *p* subshells are particularly stable, configurations with filled *d* (and *f*) subshells are also especially stable. As a result, the electronic configuration of copper is not what we might naively predict. Analysis of the atomic spectrum of gaseous copper indicates that (Ar)$3d^{10}4s^1$ is a lower energy configuration than (Ar)$3d^94s^2$.

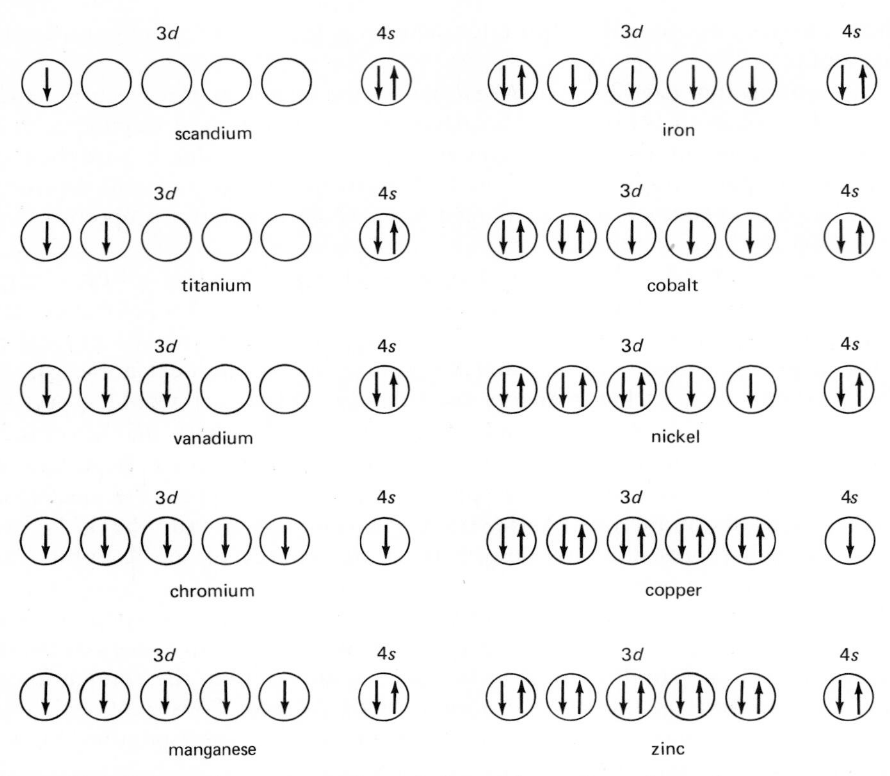

Figure 20.3. Schematic representation of electronic configurations of gaseous atoms of elements scandium through zinc. The argon core configuration common to all elements with atomic number greater than 18 is not shown.

Zinc atoms are diamagnetic with an electronic configuration (Ar)$3d^{10}4s^2$.

Figure 20.3 is a schematic representation of the electronic configuration of atoms of the first row transition elements. These configurations, along with those of all elements in the periodic table, are given in Appendix V.

With zinc the $3d$ and $4s$ subshells have been completely filled in our imaginary building process. The "last" electron in the next element, gallium, with $Z = 31$, must go into the next lowest energy orbital. The electronic configuration of gallium is therefore (Ar)$3d^{10}4s^24p^1$. Elements immediately following gallium in the periodic table each have one additional $4p$ electron, until we come to krypton with configuration (Ar)$3d^{10}4s^24p^6$.

All elements with atomic numbers greater than 36 have a kryptonlike core of electrons, represented by (Kr). Thus the electronic configuration of rubidium is (Kr)$5s^1$ and that of strontium is (Kr)$5s^2$.

The arrangement of elements in the periodic table suggests that the $4d$ orbitals are at lower energy than the $5p$ orbitals, which has been confirmed by analyses of atomic spectra. Thus the elements yttrium ($Z = 39$) through cadmium ($Z = 48$) are a second row transition metal series. The configuration of yttrium is (Kr)$4d^15s^2$, and that of cadmium is (Kr)$4d^{10}5s^2$. Configurations of the atoms of the second row transition metal series are not, however, always similar to those of the elements immediately above them in the periodic table. For example, the configuration of palladium is $(Kr)4d^{10}5s^0$, whereas that of nickel is $(Ar)3d^84s^2$. In the heavier elements, orbital energies are quite close together and inversions in the "expected" order of filling are fairly common.

The $5p$ orbitals start to fill with indium ($Z = 49$). Each of the elements immediately following indium contains one additional $5p$ electron until we come to xenon with configuration $(Kr)4d^{10}5s^25p^6$.

After cesium with a $(Xe)6s^1$ configuration and barium with a $(Xe)6s^2$ configuration, we have lanthanum with one electron in a $5d$ orbital. Thus lanthanum can be regarded as the first element in the third series of transition elements. Following lanthanum is a group of fourteen elements in which the $4f$ orbitals ($n = 4, l = 3$) are being filled. These 15 elements (lanthanum with $Z = 57$ through lutetium with $Z = 71$) have very similar chemical properties and are generally called **rare earth** metals or **lanthanides.**

Following lutetium, the $5d$ orbitals begin to fill with hafnium ($Z = 72$) and are completely filled at mercury ($Z = 80$). Then the $6p$ orbitals begin to fill at $Z = 81$ (thallium) and are completely filled at the next noble gas (radon, $Z = 86$).

In going across any row in the periodic table there is a general decrease in covalent or ionic radii. This "shrinkage" arises from incomplete shielding of increasing nuclear charge, as discussed in Chapter 7. The radius decrease among the rare earths, the so-called **lanthanide contraction,** has important chemical consequence for the post-lanthanide elements. As we have seen, chemical properties of metal ions depend on both size and charge. Because of the lanthanide contraction, the radii and chemical properties of the third row transition metals are very similar to those of the second row transition elements immediately above them in the periodic table. For example, tungsten is more similar to molybdenum than molybdenum is similar to chromium.

After francium with a (Rn) $7s^1$ configuration and radium with a (Rn) $7s^2$ configuration, we have actinium, which has one electron in a $6d$ orbital and starts a fourth series of transition elements. In the fourteen elements following actinium the $5f$ orbitals are being filled. These 15 elements (actinium with $Z = 89$ through lawrencium with $Z = 103$) are generally called **actinides.** The heavier actinides are all man-made elements and are radioactive with short half-lives.

With element 104 the fourth row of transition elements is resumed. There is a disagreement between Russian and American scientists over isolation and naming of element 104. The two proposed names are kurchatovium (Russian) and rutherfordium (American). Isotopes of element 105 (with the proposed name of hahnium) have also been obtained by the American scientists at Berkeley under the direction of A. Ghiorso.

Oxidation States of Transition Metals

In this section we discuss some of the important compounds and ions of the transition metals, with emphasis on the first row transition elements. Many of the chemical properties of these compounds and ions are concisely summarized by various equilibrium constants and reduction potentials. Formulas and chemical properties of several ions are discussed in terms of electronic configurations.

Scandium, Yttrium, and Lanthanum

The +3 oxidation state is the only chemically important one for these elements. The M^{3+} ions all have the electronic configurations of the preceding noble gases. Their compounds are all diamagnetic and colorless. The metals are highly electropositive and are generally obtained by electrolysis of the molten chlorides. For all three elements $\mathscr{E}^0$ for the half-reaction

$$M^{3+}(aq) + 3\,e^- \rightleftharpoons M(c)$$

is more negative than −2.0 volts.

With the exception of the fluorides, the halides of Sc^{3+}, Y^{3+}, and La^{3+} are quite soluble in water. Fluorides, hydroxides, carbonates, and oxalates of these metals are only slightly soluble in water. Except for the hydroxides, these compounds dissolve in the presence of excess precipitating agent to form negatively charged complex ions. For example, ScF_3 dissolves in excess $F^-(aq)$ to form $ScF_6^{3-}(aq)$. Evaporation of water from solutions containing excess precipitating agent yields such compounds as $K_3ScF_6(c)$.

Titanium

Titanium is distributed widely, amounting to about 0.6% (by mass) of the earth's crust. One of the most common ores is rutile, TiO_2, which is frequently colored brown or black by impurities. Pure TiO_2 is much used as a white pigment in paints and cosmetics. Free titanium is usually obtained by passing chlorine over a hot mixture of TiO_2 and carbon to yield titanium tetrachloride, $TiCl_4$:

$$TiO_2 + C + 2\,Cl_2 \rightarrow TiCl_4 + CO_2 \qquad (20.1)$$

The resultant $TiCl_4$ is then reduced with magnesium. Because of its strength, low density, and resistance to corrosion, metallic titanium is used in high speed aircraft and is also used to improve the properties of certain steels.

The most common oxidation state of titanium is +4. Solid compounds containing Ti^{4+} are colorless and diamagnetic, but there is no evidence for existence of $Ti^{4+}(aq)$ ions. Addition of most compounds containing Ti^{4+} to alkaline, neutral, or weakly acidic solutions results in precipitation of TiO_2. A similar tendency to hydrolyze to an oxide, hydroxide, or oxyion is common to other species in +4 or higher oxidation states. Reasonably stable complex ions involving titanium in the +4 oxidation state are formed in the presence of large concentrations of various anions. For example, $(NH_4)_2TiCl_6$ is precipitated when NH_4Cl is added to a solution of $TiCl_4$ in concentrated HCl.

Compounds containing titanium in the +2 and +3 oxidation states are also known. These compounds are colored and paramagnetic owing to the presence of unpaired *d* electrons. The $Ti^{2+}(aq)$ ion is a strong, and generally rapid, reducing agent that readily reduces $H^+(aq)$ to $H_2(g)$. For the half-reaction we have

$$Ti^{3+}(aq) + e^- \rightleftharpoons Ti^{2+}(aq) \qquad \mathscr{E}^0 = \sim -2.3 \text{ volt}$$

Although $Ti^{3+}(aq)$ is a moderately strong reducing agent, there is no meaningful Ti^{4+}/Ti^{3+} half-reaction or potential. Oxidation of $Ti^{3+}(aq)$ yields a precipitate of TiO_2 or various complex ions containing titanium in the +4 oxidation state.

Vanadium

The element vanadium is relatively rare and difficult to obtain as a pure metal. Vanadium is used as a hardening additive in steels for high-speed tools and valve springs. It is generally prepared as ferrovanadium (solid solution of Fe and V) by high temperature reduction of mixed iron and vanadium oxides with carbon.

In its common compounds vanadium exists in +2, +3, +4, and +5 oxidation states. These correspond to removal of the 4*s* electrons and then successive removal of the 3*d* electrons. Many of the compounds are characteristically colored. We have the following potentials (acidic solutions):

$$V^{2+}(aq) + 2\,e^- \rightleftharpoons V(c) \qquad \mathscr{E}^0 = -1.13 \text{ volt}$$
$$V^{3+}(aq) + e^- \rightleftharpoons V^{2+}(aq) \qquad \mathscr{E}^0 = -0.26 \text{ volt}$$
$$VO^{2+}(aq) + 2\,H^+(aq) + e^- \rightleftharpoons V^{3+}(aq) + H_2O(liq) \qquad \mathscr{E}^0 = 0.34 \text{ volt}$$
$$VO_2^+(aq) + 2\,H^+(aq) + e^- \rightleftharpoons VO^{2+}(aq) + H_2O(liq) \qquad \mathscr{E}^0 = 1.00 \text{ volt}$$

These ions are commonly named as follows:

V^{2+}	vanadous
V^{3+}	vanadic (green)
VO^{2+}	vanadyl (blue)
VO_2^+	pervanadyl

The principal species in acidic solution (pH $< \sim 2$) of vanadium in the +5 oxidation state is the pervanadyl ion represented above by $VO_2^+(aq)$. In slightly less acid solutions with pH up to about 6.5, the principal species appear to be various "decavanadates," represented by such formulas as $H_2V_{10}O_{28}^{4-}(aq)$. In approximately neutral solutions the principal species appear to be "metavanadates," which are represented by $V_4O_{12}^{4-}(aq)$ and $V_3O_9^{3-}(aq)$. The orthovanadate ion $VO_4^{3-}(aq)$ is the principal species of V(V) in strongly alkaline solutions.

One of the most important compounds of vanadium is V_2O_5, which is used as a catalyst in various oxidation reactions such as conversion of SO_2 to SO_3 in the manufacture of sulfuric acid.

Chromium

Chromium is one of the less abundant metals ($\sim$0.04% of the earth's crust), but it finds extensive use in steel manufacture and in metal plating. Its principal ore is chromite, which can be represented by $FeCr_2O_4$. This ore is often reduced with carbon to form ferrochromium (solid solution of Cr and Fe). Small concentrations of chromium in iron lead to hard, tough steels; larger concentrations ($\sim$14%) give stainless steel with high corrosion resistance. Because metallic chromium takes a high polish and is corrosion resistant, it is widely used as a plating material. Plating is usually accomplished by electrolyzing the object in a bath of $Na_2Cr_2O_7$ in $H_2SO_4(aq)$.

The most important oxidation states of chromium are +2 (chromous), +3 (chromic), and +6. All compounds of chromium are colored, and reactions are generally accompanied by color changes. Chromium forms a large number of complex ions, some of which are discussed later in this chapter.

We begin our discussion of compounds and ions of chromium with the +2 oxidation state, for which we have the following potentials:

$$Cr^{2+}(aq) + 2\,e^- \rightleftharpoons Cr(c) \qquad \mathscr{E}^0 = -0.9 \text{ volt}$$
$$Cr^{3+}(aq) + e^- \rightleftharpoons Cr^{2+}(aq) \qquad \mathscr{E}^0 = -0.4 \text{ volt}$$

The second of these potentials shows that $Cr^{2+}(aq)$ ions act as a fairly strong reducing agent (Cr^{2+} is easily oxidized). In fact, oxygen in air readily oxidizes blue $Cr^{2+}(aq)$ to violet $Cr^{3+}(aq)$. The potential also shows that $Cr^{2+}(aq)$ can reduce $H^+(aq)$ to $H_2(g)$ while being oxidized to $Cr^{3+}(aq)$, but this reaction is slow.

Addition of base to chromous solutions precipitates yellow-brown chromous hydroxide, $Cr(OH)_2$, which is easily oxidized by oxygen in air to green chromic hydroxide, $Cr(OH)_3$, sometimes represented as a hydrated oxide, $Cr_2O_3 \cdot nH_2O$. Removing water by heating yields inert green Cr_2O_3, which is used as a pigment called chrome green.

Since $Cr(OH)_2$ is soluble in acidic solutions but not in excess base, we call it a basic compound; $Cr(OH)_3$ is soluble in acid or in excess base and is called **amphoteric.** Addition of base to solutions containing chromic ions first precipitates $Cr(OH)_3$, which then reacts with more base to dissolve:

$$Cr(OH)_3(s) + OH^-(aq) \rightleftharpoons Cr(OH)_4^-(aq) \qquad K \cong 10^{-2} \quad (20.2)$$

The chromite ion is not well characterized and might be represented by $CrO_2^-(aq)$ or by $Cr(OH)_4^-(aq)$ as above. The oxide containing chromium in the +6 state is distinctly acidic.

Increasing acidity of oxides with increasing oxidation state of the central element is common. For example, HOCl, HNO_2, and H_2SO_3 are all weak acids, whereas $HClO_4$, HNO_3, and H_2SO_4 are all stronger acids.

Solutions of chromic ion are slightly acidic due to a hydrolysis reaction that may be represented in either of the following ways:

$$Cr^{3+}(aq) + H_2O(liq) \rightleftharpoons CrOH^{2+}(aq) + H^+(aq) \qquad (20.3)$$

$$Cr(H_2O)_6^{3+}(aq) \rightleftharpoons Cr(H_2O)_5OH^{2+}(aq) + H^+(aq) \qquad (20.4)$$

Equations (20.3) and (20.4) differ in that we have represented the hydration of Cr^{3+} more explicitly in the second equation than in the first. In either case we have $K \cong 10^{-4}$.

Some compounds containing chromium in the +3 oxidation state dissolve in water or aqueous acid to form violet solutions containing chromic ions that we represent by $Cr(H_2O)_6^{3+}(aq)$. Addition of chloride ions to these solutions yields first the light green ion $Cr(H_2O)_5Cl^{2+}(aq)$ and then the dark green ion $Cr(H_2O)_4Cl_2^+(aq)$. Addition of excess $NH_3(aq)$ to chromic solution yields the yellow complex ion $Cr(NH_3)_6^{3+}(aq)$. With excess cyanide ions the yellow $Cr(CN)_6^{3-}(aq)$ is formed. Solid compounds containing these and many other complex ions can be obtained from solution. All of these complex ions have six anions or polar molecules (sometimes called **ligands**) in an octahedral arrangement around the central Cr^{3+} ion. Because many reactions involving these complexes are slow, they have been extensively studied.

In the +6 oxidation state, corresponding to complete removal of the 3*d* and

4*s* electrons, we have CrO_3 and many compounds containing oxyanions; of these the most common are dichromate, $Cr_2O_7^{2-}$, and chromate, CrO_4^{2-}. Some common salts containing these anions are $K_2Cr_2O_7$, K_2CrO_4, and a number of chromates that are only slightly soluble in water as indicated by the following solubility products (25 °C):

$$Ag_2CrO_4(c) \rightleftharpoons 2\,Ag^+(aq) + CrO_4^{2-}(aq) \qquad K_{sp} = 1 \times 10^{-12}$$
$$PbCrO_4(c) \rightleftharpoons Pb^{2+}(aq) + CrO_4^{2-}(aq) \qquad K_{sp} = 2 \times 10^{-16}$$
$$BaCrO_4(c) \rightleftharpoons Ba^{2+}(aq) + CrO_4^{2-}(aq) \qquad K_{sp} = 1 \times 10^{-10}$$

The principal Cr(VI) species in aqueous solutions are chromate ion represented by $CrO_4^{2-}(aq)$, dichromate ion represented by $Cr_2O_7^{2-}(aq)$, and hydrogen chromate (sometimes called bichromate) ion represented by $HCrO_4^-(aq)$. Relationships between these ions at 25 °C are summarized by the following reaction equations and equilibrium constants:

$$HCrO_4^-(aq) \rightleftharpoons H^+(aq) + CrO_4^{2-}(aq) \qquad K = 3.4 \times 10^{-7}$$
$$2\,HCrO_4^-(aq) \rightleftharpoons Cr_2O_7^{2-}(aq) + H_2O(liq) \qquad K = 34$$

According to these equilibrium constants, $CrO_4^{2-}(aq)$ is the principal species in alkaline solution. In acidic solution in which the total Cr(VI) concentration is low, $HCrO_4^-(aq)$ is the principal species; $Cr_2O_7^{2-}(aq)$ becomes the principal species at higher total concentration of Cr(VI).

Acidic solutions of Cr(VI) are strong oxidizing agents as shown by the standard potential:

$$Cr_2O_7^{2-}(aq) + 14\,H^+(aq) + 6\,e^- \rightleftharpoons 2\,Cr^{3+}(aq) + 7\,H_2O(liq) \qquad \mathcal{E}^0 = 1.3 \text{ volt}$$

This potential also shows that a very strong oxidizing agent, such as $S_2O_8^{2-}(aq)$, is required to oxidize Cr(III) to Cr(VI) in acidic solution. On the other hand, $CrO_4^{2-}(aq)$ in alkaline solution is a weak oxidizing agent, and Cr(III) in the form of chromite ions or precipitated chromic hydroxide is easily oxidized to $CrO_4^{2-}(aq)$.

When solutions of dichromates are made very acidic with concentrated sulfuric acid, red needles of CrO_3 are precipitated. Sulfuric acid solutions of Cr(VI) are powerful oxidizing agents that are widely used in organic chemistry, and also as a "cleaning solution" for laboratory glassware. The cleaning action is largely due to oxidation of grease and organic residues.

Manganese

Managanese has a low natural abundance (0.08% of the earth's crust) and is found chiefly as pyrolusite, MnO_2. Because the major use of manganese is as a steel additive, mixed oxides of iron and manganese are often reduced together in a blast furnace. In various compounds and ions, manganese has oxidation states of +2, +3, +4, +6, and +7. Most of the compounds are colored and paramagnetic.

Much of the aqueous chemistry of manganese is summarized by the potential diagrams shown in Figure 20.4. Here we discuss some reactions of compounds and ions of manganese in relation to these potentials.

Aqueous permanganate ions, generally obtained by dissolving $KMnO_4(c)$,

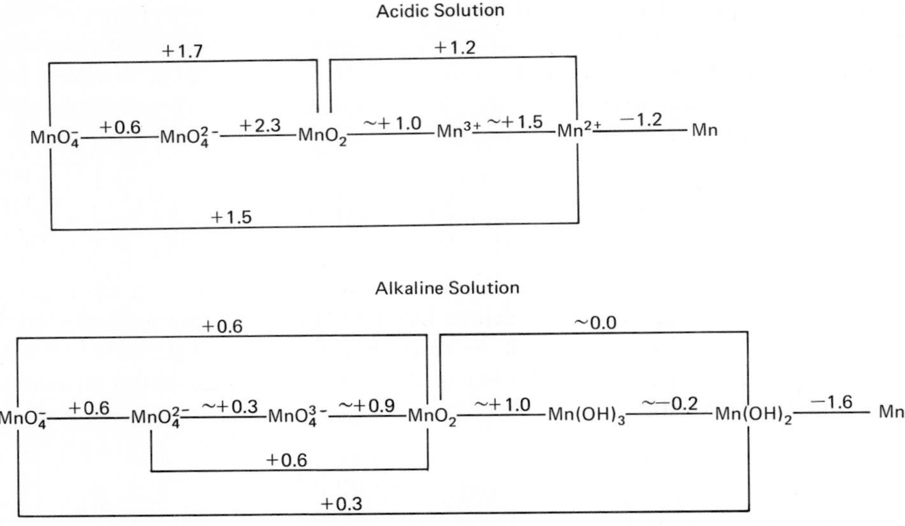

Figure 20.4. Standard potential diagram for manganese compounds and ions in acidic and alkaline aqueous solutions at 25 °C.

contain manganese in the +7 oxidation state and act as a powerful oxidizing agent in acidic solution, ordinarily being reduced to Mn^{2+}(aq). Because of the color change from the deep violet of MnO_4^-(aq) to pink of Mn^{2+}(aq), these reactions are easily followed. In alkaline solutions, permanganate is a less strong oxidizing agent, generally being reduced only to MnO_2(c).

Manganese in the +6 oxidation state exists as the manganate ion, MnO_4^{2-}. (Note the difference between manganate and permanganate.) Manganate ions are highly unstable in acidic solution because they can oxidize water and also disproportionate to MnO_2(c) and MnO_4^-(aq). In alkaline solutions, however, manganate ions are stable.

The MnO_4^{3-}(aq) ion is known only in alkaline solution, where it is unstable with respect to disproportionation as shown by the standard potential diagram.

Manganese dioxide is a strong oxidizing agent in acidic solutions. The largest use of MnO_2 is as the oxidizing component of ordinary dry cells used in flashlights. These dry cells, which are a convenient source of small electric currents at low voltages, consist of a zinc can containing a graphite rod that is surrounded by a moist paste of MnO_2, $ZnCl_2$, and NH_4Cl. Although the electrode reactions that occur when a dry cell delivers current are complicated, it is clear that the zinc is oxidized at the anode and MnO_2 is reduced at the cathode.

The ratio of elements in MnO_2 is never exactly one to two, and manganese dioxide is frequently called a nonstoichiometric compound. Nonstoichiometric solids are quite common among transition metal oxides and sulfides.

The Mn^{3+}(aq) ion is a sufficiently strong oxidizing agent to oxidize water in acidic solution, where it is also unstable with respect to disproportionation to Mn^{2+}(aq) and MnO_2(c). In alkaline solution various slightly soluble compounds of Mn(III), such as $Mn(OH)_3$, and some complex ions are reasonably stable.

Most salts of Mn(II) are quite soluble in water, but MnS and $Mn(OH)_2$ are slightly soluble and can be precipitated. Very powerful oxidizing agents are required to oxidize Mn^{2+}(aq) ions in acidic solutions, but $Mn(OH)_2$ is relatively easy to oxidize in alkaline solutions.

The increasing acidity of oxycompounds as the charge density of the central atom increases was discussed in Chapter 10 and is well illustrated with vanadium, chromium, and manganese compounds. Oxides of these elements in low oxidation states are basic. In higher oxidation states, these metals form oxyanions derived from acidic oxides. The oxides of the intermediate oxidation states are generally amphoteric.

Iron

Iron constitutes about 5% of the earth's crust, rånking fourth in abundance after oxygen, silicon, and aluminum. Because of its abundance, the relative ease of obtaining the metal from its ores, and its many useful properties (particularly when impure), iron may be regarded as the most important metal in our society.

On the basis of a variety of indirect evidence (including study of earthquakes), it is believed that the earth's core is largely molten iron with lesser amounts of cobalt and nickel. Metallic meteors are generally over 90% iron, indicating that iron is also abundant throughout the solar system.

The principal oxidation states of iron are +2 (ferrous) and +3 (ferric). The less common ferrate ion, FeO_4^{2-}, containing iron in the +6 oxidation state, is a powerful oxidizing agent. Apparent fractional oxidation states are best considered as mixtures of two integral oxidation states. For example, in Fe_3O_4 two thirds of the iron is Fe^{3+} and one third is Fe^{2+}.

We write the following standard potentials for acidic solutions:

$$Fe^{2+}(aq) + 2\,e^- \rightleftharpoons Fe(c) \qquad \mathscr{E}^0 = -0.47 \text{ volt}$$
$$Fe^{3+}(aq) + e^- \rightleftharpoons Fe^{2+}(aq) \qquad \mathscr{E}^0 = +0.77 \text{ volt}$$

Here it might be noted that the potential of the Fe^{2+}/Fe half-reaction was long thought to be −0.44 volt. More recent investigations have led to −0.41 volt and −0.47 volt, and we have chosen the latter as apparently being the "best."

These potentials show that metallic iron is a moderately strong reducing agent. Pale green Fe^{2+}(aq) ions are oxidized to the +3 state by oxygen in air or by other moderately strong oxidizing agents such as Br_2. Ferric ions are moderately strong oxidizing agents, able to oxidize I^-(aq) to I_2 but not able to oxidize Br^-(aq) to Br_2.

Hydrolysis of Fe^{2+}(aq) ions to $FeOH^+$(aq) and H^+(aq) is very slight, as shown by the small value for the equilibrium constant for the reaction (see problem 25, Chapter 14). Ferric ions, with a greater charge density (smaller size, higher charge) than ferrous ions, are hydrolyzed to a much greater extent. The first step in the hydrolysis may be written as

$$Fe^{3+}(aq) + H_2O(liq) \rightleftharpoons FeOH^{2+}(aq) + H^+(aq) \qquad K \cong 5 \times 10^{-3}$$

Further hydrolysis results in a red-brown gelatinous precipitate, which is commonly called ferric hydroxide but is best described as hydrous ferric oxide, $Fe_2O_3 \cdot nH_2O$.

The gradual conversion of free metals to compounds is often called **corrosion.** In the special case of iron, the corrosion process is called **rusting.** Rusting is a serious economic problem; it has been estimated that about 10% of the iron produced each year goes to replace that damaged by rusting.

Most rusting of iron occurs in the presence of both water and oxygen. If the water is acidic, rusting proceeds more rapidly. In dry air or oxygen-free water, rusting is usually very slow. Other conditions that expedite rusting are strains in the metal, presence of less active metals (metals that are harder to oxidize), presence of electrolytes in the water, and the presence of rust itself.

Although mechanisms of corrosion reactions are incompletely understood, some means of preventing or at least inhibiting rusting are known. Covering the iron with grease or paint effectively keeps away the reactants, oxygen and water. Electroplating with other metals that are less subject to corrosion is common and useful. Chromium plated iron and steel is familiar as the "chrome" on cars. Zinc plating, also called galvanizing, is widely used for pails and roofing. Tin plating as in some "tin cans" is also familiar.

Another method of inhibiting rust formation is called "cathodic protection." This name is related to the description of rusting as an electrochemical process in which iron functions as an anode that is oxidized. Anything that tends to make iron cathodic (where reduction rather than oxidation takes place) tends to inhibit rusting and is therefore called cathodic protection. Rusting of steel ships in salt water is a particularly serious problem, but it can be significantly reduced by strapping "sacrificial" blocks of magnesium to the hulls. These magnesium blocks preferentially oxidize (as anodes) but are easily replaced. Iron pipes may similarly be protected by attaching strips of zinc or magnesium to them. Note that cathodic protection does not stop corrosion; rather, it transfers attack away from the structurally important metal to an easily replaceable substitute.

Both Fe^{2+} and Fe^{3+} form a variety of stable complexes in aqueous solution and in the solid state. Examples are $Fe(CN)_6^{4-}$ (ferrocyanide, containing iron in the +2 state) and $Fe(CN)_6^{3-}$ (ferricyanide, containing iron in the +3 state) and the corresponding potassium salts $K_4Fe(CN)_6$ and $K_3Fe(CN)_6$. Both complex ions have the six cyanide ions arranged octahedrally around the central iron. These complex ions are so stable that their solutions show virtually none of the properties of iron or cyanide ions. Solutions and salts containing ferricyanide are paramagnetic with one unpaired electron per iron, whereas ferrocyanide species are diamagnetic.

One of the most important complexes of Fe^{2+} is hemoglobin, the oxygen carrying protein in blood. This molecule consists of a protein, **globin,** with four **heme** units attached to it. Heme has the basic porphyrin structure illustrated in Figure 20.5. The Fe^{2+} ion in this complex is in an octahedral environment with one vacant site, which can be occupied by oxygen. The details of binding oxygen to heme and subsequent release in the body are fairly well understood but are far too complex to review here.

It is also known that carbon monoxide forms a more stable complex with heme than does oxygen. Thus carbon monoxide reduces the oxygen carrying capacity of blood. In this way the susceptibility of man to carbon monoxide poisoning is readily explained.

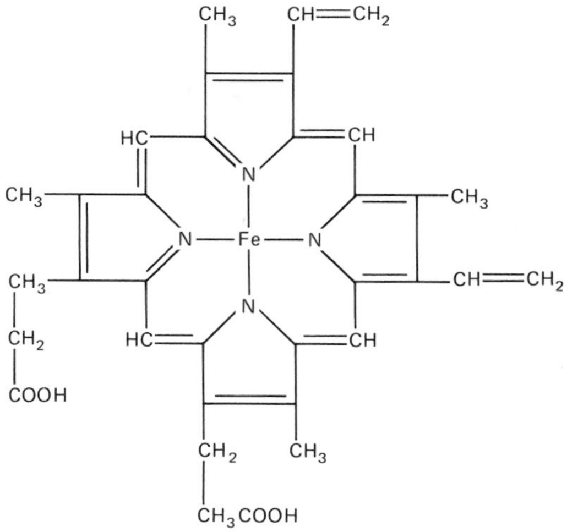

Figure 20.5. The iron porphyrin complex with the side chains shown here is called a **heme** group. Four heme groups combine with one globin molecule, a protein, to form **hemoglobin.**

Cobalt

Cobalt is quite rare (about 0.002% of earth's crust) and is found only in the combined state, mostly with arsenic and sulfur. The metal is used in corrosion-resistant alloys and in alloys for permanent magnets.

The important oxidation states of cobalt are +2 and +3, for which we have the following standard potentials in acidic solution:

$$Co^{2+}(aq) + e^- \rightleftharpoons Co(c) \qquad \mathscr{E}^0 = -0.28 \text{ volt}$$
$$Co^{3+}(aq) + e^- \rightleftharpoons Co^{2+}(aq) \qquad \mathscr{E}^0 = +1.9 \text{ volt}$$

This second potential shows that $Co^{3+}(aq)$ is a very strong oxidizing agent, strong enough to oxidize water.

Co(II) forms many complex compounds and ions with a variety of stereochemical arrangements. Octahedral and tetrahedral arrangements about the central cobalt are most common. Many of these cobaltous complexes are easily oxidized to the +3 state. For example, Co(II) forms a complex in the presence of excess cyanide that is able to reduce water to H_2 while being oxidized to $Co(CN)_6^{3-}$. A number of other complexes and slightly soluble compounds of Co(III) are also reasonably stable as compared to the $Co^{3+}(aq)$ ion. Nearly all Co(III) complexes are octahedral, although in a few solid state complexes different geometries are observed.

An important complex of Co^{3+} that occurs in nature is vitamin B_{12}. This enzyme (a biological catalyst) contains the cobalt ion in a porphyrin-like ring.

Nickel

Nickel is not very abundant in the earth's crust (0.01%) but is believed to be a major constituent of the earth's core. Nickel ores are principally sulfides and silicates mixed with other metals. Metallic nickel is used in various alloys, as a coinage metal, and as a catalyst in hydrogenation reactions.

Although nickel refining is a complicated process, it can be summarized as follows: In general, the ore is transformed to NiS, which is roasted in air to remove sulfur as SO_2 (sometimes a serious pollutant) and yield NiO. The oxide is then reduced to the metal with carbon or water gas ($CO + H_2$). Very pure nickel is produced by electrolysis or through formation and subsequent decomposition of gaseous nickel tetracarbonyl, $Ni(CO)_4$. In the Mond process, CO is passed over impure metal at ~60°C to form $Ni(CO)_4(g)$. This $Ni(CO)_4$ is then decomposed at ~200°C to yield CO, which is recycled, and very pure nickel. In $Ni(CO)_4$, nickel is assigned a zero oxidation state.

For the Ni^{2+}/Ni half-reaction we have

$$Ni^{2+}(aq) + 2\,e^- \rightleftharpoons Ni(c) \qquad \mathscr{E}^0 = -0.24 \text{ volt}$$

Metallic nickel is thus a mild reducing agent.

Nickel(II) forms a variety of complex compounds and ions. Some of these have octahedral configurations, some are square planar, and some are tetrahedral. No fully characterized compounds or ions containing nickel in oxidation states higher than +2 are known, but one such compound or mixture is of some importance.

Addition of NaOH to a solution of $NiCl_2$ or other salt of Ni(II) causes light green $Ni(OH)_2$ to precipitate. Electrochemical oxidation of this $Ni(OH)_2$ yields a black powder containing nickel in oxidation state greater than +2, variously represented as $Ni(OH)_3$, $Ni_2O_3 \cdot nH_2O$, or NiO_2. This material is used as one component of the Edison cell.

The Edison or nickel-iron battery uses KOH as the electrolyte and is based on a reaction that we represent by equation (20.5).

$$Fe + 2\,Ni(OH)_3 \underset{\text{charge}}{\overset{\text{discharge}}{\rightleftharpoons}} Fe(OH)_2 + 2\,Ni(OH)_2 \qquad (20.5)$$

The battery produces about 1.3 volt and has the advantage that OH^- produced at one electrode is consumed at the other electrode. In contrast to the lead storage battery and several other batteries, the output voltage of the Edison cell does not change significantly as it is discharged.

Copper, Silver, and Gold

Unlike the other metals we have discussed, copper, silver, and gold are sometimes found in nature as uncombined metals. Consequently, these metals have been known to man much longer than more reactive metals. They are all rare (Cu, 10^{-4}%; Ag, 10^{-8}%; Au, 10^{-9}% of the earth's crust), but some deposits are concentrated and relatively easily worked. Early use of these metals was in coins, hence the term **coinage metals.** The principal use of copper today is in the electrical industry. Considerable quantities of silver halides are used in photography.

These elements all have one electron in an *s* subshell outside a completed *d* subshell. We might therefore expect the +1 oxidation state to be most stable, but this is generally true only for silver. All of these elements have higher oxidation states, indicating that the *ns* and $(n-1)d$ orbitals are close in energy, as shown in Figure 20.2. Although +1, +2, and +3 oxidation states are known for copper, only the two lower states are of general importance. Silver exhibits

+2 and +3 oxidation states as well as the more stable +1 state. The oxidation states of gold are +1 and +3, the latter generally being most stable.

Many half-reaction potentials are known for copper compounds and ions. Some of these are summarized in the following potential diagram:

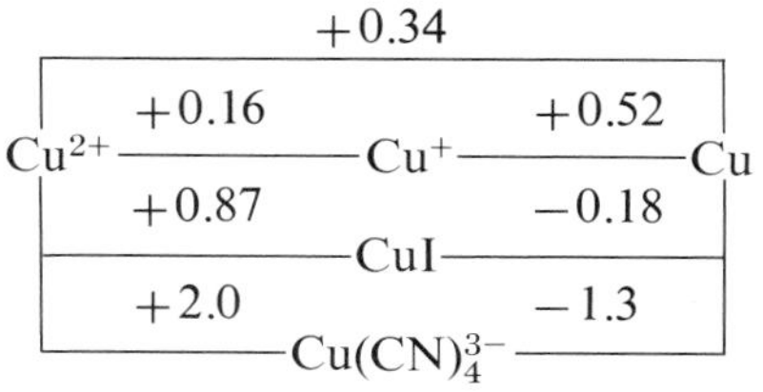

$$Cu^{2+} \xrightarrow{+0.16} Cu^{+} \xrightarrow{+0.52} Cu \qquad (Cu^{2+} \xrightarrow{+0.34} Cu)$$

$$Cu^{2+} \xrightarrow{+0.87} CuI \xrightarrow{-0.18} Cu$$

$$Cu^{2+} \xrightarrow{+2.0} Cu(CN)_4^{3-} \xrightarrow{-1.3} Cu$$

These potentials show that metallic copper will not liberate $H_2(g)$ from 1 M $H^+(aq)$. We can also see that the $Cu^+(aq)$ ion is unstable with respect to disproportionation to $Cu^{2+}(aq)$ and Cu(c) as in

$$2\ Cu^+(aq) \rightleftharpoons Cu(c) + Cu^{2+}(aq) \qquad K = 1.2 \times 10^6$$

Therefore $Cu^+(aq)$ ions can be present in aqueous solution only at very low concentrations.

The +1 oxidation state of copper is stabilized by reagents that lead to very stable complex ions or slightly soluble compounds. Examples are $CN^-(aq)$, which in excess yields the very stable $Cu(CN)_4^{3-}(aq)$ complex ion, and $I^-(aq)$, which leads to precipitation of the slightly soluble ($K_{sp} = 1.3 \times 10^{-12}$) CuI(c). The $Cu(CN)_4^{3-}/Cu$ potential shows that copper is easily oxidized (a strong reducing agent) in presence of excess $CN^-(aq)$.

The Cu^{2+}/CuI potential shows that $Cu^{2+}(aq)$ is a strong oxidizing agent in the presence of $I^-(aq)$. In fact, $Cu^{2+}(aq)$ is able to oxidize the $I^-(aq)$ ions so that the net reaction that occurs on mixing $Cu^{2+}(aq)$ with excess $I^-(aq)$ is

$$2\ Cu^{2+}(aq) + 5\ I^-(aq) \rightleftharpoons 2\ CuI(c) + I_3^-(aq) \qquad (20.6)$$

Here we have represented the oxidation product of $I^-(aq)$ as triiodide ion $I_3^-(aq)$, which is a combination of $I_2 + I^-$. The reaction above, followed by titration of the $I_3^-(aq)$ with thiosulfate solution, is the basis for a useful quantitative analysis procedure for copper.

In contrast to $Cu^+(aq)$, both CuI(c) and $Cu(CN)_4^{3-}(aq)$ are stable to disproportionation, as indicated by the large negative potentials for these reactions.

Aqueous solutions of cupric salts are light blue due to the presence of $Cu^{2+}(aq)$. Addition of aqueous ammonia first precipitates light blue $Cu(OH)_2$, which dissolves in the presence of excess ammonia because of formation of various complex ions such as $Cu(NH_3)_4^{2+}(aq)$ that are dark blue.

One of the best known compounds of copper is cupric sulfate pentahydrate, $CuSO_4 \cdot 5H_2O$, sometimes called blue vitriol. In this compound each Cu^{2+} ion is surrounded by four water molecules in a plane, with the two remaining octahedral sites occupied by oxygen atoms of neighboring sulfate ions. The fifth water molecule is hydrogen bonded between a second sulfate oxygen and a water molecule in the plane. Because $Cu^{2+}(aq)$ is toxic to many lower organisms such as algae, blue vitriol is used extensively in water treatment as a germicide and fungicide.

The standard reduction potential for silver is

$$Ag^{+}(aq) + e^{-} \rightleftharpoons Ag(c) \qquad \mathscr{E}^{0} = +0.80 \text{ volt}$$

which indicates that oxidation of Ag(c) to $Ag^{+}(aq)$ requires strong oxidizing agents (such as nitric acid). Conversely, $Ag^{+}(aq)$ ions are themselves fairly strong oxidizing agents. Many Ag^{+} salts have low solubility in water; the only common soluble salt is silver nitrate, $AgNO_3$. The trend in K_{sp} values and solubilities of the silver halides was discussed in Chapter 17. Two common, stable complexes of silver are $Ag(NH_3)_2^{+}$ and $Ag(CN)_2^{-}$, both of which are linear.

One of the major uses of silver today is in photography. Except for silver fluoride, all of the silver halides are sufficiently light sensitive to be used in photographic emulsions. The basic steps in the photographic process may be outlined as follows. Photographic film commonly consists of a dispersion of AgBr in gelatin. Exposure to light results in formation of an invisible **latent image.** This latent image consists of tiny AgBr crystals that have been **activated,** the extent of activation depending on the intensity of incident light. Activated crystals are more susceptible to reduction to elemental silver by mild reducing agents in a developing solution than nonactivated crystals. Where the light was most intense, there is the greatest amount of elemental silver, making the film blackest at these spots. Since excess AgBr would also turn black on further exposure to light, it must be removed. The film is therefore **fixed** by washing it with **hypo** (sodium thiosulfate, $Na_2S_2O_3$) solution, which dissolves the nonactivated AgBr. The result is a fixed negative image of the exposure. A positive image is then obtained by shining light through the negative onto another photographic emulsion, which is developed and fixed.

Although AgBr(c) has low solubility in water ($K_{sp} = 5.0 \times 10^{-13}$), it dissolves readily in hypo because of formation of a stable complex ion with thiosulfate:

$$AgBr(c) + 2\, S_2O_3^{2-}(aq) \rightleftharpoons Ag(S_2O_3)_2^{3-}(aq) + Br^{-}(aq) \qquad (20.7)$$

Another recent application of the photosensitivity of silver halides is in "photochromic" glass. This glass can be used to make eyeglasses that darken in sunlight but become clear again indoors. Photochromic glass is transparent to visible light but contains minute silver chloride crystals that decompose when exposed to ultraviolet light:

$$AgCl + h\nu \rightleftharpoons Ag + Cl \qquad (20.8)$$

The metallic silver particles cause the darkening. Since the silver and chlorine atoms cannot diffuse away from each other through the glass structure, silver chloride reforms with a release of energy when the glass is no longer exposed to ultraviolet light.

Some half-reaction potentials for gold are summarized in the potential diagram:

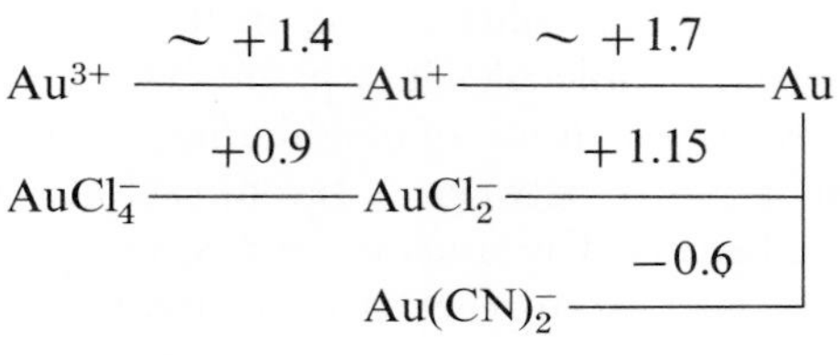

$$Au^{3+} \xrightarrow{\sim +1.4} Au^{+} \xrightarrow{\sim +1.7} Au$$

$$AuCl_4^{-} \xrightarrow{+0.9} AuCl_2^{-} \xrightarrow{+1.15} Au$$

$$Au(CN)_2^{-} \xrightarrow{-0.6} Au$$

These potentials show that $Au^+(aq)$ is unstable with respect to disproportionation. Gold is generally very difficult to oxidize (a weak reducing agent), except in cyanide solution where the stable complex ion $Au(CN)_2^-(aq)$ is formed.

Zinc, Cadmium, and Mercury

The elements of the zinc subgroup are slightly more abundant than the elements of the copper subgroup. The characteristic oxidation state for the zinc subgroup elements is +2, although there are a number of compounds containing mercury in the +1 oxidation state. No oxidation states greater than +2 are stable.

The most common ores of these metals are sulfides. Zinc and cadmiun sulfides, which commonly occur together, are roasted in air to form oxides and sulfur dioxide. The metal oxides are then reduced at high temperature with carbon. Cadmium, being more volatile than zinc, distills off first. Roasting mercuric sulfide (HgS) in air yields mercury directly; the metal is then purified by distillation.

The relatively low melting and boiling points of the zinc subgroup metals indicate correspondingly weak attractive forces between atoms. The exceptionally low melting point of mercury (−38.9°C) makes it the only liquid metal at room temperature. (Gallium and cesium melt slightly above "normal" room temperature.)

Although mercury has higher electrical resistance than any other common metal, its fluidity is such an advantage that is is commonly used in electrical switches and other electrical applications. In spite of its well known toxicity, the low volatility, high density, and resistance to air oxidation make mercury an extremely useful liquid for many kinds of laboratory work.

Because of their toxicity to lower forms of life, various compounds of mercury have long been used as fungicides (for example in paper manufacture and seed grain treatment). Despite the value of these uses, considerable damage to parts of the environment has occurred due to accumulation of mercury residues.

Both zinc and cadmium are widely used to protect iron and steel against corrosion. Although zinc can be toxic, small quantities are apparently necessary for proper human development and reproduction. Cadmium is also toxic to humans.

Zinc is a strong reducing agent in acidic solution, as shown by

$$Zn^{2+}(aq) + 2\,e^- \rightleftharpoons Zn(c) \qquad \mathscr{E}^0 = -0.76 \text{ volt}$$

Aqueous Zn^{2+} ions are colorless and diamagnetic, consistent with an electronic configuration of $(Ar)3d^{10}$. Adding base to solutions of zinc salts precipitates white zinc hydroxide, $Zn(OH)_2$, which is amphoteric and dissolves in excess hydroxide. The zincate ion that is formed in basic solution is not well characterized; it can be represented by $ZnO_2^{2-}(aq)$ or by $Zn(OH)_4^{2-}(aq)$. Zinc is also a strong reducing agent in alkaline solution, as indicated by

$$ZnO_2^{2-}(aq) + 2\,H_2O(liq) + 2\,e^- \rightleftharpoons Zn(c) + 4\,OH^-(aq) \qquad \mathscr{E}^0 = -1.19 \text{ volt}$$

Zinc forms a number of complex ions, most of which are four coordinate and tetrahedral. Examples are $Zn(NH_3)_4^{2+}$ and $Zn(CN)_4^{2-}$. Both of these complexes are sufficiently stable that $Zn(OH)_2$ dissolves in an excess of either aqueous ammonia or cyanide. Because four-coordinate complexes are so common in zinc chemistry, it may be most realistic to represent the zincate ion by $Zn(OH)_4^{2-}$ rather than as ZnO_2^{2-}, but this makes no difference to the potential above or to equilibrium constant calculations.

Many of the chemical properties of zinc and cadmium are very similar. The aqueous Cd^{2+} ion is colorless and diamagnetic, but is more toxic to man than Zn^{2+}. $Cd(OH)_2$ is easily precipitated, but it is much less soluble in excess hydroxide than is $Zn(OH)_2$. Although $Cd(OH)_2$ does not dissolve appreciably in excess hydroxide, it does dissolve in excess cyanide or ammonia owing to formation of complex ions. Both $Cd(CN)_4^{2-}$ and $Cd(NH_3)_4^{2+}$ are diamagnetic and tetrahedral.

Unlike zinc and cadmium, mercury exhibits a +1 (mercurous) oxidation state as well as the +2 (mercuric) state. Gaseous mercurous ions have an electronic configuration of $(Xe)4f^{14}5d^{10}6s^1$; therefore, we might expect mercurous compounds and the aqueous mercurous ion to be paramagnetic. They are, however, diamagnetic. Mercurous compounds do not contain discrete Hg^+ ions, but rather the double ion Hg_2^{2+}, in which two Hg^+ are joined by a covalent σ bond.

Additional evidence for Hg_2^{2+} comes from studies of the equilibrium involving mercury, mercurous ions, and mercuric ions in aqueous solution. The equilibrium reaction and corresponding equilibrium constant might be expressed in terms of $Hg^+(aq)$ or in terms of $Hg_2^{2+}(aq)$ as follows:

$$Hg(liq) + Hg^{2+}(aq) \rightleftharpoons 2\,Hg^+(aq) \qquad K' = [Hg^+]^2/[Hg^{2+}]$$
$$Hg(liq) + Hg^{2+}(aq) \rightleftharpoons Hg_2^{2+}(aq) \qquad K = [Hg_2^{2+}]/[Hg^{2+}]$$

Calculations with experimental data lead to a K that is constant, but to K' values that vary markedly with concentration. We therefore conclude that the reaction equation involving $Hg_2^{2+}(aq)$ ions is the most realistic representation. At 25°C, we have $K = 81$.

The mercurous halides are slightly soluble, as shown by the following:

$$Hg_2Cl_2(c) \rightleftharpoons Hg_2^{2+}(aq) + 2\,Cl^-(aq) \qquad K_{sp} = 1.4 \times 10^{-18}$$
$$Hg_2Br_2(c) \rightleftharpoons Hg_2^{2+}(aq) + 2\,Br^-(aq) \qquad K_{sp} = 6.3 \times 10^{-23}$$
$$Hg_2I_2(c) \rightleftharpoons Hg_2^{2+}(aq) + 2\,I^-(aq) \qquad K_{sp} = 5.0 \times 10^{-29}$$

Mercurous fluoride, Hg_2F_2, is hydrolyzed by water to yield a mixture of HgO and Hg. Note the similarities in the solubilities of the mercurous halides and silver halides.

Mercurous chloride, also called *calomel,* has been used in medicine because of its stimulating action on the liver and other secretive organs. The calomel electrode, a mixture of mercury and mercurous chloride in contact with a solution containing chloride ions, is widely used as a reference electrode for cell measurements. The half-reaction and standard potential are as follows:

$$Hg_2Cl_2(c) + 2\,e^- \rightleftharpoons Hg(liq) + 2\,Cl^-(aq) \qquad \mathcal{E}^0 = 0.268 \text{ volt}$$

Unlike the aqueous silver ion, $Hg_2^{2+}(aq)$ forms few soluble complex species. A convenient way to distinguish between AgCl(c) and $Hg_2Cl_2(c)$, both of which may be obtained as white precipitates, is by addition of aqueous ammonia. Silver chloride dissolves because of $Ag(NH_3)_2^+(aq)$ formation, but reaction of ammonia with mercurous chloride produces a dark gray precipitate as represented by equation (20.8):

$$Hg_2Cl_2(c) + 2\ NH_3(aq) \rightarrow HgNH_2Cl(c) + Hg(liq) + NH_4^+(aq) + Cl^-(aq) \quad (20.8)$$

The $HgNH_2Cl$ is white, but the dark finely divided mercury causes the precipitate to appear dark gray.

Some half-reactions and potentials relating Hg(liq), $Hg_2^{2+}(aq)$, and $Hg^{2+}(aq)$ are the following:

$$Hg_2^{2+}(aq) + 2\ e^- \rightleftharpoons 2\ Hg(liq) \qquad \mathscr{E}^0 = +0.80 \text{ volt}$$
$$Hg^{2+}(aq) + 2\ e^- \rightleftharpoons Hg(liq) \qquad \mathscr{E}^0 = +0.85 \text{ volt}$$
$$2\ Hg^{2+}(aq) + 2\ e^- \rightleftharpoons Hg_2^{2+}(aq) \qquad \mathscr{E}^0 = +0.91 \text{ volt}$$

The first potential above shows that a reasonably strong oxidizing agent is required to oxidize Hg(liq) to $Hg_2^{2+}(aq)$. Because oxidizing agents that are capable of oxidizing Hg(liq) to $Hg_2^{2+}(aq)$ are also able to oxidize $Hg_2^{2+}(aq)$ to $Hg^{2+}(aq)$, an excess of oxidizing agent leads to complete oxidation of mercury to the +2 state.

The mercuric halides are reasonably soluble in water, but are not "typical" salts. As we have seen, most salts dissolve in water to yield aqueous ions. But the mercuric halides are weak electrolytes, being only slightly dissociated into ions. In the presence of excess halide ion, such species as $HgX_4^{2-}(aq)$ are stable.

For HgS we have

$$HgS(c) \rightleftharpoons Hg^{2+}(aq) + S^{2-}(aq) \qquad K_{sp} = 10^{-52}$$

Compounds such as HgS and AgCl are frequently described as being "insoluble" in water. Although this is a convenient expression, it is not strictly true. A truly insoluble salt would have $K_{sp} = 0$ and corresponding $\Delta G^0 = +\infty$. However, the solubility of HgS is *very* small, and it is adequately descriptive to call it insoluble.

One common way to dissolve "insoluble" salts is to add some reagent that will react with one of the product ions and thereby cause the equilibrium to be shifted to the right. For example, silver chloride dissolves in ammonia because of the formation of $Ag(NH_3)_2^+(aq)$. Similarly, $BaCrO_4$ and other chromates are soluble in acidic solutions because chromate ions react with $H^+(aq)$ to yield dichromate ($Cr_2O_7^{2-}$) ions. Sulfides dissolve in strong acid because sulfide ions react with $H^+(aq)$ to form $HS^-(aq)$ and H_2S. Mercuric sulfide is best dissolved by aqua regia (mixed HCl and NHO_3). The nitrate ions oxidize S^{2-} and the chloride ions lead to formation of stable $HgCl_4^{2-}(aq)$, thus shifting the equilibrium to the right and causing HgS to dissolve.

Coordination Chemistry—General

We picture a bond being formed in most covalent molecules through sharing of electrons. In homonuclear molecules, such as H_2 or F_2, we imagine equal sharing of the pair of bonding electrons. For heteronuclear molecules, such as HCl or CF_4, we visualize unequal sharing but still consider that one electron in each bonding pair comes from each of the bonded atoms.

There is another class of compounds, called **coordination compounds,** in which both electrons for the covalent bond are supplied by a donor atom. The resultant bond is sometimes called a **coordinate covalent bond.** For example, in the reaction between gaseous trifluoroborane (BF_3) and ammonia,

$$BF_3(g) + :NH_3(g) \rightleftharpoons F_3B:NH_3(c) \tag{20.9}$$

both electrons for the B—N bond come from nitrogen. In Lewis acid-base terminology, BF_3 is an acid (electron pair acceptor) and NH_3 is a base (electron pair donor). Similarly, in formation of phosphonium bromide,

$$PH_3(g) + HBr(g) \rightleftharpoons PH_4^+Br^-(c) \tag{20.10}$$

HBr is an acid and PH_3 is a base. In formation of the hexafluoroarsenate ion, AsF_6^-,

$$AsF_5(g) + NaF(c) \rightleftharpoons Na^+AsF_6^-(c) \tag{20.11}$$

arsenic pentafluoride is an acid and fluoride ion is a base.

Because of the broadness of the Lewis acid-base definitions, it is difficult to describe classes of compounds as fitting either category. In general, the central atom of a Lewis acid must either have less than an octet of electrons (BF_3, $AlCl_3$) or be capable of expanding its octet (AsF_5, PCl_5, BrF_3). The central atom of a Lewis base must have at least one pair of unshared electrons (NH_3, H_2O, I^-). According to these broad definitions we might classify HCl as a base, whereas we generally consider it an acid. The ultimate criterion must be the availability of a vacant orbital in a Lewis acid and the availability of an electron pair in a Lewis base for the particular reaction under consideration.

Coordination Chemistry—Transition Metals

Several complex ions involving transition metals have already been mentioned. These complexes may be considered as coordination compounds in the Lewis acid-base classification. The transition metals have vacant orbitals and can function as Lewis acids. The ligands (anions or polar molecules coordinated to the transition metal) are Lewis bases.

Among the most significant aspects of transition metal complexes are their geometries, magnetic properties, and colors. Any successful bonding theory concerning these complexes must explain the stereochemical arrangements, numbers of unpaired electrons, and absorptions in the ultraviolet-visible region of the spectrum. Several bonding theories have been proposed with varying degrees of success in accounting for these properties. Before considering four of these theories, we first clarify some terminology.

Ligands that occupy one coordination site of the metal are called **unidentate** (one tooth). **Polydentate** ligands contain more than one donor atom and may occupy more than one coordination site. For example, ethylenediamine (en) is a bidentate ligand; nitrilotriacetate (NTA) is tetradentate; and ethylenediaminetetraacetate (EDTA) is hexadentate. Any ligand that bridges more than one coordination site, thus forming one or more rings, is called a **chelating** agent. Some common ligands, with the formulas and names, are shown in Table 20.1. Note that names of anionic ligands end in **-o.** Except for water, ammonia, and carbon

Table 20.1. Names and Formulas of Some Common Ligands*

Name	*Formula*	*Name*	*Formula*
Fluoro	F^-	Thiosulfato	$S_2O_3^{2-}$
Chloro	Cl^-	Aqua	H_2O
Hydroxo	OH^-	Ammine	NH_3
Cyano	CN^-	Carbonyl	CO
Nitro	NO_2^-	Oxalato (ox)	$^-O_2CCO_2^-$ (↗ ↖)

Name	*Formula*
Ethylenediamine (en)	$H_2NCH_2CH_2NH_2$ (↗ ↖)
Nitrilotriacetato (NTA)	$\leftarrow N(CH_2CO_2^- \searrow)(CH_2CO_2^- \rightarrow)(CH_2CO_2^- \nearrow)$
Ethylenediaminetetraacetato (EDTA)	$(\swarrow {}^-O_2CCH_2)(\nwarrow {}^-O_2CCH_2)\overset{\nearrow}{N}CH_2CH_2\overset{\nwarrow}{N}(CH_2CO_2^- \searrow)(CH_2CO_2^- \nearrow)$

*In polydentate ligands arrows are used to indicate the coordination sites.

monoxide, the names of neutral ligands are not changed. Also, note the spelling of **ammine** for NH_3.

Now we consider some of the rules of nomenclature and their application to a few compounds. If the complex is an anion, the ending **-ate** is used. If the complex is a cation or is neutral, the metal name is not changed. In either case, the oxidation state of the metal is indicated by Roman numerals in parentheses. Ligands are listed in alphabetical order regardless of the number of each. The number of simple ligands is indicated by the prefixes di, tri, tetra, penta, and hexa (from 2 to 6). For ligand names that already contain a numerical designation, such as ethylene*di*amine, the prefixes bis, tris, tetrakis (2 to 4) are used. Some examples of application of the above rules follow:

$Co(NH_3)_6^{3+}$	hexaamminecobalt(III) ion
$Fe(CO)_5$	pentacarbonyliron(0)
$K_3Fe(CN)_6$	potassium hexacyanoferrate(III)
$Cr(H_2O)_5OH^{2+}$	pentaaquahydroxochromium(III) ion
$Co(en)_2Cl_2^+$	dichlorobis(ethylenediamine)cobalt(III) ion
$[Pt(NH_3)_4][PtCl_6]$	tetraammineplatinum(II) hexachloroplatinate(IV)

Common names, such as potassium ferricyanide for the third compound above, are still frequently used.

Magnetic properties of solids can be determined with a magnetic (Gouy)

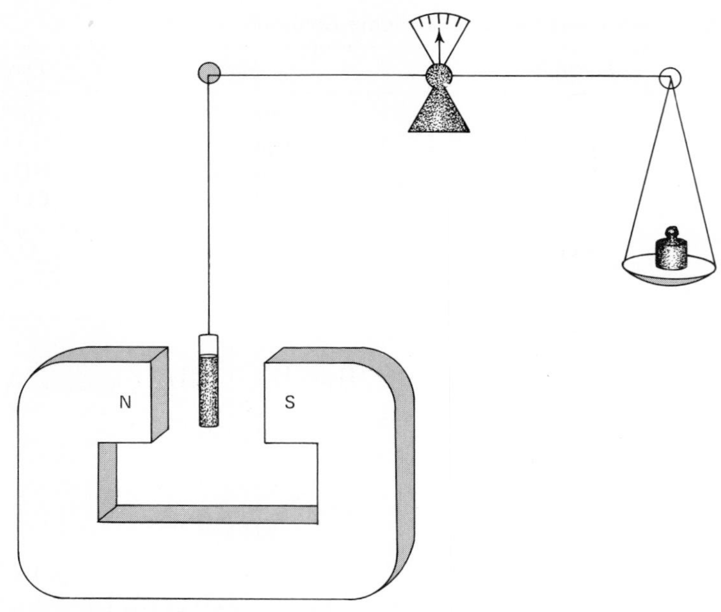

Figure 20.6. Magnetic balance. Because of repulsions, diamagnetic substances weigh less in the magnetic field than out of it. Paramagnetic substances are attracted into the magnetic field and thus weigh more.

balance, shown schematically in Figure 20.6. **Diamagnetic** substances, with no unpaired electrons, are slightly repelled by the magnetic field. **Paramagnetic** substances are attracted into the magnetic field, the amount of attraction being related to the number of unpaired electrons. The magnetic field aligns the unpaired electron spins and magnetizes the substance. When the magnetic field is shut off or removed, paramagnetic substances lose their magnetism. Substances such as metallic iron and cobalt that retain their magnetism are called **ferromagnetic.**

Paramagnetism in general depends on both the intrinsic spin and the orbital motions of unpaired electrons. For first row transition elements, however, the orbital contributions to the magnetic moment are effectively "quenched." Thus many experimental magnetic moments agree quite well with values calculated from

$$\text{"spin only" magnetic moment} = \sqrt{n(n+2)}$$

in which n is the number of unpaired electrons and the units are Bohr magnetons (BM). For example, a substance containing a V^{2+} ion has three unpaired d electrons and is calculated to have a magnetic moment of 3.87 BM; experimental values are 3.80 to 3.90 BM. Compounds of Ni^{2+} have experimental magnetic moments of 2.8 to 3.5 BM compared to the calculated value of 2.83 BM for two unpaired d electrons. Interpretation of magnetic properties of compounds of second and third row transition elements is more complicated because of appreciable orbital contributions.

Early Views of Bonding

The postulates about coordination compounds put forward by Swiss chemist Alfred Werner show excellent reasoning from relatively simple chemical information. When we realize that his proposals (1893) came before the discovery of the electron, which is the basis for all modern theories of bonding, his contributions are astounding.

Werner suggested that each metal has two kinds of valences: primary and secondary. Primary valences are nondirectional, but secondary valences have directional properties. Thus in $[Cr(NH_3)_6]Cl_3$ the secondary valences are satisfied by ammonia molecules at the corners of a regular octahedron; the three chlorides satisfy primary valences. In solutions of this compound all three Cl^- are readily precipitated by silver nitrate. In $[Cr(NH_3)_5Cl]Cl_2$ two chlorides satisfy primary valences and one chloride satisfies both a primary and a secondary valence. This third chloride is firmly bound to chromium and is therefore not readily precipitated from solution by silver nitrate.

G. N. Lewis (1916) pictured Werner's secondary valence in terms of a coordinate covalent bond. Sidgwick extended these ideas to view coordination of ligands as the means by which transition metals could achieve a noble gas configuration. He proposed that the **effective atomic number** (EAN) of a metal should be calculated as the total number of electrons of the metal atom or ion plus those shared with it through coordination. Thus in $Co(NH_3)_6^{3+}$, Co(III) has $27 - 3 = 24$ electrons plus two electrons from each NH_3 for a total of 36 electrons, the configuration of krypton.

The EAN concept is also consistent with the formulas of several metal carbonyls in which the metal is considered to be in the zero oxidation state. For example, the metals in $Cr(CO)_6$, $Fe(CO)_5$, and $Ni(CO)_4$ all have EAN values of 36. Although the EAN approach "works" well for many coordination compounds, it often fails and can therefore be taken only as a rough guide. Examples of its "failure" are four-coordinate and six-coordinate complexes of Ni^{2+} with EAN values of 34 and 38.

Valence Bond Theory

Much of the valence bond treatment of coordination compounds was developed by Linus Pauling. Because it is easily visualized, this approach was useful and popular in the 1930s but has now been largely replaced by the more quantitative ligand field and molecular orbital theories.

In valence bond theory of coordination compounds, bonds are pictured as resulting from electron donation by ligands to vacant hybrid orbitals of the metal. These hybrid metal orbitals are directed in space and may therefore be the basis for explanation of the geometries of various complexes. The number of vacant metal orbitals determines the coordination number of the metal.

Some common hybrid orbitals are listed in Table 20.2, with geometries and examples.

As an illustration of application of valence bond theory, consider the octahedral complexes $Fe(H_2O)_6^{2+}$ and $Fe(CN)_6^{4-}$. The aqua complex has a paramagnetism

Table 20.2. Some Common Hybrid Orbitals

Orbital	*Geometry*	*Examples*
sp	Linear	$Ag(NH_3)_2^+$, $CuCl_2^-$
sp^3	Tetrahedral	$Ni(CO)_4$, $Zn(NH_3)_4^{2+}$
dsp^2	Square planar	$Ni(CN)_4^{2-}$, $Pt(NH_3)_4^{2+}$
dsp^3	Trigonal bypyramidal	$Fe(CO)_5$, $CuCl_5^{3-}$
d^2sp^3	Octahedral	$Co(NH_3)_6^{3+}$, $Fe(CN)_6^{3-}$
sp^3d^2	Octahedral	$Fe(H_2O)_6^{2+}$, $Zn(NH_3)_6^{2+}$

corresponding to four unpaired electrons, whereas the cyano complex is diamagnetic.

We begin by assuming that the two 4*s* electrons of iron are lost in forming the free ferrous ion. The electronic configuration of Fe^{2+} beyond the (Ar) core is therefore represented by

3*d*	4*s*	4*p*
⇅ ↑ ↑ ↑ ↑	○	○ ○ ○

Consistent with Hund's rule, the six *d* electrons are distributed in all five of the 3*d* orbitals.

With six ligands forming covalent bonds, there are two possible ways to form an octahedral complex with Fe^{2+}: an "inner orbital" complex using 3*d* orbitals, or an "outer orbital" complex using 4*d* orbitals. We conclude from the observed magnetic properties that $Fe(H_2O)_6^{2+}$ is an "outer orbital" complex and that $Fe(CN)_6^{4-}$ is an "inner orbital" complex as indicated by the following representations:

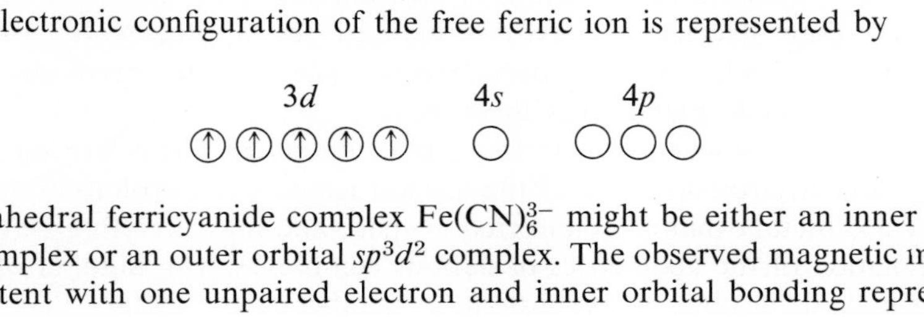

	3*d*	4*s*	4*p*	4*d*	
$Fe(H_2O)_6^{2+}$	⇅ ↑ ↑ ↑ ↑	[⇅	⇅ ⇅ ⇅	⇅ ⇅]	sp^3d^2
$Fe(CN)_6^{4-}$	⇅ ⇅ ⇅ [⇅ ⇅	⇅	⇅ ⇅ ⇅]		d^2sp^3

The electronic configuration of the free ferric ion is represented by

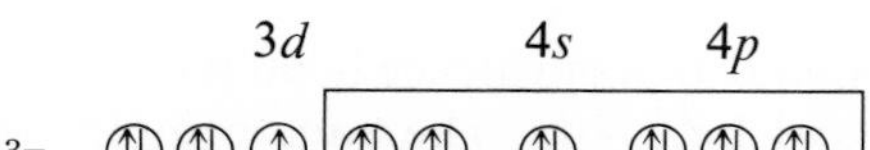

3*d*	4*s*	4*p*
↑ ↑ ↑ ↑ ↑	○	○ ○ ○

The octahedral ferricyanide complex $Fe(CN)_6^{3-}$ might be either an inner orbital d^2sp^3 complex or an outer orbital sp^3d^2 complex. The observed magnetic moment is consistent with one unpaired electron and inner orbital bonding represented by

	3*d*	4*s*	4*p*	
$Fe(CN)^{3-}$	⇅ ⇅ ↑ [⇅ ⇅	⇅	⇅ ⇅ ⇅]	d^2sp^3

If this complex had outer orbital sp^3d^2 bonding, there would be five unpaired electrons.

Ligands such as CN^- and CO generally form inner orbital complexes, whereas ligands such as H_2O and halide ions generally form outer orbital complexes.

Although valence bond theory leads to satisfactory explanations for the geometries and magnetic properties of many complexes, various difficulties have led to its being largely replaced by other theories that are discussed in the following sections.

One shortcoming of valence bond theory is the lack of any obvious means of predicting whether four-coordinate complexes of some metals are square planar or tetrahedral. For example, Cu(II) has a (Ar)$3d^9$ electronic configuration. Either square planar or tetrahedral complexes would have one unpaired electron as indicated in the following diagrams:

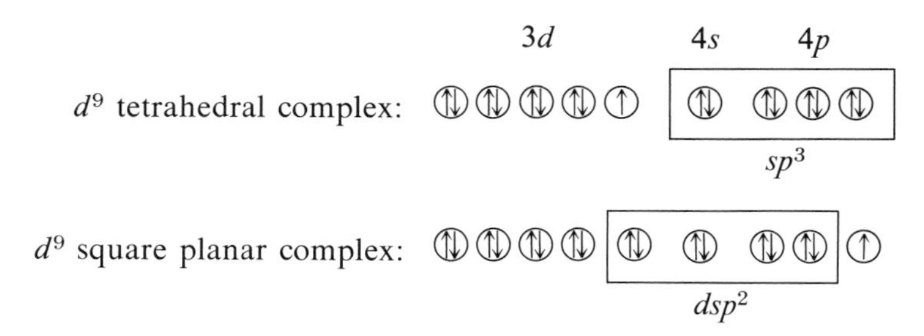

Complexes of Cu(II) with both geometries are known.

Another shortcoming of valence bond theory is that it is not conveniently related to the colors of complexes, which are due to absorption of light in the visible region of the spectrum.

Valence bond theory can be extended to cover these and most other problems, but in the process the chief attraction of the theory, its simplicity, is lost.

Crystal Field Theory

In our previous discussions of energies of orbitals we have referred only to isolated atoms. In crystal field theory (CFT), which became popular in the 1950s, the influence of neighbors on the *d* orbitals of the central metal ion or atom is important.

We start by considering electrostatic interactions. To minimize repulsions, we expect six ligands to be located at the corners of a regular octahedron, consistent with observed geometries of most hexacoordinated complexes. But the presence of these ligands creates an electrostatic field (or crystal field) that affects the energies of the *d* orbitals of the metal.

Because of their different orientations in space (see Figure 20.2), the metal *d* orbitals are not all affected the same way by this octahedral crystal field. The d_{z^2} and $d_{x^2-y^2}$ orbitals have lobes pointing directly at the incoming ligands. Electrons in these orbitals will be repelled to a higher energy state relative to the other three orbitals. The d_{xy}, d_{yz}, and d_{xz} orbitals have lobes oriented between the ligands so that electrons in these orbitals are more stable. The net result is that the five metal *d* orbitals, which were degenerate (of equal energy) in the

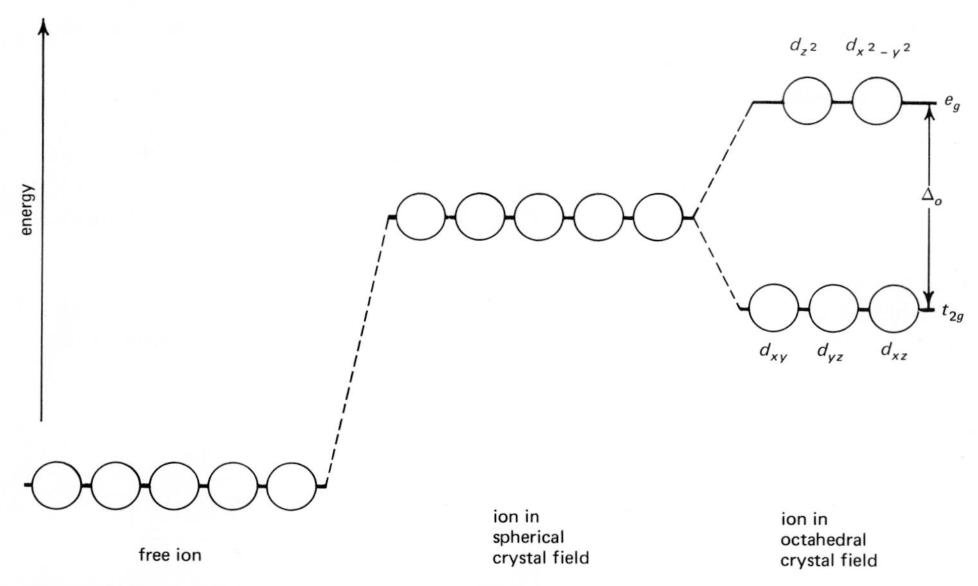

Figure 20.7. Energy level diagram of metal *d* orbitals in octahedral crystal field. The crystal field splitting energy, Δ_o, arises because of the different spatial orientations of the *d* orbitals. The subscript *o* designates the octahedral case.

free ion, are split into two energy levels in the octahedral crystal field. The separation between these two levels is called the **crystal field splitting energy,** Δ_o, as shown in Figure 20.7. The subscript o designates the octahedral case. The upper level ($d_{x^2-y^2}$ and d_{z^2}) is called the e_g level, the lower level (d_{xy}, d_{yz}, d_{xz}) the t_{2g} level. (The e_g and t_{2g} symbols come from group theory, but for our purposes may be considered simply as labels.)

Values of Δ_o can be obtained from electronic spectra. Consider the hexaaquatitanium(III) ion, $Ti(H_2O)_6^{3+}$, which has one *d* electron. Solutions of $Ti(H_2O)_6^{3+}$ are red-violet because they absorb yellow light. The absorption at 5000 Å corresponds to an energy of 57 kcal mole^{-1}. In crystal field terminology, this corresponds to an electronic transition between the t_{2g} and e_g orbitals. The crystal field splitting energy for $Ti(H_2O)_6^{3+}$ is thus 57 kcal mole^{-1}. More complicated analysis is required to obtain Δ_o values for complexes involving more than one *d* electron, but reasonable values are obtained.

Interconversion among the different units used in various branches of spectroscopy is frequently necessary and is illustrated in the following example problem.

Example Problem 20.1. From spectroscopic data the crystal field splitting energy, Δ_o, for Co(III) in the $Co(NH_3)_6^{3+}$ complex is 2.30×10^4 cm^{-1}. Express this value in the "traditional" units of Å, mμ, sec^{-1}, and kcal mole^{-1}, and also in the SI units of nm, Hz, and kJ mole^{-1}.

To convert cm^{-1}, which is a unit of reciprocal wavelength, to Å, we write

$$\left(\frac{1}{2.30 \times 10^4 \text{ cm}^{-1}}\right)\left(\frac{1 \text{ Å}}{1.0 \times 10^{-8} \text{ cm}}\right) = 4350 \text{ Å}$$

The splitting energy is thus in the violet region of the visible spectrum.

The nanometer (nm) unit, which is another way of expressing wavelength, is identical with the previously popular millimicron (mμ) unit. Since

$$1 \text{ m}\mu = 1 \text{ nm} = 1.0 \times 10^{-9} \text{ m} = 10 \text{ Å}$$

we note that

$$4350 \text{ Å} = 435 \text{ m}\mu = 435 \text{ nm}$$

The SI frequency unit of hertz (Hz) is identical with the traditional sec^{-1} unit. Since $E = h\nu$ and $1/\lambda = \nu/c$, both reciprocal wavelength and frequency units are directly proportional to energy. From $\nu = c\lambda^{-1}$, we calculate

$$(2.30 \times 10^4 \text{ cm}^{-1})\,(3.00 \times 10^{10} \text{ cm sec}^{-1}) = 6.90 \times 10^{14} \text{ sec}^{-1}$$
$$= 6.90 \times 10^{14} \text{ Hz}$$

To convert to kcal and kJ, the following conversion factors are useful:

$$1 \text{ kcal mole}^{-1} = 350 \text{ cm}^{-1} \text{ and } 1 \text{ kcal mole}^{-1} = 4.184 \text{ kJ mole}^{-1}$$

Therefore

$$(2.30 \times 10^4 \text{ cm}^{-1})\left(\frac{1 \text{ kcal mole}^{-1}}{350 \text{ cm}^{-1}}\right) = 65.7 \text{ kcal mole}^{-1}$$

$$(65.7 \text{ kcal mole}^{-1})\left(\frac{4.18 \text{ kJ mole}^{-1}}{1 \text{ kcal mole}^{-1}}\right) = 275 \text{ kJ mole}^{-1}$$

We thus see that the splitting energy for this cobalt complex is of the same order of magnitude as many bond energies. Values of Δ_o for first row transition metals range between 20 and 70 kcal mole^{-1}. Values of Δ_o for second and third row transition metals are generally greater, often as high as 100 kcal mole^{-1}. ■

It is important to realize that Δ_o depends upon both the central metal ion and the particular ligand. Ligands that produce the strongest electrostatic fields should also produce the greatest splitting of metal *d* orbitals. From studies of spectra of transition metal complexes the ability of some common ligands to split the *d* orbitals, called the **spectrochemical series,** is in the following order:

$$\text{CO, CN}^- > \text{en} > \text{NH}_3 > \text{H}_2\text{O} > \text{OH}^-,\ \text{F}^- > \text{Cl}^- > \text{Br}^- > \text{I}^-$$

strong field ligands	intermediate field ligands	weak field ligands

Although we might expect F^- with its higher charge density to produce a stronger field than I^-, it is clear that electrostatic interactions are not the entire explanation. Carbon monoxide, a neutral ligand, produces one of the strongest ligand fields. We shall return to this point later in our discussion.

As an example of the application of CFT, we consider the $Fe(H_2O)_6^{3+}$ and $Fe(CN)_6^{3-}$ complexes, both d^5 cases. Cyanide produces a greater crystal field splitting than does water. The $Fe(H_2O)_6^{3+}$ complex would achieve the lowest energy state by having one electron in each of the five *d* orbitals. In $Fe(CN)_6^{3-}$, there is a saving of energy by placing all electrons in the t_{2g} orbitals even though it "costs" energy to pair electrons. A schematic representation of the energy levels for the two iron complexes consistent with observed magnetic moments is shown in Figure 20.8.

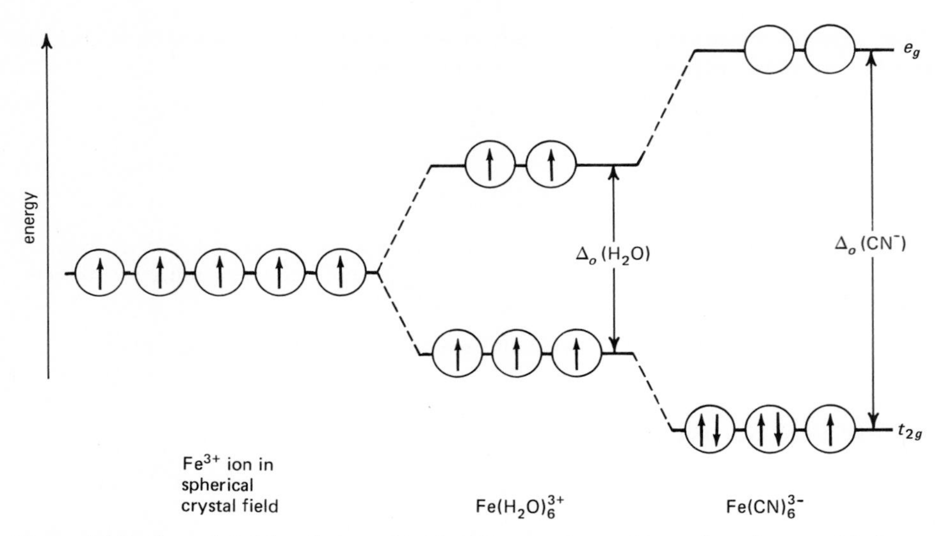

Figure 20.8. Crystal field splittings for Fe(III) complexes. Note that the e_g orbitals are vacant in the cyano complex, and that $\Delta_o(CN^-)$ is greater than $\Delta_o(H_2O)$.

With crystal field theory it is not only possible to explain magnetic properties but to predict them as well. Knowing that CN^- is a strong field ligand, we would expect it to produce a low spin complex. A weak field ligand such as H_2O, which would cause less splitting of the *d* orbitals, should produce a high spin complex. Geometries are also easily explained. In fairness, we should note that it is also possible to explain and predict magnetic properties for most complexes using valence bond theory. Perhaps the most significant advantage of CFT over valence bond theory is that colors of transition metal complexes are readily explained on the basis of transitions between the t_{2g} and e_g orbitals.

One limitation of crystal field theory is that there is no simple explanation of the relative ability of ligands to split metal *d* orbitals. An extension of CFT is **ligand field theory,** in which the possibility of π bonding is also considered. If metal *d* orbitals can overlap with ligand *p* or *d* orbitals to form a π bond, a more stable complex is formed, and the crystal field splitting energy increases. The ligands that produce the strongest crystal fields are all capable of π bond formation. Thus ligand field theory allows for covalent bonding in an originally pure ionic model, making it an approach towards the more general molecular orbital theory. As discussed in Chapter 8, molecular orbital theory can include the extremes of ionic and covalent bonding.

Before proceeding to a discussion of molecular orbital treatment of transition metal complexes, we consider two nonoctahedral crystal field examples.

It is convenient to view a square planar complex as formed from an octahedral complex in which the ligands on the *z* axis are moved to infinity. With the complex in the *xy* plane, the $d_{x^2-y^2}$ orbital has all four lobes pointing directly at ligands and is least stable. The d_{xy} orbital has lobes oriented between the ligands and is next highest in energy. The other three orbitals are out of the plane of the ligands and are most stable; the d_{z^2} orbital has a component in the *xy* plane and is less stable than the d_{xz} or d_{yz} orbitals. The arrangement of ligands and splitting pattern is shown in Figure 20.9.

It is easiest to visualize a tetrahedral complex with ligands on the opposite corners of a cube as shown in Figure 20.9. The t_{2g} orbitals (d_{xy}, d_{yz}, d_{xz}) have lobes that point to the midpoints of the edges of the cube. The lobes of the e_g orbitals are directed at the cube faces. The lobes of the t_{2g} orbitals are thus in closer proximity to the ligands, and electrons in these orbitals are less stable than those in the e_g orbitals. The energy separation for the tetrahedral case, Δ_t, is approximately half that for the octahedral case.

Because formation of six rather than four bonds leads to a greater total bond energy, the octahedral arrangement is intrinsically more stable than square planar. Consistent with predictions from CFT, square planar complexes are most common

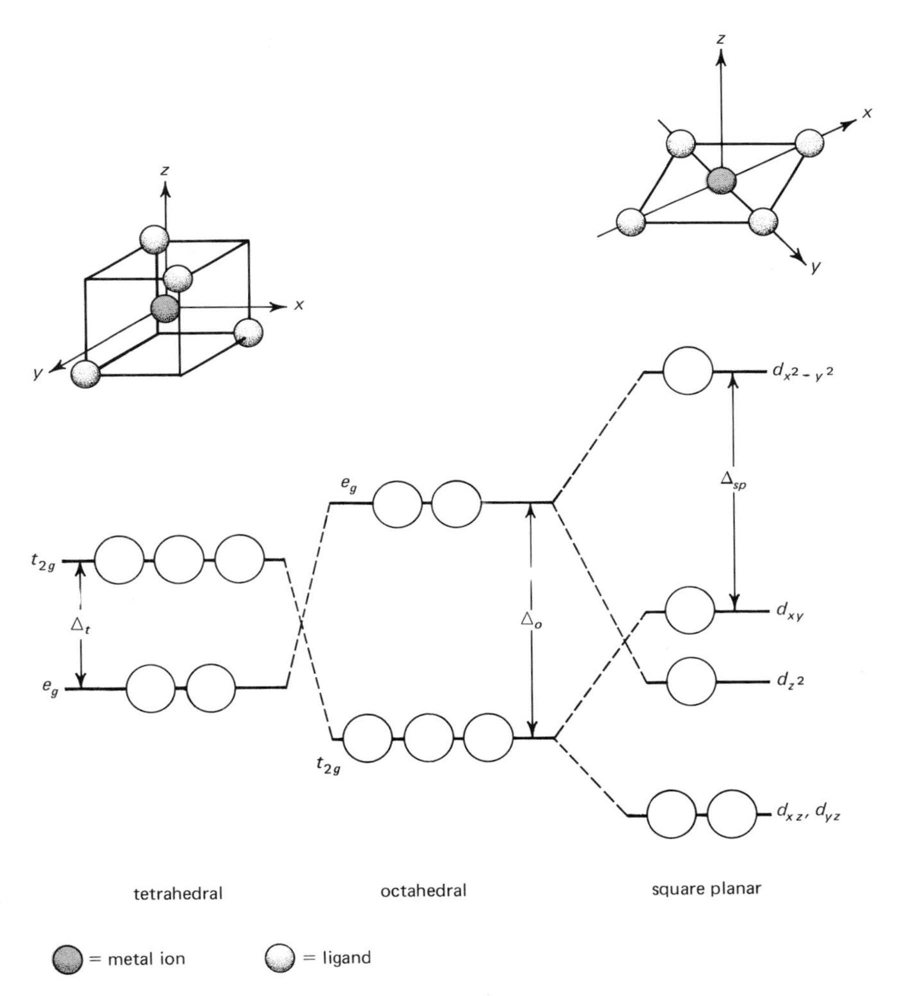

Figure 20.9. Relative energy levels for metal *d* orbitals in tetrahedral, octahedral, and square planar crystal fields.

Table 20.3. Representative Values of Crystal Field Splitting Energies

Octahedral complexes	Δ_o, cm^{-1}	*Tetrahedral complexes*	Δ_t, cm^{-1}
$Ti(H_2O)_6^{3+}$	20,300	VCl_4	9,000
TiF_6^{3-}	17,000	$CoCl_4^{2-}$	3,300
$Cr(H_2O)_6^{3+}$	17,400	$CoBr_4^{2-}$	2,900
$Cr(NH_3)_6^{3+}$	21,600	CoI_4^{2-}	2,700
$Cr(CN)_6^{3-}$	26,600		
$Cr(CO)_6$	34,150		
$Fe(H_2O)_6^{2+}$	10,400		
$Fe(CN)_6^{4-}$	33,800	*Square planar complexes*	Δ_{sp}, cm^{-1}
$Fe(H_2O)_6^{3+}$	13,700		
CoF_6^{3-}	13,000	$PdCl_4^{2-}$	23,600
$Co(H_2O)_6^{3+}$	18,200	$PtCl_4^{2-}$	29,700
$Co(NH_3)_6^{3+}$	22,900		
$Co(CN)_6^{3-}$	34,800		
$Co(H_2O)_6^{2+}$	9,300		

for d^8 configurations (Ni^{2+}, Pd^{2+}, Pt^{2+}). By placing the seventh and eighth electrons in the d_{xy} orbital in a square planar configuration, a lower energy state is achieved than by using a relatively high energy e_g orbital in an octahedral arrangement. Although it appears from Figure 20.9 that square planar coordination would produce a lower energy state for d^1 to d^6 cases, the extra bond energy for two additional ligands causes octahedral coordination to predominate.

Tetrahedral coordination is rare, also consistent with predictions of crystal field theory. Tetrahedral complexes not only have fewer bonds than octahedral complexes but the stabilization achieved by virtue of splitting the d orbitals is also less.

Some representative values for crystal field splitting energies are given in Table 20.3. These are experimental values obtained from ultraviolet-visible spectra. Note the relative values of Δ for octahedral, square planar, and tetrahedral complexes. Also note the relative ability of different ligands to split the d orbitals of any given metal and the relative splitting for metals in different oxidation states.

Molecular Orbital Theory

For a simplified molecular orbital treatment of coordination compounds, we consider that each ligand in an octahedral complex has a σ orbital available for bond formation. These six ligand orbitals can combine with the six metal orbitals that have components along the principal coordinate axes: s, p_x, p_y, p_z, $d_{x^2-y^2}$ and d_{z^2}. Note that these are the same metal orbitals used to form the d^2sp^3 hybrid orbitals in valence bond theory. The d_{xy}, d_{yz}, and d_{xz} orbitals have the wrong symmetry to form σ orbitals and therefore become nonbonding orbitals.

A resulting energy level diagram for hexafluorocobaltate(III) is shown in Figure 20.10. Six ligand orbitals plus six metal orbitals form six bonding and six antibonding molecular orbitals. The six bonding orbitals are filled, corresponding to six metal-ligand bonds. The remaining six electrons are distributed among the nonbonding t_{2g} orbitals and antibonding σ_d^* orbitals. The remaining four anti-

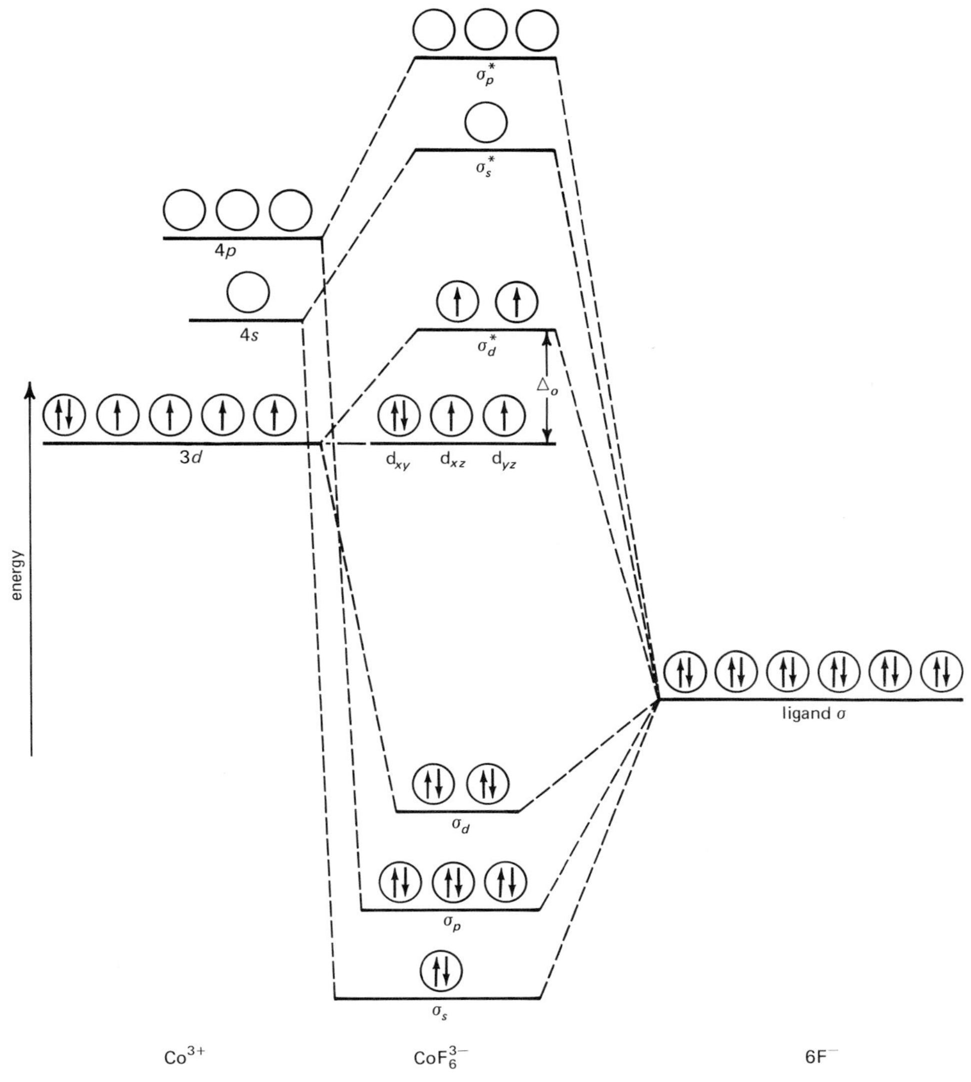

Figure 20.10. Molecular orbital energy diagram for octahedral coordination. CoF_6^{3-} represents a high spin case where Δ_o is relatively small.

bonding orbitals are relatively high in energy and are only of interest for excited states.

Placement of electrons in the t_{2g} and σ_d^* orbitals is quite analogous to crystal field filling of t_{2g} and e_g orbitals. If Δ_o is small, the lowest energy state is achieved by placing two electrons in the antibonding orbitals, a $t_{2g}^4\,\sigma_d^{*2}$ configuration. When Δ_o is large, as in $Co(NH_3)_6^{3+}$, the t_{2g}^6 low spin configuration represents the lowest energy state. However, the explanation for splitting in the two theories is quite different. Crystal field splitting arises from electrostatic repulsions of *d* electrons by the ligands. In molecular orbital theory the splitting is mostly attributed to

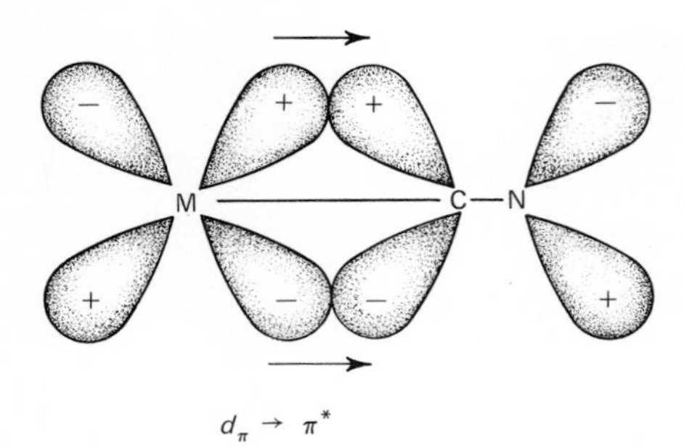

Figure 20.11. Formation of ligand-metal π bond between filled metal orbital and vacant ligand antibonding orbital. Increasing electron density in a π^* orbital of CN^- causes a weakening of the C—N bond. In effect, both the metal-carbon bond and carbon-nitrogen bond approach double bond characteristics.

covalent bonding. The greater the overlap of e_g orbitals with ligand orbitals, the higher in energy will be the corresponding σ_d^* orbitals.

The influence of π bonding on the stability of metal complexes and magnitude of Δ_o is readily explained with molecular orbital theory. We shall only attempt a qualitative explanation here by considering hexacyanocobaltate(III). As discussed in Chapter 18, the cyanide ion has a triple bond. It is the lone pair on carbon in a σ orbital that overlaps with a vacant metal orbital to form a bonding molecular orbital for the complex. The symmetry of the empty antibonding π^* orbitals of CN^- makes them suitable for overlap with filled t_{2g} orbitals of the metal. As shown in Figure 20.11, a metal-ligand π bond (designated $d_\pi \rightarrow \pi^*$) is formed. The metal-carbon bond thus has partial double bond character.

The large crystal fields caused by ligands that can π bond, such as CN^- and CO, can then be explained. The t_{2g} orbitals will no longer be nonbonding but will form π and π^* molecular orbitals with the ligands. These π orbitals will be lower in energy than the atomic d orbitals. By looking at Figure 20.10 we can see that any process that lowers the energy of the t_{2g} orbitals must increase Δ_o.

In summary, we should regard all three of these theories as only good approximations. They provide an excellent example of how the same physical facts can be explained by different and seemingly contradictory assumptions. All the theories can be used to account for many features of transition metal complexes. The most versatile and perhaps the most realistic is molecular orbital theory; unfortunately, it is also the most complicated and difficult to visualize. Currently the most successful is ligand field theory, an amalgamation of crystal field and molecular orbital theories.

Isomerism in Metal Complexes

Another significant aspect of transition metal complexes is **isomerism,** the existence of compounds with the same formula but different structures. Isomers generally differ in their chemical and physical properties and can often be isolated if they react slowly. Complexes that react rapidly generally rearrange in solution to yield only the most stable isomer. There are several different kinds of isomerism, but we will consider only three of the more important types.

Ionization isomers are complexes that yield different ions in solution and are common among octahedral, square planar, and tetrahedral complexes. Ions that are outside the coordination sphere of the metal react rapidly, whereas ions that are inside the coordination sphere react slowly.

A good example of ionization isomerism is the following pair of octahedral Co(III) complexes:

$[Co(NH_3)_5Br]SO_4$ pentaamminebromocobalt(III) sulfate violet

and

$[Co(NH_3)_5SO_4]Br$ pentaamminesulfatocobalt(III) bromide red

Conductivity data indicate that both isomers form two ions in aqueous solution. An aqueous solution of the violet isomer immediately yields a precipitate of $BaSO_4$ on addition of Ba^{2+}(aq) but does not yield a precipitate on addition of dilute Ag^{+}(aq). (More concentrated solutions of Ag^{+}(aq) would cause precipitation of white Ag_2SO_4.) A solution of the red isomer immediately yields pale yellow AgBr on addition of Ag^{+}(aq) but no precipitate with Ba^{2+}(aq).

In some metal complexes the ligands may occupy different positions around the central atom, resulting in **geometrical isomers.** The particular ligands may be next to one another (*cis*) or opposite each other (*trans*). Since all positions are equivalent in tetrahedral complexes, geometrical isomerism is not possible for this geometry.

Numerous examples of geometrical isomers are known for square planar complexes. Good examples are *cis*- and *trans*-diamminedichloroplatinum(II):

$$\begin{array}{ccc} Cl & & NH_3 \\ & Pt & \\ Cl & & NH_3 \end{array} \qquad \begin{array}{ccc} Cl & & NH_3 \\ & Pt & \\ H_3N & & Cl \end{array}$$

cis *trans*

Because it is not symmetrical, the *cis* isomer has a dipole moment; the *trans* isomer, which is symmetrical, does not.

Geometrial isomerism is also common among octahedral complexes. Examples are *cis*- and *trans*-tetraamminedichlorocobalt(III) and *cis*- and *trans*-dichlorobis (ethylenediamine)cobalt(III) ions, which are shown in Figure 20.12.

A third type of isomerism appears with *cis*-$[Co(en)_2Cl_2]^+$. Two forms of a molecule or ion that do not have superimposable mirror images are called **optical isomers.** They bear the same relationship to each other as an object and its mirror image, or as a right hand and left hand. The properties of optical isomers are nearly identical in all respects, with the exceptions of speed of reaction with other optical isomers and behavior toward plane-polarized light. Light that has passed through a polarizer, such as a polaroid filter, consists of waves that vibrate in a single plane. When this polarized light is passed through an optically active substance, it emerges vibrating in a different plane, which can be detected with a device called a polarimeter.

Many octahedral complexes, particularly those formed with chelating agents, exhibit optical activity. Square planar complexes rarely show optical activity, because the plane containing the metal and ligands divides the complex into two symmetrical parts. The only major exceptions are square planar complexes containing ligands that are themselves optically active.

If all four ligands of a tetrahedral complex are different, the complex will not

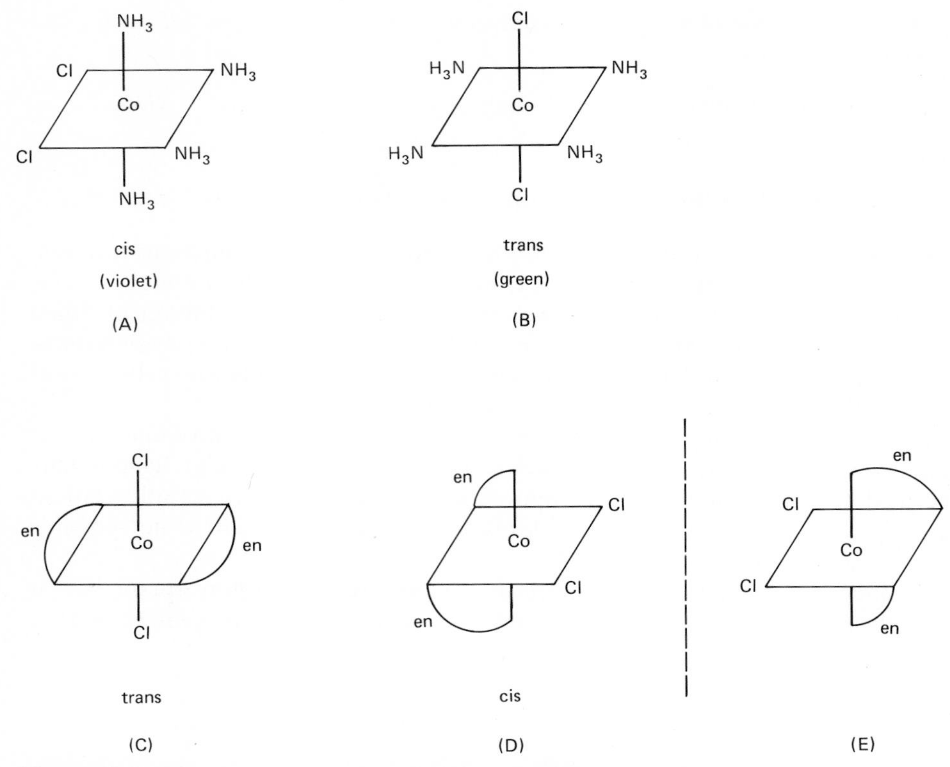

Figure 20.12. The geometrical isomers of tetraamminedichlorocobalt(III) are represented by (A) and (B). Geometrical isomers of dichlorobis(ethylenediamine)cobalt(III) are represented by (C) and (D). Because the mirror image (E) can not be rotated to yield a superimposable configuration with (D), the cis complex is optically active.

have a mirror plane, and optical activity is possible. Tetrahedral ligands exchange rapidly, however, and these optical isomers can rarely be isolated.

Optical activity is also common among organic compounds, particularly naturally occurring substances such as proteins and sugars.

References

Ameen, J. G., and H. F. Durfee: The structure of metal carbonyls, *J. Chem. Educ.*, **48:**372 (1971).

Basolo, F. A., and R. Johnson: *Coordination Chemistry*. W. A. Benjamin, Inc., New York, 1964 (paperback).

Canham, G. W. R., and A. B. P. Lever: Bioinorganic chemistry: simple models of iron sites in some biological systems. *J. Chem. Educ.*, **49:**656 (1972).

Cotton, F. A.: Ligand field theory. *J. Chem. Educ.*, **41:**9 (1964) (AC_3 resource paper).

Huheey, J. E., and C. L. Huheey: Anomalous properties of elements that follow long periods of elements. *J. Chem. Educ.*, **49:**227 (1972).

Kettle, S. F. A.: *Coordination Compounds*. Appleton-Century-Crofts, New York, 1970 (paperback).

King, H. C. A.: Preparation and properties of a series of cobalt(II) complexes. *J. Chem. Educ.*, **48**:482 (1971).
Larsen, E. M.: *Transitional Elements.* W. A. Benjamin, Inc., New York, 1965 (paperback).
Moeller, T.: *The Chemistry of the Lanthanides.* Reinhold Publishing Corp., New York, 1963 (paperback).
Moeller, T.: Periodicity and the lanthanides and actinides. *J. Chem. Educ.*, **47**:417 (1970).
Murmann, R. K.: *Inorganic Complex Compounds.* Reinhold Publishing Corp., New York, 1964 (paperback).
Orgel, L. E.: *An Introduction to Transition Metal Chemistry: Ligand Field Theory.* Methuen, London, 1966.
Pauling, L.: Valence bond theory in coordination chemistry. *J. Chem. Educ.*, **39**:461 (1962).
Plumb, R. C., L. E. Strong, and J. Blazyk: Thanksgiving dinner and transition metal complexes. *J. Chem. Educ.*, **48**:265 (1971).
Seaborg, G. T.: *Man-Made Transuranium Elements.* Prentice-Hall, Inc., Englewood Cliffs, N.J., 1963 (paperback).
Seaborg, G. T.: Prospects for further considerable extension of the periodic table. *J. Chem. Educ.*, **46**:626 (1969).

Problems

1. Predict the atomic number for each of the following: (a) noble gas after radon, (b) first element of third inner transition series, (c) first element of fifth transition element series.

2. Would an experimental investigation of the paramagnetism of gaseous copper atoms help to distinguish between the $(Ar)3d^{10}4s^1$ and $(Ar)3d^94s^2$ configurations? Explain.

3. Various mnemonic aids have been devised to assist students in remembering relative orbital energies. One of these is the following, in which one starts at the top with the 1*s* orbital and then follows the arrows.

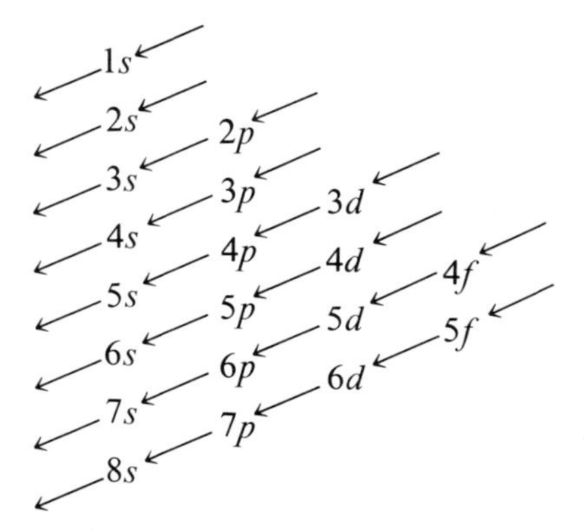

It is important to realize that these mnemonic aids are only rough guides and that there are inversions in the expected order, particularly among the heavier elements where orbital energies are close together. Predict the electronic configurations of the following elements:

Ca V Cu Mo Cd La Pt Hg Gd Cf

Compare your predictions with Appendix V.

4. (a) If only elements with partially filled *d* orbitals were included as transition elements, which elements of atomic numbers 21 through 30, and atomic numbers 39 through 48 would be excluded?

(b) If only elements with partially filled *f* orbitals were included as inner transition elements, which of the lanthanides and actinides would be excluded?

5. Predict the number of unpaired electrons for each of the following gaseous ions: Mo^{3+}, Pt^{2+}, Ce^{3+}, U^{3+}.

6. Draw representations of the $d_{z^2-x^2}$ and $d_{z^2-y^2}$ orbitals and show qualitatively that a linear combination of these orbitals produces the d_{z^2} orbital shown in Figure 20.1.

7. The ionic radii of Y^{3+}(0.93 Å) and La^{3+}(1.15 Å) are quite different from each other whereas the ionic radii of Zr^{4+}(0.80 Å) and Hf^{4+}(0.81 Å) are almost identical. Explain.

8. Write the anode, cathode, and cell equations for electrolysis of molten $LaCl_3$. How much elemental lanthanum can be obtained by a steady current of 3.0 amp for 1.0 hr?

9. Consider the reduction potentials in acidic solution summarized in the following diagram:

$$VO_2^+ \xrightarrow{+1.00} VO^{2+} \xrightarrow{+0.34} V^{3+} \xrightarrow{+0.26} V^{2+} \xrightarrow{-1.13} V$$

If metallic zinc is added to an acidic solution containing VO_2^+(aq) ions, what vanadium species will result? Write a balanced equation for the reaction.

10. Predict whether UO_2 or UO_3 is more acidic. Which is more soluble in aqueous sodium hydroxide?

11. Write a balanced equation for the reduction of chromite, $FeCr_2O_4$, with carbon. Ignoring purification problems and possible formation of carbides, how many tons of chromium are available from 100 tons of pure chromite ore?

12. Write a balanced equation for oxidation of Fe^{2+}(aq) by oxygen in acidic solution. Use $\mathscr{E}^0$ values from Appendix IV to calculate ΔG^0 for this reaction.

13. By considering the following equilibria:

$$HCrO_4^-(aq) \rightleftharpoons H^+(aq) + CrO_4^{2-}(aq) \qquad K = 3.4 \times 10^{-7}$$
$$2\,HCrO_4^-(aq) \rightleftharpoons Cr_2O_7^{2-}(aq) + H_2O \qquad K = 34$$

and LeChatelier's principle, we predict correctly that at low pH and low concentrations of Cr(VI) the $HCrO_4^-$(aq) ion is most stable. State the conditions of pH and Cr(VI) concentration at which the CrO_4^{2-}(aq) and $Cr_2O_7^{2-}$(aq) ions are most stable.

14. Many chromates, such as $BaCrO_4$, are only slightly soluble in water. Using the data in problem 13, predict the effect of acids on the solubility of $BaCrO_4$.

15. Using standard reduction potentials from Appendix IV, decide which of the following oxidations are possible with $Cr_2O_7^{2-}$ in aqueous acid: (a) Br^- to Br_2, (b) Cl_2 to HOCl, (c) SO_4^{2-} to $S_2O_8^{2-}$, (d) MnO_2 to MnO_4^-, (e) Hg to Hg^{2+}.

16. Should the object to be "chrome-plated" be made the anode or cathode in a $Na_2Cr_2O_7/H_2SO_4$ bath? Explain.

17. Since manganese forms a carbide, reduction of MnO_2 with carbon is generally avoided. An alternative reaction is reduction with powdered aluminum. Write the balanced equation for this reaction and calculate ΔH^0. This reaction can be used to obtain several metals in high purity and is called the Goldschmidt reaction. It owes its success to the great thermodynamic stability of aluminum oxide.

18. Using $\mathscr{E}^0$ values from Figure 20.4 calculate the equilibrium constant at 25°C for disproportionation of Mn^{3+}(aq) ions.

19. Choose the most acidic oxide in each of the following sets:

(a) Mo_2O_3, MoO_2, Mo_2O_5, MoO_3

(b) VO, V_2O_3, VO_2, V_2O_5

(c) FeO, Fe_2O_3

(d) WO_2, W_2O_5, WO_3

20. If the tin plating on an iron can is punctured, rusting is more rapid than if the tin were not present. Consider reduction potentials and suggest an explanation.

21. What happens to the pH of a solution containing ferrous ions when air is bubbled through it? Write a reaction equation to justify your answer.

22. Assuming that nickel tetracarbonyl behaves as an ideal gas, calculate its density at 60°C and 1.0 atm pressure.

23. Write balanced equations for the following reactions. (It may be necessary to look up solubilities and other data.)

(a) Roasting of NiS and subsequent conversion of NiO to Ni with water gas
(b) Formation of CrO_3 from an aqueous solution containing Cr^{2+} ions
(c) Air oxidation of free Ag to $Ag(CN)_2^-$ in presence of alkaline cyanide solutions
(d) Dissolving Au in aqua regia.

24. Use the half-reaction potentials in the text to calculate the K_{sp} for CuI.

25. Although roasting sulfide ores usually produces a metal oxide, free mercury can be obtained directly in this way. Suggest an explanation.

26. From museum samples it appears that the concentrations of mercury in fish have been relatively constant (at 0.1 part per million or less) for over a century. Why has mercury pollution only recently become a "problem" to man? We should note that these concentrations are found in fish caught in open waters and are due to natural sources of mercury such as weathering of rocks. In waters where industrial activities have led to higher than "normal" mercury concentrations, such as Lake Erie and Minamata Bay in Japan, fish show much higher levels of mercury, often greater than 1 ppm.

27. One of the oldest ways of separating transition metals involves treatment with H_2S at different $H^+(aq)$ concentrations. Controlling the pH controls the sulfide ion concentration, as may be judged by the following equilibria at 25°C:

$$H_2S(aq) \rightleftharpoons H^+(aq) + HS^-(aq) \qquad K_1 = 1 \times 10^{-7}$$
$$HS^-(aq) \rightleftharpoons H^+(aq) + S^{2-}(aq) \qquad K_2 = 1 \times 10^{-13}$$

Combining K_1 and K_2, we obtain

$$K_1 \times K_2 = 1 \times 10^{-20} = \frac{[H^+][HS^-]}{[H_2S]} \times \frac{[H^+][S^{2-}]}{[HS^-]} = \frac{[H^+]^2[S^{2-}]}{[H_2S]}$$

If we restrict ourselves to saturated solutions at 1 atm $H_2S(g)$, the concentration of $H_2S(aq)$ is 0.1 *M* and our expression becomes

$$1.0 \times 10^{-20} = \frac{[H^+]^2[S^{2-}]}{0.1}$$

which is rearranged to

$$[S^{2-}] = \frac{1.0 \times 10^{-21}}{[H^+]^2}$$

(a) At what pH is $[S^{2-}] = 1.0 \times 10^{-15}$ *M*?
(b) Calculate $[S^{2-}]$ when pH = 4.2.

28. Calculate the concentration of Zn^{2+} ions in a solution buffered at pH = 6 that is saturated with H_2S. For zinc sulfide, $K_{sp} = 1 \times 10^{-20} = [Zn^{2+}][S^{2-}]$. Remember that the concentrations of all species must be such that all relevant equilibrium constant expressions are satisfied, in this case K_{sp} and $[S^{2-}] = 1 \times 10^{-21}/[H^+]^2$.

29. By calculations, show that Pb^{2+} and Zn^{2+} are effectively separated by making $[H^+] = 0.3$ *M* and then saturating with H_2S. For PbS, $K_{sp} = 1 \times 10^{-28} = [Pb^{2+}][S^{2-}]$.

30. Calculate the concentration of Hg^{2+} ions in a 1.0 *M* H^+ solution that is saturated with H_2S. For HgS, $K_{sp} = 1 \times 10^{-52} = [Hg^{2+}][S^{2-}]$.

31. Explain why addition of a solution of aqueous mercuric nitrate to an unbuffered solution of H_2S increases the H^+ ion concentration.

32. Give systematic names for each of the following: $Ag(NH_3)_2^+$, FeF_6^{3-}, $Co(en)_3^{3+}$, $Cr(CO)_6$, $[Cr(C_2O_4)_2(H_2O)_2]^-$, $[Co(NO_2)_3(NH_3)_3]$.

33. Predict the spin-only magnetic moment for each of the following: Mn^{2+}, $Zn(CN)_4^{2-}$, $Cu(NH_3)_4^{2+}$, $CrCl_2(H_2O)_4^+$.

34. Calculate the "effective atomic number" of the metal in each of the complexes in problem 32 and compare with that of the appropriate noble gas.

35. What atomic orbitals would be used to form a set of dsp^2 hybrid orbitals with lobes pointing toward the corners of a square.

36. Predict the geometry of the following ions: NiF_6^{4-} (two unpaired electrons); $AuCl_4^-$ (diamagnetic); CrF_6^{4-} (four unpaired electrons); $FeCl_4^-$ (five unpaired electrons).

37. Account for the fact that there is one form of the $Co(NH_3)_6^{3+}$ complex but two forms of the $Co(NH_3)_4Cl_2^+$ complex. All forms are octahedral.

38. There are three different forms of the compound $CrCl_3 \cdot 6H_2O$. Aqueous $AgNO_3$ precipitates one third of the chloride in the dark green form, two thirds in the light green form, and all in the violet form. The dark green compound easily loses one third of its water and the light green compound easily loses one sixth of its water. Write structural formulas for each compound.

39. The spectroscopically determined ligand field splitting energy for $Cr(NCS)_6^{3-}$ is 17,800 cm^{-1}. Using data in Table 20.3, place NCS^- (isothiocyanate) in the spectrochemical series.

40. Which complex in each of the following sets should have the larger Δ value:
(a) $Fe(CN)_6^{4-}$ or $Fe(CN)_6^{3-}$?
(b) PtI_4^{2-} or $PtCl_4^{2-}$?
(c) $Co(NH_3)_6^{3+}$ or $Co(NH_3)_6^{2+}$?
(d) $Cr(CO)_6$ or $Cr(en)_3^{3+}$?

41. Draw a molecular orbital energy level diagram for the $Co(CN)_6^{3-}$ complex and show the electron population of orbitals. Will the carbon-nitrogen bond energy in the complex be greater or less than that in HCN? Explain.

42. Suggest an explanation for the fact that $Zn(NH_3)_2Cl_2$ exists in only one form, but two forms of $Pt(NH_3)_2Br_2$ are observed.

43. Draw structures for all the geometrical isomers of $Co(en)Cl_2Br_2^-$. Are any of these isomers optically active?

44. A 0.412 g sample of brass, an alloy of copper and zinc, was dissolved in dilute nitric acid. Excess KI was added to precipitate CuI. If 38.4 ml of 0.100 *M* $Na_2S_2O_3$ solution were required to react with the resultant $I_3^-(aq)$, calculate the percentage of copper in the sample. The reaction equations are:

$$2\,Cu^{2+}(aq) + 5\,I^-(aq) \rightleftharpoons 2\,CuI(c) + I_3^-(aq)$$
$$I_3^-(aq) + 2\,S_2O_3^{2-}(aq) \rightleftharpoons S_4O_6^{2-}(aq) + 3\,I^-(aq)$$

21

SOLIDS

Introduction

Consideration of the regular forms of crystals led Hooke (1665) and Haüy (1784) to propose that crystals consist of small particles packed together in orderly ways. Various workers deduced from the small compressibilities of solids that these small particles must be packed together rather closely. Because solids are rigid and do not flow easily, it was also concluded long ago that the small particles do not slide over one another in crystals as easily as they do in liquids. It was even possible to deduce something of the arrangement of atoms in crystals from considerations of their external forms, but detailed investigations of structures of solids were impossible before the discovery of x rays.

One of the simpler principles of x-ray investigation of crystal structures is discussed briefly in this chapter, and the structures of a few solids are illustrated. Some of the factors that determine crystal structures of ionic crystals are also discussed.

It is possible to use our knowledge of crystal structures in several ways. One way that is illustrated in this chapter is in the calculation of lattice energies and enthalpies.

Part of this chapter is concerned with the properties and production of various metals. Another part is devoted to discussion of some synthetic polymers. In connection with our discussion of polymers, we also develop a quantitative treatment of osmotic pressure. Our immediate concern is with the relationship between osmotic properties and molecular weights of polymers, but the same principles are readily extended to a variety of other problems, such as the effects of various solute ions on transport of water in biological systems.

In this chapter we also have an introduction to the principles of statistical thermodynamics, with application to heat capacities of solids and the third law of thermodynamics.

Structures of Crystals

Diffraction patterns are produced whenever light passes through or is reflected by a periodic or regularly repeating structure. One example is illustrated in Figure 21.1. Another example is a piece of glass, called a diffraction grating, on which many lines have been scratched at regular intervals.

Consider a beam of x rays of wavelength λ impinging on two parallel planes of atoms in a crystal, as pictured in Figure 21.2. We are concerned with the x rays that are scattered or reflected by the atoms in these two layers. Maximum intensity of the reflected x rays will be observed at angles θ where waves reflected from different planes are in phase with each other, which occurs when the differences in path lengths of the x rays are equal to an integral number of wavelengths, expressed as $n\lambda$. Because the difference in path lengths is DEF (Figure 21.2), we have

$$\text{distance } DEF = 2\,DE = n\lambda \tag{21.1}$$

From trigonometry we have

$$\sin\theta = DE/d$$

and then

$$DE = d\sin\theta \tag{21.2}$$

Substituting (21.2) in (21.1) gives the Bragg equation

$$2d\sin\theta = n\lambda \tag{21.3}$$

This equation permits evaluation of the distance d between planes of atoms in a crystal and can ultimately lead to knowledge of the structure of the crystals.

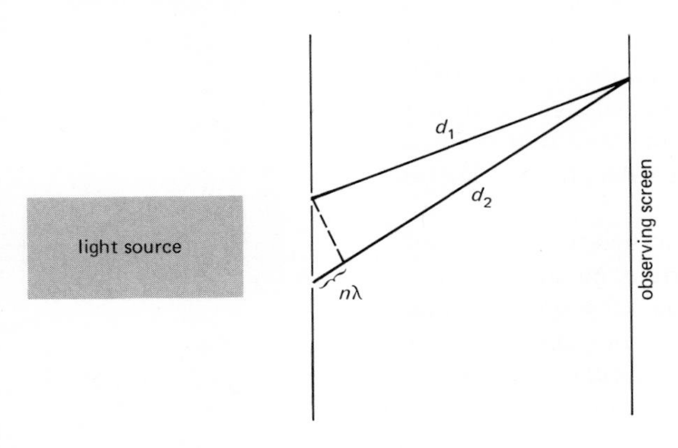

Figure 21.1. Illustration of a simple two-slit diffraction apparatus. The intensity of light on the observing screen is a maximum at those points where the difference in path lengths denoted by d_1 and d_2 is equal to $n\lambda$, in which n is any integer and λ is the wavelength of the light.

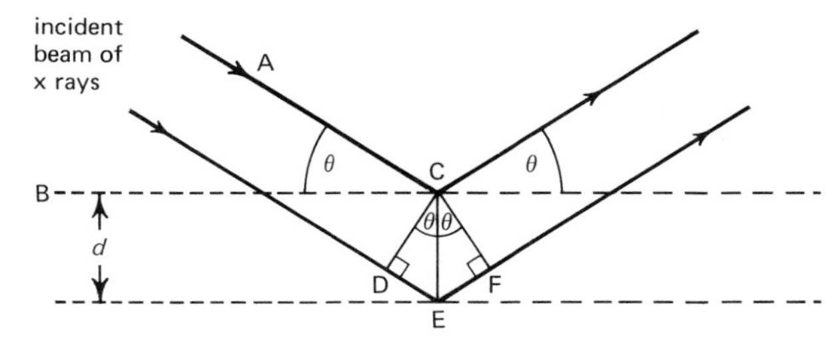

Figure 21.2. Schematic illustration of reflection of x rays by two parallel planes of atoms in a crystal. The equal angles of incidence and reflection of the x rays are denoted by θ. See problem 11 at the end of this chapter for proof that angles DCE and ECF are also equal to θ.

Example Problem 21.1. X rays made by bombarding a copper target with electrons have wavelength of 1.540 Å. These x rays are reflected with maximum intensities at angles of 14°11′, 29°21′, and 47°18′ from a crystal of KCl. Calculate the distance between planes of ions in KCl.

We write equation (21.3) as

$$d = n\lambda/2 \sin \theta$$

and insert $\lambda = 1.540$ Å, $n = 1$, and $\theta = 14°11'$ to obtain

$$d = \frac{(1)(1.540 \text{ Å})}{2 \sin (14°11')} = \frac{1.540 \text{ Å}}{(2)(0.245)} = 3.146 \text{ Å}$$

as the distance between planes of ions.

Similarly, we can use $n = 2$ with $\theta = 29°21'$ and $n = 3$ with $\theta = 47°18'$ to obtain the same value of d. ■

Most of the alkali metal halides and a number of other substances have the sodium chloride or rock salt crystal structure, which is illustrated in different ways in Figures 21.3 and 21.4. Another kind of crystal structure, that of cesium chloride, is illustrated in Figure 21.5.

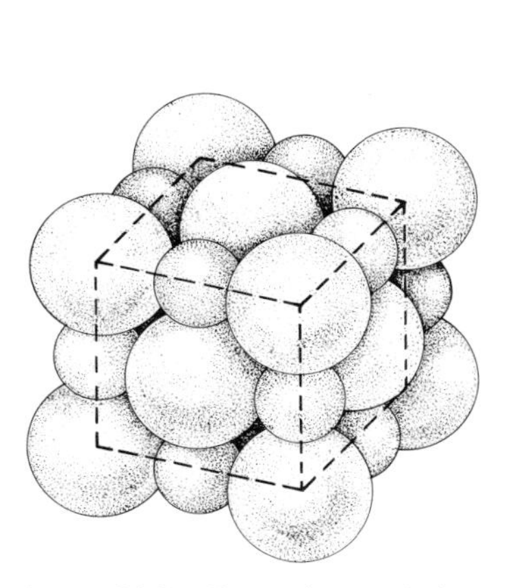

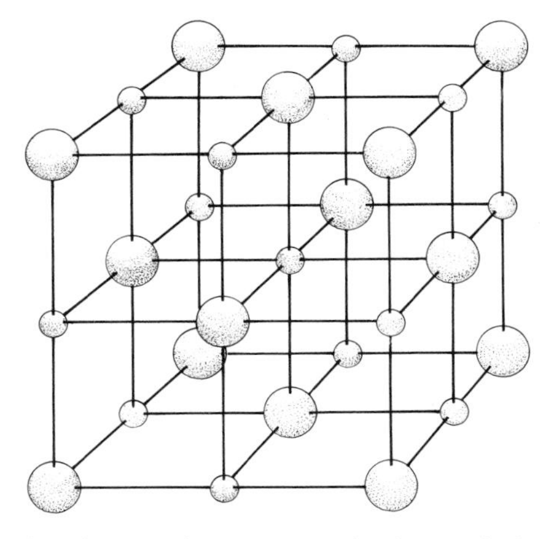

Figure 21.3. Illustrations of the NaCl or rock salt crystal structure. The large circles represent Cl^- ions and the small circles represent Na^+ ions. The illustration on the left is most realistic, while that on the right shows more clearly the arrangement of ions.

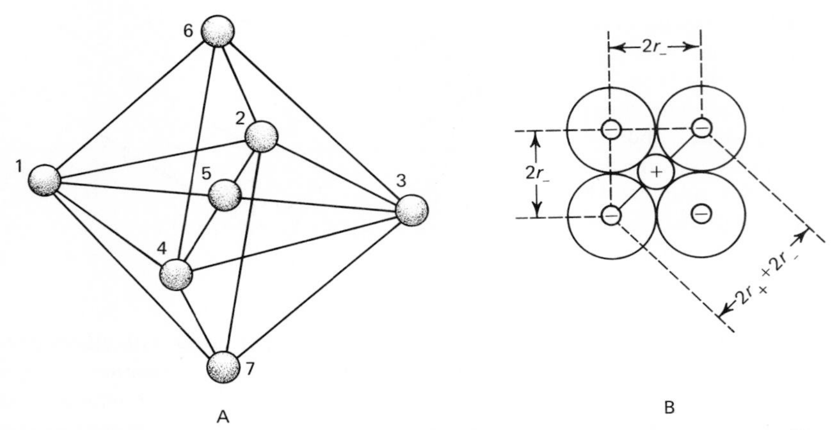

Figure 21.4. Illustration of octahedral coordination in the rock salt structure. Drawing (A) is a part of the structure shown in Figure 21.3 and shows that each ion is surrounded by six oppositely charged ions. This octahedral configuration is the same as in coordination compounds with d^2sp^3 bonding. Drawing (B) shows a top view of arrangement of anions in one plane around a cation. The four anions in (B) correspond to ions 1, 2, 3, and 4 in (A).

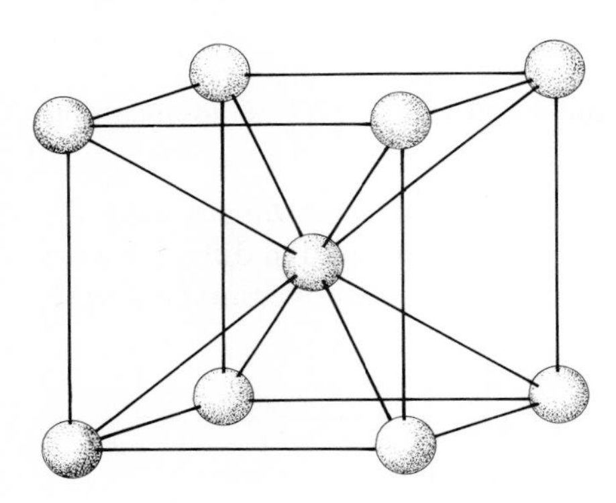

Figure 21.5. Schematic illustration of the CsCl (body centered cubic) structure with coordination number 8.

Ionic Radii and Radius Ratios

In order to approach understanding of the factors that lead to formation of any particular crystal structure, let us consider bringing anions from a great distance to rest around a cation. After the first anion comes to rest at its equilibrium position, each successive anion settles in the particular location where there is the most favorable balance of attraction to the cation and repulsion from other anions. Similar considerations apply to addition of cations to a group of one anion and several cations. A reasonably realistic model for such systems of ions is obtained when we picture each ion as a hard sphere of some specified radius bearing some specified charge. With this model it is possible to carry out useful calculations of properties of real crystals.

On the basis of the idea that it is nearly realistic to picture an ionic crystal as a collection of hard spheres, it has been possible for Pauling and others to

Table 21.1. Ionic Radii (Å)

Alkali metals	*r*	*Alkaline earths*	*r*	*Aluminum group*	*r*
Li^{+}	0.60	Be^{2+}	0.31	Al^{3+}	0.50
Na^{+}	0.95	Mg^{2+}	0.65	Ga^{3+}	0.62
K^{+}	1.33	Ca^{2+}	0.99	In^{3+}	0.81
Rb^{+}	1.48	Sr^{2+}	1.13	Tl^{3+}	0.95
Cs^{+}	1.69	Ba^{2+}	1.35	Tl^{+}	1.40

Oxygen group	*r*	*Halides*	*r*
O^{2-}	1.40	F^{-}	1.36
S^{2-}	1.84	Cl^{-}	1.81
Se^{2-}	1.98	Br^{-}	1.95
Te^{2-}	2.21	I^{-}	2.16

Transition metals

Ion	Sc^{3+}	Ti^{2+}	V^{2+}	Cr^{2+}	Mn^{2+}	Fe^{2+}	Co^{2+}	Ni^{2+}	Cu^{+}	Zn^{2+}
r	0.81	0.90	0.88	0.84	0.80	0.76	0.74	0.72	0.95	0.74
Ion		Ti^{4+}		Cr^{3+}		Fe^{3+}			Cu^{2+}	Cd^{2+}
r		0.68		0.69		0.64			0.69	0.94
Ion									Ag^{+}	Hg^{2+}
r									1.26	1.10

deduce radii of a number of ions and to derive relationships between radius ratios and the most likely crystal structures for various compounds. Because the electron cloud around the "exterior" of an ion is not as definite as the surface of an imaginary hard sphere, ionic radii are not entirely definite quantities and deductions concerning radius ratios can only be regarded as useful guides rather than as completely reliable proofs. A number of ionic radii are listed in Table 21.1. Although inexact both numerically and in terms of the indistinct boundaries of real ions, we shall see that these radii are useful.

Now we use Figure 21.4B as the basis for an illustrative calculation concerning radius ratios and crystal structures. The ions pictured as circles in two dimensions (or as spheres in three dimensions) in Figure 21.4B are just the right size to permit maximum ion-ion contact. If the cation were smaller or the anions larger, the hole between anions would be too large for the cation. We therefore expect that when the radius ratio r_+/r_- is less than the value associated with Figure 21.4B, the sodium chloride structure will be less stable than some other structure.

The value of the limiting radius ratio r_+/r_- can be derived by applying the Pythagorean theorem to Figure 21.4B as follows:

$$(2r_+ + 2r_-)^2 = (2r_-)^2 + (2r_-)^2$$

We expand the squared term on the left and combine terms on the right to obtain

$$4(r_+)^2 + 8(r_+r_-) + 4(r_-)^2 = 8(r_-)^2$$

and thence

$$4(r_+)^2 + 8(r_+r_-) - 4(r_-)^2 = 0$$

Now we divide by $4(r_-)^2$ to obtain

$$\left(\frac{r_+}{r_-}\right)^2 + 2\left(\frac{r_+}{r_-}\right) - 1 = 0.$$

Application of the quadratic equation leads to

$$\left(\frac{r_+}{r_-}\right) = 0.414$$

for the desired limiting radius ratio. Whenever (r_+/r_-) is less than 0.41, we should expect the NaCl crystal structure (octahedral configuration and coordination number 6) to be unstable.

The CsCl crystal structure, also called body centered cubic, is illustrated in Figure 21.5. The arrangement of anions about cations in this structure is cubic and the coordination number is 8. Geometrical considerations similar to those applied to the rock salt structure lead to

$$\left(\frac{r_+}{r_-}\right) = 0.732$$

for the smallest cation/anion radius ratio that does not result in overcrowding of the anions in the body centered cubic structure.

Just as eight anions cannot fit around a cation when r_+/r_- is less than 0.73, six anions cannot fit around a cation when r_+/r_- is less than 0.41. Crystals of the MX type with r_+/r_- less than 0.41 are therefore expected to have a structure with fewer than six anions about each cation. A structure of this type is the zinc blende structure pictured in Figure 21.6, in which the coordination number is 4.

Our considerations of ionic crystals and radius ratios are based on a model in which ions are pictured as hard spheres with no specific directional forces between ions. Real crystals consist of real ions that are only approximately like hard spheres. Furthermore, some "ionic" crystals are really partly covalent so that

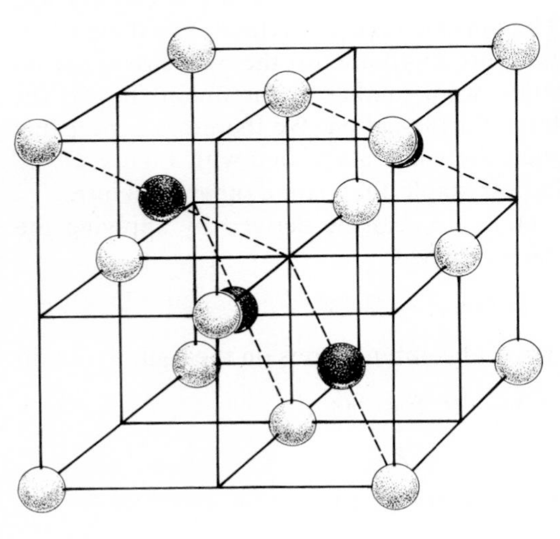

Figure 21.6. Schematic illustration of the zinc blende (ZnS) structure in which each ion is surrounded tetrahedrally by four ions of opposite charge.

Table 21.2. Radius Ratios and Crystal Structures

CsCl structure 8 coordinated		*NaCl structure 6 coordinated*		*ZnS structure 4 coordinated*	
Compound	r_+/r_-	*Compound*	r_+/r_-	*Compound*	r_+/r_-
CsCl	0.93	KCl	0.73	ZnS	0.40
CsBr	0.87	CaO	0.71	MgTe*	0.29
CsI	0.77	NaCl	0.52	BeO*	0.21
		MgO	0.46	BeS	0.17
		NaI	0.44		
		LiF	0.44		

*These crystals actually have the wurtzite structure, which also has four anions about each cation.

there are specific directional forces as well as nondirectional coulombic forces to be considered. Because of the inadequacies of our model, we should expect to find exceptions to our predictions based on radius ratios. Nevertheless, these predictions do work reasonably well and provide a useful guide to crystal structures.

We may summarize the radius ratio (r_+/r_-) generalizations already given as follows:

$$\begin{array}{ccccc} \text{body centered cubic} & \overset{0.73}{\longleftrightarrow} & \text{simple cubic} & \overset{0.41}{\longleftrightarrow} & \text{zinc blende} \\ \text{8 coordinated} & & \text{6 coordinated} & & \text{4 coordinated} \end{array}$$

In Table 21.2 we list a number of compounds with structures that are in accord with our radius ratio rules. There are, however, other compounds with structures that are not in accord with the radius ratio rules. For example, TlBr has the CsCl structure although its $r_+/r_- = 0.72$ suggests that it "ought" to have the NaCl structure. Both LiCl and RbCl have the NaCl structure, although radius ratios suggest that LiCl ($r_+/r_- = 0.33$) ought to have the ZnS structure (or another 4 coordinated structure) while RbCl ($r_+/r_- = 0.82$) ought to have the CsCl structure. Our last example of this sort is AgI with $r_+/r_- = 0.58$, which actually has the ZnS structure rather than the predicted NaCl structure.

Types of Solids

Just as our attempts to classify bonding in terms of the extremes of pure ionic or pure covalent are inexact but useful, similar classifications of solids as **ionic, covalent, molecular,** or **metallic** according to the general criteria summarized in Table 21.3 are inexact but useful.

The lattice positions in covalent crystals are occupied by atoms that share electrons with their neighbors. Because covalent bonds extend in fixed directions (fixed bond angles) from each atom, a covalent crystal can be considered to be a giant molecule. Examples of covalent crystals are diamond, graphite, silica, and boron nitride.

The structure of diamond is illustrated in Figure 21.7. Each carbon atom is bonded strongly to four neighboring atoms, which are located at the corners of

Table 21.3. Some Classifications of Solids

	Ionic	*Covalent*	*Molecular*	*Metallic*
Units that occupy crystal lattice sites	Positive and negative ions	Atoms	Molecules	Positive ions in a sea of electrons
Binding force	Electrostatic attraction	Shared electrons	Van der Waals forces; Also dipole-dipole forces in some molecular crystals	Attraction between positive ions and surrounding negative electrons; see text for other explanation
Physical properties	High melting points and brittle	High melting points, hard	Low melting and boiling points, soft	Moderate to high melting points, hard or soft
Electrical properties	Poor conductors in solid state, good conductors when melted or dissolved in water	Poor conductors in solid and liquid states	Poor conductors in solid and liquid states	Good conductors in solid and liquid states
Examples	NaCl MgO CaF_2 K_2SO_4	Diamond Quartz (SiO_2)	Argon Benzene CO_2	Fe Cu Na Hg

a regular tetrahedron. Because each atom at the corner of one tetrahedron is at the center of another tetrahedron, the lattice pictured in Figure 21.7 continues indefinitely in three dimensions. Diamonds have a high melting point (~3500°C) and are very hard because the positions of the carbon atoms are rigidly defined by strong covalent bonds. Sharing of four electrons by each carbon atom fills all the $n = 2$ orbitals. Because the bonding electron pairs are localized between specific pairs of atoms, diamonds are electrical insulators.

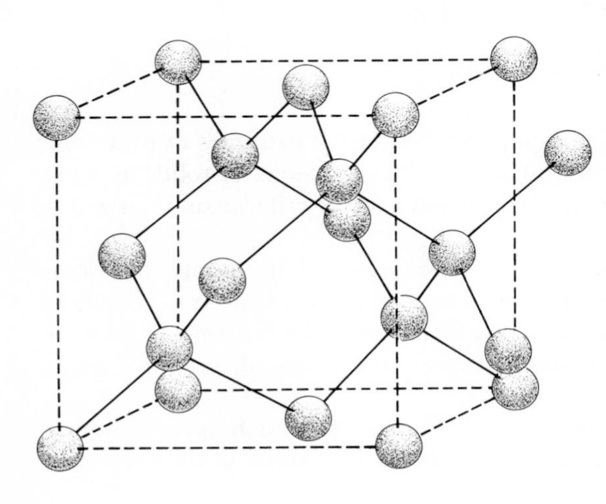

Figure 21.7. Illustration of the cubic unit cell of diamond. Each carbon atom in this structure is surrounded tetrahedrally by four other carbon atoms.

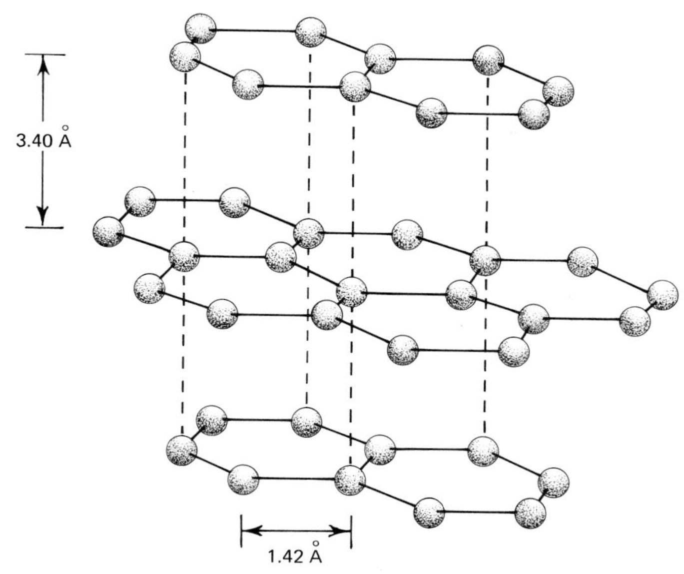

Figure 21.8. Illustration of the layer structure of graphite. Electrical conductivity in graphite is principally in the two dimensions between layers of carbon atoms, not across layers.

Each carbon atom in graphite is bonded to three other atoms at a distance of 1.42 Å. These bonded atoms are arranged in sheets that are 3.40 Å apart, as shown in Figure 21.8. The distance between sheets is too great for covalent bonding, so the sheets are held to one another only by weak van der Waals forces. Because the sheets can slide over one another easily, graphite is soft and slippery. On the other hand, the high melting point is due to the strong bonds between atoms in the same sheet. Bonding in graphite may be regarded as a resonance hybrid of the bonding structures shown in Figure 21.9. Because some of the bonding electrons are not localized between any two atoms, graphite conducts electricity considerably better than does diamond.

Boron nitride, BN, is obtained as a slippery white solid by reaction of hot boron with ammonia or nitrogen. The structure of this compound is similar to the layer structure of graphite. Graphite-like BN is converted at high temperature and

Figure 21.9. Illustration of one of the three valence bond resonance structure representations of bonding in graphite.

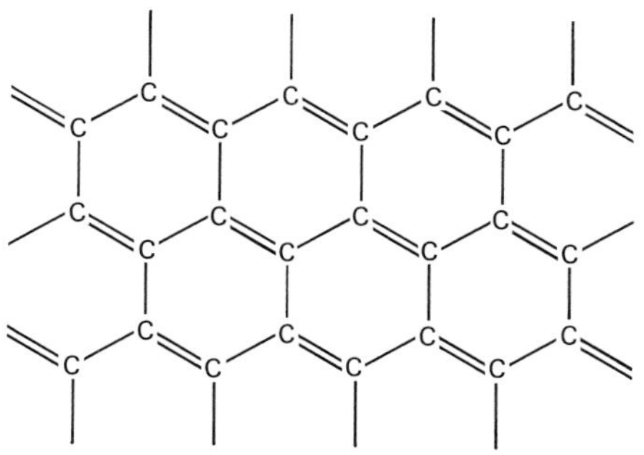

pressure to a cubic form with a diamond-like structure. Cubic BN is so hard that it can scratch diamonds.

Silicon dioxide, commonly called silica, occurs in three important crystalline forms. Quartz is the stable form at low temperatures, with other forms being more stable at higher temperatures as indicated by the following:

$$\xrightarrow{\text{quartz}} 870\,^\circ\text{C} \xleftrightarrow{\text{tridymite}} 1470\,^\circ\text{C} \xleftrightarrow{\text{cristobalite}} 1710\,^\circ\text{C}.$$

These three forms of silica are all built up of SiO_4 tetrahedra linked together so that every oxygen atom is a part of two tetrahedra, thus giving the composition SiO_2. The arrangements of tetrahedra are different in the three different crystal forms.

When molten SiO_2 is cooled, a glassy substance called vitreous silica is sometimes obtained rather than one of the crystalline silicas. X-ray studies of vitreous silica have shown that the oxygen atoms are arranged tetrahedrally about the silicon atoms, but the tetrahedra are not arranged as regularly as they are in crystalline silica. This short range order and long range disorder of vitreous silica is characteristic of other glassy substances and is also characteristic of liquids. Glasses may be considered to be a special kind of supercooled liquid that flows slowly and has considerable mechanical strength.

Common "glass" is made by fusing SiO_2, usually obtained as sand, with various oxides or carbonates of which CaO and Na_2CO_3 are most often used. Some glasses also contain B_2O_3, Al_2O_3, or still other oxides.

Silicon has never been found in nature as the free element, but the dioxide and numerous silicates constitute about 87% of the earth's crust.

A few mineral silicates have structures in which there are discrete orthosilicate ions, SiO_4^{4-}. Examples are willemite (Zn_2SiO_4) and forsterite (Mg_2SiO_4) in which SiO_4^{4-} tetrahedra are arranged so that each M^{2+} ion is surrounded octahedrally by six oxygens.

More numerous than the orthosilicates are minerals with larger silicate units. Substances containing $Si_2O_7^{6-}$, $Si_3O_9^{6-}$, and $Si_6O_{18}^{12-}$ silicate anions are known. Both the $Si_3O_9^{6-}$ and $Si_6O_{18}^{12-}$ ions have cyclic structures containing —Si—O—Si— bonds. Emeralds are beryl crystals ($Be_3Al_2Si_6O_{18}$) that have a green color due to presence of Cr^{3+} ions in small concentrations.

There are also large chainlike silicate anions, all containing —O—Si—O—Si—O— bonding systems. Pyroxenes are single stranded chains of composition $(SiO_3^{2-})_n$ and amphiboles are double stranded chains of composition $(Si_4O_{11}^{6-})_n$. Various asbestos minerals are amphiboles.

The SiO_4 tetrahedra in silicates sometimes form sheet-like networks, with the simplest formula of the resulting anions being $(Si_2O_5^{2-})_n$. The micas are a common class of minerals having such sheet-like silicate anions. These substances are cleaved easily into thin sheets.

The feldspars and zeolites are important aluminosilicates that have three dimensional framework structures. Feldspars such as orthoclase ($KAlSi_3O_8$) and anorthite ($CaAl_2Si_2O_8$) are the major constitutents of igneous rocks. Various synthetic and naturally occurring zeolites are used as ion exchangers and "molecular sieves." For example, some molecular sieve zeolites absorb straight chain hydrocarbons but not branched chain or cyclic hydrocarbons.

We will now briefly consider molecular crystals, which have been described in Table 21.3 as being held together by van der Waals forces and sometimes also by dipole-dipole forces. Nonpolar molecules such as H_2, N_2, CH_4, and CCl_4 form molecular crystals in which relatively weak van der Waals forces loosely bind the molecules together. Dipole-dipole attractive forces (see problem 12) are also important in the crystals of such polar molecules as HCl, BrCl, SO_2, and many others.

Metals

Characteristics common to all metals are their high electrical and thermal conductivities as compared to nonmetals. As long ago as 1916 H. A. Lorentz was able to summarize a theory that qualitatively accounted for these characteristic properties and also for the large enthalpies of vaporization of most metals. This theory was based on a model of a metal as an arrangement of metal ions at lattice points with nearly free electrons in the space between ions. Motion of these electrons was regarded as responsible for conduction of both electricity and heat. This free electron theory, especially as developed since 1927 by application of quantum theory, successfully accounts for many of the properties of most metals.

The free electron theory, however, does not explain adequately why some elements are good conductors of electricity, with resistances that *increase* approximately linearly with increasing temperature, while other elements have much larger resistances that *decrease* approximately exponentially with increasing temperature. Substances of this latter sort are called **semiconductors.** Still other elements have such high resistances in the solid state that they are classed as insulators. Resistances of the best insulators and best conductors differ by a factor of about 10^{30}, which is one of the widest ranges of any observed property of matter.

Another theory, largely worked out by Pauling, pictures the attractive forces between atoms in a metal as being due to covalent bonds. The state of the bonding electrons must be represented as a resonance hybrid of the various resonance forms that can be drawn.

The most natural and probably the best approach to a theory of metals is similar to the molecular orbital theory of bonding in covalent molecules. This approach leads to **band theory** in which electrons are thought of as belonging to the whole crystal rather than to individual atoms or to pairs of atoms. According to this theory, energy levels are grouped into bands, and each atomic orbital contributes one energy level to a band.

The band theory picture of the electrons in a crystal of sodium is illustrated in Figure 21.10. There are just enough levels in the lower bands to accomodate the lower energy electrons, which do not ordinarily contribute to the electrical conductivity. The higher energy band labelled 3*s* is only half filled. Some of the 3*s* electrons are in energy levels in the dotted portion of this band and are easily able to move through the crystal when influenced by an electric field. As the temperature of the metal is increased, vibrations of the metal ions about the lattice points increase in amplitude so that the increasingly irregular array of ions interferes more and more with the flow of electrons. The electrical resistance therefore increases with increasing temperature.

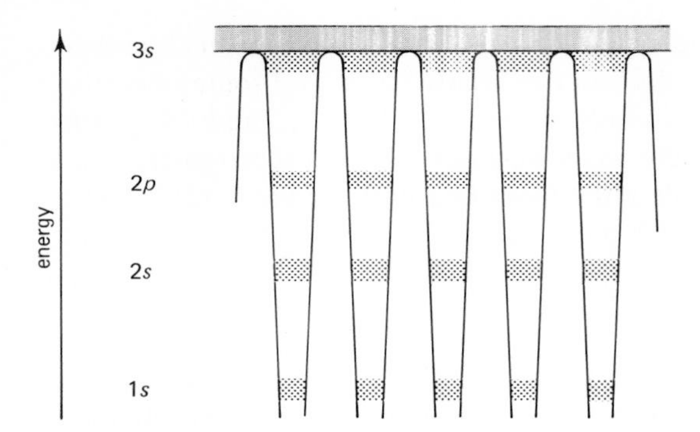

Figure 21.10. Illustration of the electron energy bands in a crystal of sodium. The 1*s*, 2*s*, and 2*p* bands are full, as is the dotted portion of the 3*s* band. The gray portion of the 3*s* band is easily reached by electrons from the bottom half of the band, and it is electrons in this "conduction band" that account for the high conductivity (low resistance) of sodium.

Another way of picturing the energy bands in metals is illustrated in Figure 21.11, where similar diagrams are also given for insulators and semiconductors. We see in this picture that metals are characterized by a low energy conduction band so that many electrons are "ready" to move under the influence of an electric field. Metals therefore have very low electrical resistances.

Insulators are characterized by large energy gaps between the valence and conduction bands. In these substances there are practically no electrons to carry a current, and the electrical resistance is very high.

Our illustration (Figure 21.11) also shows that the energy gap between the highest filled valence band and the conduction band in an intrinsic semiconductor is small enough that an appreciable number of electrons are raised into the conduction band by thermal excitation. Excitation of electrons to the conduction band leaves "positive holes" in the valence band. Both positive holes in the valence band and negative electrons in the conduction band can move under the influence of an electric field and contribute to the electrical conductivity, which is intermediate between conductivities of metals and insulators.

Because insulators are substances with "large" energy gaps between valence and conduction bands while semiconductors have "small" gaps between these bands, there is no clear distinction between insulators and semiconductors.

Before returning to considerations of metals, we point out that there are very useful "impurity semiconductors" that are made by introducing minute quantities of selected impurities into otherwise pure material. For example, a trace of arsenic (five valence electrons) or gallium (three valence electrons) in germanium (four valence electrons) yields useful impurity semiconductor material. The charge carriers associated with arsenic in germanium are electrons, while those associated with gallium in germanium are positive holes.

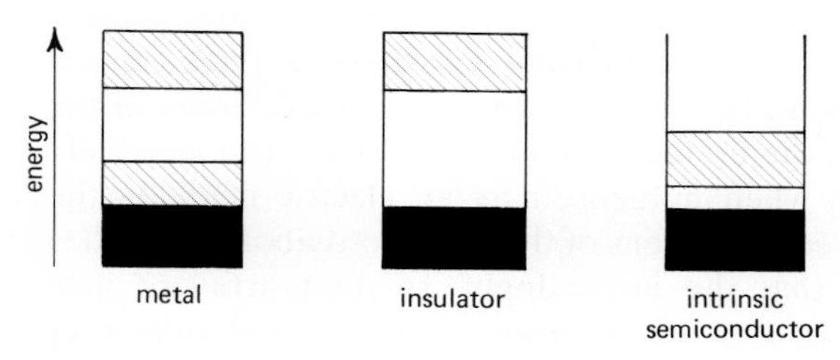

Figure 21.11. In this schematic illustration of the band model of solids the black areas represent valence or bonding bands, white areas represent forbidden energy gaps, and gray areas represent conduction bands.

Now, as the concluding part of our discussion of metals, we consider processes by which we obtain metals from their ores.

In a general way, metallurgical processes begin with a relatively large body of ore that contains ultimately useful metal mixed or chemically combined with other material from which the metal must be separated. A general equation for treatment of the ore is

ore (metal that is mixed or combined with "waste") → metal + "waste" (21.4)

Because separated metal and "waste" have lower entropy than the mixed-up metal and "waste" in ore, we see that ΔS is negative for the part of the metallurgical process represented by the equation above. But we also know that the second law of thermodynamics absolutely requires that the *total* ΔS for any real process must be positive. There must, therefore, be some other accompanying process or reaction with sufficiently large positive ΔS that the *total* ΔS becomes positive. This other process is typically the burning of coal or some other related fuel-consuming reaction.

When we work with low grade ore (relatively lots of "waste" compared to metal), the entropy loss associated with (21.4) is larger than when we work with high grade ore. The required entropy increase in the accompanying reactions or processes must therefore be greater for processing low grade ore than for similar processing of high grade ore. It therefore inevitably requires more fuel consumption to recover metal from low grade ore than from high grade ore. This requirement has an important bearing on future needs for energy and is also pertinent to considerations of various recycling processes that are designed to permit reuse of various materials.

As the world moves toward exhaustion of readily available high grade ores of most metals, we are increasingly faced with two related problems: (1) obtaining metals from low grade ores and (2) recycling various metal wastes. Both approaches will be necessary, and a fundamental problem associated with each is the large amount of energy that must be expended to obtain useful material. It is partly for this reason that efficiency is an ecological necessity as well as economically desirable. Many of the principles of thermodynamics and electrochemistry already discussed in this book are fundamental to achieving the kinds of efficient processes that are needed. And, as we shall see in the next chapter, the principles of chemical kinetics are also important in this connection.

Now, as an important specific example, let us consider some of the processes involved in recovery of iron from its ores. The crust of the earth contains huge quantities of iron, but much of this iron is tied up in complex silicates that are not presently practical sources of iron. However, weathering of these silicates has in some places led to formation of the useful iron ores listed in Table 21.4.

The basic chemical process in production of metallic iron from iron ore is reduction of the iron in a compound (+2 or +3 oxidation state) to the zero oxidation state of the elementary metal. This reduction is commonly carried out at high temperature, using coal and carbon monoxide as reducing agents. The carbon monoxide is produced by burning coal, a reaction which also provides

Table 21.4. Iron Ores

Name	*Formula*
Hematite	Fe_2O_3
Magnetite	Fe_3O_4
Siderite	$FeCO_3$
Limonite	$Fe_2O_3 \cdot xH_2O$

the heat required for high temperature reaction. Equations for production of carbon monoxide and reduction of hematite by carbon monoxide are

$$2\ C(coal) + O_2 \rightarrow 2\ CO + heat$$

and

$$3\ CO + Fe_2O_3 \rightarrow 2\ Fe + 3\ CO_2$$

A complicating factor in the metallurgical reactions derives from the presence of various silicates in the ore. Addition of $CaCO_3$ in the form of limestone to such ores yields complex silicates called slags, which can be separated from the iron produced by reduction.

The molten iron that is produced as described above usually contains a few per cent of carbon and a few other impurities. Cooling this material slowly commonly yields "gray cast iron," which consists of nearly pure iron interspersed with tiny flakes of graphite. More rapid cooling of the impure iron can yield "white cast iron," in which the carbon is in the form of Fe_3C.

Purification of cast iron leads to iron (called steel) that ordinarily contains 0.1–1.5% carbon. Many special purpose steels are then made by adding various metals to the molten steel issuing from the cast iron purification procedure. One stainless steel contains 8% nickel and 18% chromium. Other steel alloys may contain vanadium, manganese, cobalt, molybdenum, tungsten, or other metallic elements.

Lattice Energies and Enthalpies

In several earlier chapters we have made use of lattice energies and lattice enthalpies in understanding various ΔH^0 values and thence in gaining understanding of chemical properties of many substances. Now we consider this subject in a way that will tell us how lattice energies and enthalpies depend on charges and sizes of the ions.

The electrostatic potential energy of a pair of ions of charges $z_i e$ and $z_j e$ (z's may be ± 1, ± 2, etc., and e is the charge of an electron) at distance r from each other is

$$\frac{z_i z_j e^2}{r}$$

For oppositely charged ions $z_i z_j$ is negative, corresponding to attraction. For ions that have charges of the same sign, $z_i z_j$ is positive, corresponding to repulsion.

At very small distances the electron clouds of ions overlap, which results in

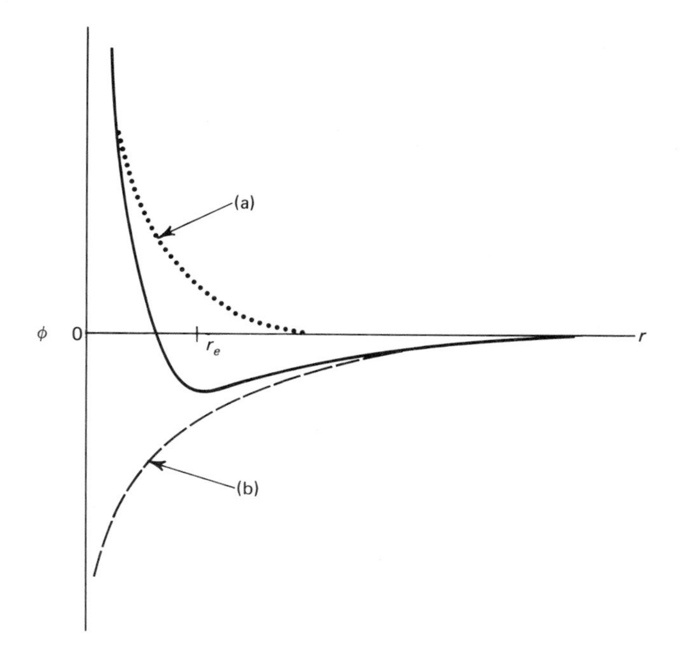

Figure 21.12. Potential energy diagram based on equation (21.5). The dotted line (a) represents the repulsive energy (be^2/r^n) and the dashed line (b) represents the attractive energy between two oppositely charged ions ($z_i z_j e^2/r$). The solid line represents the net potential energy, which is a sum of the repulsive and attractive terms as in equation (21.5). The minimum in the potential energy curve ($d\phi/dr = 0$) corresponds to the equilibrium distance r_e.

a repulsion energy that Born represented by

$$\frac{be^2}{r^n}$$

in which n is ordinarily taken to be about 9. Because of the relatively large value of n, which can be evaluated by way of calculations based on compressibility data, this repulsive energy term is significant only for small distances.

Combination of the electrostatic and overlap terms above gives

$$\phi_{ij} = \frac{z_i z_j e^2}{r} + \frac{be^2}{r^n} \tag{21.5}$$

for the mutual potential energy of two ions. Figure 21.12 illustrates how the two terms combine for ions of opposite charge.

The total potential energy of an ionic crystal of MX type ($z_i = -z_j = z$) can be obtained as the sum of the ϕ_{ij} terms for every pair of ions in the crystal. Because all of the interionic distances are geometrically related to the distance r between neighboring oppositely charged ions, the total potential energy of a mole of MX crystal can be written as

$$\phi = -\frac{NMz^2e^2}{r} + \frac{Be^2}{r^n} \tag{21.6}$$

The dimensionless **Madelung constant** M depends on the arrangement of ions (simple cubic, body centered cubic, etc.) and can be evaluated by summing series as explained in references cited at the end of this chapter. The constant B is related to the Born repulsive constant b, and N is Avogadro's number.

A crystal is most stable when the interionic distance r corresponds to the minimum in the potential energy curve, as represented by r_e in Figure 21.12. Making use of the fact that $d\phi/dr = 0$ at the equilibrium distance r_e (see problem 17 at the end of this chapter), it is possible to show that

$$B = \frac{NMz^2(r_e)^{n-1}}{n} \tag{21.7}$$

Substitution of this value for B in (21.6) and rearrangement gives

$$\phi = -\frac{NMz^2e^2}{r_e}\left(1 - \frac{1}{n}\right) \tag{21.8}$$

The ϕ in (21.8) can be identified with ΔE (very nearly equal to ΔH^0) for the reaction represented by

$$M^+(g) + X^-(g) \rightarrow MX(c)$$

and can be called the lattice energy (or enthalpy).

Example Problem 21.2. Compare ϕ from equation (21.8) with ΔH^0 for the reaction

$$Na^+(g) + Cl^-(g) \rightarrow NaCl(c)$$

For NaCl we have $M = 1.748$, $r_e = 2.814$ Å, and $n = 9$. Appropriate substitution in (21.8) gives

$$\phi = -\frac{(6.02 \times 10^{23} \text{ mole}^{-1})(1.748)(1)^2(4.80 \times 10^{-10} \text{ esu})^2}{(2.814 \times 10^{-8} \text{ cm})}\left(1 - \frac{1}{9}\right)$$

The dimensions of the esu are $g^{\frac{1}{2}}$ $cm^{\frac{3}{2}}$ sec^{-1} and those of the erg are g cm^2 sec^{-2} so that we have

$$\begin{aligned}\phi &= -76.6 \times 10^{11} \text{ ergs mole}^{-1}\\ &= -76.6 \times 10^{4} \text{ joules mole}^{-1}\\ &= -183 \text{ kcal mole}^{-1}\end{aligned}$$

(Readers should make sure that they understand the units used above and can convert from ergs to joules to calories.)

From ΔH_f^0 values we calculate $\Delta H^0 = -185$ kcal $mole^{-1}$.

We conclude that equation (21.8) gives a good approximation to the lattice energy (or enthalpy) of NaCl. ■

We see from equation (21.8) that a compound that consists of M^{2+} and X^{2-} ions will have four times the lattice energy of a compound of M^+:X^- type with the same r_e and M values.

Because various authors define M differently in terms of z and r_e values, it is necessary to be careful in using M values from different publications.

Statistical Thermodynamics, Heat Capacities, and the Third Law of Thermodynamics

We speak of atomic theories that do not involve recognition of the discrete nature of energy and that do not contain Planck's constant as "classical" theories. In classical theory a solid is pictured as consisting of atoms that vibrate about certain fixed points called lattice points. The vibrational energy of a classical oscillator is equal to kT, where k is the Boltzmann constant, which has been identified as the gas constant for one molecule ($k = R/N$). Because there are N atoms in a mole of monatomic solid and each atom vibrates in three dimensions, we have $E = 3NkT = 3RT$ for the vibrational energy of a mole of classical monatomic solid. Remembering that $C_v = dE/dT$ (equations 11.16 and 11.18), we obtain $C_v = 3R = 3 \times 1.987$ cal deg^{-1} mole^{-1}.

Calorimetric measurements on solids ordinarily lead to heat capacities at constant pressure (C_p). The difference between C_p and C_v is negligible at low temperatures and is still relatively small at high temperatures.

The prediction of classical theory that heat capacities of monatomic solids should be about 6 cal deg^{-1} mole^{-1} at all temperatures is in rough agreement with experimental results at high temperatures (room temperature is high enough for heavy elements, but not for light ones). But heat capacities of all solids are *much less* than $3R$ at low temperatures. Therefore something is clearly wrong with the classical theory.

Einstein was the first to recognize (in 1905) the relationship between quantum theory and heat capacities. We now present some statistical thermodynamics along with Einstein's applications of these ideas to the problem of heat capacities.

The starting point for the statistical thermodynamics of present interest is the Boltzmann equation, which describes the most probable distribution of atoms among the various available energy states. Rigorous derivation of the Boltzmann equation is beyond the scope of this book, but a simple derivation suggested by Giauque is easily given.

Consider the equilibrium state for some atomic species that can be in either of two energy states, indicated by subscripts i and j. The "reaction equation" for the equilibrium and the corresponding "equilibrium constant" expression are

$$j \rightleftharpoons i \qquad \text{and} \qquad K = n_i/n_j$$

in which n_i and n_j represent the numbers of atoms with molar energies E_i and E_j. Corresponding energies per atom are represented by the Greek letter epsilon (ε) and are obtained by dividing E_i or E_j by Avogadro's number.

Combining the equilibrium constant expression with

$$\Delta G^0 = -RT \ln K = \Delta H^0 - T\,\Delta S^0$$

gives

$$-RT \ln (n_i/n_j) = \Delta H^0 - T\,\Delta S^0$$

Because equal volumes are available for atoms in each state, $\Delta H^0 = \Delta E^0 = E_i - E_j$ and $\Delta S^0 = 0$ (see equation 12.8). The equation above then becomes

$$-RT \ln (n_i/n_j) = E_i - E_j$$

Dividing by Avogadro's number and remembering $R/N = k$ and $E/N = \varepsilon$ gives

$$-kT \ln (n_i/n_j) = \varepsilon_i - \varepsilon_j$$

which is rearranged to

$$\frac{n_i}{n_j} = e^{-(\varepsilon_i - \varepsilon_j)/kT}$$

As previously explained in connection with the first law of thermodynamics, we do not know energies on an absolute basis but are free to choose any convenient "sea level" for our reference state to which we assign a value of zero energy. In this case it is most convenient to let ε_j represent the energy of the lowest energy level or state and take this state or level to be the reference state that is assigned zero energy. Thus we replace n_j by n_0 and ε_j by 0 in the equation above and obtain

$$n_i = n_0 e^{-\varepsilon_i/kT} \tag{21.9}$$

for the Boltzmann equation in its usual form.

The total energy of a system is given by the sum of a series of terms, each of which is the number of atoms with a particular energy times that energy. The equation is

$$E = n_0\varepsilon_0 + n_1\varepsilon_1 + n_2\varepsilon_2 + \cdots$$

in which $n_0\varepsilon_0 = 0$ because $\varepsilon_0 = 0$. Substituting the appropriate Boltzmann expressions (21.9) for $n_1, n_2 \ldots$ gives

$$\begin{aligned} E &= n_0\varepsilon_1 e^{-\varepsilon_1/kT} + n_0\varepsilon_2 e^{-\varepsilon_2/kT} + \cdots \\ &= n_0 \Sigma \varepsilon_i e^{-\varepsilon_i/kT} \end{aligned} \tag{21.10}$$

In order to make use of (21.10) we must evaluate n_0, which we now begin by expressing the total number of atoms N as the sum of the number of atoms in each energy state:

$$N = n_0 + n_1 + n_2 + \cdots = \Sigma n_i \tag{21.11}$$

Substituting the appropriate Boltzmann expressions (21.9) now gives

$$N = n_0 + n_0 e^{-\varepsilon_1/kT} + n_0 e^{-\varepsilon_2/kT} + \cdots$$

and then factoring leads to

$$N = n_0(1 + e^{-\varepsilon_1/kT} + e^{-\varepsilon_2/kT} + \cdots)$$

This equation is easily solved for the desired

$$n_0 = N/(1 + e^{-\varepsilon_1/kT} + e^{-\varepsilon_2/kT} + \cdots) \tag{21.12}$$

Because the term in the denominator of (21.12) occurs frequently in statistical thermodynamics and is awkward to write out, it is usually given a symbol and name of its own. We use the name **partition function** and represent it by Z. Remembering that $\varepsilon_0 = 0$ and that $e^{-0} = 1$, we see that Z can be written as

$$Z = \Sigma e^{-\varepsilon_i/kT} \tag{21.13}$$

Thus we now have

$$n_0 = N/Z = N/\Sigma e^{-\varepsilon_i/kT} \tag{21.14}$$

We combine our equations for E (21.10) and n_0 (21.14) to obtain

$$E = N\Sigma\varepsilon_i e^{-\varepsilon_i/kT}/\Sigma e^{-\varepsilon_i/kT} \tag{21.15}$$

Inspection of (21.15) while remembering the formula for differentiation of an exponential (see Table I.1, Appendix I) suggests that we evaluate dZ/dT. Carrying out this differentiation (readers should do this for practice) gives

$$kT^2\frac{dZ}{dT} = \Sigma\varepsilon_i e^{-\varepsilon_i/kT} \tag{21.16}$$

Now we can simplify equation (21.15) by substituting Z for the denominator and $kT^2(dZ/dT)$ for the summation in the numerator to obtain

$$E = \frac{NkT^2}{Z}\left(\frac{dZ}{dT}\right)$$

Finally, remembering that du/u = d ln u leads to

$$E = NkT^2\left(\frac{d\ln Z}{dT}\right) = RT^2\left(\frac{d\ln Z}{dT}\right) \tag{21.17}$$

As we shall see, Z, $\ln Z$, and $(d\ln Z/dT)$ can be evaluated. Therefore equation (21.17) can be used to calculate the energy of a system of interest. But we are most interested in the heat capacity C_v, which is obtained as $C_v = dE/dT$:

$$C_v = R\left[T^2\left(\frac{d^2\ln Z}{dT^2}\right) + 2T\left(\frac{d\ln Z}{dT}\right)\right] \tag{21.18}$$

Now we are ready to follow Einstein in evaluating the partition function Z_{vib} for vibration of the atoms in a monatomic solid. Einstein pictured the atoms as vibrating with some characteristic frequency ν. According to the Planck equation, which correctly accounted for the radiation emitted by hot solids, the only allowed vibrational energies are given by

$$\varepsilon = \upsilon h\nu \tag{21.19}$$

where υ is an integer called the vibrational quantum number. Thus we write the partition function for vibration as

$$Z_{vib} = 1 + e^{-h\nu/kT} + e^{-2h\nu/kT} + \cdots \tag{21.20}$$

Multiplying both sides of (21.20) by $e^{-h\nu/kT}$ gives

$$Z_{vib}e^{-h\nu/kT} = e^{-h\nu/kT} + e^{-2h\nu/kT} + \cdots \tag{21.21}$$

We subtract (21.21) from (21.20) to obtain

$$Z_{vib} - Z_{vib}e^{-h\nu/kT} = 1$$

and rearrange to

$$Z_{vib} = \frac{1}{1 - e^{-h\nu/kT}} \tag{21.22}$$

To use this concise expression for Z_{vib}, we take the natural logarithm to obtain

$$\ln Z_{vib} = -\ln(1 - e^{-h\nu/kT}) \tag{21.23}$$

Differentiation of (21.13) gives

$$\frac{d\ln Z}{dT} = \left(\frac{h\nu}{kT^2}\right)\left(\frac{e^{-h\nu/kT}}{1 - e^{-h\nu/kT}}\right) = \left(\frac{h\nu}{kT^2}\right)\left(\frac{1}{e^{h\nu/kT} - 1}\right) \tag{21.24}$$

Now we substitute (21.24) into (21.17) to obtain

$$E_{\text{vib}} = \frac{Nh\nu}{e^{h\nu/kT} - 1} \tag{21.25}$$

We can obtain C_v in either of two ways: (1) differentiate (21.25) with respect to temperature or (2) differentiate (21.24) with respect to temperature to obtain the second derivative ($d^2 \ln Z/dT^2$) and then substitute this and (21.24) into (21.18). Both procedures lead to

$$C_v(\text{vib}) = \frac{R(h\nu/kT)^2\, e^{h\nu/kT}}{(e^{h\nu/kT} - 1)^2} \tag{21.26}$$

Equation (21.25) gives the vibrational energy of one mole of oscillators, each oscillating in one dimension. Similarly, (21.26) gives the vibrational heat capacity of one mole of oscillators, each oscillating in one dimension. Because the atoms in a solid are vibrating in three dimensions, we must multiply each equation by 3 to obtain the molar quantities for a monatomic solid. We now multiply (21.26) by 3 and also let $h\nu/kT$ be represented by x, which gives us

$$C_v = \frac{3Rx^2e^x}{(e^x - 1)^2} \tag{21.27}$$

as a convenient form of the Einstein equation for the heat capacity of a monatomic solid.

At low temperature $h\nu/kT = x$ is large so that $e^x - 1 \cong e^x$, which gives

$$C_v = \frac{3Rx^2e^x}{(e^x)^2} = \frac{3Rx^2}{e^x} \quad \text{(low temp. limit)}$$

As T approaches zero so that x increases, e^x increases much more than does x^2. We therefore see that C_v approaches zero as T approaches 0°K. Because heat capacities do approach zero at very low temperatures, the Einstein equation is at least qualitatively satisfactory where the classical theory was entirely unsatisfactory.

At high temperatures $h\nu/kT = x$ is very small, which suggests that e^x in (21.27) can be usefully approximated by the first part of the series (see Table 1.1)

$$e^x = 1 + x + \cdots$$

Substitution in (21.27) now leads to

$$C_v = \frac{3Rx^2(1)}{(1 + x - 1)^2} = 3R \quad \text{(high temp. limit)}$$

for the heat capacity of an Einstein solid at high temperature, which is in exact agreement with classical theory and in reasonable agreement with results of measurements.

Although the Einstein theory was a great qualitative success, detailed comparison of heat capacities calculated from equation (21.27) with experimental results showed that the Einstein model was oversimplified. In 1912 Debye improved the Einstein theory by recognizing that interactions of the atoms should result in a distribution of vibration frequencies. He then deduced the form of this distribution of frequencies and evaluated the corresponding partition function, which led to an equation for C_v that was in much better agreement with results of laboratory measurement.

Einstein's combination of quantum theory with the older statistical thermodynamics of Boltzmann, Gibbs, and others provided clear indication of both the general validity and the importance of Planck's quantum postulate. Einstein's work also led to later development by Fowler, Giauque, Tolman, and others of useful methods for calculating heat capacities and entropies of gases from knowledge (derived from spectroscopy) of the vibrational and rotational properties of their molecules. These calculations of entropies have direct bearing on the third law of thermodynamics, which we now discuss.

We may obtain a simple statistical feeling for the third law of thermodynamics by combining the Boltzmann equation (21.9) with the idea that entropy is a measure of the disorder or randomness in a system. According to the Boltzmann equation, all atoms or molecules will be in the lowest energy state at $T = 0$ because $e^{-\varepsilon_i/0} = 0$ except when $\varepsilon_i = 0$.* When all particles are in the same energy state, randomness and entropy become zero, which can be regarded as an informal statement of the third law.

For our purposes, the third law of thermodynamics can be stated as follows: *Entropies of pure crystalline solids may be taken to be equal to zero at 0° K.* Certain restrictions on this statement are given in more advanced works.

Experimental evidence for the validity of the above formulation of the third law was first provided by the work of Richards, Nernst, Planck, Haber, and several who worked with Lewis.

One of the important consequences of the third law is that it permits us to calculate entropies from the results of heat capacity measurements, which may then be combined with ΔH^0 values to yield ΔG^0 values and thence equilibrium constants. Thus the third law makes it possible to obtain equilibrium constants for reactions that have never been investigated.

In an earlier discussion of temperature and entropy we developed equation (12.19), which we now write as

$$dS = \frac{C_p}{T} dT \tag{21.28}$$

Integration of this equation gives

$$S_T - S_0 = \int_0^T \frac{C_p}{T} dT \tag{21.29}$$

The third law tells us that $S_0 = 0$ and the integral on the right side can be evaluated as the area under the line in a graph of C_p/T against T, thus yielding the entropy of the solid at temperature T.

*With x finite in $e^{-x/y}$, as y approaches zero, x/y approaches infinity. By definition, $e^{-\infty} = 0$. If x is zero and y approaches zero, x/y also approaches zero. By definition, $e^0 = 1$. Hence, from either analysis we see that Boltzmann statistics predict that all particles are in the ground energy state at 0°K.

If the substance under investigation is a gas at the temperature of interest, the entropy is evaluated as follows:

$$S_T^0 = \int_0^{T_m} \frac{C_p(\text{solid})}{T} dT + \frac{\Delta H_m}{T_m} + \int_{T_m}^{T_b} \frac{C_p(\text{liq})}{T} dT + \frac{\Delta H_b}{T_b} + \int_{T_b}^{T} \frac{C_p(\text{gas})}{T} dT \quad (21.30)$$

In (21.30) ΔH_m and ΔH_b refer to enthalpies of melting and boiling and T_m and T_b refer to the normal melting and boiling temperatures. Entropies of gases calculated from thermal data according to (21.30) have been found by Giauque and others to be in excellent agreement with entropies calculated by means of statistical thermodynamics applied to rotational and vibrational data obtained spectroscopically, thus providing support for the original assignment of $S_0 = 0$.

In our earlier discussions of enthalpies of reaction, we have often worked in terms of standard enthalpies of formation represented by ΔH_f^0. But for similar calculations with entropies it is unnecessary to work so often with entropies of formation represented by ΔS_f^0. Instead, we can and usually do work with the entropies themselves, frequently as obtained from application of equations (21.29) and (21.30) to thermal data. The standard entropy change for the chemicals in a reaction can be calculated as

$$\Delta S^0_{\text{reaction}} = \Sigma S^0_{\text{products}} - \Sigma S^0_{\text{reactants}} \quad (21.31)$$

We illustrate some applications of the third law with the following Example Problems.

Example Problem 21.3. Calculate ΔG_f^0 for gaseous methanol (methyl alcohol, CH_3OH) at 298°K from ΔH_f^0 and S^0 values listed in Table 21.5.

The standard free energy of formation of $CH_3OH(g)$ has been defined as ΔG^0 for the reaction

$$C(gr) + \tfrac{1}{2} O_2(g) + 2\,H_2(g) \rightarrow CH_3OH(g) \quad (21.32)$$

The ΔH^0 for this reaction is equal to the ΔH_f^0 for $CH_3OH(g)$ and we calculate ΔS^0 by combining the S^0 values in Table 21.5 with equation (21.31) to obtain

$$\Delta S^0 = 57.3 - [1.4 + \frac{49.0}{2} + 2(31.2)] = -31.0 \text{ cal deg}^{-1} \text{ mole}^{-1}$$

Table 21.5. Thermodynamic Data*

Substance	ΔH_f^0 *(kcal mole^{-1})*	S^0 *(cal deg^{-1} mole^{-1})*
$CH_3OH(g)$	−47.96	57.3
C (graphite)	0	1.4
$O_2(g)$	0	49.0
$H_2(g)$	0	31.2

*The ΔH_f^0 for $CH_3OH(g)$ is derived from its enthalpy of combustion. Entropies of $CH_3OH(g)$, $O_2(g)$, and $H_2(g)$ have been obtained both by application of equation (21.30) to thermal data and by way of statistical calculations. The entropy of graphite comes from application of equation (21.29) to heat capacity data.

Substituting $\Delta H^0 = -47{,}960$ cal mole^{-1} and $\Delta S^0 = -31.0$ cal deg^{-1} mole^{-1} in $\Delta G^0 = \Delta H^0 - T\,\Delta S^0$ gives us

$$\Delta G^0 = -47{,}960 - 298(-31.0) = -38{,}720 \text{ cal mole}^{-1} \qquad (21.33)$$

for reaction (21.32). Therefore we have $\Delta G_f^0 = -38.72$ kcal mole^{-1} for $CH_3OH(g)$. ■

Note that ΔH_f^0 and ΔH^0 values are usually expressed in kcal mole^{-1} while S^0 and ΔS^0 values are usually expressed in cal deg mole^{-1}. When combining ΔH^0 and $T\,\Delta S^0$ values in $\Delta G^0 = \Delta H^0 - T\,\Delta S^0$ it is necessary to use the same units throughout, which is why we expressed ΔH^0 in cal mole^{-1} in equation (21.33).

Calculations like the one we have just completed are the source of a large number of ΔG_f^0 values, which are very important in connection with calculation of ΔG^0 values and thence equilibrium constants for a huge number of chemical reactions.

Example Problem 21.4. Use thermodynamic data from Appendix II to investigate the possibility of making methyl alcohol (also called methanol) from carbon monoxide and hydrogen.

An equation for the reaction to be considered is

$$CO(g) + 2\,H_2(g) \rightleftharpoons CH_3OH(g) \qquad (21.34)$$

We remember that $\Delta G_f^0 = 0$ for $H_2(g)$ because of our earlier choice of reference or "sea level" for free energies and enthalpies and we look up ΔG_f^0 values for $CH_3OH(g)$ and $CO(g)$. Substitution of these values in equation (13.23) gives

$$\Delta G^0 = -38.72 - (32.78) = -5.94 \text{ kcal mole}^{-1} = -5940 \text{ cal mole}^{-1}$$

for reaction (21.34). Using this value with $\Delta G^0 = -RT \ln K$ we find

$$-5940 = -1.987 \times 298 \times 2.303 \log K$$

and $K = 2.24 \times 10^4$

This equilibrium constant is reasonably large and we conclude that it is *possible* to obtain a good yield of gaseous methanol from CO and H_2. On the basis of thermodynamics, we are unable to say whether the rate of reaction will be favorable or not.

From ΔH_f^0 values we calculate that $\Delta H^0 = -21.54$ kcal mole^{-1} for reaction (21.34). The equilibrium constant for this reaction therefore decreases with increasing temperature, and its values at temperatures other than 298°K can be calculated by way of the van't Hoff equation as in Chpater 13.

We also note that S^0 values in Appendix II lead to $\Delta S^0 = -52.3$ cal deg^{-1} mole^{-1} for reaction (21.34). Combination of this ΔS^0 with the ΔH^0 given above leads to $\Delta G^0 = -5.94$ kcal mole^{-1}, exactly as already calculated from ΔG_f^0 values. ■

Polymers

Polymer is a name given to very large molecules, generally consisting of recognizable repeating units. Silica and diamond crystals may be considered to be polymers, but the word is more commonly applied to such substances as polypeptides, nylon, rubber, polyethylene, and others discussed in this chapter.

We now consider some chemical reactions by which synthetic polymers are prepared. These reactions have been classified by W. H. Carothers as **condensation polymerization** and **addition polymerization.**

A familiar condensation reaction in organic chemistry is that of an alcohol with a carboxylic acid to yield an ester:

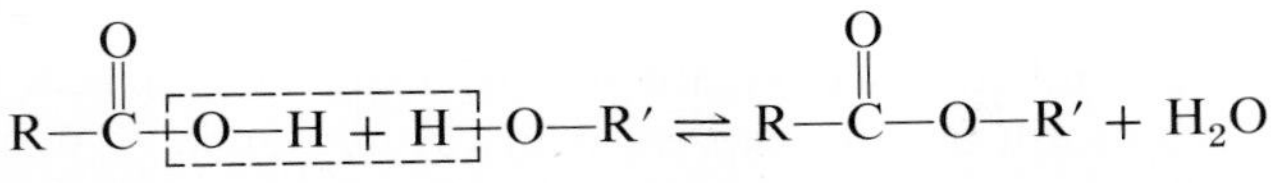

$$\mathrm{R{-}\overset{\displaystyle O}{\overset{\|}{C}}{-}\boxed{O{-}H + H}{-}O{-}R' \rightleftharpoons R{-}\overset{\displaystyle O}{\overset{\|}{C}}{-}O{-}R' + H_2O}$$

Now suppose that this same kind of reaction is carried out with a carboxylic acid that has COOH groups at each end and an alcohol that has OH groups at each end:

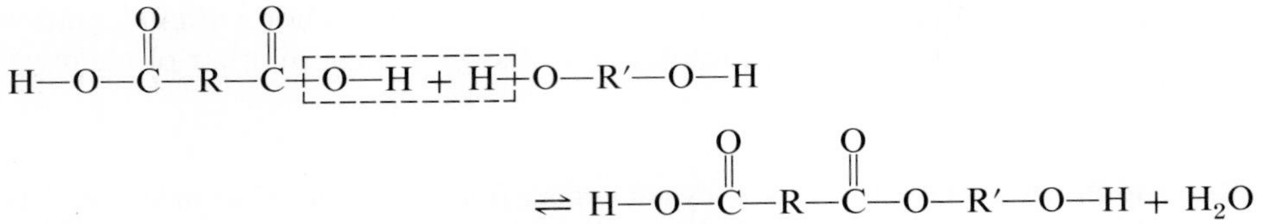

$$\mathrm{H{-}O{-}\overset{\displaystyle O}{\overset{\|}{C}}{-}R{-}\overset{\displaystyle O}{\overset{\|}{C}}{-}\boxed{O{-}H + H}{-}O{-}R'{-}O{-}H}$$

$$\mathrm{\rightleftharpoons H{-}O{-}\overset{\displaystyle O}{\overset{\|}{C}}{-}R{-}\overset{\displaystyle O}{\overset{\|}{C}}{-}O{-}R'{-}O{-}H + H_2O}$$

The ester produced by this reaction can react with another alcohol at one end and with another carboxylic acid at the other end to form a larger molecule, which can continue to react in the same way to form finally a very large molecule called a polymer—in this case a polyester. Polyesters are used in a variety of ways; alkyd paints, Dacron fiber fabrics, and Mylar films are examples.

Carboxylic acids also undergo condensation reactions with amines as indicated by

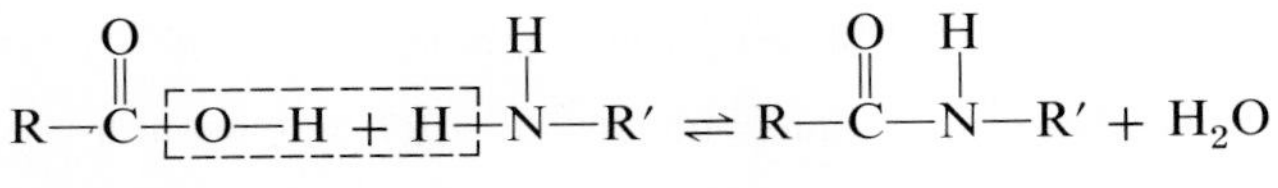

$$\mathrm{R{-}\overset{\displaystyle O}{\overset{\|}{C}}{-}\boxed{O{-}H + H}{-}\overset{\displaystyle H}{\overset{|}{N}}{-}R' \rightleftharpoons R{-}\overset{\displaystyle O}{\overset{\|}{C}}{-}\overset{\displaystyle H}{\overset{|}{N}}{-}R' + H_2O}$$

in which an amide is formed. Similar reactions of acids containing two or more COOH groups with amines containing two or more NH_2 groups lead to very large molecules called polyamides. Various nylons are the most important polyamides in industry. One indication of the importance of nylon (used in tire cord, clothing, plastic substitutes for metals, fishing line, tooth brushes, etc.) is that about 1.4×10^9 lb of it were produced in the United States alone in 1969. The polypeptides and proteins of great biological importance are also polyamides.

Addition polymerization consists of chain reactions that result in combination of many monomers to form large polymeric molecules. An important example is the polymerization of ethylene to form polyethylene with a structure that can be represented by

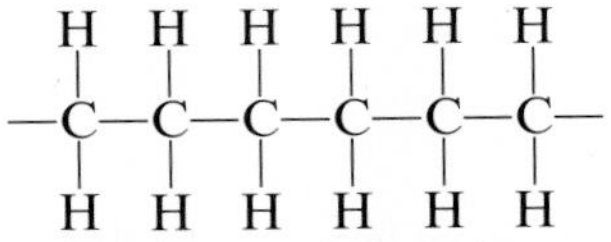

$$\mathrm{{-}\underset{\displaystyle H}{\underset{|}{\overset{\displaystyle H}{\overset{|}{C}}}}{-}\underset{\displaystyle H}{\underset{|}{\overset{\displaystyle H}{\overset{|}{C}}}}{-}\underset{\displaystyle H}{\underset{|}{\overset{\displaystyle H}{\overset{|}{C}}}}{-}\underset{\displaystyle H}{\underset{|}{\overset{\displaystyle H}{\overset{|}{C}}}}{-}\underset{\displaystyle H}{\underset{|}{\overset{\displaystyle H}{\overset{|}{C}}}}{-}\underset{\displaystyle H}{\underset{|}{\overset{\displaystyle H}{\overset{|}{C}}}}{-}}$$

Many addition polymers can be represented as combinations of molecules of type

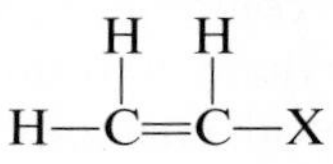

$$\mathrm{H{-}\overset{\displaystyle H}{\overset{|}{C}}{=}\overset{\displaystyle H}{\overset{|}{C}}{-}X}$$

When X = H, we have ethylene as monomer and polyethylene as the resulting polymer. Another important polymer of this type is polystyrene in which X represents

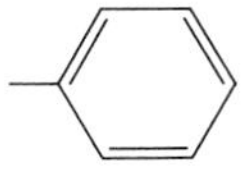

Network or **crosslinked** polymers are formed by condensation polymerization of monomers containing more than two functional groups. For example, condensation polymerization of glycerine (also called glycerol)

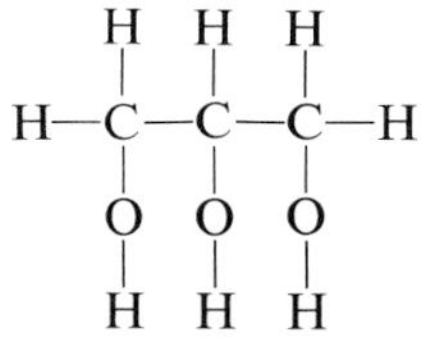

$$\begin{array}{ccccccc} & & H & & H & & H & \\ & & | & & | & & | & \\ H & — & C & — & C & — & C & — & H \\ & & | & & | & & | & \\ & & O & & O & & O & \\ & & | & & | & & | & \\ & & H & & H & & H & \end{array}$$

with a dicarboxylic or tricarboxylic acid yields a crosslinked polymer. Some polymers are made by having several kinds of monomers in the reaction. The resulting **copolymers** often contain branched chains and crosslinking.

Polymeric materials may be fabricated into useful shapes by application of heat and pressure. Crosslinking during fabrication yields material that is called **thermosetting** because it loses the ability to flow at high temperature after the original heat treatment that initiated crosslinking. On the other hand, linear polymers that remain linear during heat treatment are called **thermoplastic** because they retain the ability to flow at high temperature.

There are very many interesting and important problems in polymer chemistry (and biology and physics). Examples include such diverse topics as x-ray diffraction by partly crystalline solid polymers, dielectric properties of polymers that are used for electrical insulation, tensile strengths of fibers made from polymers, chemical reactions involved in dyeing of polymers, synthesis of polymers that will decompose in sunlight, rates of polymerization reactions, optical properties of thin films made of polymers, configurations of many biological polymers, and on and on. But in the remainder of this chapter we will focus on only one property of polymers—molecular weights and their determinations.

Readers may have anticipated the well-established fact that polymeric molecules of the same "kind" do not all have the same molecular weight. A sample of polymer is likely to contain similar molecules that have a range of molecular weights. Various samples may have either narrow or wide ranges of molecular weights. Various methods of determining molecular weights average the various molecular weights in a sample in different ways that will not concern us. We shall speak simply of "the" molecular weight of a polymer when we mean an average molecular weight.

In our earlier studies we have seen that the molecular weight of a substance can be calculated from the density of the substance in the gaseous state at some specified temperature and pressure. Because high molecular weight polymers cannot be obtained as gases (they decompose), this method does not presently concern us.

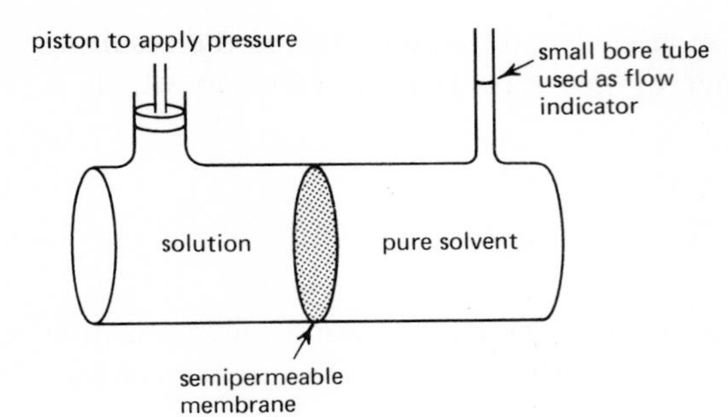

Figure 21.13. Schematic illustration of an apparatus for measuring osmotic pressure. The osmotic pressure is the applied pressure necessary to keep solvent from flowing through the semipermeable membrane.

We have also seen that molecular weights can be obtained from freezing point depressions of dilute solutions. Let us consider application of this method to a polymer with molecular weight of about 100,000 g mole^{-1}. A dilute solution of this polymer might contain 1.0 g of polymer and 100 g of water so that the concentration of polymer is 1.0×10^{-4} *m*. By equation (4.9) we calculate that the freezing point depression is 1.86×10^{-4} degree. A similar 1% solution of polymer having molecular weight of about 200,000 g mole^{-1} would have freezing point depression of 0.93×10^{-4} degree. These freezing point depressions are so small that it is difficult to distinguish between them, and we certainly could not expect to obtain reliable molecular weights of such large molecules by way of these measurements.

An obvious question: Why not use more concentrated solutions so that freezing point depressions would be larger? There are two reasons. (1) Many polymers are not very soluble. (2) Equation (4.9) is valid only for very dilute solutions.

Calculations similar to the ones above show that neither boiling point elevations nor vapor pressure lowering measurements are useful methods for determination of molecular weights of polymers.

A method of molecular weight determination that is well suited to polymers is based on measurement of osmotic pressure, frequently in an apparatus like that pictured in Figure 21.13. In this apparatus, the semipermeable membrane permits flow of solvent but not of the large solute molecules. Because the escaping tendency of solvent is greater from pure solvent than from solution, solvent tends to flow from the pure solvent through the membrane to the solution. This flow can be prevented by applying pressure to the solution. The applied pressure that is just sufficient to prevent flow of solvent through the membrane is called the osmotic pressure and given the symbol π. Our approach to quantitative treatment of osmotic pressures is to relate the free energy decrease of solvent associated with adding solute to form the solution to the free energy increase associated with the applied pressure.

The lowering of free energy (per mole of solvent) that results from adding solute to pure solvent can be expressed in terms of the lowering of the vapor pressure of solvent as

$$\Delta G = RT \ln \frac{P}{P^0} \quad \text{(taken from equation 13.25)}$$

in which P and P^0 represent the solvent vapor pressure over the solution and over the pure solvent.

The free energy increase (again per mole of solvent) due to the applied pressure π is obtained by integrating (13.24) as follows:

$$\int dG = \int_0^{\pi} V_i\, dP = V_1 \int_0^{\pi} dP$$

$$\Delta G = V_1 \pi$$

Here we have used V_1 to represent the volume of one mole of solvent, which is regarded as incompressible so that it can be treated as constant.

Because the applied pressure π was chosen to maintain equilibrium, the net change in free energy of solvent must be zero as expressed by

$$\Sigma \Delta G = 0 = V_1 \pi + RT \ln \frac{P}{P^0}$$

or

$$\pi V_1 = -RT \ln \frac{P}{P^0}$$

Raoult's law applies to the solvent in all dilute solutions, which permits replacement of the ratio P/P^0 by the mole fraction of solvent X_1. The result is

$$\pi V_1 = -RT \ln X_1 = -RT \ln (1 - X_2) \tag{21.35}$$

in which we have made use of $X_1 + X_2 = 1$ and thence $X_1 = 1 - X_2$. Although equation (21.35) can be used to relate composition of solution and osmotic pressure to molecular weight as in problem 22 at the end of this chapter, this is more commonly done by way of the related equation (see problem 24)

$$\pi = cRT \tag{21.36}$$

in which c represents the molarity of the solution. Application of (21.36) is illustrated in the following Example Problem.

Example Problem 21.5. A sample of polymer weighing 2.50 g is dissolved in 100 cm^3 of benzene at 298°K. The osmotic pressure π of the resulting solution is 2.51×10^{-3} atm. What is the average molecular weight of the polymer?

We rearrange (21.36) to obtain $c = \pi/RT$ and thence calculate $c = 1.03 \times 10^{-4}$ M. Making use of the definition of molarity ($M = n/V$) we calculate that $n = 1.03 \times 10^{-5}$ mole of polymer. Finally, we combine the mass of polymer with the now known number of moles to obtain its average molecular weight as 2.43×10^5 g mole^{-1}.

Although the step-wise procedure outlined above is entirely correct, it is sometimes more convenient to combine all of the steps to obtain a final equation before undertaking numerical calculations. To do so we express the concentration in terms of number of moles of polymer and volume of solution as $c = n/V$ and the number of moles of polymer as $n = w/W$ in which w and W represent weight and molecular weight of polymer. Thus we have $c = w/WV$ to substitute in (21.36) to obtain $\pi = wRT/WV$. Now we do the calculation all at once as follows:

$$W = \frac{wRT}{\pi V} = \frac{(2.50\text{ g})(0.082\text{ liter atm deg}^{-1}\text{ mole}^{-1})(298\text{ deg})}{(2.51 \times 10^{-3}\text{ atm})(0.100\text{ liter})}$$

$$= 2.43 \times 10^5\text{ g mole}^{-1}$$

Note the choice of value and units for R in this calculation. ■

References

Azbel M. A. Ya., M. I. Kaganov, and I. M. Lifshitz: Conduction electrons in metals. *Sci. Amer.*, **228:**88 (Jan. 1973).

Beveridge, D. L., and B. J. Bulkin: Descriptive crystal orbital theory of conduction in diamond and graphite. *J. Chem. Educ.*, **48:**587 (1971).

Billmeyer, F. W.: *Textbook of Polymer Science,* 2nd ed. J. Wiley and Sons, Inc., New York, 1971.

Hakala, R. W.: A new derivation of the Boltzmann distribution law. *J. Chem. Educ.*, **39:**10 (1961).

Hecht, C. E.: Desalination of water by reverse osmosis. *J. Chem. Educ.,* **44:**53 (1967).

Kapecki, J. A.: An introduction to x-ray structure determination. *J. Chem. Educ.* **49:**231 (1972).

Melrose, M. P.: Statistical mechanics and the third law. *J. Chem. Educ.*, **47:**283 (1970).

Mergerison, D., and G. C. East: *An Introduction to Polymer Chemistry*. Pergamon Press, London, 1967 (paperback).

Moore, W. J.: *Seven Solid States: An Introduction to the Chemistry and Physics of Solids.* W. A. Benjamin, Inc., New York, 1967.

O'Driscoll, K. F.: *The Nature and Chemistry of High Polymers.* Reinhold Publishing Corp., New York, 1964 (paperback).

Quane, D.: Crystal lattice energy and the Madelung constant. *J. Chem. Educ.*, **47:**396 (1970).

Suess, M. J.: Reverse osmosis. *J. Chem. Educ.*, **48:**190 (1971).

Weller, P. F.: An analogy for elementary band theory concepts in solids. *J. Chem. Educ.*, **44:**391 (1967).

Weller, P. F.: An introduction to principles of the solid state. *J. Chem. Educ.*, **47:**501 (1970).

Weller, P. F.: An introduction to principles of the solid state: extrinsic semiconductors. *J. Chem. Educ.*, **48:**831 (1971).

Problems

1. The face centered cubic structure is found for many metals (for example, the coinage metals) and is a simple cubic structure with an added atom or ion in the center of each face. Sketch this structure.
2. Most anions are larger than most cations. For example, the radii of Cl^- and K^+, which have the same number of electrons, are about 1.81 Å and 1.33 Å. Explain why the radius of Cl^- is larger than that of K^+.
3. The density of diamond is greater than that of graphite, as predicted from the knowledge that graphite has a more open structure than diamond. Is the synthesis of diamond from graphite favored by high pressure or by low pressure? What equation would you use for a quantitative calculation of the effect of pressure on the equilibrium between graphite and diamond?
4. Predict the sign of ΔV for the reaction

$$\text{BN (graphite structure)} \rightleftharpoons \text{BN (diamond structure)}$$

   Why is the diamond structure favored by high pressure?
5. The bond angles in diamond and methane are the same. Suggest an explanation.
6. Predict whether a pure metal or an alloy of that metal with a small concentration of another metal would have the higher electrical conductivity. Would the pure metal or the alloy be expected to have the greater thermal conductivity? Explain the basis for your predictions.

7. Many years ago Wiedemann and Franz observed that the ratio of thermal conductivity to electrical conductivity of metals is proportional to the temperature as in

$$\frac{K}{\sigma} = LT$$

in which K and σ represent thermal and electrical conductivities and L is nearly the same constant for all metals. Explain why the form of this equation and the approximate constancy of L are not surprising. What are your chances of finding a metal that is a good conductor of electricity and a poor conductor of heat?
8. Why do solids give sharper x-ray diffraction patterns than do liquids?
9. Why might two elements have the same molar volumes but different atomic radii? Or, why might they have the same atomic radii but different molar volumes?
10. Zone melting is now used as a method for purifying materials that are already nearly pure, such as semiconductors for use in transistors. The procedure may be summarized as follows. A heater placed around one end of a bar of the material is used to melt a small zone of the material. The heater is then moved slowly down the bar so that the molten zone moves from one end of the bar to the other. The process can be repeated several times. The impurities accumulate in the molten zone and the solid material in back of the heater is purified, thus gradually leading to pure material at one end of the bar with impurities concentrated at the other end. Why do the impurities accumulate in the molten zone?
11. Prove that angles DCE and ECF are equal to θ in Figure 21.2.
Hint: Note that $\angle ACD = \angle BCE = 90°$ and that 90° angles can be expressed as sums of two smaller angles.
12. It has been stated in the text that polar molecules (such as BrCl) are attracted to each other by dipole-dipole forces. Since it is not obvious that dipole-dipole forces should be attractive rather than repulsive, consider one attack on this problem by way of calculations based on the arrangement of molecules as in Figure 21.14.

The potential energy ϕ is given by the product of the interacting charges divided by the distance of separation. Thus the mutual potential energy of the atoms labeled (1) and (3) is

$$\phi_{13} = +e^2/R \quad \text{(like charges repel, } \phi \text{ positive)}$$

and for atoms labeled (1) and (4) is

$$\phi_{14} = -e^2/(R + r) \quad \text{(unlike charges attract, } \phi \text{ negative)}$$

Write down appropriate expressions for ϕ_{23} and ϕ_{24}. Then show that the sum of $\phi_{13} + \phi_{14} + \phi_{23} + \phi_{24}$ gives the total potential energy in the form

$$\phi = e^2\left(\frac{2}{R} - \frac{2R}{(R + r)(R - r)}\right)$$

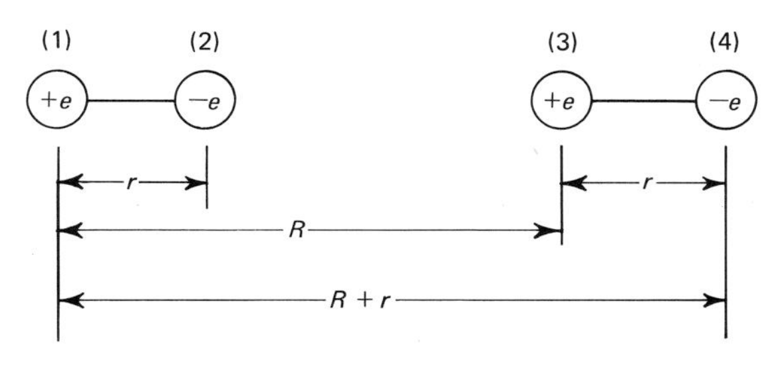

Figure 21.14. Illustration of polar molecules. See problem 12.

Finally, show that the term on the right inside the brackets in the equation above is larger than $2/R$ so that ϕ is negative, corresponding to net attraction between these dipoles.

13. Production of titanium from the common ore rutile has been carried out by means of the following series of reactions:
(i) The ore, largely TiO_2, is heated with carbon and chlorine to give $TiCl_4(g)$ and $CO_2(g)$.
(ii) The $TiCl_4$ is reduced at high temperature by magnesium.
(iii) The $MgCl_2$ that is produced and excess Mg are separated from the metallic titanium by treatment with dilute hydrochloric acid, which does not attack titanium.
(a) Write balanced equations for reactions described above.
(b) What is the maximum amount of titanium that can be obtained from 1.0 ton of TiO_2?
(c) What are the minimum amounts of carbon, chlorine, magnesium, and hydrochloric acid that must be used for full recovery of titanium from 1.0 ton of TiO_2?

14. A number of alloys of bismuth expand on freezing and are therefore useful in making castings. Predict the effect of increasing pressure on freezing points of these alloys.

15. Compositions of some important alloys of copper with other metals are listed in Table 21.6. Devise *chemical* tests that would enable you to distinguish these alloys from each other. Use information in Chapter 20, Appendices III and IV, or from a Handbook.

Table 21.6. Compositions of Useful Copper Alloys

Alloy	*Composition, %*
Brass	60–90 Cu, 10–40 Zn
Bronze	80 Cu, 15 Sn, 5 Zn
Bell metal	78 Cu, 22 Sn
Constantan	60 Cu, 40 Ni
Manganin	82 Cu, 15 Mn, 3 Ni
German silver	52–60 Cu, 25 Zn, 15–23 Ni

16. An aluminum rod will dissolve more rapidly in aqueous sodium hydroxide than in hydrochloric acid. Similarly, aluminum oxide dissolves more rapidly in base than in acid. How does the second observation account for the first?

17. Differentiate equation (21.6) with respect to r to obtain

$$\frac{d\phi}{dr} = \frac{NMz^2e^2}{r^2} + \frac{nBe^2}{r^{n+1}}$$

At the minimum of the ϕ against r curve, we have $d\phi/dr = 0$ and $r = r_e$. Go through the missing steps to obtain equations (21.7) and (21.8).

18. It has been stated in the text that the value of n in equation (21.8) is about 9 for NaCl. Repeat the calculation of Example Problem 21.2 with $n = \infty$ and $n = 10$ and then decide whether the exact value of n is of much importance with respect to the lattice enthalpy.

19. Suppose that the experiment described in Example Problem 21.1 had been carried out with x rays having wavelength of 2.290 Å. At what angles would these x rays be reflected with maximum intensity?

20. Draw a structural formula for the polymer formed from vinyl chloride, $H_2C{=}CHCl$.

21. What is the mass of one molecule of a polymer with a molecular weight of 1,000,000 g mole^{-1}? Assuming that the polymer in question has density of about 1.0 g cm^{-3} and that a molecule is nearly spherical, calculate the radius of one molecule. Could you see that molecule? Estimate the molecular weight of the smallest molecule of this polymer that could be seen by the unaided eye.

22. On the basis of the van't Hoff equation (13.48) or the Boltzmann equation (21.9), predict the effect of increasing temperature on the electrical conductance and resistance of an intrinsic semiconductor.

23. Use equation (21.35) to obtain the molecular weight of the polymer discussed in Example Problem 21.5. The procedure to be used in this calculation is similar to that in Example Problem 21.5.

24. For small X_2 corresponding to dilute solution, it is a good approximation (see Table 1.1) to take

$$\ln(1 - X_2) \cong -X_2$$

in equation (21.35). Now carry out a detailed derivation of equation (21.36). *Hint:* See equation (4.1) and remember that $\Sigma n_i \cong n_1$ for dilute solutions.

25. Using data in Appendix II, calculate the equilibrium constant at 298°K for the reaction

$$CH_2CH_2(g) + H_2O(g) \rightleftharpoons CH_3CH_2OH(g)$$

What is the effect of increasing temperature on the equilibrium constant for this reaction? How could you calculate the equilibrium constant at some temperature higher than 298°K? What conditions of pressure and temperature are suggested by thermodynamics (LeChatelier's principle) for maximum equilibrium conversion of ethylene to ethyl alcohol?

26. Standard free energies of formation of many organic compounds are derived from ΔH_f^0 values and S^0 values. Describe in detail the experiments and calculations leading to ΔH_f^0 and S^0 for gaseous acetylene. Then use the ΔH_f^0 and S^0 values in Appendix II to derive ΔG_f^0 for C_2H_2, showing all calculations.

27. Heat capacities of pure substances are positive at all temperatures above 0°K. Entropies of pure substances at constant pressure always increase with increasing temperature. Show that the second statement follows from the first.

28. Dissolving KCl(c) in water is an endothermic process, with $\Delta H = +4$ kcal mole^{-1}. What can you say about relative magnitudes of the lattice enthalpy and hydration enthalpy of this substance?

29. Estimate the thermodynamic stability of the hypothetical ionic compound ArCl. First, estimate the radius of the Ar^+ ion and then calculate an approximate lattice enthalpy for ArCl, which you may assume to have the NaCl structure.

30. Planck's quantum theory led to $\varepsilon = vh\nu$ for vibrational energies, as in equation (21.19). Modern quantum theory leads to the similar equation

$$\varepsilon = (v + \tfrac{1}{2})h\nu$$

Explain why you anticipate that use of this equation as in equations (21.20–21.25) will lead to the same heat capacity expression (21.26) that Einstein found. Those who cannot see or question the verbal explanation should carry out the various mathematical operations to prove this point.

31. Calculate the osmotic pressure of an aqueous solution that is 0.010 M at 300°K.

32. The tallest trees known (coastal redwoods of California) are about 350 ft tall. It is osmotic pressure that supplies water to the topmost branches of these trees. Estimate the concentration of solute in "tree-water" that is sufficient to account for this osmotic pressure. State clearly all assumptions and approximations on which your estimate is based.

33. Debye's improved version of Einstein's theory of heat capacities of solids gives

$$C_p = DT^3$$

for very low temperatures. Suppose that measurements give $C_p = A$ for some solid at 10°K. Calculate the entropy of this solid (in terms of A) at 10°K.

34. At high temperatures the partition function for vibration of a diatomic molecule is approximated by $Z = T/a$ in which a is a constant that depends only on the molecule under consideration. Calculate the contributions of vibration to the energy and heat capacity of a diatomic gas at high temperatures.

35. The radius of Na^+ is almost the same as that of Ca^{2+}, and the radius of Cl^- is almost the same as that of S^{2-}. Taking the lattice enthalpy of NaCl to be 185 kcal mole^{-1} and assuming that NaCl and CaS have the same crystal structure, make a quick estimate of the lattice enthalpy of CaS.

36. There is current interest in development of polymers (plastics) that will decompose in sunlight, with the intention of freeing the world of its ever increasing stock of plastic rubbish. Like many problems, this one is complicated and has many facets. Think about and then write out your thoughts on the following:

(a) What kinds of tests should be made to establish the presence or absence of both short term and long term problems associated with use of all new plastics, including those that decompose in sunlight.

(b) Plastics that degrade *safely* in our atmosphere are surely better than those that clutter up our planet. But each piece of plastic that is used and discarded represents valuable resources that we cannot use again. What are these resources?

(c) Because we should conserve the resources that go into manufacture of plastics, maybe we should use returnable plastic bottles or other returnable containers. What are the problems associated with reuse of plastic bottles and the like?

(d) The world contains *huge* quantities of silica and the other components of glass. To what extent should we decrease the use of plastics and increase the use of glass?

22

CHEMICAL KINETICS

Introduction

In chemical kinetics we are concerned with rates of reactions and with using experimental information about reaction rates to obtain knowledge about mechanisms or pathways of chemical reactions.

We begin our study of chemical kinetics by considering some experimental methods for investigating reaction rates and then show how reaction rate data are summarized by useful equations. Then we turn to consideration of some theories of chemical kinetics and to the relationship between reaction rate and reaction mechanism.

Many reactions with large negative ΔG^0 (large equilibrium constant K) proceed slowly. Other reactions that are less favored thermodynamically (smaller K) may proceed rapidly. Rates of many reactions can be dramatically increased or decreased by substances called catalysts and inhibitors, which do not affect the position of equilibrium. Thus there is no general relationship between ΔG^0 or K and the rate of the reaction. Thermodynamics yields a description of the *equilibrium* state of a system, but tells us nothing about the *rate* of approach to equilibrium. But we shall see that thermodynamic considerations are sometimes important in connection with some of the steps that make up the mechanisms of reactions, and we shall also see that rate measurements can sometimes lead to knowledge of the equilibrium constant of a reaction.

Combinations of thermodynamic and kinetic considerations have led chemical and related industries to some of their most useful processes. Manufacture of ammonia and nitric acid from atmospheric nitrogen and reforming of petroleum are important industrial examples of applications of combinations of thermodynamics and kinetics.

Experiments in Chemical Kinetics

Most experiments in chemical kinetics are intended to yield quantitative information about the rates of chemical reactions. The measurements that are made depend on the properties of reactants and products and the time available for measurements. Specific examples are discussed below.

Rates of disintegration of radioactive atoms were discussed briefly in Chapter 2. These disintegration processes or "reactions" may be represented in general by

$$\text{A} \rightarrow \text{B} + \text{particle} \tag{22.1}$$

in which A, B, and particle represent the initial radioactive atom, the atomic product of the disintegration, and the emitted particle (α particle, electron called a β^- particle, etc.). Gamma or x rays should be included in (22.1) for many radioactive disintegrations.

Investigation of the rate of radioactive disintegration might take the form of measurements of the amounts of A or B present at different times, but the usual procedure is to "count" the emitted particles with a suitable instrument such as a Geiger counter. The data recorded in the investigator's notebook could be the numbers of counts registered on the detecting instrument at the end of each minute. Then the investigator could easily calculate the number of particles counted in any particular minute. After this calculation has been carried out for each minute from beginning to end of the experiment, the investigator might summarize his results in a table in which one column is headed "Disintegrations per Minute" and the other "Time." Zero time would ordinarily be the time when the counting instrument began counting particles. Heading a column "Disintegrations per Minute" when it was emitted particles that were actually counted implies a one-to-one correspondence between emitted particles and disintegrations, which should be verified by some kind of analysis.

When the investigator has completed the table of results and perhaps displayed them on a graph, the next step is often mathematical representation of the results. Various theoretical principles are an important help in finding an equation that will represent the experimental results and are then essential for finding out about the mechanism of the reaction.

The discussion above implies a greater separation of theory and experiment than is common or desirable in most investigations or applications of chemical kinetics. Someone's theory (possibly only half formed) or need for information to be used in some way not directly connected with the kinetics experiments often provides the basis for deciding to undertake an investigation. Then some combination of theory, experience, and intuition suggests the important quantities to be measured and finally tabulated or graphed or represented by an equation.

Nevertheless, we shall find it convenient to continue an artificial separation of experiment from mathematics and theory while we discuss a few more measurements in chemical kinetics.

Investigations of rates of most chemical reactions usually reduce to quantitative determinations at different times of the amounts or concentrations of one or more of the species involved in the reaction. Most of the analytical methods used are conveniently classed as either chemical or physical.

One way of investigating chemical reaction kinetics is to make a large batch of the reaction mixture and remove small samples for analysis at known time intervals. If the total time for sample removal and analysis is fast compared to the reaction rate, this procedure effectively gives the concentration or amount of chemical at the time the sample was removed from the reaction vessel. More often, however, the analytical procedure is relatively slow so that the reaction must be stopped (or nearly stopped) by a sudden change, such as lowering the temperature, removing a reactant or catalyst, or adding an inhibitor.

Physical methods of analysis are often best for investigations of reaction rates. One such method is based on pressure measurements for reactions involving one or more gases. Solutions may be analyzed by various optical or electrical methods involving measurements of absorption of light of certain wavelengths, rotation of the plane of polarized light, refractive index, electrical conductivity, or dielectric constant. Still other kinetic investigations have been based on measurements of the heat evolved by a reaction or the volume change accompanying reaction in solution. Some of these physical methods have the advantage of being almost instantaneous so that nearly continuous determinations can be made without removing samples from the reaction vessel. These methods also make possible the investigation of very rapid reactions.

Reaction Rate Equations

To illustrate deduction of a rate equation from experimental rate data, we consider decomposition of N_2O_5 in carbon tetrachloride as represented by the balanced reaction equation

$$2\,N_2O_5(\text{in }CCl_4) \rightarrow 2\,N_2O_4(\text{in }CCl_4) + O_2(g)$$

Eyring and Daniels measured the volumes of oxygen produced at constant temperature and pressure as the reaction proceeded. From these data the number of moles of oxygen produced and then the number of moles of N_2O_5 decomposed were calculated. Since the initial concentration of N_2O_5 was known, knowledge of the number of moles decomposed at various times permitted calculation of the concentrations of N_2O_5 remaining at these times, as shown in Table 22.1.

Our task now is to find or develop an equation that will describe the concentration of N_2O_5 as a function of time. Because there is no general and reliable method for predicting the form of the desired equation, we must in this case and many others proceed partly by trying and testing various equations.

In order to obtain an equation that is worth testing, let us tentatively assume that the rate of disappearance of N_2O_5 is proportional to the concentration of

Table 22.1. Rate of Decomposition of N_2O_5 in CCl_4 at 45°C

Time (sec)	*Molarity of* N_2O_5
0	2.33
319	1.91
526	1.67
867	1.36
1198	1.11
1877	0.72
2315	0.55
3144	0.34

N_2O_5 as expressed by the differential equation

$$-\frac{d[N_2O_5]}{dt} = k[N_2O_5] \tag{22.2}$$

in which k is a proportionality constant usually called the **rate constant.** We now integrate this equation as an indefinite integral:

$$-\int \frac{d[N_2O_5]}{[N_2O_5]} = k \int dt$$

$$-\ln [N_2O_5] = kt + c$$

$$\ln [N_2O_5] = -kt - c \tag{22.3}$$

Equation (22.3) is of the familiar $y = mx + b$ form. We therefore know that a graph of $\ln [N_2O_5]$ against t should give a straight line of slope $-k$ IF our original assumption of (22.2) is correct. Figure 22.1 shows a graph of $\log [N_2O_5]$ against

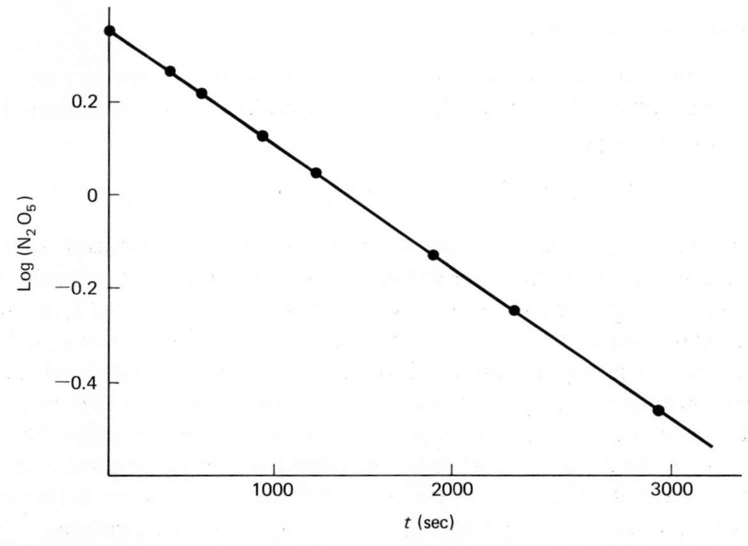

Figure 22.1. Graph of rate data for decomposition of N_2O_5 in CCl_4 solution. Plotting $\log (N_2O_5)$ versus t was suggested by equation (22.3).

t, based on the data in Table 22.1. The line is straight and has slope of $-2.7 \times 10^{-4}\ \text{sec}^{-1}$ from which we calculate that $k = -(2.303)(-2.7 \times 10^{-4}\ \text{sec}^{-1}) = 6.2 \times 10^{-4}\ \text{sec}^{-1}$. Because Figure 22.1 shows that equation (22.3) is consistent with the experimental rate data, we say that we have a satisfactory rate equation for this reaction.

Note that we can evaluate c in equation (22.3) from the intercept of the line in Figure 22.1. Or we can set $t = 0$ in (22.3) to obtain $c = -\ln [N_2O_5]_0$ so that we have

$$\ln [N_2O_5] = -kt + \ln [N_2O_5]_0$$

or

$$\ln \frac{[N_2O_5]}{[N_2O_5]_0} = -kt \tag{22.4}$$

Equation (22.4) could be tested by plotting $\ln ([N_2O_5]/[N_2O_5]_0)$ against t. In this case, we obtain a straight line through the origin (zero intercept) with slope equal to $-k$.

It should also be noted that we can obtain (22.4) by integrating (22.2) as a definite integral.

Readers may well wonder why we began with equation (22.2) rather than $-d[N_2O_5]/dT = k[N_2O_5]^2$, which is just as reasonable for a trial equation. Had we started with this latter equation, we would have integrated it and arranged the result in such fashion that we could test it graphically as shown later in this chapter. In this case, we would find a curved line rather than a straight line (see problem 18) and would be forced to conclude that this equation is unsatisfactory. In general, we would find that all equations tested are unsatisfactory until we cleverly or accidentally hit on (22.2) and thence (22.3) or (22.4).

Reactions with rates that are described by equations like (22.2–22.4) are called **first-order** reactions and are said to follow or be described by a first-order rate equation. *Note that the order of a reaction or rate equation is not necessarily related to coefficients in the balanced equation for the reaction.*

The **half-life** of a chemical reaction is the time required for half of the total possible reaction to occur. For decomposition of N_2O_5 this means that $[N_2O_5]/[N_2O_5]_0 = \frac{1}{2} = 0.5$. Substitution in (22.4) leads to

$$\ln (0.5) = -kt_{\frac{1}{2}}$$

and thence

$$k = 0.693/t_{\frac{1}{2}} \qquad \text{or} \qquad t_{\frac{1}{2}} = 0.693/k \tag{22.5}$$

A general characteristic of first-order reactions is that the time it takes for the concentration to be halved is independent of the initial concentration, which gives us an easy way to recognize a first-order reaction. Consider the data in Table 22.1 in this connection. We see that it takes about 1100 sec for the concentration of N_2O_5 to go from 2.33 *M* to 1.16 *M*. It also takes about 1100 sec for the concentration of N_2O_5 to go from 1.11 *M* to 0.55 *M*. This kind of observation suggests equation (22.2), which should then be tested by the procedure leading to Figure 22.1.

Many reactions have rates that cannot be described by first-order rate equations. One example is the high temperature decomposition of hydrogen iodide represented by

$$2\ \mathrm{HI(g)} \rightarrow \mathrm{H_2(g)} + \mathrm{I_2(g)} \tag{22.6}$$

Although a complete investigation of the kinetics of reaction (22.6) must also be concerned with the reverse reaction of H_2 and I_2 to give HI, this reverse reaction can be ignored in the early stages of the decomposition of HI that we are now discussing.

Kistiakowsky investigated the rate of decomposition of HI by heating bulbs containing HI in a molten lead constant temperature bath. Reaction in a particular bulb could be effectively stopped at any time by removing the bulb and rapidly cooling it. Then analysis of the contents gave the composition at the time the bulb was cooled.

It might first be assumed that the rate of decomposition of HI is proportional to the concentration of HI as in

$$-d[\mathrm{HI}]/dt = k[\mathrm{HI}] \tag{22.7}$$

so that integration leads to

$$\ln [\mathrm{HI}] = -kt - c \tag{22.8}$$

In this case, a graph of ln [HI] against t yields a distinctly curved line.

We might also write the rate equation in the form

$$\ln ([\mathrm{HI}]/[\mathrm{HI}]_0) = -kt \tag{22.9}$$

A graph of $\ln ([\mathrm{HI}]/[\mathrm{HI}]_0)$ against t yields a curved line rather than the straight line expected if we are working with a rate equation appropriate to this reaction. We must conclude that equations (22.7–22.9) do not represent the rate of decomposition of HI.

A reasonable second try is the **second-order** rate equation

$$-\frac{d[\mathrm{HI}]}{dt} = k[\mathrm{HI}]^2 \tag{22.10}$$

in which $[\mathrm{HI}]^2$ represents the square of the concentration (usually expressed in moles per liter) of HI. Now we integrate (22.10) as follows:

$$-\int \frac{d[\mathrm{HI}]}{[\mathrm{HI}]^2} = -\int [\mathrm{HI}]^{-2}\, d[\mathrm{HI}] = k \int dt$$

$$\frac{1}{[\mathrm{HI}]} = kt + c \tag{22.11}$$

A test of these equations is made by plotting $1/[\mathrm{HI}]$ against t, which is found to yield a straight line. The slope of the line is equal to the rate constant k.

The original second-order rate equation (22.10) can also be integrated as a definite integral from $t = 0$ and $[\mathrm{HI}]_0$ to time t and $[\mathrm{HI}]_t$ to obtain

$$\frac{1}{[\mathrm{HI}]_t} - \frac{1}{[\mathrm{HI}]_0} = kt \tag{22.12}$$

Table 22.2. Rate Data for Reaction (22.13) at 25°C

First Experiment:

Initial concentration of $S_2O_8^{2-}(aq) = 0.00016\ M$
Initial concentration of $I^-(aq) = 0.024\ M$

Time (min)	$[S_2O_8^{2-}]$
0	0.00016 M
7	0.00012 M
14	0.00009 M

Second Experiment:

Initial concentration of $S_2O_8^{2-}(aq) = 0.00016\ M$
Initial concentration of $I^-(aq) = 0.012\ M$

The solvent for this second experiment was 0.012 M KNO_3 rather than water as in the first experiment. This concentration of KNO_3 was used to make the total concentration of ions in the solution the same in both experiments and thereby minimize effect of activity coefficients.

Time (min)	$[S_2O_8^{2-}]$
0	0.00016 M
14	0.00012 M
28	0.00009 M

Equation (22.12) shows that the constant in (22.11) has the value $1/[HI]_0$, which might also have been obtained by setting $t = 0$ in (22.11). Equation (22.12) is easily solved for k so that each value of $[HI]_t$ with $[HI]_0$ permits calculation of an independent value of k. The values so obtained are found to be satisfactorily constant, again providing evidence that the second-order rate equations (22.10–22.12) are appropriate for this reaction. The same k value is obtained from this calculation as was obtained from the graph described in the paragraph above.

Now let us consider rate data for the reaction of peroxydisulfate (sometimes called persulfate) ions with iodide ions in dilute aqueous solution for which we have the balanced reaction equation:

$$S_2O_8^{2-}(aq) + 2\ I^-(aq) \rightarrow 2\ SO_4^{2-}(aq) + I_2(aq) \tag{22.13}$$

Some experimental data are given in Table 22.2. We will use these data to find the form of the rate equation and to evaluate the rate constant as follows.

First of all, we note that $[I^-] \gg [S_2O_8^{2-}]$ for both experiments so that relatively little I^- reacts. The $[I^-]$ in each experiment is therefore nearly constant at its initial value, which we represent by $[I^-]_0$.

The data from the first experiment show that one fourth of the peroxydisulfate ions react in each 7 minute period. Our previous discussion therefore suggests (see problem 19 for proof) that the reaction is first-order in $[S_2O_8^{2-}]$. Since cutting the $I^-(aq)$ concentration in half in the second experiment doubles the time required (cuts rate in half) for one fourth of the $S_2O_8^{2-}(aq)$ to react, we must also conclude that the reaction is first-order in $[I^-]$. The rate law is now written as

$$\frac{d[S_2O_8^{2-}]}{dt} = k[I^-][S_2O_8^{2-}] \tag{22.14}$$

We set $[I^-] = [I^-]_0$ before rearranging and integrating

$$-\int_0^t \frac{d[S_2O_8^{2-}]}{[S_2O_8^{2-}]} = k[I^-]_0 \int_0^t dt$$

$$-\ln \frac{[S_2O_8^{2-}]_t}{[S_2O_8^{2-}]_0} = k[I^-]_0\, t \tag{22.15}$$

$$k = \frac{2.303}{[I^-]_0\, t} \log \frac{[S_2O_8^{2-}]_0}{[S_2O_8^{2-}]_t} \tag{22.16}$$

Note that we have obtained (22.16) from (22.15) by changing the sign and inverting the log term.

Inserting numerical values in (22.16) from the first experiment now gives

$$k = \frac{2.303}{0.024 \times 7.0} \log \frac{0.00016}{0.00012} = 1.70 \text{ liter mole}^{-1} \text{ min}^{-1}$$

Readers should verify that the units for k are liters per mole per minute as written above and that the other data in Table 22.2 also lead to the same value for k, thus confirming the rate equation (22.14) on which (22.16) is based.

Example Problem 22.1. Calculate the time at which $[S_2O_8^{2-}] = 0.00008\ M$ in the Second Experiment (Table 22.2). Also calculate $[S_2O_8^{2-}]$ at $t = 60$ min in the First Experiment (Table 22.2).

We solve (22.15) for t to obtain (note change of sign and inversion of log term)

$$t = \frac{2.303}{k[I^-]_0} \log \frac{[S_2O_8^{2-}]_0}{[S_2O_8^{2-}]_t}$$

Insertion of numerical values gives

$$t = \frac{2.303}{1.70 \times 0.012} \log 2 = 34 \text{ min}$$

Readers should verify that units of t are indeed minutes.

For our second calculation we have the following:

$$\log \frac{[S_2O_8^{2-}]_t}{[S_2O_8^{2-}]_0} = -\frac{1.70 \times 0.024 \times 60}{2.303} = -2.45 = 0.55 - 3$$

$$\frac{[S_2O_8^{2-}]_t}{[S_2O_8^{2-}]_0} = 3.6 \times 10^{-3}$$

$$[S_2O_8^{2-}]_t = (3.6 \times 10^{-3})(1.6 \times 10^{-4}) = 5.8 \times 10^{-7}\ M \quad \blacksquare$$

Now let us consider the oxidation of iodide ions by hypochlorite ions in alkaline solution as investigated by Chia and Connick:

$$I^-(aq) + OCl^-(aq) \rightarrow OI^-(aq) + Cl^-(aq) \tag{22.17}$$

We might guess, partly by analogy with reaction (22.13), that the kinetics of this reaction will be in accord with

$$+\frac{d[OI^-]}{dt} = k[I^-][OCl^-] \tag{22.18}$$

Note that equation (22.18) is expressed in terms of the rate of formation of hypoiodite. Measurements of the rate of reaction in 1.00 *M* NaOH at 25°C led to results that were consistent with this rate law and $k = 61$ liter mole^{-1} sec^{-1}. Another series of measurements in 0.50 *M* NaOH also led to results consistent with (22.18), but in this case gave $k = 120$ liter mole^{-1} sec^{-1}. Still other measurements in 0.25 *M* NaOH led to $k = 230$ liter mole^{-1} sec^{-1}.

It is now apparent that our first guess was incomplete. The reaction rate depends on the concentration of hydroxide ions in a way that is consistent with

$$+\frac{d[OI^-]}{dt} = k\frac{[I^-][OCl^-]}{[OH^-]} \tag{22.19}$$

and with an average $k = 60$ sec^{-1}. Thus we see that rate equations sometimes involve chemical species that are neither consumed nor produced in the net reaction.

There are many important examples of reactions with rates that are drastically affected by substances that do not appear in the net balanced equation for the reaction. In general substances that increase the rate of reaction are called **catalysts,** and substances that decrease the rate are called **inhibitors.** Neither catalysts nor inhibitors affect equilibrium constants—they only change the rate of approach to equilibrium.

The Haber synthesis of NH_3 from N_2 and H_2 is an extremely important reaction that proceeds at a useful rate only in the presence of certain catalysts. Oxidation of ammonia to nitric acid is also a catalytic reaction. Another important catalytic oxidation is reaction of SO_2 with O_2 to yield SO_3 and thence sulfuric acid. Enzymes are essential catalysts for many biological reactions, and are also used in some industrial processes. Many addition polymerization reactions are only possible in presence of catalysts.

Reaction Rate Theory

The first reasonably satisfactory theory of reaction rates was largely originated by van't Hoff and Arrhenius in the years 1884–1889. In this theory and in later work along the same lines by others, it was postulated that the rate of reaction between molecules A and B in the gas phase is proportional to the number of collisions per unit time of A molecules with B molecules. The kinetic theory of gases shows that the rate of collision of A molecules with B molecules is proportional to the product of their concentrations, thus explaining the form of the second-order rate equation

$$-\frac{d[A]}{dt} = k_r[A][B]$$

Similarly, the second-order reaction of A molecules with A molecules is explained on the basis of bimolecular collisions, consistent with the rate equation

$$-\frac{d[\mathrm{A}]}{dt} = k_r[\mathrm{A}]^2$$

Here we have used k_r for the rate constant to distinguish it from the Boltzmann constant k.

Arrhenius and others have postulated that only those collisions having energy greater than some critical amount called the **activation energy** can be effective in leading to reaction. According to the quantitative kinetic theory of gases based on the Boltzmann equation (21.9), the fraction of molecules having energy greater than some energy ε_{a} is proportional to the Boltzmann factor $e^{-\varepsilon_{\mathrm{a}}/kT}$ in which ε_{a} represents the activation energy per molecule. Since it is more convenient to deal with energies per mole than energies per molecule, we often say that the fraction of molecules having energies greater than the activation energy is proportional to $e^{-\Delta E_{\mathrm{a}}/RT}$ in which ΔE_{a} represents the activation energy per mole. The rate constant is therefore proportional to $e^{-\Delta E_{\mathrm{a}}/RT}$ as in

$$k_r = \mathrm{A}e^{-\Delta E_{\mathrm{a}}/RT} \tag{22.20}$$

or

$$\ln k_{\mathrm{r}} = -\frac{\Delta E_{\mathrm{a}}}{RT} + \ln \mathrm{A} \tag{22.21}$$

Graphs of $\ln k_r$ against $1/T$ do give straight lines consistent with equation (22.21) and therefore in accord with the collision theory. The proportionality constant A is sometimes called the Arrhenius factor or pre-exponential factor.

The idea of an activation energy is illustrated in Figure 22.2 as a kind of barrier the reactant molecules have to hurdle to become products. This same idea has been illustrated in Figures 13.2–13.5, which were concerned with equilibrium distributions of idealized Mexican jumping beans.

As a result of many experimental and theoretical investigations of various reaction rates, a new approach called the **transition state theory** was developed by Evans, Polanyi, Wynne-Jones, and especially Eyring. Collision theories may be included in this newer transition state theory as special cases.

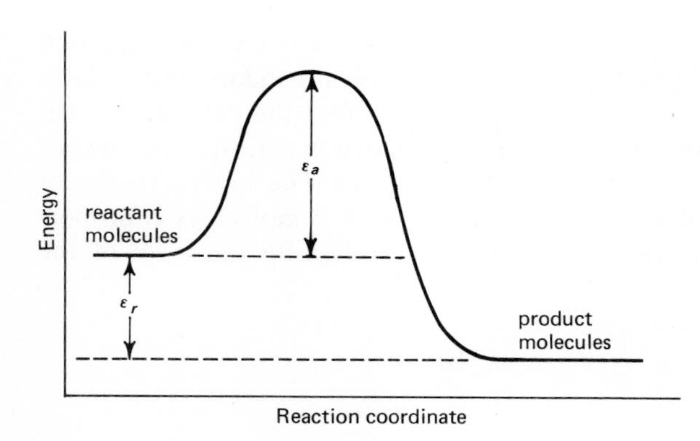

Figure 22.2. The activation energy per molecule is represented by ε_a. For the molar activation energy (ΔE_a) we have $N\varepsilon_a$. The energy difference between product and reactant is ε_r per molecule or $N\varepsilon_r$ per mole.

The transition state theory is specifically concerned with the "species" in the reaction process that corresponds to the maximum energy in the diagram in Figure 22.2. For example, consider a reaction

$$A + B \rightleftharpoons \text{products} \tag{22.22}$$

with rate described by a second-order rate equation. The species called the **activated complex** that corresponds to the maximum energy in Figure 22.2 has composition AB and is represented by $AB^{\ddagger}$. Equation (22.22) may therefore be written

$$A + B \rightleftharpoons AB^{\ddagger} \rightleftharpoons \text{products} \tag{22.23}$$

in which we have used $\rightleftharpoons$ to indicate reversible reactions for reasons discussed later.

In this theory the activated complex is considered to behave like other molecules in that it may move, rotate, and vibrate. Decomposition of $AB^{\ddagger}$ to reactants or products occurs when a particular vibration breaks the complex. The rate of reaction is therefore supposed to depend on the concentration of $AB^{\ddagger}$ and the frequency with which $AB^{\ddagger}$ breaks up to give products.

It is postulated in the transition state theory that the reactants are in thermodynamic equilibrium with the activated complex, as indicated by

$$A + B \rightleftharpoons AB^{\ddagger}$$

and

$$K^{\ddagger} = \frac{[AB^{\ddagger}]}{[A][B]} \tag{22.24}$$

where $K^{\ddagger}$ is an equilibrium constant.

Calculations based on the transition state model lead to $\sim kT/h$ (k is the Boltzmann constant and h is the Planck constant) as the frequency of the vibration that leads to decomposition of the activated complex. Letting κ (kappa) represent the probability that the vibration will lead to products rather than reactants, the rate of decomposition of the activated complex to give products is taken to be $\kappa kT/h$, in which κ is necessarily between 0 and 1 and is sometimes called a transmission factor.

Since the rate of reaction is given by the concentration of activated complex times the frequency of decomposition to products, we write

$$\text{reaction rate} = -\frac{d[A]}{dt} = \frac{\kappa kT}{h}[AB^{\ddagger}]$$

Now we solve (22.24) for $[AB^{\ddagger}] = K^{\ddagger}[A][B]$ and substitute in this equation to obtain

$$\text{reaction rate} = -\frac{d[A]}{dt} = \frac{\kappa kT}{h}K^{\ddagger}[A][B] \tag{22.25}$$

Comparison of (22.25) with the second-order rate equation

$$-\frac{d[A]}{dt} = k_r[A][B]$$

shows that we can identify the rate constant as

$$k_r = \frac{\kappa k T}{h} K^\ddagger \tag{22.26}$$

Application of thermodynamics to the equilibrium constant $K^\ddagger$ gives

$$-RT \ln K^\ddagger = \Delta H^\ddagger - T\Delta S^\ddagger$$

This equation can be rearranged to

$$K^\ddagger = e^{-\Delta H^\ddagger/RT} e^{\Delta S^\ddagger/R} \tag{22.27}$$

Substituting the right hand side of (22.27) for $K^\ddagger$ in (22.26) gives

$$k_r = \frac{\kappa k T}{h} e^{-\Delta H^\ddagger/RT} e^{\Delta S^\ddagger/R}$$

and thence

$$\ln k_r = -\frac{\Delta H^\ddagger}{RT} + \left[\frac{\Delta S^\ddagger}{R} + \ln\left(\frac{\kappa k T}{h}\right)\right] \tag{22.28}$$

Because the term in brackets changes very much less rapidly with changing temperature than does the term $\Delta H^\ddagger/RT$, we expect from (22.28) that a graph of $\ln k_r$ (or $\log k_r$) against $1/T$ should give a (very nearly) straight line with slope $-\Delta H^\ddagger/R$ (or $-\Delta H^\ddagger/2.3R$). Such graphs in general *do* give straight or nearly straight lines.

The quantity $\Delta H^\ddagger$ is called the **enthalpy of activation** and has physical significance similar to that of ΔE_a in Figure 22.2, except that the ordinate should now be enthalpy rather than energy. The difference between $\Delta H^\ddagger$ and ΔE_a is small for reactions taking place in solution and is calculated from ΔnRT for ideal gas reactions.

For most purposes we use the relationship between the rate constant k_r and the enthalpy of activation $\Delta H^\ddagger$ as either

$$\ln k_r = -\frac{\Delta H^\ddagger}{RT} + C \tag{22.29}$$

or

$$\frac{d \ln k_r}{dT} = \frac{\Delta H^\ddagger}{RT^2} \tag{22.30}$$

We may also combine two equations of the form of (22.29) or integrate (22.30) as a definite integral to obtain an equation of the form of (13.41) that was obtained by integration of $d \ln K/dT = \Delta H^0/RT^2$. The most common method of obtaining $\Delta H^\ddagger$ values is based on evaluation of the slope of a graph of $\ln k_r$ (or $\log k_r$) against $1/T$ as suggested by equation (22.29).

Mechanisms of Chemical Reactions

A complete account of the mechanism of a reaction would be something like a (slow motion) moving picture of the reactant atoms or molecules coming together, possibly in a complicated sequence of steps, breaking and forming bonds, and finally becoming products of the reaction. By "reaction mechanism" we mean the series of steps that lead to the overall reaction represented by the balanced net reaction equation. A *complete* reaction mechanism would also include a detailed description of the bonds that are broken and formed and the various rearrangements involved in formation and decomposition of the activated complex. In this book, however, we are limited to considering the sequence of reaction steps that lead to the overall reaction.

We now consider the rate equation for a particular reaction in relation to the reaction mechanism, and use this reaction to illustrate some definitions and useful procedures.

Decomposition of ozone to oxygen as represented by

$$2\,O_3(g) \rightarrow 3\,O_2(g)$$

has been reported to occur at a rate expressed by the equation

$$-\frac{d[O_3]}{dt} = \frac{k_r[O_3]^2}{[O_2]} \tag{22.31}$$

As we shall demonstrate, a reaction mechanism that is consistent with this rate equation is the following:

$$O_3 \underset{\text{equilibrium}}{\overset{\text{rapid}}{\rightleftharpoons}} O_2 + O \tag{22.32}$$

$$O + O_3 \xrightarrow{\text{rate determining step}} 2\,O_2 \tag{22.33}$$

Note that the sum of the steps in this postulated mechanism gives the overall chemical change represented by the balanced equation for the reaction.

The idea of a relatively slow rate determining step as in (22.33) is a common one in many fields. For example, the rate of traffic movement on a highway system may be limited by the slow rate of movement through one "bottleneck" such as a bridge or tunnel. Similarly, rate of manufacture of some instrument may be effectively limited and therefore determined by the rate at which some slow step in an otherwise fast procedure can be carried out. A chemical reaction may proceed by a more or less complicated sequence of steps. If one step is slow and all others are fast, it is the rate of the slow step that effectively limits or determines the rate of the reaction. We shall see that the form of the reaction rate equation in combination with the idea of the activated complex in transition state theory gives us a useful clue to the reaction mechanism.

Because we have postulated that the slow or rate determining step is represented by equation (22.33), we now write the rate of overall reaction as

$$-\frac{d[O_3]}{dt} = k_a[O][O_3] \tag{22.34}$$

in which we have used k_a to represent the rate constant for this step in the reaction. This rate equation contains [O] and is therefore not comparable to the rate equation (22.31) that was derived from results of experiments. We therefore write out the equilibrium constant expression for the reaction represented by (22.32) as

$$K = \frac{[O_2][O]}{[O_3]}$$

and solve for

$$[O] = \frac{K[O_3]}{[O_2]}$$

to substitute in (22.34) to obtain

$$-\frac{d[O_3]}{dt} = \frac{k_a K[O_3]^2}{[O_2]} \qquad (22.35)$$

Equations (22.31) and (22.35) lead to the rate constant k_r evaluated from the kinetic data being equal to $k_a K$.

Because equations (22.31) and (22.35) are of the same form, we can say that the proposed mechanism is consistent with the kinetic data as expressed by the rate equation. Since it is possible that some other proposed mechanism can also lead to a rate equation of the right form, we have not proven the correctness of the proposed mechanism. Proof (or disproof) of the proposed mechanism could take the form of determination or calculation of K coupled with independent evaluation of k_a to find out if k_r really does equal $k_a K$.

Although it is possible to deduce a unique rate equation from a proposed mechanism, one cannot uniquely determine a mechanism from a rate equation. It is therefore possible to disprove a proposed mechanism by showing that it leads to a rate equation of form different than that derived from experiment. Because complete proof of a mechanism is very difficult, we usually stop with a chemically reasonable mechanism that can be shown to lead to a rate equation of the same form as the rate equation derived directly from experiment.

The slow step in the postulated mechanism (equation 22.33) involves a total of four atoms of oxygen. Thus we may say that the activated complex in the decomposition of ozone has the composition O_4. It should be clear that it will be a great help in deducing a reasonable mechanism from a rate equation if we have some easy way of knowing the overall composition of the activated complex. On the basis of the one example we have been discussing (and other examples that could be cited) and the form of K in (22.24) we can now give the desired rule as follows:

The atoms involved in the slow step of a reaction can be evaluated as the sum of those represented in the numerator of the rate equation minus those represented in the denominator of the rate equation.

Example Problem 22.2. Burns and Dainton have investigated the kinetics of the oxidation of CO by Cl_2 to yield phosgene as in

$$CO(g) + Cl_2(g) \rightarrow COCl_2(g)$$

and found their experimental results to be represented by the rate equation

$$\frac{d[COCl_2]}{dt} = k_r[Cl_2]^{\frac{3}{2}}[CO]$$

Postulate a mechanism that is consistent with this rate equation.

According to the rule above, the activated complex in this reaction has overall composition Cl_3CO. A reasonable suggestion for the slow step in this reaction is therefore

$$COCl(g) + Cl_2(g) \rightarrow COCl_2(g) + Cl(g)$$

This equation alone does not constitute a mechanism; we must now postulate reasonable rapid steps to go with this slow step.

A mechanism that has been suggested for this reaction is as follows:

$$Cl_2(g) \xrightleftharpoons[K_1]{\text{rapid equilibrium}} 2\ Cl(g)$$

$$Cl(g) + CO(g) \xrightleftharpoons[K_2]{\text{rapid equilibrium}} COCl(g)$$

$$COCl(g) + Cl_2(g) \xrightarrow[k_a]{\text{rate determining}} COCl_2(g) + Cl(g)$$

Now we show that this mechanism is indeed consistent with the experimental rate equation.

On the basis of the slow step in this mechanism, we first write

$$\frac{d[COCl_2]}{dt} = k_a[COCl][Cl_2] \tag{22.36}$$

The equilibrium constant expression for formation of COCl is written down and solved for

$$[COCl] = K_2[Cl][CO]$$

We similarly obtain

$$[Cl] = (K_1)^{\frac{1}{2}}[Cl_2]^{\frac{1}{2}}$$

and substitute in the expression above to obtain

$$[COCl] = K_2(K_1)^{\frac{1}{2}}[Cl_2]^{\frac{1}{2}}[CO] \tag{22.37}$$

Finally, substitution of (22.37) in (22.36) gives

$$\frac{d[COCl_2]}{dt} = k_a K_2(K_1)^{\frac{1}{2}}[Cl_2]^{\frac{3}{2}}[CO] \tag{22.38}$$

Now we identify $k_a K_2(K_1)^{\frac{1}{2}}$ in (22.38) with the experimental rate constant k_r and see that the proposed mechanism is consistent with the kinetic evidence. Have we proved this mechanism? NO! We have, however, shown that this proposed mechanism is consistent with the experimental rate equation and therefore may be a good representation of the true reaction mechanism. Further, we might progress toward proof of this mechanism by detecting the intermediate COCl or by separate evaluation of K_1, K_2, and k_a. ■

Example Problem 22.3. Propose a mechanism for the reaction

$$I^-(aq) + OCl^-(aq) \rightarrow OI^-(aq) + Cl^-(aq)$$

that is consistent with the experimental rate equation

$$+\frac{d[OI^-]}{dt} = k_r \frac{[I^-][OCl^-]}{[OH^-]}$$

At first thought we are unable to apply the rule that relates the form of the rate equation to the composition of the activated complex because we cannot carry out the subtraction IOCl − OH. But we should remember that $[I^-]$ represents the concentration of $I^-(aq)$ ions—that is, the concentration of hydrated ions in solution. Now, recognizing that there are unspecified and usually unknown numbers of water molecules associated with all aqueous species, we add one water to the composition indicated by the numerator and carry out the subtraction $IOClH_2O - OH = IOClH$ to obtain information about the composition of the activated complex. As with other aqueous species, when we say that the activated complex is IOClH(aq), we really mean $IOClH \cdot nH_2O$ in solution, with n usually unknown.

A reasonable slow step involving the atoms represented by IOClH is

$$I^-(aq) + HOCl(aq) \rightarrow HOI(aq) + Cl^-(aq)$$

Both HOCl(aq) and HOI(aq) are known. Both are weak acids that exist mostly as $OX^-(aq)$ ions in alkaline solution, consistent with the way we have written the net equation for the overall reaction, but both can exist in small concentrations in alkaline solutions. Now we must put this proposed rate determining reaction together with possible fast reactions that will lead to the overall observed reaction.

Here is one combination of reactions that appears to be consistent with the information we have about the reaction:

$$OCl^-(aq) + H_2O(liq) \rightleftharpoons HOCl(aq) + OH^-(aq) \qquad \text{rapid equilibrium}$$
$$I^-(aq) + HOCl(aq) \rightarrow HOI(aq) + Cl^-(aq) \qquad \text{rate determining}$$
$$OH^-(aq) + HOI(aq) \rightleftharpoons H_2O(liq) + OI^-(aq) \qquad \text{rapid equilibrium}$$

Because conversion of HOI(aq) to $OI^-(aq)$ by the third reaction is rapid, we can write

$$\frac{d[OI^-]}{dt} = k_a[I^-][HOCl]$$

in which k_a represents the rate constant for the proposed rate determining step. Now we write down the equilibrium constant expression for the proposed first rapid equilibrium and solve for

$$[HOCl] = K_1\frac{[OCl^-]}{[OH^-]}$$

and substitute this in our rate equation above to obtain

$$\frac{d[OI^-]}{dt} = k_aK_1\frac{[I^-][OCl^-]}{[OH^-]}$$

This rate equation is of the same form as the experimental rate equation and leads to $k_r = k_aK_1$. Readers may see (possibly after reviewing in Chapter 14) that $K_1 = K_w/K_{HOCl}$ so that we can obtain a numerical value for K_1 from information already available.

Another "reasonable" mechanism that is consistent with the experimental rate equation is

$$OCl^-(aq) + H_2O(liq) \rightleftharpoons HOCl(aq) + OH^-(aq) \qquad \text{rapid equilibrium}$$
$$I^-(aq) + HOCl(aq) \rightarrow ICl(aq) + OH^-(aq) \qquad \text{rate determining}$$
$$ICl(aq) + 2\,OH^-(aq) \rightleftharpoons OI^-(aq) + Cl^-(aq) + H_2O(liq) \qquad \text{rapid equilibrium}$$

This second proposed mechanism differs from the first only in what is supposed to happen after formation of the activated complex from I^- and HOCl (and $n\,H_2O$). Our information from kinetic data is insufficient to tell us what happens after the activated complex is formed. ■

Example Problem 22.4. Many organic halogen compounds react with water (are "hydrolyzed") to yield the corresponding alcohol, as indicated by the reaction equation

$$RCl(aq) + H_2O(liq) \rightarrow ROH(aq) + H^+(aq) + Cl^-(aq)$$

Investigations of rates of a number of such reactions have led to rate equations of the form

$$-\frac{d[RCl]}{dt} = +\frac{d[ROH]}{dt} = \frac{k_a[RCl]}{1 + k_b[Cl^-]}$$

Because the concentration of water is effectively constant when these reactions are carried out in aqueous solution, $[H_2O]$ does not appear in the rate equation, and we have no information from this source about hydration of the various species involved in the reaction.

Propose a "reasonable" mechanism that is consistent with the rate equation above.

At very low $[Cl^-]$ when $1 + k_b[Cl^-] \cong 1$, the activated complex has composition RCl. For larger $[Cl^-]$ when $1 + k_b[Cl^-] \cong k_b[Cl^-]$, the composition of activated complex is $RCl - Cl^- = R^+$. A mechanism that is consistent with RCl and R^+ as activated complexes is

$$RCl(aq) \underset{k_2}{\overset{k_1}{\rightleftharpoons}} R^+(aq) + Cl^-(aq) \tag{22.39}$$

$$R^+(aq) + H_2O(liq) \xrightarrow{k_3} ROH(aq) + H^+(aq) \tag{22.40}$$

To simplify the mathematics of finding the rate equation associated with the proposed mechanism, we assume that the concentration of $R^+(aq)$ reaches a *steady state* value so that $d[R^+]/dt = 0$. This steady state treatment is justified if $[R^+]$ is sufficiently small (as in this case) and may be justified in certain other cases.

The rate of appearance of $R^+(aq)$ due to the forward part of (22.39) is

$$\frac{d[R^+]}{dt} = k_1[RCl]$$

The rates of disappearance of $R^+(aq)$ due to the backward part of (22.39) and to (22.40) are

$$-\frac{d[R^+]}{dt} = k_2[R^+][Cl^-]$$

and

$$-\frac{d[R^+]}{dt} = k_3[H_2O][R^+] = k_3'[R^+]$$

In the proposed steady state the net $d[R^+]/dt = 0$. We therefore combine the three equations above to obtain

$$k_1[RCl] - k_2[R^+][Cl^-] - k_3'[R^+] = 0$$

This equation is rearranged and solved for

$$[R^+] = \frac{k_1[RCl]}{k_3' + k_2[Cl^-]} \tag{22.41}$$

According to (22.40), the rate of appearance of alcohol ROH is

$$\frac{d[ROH]}{dt} = k_3[H_2O][R^+] = k_3'[R^+]$$

We substitute (22.41) in this expression to obtain

$$\frac{d[\text{ROH}]}{dt} = \frac{k_3' k_1[\text{RCl}]}{k_3' + k_2[\text{Cl}^-]} = \frac{k_1[\text{RCl}]}{1 + (k_2/k_3')[\text{Cl}^-]} \quad \blacksquare$$

This rate equation is of the same form as the experimental rate equation and permits identification of k_1 with k_a and k_2/k_3' with k_b.

Reaction Rates and Equilibrium

We shall begin our considerations of the relationship between rates and equilibria by looking at one specific case for which the equations are simple:

$$\text{A} + \text{B} \rightleftharpoons \text{C} + \text{D}$$

Suppose that the rate of the forward reaction has been investigated under conditions such that the reverse or backward rate is negligible and that the rate of the forward reaction is described by

$$-\frac{d[\text{A}]}{dt} = k_{\text{f}}[\text{A}][\text{B}] \tag{22.42}$$

This rate equation shows that the composition of the activated complex is the sum of the compositions of A and B.

Now suppose that the rate of the reverse or backward reaction has been investigated under such conditions that the forward reaction is negligible and that the rate of this backward reaction is described by

$$+\frac{d[\text{A}]}{dt} = k_{\text{b}}[\text{C}][\text{D}] \tag{22.43}$$

The composition of the activated complex for this backward reaction is the sum of the compositions of C and D, which is the same composition as the activated complex for the forward reaction. As we shall see later, the activated complex *must* have the same composition for forward and backward reactions.

We now assume that the rate equations (22.42) and (22.43) adequately represent rates of forward and backward reactions over the whole range of reaction. Knowing that the forward rate equals the reverse rate at equilibrium, we equate the right sides of (22.42) and (22.43) to obtain

$$k_{\text{f}}[\text{A}][\text{B}] = k_{\text{b}}[\text{C}][\text{D}]$$

and rearrange to

$$\frac{k_{\text{f}}}{k_{\text{b}}} = \frac{[\text{C}][\text{D}]}{[\text{A}][\text{B}]}$$

Recognizing that [C][D]/[A][B] is the equilibrium constant expression for the reaction, we write

$$\frac{k_{\text{f}}}{k_{\text{b}}} = K \tag{22.44}$$

Although we have derived equation (22.44) for a very special case, this important result is quite general.

The general validity of (22.44) and the statement that the activated complex *must* have the same composition for forward and backward reactions are based on the **principle of microscopic reversibility,** which has also been called the principle of detailed balancing, the law of entire equilibrium, and the law of reversibility to the last detail. In 1925 G. N. Lewis expressed this statement as follows: *Corresponding to every individual process there is a reverse process, and in a state of equilibrium the average rate of every process is equal to the average rate of its reverse process.* A more recent statement by E. L. King is: *A pathway for a forward reaction is also a pathway for the reverse reaction. In a system at equilibrium, not only is the total rate of the forward reaction equal to the total rate of the reverse reaction, but also the forward rate by each pathway is equal to the reverse rate by the same pathway.*

We illustrate this principle by way of an illustration developed by G. M. Fleck. Consider a fast gas phase reaction

$$X(g) \rightarrow Y(g) + Z(g) \tag{22.45}$$

and the slow reverse reaction

$$Y(g) + Z(g) \rightarrow X(g) \tag{22.46}$$

Suppose that a catalyst D is found to increase the rate of this reverse reaction without having any effect on the rate of the forward reaction.

We can now imagine placing chemical X and catalyst D in the piston-cylinder apparatus pictured in Figure 22.3. As fast reaction (22.45) proceeds, the number of moles of gas increases and pushes the piston upwards. The movement of the piston permits D to catalyse the reverse reaction (22.46) and thereby decrease the number of moles of gas, which allows the external atmospheric pressure to push the piston downward and close the box. When the box is closed so that D cannot catalyze reaction (22.46), the rapid reaction (22.45) takes over again. The result is that the piston goes up and down. Now we need only connect the

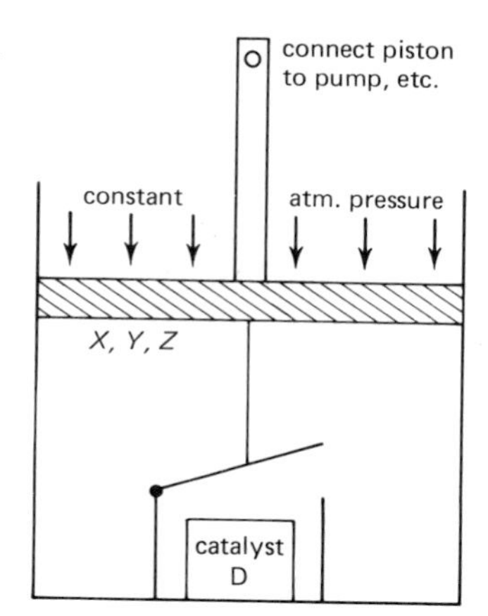

Figure 22.3. Schematic illustration of a useful perpetual motion machine based on reactions (22.45) and (22.46) and the hypothetical (impossible!) catalyst D.

piston to a pump and we have achieved a perpetual motion machine that will do useful work forever without any input of energy.

On the basis of our earlier discussions of thermodynamics, we are confident that this proposed machine is impossible—too good to be true. We must therefore conclude that the catalyst D cannot function as we have described it above. More particularly, we conclude that the equilibrium constant cannot be affected by the catalyst and that the catalyst must therefore have similar effects on both the forward and reverse reactions. The result of all this is the principle of microscopic reversibility.

According to the principle of microscopic reversibility, there can be no reactions of the sort we have often written with a single arrow $\rightarrow$. Instead, each of these reactions must be reversible as indicated for many reactions by $\rightleftharpoons$. Our practical justification for writing $\rightarrow$ and considering only the forward parts of many reactions has been that under the conditions of interest only the forward reaction occurred to an appreciable extent, or we had chosen to focus our attention only on this part of a complete reversible reaction.

Now we consider application of (22.44) to specific reactions.

Nitrous acid decomposes in acidic solution as indicated by

$$3\,HNO_2(aq) \rightleftharpoons H^+(aq) + NO_3^-(aq) + 2\,NO(g) + H_2O(liq) \tag{22.47}$$

Measurements on this reaction under different conditions have shown that the forward rate is given by $k_f[HNO_2]^4/(P_{NO})^2$ and that the backward rate is given by $k_b[H^+][NO_3^-][HNO_2]$. Thus the net rate of reaction can be expressed as

$$\frac{d[NO_3^-]}{dt} = \frac{k_f[HNO_2]^4}{(P_{NO})^2} - k_b[H^+][NO_3^-][HNO_2] \tag{22.48}$$

At equilibrium the net rate of reaction is zero, so that we can set $d[NO_3^-]/dt = 0$ and rearrange (22.48) to

$$\frac{k_f}{k_b} = \frac{[H^+][NO_3^-](P_{NO})^2}{[HNO_2]^3}$$

Again we may identify k_f/k_b with the equilibrium constant K.

One of the earliest and most convincing demonstrations of the validity of $k_f/k_b = K$ was given by van't Hoff who made use of experimental results obtained by Knoblauch in 1897. Knoblauch had investigated the rate of esterification of acetic acid by ethyl alcohol and also the rate of hydrolysis of ethyl acetate. The results permitted van't Hoff to calculate values of k_f and k_b for

$$H_3C{-}\overset{\overset{\displaystyle O}{\|}}{C}{-}O{-}H(aq) + H{-}O{-}CH_2{-}CH_3(aq) \underset{k_b}{\overset{k_f}{\rightleftharpoons}} H_3C{-}\overset{\overset{\displaystyle O}{\|}}{C}{-}O{-}CH_2{-}CH_3(aq) + H_2O(liq)$$

The value of $K = k_f/k_b$ that he then calculated agreed with the value based on direct evaluation from the compositions of solutions at equilibrium.

References

Arnot, C. L., Activated complex theory of bimolecular gas reactions. *J. Chem. Educ.*, **49**:480 (1972).

Back, M. A., and K. J. Laidler: *Chemical Kinetics.* Pergamon Press, Elmsford, N.Y., 1967 (paperback).

Birk, J. P., Mechanistic implications and ambiguities of rate laws, *J. Chem. Educ.*, **47**:805 (1970).

Chesick, J. P., and A. Patterson, Jr.: Determination of reaction rates with an a.c. conductivity bridge (a student experiment). *J. Chem. Educ.*, **37**:242 (1960).

Clarke, J. R., The kinetics of the bromate-bromide reaction. *J. Chem. Educ.*, **47**:775 (1970).

Edwards, J. O., *Inorganic Reaction Mechanisms: An Introduction.* W. A. Benjamin, Inc., New York, 1964 (paperback).

Edwards, J. O., Reaction rates for practical chemists. *J. Chem. Educ.*, **34**:47 (1957).

Eyring, H., and E. M. Eyring: *Modern Chemical Kinetics.* Van Nostrand Reinhold, New York, 1963 (paperback).

Finlayson, M. E., and D. G. Lee: Oxidation of ethanol by chromium(VI). *J. Chem. Educ.*, **48**:473 (1971).

Fleck, G. M.: *Chemical Reaction Mechanisms.* Holt, Rinehart and Winston, Inc., New York, 1971.

Hammes, G. G., and L. E. Erickson: Kinetic studies of systems at equilibrium. *J. Chem. Educ.*, **35**:611 (1958).

Harris, G. M.: *Chemical Kinetics.* Heath, Boston, 1966 (paperback).

Hedrick, C. E.: Formation of the chromium-EDTA complex (an undergraduate kinetics experiment). *J. Chem. Educ.*, **47**:231 (1970).

Herbrandson, H. F., The hydrolysis of t-butyl chloride. *J. Chem. Educ.*, **48**:706 (1971).

Herman, I. J., and A. Lifshitz: The preparation and kinetics of aquation of bromopentaaquochromium(III). *J. Chem. Educ.*, **47**:231 (1970).

King, E. L.: *How Chemical Reactions Occur.* W. A. Benjamin, Inc., New York, 1963 (paperback).

Liu, M. T. H., Kinetics of the oxidation of benzyl alcohol. *J. Chem. Educ.*, **48**:702 (1971).

Problems

1. Rate measurements have shown that the decomposition of $SO_2Cl_2(g)$ to $SO_2(g)$ and $Cl_2(g)$ is first-order, with $k_r = 2 \times 10^{-5}\ \text{sec}^{-1}$ at 320°C. What fraction of a sample of SO_2Cl_2 maintained at 320°C will decompose in 1.0 hr? How long would it take for half of a sample of $SO_2Cl_2(g)$ to decompose at 320°C?
2. The first-order decomposition of cyclobutane to yield ethylene has been investigated over a range of temperatures and it has been found that the first-order rate constant can be expressed as $k_r = Ae^{-31,500/T}$. Evaluate $\Delta H^\ddagger$ for this reaction.
3. An increase of 10°C (in the neighborhood of room temperature) approximately doubles the rates of many reactions. Calculate an approximate $\Delta H^\ddagger$ for such a reaction.
4. If a reaction goes twice as fast at 10°C as at 0°C, by what factor is the rate increased as the temperature is raised by 100°C to 110°C? Evaluate $\Delta H^\ddagger$ for this reaction.
5. Radioactive decay as represented generally by equation (22.1) has been established to be a first-order process. The half-life for radioactive decay of $^{64}_{29}Cu$ by way of β^- emission has been found to be 12.8 hours. What is the value (number, with units) for the rate constant for this process?
6. The half-life for radioactive decay of $^{90}_{38}Sr$ is 20 years. How long does it take for 90% of the radioactivity in a sample of $^{90}_{38}Sr$ to disappear? Why should anyone care about the results of this calculation?

7. It has been suggested that the mechanism of reaction (22.47) is as follows:

$$2\,HNO_2(aq) \rightleftharpoons NO(g) + NO_2(aq) + H_2O(liq) \quad (22.49)$$
$$2\,NO_2(aq) \rightleftharpoons N_2O_4(aq) \quad (22.50)$$
$$N_2O_4(aq) + H_2O(liq) \rightleftharpoons HNO_2(aq) + H^+(aq) + NO_3^-(aq) \quad (22.51)$$

It is postulated that reactions (22.49) and (22.50) are rapid equilibria with equilibrium constants K_1 and K_2, and that (22.51) is rate determining with forward and backward rate constants represented by k_1 and k_{-1} respectively.
(a) What can you say about the formula of the activated complex for this reaction?
(b) Show that the proposed mechanism is (or is not) consistent with the rate equation (22.48). (*Hint:* it may be easiest to begin by considering the backwards reaction first.)

8. An important reaction in analytical chemistry is

$$I_3^-(aq) + 2\,S_2O_3^{2-}(aq) \rightarrow S_4O_6^{2-}(aq) + 3\,I^-(aq).$$

It has been suggested that the mechanism for this reaction is

$$S_2O_3^{2-}(aq) + I_3^-(aq) \rightleftharpoons S_2O_3I^-(aq) + 2\,I^-(aq)$$
$$S_2O_3I^-(aq) + S_2O_3^{2-}(aq) \rightarrow S_4O_6^{2-}(aq) + I^-(aq)$$

The first reaction is supposed to be a rapid equilibrium, while the second is supposed to be rate determining.
On the basis of this proposed mechanism, derive an equation for $-d[I_3^-]/dt$.

9. It has been reported that the rate of the reaction

$$S_2O_8^{2-}(aq) + 2\,VO^{2+}(aq) + 4\,H_2O(liq) \rightarrow 2\,HVO_3(aq) + 2\,SO_4^{2-}(aq) + 6\,H^+(aq)$$

in the presence of $Ag^+(aq)$ catalyst is accurately represented by

$$-\frac{d[S_2O_8^{2-}]}{dt} = k_r[S_2O_8^{2-}][Ag^+]$$

It has also been reported that the rate of the reaction

$$3\,S_2O_8^{2-}(aq) + 2\,Cr^{3+}(aq) + 7\,H_2O(liq) \rightarrow Cr_2O_7^{2-}(aq) + 6\,SO_4^{2-}(aq) + 14\,H^+(aq)$$

is represented by the same rate equation when $Ag^+(aq)$ is present as catalyst. Suggest mechanisms for these reactions. (*Hint:* silver in oxidation state > 1 is known to be a rapid oxidizing agent.)

10. In many reactions between a solid and a gas (or solution), a solid product is formed that covers the surface. The resulting protective covering causes the reaction rate to decrease as reaction proceeds because the protective coating becomes thicker and more protective. It seems reasonable to propose that the rate equation for such a reaction might be

$$\frac{dA}{dt} = \frac{k}{A}$$

in which A represents the extent of reaction, which is proportional to coating thickness. The rate of oxidation of metallic nickel by O_2 has been described by

$$A^2 = k_r t$$

Is this experimental rate equation consistent with the above proposed differential rate equation for solid-gas reactions?

11. Suppose that both forward and backward rates of some reaction have been extensively investigated so that numerical values of both k_f and k_b are available at several temperatures. How can you use these values in calculating ΔH^0 for the reaction? (*Hint:* Consider Figure 22. 4.)

12. Suppose that $\Delta H_f^‡ = A$ and $\Delta H_b^‡ = B$ for some uncatalyzed reaction (where subscripts f, b, and r represent forward, backward, and reaction). Express ΔH^0 for the reaction in terms of A and B. Now suppose that some catalyst is found that lowers $\Delta H_f^‡$ to $A/2$. What is the value of $\Delta H_b^‡$ for the catalyzed reaction? How do you know that ΔH_r^0 for the reaction is the same for the catalyzed reaction as for the uncatalyzed reaction? (See Figure 22.4.)

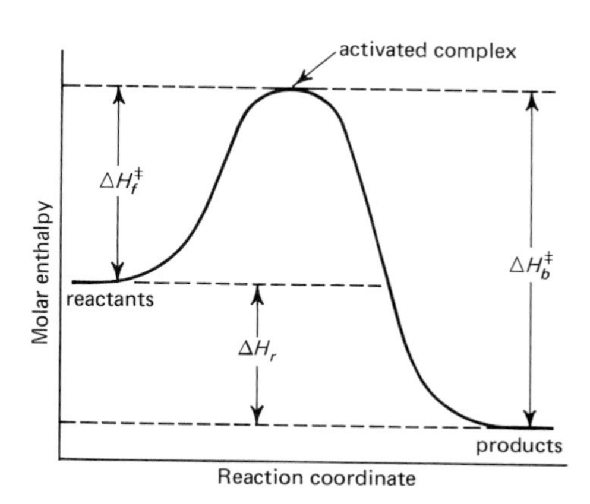

Figure 22.4. Illustration showing relationship between $\Delta H^‡$ for forward and backward reactions and ΔH of reaction. (Subscripts f = forward, b = backward, r = reaction.)

13. For the reaction

$$H_2PO_2^-(aq) + OH^-(aq) \rightarrow HPO_3^{2-}(aq) + H_2(g)$$

we have

$$-\frac{d[H_2PO_2^-]}{dt} = k_r[H_2PO_2^-][OH^-]^2$$

Suggest a mechanism that is consistent with this rate equation.

14. In 1916 Langmuir applied the idea of equal forward and reverse rates at equilibrium to adsorption of gases on solid surfaces. He considered a solid surface to have a large number of identical adsorption sites that can be occupied by adsorbed molecules, and also assumed that the presence of an adsorbed molecule on one site has no effect on adsorption at neighboring sites. He represented the fraction of occupied sites by θ.

Because the rate of collision of gas molecules with the surface is proportional to concentration or pressure of the gas and the chance of a colliding molecule hitting an empty site is proportional to $(1 - \theta)$, the rate of adsorption can be expressed as $k_aP(1 - \theta)$. The rate of desorption is given by $k_d\theta$ in which k_d represents the rate of desorption from completely covered surface.

(a) Equate the two rates and show that

$$\theta = \frac{bP}{1 + bP}$$

in which $b = k_a/k_d$.

(b) Since the amount of gas adsorbed is proportional to θ, we also have

$$A = \frac{abP}{1 + bP}$$

in which A represents the amount of gas adsorbed and a is a proportionality constant. Rearrange this equation to show that a graph of $1/A$ against $1/P$ should be linear. Explain how to evaluate a and b graphically.

15. Chemists often assume that lowering the temperature at which a reaction is carried out will increase the proportion of the main reaction product and decrease the proportion of by-products produced by way of side reactions. Provide kinetic justification for this assumption in terms of a ratio of rate constants on the basis of the following generalities. (i) Rate constants can be expressed in terms of equation (22.29). (ii) Values of C for main reaction and side reaction are often about the same. (iii) $\Delta H^{\ddagger}$ for the main reaction is commonly smaller than $\Delta H^{\ddagger}$ for the side reaction. Note that $\Delta H^{\ddagger}_{m} < \Delta H^{\ddagger}_{s}$ (m for main reaction and s for side reaction) is consistent with the idea that $k_m > k_s$ and our labels of "main" and "side" for the two reactions.

16. Halpern, Legare, and Lumry have made measurements leading to equilibrium constants and rate constants at different temperatures for the reaction

$$Fe(DMP)_3^{2+}(aq) + IrCl_6^{2-}(aq) \rightleftharpoons Fe(DMP)_3^{3+}(aq) + IrCl_6^{3-}(aq)$$

in which DMP represents a large organic molecule bound to iron. Their results lead to (at 18°C) the following: $K = 0.40$, $\Delta H^0 = -5.3$ kcal mole^{-1}, $k_f = 1.05 \times 10^9\ M^{-1}$ sec^{-1}, $k_b = 2.5 \times 10^9\ M^{-1}$ sec^{-1}, $\Delta H^{\ddagger}_{f} = 0.5$ kcal mole^{-1}, and $\Delta H^{\ddagger}_{b} = 6$ kcal mole^{-1}.

Carry out calculations to show that the kinetic results are (or are not) in reasonable agreement with the equilibrium results.

17. Rates of the reaction

$$Tl^{3+}(aq) + 2\ Fe^{2+}(aq) \rightarrow Tl^{+}(aq) + 2\ Fe^{3+}(aq)$$

have been investigated several times as follows.

(a) The first investigations led to a rate equation of the form

$$\frac{d[Tl^{+}]}{dt} = k_r[Tl^{3+}][Fe^{2+}] \tag{22.52}$$

Are either or both of the following mechanisms consistent with this rate equation?

$$\begin{cases} Fe^{2+}(aq) + Tl^{3+}(aq) \rightarrow Fe^{3+}(aq) + Tl^{2+}(aq) & \text{slow} \\ Fe^{2+}(aq) + Tl^{2+}(aq) \rightarrow Fe^{3+}(aq) + Tl^{+}(aq) & \text{fast} \end{cases}$$

$$\begin{cases} Fe^{2+}(aq) + Tl^{3+}(aq) \rightarrow Fe^{4+}(aq) + Tl^{+}(aq) & \text{slow} \\ Fe^{2+}(aq) + Fe^{4+}(aq) \rightarrow 2\ Fe^{3+}(aq) & \text{fast} \end{cases}$$

(b) More extensive investigations by Ashurst and Higginson showed that (22.52) adequately represents the *initial* rate of reaction, but that a more complicated equation is required to describe the rate of the reaction as it proceeds further or is carried out in the presence of a substantial initial concentration of $Fe^{3+}(aq)$. This rate equation can be written

$$\frac{d[Tl^{+}]}{dt} = \frac{k_1 k_3 [Tl^{3+}][Fe^{2+}]^2}{k_2[Fe^{3+}] + k_3[Fe^{2+}]} \tag{22.53}$$

Show that this rate equation is consistent with the following postulated mechanism:

$$Tl^{3+}(aq) + Fe^{2+}(aq) \underset{k_2}{\overset{k_1}{\rightleftharpoons}} Tl^{2+}(aq) + Fe^{3+}(aq)$$

$$Tl^{2+}(aq) + Fe^{2+}(aq) \xrightarrow{k_3} Tl^{+}(aq) + Fe^{3+}(aq)$$

Both k_2 and k_3 are supposedly much larger than k_1 so that a small steady state of $Tl^{2+}(aq)$ is obtained.

(c) Explain why (22.53) is equivalent to (22.52) for the initial rate of reaction.

(d) Are the rate equation (22.53) and the proposed mechanism in (b) consistent with the experimental observations that a large initial concentration of $Fe^{3+}(aq)$ makes the reaction slower while $Tl^{+}(aq)$ has no effect on the rate?

18. In the absence of any other information, it is reasonable to wonder if the rate of decomposition of N_2O_5 might be expressed by $-d[N_2O_5]/dt = k[N_2O_5]^2$. Integrate this equation to obtain an equation similar to (22.11) and then test by making an appropriate graph based on the data in Table 22.1. Then either evaluate the constant of integration by setting $t = 0$ or carry out a definite integration to obtain an equation like (22.12). Solve this equation for k and use all of the data in Table 22.1 for several evaluations of k. If this k is not (approximately) constant, we can conclude that the second-order rate equation above is not applicable to this reaction.

19. Make a graph based on an equation like (22.3) or do numerical calculations based on an equation like (22.4) to show that reaction (22.13) is first-order with respect to $[S_2O_8^{2-}]$.

20. Crowell and Hammett have investigated the rate of reaction of propyl bromide with thiosulfate in aqueous alcohol solution for which we write the balanced equation

$$C_3H_7Br + S_2O_3^{2-} \rightarrow C_3H_7S_2O_3^- + Br^-$$

Their experimental results could be described by the equation

$$-\frac{d[\mathrm{PrBr}]}{dt} = k_r[\mathrm{PrBr}][S_2O_3^{2-}]$$

with $k_r = 1.6 \times 10^{-3}$ liter mole^{-1} sec^{-1}.

(a) Some people would write $k_r = 1.6 \times 10^{-3}\ M^{-1}$ sec^{-1} for this reaction. Are these units consistent with those given above?

(b) Express the rate constant for this reaction in terms of minutes.

(c) Suppose that initial concentrations for this reaction are 0.10 M for $S_2O_3^{2-}$ and 0.002 M for C_3H_7Br. How long will it take for half of the C_3H_7Br to react? Note that the concentration of $S_2O_3^{2-}$ is very nearly constant throughout the experiment.

21. A common method for quantitative determination of hydrogen peroxide in aqueous solution is based on titration with permanganate solution and the reaction

$$2\ MnO_4^-(aq) + 5\ H_2O_2(aq) + 6\ H^+(aq) \rightarrow 2\ Mn^{2+}(aq) + 5\ O_2(g) + 8\ H_2O(liq)$$

When the first drop of MnO_4^- is added to begin a titration, the pink color persists long enough that an impatient experimenter might conclude that the solution contains no H_2O_2. Thus we see that the reaction represented above is slow. But the pink color due to the second drop of MnO_4^- disappears quicker, and by the time a few more drops have been added, the reaction appears to be practically instantaneous. Can you explain these observations?

Appendix I

ELEMENTARY CALCULUS USED IN GENERAL CHEMISTRY

Only elementary differential and integral calculus is used in this book. The mathematical operations involved are simple and easily learned from the brief discussion that follows and examples in several chapters. On the other hand, the meaning of these operations and the insight that directs their application to chemical problems is more difficult to grasp and requires special attention. In the following discussion we summarize some of the relevant operations of calculus and illustrate the meanings of these operations. Detailed discussions of the background reasoning that leads to the use of calculus in chemistry are given in connection with specific applications in Chapters 11, 12, 13, 16, 21, and 22.

We begin by considering the slope at point P of the line in Figure I.1. The dashed line through point P is the tangent to the curve at this point and has

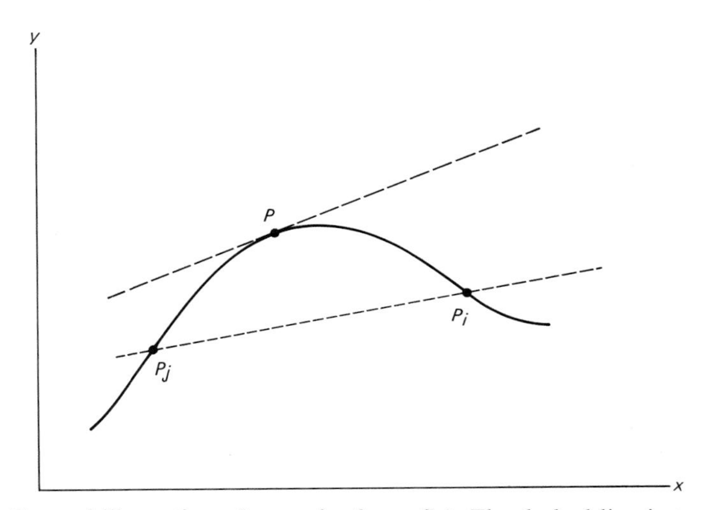

Figure I.1. General illustration of a graph of $y = f(x)$. The dashed line is tangent to the curve at point P and has the same slope as the curve at this point. The dotted line will approach the dashed line and therefore will have slope that approaches that of the curve at point P as the points P_i and P_j move closer to point P.

the same slope that the curve has at this point. The dotted line in Figure I.1 clearly has a different slope. But as the points P_i and P_j that define the dotted line move closer and closer to point P, the dotted line moves closer and closer to the dashed line, and the slope of the dotted line more and more closely approximates the slope of the dashed line. We thus see that the desired slope of the dashed line can be obtained by calculating the slope of the dotted line when its two defining points P_i and P_j are very close to point P.

In order to start the suggested calculation, we write (as in Figure I.2)

$$m = \frac{y_i - y_j}{x_i - x_j} \tag{I.1}$$

for the slope of the dotted line. This equation may also be written as

$$m = \frac{\Delta y}{\Delta x} \tag{I.2}$$

in which Δ indicates the increment or change or difference in the following quantity. For a straight line (a line having constant slope), we may choose any points P_i and P_j on the line and then calculate the slope correctly by means of either equation (I.1) or (I.2). However, according to the discussion in the paragraph above, we can apply equations (I.1) and (I.2) to calculate the slope of a curved line (one that has different slopes at different points) only when the points P_i and P_j on the line are chosen very close to the point P of interest so that the differences in y and x (represented by Δy and Δx) are both very small.

Instead of Δ in equation (I.2), we now use the letter d specifically to indicate infinitesimal increments in x and y, corresponding to P_i and P_j very close together. The resulting equation for the slope is

$$m = \frac{dy}{dx} \tag{I.3}$$

Using the language of calculus, we say that the derivative of y with respect to x (or the ratio of the differentials of y and x) is equal to the desired slope at any point. The slope has physical significance as a rate of change of y with changing x. For example, letting D represent distance of an automobile from some starting point and t represent time, dD/dt is the slope of the curve in a graph of D versus t and is also the rate of change of position (the velocity) of the automobile.

Finding the maximum or minimum of a function is a useful application of differential calculus. It is seen in Figure I.2 that the slope of a curve is zero at a maximum or a minimum. Differentiating the equation for a curve to obtain the derivative dy/dx and then setting the derivative equal to zero permits location of a minimum or maximum on a curve.

The details of development of procedures for obtaining derivatives of various functions are thoroughly described in texts concerned with calculus, including those cited at the end of Chapter 1. Here we need only write down the results of these developments. For this purpose some important formulas for differentiation are summarized in Table I.1.

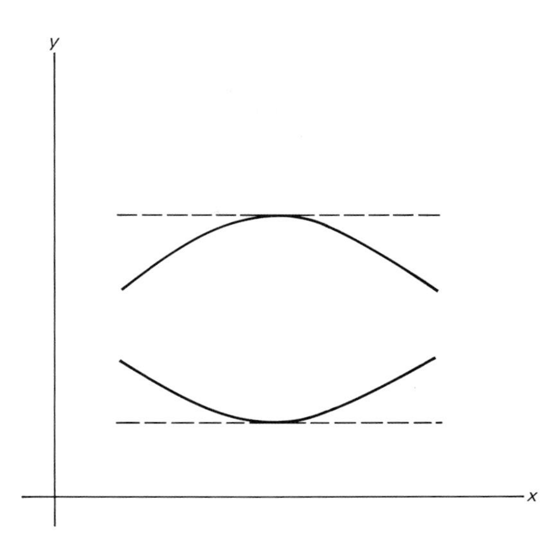

Figure I.2. A curve with a maximum and a curve with a minimum. The tangents, represented by dashed lines, have zero slope, showing that $dy/dx = 0$ at a maximum or a minimum.
If there is any doubt as to whether a given point (where $dy/dx = 0$) is a minimum or a maximum, the derivative dy/dx is itself differentiated to obtain the second derivative, d^2y/dx^2. If d^2y/dx^2 is positive, the point is a minimum; if d^2y/dx^2 is negative, the point is a maximum.
A horizontal point of inflection also has $dy/dx = 0$, but it has $d^2y/dx^2 = 0$ to distinguish it from a minimum or maximum.

Integration is the reverse of differentiation, just as squaring a number is the reverse of calculating a square root. Since differentiation of x^n/n gives x^{n-1}, integration of $x^{n-1}dx$ gives x^n/n. In general, rules for integration may be deduced from the rules for differentiation.

To understand the meaning of integration, consider an automobile accelerating from rest to some final constant velocity, as illustrated in Figure I.3. Calculation of the distance D traveled in a time interval $\Delta t = t_2 - t_1$ presents no difficulty if the velocity is constant during the time interval in question. In this case,

$$D = v(t_2 - t_1) \tag{I.4}$$

The product $v(t_2 - t_1)$ is equal to the shaded area in Figure I.3 and could have been determined from the graph.

For calculating the distance traveled in a time interval where v is changing, we must modify our use of equation (I.4). Suppose that the velocity changes linearly with time, as shown in Figure I.4. In this case we calculate the distance traveled as the average velocity times the time,

$$D = \frac{v_1 + v_2}{2}(t_2 - t_1) \tag{I.5}$$

The average velocity is the velocity at the midpoint of the time interval, so we can also calculate the distance traveled as

$$D = v_m(t_2 - t_1) = v_m\Delta t \tag{I.6}$$

where v_m represents the velocity at the midpoint of the time interval. The distance calculated by either (I.5) or (I.6) is equal to the shaded area in Figure I.4 and could be determined from the graph.

To calculate the distance traveled in a time interval in which the velocity is changing nonlinearly with time, we might divide the time interval of interest into

Table I.1. Formulas for Differentiation

(In all of the following, ln refers to natural logarithms with base 2.718 . . . , u and v represent functions of x, and c represents a constant.)

Function	*Derivative*	*Differential*
$y = c$	$\frac{dy}{dx} = \frac{d(c)}{dx} = 0$	$dy = d(c) = 0$
$y = cu$	$\frac{dy}{dx} = \frac{d(cu)}{dx} = c\frac{du}{dx}$	$dy = d(cu) = cdu$
$y = x^n$	$\frac{dy}{dx} = \frac{d(x^n)}{dx} = nx^{n-1}$	$dy = d(x^n) = nx^{n-1}\,dx$
$y = u^n$	$\frac{dy}{dx} = \frac{d(u^n)}{dx} = nu^{n-1}\frac{du}{dx}$	$dy = d(u^n) = nu^{n-1}\,du$
$y = uv$	$\frac{dy}{dx} = \frac{d(uv)}{dx} = u\frac{dv}{dx} + v\frac{du}{dx}$	$dy = d(uv) = u\,dv + v\,du$
$y = \frac{u}{v}$	$\frac{dy}{dx} = \frac{d(u/v)}{dx} = \frac{v(du/dx) - u(dv/dx)}{v^2}$	$dy = d(u/v) = \frac{v\,du - u\,dv}{v^2}$
$y = e^x$	$\frac{dy}{dx} = \frac{d(e^x)}{dx} = e^x$	$dy = d(e^x) = e^x\,dx$
$y = e^u$	$\frac{dy}{dx} = \frac{d(e^u)}{dx} = e^u\frac{du}{dx}$	$dy = d(e^u) = e^u\,du$
$y = \ln x$	$\frac{dy}{dx} = \frac{d(\ln x)}{dx} = \frac{1}{x}$	$dy = d(\ln x) = \frac{dx}{x}$
$y = \ln u$	$\frac{dy}{dx} = \frac{d(\ln u)}{dx} = \frac{1}{u}\frac{du}{dx}$	$dy = d(\ln u) = \frac{du}{u}$

Natural logarithms (base $e = 2.718$) and common logarithms (base 10) are related according to

$$\ln u = 2.303 \log u.$$

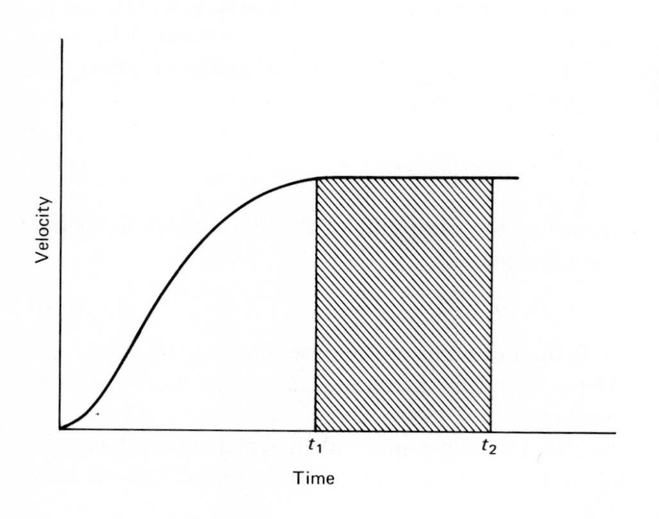

Figure I.3. Graph of the velocity of a car, initially at rest, that accelerates to a final constant velocity.

Figure I.4. Graph of the velocity of a car with velocity that is increasing linearly with time.

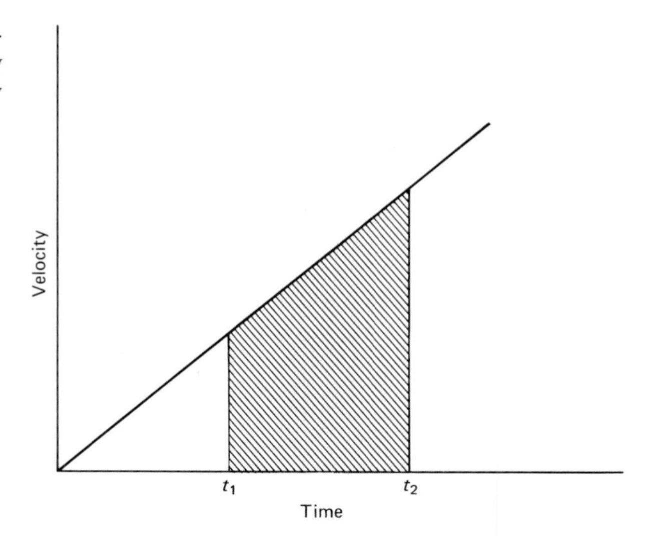

many smaller time intervals. Within any of these small time intervals, the velocity-time curve is nearly linear, so that it is a good approximation to apply equation (I.6) to calculating the distance traveled during each small time interval. Then we could sum all these distances to obtain the distance traveled in the entire time interval, as summarized by

$$D = \Sigma v_m \Delta t \tag{I.7}$$

where Σ is used as a summation sign. This procedure is equivalent to determining the area of all the rectangles pictured in Figure I.5, so we might also have found the desired distance by measuring the area under the part of the curve from t_1 to t_2.

The distance calculated from equation (I.7) is not exactly correct because the velocities at the midpoints of the time intervals are not exactly equal to the average velocities in the corresponding time intervals. The magnitude of the discrepancy can be diminished merely by choosing smaller time intervals for use with equation (I.7).

If we choose the time interval to be infinitesimally small, we write equation (I.7) as

$$\text{area} = D = \int v\, dt \tag{I.8}$$

where the $\int$ is used as an integral sign meaning a sum of an infinite number of $v\,dt$ areas extending over the time interval of interest. In order to carry out the desired integration, it is necessary to know v as a function of t.

Further discussion of the meaning and use of integration follows illustration of some of the integration formulas that are summarized in Table I.2.

Suppose that chemical considerations lead to the equation

$$\frac{dy}{dx} = 6.8 \tag{I.9}$$

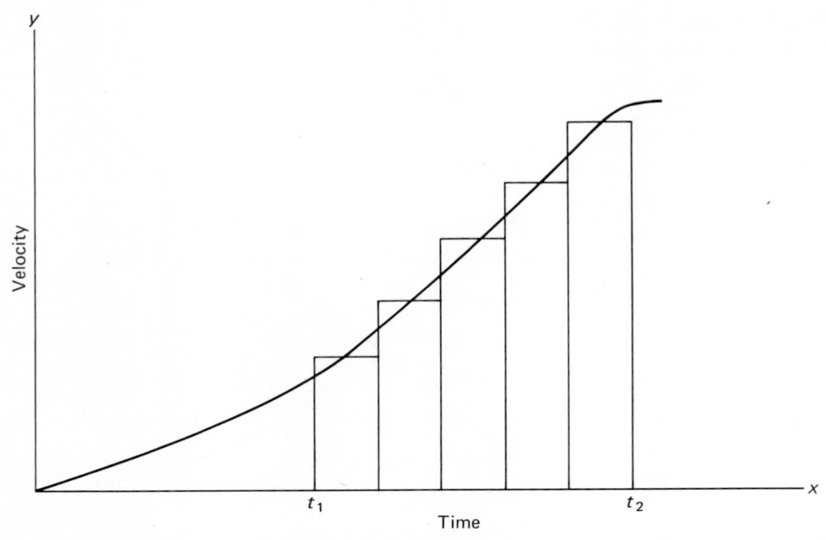

Figure I.5. Illustration of application of equation (I.7) to calculation of the distance traveled by a car in a time interval in which the car's velocity is neither constant nor changing linearly with time.

and we are faced with the problem of integrating to obtain an ordinary equation that does not contain differentials. Rearrangement gives

$$dy = 6.8\,dx$$

and then

$$\int dy = 6.8 \int dx$$

Integrating according to Table I.2 gives us

$$y = 6.8x + c \tag{I.10}$$

where c is a constant of integration that can be evaluated with the aid of specific information about y and x. For instance, if $y = 18.8$ when $x = 2.0$,

$$c = y - 6.8x = 18.8 - 13.6 = 5.2$$

Table I.2. Integrals*

$\int du = u + c$	$\int u^n\,du = \dfrac{u^{n+1}}{n+1} + c$	(when $n \neq -1$)
$\int \dfrac{du}{u} = \ln u + c$	$\int e^u\,du = e^u + c$	
$\int a^u\,du = \dfrac{a^u}{\ln a} + c$	$\int e^{au}\,du = \dfrac{e^{au}}{a} + c$	

*Letters a and c represent constants.

so that we can write equation (I.10) as

$$y = 6.8x + 5.2.$$

Since differentiation of equation (I.10) with respect to x gives $dy/dx = 6.8$, regardless of the value assigned to c, there is an infinite number of equations that are solutions of equation (I.9). There is, however, only one equation that is consistent with both (I.9) and the condition that $y = 18.8$ when $x = 2.0$.

Now we consider the distance traveled by an automobile with velocity given by

$$v = at + bt^2 \tag{I.11}$$

where a and b represent known numerical constants so that v can be calculated at any time t. Following equation (I.8), we calculate the distance traveled from zero time to time t as

$$\begin{aligned} D &= \int v\,dt \\ &= \int (at + bt^2)dt \\ &= a\int t\,dt + b\int t^2\,dt \\ &= \frac{at^2}{2} + \frac{bt^3}{3} + c. \end{aligned} \tag{I.12}$$

Since we know that $D = 0$ at $t = 0$, we see that the constant of integration equals zero for this problem and that we can calculate the distance D traveled at any time t from

$$D = \frac{at^2}{2} + \frac{bt^3}{3} \tag{I.13}$$

Differentiation of (I.12) with respect to t gives

$$\frac{dD}{dt} = v = at + bt^2$$

as it must, since integration is the reverse of differentiation.

Integration leading to a constant of integration, c, as in (I.10) and (I.12), is called indefinite integration. It is sometimes more convenient to work with what are called definite integrals, as in the case of finding the distance an automobile travels in the time interval between t_1 and t_2, which is illustrated in Figure I.5. Following (I.8), we write

$$D = \int_{t_1}^{t_2} v\,dt$$

and for the case where

$$v = at + bt^2$$

we have

$$D = \int_{t_1}^{t_2} (at + bt^2)\, dt$$

Integration gives

$$D = \left[\frac{at^2}{2} + \frac{bt^3}{3}\right]_{t_1}^{t_2}$$

where $[\ \ ]_{t_1}^{t_2}$ means that we have integrated between the limits t_1 and t_2 and are currently concerned with the value of D that results from subtracting the value of D corresponding to t_1 from the value of D corresponding to t_2.
Thus, we obtain

$$D = \frac{a}{2}t_2^2 + \frac{b}{3}t_2^3 - \frac{a}{2}t_1^2 - \frac{b}{3}t_1^3$$

$$= \frac{a}{2}(t_2^2 - t_1^2) + \frac{b}{3}(t_2^3 - t_1^3) \qquad \text{(I.14)}$$

Inserting the chosen values of t_1 and t_2 in (I.14) permits us to calculate the distance traveled in the time between t_1 and t_2. Note that $(t_2^2 - t_1^2) \neq (t_2 - t_1)^2$.
Taking the special case of $t_1 = 0$, we obtain from (I.14)

$$D = \frac{a}{2}t_2^2 + \frac{b}{3}t_2^3 \qquad \text{(I.15)}$$

Equations (I.13) and (I.15) give the same value for D at any time t, showing that the same result can be obtained whether we do a definite integration between limits or an indefinite integration with a constant of integration to be evaluated from available information. The choice of integration procedure is often made on the basis of convenience with respect to following numerical calculation.

Appendix II

THERMODYNAMIC DATA

(All of these values, mostly taken from the *Technical Notes* 270 of the National Bureau of Standards, are for substances at 298°K.)

Substance	ΔH_f^0 (*kcal mole*$^{-1}$)	ΔG_f^0 (*kcal mole*$^{-1}$)	S^0 (*cal deg*$^{-1}$ *mole*$^{-1}$)
Noble Gases			
He(g)	0	0	30.124
He^+(g)	568.459		
Ne(g)	0	0	34.947
Ne^+(g)	498.77		
Ar(g)	0	0	36.982
Ar^+(g)	364.90		
Kr(g)	0	0	39.190
Kr^+(g)	324.32		
Xe(g)	0	0	42.529
Xe^+(g)	281.20		
XeF_4(c)	−62.5		
Oxygen and Hydrogen			
O_2(g)	0	0	49.003
O(g)	59.553	55.389	38.467
O^+(g)	375.070		
O^-(g)	24.29		
O_2^+(g)	281.48		
O_3(g)	34.1	39.0	57.08
H_2(g)	0	0	31.208
H(g)	52.095	48.581	27.391
H^+(g)	367.161		
H^+(aq)	0	0	0
H^-(g)	33.39		
H_2^+(g)	357.23		
OH^-(aq)	−54.970	−37.594	−2.57
H_2O(liq)	−68.315	−56.687	16.71
H_2O(g)	−57.796	−54.634	45.104
H_2O_2(aq)	−45.69	−32.05	34.4
HO_2^-(aq)	−38.32	−16.1	5.7
Halogens			
F_2(g)	0	0	48.44
F(g)	18.88	14.80	37.917
F^-(g)	−64.7		
F^-(aq)	−79.50	−66.64	−3.3
HF(aq)	−76.50	−70.95	21.2
HF(g)	−64.8	−65.3	41.508
Cl_2(g)	0	0	53.288
Cl(g)	29.082	25.262	39.457
Cl^-(g)	−58.8		
Cl^-(aq)	−39.952	−31.372	13.5
HCl(g)	−22.062	−22.777	44.646
HOCl(aq)	−28.9	−19.1	34
OCl^-(aq)	−25.6	−8.8	10
Br_2(liq)	0	0	36.384
Br_2(g)	7.387	0.751	58.641
Br(g)	26.741	19.701	41.805
Br^-(g)	−55.9		
Br^-(aq)	−29.05	−24.85	19.7
HBr(g)	−8.70	−12.77	47.463
I_2(c)	0	0	27.757

Substance	ΔH_f^0 (*kcal mole*$^{-1}$)	ΔG_f^0 (*kcal mole*$^{-1}$)	S^0 (*cal deg*$^{-1}$ *mole*$^{-1}$)
I_2(g)	14.923	4.627	62.28
I(g)	25.535	16.798	43.184
I^-(g)	−47.0		
I^-(aq)	−13.19	−12.33	26.6
HI(g)	6.33	0.41	49.351
Sulfur			
S(rhombic)	0	0	7.60
S(monoclinic)	0.08		
S(g)	66.636	59.951	40.084
S_2(g)	30.68	18.96	54.51
S^{2-}(aq)	7.9	20.5	−3.5
HS^-(aq)	−4.2	2.88	15.0
H_2S(aq)	−9.5	−6.66	29
H_2S(g)	−4.93	−8.02	49.16
SO_2(g)	−70.944	−71.748	59.30
SO_3(g)	−94.58	−88.69	61.35
SO_4^{2-}(aq)	−217.32	−177.97	4.8
HSO_4^-(aq)	−212.08	−180.69	31.5
Nitrogen			
N_2(g)	0	0	45.77
N_2^+(g)	436		
N(g)	112.979	108.886	36.613
N_2O(g)	19.61	24.90	52.52
NO(g)	21.57	20.69	50.347
NO_2(g)	7.93	12.26	57.35
N_2O_4(g)	2.19	23.38	72.70
NO_2^-(aq)	−25.0	−8.9	33.5
NO_3^-(aq)	−49.56	−26.61	35.0
NH_3(g)	−11.02	−3.94	45.97
NH_3(aq)	−19.19	−6.35	26.6
NH_4^+(aq)	−31.67	−18.97	27.1
NH_4Cl(c)	−75.15	−48.51	22.6
NH_4NO_2(c)	−61.3		
NH_4NO_3(c)	−87.37	−43.98	36.11
HNO_2(aq)	−28.5	−13.3	36.5
Carbon			
C(graphite)	0	0	1.372
C(diamond)	0.4533	0.6930	0.568
C(g)	171.291	160.442	37.7597
CO(g)	−26.416	−32.780	47.219
CO_2(g)	−94.051	−94.254	51.06
CO_3^{2-}(aq)	−161.84	−126.17	−13.6
HCO_3^-(aq)	−165.39	−140.26	21.8
CH_4(g)	−17.88	−12.13	44.492
C_2H_2(g)	54.19	50.00	48.00
C_2H_4(g)	12.49	16.28	52.45
C_2H_6(g)	−20.24	−7.86	54.85
C_3H_4(g, propyne)	44.32	46.47	59.30
C_3H_6(g, propylene)	4.88	14.99	63.80
C_3H_6(g, cyclopropane)	12.74	24.95	56.75

Substance	ΔH_f^0 (kcal mole^{-1})	ΔG_f^0 (kcal mole^{-1})	S^0 (cal deg^{-1} mole^{-1})
C_3H_8(g, propane)	−24.82	−5.61	64.51
C_4H_{10}(g, *n*-butane)	−30.15	−4.10	74.12
C_4H_{10}(g, isobutane)	−32.15	−4.99	70.42
C_4H_8(g, cyclobutane)	6.37	26.30	63.43
C_6H_6(liq, benzene)	11.72	29.72	41.41
C_6H_6(g, benzene)	19.81	30.99	64.34
C_8H_{18}(liq, *n*-octane)	−59.78		
C_8H_{18}(g, *n*-octane)	−49.86		
CH_3OH(liq)	−57.04	−39.76	30.3
CH_3OH(g)	−47.96	−38.72	57.29
C_2H_5OH(liq)	−66.37	−41.80	38.4
C_2H_5OH(g)	−56.19	−40.29	67.54
CF_4(g)	−221	−210	62.50
CCl_4(liq)	−32.37	−15.60	51.72
CCl_4(g)	−24.6	−14.49	74.03
CN^-(aq)	36.0	41.2	22.5
HCN(aq)	25.6	28.6	29.8
HCN(g)	32.3	29.8	48.20
$CO(NH_2)_2$(c, urea)	−79.56	−47.04	25.0
NH_4CNO(c)	−72.75	~−40	~25
Transition metals			
Zn(c)	0	0	9.95
Zn(g)	31.245	22.748	38.450
Zn^{2+}(g)	665.09		
Zn^{2+}(aq)	−36.78	−35.14	−26.8
Cd(c)	0	0	12.37
Cd(g)	26.77	18.51	40.066
Cd^{2+}(g)	627.04		
Cd^{2+}(aq)	−18.14	−18.542	−17.5
Hg(liq)	0	0	18.17
Hg(g)	14.655	7.613	41.79
Hg^{2+}(g)	690.83		
Hg^{2+}(aq)	40.9	39.30	−7.7
Hg_2^{2+}(aq)	41.2	36.70	20.2
$HgCl_2$(c)	−53.6	−42.7	34.9
$HgBr_2$(c)	−40.8	−36.6	41
HgI_2(c)	−25.2	−24.3	43
Hg_2Cl_2(c)	−63.39	−50.377	46.0
Hg_2Br_2(c)	−49.45	−43.278	52
Hg_2I_2(c)	−29.00	−26.53	55.8
HgS(c, red)	−13.9	−12.1	19.7
HgS(c, black)	−12.8	−11.4	21.1
Hg_2SO_4(c)	−177.61	−149.589	47.96
Cu(c)	0	0	7.923
Cu(g)	80.86	71.37	39.74
Cu^+(g)	260.513		
Cu^+(aq)	17.13	11.95	9.7
Cu^{2+}(g)	729.93		
Cu^{2+}(aq)	15.48	15.66	−23.8
CuCl(c)	−32.8	−28.65	20.6
CuI(c)	−16.2	−16.6	23.1
Ag(c)	0	0	10.17

Substance	ΔH_f^0 (*kcal mole*$^{-1}$)	ΔG_f^0 (*kcal mole*$^{-1}$)	S^0 (*cal deg*$^{-1}$ *mole*$^{-1}$)
$Ag(g)$	68.01	58.72	41.321
$Ag^+(g)$	243.59		
$Ag^+(aq)$	25.234	18.433	17.37
$AgCl(c)$	−30.370	−26.244	23.0
$AgBr(c)$	−23.99	−23.16	25.6
$AgI(c)$	−14.78	−15.82	27.6
$Ag(NH_3)_2^+(aq)$	−26.60	−4.12	58.6
$Ag(CN)_2^-(aq)$	64.6	73.0	46
$Ag_2CrO_4(c)$	−174.89	−153.40	52.0
$Mn(c)$	0	0	7.65
$Mn^{2+}(aq)$	−52.76	−54.5	−17.6
$MnO_2(c)$	−124.29	−111.18	12.68
$MnO_4^-(aq)$	−129.4	−106.9	45.7
$Cr(c)$	0	0	5.68
$Cr_2O_3(c)$	−272.4	−252.9	19.4
$CrO_3(c)$	−140.9		
$CrO_4^{2-}(aq)$	−210.60	−173.96	12.00
$HCrO_4^-(aq)$	−209.9	−182.8	44.0
$Cr_2O_7^{2-}(aq)$	−356.2	−311.0	62.6
$Mo(c)$	0	0	6.85
$MoO_2(c)$	−140.76	−127.40	11.06
$MoO_3(c)$	−178.08	−159.66	18.58
$MoS_2(c)$	−56.2	−54.0	14.96
Nontransition metals			
$Pb(c)$	0	0	15.49
$Pb(g)$	46.6	38.7	41.889
$Pb^{2+}(g)$	567.25		
$Pb^{2+}(aq)$	−0.4	−5.83	2.5
$PbO_2(c)$	−66.3	−51.95	16.4
$PbSO_4(c)$	−219.87	−194.36	35.51
$Al(c)$	0	0	6.77
$Al^{3+}(aq)$	−127	−116	−76.9
$Al_2O_3(c)$	−400.5	−378.2	12.17
Alkaline earth metals			
$Mg(c)$	0	0	7.81
$Mg(g)$	35.30	27.04	35.502
$Mg^+(g)$	213.100		
$Mg^{2+}(g)$	561.299		
$Mg^{2+}(aq)$	−111.58	−108.7	−33.0
$MgO(c)$	−143.81	−136.10	6.44
$Mg(OH)_2(c)$	−220.97	−199.23	15.10
$MgF_2(c)$	−268.5	−255.8	13.68
$MgCl_2(c)$	−153.28	−141.45	21.42
$MgCl_2 \cdot 6H_2O(c)$	−597.28	−505.49	87.5
$Mg_3N_2(c)$	−110.1		
$MgSO_4(c)$	−307.1	−279.8	21.9
$MgSO_4 \cdot 7H_2O(c)$	−809.92	−686.4	89
$Ca(c)$	0	0	9.90
$Ca(g)$	42.6	34.5	36.992
$Ca^+(g)$	185.05		
$Ca^{2+}(g)$	460.29		

Substance	ΔH_f^0 (*kcal mole*$^{-1}$)	ΔG_f^0 (*kcal mole*$^{-1}$)	S^0 (*cal deg*$^{-1}$ *mole*$^{-1}$)
Ca^{2+}(aq)	−129.74	−132.30	−12.7
CaO(c)	−151.79	−144.37	9.50
$Ca(OH)_2$(c)	−235.68	−214.76	19.93
CaCl(g)	−23.4	−29.7	57.70
CaF_2(c)	−291.5	−279.0	16.46
$CaCl_2$(c)	−190.2	−178.8	25.0
$CaCl_2 \cdot 6H_2O$(c)	−623.3		
$CaCO_3$(c, calcite)	−288.46	−269.80	22.2
$CaCO_3$(c, aragonite)	−288.51	−269.55	21.2
CaC_2(c)	−14.3	−15.5	16.72
Sr(c)	0	0	12.5
Sr(g)	39.3	31.3	39.32
Sr^{2+}(g)	427.96		
Sr^{2+}(aq)	−130.45	−133.71	−7.8
Ba(c)	0	0	15.0
Ba(g)	43	35	40.663
Ba^{+}(g)	164.67		
Ba^{2+}(g)	396.86		
Ba^{2+}(aq)	−128.50	−134.02	2.3
$BaCl_2$(c)	−205.2	−193.7	29.56
$BaSO_4$(c)	−352.1	−325.6	31.6
$BaCO_3$(c)	−290.7	−271.9	26.8
Alkali metals			
Na(c)	0	0	12.24
Na(g)	25.98	18.67	36.715
Na^{+}(g)	146.015		
Na^{+}(aq)	−57.39	−62.59	14.10
NaCl(c)	−98.27	−91.82	17.30
K(c)	0	0	15.34
K(g)	21.51	14.62	38.296
K^{+}(g)	123.07		
K^{+}(aq)	−60.32	−67.70	24.5
K^{2+}(g)	858.15		
KCl(c)	−104.38	−97.79	19.76
KBr(c)	−94.16	−91.05	23.1

Appendix III

EQUILIBRIUM CONSTANTS FOR AQUEOUS SOLUTIONS AT 298°K

$H_2O(liq) \rightleftharpoons H^+(aq) + OH^-(aq)$	$K_w = 1.0 \times 10^{-14}$
$HF(aq) \rightleftharpoons H^+(aq) + F^-(aq)$	$K = 7 \times 10^{-4}$
$HOCl(aq) \rightleftharpoons H^+(aq) + OCl^-(aq)$	$K = 2.8 \times 10^{-8}$
$HClO_2(aq) \rightleftharpoons H^+(aq) + ClO_2^-(aq)$	$K = 1 \times 10^{-2}$
$HIO_3(aq) \rightleftharpoons H^+(aq) + IO_3^-(aq)$	$K = 0.16$
$H_5IO_6(aq) \rightleftharpoons H^+(aq) + H_4IO_6^-(aq)$	$K = 5 \times 10^{-4}$
$H_4IO_6^-(aq) \rightleftharpoons H^+(aq) + H_3IO_6^{2-}(aq)$	$K = 10^{-7}$
$H_4IO_6^-(aq) \rightleftharpoons IO_4^-(aq) + 2\ H_2O(liq)$	$K = 40$
$NH_4^+(aq) \rightleftharpoons H^+(aq) + NH_3(aq)$	$K = 5.6 \times 10^{-10}$
$NH_3(aq) + H_2O(liq) \rightleftharpoons NH_4^+(aq) + OH^-(aq)$	$K = 1.8 \times 10^{-5}$
$HCN(aq) \rightleftharpoons H^+(aq) + CN^-(aq)$	$K = 6 \times 10^{-10}$
$HNO_2(aq) \rightleftharpoons H^+(aq) + NO_2^-(aq)$	$K = 5.9 \times 10^{-4}$
$NH_2OH(aq) + H_2O(liq) \rightleftharpoons NH_3OH^+(aq) + OH^-(aq)$	$K = 1 \times 10^{-8}$
$H_2S(aq) \rightleftharpoons H^+(aq) + HS^-(aq)$	$K = 1.0 \times 10^{-7}$
$HS^-(aq) \rightleftharpoons H^+(aq) + S^{2-}(aq)$	$K = 1 \times 10^{-13}$
$HSO_3^-(aq) \rightleftharpoons H^+(aq) + SO_3^{2-}(aq)$	$K = 6 \times 10^{-8}$
$HSO_4^-(aq) \rightleftharpoons H^+(aq) + SO_4^{2-}(aq)$	$K = 1.0 \times 10^{-2}$
$H_3PO_3(aq) \rightleftharpoons H^+(aq) + H_2PO_3^-(aq)$	$K = 1.6 \times 10^{-2}$
$H_2PO_3^-(aq) \rightleftharpoons H^+(aq) + HPO_3^{2-}(aq)$	$K = 7 \times 10^{-7}$
$H_3PO_4(aq) \rightleftharpoons H^+(aq) + H_2PO_4^-(aq)$	$K = 7.1 \times 10^{-3}$
$H_2PO_4^-(aq) \rightleftharpoons H^+(aq) + HPO_4^{2-}(aq)$	$K = 6.3 \times 10^{-8}$
$HPO_4^{2-}(aq) \rightleftharpoons H^+(aq) + PO_4^{3-}(aq)$	$K = 5 \times 10^{-13}$
$CO_2(aq) + H_2O(liq) \rightleftharpoons H^+(aq) + HCO_3^-(aq)$	$K = 4.3 \times 10^{-7}$
$HCO_3^-(aq) \rightleftharpoons H^+(aq) + CO_3^{2-}(aq)$	$K = 4.7 \times 10^{-11}$
CH_3CO_2H(aq, acetic acid) $\rightleftharpoons H^+(aq) + CH_3CO_2^-(aq)$	$K = 1.8 \times 10^{-5}$
$C_6H_5CO_2H$(aq, benzoic acid) $\rightleftharpoons H^+(aq) + C_6H_5CO_2^-(aq)$	$K = 6.5 \times 10^{-5}$
C_6H_5OH(aq, phenol) $\rightleftharpoons H^+(aq) + C_6H_5O^-(aq)$	$K = 1.1 \times 10^{-10}$

Reaction	Constant
$Fe^{3+}(aq) + H_2O(liq) \rightleftharpoons H^+(aq) + FeOH^{2+}(aq)$	$K = 5 \times 10^{-3}$
$HCrO_4^-(aq) \rightleftharpoons H^+(aq) + CrO_4^{2-}(aq)$	$K = 3.4 \times 10^{-7}$
$2\ HCrO_4^-(aq) \rightleftharpoons Cr_2O_7^{2-}(aq) + H_2O(liq)$	$K = 34$
$Ag^+(aq) + 2\ NH_3(aq) \rightleftharpoons Ag(NH_3)_2^+(aq)$	$K = 1.7 \times 10^{7}$
$Ag^+(aq) + 2\ CN^-(aq) \rightleftharpoons Ag(CN)_2^-(aq)$	$K = 2.8 \times 10^{20}$
$AgCl(c) \rightleftharpoons Ag^+(aq) + Cl^-(aq)$	$K_{sp} = 1.76 \times 10^{-10}$
$AgBr(c) \rightleftharpoons Ag^+(aq) + Br^-(aq)$	$K_{sp} = 5.3 \times 10^{-13}$
$AgI(c) \rightleftharpoons Ag^+(aq) + I^-(aq)$	$K_{sp} = 8.5 \times 10^{-17}$
$Ag_2CrO_4(c) \rightleftharpoons 2\ Ag^+(aq) + CrO_4^{2-}(aq)$	$K_{sp} = 1.2 \times 10^{-12}$
$MgF_2(c) \rightleftharpoons Mg^{2+}(aq) + 2\ F^-(aq)$	$K_{sp} = 7 \times 10^{-11}$
$CaF_2(c) \rightleftharpoons Ca^{2+}(aq) + 2\ F^-(aq)$	$K_{sp} = 1.4 \times 10^{-10}$
$BaSO_4(c) \rightleftharpoons Ba^{2+}(aq) + SO_4^{2-}(aq)$	$K_{sp} = 1.05 \times 10^{-10}$
$BaCrO_4(c) \rightleftharpoons Ba^{2+}(aq) + CrO_4^{2-}(aq)$	$K_{sp} = 1 \times 10^{-10}$
$PbSO_4(c) \rightleftharpoons Pb^{2+}(aq) + SO_4^{2-}(aq)$	$K_{sp} = 1.8 \times 10^{-8}$
$PbCrO_4(c) \rightleftharpoons Pb^{2+}(aq) + CrO_4^{2-}(aq)$	$K_{sp} = 2 \times 10^{-16}$
$CaCO_3(c) \rightleftharpoons Ca^{2+}(aq) + CO_3^{2-}(aq)$	$K_{sp} = 1.4 \times 10^{-10}$
$SrCO_3(c) \rightleftharpoons Sr^{2+}(aq) + CO_3^{2-}(aq)$	$K_{sp} = 5.6 \times 10^{-10}$
$Hg_2Cl_2(c) \rightleftharpoons Hg_2^{2+}(aq) + 2\ Cl^-(aq)$	$K_{sp} = 1.4 \times 10^{-18}$
$Hg_2Br_2(c) \rightleftharpoons Hg_2^{2+}(aq) + 2\ Br^-(aq)$	$K_{sp} = 6.3 \times 10^{-23}$
$Hg_2I_2(c) \rightleftharpoons Hg_2^{2+}(aq) + 2\ I^-(aq)$	$K_{sp} = 5.0 \times 10^{-29}$
$Hg(liq) + Hg^{2+}(aq) \rightleftharpoons Hg_2^{2+}(aq)$	$K = 81$
$2\ Cu^+(aq) \rightleftharpoons Cu(c) + Cu^{2+}(aq)$	$K = 1.2 \times 10^{6}$
$CuI(c) \rightleftharpoons Cu^+(aq) + I^-(aq)$	$K_{sp} = 1.3 \times 10^{-12}$
$HgS(ppct) \rightleftharpoons Hg^{2+}(aq) + S^{2-}(aq)$	$K_{sp} = 10^{-52}$
$ZnS(ppct) \rightleftharpoons Zn^{2+}(aq) + S^{2-}(aq)$	$K_{sp} = 10^{-20}$
$CuS(ppct) \rightleftharpoons Cu^{2+}(aq) + S^{2-}(aq)$	$K_{sp} = 10^{-36}$
$Cu_2S(ppct) \rightleftharpoons 2\ Cu^+(aq) + S^{2-}(aq)$	$K_{sp} = 10^{-48}$
$MnS(ppct) \rightleftharpoons Mn^{2+}(aq) + S^{2-}(aq)$	$K_{sp} = 10^{-14}$
$PbS(ppct) \rightleftharpoons Pb^{2+}(aq) + S^{2-}(aq)$	$K_{sp} = 10^{-28}$
$Ag_2S(ppct) \rightleftharpoons 2\ Ag^+(aq) + S^{2-}(aq)$	$K_{sp} = 10^{-50}$
$CdS(ppct) \rightleftharpoons Cd^{2+}(aq) + S^{2-}(aq)$	$K_{sp} = 10^{-28}$
$FeS(ppct) \rightleftharpoons Fe^{2+}(aq) + S^{2-}(aq)$	$K_{sp} = 10^{-17}$
$Mg(OH)_2(c) \rightleftharpoons Mg^{2+}(aq) + 2\ OH^-(aq)$	$K_{sp} = 6 \times 10^{-12}$
$Ca(OH)_2(c) \rightleftharpoons Ca^{2+}(aq) + 2\ OH^-(aq)$	$K_{sp} = 5 \times 10^{-6}$
$Cu(OH)_2(c) \rightleftharpoons Cu^{2+}(aq) + 2\ OH^-(aq)$	$K_{sp} = 5 \times 10^{-20}$
$Cu_2O(c) + H_2O(liq) \rightleftharpoons 2\ Cu^+(aq) + 2\ OH^-(aq)$	$K = 1 \times 10^{-30}$
$Ag_2O(c) + H_2O(liq) \rightleftharpoons 2\ Ag^+(aq) + 2\ OH^-(aq)$	$K = 4 \times 10^{-16}$
$Zn(OH)_2(c) \rightleftharpoons Zn^{2+}(aq) + 2\ OH^-(aq)$	$K_{sp} = 6 \times 10^{-17}$
$Cd(OH)_2(c) \rightleftharpoons Cd^{2+}(aq) + 2\ OH^-(aq)$	$K_{sp} = 5 \times 10^{-15}$

$Mn(OH)_2(c) \rightleftharpoons Mn^{2+}(aq) + 2\,OH^-(aq)$ $K_{sp} = 2 \times 10^{-13}$
$Fe(OH)_3(ppct) \rightleftharpoons Fe^{3+}(aq) + 3\,OH^-(aq)$ $K_{sp} = 10^{-39}$
$Co(OH)_2(ppct) \rightleftharpoons Co^{2+}(aq) + 2\,OH^-(aq)$ $K_{sp} = 2 \times 10^{-16}$
$Ni(OH)_2(ppct) \rightleftharpoons Ni^{2+}(aq) + 2\,OH^-(aq)$ $K_{sp} = 6 \times 10^{-16}$
$Al(OH)_3(ppct) \rightleftharpoons Al^{3+}(aq) + 3\,OH^-(aq)$ $K_{sp} = 10^{-33}$

Appendix **IV**

STANDARD STATE HALF-REACTION POTENTIALS FOR AQUEOUS SOLUTIONS AT 298°K

Hydrogen and Oxygen	$\mathscr{E}^0$ (volts)
$2\ H^+(aq) + 2\ e^- \rightleftharpoons H_2(g)$	0.000
$2\ H^+(aq,\ 10^{-7}\ M) + 2\ e^- \rightleftharpoons H_2(g)$	−0.414
$2\ H_2O(liq) + 2\ e^- \rightleftharpoons H_2(g) + 2\ OH^-(aq)$	−0.828
$O_2(g) + 4\ H^+(aq) + 4\ e^- \rightleftharpoons 2\ H_2O(liq)$	+1.229
$O_2(g) + 4\ H^+(aq,\ 10^{-7}\ M) + 4\ e^- \rightleftharpoons 2\ H_2O(liq)$	+0.815
$O_2(g) + 2\ H_2O(liq) + 4\ e^- \rightleftharpoons 4\ OH^-(aq)$	+0.401
$O_2(g) + 2\ H^+(aq) + 2\ e^- \rightleftharpoons H_2O_2(aq)$	+0.695
$O_2(g) + H_2O(liq) + 2\ e^- \rightleftharpoons HO_2^-(aq) + OH^-(aq)$	−0.065
$H_2O_2(aq) + 2\ H^+(aq) + 2\ e^- \rightleftharpoons 2\ H_2O(liq)$	+1.763
$HO_2^-(aq) + H_2O(liq) + 2\ e^- \rightleftharpoons 3\ OH^-(aq)$	+0.867
$O_3(g) + 2\ H^+(aq) + 2\ e^- \rightleftharpoons O_2(g) + H_2O(liq)$	+2.07
$O_3(g) + H_2O(liq) + 2\ e^- \rightleftharpoons O_2(g) + 2\ OH^-(aq)$	+1.25

$O_2(g)$ to $H_2O(liq)$: +1.229

$$O_3(g) \overset{+2.07}{\text{———}} O_2(g) \overset{+0.695}{\text{———}} H_2O_2(aq) \overset{+1.763}{\text{———}} H_2O(liq)$$

Acidic solution

$O_2(g)$ to $OH^-(aq)$: +0.401

$$O_3(g) \overset{+1.25}{\text{———}} O_2(g) \overset{-0.065}{\text{———}} HO_2^-(aq) \overset{+0.867}{\text{———}} OH^-(aq)$$

Alkaline solution

Halogens	$\mathscr{E}^0$ (volts)
$F_2(g) + 2\ e^- \rightleftharpoons 2\ F^-(aq)$	+2.890
$F_2(g) + 2\ H^+(aq) + 2\ e^- \rightleftharpoons 2\ HF(aq)$	+3.077
$Cl_2(g) + 2\ e^- \rightleftharpoons 2\ Cl^-(aq)$	+1.360
$2\ HOCl(aq) + 2\ H^+(aq) + 2\ e^- \rightleftharpoons Cl_2(g) + 2\ H_2O(liq)$	+1.630

$2\ OCl^-(aq) + 2\ H_2O(liq) + 2\ e^- \rightleftharpoons Cl_2(g) + 4\ OH^-(aq)$	+0.421
$HOCl(aq) + H^+(aq) + 2\ e^- \rightleftharpoons Cl^-(aq) + H_2O(liq)$	+1.495
$OCl^-(aq) + H_2O(liq) + 2\ e^- \rightleftharpoons Cl^-(aq) + 2\ OH^-(aq)$	+0.891
$HClO_2(aq) + 2\ H^+(aq) + 2\ e^- \rightleftharpoons HOCl(aq) + H_2O(liq)$	+1.674
$ClO_2^-(aq) + H_2O(liq) + 2\ e^- \rightleftharpoons OCl^-(aq) + 2\ OH^-(aq)$	+0.681
$ClO_3^-(aq) + 3\ H^+(aq) + 2\ e^- \rightleftharpoons HClO_2(aq) + H_2O(liq)$	+1.181
$ClO_3^-(aq) + H_2O(liq) + 2\ e^- \rightleftharpoons ClO_2^-(aq) + 2\ OH^-(aq)$	+0.295
$ClO_2(g) + H^+(aq) + e^- \rightleftharpoons HClO_2(aq)$	+1.188
$ClO_2(g) + e^- \rightleftharpoons ClO_2^-(aq)$	+1.071
$ClO_3^-(aq) + 2\ H^+(aq) + e^- \rightleftharpoons ClO_2(g) + H_2O(liq)$	+1.175
$ClO_3^-(aq) + H_2O(liq) + e^- \rightleftharpoons ClO_2(g) + 2\ OH^-(aq)$	−0.481
$2\ ClO_3^-(aq) + 12\ H^+(aq) + 10\ e^- \rightleftharpoons Cl_2(g) + 6\ H_2O(liq)$	+1.468
$ClO_3^-(aq) + 2\ H_2O(liq) + 4\ e^- \rightleftharpoons OCl^-(aq) + 4\ OH^-(aq)$	+0.489
$ClO_3^-(aq) + 3\ H_2O(liq) + 6\ e^- \rightleftharpoons Cl^-(aq) + 6\ OH^-(aq)$	+0.622
$ClO_4^-(aq) + 2\ H^+(aq) + 2\ e^- \rightleftharpoons ClO_3^-(aq) + H_2O(liq)$	+1.202
$ClO_4^-(aq) + H_2O(liq) + 2\ e^- \rightleftharpoons ClO_3^-(aq) + 2\ OH^-(aq)$	+0.374

Acidic solution

$$ClO_4^- \xrightarrow{+1.20} ClO_3^- \xrightarrow{+1.18} HClO_2 \xrightarrow{+1.67} HOCl \xrightarrow{+1.63} Cl_2 \xrightarrow{+1.36} Cl^-$$

$$ClO_3^- \xrightarrow{+1.18} ClO_2 \xrightarrow{+1.19} HClO_2 \qquad HOCl \xrightarrow{+1.49} Cl^- \qquad ClO_3^- \xrightarrow{+1.47} Cl_2$$

Alkaline solution

$$ClO_4^- \xrightarrow{+0.37} ClO_3^- \xrightarrow{+0.29} ClO_2^- \xrightarrow{+0.68} OCl^- \xrightarrow{+0.42} Cl_2 \xrightarrow{+1.36} Cl^-$$

$$ClO_3^- \xrightarrow{-0.48} ClO_2 \xrightarrow{+1.07} ClO_2^- \qquad ClO_3^- \xrightarrow{+0.49} OCl^- \qquad OCl^- \xrightarrow{+0.89} Cl^- \qquad ClO_3^- \xrightarrow{+0.62} Cl^-$$

Acidic solution

$$BrO_3^- \xrightarrow{+1.45} HOBr \xrightarrow{+1.60} Br_2 \xrightarrow{+1.08} Br^- \qquad BrO_3^- \xrightarrow{+1.48} Br_2$$

Alkaline solution

$$BrO_3^- \xrightarrow{+0.49} OBr^- \xrightarrow{+0.46} Br_2 \xrightarrow{+1.08} Br^- \qquad BrO_3^- \xrightarrow{+0.48} Br_2 \qquad BrO_3^- \xrightarrow{+0.58} Br^-$$

Acidic solution

$$H_5IO_6 \xrightarrow{\sim +1.7} IO_3^- \xrightarrow{+1.15} HOI \xrightarrow{+1.43} I_2(c) \xrightarrow{+0.54} I^- \qquad IO_3^- \xrightarrow{+1.21} I_2(c)$$

Alkaline solution

$$H_3IO_6^{2-} \xrightarrow{\sim +0.7} IO_3^- \xrightarrow{+0.17} OI^- \xrightarrow{+0.40} I_2(c) \xrightarrow{+0.54} I^- \qquad IO_3^- \xrightarrow{+0.27} I^-$$

Sulfur

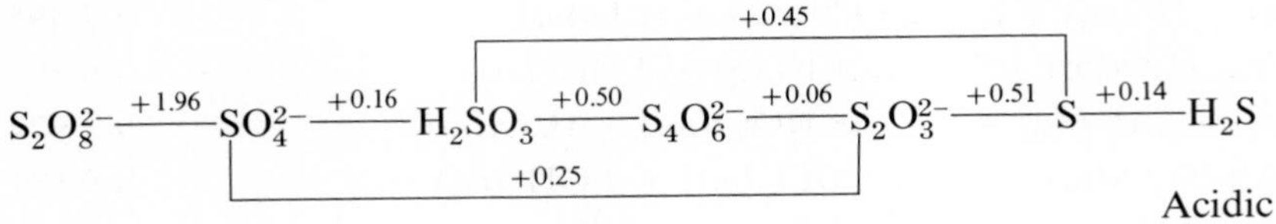

$$S_2O_8^{2-} \xrightarrow{+1.96} SO_4^{2-} \xrightarrow{+0.16} H_2SO_3 \xrightarrow{+0.50} S_4O_6^{2-} \xrightarrow{+0.06} S_2O_3^{2-} \xrightarrow{+0.51} S \xrightarrow{+0.14} H_2S$$

$H_2SO_3 \xrightarrow{+0.45} S$; $SO_4^{2-} \xrightarrow{+0.25} S_2O_3^{2-}$

Acidic solution

$$S_2O_8^{2-} \xrightarrow{+1.96} SO_4^{2-} \xrightarrow{-0.94} SO_3^{2-} \xrightarrow{-0.6} S_2O_3^{2-} \xrightarrow{-0.7} S \xrightarrow{-0.44} S^{2-}$$

$SO_3^{2-} \xrightarrow{-0.66} S$

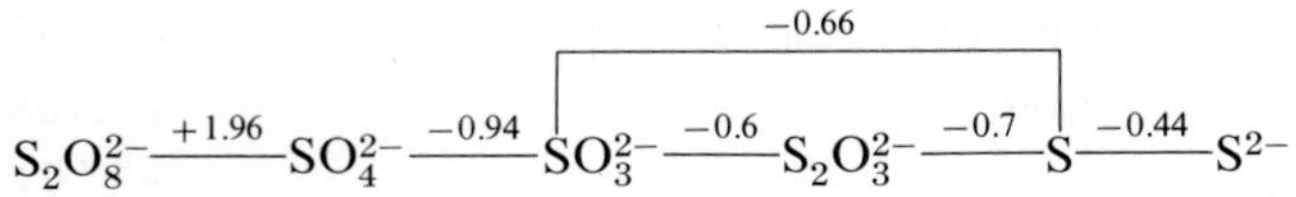

Alkaline solution

Nitrogen

See Table 18.1 and Figure 18.2.

Arsenic

$$H_3AsO_4(aq) + 2\,H^+(aq) + 2\,e^- \rightleftharpoons HAsO_2(aq) + 2\,H_2O(liq) \qquad \mathcal{E}^0 = +0.58 \text{ volt}$$

Tin

$$Sn(IV, aq\ HCl) \xrightarrow{+0.15} Sn^{2+} \xrightarrow{-0.14} Sn$$

Aluminum

$$Al^{3+}(aq) + 3\,e^- \rightleftharpoons Al(c) \qquad \mathcal{E}^0 = -1.68 \text{ volts}$$

$$Al(OH)_3(c) + 3\,e^- \rightleftharpoons Al(c) + 3\,OH^-(aq) \qquad \mathcal{E}^0 = -2.3 \text{ volts}$$

$$AlO_2^-(aq) + 2\,H_2O(liq) + 3\,e^- \rightleftharpoons Al(c) + 4\,OH^-(aq) \qquad \mathcal{E}^0 = -2.3 \text{ volts}$$

Lead

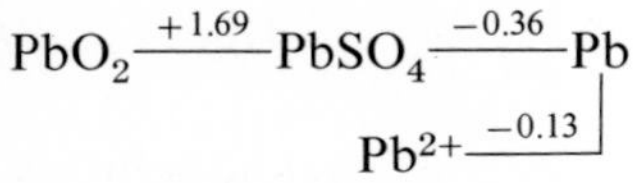

$$PbO_2 \xrightarrow{+1.69} PbSO_4 \xrightarrow{-0.36} Pb$$

$Pb^{2+} \xrightarrow{-0.13} Pb$

Acidic solution

Vanadium

$$VO_2^+ \xrightarrow{+1.00} VO^{2+} \xrightarrow{+0.34} V^{3+} \xrightarrow{-0.26} V^{2+} \xrightarrow{-1.13} V$$

Acidic solution

Chromium

$$Cr_2O_7^{2-} \xrightarrow{+1.3} Cr^{3+} \xrightarrow{-0.4} Cr^{2+} \xrightarrow{-0.9} Cr$$

Acidic solution

$$CrO_4^{2-} \xrightarrow{\sim -0.1} Cr(OH)_3 \xrightarrow{\sim -1.2} Cr(OH)_2 \xrightarrow{\sim -1.5} Cr$$

Alkaline solution

Manganese

See Figure 20.4.

Iron

$Fe^{3+} \xrightarrow{+0.77} Fe^{2+} \xrightarrow{-0.47} Fe$ Acidic solution

$Fe(OH)_3 \xrightarrow{-0.55} Fe(OH)_2 \xrightarrow{-0.92} Fe$ Alkaline solution

$Fe(CN)_6^{3-}(aq) + e^- \rightleftharpoons Fe(CN)_6^{4-}(aq)$ $\mathscr{E}^0 = +0.36$ volt

Cobalt

$Co^{3+} \xrightarrow{+1.9} Co^{2+} \xrightarrow{-0.28} Co$ Acidic solution

$Co(OH)_3 \xrightarrow{+0.1} Co(OH)_2 \xrightarrow{-0.7} Co$ Alkaline solution

Nickel

$Ni^{2+} \xrightarrow{-0.24} Ni$ Acidic solution

$Ni(OH)_3 \xrightarrow{\sim +0.5} Ni(OH)_2 \xrightarrow{-0.7} Ni$ Alkaline solution

Copper

$Cu(OH)_2 \xrightarrow{-0.10} Cu_2O \xrightarrow{-0.36} Cu$ Alkaline solution

Silver

$AgO^+ \xrightarrow{\sim +2.1} Ag^{2+} \xrightarrow{\sim +2.0} Ag^+ \xrightarrow{+0.799} Ag$

$Ag_2O_3 \xrightarrow{+1.71} Ag^{2+}$

$Ag_2O_3 \xrightarrow{+1.76} Ag^+$

$AgCl \xrightarrow{+0.222} Ag$

$AgBr \xrightarrow{+0.071} Ag$

$AgI \xrightarrow{-0.152} Ag$

Acidic solution

$$Ag_2O_3 \xrightarrow{+0.88} AgO \xrightarrow{+0.60} Ag_2O \xrightarrow{+0.34} Ag$$

$Ag_2O_3 \rightarrow Ag_2O$: $+0.74$

$Ag(NH_3)_2^+ \xrightarrow{+0.37} Ag$

$Ag(CN)_2^- \xrightarrow{-0.41} Ag$

Alkaline solution

Gold

$$Au^{3+} \xrightarrow{\sim+1.4} Au^+ \xrightarrow{\sim+1.7} Au$$

$AuCl_4^- \xrightarrow{+0.9} AuCl_2^- \xrightarrow{+1.15} Au$

$Au(CN)_2^- \xrightarrow{-0.6} Au$

Acidic solution

Zinc

	$\mathscr{E}^0$ (volts)
$Zn^{2+}(aq) + 2\,e^- \rightleftharpoons Zn(c)$	-0.762
$Zn(OH)_2(c) + 2\,e^- \rightleftharpoons Zn(c) + 2\,OH^-(aq)$	-1.24
$ZnO_2^{2-}(aq) + 2\,H_2O(liq) + 2\,e^- \rightleftharpoons Zn(c) + 4\,OH^-(aq)$	-1.19
$Zn(NH_3)_4^{2+}(aq) + 2\,e^- \rightleftharpoons Zn(c) + 4\,NH_3(aq)$	-1.01
$Zn(CN)_4^{2-}(aq) + 2\,e^- \rightleftharpoons Zn(c) + 4\,CN^-(aq)$	-1.26

Cadmium

$Cd^{2+}(aq) + 2\,e^- \rightleftharpoons Cd(c)$	-0.402
$Cd(OH)_2(c) + 2\,e^- \rightleftharpoons Cd(c) + 2\,OH^-(aq)$	-0.82
$Cd(NH_3)_4^{2+}(aq) + 2\,e^- \rightleftharpoons Cd(c) + 4\,NH_3(aq)$	-0.62
$Cd(CN)_4^{2-}(aq) + 2\,e^- \rightleftharpoons Cd(c) + 4\,CN^-(aq)$	-0.94

Mercury

$$Hg^{2+} \xrightarrow{+0.91} Hg_2^{2+} \xrightarrow{+0.80} Hg$$

$Hg^{2+} \rightarrow Hg$: $+0.85$

$Hg_2Cl_2 \xrightarrow{+0.268} Hg$

$Hg_2Br_2 \xrightarrow{+0.139} Hg$

$Hg_2I_2 \xrightarrow{-0.041} Hg$

$Hg_2SO_4 \xrightarrow{+0.615} Hg$

Acidic solution

	$\mathscr{E}^0$ (volts)
$HgO(c) + H_2O(liq) + 2\,e^- \rightleftharpoons Hg(liq) + 2\,OH^-(aq)$	$+0.098$

Alkaline Earth and Alkali Metals

$Mg^{2+}(aq) + 2\,e^- \rightleftharpoons Mg(c)$	−2.36
$Mg(OH)_2(c) + 2\,e^- \rightleftharpoons Mg(c) + 2\,OH^-(aq)$	−2.69
$Ca^{2+}(aq) + 2\,e^- \rightleftharpoons Ca(c)$	−2.87
$Sr^{2+}(aq) + 2\,e^- \rightleftharpoons Sr(c)$	−2.90
$Ba^{2+}(aq) + 2\,e^- \rightleftharpoons Ba(c)$	−2.91
$Ra^{2+}(aq) + 2\,e^- \rightleftharpoons Ra(c)$	−2.91
$Na^+(aq) + e^- \rightleftharpoons Na(c)$	−2.71
$K^+(aq) + e^- \rightleftharpoons K(c)$	−2.94

Appendix **V**

ELECTRONIC CONFIGURATIONS OF ATOMS
(Lowest energy states for gaseous atoms)

Z	*Atom*	*Orbital electronic configuration*	*Z*	*Atom*	*Orbital electronic configuration*
1	H	$1s^1$	53	I	(Kr)$4d^{10}5s^25p^5$
2	He	$1s^2$	54	Xe	(Kr)$4d^{10}5s^25p^6$
3	Li	(He)$2s^1$	55	Cs	(Xe)$6s^1$
4	Be	(He)$2s^2$	56	Ba	(Xe)$6s^2$
5	B	(He)$2s^22p$	57	La	(Xe)$5d^16s^2$
6	C	(He)$2s^22p^2$	58	Ce	(Xe)$4f^15d^16s^2$
7	N	(He)$2s^22p^3$	59	Pr	(Xe)$4f^36s^2$
8	O	(He)$2s^22p^4$	60	Nd	(Xe)$4f^46s^2$
9	F	(He)$2s^22p^5$	61	Pm	(Xe)$4f^56s^2$
10	Ne	(He)$2s^22p^6$	62	Sm	(Xe)$4f^66s^2$
11	Na	(Ne)$3s^1$	63	Eu	(Xe)$4f^76s^2$
12	Mg	(Ne)$3s^2$	64	Gd	(Xe)$4f^75d^16s^2$
13	Al	(Ne)$3s^23p$	65	Tb	(Xe)$4f^96s^2$
14	Si	(Ne)$3s^23p^2$	66	Dy	(Xe)$4f^{10}6s^2$
15	P	(Ne)$3s^23p^3$	67	Ho	(Xe)$4f^{11}6s^2$
16	S	(Ne)$3s^23p^4$	68	Er	(Xe)$4f^{12}6s^2$
17	Cl	(Ne)$3s^23p^5$	69	Tm	(Xe)$4f^{13}6s^2$
18	Ar	(Ne)$3s^23p^6$	70	Yb	(Xe)$4f^{14}6s^2$
19	K	(Ar)$4s^1$	71	Lu	(Xe)$4f^{14}5d^16s^2$
20	Ca	(Ar)$4s^2$	72	Hf	(Xe)$4f^{14}5d^26s^2$
21	Sc	(Ar)$3d^14s^2$	73	Ta	(Xe)$4f^{14}5d^36s^2$
22	Ti	(Ar)$3d^24s^2$	74	W	(Xe)$4f^{14}5d^46s^2$
23	V	(Ar)$3d^34s^2$	75	Re	(Xe)$4f^{14}5d^56s^2$
24	Cr	(Ar)$3d^54s^1$	76	Os	(Xe)$4f^{14}5d^66s^2$
25	Mn	(Ar)$3d^54s^2$	77	Ir	(Xe)$4f^{14}5d^76s^2$
26	Fe	(Ar)$3d^64s^2$	78	Pt	(Xe)$4f^{14}5d^96s^1$
27	Co	(Ar)$3d^74s^2$	79	Au	(Xe)$4f^{14}5d^{10}6s^1$
28	Ni	(Ar)$3d^84s^2$	80	Hg	(Xe)$4f^{14}5d^{10}6s^2$
29	Cu	(Ar)$3d^{10}4s^1$	81	Tl	(Xe)$4f^{14}5d^{10}6s^26p^1$
30	Zn	(Ar)$3d^{10}4s^2$	82	Pb	(Xe)$4f^{14}5d^{10}6s^26p^2$
31	Ga	(Ar)$3d^{10}4s^24p^1$	83	Bi	(Xe)$4f^{14}5d^{10}6s^26p^3$
32	Ge	(Ar)$3d^{10}4s^24p^2$	84	Po	(Xe)$4f^{14}5d^{10}6s^26p^4$
33	As	(Ar)$3d^{10}4s^24p^3$	85	At	(Xe)$4f^{14}5d^{10}6s^26p^5$
34	Se	(Ar)$3d^{10}4s^24p^4$	86	Rn	(Xe)$4f^{14}5d^{10}6s^26p^6$
35	Br	(Ar)$3d^{10}4s^24p^5$	87	Fr	(Rn)$7s^1$
36	Kr	(Ar)$3d^{10}4s^24p^6$	88	Ra	(Rn)$7s^2$
37	Rb	(Kr)$5s^1$	89	Ac	(Rn)$6d^17s^2$
38	Sr	(Kr)$5s^2$	90	Th	(Rn)$6d^27s^2$
39	Y	(Kr)$4d^15s^2$	91	Pa	(Rn)$5f^26d^17s^2$
40	Zr	(Kr)$4d^25s^2$	92	U	(Rn)$5f^36d^17s^2$
41	Nb	(Kr)$4d^45s^1$	93	Np	(Rn)$5f^46d^17s^2$
42	Mo	(Kr)$4d^55s^1$	94	Pu	(Rn)$5f^67s^2$
43	Tc	(Kr)$4d^55s^2$	95	Am	(Rn)$5f^77s^2$
44	Ru	(Kr)$4d^75s^1$	96	Cm	(Rn)$5f^76d^17s^2$
45	Rh	(Kr)$4d^85s^1$	97	Bk	(Rn)$5f^97s^2$
46	Pd	(Kr)$4d^{10}$	98	Cf	(Rn)$5f^{10}7s^2$
47	Ag	(Kr)$4d^{10}5s^1$	99	Es	(Rn)$5f^{11}7s^2$
48	Cd	(Kr)$4d^{10}5s^2$	100	Fm	(Rn)$5f^{12}7s^2$
49	In	(Kr)$4d^{10}5s^25p^1$	101	Md	(Rn)$5f^{13}7s^2$
50	Sn	(Kr)$4d^{10}5s^25p^2$	102	No	(Rn)$5f^{14}7s^2$
51	Sb	(Kr)$4d^{10}5s^25p^3$	103	Lr	(Rn)$5f^{14}6d^17s^2$
52	Te	(Kr)$4d^{10}5s^25p^4$			

Appendix VI

VAPOR PRESSURES OF WATER

Temperature, t (°C)	*Pressure, P (mm Hg)*	*Temperature, t (°C)*	*Pressure, P (mm Hg)*
−15(ice)	1.4	30	31.8
−10(ice)	2.1	40	55.3
−5(ice)	3.2	50	92.5
0	4.6	60	149.4
5	6.5	70	233.7
10	9.2	80	355.1
15	12.8	90	525.8
20	17.5	100	760.0
25	23.8	110	1074.6
		150	3570.5

ANSWERS TO PROBLEMS

Chapter 1

1. (a) $\log 3740 = 3.57$; $\ln 3740 = 8.21$;
(b) $\log 0.000489 = -3.31$; $\ln 0.000489 = -7.61$;
(c) $\log(832.1 \times 10^{-6}) = -5.08$; $\ln(832.1 \times 10^{-6}) = -11.68$;
(d) $\log(6.49 \times 10^{7}) = 7.81$; $\ln(6.49 \times 10^{7}) = 17.95$

2. (a) $\log x = 2.6037$; $x = 4.016 \times 10^{2}$;
(b) $\log x = -8.4195$; $x = 3.806 \times 10^{-9}$

3. (a) $\ln x = -2.0910$; $x = 0.123$;
(b) $\ln x = 23.456$; $x = 1.59 \times 10^{10}$

4. 4.551

5. (a) 95.5; (b) 1.23×10^{-5}; (c)2.19×10^{-7}; (d) 1.01×10^{9}

6. (a) 6.78×10^{22}; (b) 2.19×10^{3}; (c) 6.49×10^{4}

7. $x = 0.0052$

8. $x = 0.0052$

9. $x = 2 \times 10^{-9}$

10. $x = 2 \times 10^{-9}$

11. $x = 3.97 \times 10^{-4}$

12. $x = 1.32 \times 10^{-4}$

13. Intercept $= K_1$; $K_2 =$ slope/intercept

$$\frac{A - (K_1K_2/B) - K_1}{K_1} = K_3B$$

14. $\log P = -(B/2.3T) + \log A$; $B = -2.3 \times$ slope

15. $\mathrm{KE} = p^2/2m$

18. 1.27×10^{7} Btu; 3.2×10^{6} kcal

Chapter 2

1. 1.96×10^{-15}

2. (a) 19 protons and 20 neutrons;
(b) 79 protons and 118 neutrons;

(c) 11 protons and 13 neutrons;
(d) 56 protons and 85 neutrons

3. 6.9361 amu

4. 20.17 amu

5. β^-

6. $^{24}_{12}Mg$

7. $^{222}_{86}Rn$

8. 1/226 mole of He = 0.0177 g of He

9. $^{92}_{36}Kr$

10. $^{14}_{6}C$; $^{243}_{97}Bk$; $^{214}_{83}Bi$; $^{4}_{2}He$

11. $^{239}_{92}U$; $^{239}_{93}Np$; $^{239}_{94}Pu$

12. $^{22}_{11}Na \rightarrow \beta^+ + ^{22}_{10}Ne$

13. 1.662×10^{-24} g amu^{-1}; 3.61×10^{-23} g atom^{-1} of $^{19}_{9}F$; 3.6×10^{23} atoms of $^{31}_{15}P$

14. 9.58×10^4 coulomb g^{-1}, 1.20×10^4 coulomb g^{-1}

15. 5.47×10^{-4} g mole^{-1}

16. (b) $r = 1.8 \times 10^{-8}$ cm

17. 2.79×10^{21} erg penny^{-1}

Chapter 3

1. 2.41×10^{23} atoms

2. 1.99×10^{-23} g atom^{-1}

3. 45 g

4. 55 liters

5. 9.3 liters

6. 0.066%

7. 64 g mole^{-1}

8. 0.87 g liter^{-1}

9. C_4H_{10}

10. 3.2×10^{13} molecules

11. 1.15 g liter^{-1}

13. H_2 is twice as efficient as He

14. 2.6 atm

16. $(P/d) = (RT/M) + (bP/M)$

17. 1.1 cm sec^{-1}

18. 298 cal

19. 7.7×10^4 cm sec^{-1}

20. $P_{H_2} = 533$ mm Hg

21. $M = 256$ g mole^{-1}; S_8

22. 477°K

23. 1.27 atm

24. U ratio = 1.004, H_2-D_2 ratio = 1.4

25. 40.5 cm

26. 119 cm^3

27. 72.5°K
28. 71 g mole^{-1}

Chapter 4

1. 77.8 g
2. m = 0.87 mole alcohol/kg water
3. m = 0.341 mole NaCl/kg water; M = 0.339 mole NaCl/liter solution
4. P(benzene) = 42 mm Hg; P(toluene) = 24 mm Hg; P(total) = 66 mm Hg
5. 31.5 mm Hg
6. 0.517 g
7. m = 0.0167 mole impurity/kg water
8. 47 g mole^{-1}
11. m = 0.497 mole solute/kg water; 100.25°C
12. 4.6×10^3 mm Hg
13. (a) m = 83.3 mole NH_3/kg water, M = 31.0 mole NH_3/liter solution;
(b) m = 15.4 mole HCl/kg water, M = 11.7 mole HCl/liter solution;
(c) m = 3317 mole acetic acid/kg water, M = 17.6 mole acetic acid/liter solution
14. m = 16.07 mole alcohol/kg water; M = 8.63 mole alcohol/liter solution, X(alcohol) = 0.224
15. 15 ml of 1.00 M solution diluted to 100 ml final volume
16. 0.33°
17. 5.13 kg solvent × deg/moles of solute
18. 4.74°; cadmium chloride is not completely dissociated to Cd^{2+} and Cl^- ions
19. M = 256 g mole^{-1}; S_8
20. 5.0 g sucrose in 90 g of water and 7.5 g sucrose in 135 g of water
21. $C_6H_4Cl_2$
22. 93.1 g mole^{-1}
23. P(total) = 514 mm Hg, X(ethyl alcohol) = 0.282 in vapor
24. 202 g mole^{-1}
25. 100.42°C; 31.3 mm Hg

Chapter 5

1. Mutual orientation of water molecules decreases with increasing temperature
2. F(NaCl in alcohol)/F(NaCl in water) = 3.2
3. Water-alcohol mixtures have lower dielectric constants than has water
5. HCl(g) is not ionic, whereas NaCl(c) is ionic
6. H_2O_2 molecules are nonlinear with extensive hydrogen bonding in H_2O_2(liq)
7. 1.8×10^{-9}, fraction of water molecules ionized
8. Volume increases by 1.67 cm^3 mole^{-1}
9. Polar HBr molecules are attracted to each other by van der Waals forces *and* by dipole-dipole forces, whereas only van der Waals forces attract Kr atoms to each other
10. ~100°K

11. 55 million gal
13. 0.083 is fraction of iceberg above lake, 0.105 is fraction of iceberg above sea water.

Chapter 6

1. 3.8×10^{14} sec^{-1}; visible region
2. 6500 Å
3. (a) 1.0×10^{5} sec^{-1}, 6.62×10^{-22} ergs;
   (b) 1.0×10^{13} sec^{-1}, 6.62×10^{-14} ergs;
   (c) 2.0×10^{15} sec^{-1}, 1.32×10^{-12} ergs;
   (d) 3.0×10^{21} sec^{-1}, 2.0×10^{-6} ergs
4. 7.94×10^{-12} ergs photon^{-1}; 115 kcal mole^{-1}; ultraviolet region
5. $\nu_0 = 1.13 \times 10^{15}$ sec^{-1}; $\lambda = 2.66 \times 10^{-5}$ cm
6. 9.8×10^{7} cm sec^{-1}
7. 10 kcal mole^{-1} = 1.05×10^{14} sec^{-1} = 2.86×10^{4} Å = 3.50×10^{3} cm^{-1} (infrared)
   100 kcal mole^{-1} = 1.05×10^{15} sec^{-1} = 2.86×10^{3} Å = 3.50×10^{4} cm^{-1} (ultraviolet)
8. for $n = 1$, $v = 2.2 \times 10^{8}$ cm sec^{-1}, $E = 2.2 \times 10^{-11}$ ergs; $n = 2$, $v = 1.1 \times 10^{8}$ cm sec^{-1}, $E = 5.5 \times 10^{-12}$ erg
9. 6.56×10^{3} Å and 4.86×10^{3} Å
10. 13.6 eV
11. 1.5 eV
12. $r = 1.2 \times 10^{-8}$ cm; $v = 2.9 \times 10^{8}$ cm sec^{-1}
13. (a) 109,737 cm^{-1}; (b) 9.107×10^{-28} g; (c) 109,701 cm^{-1}
14. 329,031 cm^{-1}
15. 1.8×10^{8} cm sec^{-1}
16. v(electron) = 7.3×10^{8} cm sec^{-1}; v(proton) = 4.0×10^{5} cm sec^{-1}
17. 2.09×10^{-5} cm
18. 8.26×10^{-9} cm
19. 1.2×10^{-9} cm
20. (a) 1.2×10^{-21} cm (b) 1.9×10^{-4} cm
22. 5.8×10^{3} °K
23. $n = 3$
24. 7.85×10^{7} cm sec^{-1}
25. 1.08×10^{-23} Å

Chapter 7

1. See Table 7.1 for neutral atoms; Li^{+} has (He) configuration; Ca^{2+} and Cl^{-} have (Ar) configuration
2. $2n^2$
3. O^{2-} and Si^{4+} have (Ne) configuration; K^{+} has (Ar) configuration
6. Plots are not perfectly periodic
7. Actual values for boiling point and density of $GeCl_4$ are within 2% of average values obtained from $SiCl_4$ and $SnCl_4$
8. Increasing ΔH solution is result of increasing dipole-induced dipole interactions

10. Element 118 would be a noble gas; values can be obtained by extrapolation

11. The actual values for the boiling point (−33 °C) and melting point (−78 °C) are quite different from the extrapolated values. Simple predictions based on periodic properties ignore "other considerations," in this case hydrogen bonding

12. Largest radius (a) I, (b) Sr, (c) K, (d) Br, (e) Sb

13. Adding electrons to the "inner" 3*d* orbitals in the transition metals leads to more effective shielding of the increasing nuclear charge than when filling orbitals of same principle quantum number

14. HF, 1.08Å; ClF, 1.70 Å; HBr, 1.51 Å; BrCl, 2.13 Å

15. Radius decreases as Z increases. All are helium-like species, but nuclear charge is increasing.

16. (a) Be; (b) Cl; (c) Cl; (d) Sr; (e) Xe; (f) N

17. (a) Na; (b) Li; (c) Cl; (d) Rb

18. Radius is most important, if we can assume complete "shielding" of nuclear charge by inner electrons

19. (b) The first true noble gas compound was prepared this way

20. Less energy to remove 4*s* electron from Li

21. 54.4 eV

22. 1160 kJ $mole^{-1}$

24. −5.83 eV, energy absorbed on adding second electron and for overall process

26. Electronegativity difference between Cl and I is 0.5; a reference standard must be chosen

27. Electronegativity values on the Mulliken scale are ~2.6 times greater than on Pauling scale

28. Melting point greater, other properties less

Chapter 8

1. Ionic: CsBr, CaO, NaH; Covalent: PCl_3, CCl_4, ClF_3, SF_6; ClF_3 and SF_6 "violate" the octet rule

3. $Cl_2C{=}O$; H—N—N≡N, H—N=N=N, H—N≡N—N; S=C=S, H—C≡N, P≡P, $C{\equiv}C^{2-}$

4.

```
  :Ö:            :Ö:            :O:
   |              |              ||
   S      ↔       S      ↔       S
 //  \          /  \\          /   \
:O.  .Ö:      :Ö.  .O:       :Ö.  .Ö:
```

5. (a) For example; H—F, P—F, B—F; (b) For example: P—H, B—H, P—B, C—H; (c) Any other alkali halide

6. NO_3^- is isoelectronic with SO_3 and has similar structure. See problem 4 answer.

7. (b) Ethane; (c) acetylene

8. BBr_3, trigonal planar; SF_6, octahedral; NH_4^+, tetrahedral; BrF_4^-, square planar; H_3O^+, pyramidal; Cl_2O, bent

9. PCl_3, H_2S, and ClF_3 have dipole moments.

10. Dipole moment increases as the electronegativity of the halide increases

11.

```
      Ö                 Ö
    //  \      ↔      /  \\
  .O.   .Ö:         :Ö.   O.
```

12. $\mathrm{O{=}\dot{N}{-}O} \rightleftharpoons \mathrm{O_2N{-}NO_2}$

13. XeF_2 is linear and XeF_4 is square planar; neither has a dipole moment
14. $:N\vdots\vdots N:$
15. See discussion of O_2 bonding in Chapter 10.
16. In valence bond theory atoms come together and their electron clouds interact to form molecules; in molecular orbital theory the nuclei (plus inner shells) are placed in equilibrium positions and valence electrons placed in molecular orbitals encompassing the entire molecule
18. (a) In NH_3 the bond pair moments are reinforced by the lone pair moment; (b) The electronegativity of F is greater than that of Cl
19. (a) *sp*; (b) sp^3; (c) sp^2
20. (b) sp^2 in free BF_3, sp^3 in addition compound
21. (a) σ orbitals from $A(3s) \pm B(3s)$ and $A(3p_z) \pm B(3p_z)$, π orbitals from $3p_y$ and $3p_x$ orbitals of A and B; (b) See Figures 8.6 and 8.7
22. (a) Similar to Figure 8.8; (c) Triple bond
23. $N_2 \rightarrow N_2^+ + e^-$; electron lost from bonding molecular orbital; $O_2 \rightarrow O_2^+ + e^-$; electron lost from antibonding molecular orbital
24. He_2^+, S_2, and Li_2^- should exist; all would be paramagnetic
25. Going from O_2^+ to O_2 to O_2^- involves adding electrons to antibonding molecular orbitals so that bond strength decreases and bond length increases
26. NO^+ has a shorter bond length; NO is paramagnetic
27. CN^- has highest bond dissociation energy; CN is paramagnetic.
28. sp^2 hybridization implies a planar molecule and NH_3 is pyramidal
29. *KK* represents filled inner shells; $(\sigma_{1s})^2(\sigma_s^*)^2$
30. O should be negative end of molecule since it has a higher electronegativity; the dipole moment of CS should be less than that of CO. Actually C is the negative end of the CO molecule, and CS has the greater dipole moment.

Chapter 9

1. 1.23 g
2. (a) K, +1; Cl, +5; O, −2
   (b) K, +1; Cl, +7; O, −2
   (e) K, +1; Mn, +7; O, −2
   (g) Na, +1; Cr, +6; O, −2
   (h) K, +1; Cr, +6; O, −2
   (n) N, −3; H, +1
   (p) I, +5, O, −2
3. (a) Al is oxidized and is the reducing agent; Cr in Cr_2O_3 is reduced and is the oxidizing agent

$$2\,Al(c) + Cr_2O_3(c) \rightarrow Al_2O_3(c) + 2\,Cr(c)$$

   (c) Not an oxidation-reduction reaction
   (d) Cl in $KClO_3$ is reduced and is the oxidizing agent; O in $KClO_3$ is oxidized and

is the reducing agent

$$2\ KClO_3(c) \rightarrow 2\ KCl(c) + 3\ O_2(g)$$

4. 1.19×10^{-4} mole of water vapor; 3.57×10^{-3} mole of other gas

5. 67.6 g U from 100 g UF_6; 67.6 lb from 100 lb UF_6

6. 3.57 moles

7. 10 moles

8. 503 g

9. $P(\text{final}) = P(\text{initial})/3$

10. 0.0193 mole

11. 0.464 mole

12. (a) $5\ HOCl(aq) + Br_2(aq) + H_2O(liq) \rightarrow 5\ Cl^-(aq) + 2\ BrO_3^-(aq) + 7\ H^+(aq)$
Br_2 is oxidized and is the reducing agent; Cl in HOCl is reduced and is the oxidizing agent

(b) $2\ MnO_4^-(aq) + 5\ HNO_2(aq) + H^+(aq) \rightarrow 2\ Mn^{2+}(aq) + 3\ H_2O(liq) + 5\ NO_3^-(aq)$
Mn in MnO_4^- is reduced and is the oxidizing agent; N in HNO_2 is oxidized and is the reducing agent

(f) $4\ Zn(c) + NO_3^-(aq) + 10\ H^+(aq) \rightarrow NH_4^+(aq) + 4\ Zn^{2+}(aq) + 3\ H_2O(liq)$
Zn is oxidized and is the reducing agent; N in NO_3^- is reduced and is the oxidizing agent

(h) $2\ Cu^{2+}(aq) + 5\ I^-(aq) \rightarrow 2\ CuI(c) + I_3^-(aq)$
Cu^{2+} is reduced and is the oxidizing agent; the reducing agent is I^-, which is oxidized (note that I_3^- may be pictured as $I_2 + I^-$)

13. (a) $2\ MnO_4^-(aq) + 3\ NO_2^-(aq) + H_2O(liq) \rightarrow 2\ MnO_2(c) + 3\ NO_3^-(aq) + 2\ OH^-(aq)$
Mn in MnO_4^- is reduced and is the oxidizing agent; N in NO_2^- is oxidized and is the reducing agent

(c) $3\ Cl_2(aq) + 6\ OH^-(aq) \rightarrow 5\ Cl^-(aq) + ClO_3^-(aq) + 3\ H_2O(liq)$
Cl_2 is both oxidized and reduced and is therefore both oxidizing agent and reducing agent in this reaction

(f) $2\ Al(c) + OH^-(aq) + 2\ H_2O(liq) \rightarrow 2\ AlO_2^-(aq) + 3\ H_2(g)$
Al is oxidized and is the reducing agent; the H in H_2O is reduced and is the oxidizing agent

14. HOCl and HNO_2 are weak acids; nitric acid is a strong acid that exists as $H^+(aq)$ and $NO_3^-(aq)$

15. 2.5×10^{-4} mole of $H^+(aq)$ ions, or 1.5×10^{20} $H^+(aq)$ ions

16. 0.126 liter or 126 ml

17. 89.5% Mg

18. V_2O_5

19. C_4H_8

20. SnO_2

21. Atomic weight is $130.6/n$ in which n is any small integer; taking $n = 2$ leads to 65.3 g mole^{-1}, which would indicate that the metal might be zinc.

22. 101 g acetylene

23. 88.7% $BaCO_3$

24. 81.1% $BaCO_3$

25. 40.0% NaCl

26. 0.858 mole

27. 79.1% $BaCl_2$

Chapter 10

1. ${}^3_1H \rightarrow \beta^- + {}^3_2He$

2. (b) 0.29 liters

3. Since O_3 is polar and O_2 is not, dipole-dipole attractions between O_3 and H_2O are greater

4. (a) $D_2O + SO_3$; (b) $D_2O + Na$; (c) $D_2O + BBr_3$; (d) $D_2O + NaD$; (e) $Li + D_2$ (obtained from electrolysis of D_2O)

5. (b) Li releases most H_2

6. $\Delta V = +528$ liters

7. The isotopes of hydrogen have a uniquely large mass difference, and rates of chemical reactions are affected by the masses of the reacting species

8. ${}^6_3Li + {}^1_0n \rightarrow {}^3_1H + {}^4_2He$

9. Producing such low temperatures is expensive, and liquid hydrogen is dangerous to handle

10. 1.12 moles CO_2

11. (a) HCl; (b) CO; (c) H_2S; (d) H_2O

12. (b) 22.0 tons CH_3OH; (c) 19.2 tons

13. 82 g $KClO_3$

14. 3.1 liters air

15. 1.1 liters H_2O

16.

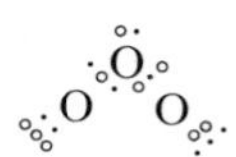

17. Use Figure 8.8 and follow aufbau rules for filling molecular orbitals; O_2^+, O_2^-, and S_2 are paramagnetic

18. (a) $H_2O(liq) + O_3(g) + 2\ I^-(aq) \rightarrow O_2(g) + 2\ OH^-(aq) + I_2(aq)$
(b) 4.9 ml

20. 6.85 atoms molecule^{-1}

21. (c) \$80, but ignoring supply and demand, transportation costs, recovery equipment costs, etc. The calculation also ignores the social costs of deterioration of health, cars, and buildings.

22. 17 *M* and 136 *m*

24. Structure similar to thiosulfate

25. SF_6, octahedral; $SeCl_4$, irregular tetrahedron; $ClSO_3^-$, tetrahedral; SO_2F_2, tetrahedral; $TeBr_2$, bent

26. 0.73 liters

27. AO is basic: BO_2 is acidic

28. 23.0% SO_2F_2

29. 3.36 Å

31. (a) H_2SeO_4; (b) HOCl; (c) CrO_3; (d) $HOSO_2F$

32. (b) 25%

Chapter 11

1. (a) -32.83 kcal mole^{-1}; (b) -42.07 kcal mole^{-1}; (c) -55.83 kcal mole^{-1}; (d) -43.7 kcal mole^{-1}

2. $\Delta H^0 = -8.61$ kcal mole^{-1}
3. $\Delta H_f^0 = -140.76$ kcal mole^{-1} for $MoO_2(c)$
4. 600 cal
5. 941.6 cal
7. 57.5°C
8. $\Delta E = 52$ cal $= 2.15$ liter atm
9. $\Delta H = 0$
10. 210.7 J $=$ 50.35 cal
11. (a) 11924 J $=$ 11.924 kJ; (b) 865 cal; (c) 62010 J; (d) 27,500 cal $=$ 115,000 J; (e) 5.65×10^{-13} ergs molecule^{-1}; (f) 0.079 kwh
12. (a) $\Delta H = +32$ kcal mole^{-1}; (b) $\Delta H = +21$ kcal mole^{-1}; (c) $\Delta H = +21$ kcal mole^{-1}; (d) $\Delta H = +53$ kcal mole^{-1}
13. Require 4100 g of H_2O, which gives 4100 ml or 4.1 liters of water; one could get plenty of exercise going to the toilet
14. $\Delta H_f^0 = -530$ kcal mole^{-1}
15. ΔH^0 (from bond energies) $= -11$ kcal/mole; true ΔH^0 is expected to be more negative than this value
16. $\Delta H_f^0 = -10.8 - 5.5n$ (expressed in kcal mole^{-1})
17. $\Delta H^0 = -57.0 - 146.3n$ (expressed in kcal mole^{-1})
19. $\Delta H^0 = +5.49$ kcal mole^{-1}
20. $\Delta H^0 = +19.0$ kcal mole^{-1}
21. ΔH^0 (from bond energies) $= +1278$ kcal mole^{-1}; ΔH^0 (first law) $= 1321$ kcal mole^{-1}; real benzene is more stable than expected on basis of simple valence bond picture, hence, one needs concept of resonance, or must turn to molecular orbital approach

Chapter 12

1. Positive
2. Positive ΔS leading to negative ΔG
3. 2.75 cal deg^{-1}
4. 7.4 kcal mole^{-1}
5. ΔS(environment) is large (positive) enough to make ΔS(total) be positive
6. ΔS_{water} is positive
7. (a) $T_f = (T_h + T_c)/2$; (b) 54.2°C
8. Since $T_f^2 = T_h T_c + (\delta^2/2)$, $T_f^2 > T_h T_c$
9. ΔS(gas) $= 0$ and T(gas) increases
10. $t_f = 35.8$°C and $\Delta S = 0.90$ cal deg^{-1}
11. $$w = RT \ln\left(\frac{V_2 + b}{V_1 - b}\right) + \left(\frac{1}{V_2} - \frac{1}{V_1}\right)$$
13. The entropy of the room decreases as its temperature is lowered; however the entropy of the heat pump, associated machinery, and the environment increases enough to make total ΔS positive
14. Temperature of the room will increase.
15. Under ordinary conditions, the kitchen will get warmer when the door of the refrigerator is left open. If the motor and exhaust are located outside the kitchen, then the room will be cooled.

Chapter 13

1. (b) $K = (P_{Br_2})(P_{HCl})^2/(P_{Cl_2})(P_{HBr})^2$
   (c) $K = (P_{NH_3})^2/(P_{N_2})(P_{H_2})^3$
2. (a) $K = 1.5 \times 10^6$; (b) $K = 1.3 \times 10^{14}$; (c) $K = 6.0 \times 10^5$; (d) $K = 2.6 \times 10^{12}$
3. (a) $\Delta H^0 = -13.64$ kcal mole^{-1}; (b) $\Delta H^0 = -19.34$ kcal mole^{-1}; (c) $\Delta H^0 = -22.04$ kcal mole^{-1}; (d) $\Delta H^0 = -23.64$ kcal mole^{-1}; all equilibrium constants in this problem decrease with increasing temperature
4. 346°K or 73°C
6. $P \cong 285$ mm Hg
7. $\Delta H_{vap} = 7.9$ kcal mole^{-1} and $T_b = 349$°K
8. $\Delta H_{vap} = -2.303RA$
9. $\log P = (4.585/T)(T - T_b)$
10. -38.15°C
11. $K = 1.82$
12. $K = 1.82$
13. $\alpha = 0.26$ and $M_{ave} = 73$ g mole^{-1}.
14. $K = 7.24 \times 10^{-5}$
15. $\Delta H = 0$; $\Delta S = nR \ln (V_2/V_1)$; and $\Delta G = nRT \ln (P_2/P_1)$
16. $n_{NH_3} = 0.315$ mole; $n_{N_2} = 1.342$ mole; $n_{H_2} = 2.043$ mole
19. High pressure, low temperature

Chapter 14

1. $[H^+] = 2.24 \times 10^{-3}$ M
2. $[H^+] = 7.6 \times 10^{-7}$ and $[OH^-] = 1.3 \times 10^{-8}$ M
3. $[H^+] = 4.7 \times 10^{-5}$ M
4. pH = 10.43
5. $Ac^-(aq) + H_2O(liq) \rightleftharpoons HAc(aq) + OH^-(aq)$; $K = 0.556 \times 10^{-9}$; pH = 9.02
6. $[H^+] = 1.62 \times 10^{-7}$ M in 1×10^{-7} M HCl; $[H^+] = 1.05 \times 10^{-7}$ M in 1×10^{-8} M HCl
7. $[H^+]$ is between 1.611×10^{-7} and 1.612×10^{-7} M
8. (a) $[H^+] = 2.4 \times 10^{-6}$ M; (b) pH = 9.22; (c) pH = 10.77
9. pH = 11.7 when $[HQ]/[Q^-] = 2$; in 1.0 M HQ solution, $[H^+] = 2 \times 10^{-13}$ M
10. pH = 4.43
11. $[OH^-] = 1.1 \times 10^{-5}$ M; $[H^+] = 9.1 \times 10^{-10}$ M; $[NH_3] = 0.020$ M; $[NH_4^+] = 0.033$ M
12. $[H^+] = 1.8 \times 10^{-5}$ M; $[Ac^-] = 0.5$ M; $[Na^+] = 0.5$ M; $[HAc] = 0.5$ M; $[OH^-] = 0.56 \times 10^{-9}$ M
13. $K_b = 4 \times 10^{-10}$, $K_a = 2.4 \times 10^{-5}$, K_a ("best") $= 2.53 \times 10^{-5}$; small discrepancy might be due to experimental error in the quoted (OH^-), to an impurity in the aniline, or to neglect of activity coefficients
14. $[H^+] = 3.5 \times 10^{-3}$ M and pH = 2.45
16. pH = $(1458/T) + 2.109$
17. pH = 2.29
18. pH = 0.31
20. $[H^+] = 1.33 \times 10^{-3}$ M in 0.1 M HAc; $[H^+] = 1.44 \times 10^{-3}$ M in impure solution;

$K_a = 2.1 \times 10^{-5}$ on basis of impure solution; $K_a = 1.81 \times 10^{-5}$ on basis of partly neutralized impure solution

22. $pK_w = pH + pOH$

23. $[Cl^-]/[I^-] = 2.08 \times 10^6$

24. 5.3×10^{-7} mole Ag_2CrO_4 dissolved in 100 ml; if $AgCrO_4^-(aq)$ is formed, real solubility will be greater than calculated here

25. $K_{sp} = 1 \times 10^{-15}$; $K_c = 5 \times 10^5$; $K_a = 5 \times 10^{-9}$

26. Graph leads to minimum solubility at $[Cl^-] \cong 3 \times 10^{-3}$ *M*; differentiation leads to minimum solubility at $[Cl^-] = 2.6 \times 10^{-3}$ *M*.

Chapter 15

1. ${}^{223}_{87}Fr \rightarrow {}^{4}_{2}He + {}^{219}_{85}At$

2. 32 g

3. (a) 0.0833 g H_2; (b) 0.0952 g H_2; (c) use CaH_2 with (mass CaH_2)/(mass H_2O) = 1.13

4. (a) 3570 cal g^{-1}; (b) 0.124 g; (c) 11,700°K

5. $CO_3^{2-}(aq) + H_2O(liq) \rightleftharpoons HCO_3^-(aq) + OH^-(aq)$

6. $Na(c) + H_2O(liq) \rightarrow Na^+(aq) + OH^-(aq) + \frac{1}{2} H_2(g)$;
$H_2(g)$ may react (explosively) with $O_2(g)$ in air to form water

7. $CaCO_3(c) \rightarrow CaO(c) + CO_2(c)$;
$CaO(c) + H_2O(liq) \rightarrow Ca(OH)_2(c)$ or $Ca^{2+}(aq) + 2\ OH^-(aq)$

8. (a) (i) $3\ Mg + N_2 \rightarrow Mg_3N_2$; (ii) $Mg_3N_2 + 6\ H_2O \rightarrow 3\ Mg(OH)_2 + 2\ NH_3$; (iii) $Mg(OH)_2 + 2\ HCl \rightarrow MgCl_2 + 2\ H_2O$; (iv) $MgCl_2 \rightarrow Mg + Cl_2$
(b) HCl for step (iii) is expensive, as is electricity for step (iv)

9. 1 mole of calcium bicarbonate is removed from solution, and 2 moles of calcium carbonate are precipitated.

10. (a) 2 moles HCl; (b) heat limestone to form CaO, which is reacted with water to give the desired $Ca(OH)_2$, 1 mole of Cl_2 is formed as by-product accompanying each mole of Mg

11. $K_{sp} = 1.05 \times 10^{-10}$

12. $K_{sp} = 1.74 \times 10^{-10}$

13. (a) Aragonite is more soluble than calcite; (b) $K_{sp} = 7.53 \times 10^{-9}$ for aragonite, $K_{sp} = 4.94 \times 10^{-9}$ for calcite; (c) 50°K; (d) yes

14. $S^0 \cong 20$ cal deg^{-1} $mole^{-1}$ for CaCl(c); $T\Delta S^0 \cong -1.5$ kcal $mole^{-1}$, which is small compared to $\Delta H^0 \cong -82$ kcal $mole^{-1}$

15. $\Delta H_f^0 \cong 144$ kcal $mole^{-1}$ for $KCl_2(c)$, which is clearly unstable

16. They react spontaneously with O_2 and H_2O

17. Increase the pH

18. Reacts with O_2 and H_2O in air; low strength, too

Chapter 16

1. 2.01 g

2. 0.592 g

3. 33.6 g

4. 424 cm^3

5. Cathode: $2\ H_2O(liq) + 2e^- \rightarrow H_2(g) + 2\ OH^-(aq)$; anode: $4\ OH^-(aq) \rightarrow O_2(g) + 2\ H_2O(liq) + 4\ e^-$
6. Ag will be in anode sludge and Zn^{2+} will be in solution
7. $K = 3.6 \times 10^{-21}$
8. $K = 3.1$; $[Ag^+] = [Fe^{2+}] = 0.044\ M$, $[Fe^{3+}] = 0.006\ M$
9. K(first reaction) $= 3.06 \times 10^{-16}$; K(second reaction) $= 1.75 \times 10^{-8}$; $K_2 = K_1^{\frac{1}{2}}$
10. (a) $\mathscr{E}^0 = 0.29$ volt; (c) Electrons flow through the external circuit from the zinc electrode to the iron electrode, with the latter being the cathode when the cell is delivering current; (d) $\mathscr{E} = 0.23$ volt when $[Zn^{2+}] = 10\,[Fe^{2+}]$
11. $[Ag^+] = 3.9 \times 10^{-9}$
12. (a) $\mathscr{E}^0 = 1.092$ volt; electrons flow from the calomel electrode through the external circuit to the chlorine electrode when the cell is delivering current
13. (a) $\mathscr{E}^0 = 0.066$ volt; (b) $2\ Ag(c) + Hg_2Br_2(c) \rightleftharpoons 2\ AgBr(c) + 2\ Hg(liq)$; (c) no
14. $K_{sp} = 1.2 \times 10^{-12}$
17. (b) $\Delta S^0 = -14.9$ cal deg^{-1} mole^{-1} and $d\mathscr{E}^0/dT$ is negative; $\mathscr{E}^0$ decreases with increasing temperature
    (c) $d\mathscr{E}^0/dt = -0.000646$ volt deg^{-1} and $\Delta H^0 = -9.57$ kcal mole^{-1}
    (d) $\Delta H^0 = -9.58$ kcal mole^{-1} from ΔH_f^0 values and -9.57 from $\Delta H^0 = \Delta G^0 + T\,\Delta S^0$
18. (a) pH $= 4.22$ and $[H^+] = 6.03 \times 10^{-5}$; (b) $K_a = 6.2 \times 10^{-5}$
20. Al is more easily oxidized than Fe, but oxide coating on Al protects the metal against further rapid oxidation
21. See nickel discussion in Chapter 20.

Chapter 17

1. The order of increasing boiling points is: CF_4, CCl_4, CBr_4, CI_4; CI_4 has the largest number of electrons per molecule, is most easily polarized, and experiences the strongest van der Waals attractions
2. 51 g I_2
3. MnO_4^-, BrO_3^-, HOCl, $S_2O_8^{2-}$, F_2, etc.
4. Only the reaction with I^- would be spontaneous
5. CaF_2 is very slightly soluble in water
6. The O—Cl—O bond angle is 110.8°; the molecule is paramagnetic
7. 72 g F_2
8. -56 kJ mole^{-1} or -13 kcal mole^{-1}
10. (a) $AsCl_3 + H_2O \rightarrow 3\ HCl + As(OH)_3$; (b) $Al_2Br_6 + 6\ H_2O \rightarrow 6\ HBr + 2\ Al(OH)_3$; (c) $BBr_3 + 3\ D_2O \rightarrow 3\ DBr + B(OD)_3$
11. (b) 0.018 moles HCl
12. 540 liters HBr
13. ΔH^0 for the first reaction is -11.17 kcal mole^{-1}; ΔH^0 for the second reaction is -12.80 kcal mole^{-1}
14. 0.69 moles NH_3
15. 3.8×10^{-5} moles AgI
16. 9.9×10^{-3} moles AgCl
17. (a) AgBr precipitates first; (b) $[Br^-] = 1.8 \times 10^{-6}\ M$
18. $\Delta H^0(KBr) = +4.7$ kcal mole^{-1}

19. Bent, an AX_2E_2 case

20. ClO_2^-, bent; Cl_2O, bent; BrO_3^-, pyramidal; HIO_4, tetrahedral; I_3^-, linear; IO_6^{5-}, octahedral

21. $K = 2.8 \times 10^{-5}$

23. $K = 7.8 \times 10^{15}$

24. 41 g

25. 7.7 liters

26. $Cl_2(g)$ liberated

27. Only HOBr will disproportionate

28. Precipitation of $BaSO_4$ helps shift equilibrium

29. (a) $HBrO_4$ a stronger acid than periodic acids, weaker than $HClO_4$; (b) HOF strongest HOX acid; (c) HAt strongest HX acid

30. HBr, HOI and HF, $HClO_2$

31. ICl_2^-, linear; BrF_3, bent—T; ClF_5, square pyramidal; BrF_6^-, pentagonal bipyramidal

32. Solubility in water is 3.2×10^{-4}

33. (a) lowest; (b) lowest; (c) greatest; (d) greatest; (e) greatest; (f) lowest

34. $T\Delta S^0$ terms are ~15% of total ΔG^0, therefore interpretations based on ΔH^0 are reasonable

35. EA (PtF_6) at least 156 kcal $mole^{-1}$

36. 103°C, but not valid to use Clausius-Clapeyron equation

37. XeO_6^{4-}, octahedral; XeO_4, tetrahedral

Chapter 18

1. (a) 1.250 g $liter^{-1}$

2. (b) 245 g NH_3 from Li, 46.6 g NH_3 from Mg

4. $\Delta G^0 = -851$ kJ $mole^{-1}$

5. NO_3^-, N_2

6. Large positive ΔS^0

9. H_2SO_4 has a high boiling point and is a dehydrating agent

10. Diffusion rate inversely proportional to $\sqrt{M}$

11. pH = 11.5

12. Yes to first part; N_2O is more favored thermodynamically

13. As T increases, K decreases

15. N_3^- and ON_3^-, which should not be very stable

16. 13.7 g

17. Ag_2NH a base in NH_3(liq)

18. (b) Haber process plus Ostwald process

19. 42.5 pounds NH_3, hydrolysis of carbonate increases pH

20. (a) linear; (b) $OSCN^-$ and SCN^- are products

21. similar to bonding in N_2

22. (d) $Na_3P_3O_9$ is used for "building" soaps

23. PBr_3, pyramidal; PF_5, trigonal bipyramidal; $POCl_3$, tetrahedral; PF_6^-, octahedral; PCl_4^+, tetrahedral

24. 0.99 tons

25. pH = 2.1
28. 146 g
29. $[H^+] = [H_2PO_4^-] = 8.1 \times 10^{-2}$ *M*; $[HPO_4^{2-}] = 7.2 \times 10^{-5}$ *M*; $[PO_4^{3-}] = 6 \times 10^{-9}$ *M*
30. $[H^+] = 1.3$ *M*
31. Arsenic acid can only oxidize I^- to I_2 and becomes stronger oxidizing agent at low pH

Chapter 19

1. $\Delta H^0 = -6.81$ kcal mole^{-1}
3. (a) $CH_3(CH_2)_3CH_3$, $(CH_3)_2CHCH_2CH_3$, $(CH_3)_4C$; (b) $CH_3(CH_2)_4CH_3$, $(CH_3)_2CH(CH_2)_2CH_3$, $(CH_3)_2CHCH(CH_3)_2$, $CH_3CH_2CH(CH_3)CH_2CH_3$, $(CH_3)_3CCH_2CH_3$
4. C_2H_6, C_2H_4, C_4H_{10}
5. 2.0×10^{-6} moles, 1.2×10^{18} molecules
6. Bond enthalpies are related to dissociation into atoms or groups of atoms; only the high energy π electrons of unsaturated compounds are involved in most reactions
7. Five alkenes plus cyclopentane
8. Average C—F bond energy in CF_4 is 117 kcal mole^{-1}
9. (a) C_8H_{18}
12. 40 kcal mole^{-1} "resonance energy"
13. (b) Two isomers from ortho- or meta- case, one isomer from para- form
14. $C_{2n}H_{n+3}$
16. (a) $C{\equiv}CH^-$ strongest base, OH^- weakest base
17. (b) $\Delta S^0 = -52.35$ cal mole^{-1} deg^{-1}, $\Delta G^0 = -5.94$ kcal mole^{-1}
18. $CH_3OC_4H_9$ (four isomers), $C_2H_5OC_3H_7$ (two isomers)
19. HCl is a stronger acid than H_2O
20. Hydrogen bonding in alcohols
21. (b) Ketones, ethers, and alcohols give negative Tollens' test
22. In valence bond theory, resonance structures; in molecular orbital theory, a delocalized π bond.
23. At 110°C, $M_{app} = 90.9$ g mole^{-1}; at 156°, $M_{app} = 71.2$ g mole^{-1}; greater dissociation of dimer with increasing *T*.
24. $\Delta S^0 = -52.3$ cal mole^{-1} deg^{-1}
25. CF_3COOH has largest K_a due to high electronegativity of F
26. Line (a) consistent with observations
27. (b) reduced dipole-dipole attractions between amino acid and water at isoelectric point
28. 24 possible peptides
29. glucose
30. ~2200
33. 91.1% cystine

Chapter 20

1. (a) 118; (b) 122; (c) 121
2. No

4. (a) Cu, Zn, Pd, Ag, and Cd have filled *d* orbitals; (b) La, Yb, Lu, Ac, Th, No, and Lw would be excluded

5. Mo^{3+}, 3; Pt^{2+}, 2; Ce^{3+}, 1; U^{3+}, 3

7. Lanthanide contraction

8. Cathode: $La^{3+} + 3\,e^- \rightarrow La$; anode: $3\,Cl^- \rightarrow \frac{3}{2}\,Cl_2 + 3\,e^-$; 5.2 g La

9. $5\,Zn(c) + 8\,H^+(aq) + 2\,VO_2^+(aq) \rightarrow 5\,Zn^{2+}(aq) + 2\,V(c) + 4\,H_2O(liq)$

10. UO_3 more acidic

11. 46.5 tons Cr

12. $4\,Fe^{2+}(aq) + O_2(g) + 4\,H^+(aq) \rightarrow 4\,Fe^{3+}(aq) + 2\,H_2O(liq)$; $\Delta G^0 = -178$ kJ mole^{-1}

13. See chromium discussion in text

14. Solubility of chromates increases as pH decreases due to formation of $Cr_2O_7^{2-}(aq)$

15. (a) Yes; (b) no; (c) no; (d) no; (e) yes

16. Cathode

17. $\Delta H^0 = -428$ kcal mole^{-1}

18. $K = 2.8 \times 10^8$

19. (a) MoO_3; (b) V_2O_5; (c) Fe_2O_3; (d) WO_3

20. Fe becomes anode, where oxidation occurs, in presence of Sn

21. pH increases

22. $d = 6.3$ g liter^{-1}

23. (b) $2\,S_2O_8^{2-}(aq) + Cr^{2+}(aq) + 3\,H_2O(liq) \rightarrow 4\,SO_4^{2-}(aq) + CrO_3(c) + 6\,H^+(aq)$
(c) $4\,Ag(c) + 8\,CN^-(aq) + O_2(g) + 2\,H_2O(liq) \rightarrow 4\,Ag(CN)_2^-(aq) + 4\,OH^-(aq)$
(d) $4\,H^+(aq) + 4\,Cl^-(aq) + Au(c) + NO_3^-(aq) \rightarrow AuCl_4^-(aq) + NO(g) + 2\,H_2O(liq)$

24. $K_{sp} = 1 \times 10^{-12}$

26. More sensitive analytical methods permit detection of lower concentrations

27. (a) pH = 3; (b) $[S^{2-}] = 3 \times 10^{-13}\,M$

28. $[Zn^{2+}] = 1 \times 10^{-11}\,M$

29. $[Pb^{2+}] \sim 10^{-8}\,M$ but $[Zn^{2+}] \sim 0.9\,M$

30. $[Hg^{2+}] \sim 1 \times 10^{-31}\,M$

32. Diamminesilver(I) ion; hexafluoroferrate(III) ion; tris(ethylenediamine)cobalt(III) ion; hexacarbonylchromium (0); diaquadioxalatochromate(III) ion; triamminetrinitrocobalt(III)

33. 5.9 BM, 0, 1.7 BM, 3.9 BM

34. Only $Co(en)_3^{3+}$, $Cr(CO)_6$, and $Co(NO_2)_3(NH)_3$ have an EAN of a noble gas

35. s, p_x, p_y, $d_{x^2-y^2}$

36. NiF_6^{4-}, octahedral; $AuCl_4^-$, square planar; CrF_6^{4-}, octahedral; $FeCl_4^-$, tetrahedral

37. The chloro complex has cis and trans isomers

39. NCS^- between H_2O and NH_3 in spectrochemical series

40. (a) $Fe(CN)_6^{3-}$; (b) $PtCl_4^{2-}$; (c) $Co(NH_3)_6^{3+}$; (d) $Cr(CO)_6$

41. C—N bond energy less than in HCN

42. The platinum complex is square planar, whereas the nickel complex is tetrahedral

43. Three geometrical isomers, one optically active

44. 59.2% Cu

Chapter 21

3. High pressure; $d\Delta G/dP$

4. ΔV is negative

6. Pure metal is expected to have higher thermal and electrical conductivities than alloy

9. Atoms packed together differently

13. (b) 0.6 ton; (c) 0.15 ton carbon, 1.8 ton chlorine, 0.61 ton magnesium, and no HCl

14. Freezing point is lowered with increasing pressure

19. $n = 1$, 21.343° or 21°21′; $n = 2$, 46.711° or 46°43′

21. Mass is 0.17×10^{-17} g molecule^{-1}, and radius of molecule is 7.4×10^{-7} cm; can see a molecule of molecular weight about 10^{12} g mole^{-1}

22. Conductance increases and resistance decreases with increasing temperature

25. $K = 26$ at 298°K; K decreases with increasing temperature; maximum yield at high pressure and low temperature (actually, the reaction is too slow at low temperature, which requires a compromise between thermodynamics and kinetics)

28. Lattice enthalpy is larger than hydration enthalpy

29. $\Delta H_f^0 \cong +124$ kcal mole^{-1} for ArCl(c), which is unstable with respect to decomposition to the elements

31. 0.246 atm

32. $\sim$0.4 M

33. $S = A/3$ at 10°K

34. $E = RT$ and $C_v = R$

35. $\sim$740 kcal mole^{-1}

Chapter 22

1. 0.93, 9.6 hr

2. 62.6 kcal mole^{-1}

3. 12.3 kcal/mole

4. $k_{110}/k_{100} = 1.45$

5. 0.054 hr^{-1} = 9×10^{-4} min^{-1}

6. 66 years

7. (a) $N_2O_4 \cdot H_2O$; (b) proposed mechanism is consistent with rate equation (22.48)

8. $-d[I_3^-]/dt = +d[S_4O_6^{2-}]/dt = kK[S_2O_3^{2-}][I_3^-]/[I^-]^2$

9. $S_2O_8^{2-}(aq) + Ag^+(aq) \rightarrow 2\ SO_4^{2-}(aq) + Ag^{3+}(aq)$, slow; $Ag^{3+}(aq) + 2\ VO^{2+}(aq) + 4\ H_2O(liq) \rightarrow 2\ HVO_3(aq) + 6\ H^+(aq) + Ag^+(aq)$, rapid

10. Equations are consistent with $k_r = 2k$

12. $\Delta H^0 = A - B$; $\Delta H_b^{\ddagger} = B - (A/2)$ for the catalyzed reaction

14. (b) $(1/A) = (1/abP) + (1/a)$; $a = 1/\text{intercept}$ and $b = \text{intercept}/\text{slope}$

16. $K = k_f/k_b = 0.42$ and $\Delta H^0 = \Delta H_f^{\ddagger} - \Delta H_b^{\ddagger} = -5.5$ kcal mole^{-1} to show reasonable agreement between results of equilibrium and kinetic measurements

17. (a) Both postulated mechanisms are consistent with the preliminary rate equation (22.52), which cannot be used as basis for choosing between these two possible mechanisms; (b) rate equation (22.53) and following mechanism are consistent with each other; (c) $[Fe^{3+}] \cong 0$ near beginning of reaction; (d) yes

20. (a) Yes; (b) $k_r = 0.096$ liter mole^{-1} min^{-1}; (c) 4330 sec

21. $Mn^{2+}(aq)$ is a catalyst for the reaction

TABLES OF LOGARITHMS

Logarithms

Natural Numbers	0	1	2	3	4	5	6	7	8	9	PROPORTIONAL PARTS								
											1	2	3	4	5	6	7	8	9
55	7404	7412	7419	7427	7435	7443	7451	7459	7466	7474	1	2	2	3	4	5	5	6	7
56	7482	7490	7497	7505	7513	7520	7528	7536	7543	7551	1	2	2	3	4	5	5	6	7
57	7559	7566	7574	7582	7589	7597	7604	7612	7619	7627	1	2	2	3	4	5	5	6	7
58	7634	7642	7649	7657	7664	7672	7679	7686	7694	7701	1	1	2	3	4	4	5	6	7
59	7709	7716	7723	7731	7738	7745	7752	7760	7767	7774	1	1	2	3	4	4	5	6	7
60	7782	7789	7796	7803	7810	7818	7825	7832	7839	7846	1	1	2	3	4	4	5	6	6
61	7853	7860	7868	7875	7882	7889	7896	7903	7910	7917	1	1	2	3	4	4	5	6	6
62	7924	7931	7938	7945	7952	7959	7966	7973	7980	7987	1	1	2	3	3	4	5	6	6
63	7993	8000	8007	8014	8021	8028	8035	8041	8048	8055	1	1	2	3	3	4	5	5	6
64	8062	8069	8075	8082	8089	8096	8102	8109	8116	8122	1	1	2	3	3	4	5	5	6
65	8129	8136	8142	8149	8156	8162	8169	8176	8182	8189	1	1	2	3	3	4	5	5	6
66	8195	8202	8209	8215	8222	8228	8235	8241	8248	8254	1	1	2	3	3	4	5	5	6
67	8261	8267	8274	8280	8287	8293	8299	8306	8312	8319	1	1	2	3	3	4	5	5	6
68	8325	8331	8338	8344	8351	8357	8363	8370	8376	8382	1	1	2	3	3	4	4	5	6
69	8388	8395	8401	8407	8414	8420	8426	8432	8439	8445	1	1	2	2	3	4	4	5	6
70	8451	8457	8463	8470	8476	8482	8488	8494	8500	8506	1	1	2	2	3	4	4	5	6
71	8513	8519	8525	8531	8537	8543	8549	8555	8561	8567	1	1	2	2	3	4	4	5	5
72	8573	8579	8585	8591	8597	8603	8609	8615	8621	8627	1	1	2	2	3	4	4	5	5
73	8633	8639	8645	8651	8657	8663	8669	8675	8681	8686	1	1	2	2	3	4	4	5	5
74	8692	8698	8704	8710	8716	8722	8727	8733	8739	8745	1	1	2	2	3	4	4	5	5
75	8751	8756	8762	8768	8774	8779	8785	8791	8797	8802	1	1	2	2	3	3	4	5	5
76	8808	8814	8820	8825	8831	8837	8842	8848	8854	8859	1	1	2	2	3	3	4	5	5
77	8865	8871	8876	8882	8887	8893	8899	8904	8910	8915	1	1	2	2	3	3	4	4	5
78	8921	8927	8932	8938	8943	8949	8954	8960	8965	8971	1	1	2	2	3	3	4	4	5
79	8976	8982	8987	8993	8998	9004	9009	9015	9020	9026	1	1	2	2	3	3	4	4	5
80	9031	9036	9042	9047	9053	9058	9063	9069	9074	9079	1	1	2	2	3	3	4	4	5
81	9085	9090	9096	9101	9106	9112	9117	9122	9128	9133	1	1	2	2	3	3	4	4	5
82	9138	9143	9149	9154	9159	9165	9170	9175	9180	9186	1	1	2	2	3	3	4	4	5
83	9191	9196	9201	9206	9212	9217	9222	9227	9232	9238	1	1	2	2	3	3	4	4	5
84	9243	9248	9253	9258	9263	9269	9274	9279	9284	9289	1	1	2	2	3	3	4	4	5
85	9294	9299	9304	9309	9315	9320	9325	9330	9335	9340	1	1	2	2	3	3	4	4	5
86	9345	9350	9355	9360	9365	9370	9375	9380	9385	9390	1	1	2	2	3	3	4	4	5
87	9395	9400	9405	9410	9415	9420	9425	9430	9435	9440	0	1	1	2	2	3	3	4	4
88	9445	9450	9455	9460	9465	9469	9474	9479	9484	9489	0	1	1	2	2	3	3	4	4
89	9494	9499	9504	9509	9513	9518	9523	9528	9533	9538	0	1	1	2	2	3	3	4	4
90	9542	9547	9552	9557	9562	9566	9571	9576	9581	9586	0	1	1	2	2	3	3	4	4
91	9590	9595	9600	9605	9609	9614	9619	9624	9628	9633	0	1	1	2	2	3	3	4	4
92	9638	9643	9647	9652	9657	9661	9666	9671	9675	9680	0	1	1	2	2	3	3	4	4
93	9685	9689	9694	9699	9703	9708	9713	9717	9722	9727	0	1	1	2	2	3	3	4	4
94	9731	9736	9741	9745	9750	9754	9759	9763	9768	9773	0	1	1	2	2	3	3	4	4
95	9777	9782	9786	9791	9795	9800	9805	9809	9814	9818	0	1	1	2	2	3	3	4	4
96	9823	9827	9832	9836	9841	9845	9850	9854	9859	9863	0	1	1	2	2	3	3	4	4
97	9868	9872	9877	9881	9886	9890	9894	9899	9903	9908	0	1	1	2	2	3	3	4	4
98	9912	9917	9921	9926	9930	9934	9939	9943	9948	9952	0	1	1	2	2	3	3	4	4
99	9956	9961	9965	9969	9974	9978	9983	9987	9991	9996	0	1	1	2	2	3	3	3	4

Logarithms

Natural Numbers	0	1	2	3	4	5	6	7	8	9	PROPORTIONAL PARTS								
											1	2	3	4	5	6	7	8	9
10	0000	0043	0086	0128	0170	0212	0253	0294	0334	0374	4	8	12	17	21	25	29	33	37
11	0414	0453	0492	0531	0569	0607	0645	0682	0719	0755	4	8	11	15	19	23	26	30	34
12	0792	0828	0864	0899	0934	0969	1004	1038	1072	1106	3	7	10	14	17	21	24	28	31
13	1139	1173	1206	1239	1271	1303	1335	1367	1399	1430	3	6	10	13	16	19	23	26	29
14	1461	1492	1523	1553	1584	1614	1644	1673	1703	1732	3	6	9	12	15	18	21	24	27
15	1761	1790	1818	1847	1875	1903	1931	1959	1987	2014	3	6	8	11	14	17	20	22	25
16	2041	2068	2095	2122	2148	2175	2201	2227	2253	2279	3	5	8	11	13	16	18	21	24
17	2304	2330	2355	2380	2405	2430	2455	2480	2504	2529	2	5	7	10	12	15	17	20	22
18	2553	2577	2601	2625	2648	2672	2695	2718	2742	2765	2	5	7	9	12	14	16	19	21
19	2788	2810	2833	2856	2878	2900	2923	2945	2967	2989	2	4	7	9	11	13	16	18	20
20	3010	3032	3054	3075	3096	3118	3139	3160	3181	3201	2	4	6	8	11	13	15	17	19
21	3222	3243	3263	3284	3304	3324	3345	3365	3385	3404	2	4	6	8	10	12	14	16	18
22	3424	3444	3464	3483	3502	3522	3541	3560	3579	3598	2	4	6	8	10	12	14	15	17
23	3617	3636	3655	3674	3692	3711	3729	3747	3766	3784	2	4	6	7	9	11	13	15	17
24	3802	3820	3838	3856	3874	3892	3909	3927	3945	3962	2	4	5	7	9	11	12	14	16
25	3979	3997	4014	4031	4048	4065	4082	4099	4116	4133	2	3	5	7	9	10	12	14	15
26	4150	4166	4183	4200	4216	4232	4249	4265	4281	4298	2	3	5	7	8	10	11	13	15
27	4314	4330	4346	4362	4378	4393	4409	4425	4440	4456	2	3	5	6	8	9	11	13	14
28	4472	4487	4502	4518	4533	4548	4564	4579	4594	4609	2	3	5	6	8	9	11	12	14
29	4624	4639	4654	4669	4683	4698	4713	4728	4742	4757	1	3	4	6	7	9	10	12	13
30	4771	4786	4800	4814	4829	4843	4857	4871	4886	4900	1	3	4	6	7	9	10	11	13
31	4914	4928	4942	4955	4969	4983	4997	5011	5024	5038	1	3	4	6	7	8	10	11	12
32	5051	5065	5079	5092	5105	5119	5132	5145	5159	5172	1	3	4	5	7	8	9	11	12
33	5185	5198	5211	5224	5237	5250	5263	5276	5289	5302	1	3	4	5	6	8	9	10	12
34	5315	5328	5340	5353	5366	5378	5391	5403	5416	5428	1	3	4	5	6	8	9	10	11
35	5441	5453	5465	5478	5490	5502	5514	5527	5539	5551	1	2	4	5	6	7	9	10	11
36	5563	5575	5587	5599	5611	5623	5635	5647	5658	5670	1	2	4	5	6	7	8	10	11
37	5682	5694	5705	5717	5729	5740	5752	5763	5775	5786	1	2	3	5	6	7	8	9	10
38	5798	5809	5821	5832	5843	5855	5866	5877	5888	5899	1	2	3	5	6	7	8	9	10
39	5911	5922	5933	5944	5955	5966	5977	5988	5999	6010	1	2	3	4	5	7	8	9	10
40	6021	6031	6042	6053	6064	6075	6085	6096	6107	6117	1	2	3	4	5	6	8	9	10
41	6128	6138	6149	6160	6170	6180	6191	6201	6212	6222	1	2	3	4	5	6	7	8	9
42	6232	6243	6253	6263	6274	6284	6294	6304	6314	6325	1	2	3	4	5	6	7	8	9
43	6335	6345	6355	6365	6375	6385	6395	6405	6415	6425	1	2	3	4	5	6	7	8	9
44	6435	6444	6454	6464	6474	6484	6493	6503	6513	6522	1	2	3	4	5	6	7	8	9
45	6532	6542	6551	6561	6571	6580	6590	6599	6609	6618	1	2	3	4	5	6	7	8	9
46	6628	6637	6646	6656	6665	6675	6684	6693	6702	6712	1	2	3	4	5	6	7	7	8
47	6721	6730	6739	6749	6758	6767	6776	6785	6794	6803	1	2	3	4	5	5	6	7	8
48	6812	6821	6830	6839	6848	6857	6866	6875	6884	6893	1	2	3	4	4	5	6	7	8
49	6902	6911	6920	6928	6937	6946	6955	6964	6972	6981	1	2	3	4	4	5	6	7	8
50	6990	6998	7007	7016	7024	7033	7042	7050	7059	7067	1	2	3	3	4	5	6	7	8
51	7076	7084	7093	7101	7110	7118	7126	7135	7143	7152	1	2	3	3	4	5	6	7	8
52	7160	7168	7177	7185	7193	7202	7210	7218	7226	7235	1	2	2	3	4	5	6	7	7
53	7243	7251	7259	7267	7275	7284	7292	7300	7308	7316	1	2	2	3	4	5	6	6	7
54	7324	7332	7340	7348	7356	7364	7372	7380	7388	7396	1	2	2	3	4	5	6	6	7

INDEX

PERIODIC CHART

Metals

Representative Elements

Transition Elements

Group / Period	IA	IIA	IIIB	IVB	VB	VIB	VIIB	VIII	
1	1 H 1.008								
2	3 Li 6.941	4 Be 9.012							
3	11 Na 22.99	12 Mg 24.31							
4	19 K 39.10	20 Ca 40.08	21 Sc 44.96	22 Ti 47.90	23 V 50.94	24 Cr 52.00	25 Mn 54.94	26 Fe 55.85	27 Co 58.93
5	37 Rb 85.47	38 Sr 87.62	39 Y 88.91	40 Zr 91.22	41 Nb 92.91	42 Mo 95.94	43 Tc (96.91)	44 Ru 101.1	45 Rh 102.9
6	55 Cs 132.9	56 Ba 137.3	57–71	72 Hf 178.5	73 Ta 180.9	74 W 183.9	75 Re 186.2	76 Os 190.2	77 Ir 192.2
7	87 Fr (223)	88 Ra (226)	89–103	104	105				

Inner

Lanthanide Series	57 La 138.9	58 Ce 140.1	59 Pr 140.9	60 Nd 144.2	61 Pm (145)	62 Sm 150.4
Actinide Series	89 Ac (227)	90 Th (232.0)	91 Pa (231.0)	92 U (238.0)	93 Np (237.0)	94 Pu (244)

Parenthetical values refer to the isotope of longest half-life when all isotopes are radioactive.